D0138737

							0 (18)
			III B (13)	IV B (14)	V B (15)	VI B (16)	VII B (17)	2 He 4.00260

(10)	I B (11)	II B (12)	III B (13)	IV B (14)	V B (15)	VI B (16)	VII B (17)	0 (18)
			5 B 10.81	6 C 12.011	7 N 14.0067	8 O 15.9994	9 F 18.998403	10 Ne 20.179
			13 Al 26.98154	14 Si 28.0855	15 P 30.97376	16 S 32.06	17 Cl 35.453	18 Ar 39.948
28 Ni 58.69	29 Cu 63.546	30 Zn 65.38	31 Ga 69.72	32 Ge 72.59	33 As 74.9216	34 Se 78.96	35 Br 79.904	36 Kr 83.80
46 Pd 106.42	47 Ag 107.868	48 Cd 112.41	49 In 114.82	50 Sn 118.69	51 Sb 121.75	52 Te 127.60	53 I 126.9045	54 Xe 131.29
78 Pt 195.08	79 Au 196.9665	80 Hg 200.59	81 Tl 204.383	82 Pb 207.2	83 Bi 208.9804	84 Po (209)	85 At (210)	86 Rn (222)

64 Gd 157.25	65 Tb 158.9254	66 Dy 162.50	67 Ho 164.9304	68 Er 167.26	69 Tm 168.9342	70 Yb 173.04	71 Lu 174.967
96 Cm (247)	97 Bk (247)	98 Cf (251)	99 Es (252)	100 Fm (257)	101 Md (258)	102 No (259)	103 Lr (260)

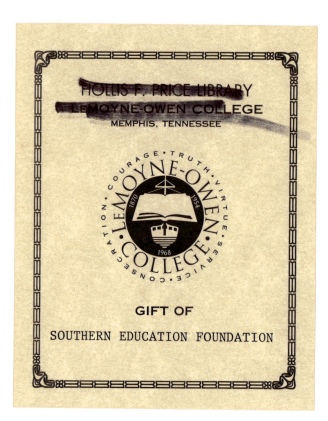

HOLLIS F. PRICE LIBRARY
LEMOYNE-OWEN COLLEGE
MEMPHIS, TENNESSEE

GIFT OF

SOUTHERN EDUCATION FOUNDATION

Concepts and Models of Inorganic Chemistry

CONCEPTS AND MODELS OF INORGANIC CHEMISTRY

2ND EDITION

Bodie E. Douglas, *University of Pittsburgh*

Darl H. McDaniel, *University of Cincinnati*

John J. Alexander, *University of Cincinnati*

John Wiley & Sons, Inc.

New York Chichester Brisbane Toronto Singapore

Hollis F. Price

Le Moyne - Owen College

Memphis, Tennessee

Withdrawn

Cover illustration: Computer-generated drawing of Carboxypeptidase A. Color coding: background α-carbon chain, blue; amino acids in the active site, red; zinc and associated atoms, white; cysteine, yellow; and inhibitor, green. (Courtesy of Dr. Robert Langridge, Computer Graphics Laboratory, University of California, San Francisco, © The Regents of the University of California.)

The American Chemical Society holds the copyright and has given permission to use the following:

Figure 3.2, p. 85 and the figure on p. 87; Table 5.5, p. 190; Figure 8.13, p. 312; Figure 8.22, p. 319; Figure 8.29, p. 326; The figure of adducts of 2,6-lutidine on p. 535 referenced to D. F. Hoeg, S. Liebman, and L. Schubert, *J. Org. Chem.* 1963, *28,* 1554.

Copyright © 1965 and 1983, by John Wiley & Sons, Inc.

All rights reserved. Published simultaneously in Canada.

Reproduction or translation of any part of
this work beyond that permitted by Sections
107 and 108 of the 1976 United States Copyright
Act without the permission of the copyright
owner is unlawful. Requests for permission
or further information should be addressed to
the Permissions Department, John Wiley & Sons.

Library of Congress Cataloging in Publication Data:

Douglas, Bodie Eugene, 1924–
   Concepts and models of inorganic chemistry.

   Includes bibliographies and indexes.
   1. Chemistry, Inorganic.   I. McDaniel, Darl
Hamilton, 1928–   II. Alexander, John J.
III. Title.

QD475.D65 1982   546   82-2606
ISBN 0-471-21984-3   AACR2

Printed in the United States of America

20 19 18 17 16 15 14 13 12

Printed and bound by the Arcata Graphics Company

Preface

The title of the book still serves as a central theme. The text should help students integrate their knowledge of chemistry, enabling them to draw on the wealth of knowledge learned in the highly compartmentalized chemistry courses. The upper level inorganic course is a fun course for many of us (instructors and students) because here many students gain an overall view of chemistry and acquire the intuitive feel of a chemist. The concepts-and-models approach should foster this goal.

The first edition of this text appeared in 1965. The developments in inorganic chemistry since then have been impressive. We have made the following additions to the text to keep abreast of these developments. We use figures more extensively than in the first edition; they have been selected with care and rendered more effectively. The treatment of bonding is more sophisticated, using molecular orbital theory (Chapter 4). Symmetry and group theory (Chapter 3) are presented for applications to bonding; with this background we are able to have a more detailed discussion of ligand field spectra (Chapter 7). Because the solid state is not given proper emphasis in inorganic courses, discussion of this topic has been expanded (Chapter 6). Reaction mechanisms have increased in importance and receive greater attention (Chapters 9 and 10). The field of organometallic chemistry has grown tremendously and requires a full chapter (Chapter 10), including reaction mechanisms. The treatment of hard and soft acids and bases was added (Chapter 12), and applications to coordination chemistry are included in the same chapter. More descriptive chemistry has been added in Chapters 13, 14, 15, and elsewhere. The importance of the emerging field of metal cluster compounds was acknowledged in the first edition by using a picture of the $Re_3Cl_{12}^{3-}$ structure on the dust jacket, and metal clusters are now recognized to be much more common than had been believed. More systematic treatments of cage and cluster compounds merit regrouping of the topics (Chapter 15). Bioinorganic chemistry (Chapter 16) has become a major component of inorganic chemistry and requires introduction.

The goal of the text is to offer a reasonable balance of material for an upper level undergraduate or first year graduate course. Even reasonable, but not comprehensive, coverage of the currently important topics in inorganic chemistry leads to a book that is too large for a one semester course. Each instructor must select topics and decide on emphasis. Material is provided for courses differing significantly in the balance between theory and descriptive chemistry. Chapters 8, 10, 13, 14, and 15 are primarily descriptive, although in Chapters 6, 9, 11, 12, and 16 we have incorporated a great deal of descriptive chemistry.

Chapter 3 presents symmetry and a brief introduction to group theory for applications to bonding and spectroscopy. The material in Chapters 3, 4, and 7 is organized so that group theory and its applications can be omitted, with the sections using group theory identified.

In a one term course we recommend coverage of the following sections as a basic "core": Sections 2.1–2.4, 3.1–3.4, 4.1–4.2, 6.1, 6.3, 7.1–7.5, 9.1–9.3.5, and 12.2–12.3. Beyond this "core", these topics can be expanded and others added to satisfy the interests of the instructor and the objectives of the course.

A Solutions Manual is available. It could prove very helpful to students if used properly.

We thank colleagues who have read parts of the manuscript and made useful suggestions, including Rex Shepherd, Darel Straub, and John Thayer. The book was improved by detailed comments on the entire manuscript by Edwin H. Abbott, William E. Hatfield, Duward F. Shriver, and Andrew Wojcicki. Finally, we wish to acknowledge the aid of Marty McDaniel in proofreading.

Bodie E. Douglas
Darl H. McDaniel
John J. Alexander

Preface to the
First Edition

This book is intended for use as a text in advanced undergraduate courses in inorganic chemistry, with physical chemistry as a prerequisite or at least a corequisite. It is also suitable for a beginning graduate course for students who have not had an advanced undergraduate course. The level of the treatment will stimulate the student to use his preparation in chemistry and physics. An inorganic chemistry course should present the challenge of modern inorganic chemistry and at the same time unify many of the principles and facts presented in earlier courses.

A textbook should serve a number of functions. The first and foremost of these is to acquaint the reader with the theory which undergirds the subject. This is particularly important in chemistry where observations in the laboratory are used to make inferences about the unseen behavior of molecular species. The chemist creates mental models to rationalize the behavior he seeks to describe, and he spends much of his time dealing with these models. A fruitful model serves not only to organize a number of observations, but also as a basis for prediction and to stimulate testing of the model by critical experiments. In this text we present some of the models and concepts of inorganic chemistry in current use. Some are models in the literal sense that they may be visualized and drawings or physical models made, as with shapes of orbitals, while some may not readily be pictured in physical space, as in the case of energy levels.

As well as providing a theoretical base for the subject, a new book in any area should keep the reader up to date. This can be done only if the new ideas appearing in the literature are selectively introduced into the text. Recent topics introduced in this text include crystal field theory, molecular orbitals, boron hydrides, and compounds of the rare gases.

A text should also organize and present effectively ideas which have proven of value for some time. The bases for concepts such as electronegativity and ionic radii are explored here.

The first four chapters present the models and concepts fundamental to inorganic chemistry. Although this book was written with a one-semester course in view, it is not expected that all chapters will necessarily be covered in one semester. Chapters V through XII may be treated as virtually independent topics with material to be selected at the discretion of the instructor. Frequent cross-references are given to facilitate such usage. Many instructors will prefer to refer students to Chapter XII (structural tools), introducing lecture material as needed for other chapters.

Some of the topics within individual chapters are included primarily to serve as a springboard for more advanced courses. For example, the treatment of spectroscopic terms given here provides the left-hand side of an Orgel diagram. The development of the right-hand side may then be carried out in a course in coordination compounds. The brief review of symmetry (Appendix C) and the discussion of the application of symmetry to molecular orbitals (Appendix D) represent extensions, which are becoming increasingly important, of the material in the text. It seemed most appropriate to separate this material from the main body of the text to provide better continuity for those instructors who do not cover this material.

Tables and figures have been used liberally throughout this book. These contain information essential to the portions of the text where they appear and are appropriately indexed. The Appendices and references included in each chapter are intended to extend the usefulness of the book beyond the classroom. It is hoped that the student will find it a useful reference source, which will continue to be of value after he has finished with it as a text.

The authors would like to thank Dr. Elmer Amma for the preparation of a treatment on molecular orbitals, Appendix D, and Dr. Alan Searcy for the example problem on the use of approximation methods in establishing half-cell emf values. We are indebted to our many friends and colleagues who read parts of the manuscript during its preparation and made numerous helpful suggestions. Among these we would like especially to thank Drs. C. H. Brubaker, Jr., T. B. Cameron, J. C. Carter, Joyce Corey, H. S. Frank, H. H. Jaffé, W. L. Jolly, L. N. Mulay, and D. K. Straub. For critically reading the entire manuscript we would like to express our gratitude to Dr. Gordon Atkinson. Finally, we would like to acknowledge the less specific, but no less important, contributions of our teachers and colleagues, who will be aware of their influence in the development of the ideas in this text.

Contents

XVI Some Aspects of Bioinorganic Chemistry 720

Appendix A Units and Physical Constants 756

Appendix B Nomenclature of Inorganic Chemistry 759

Appendix C Potential Diagrams 772

Concepts and Models of
Inorganic Chemistry

I

Atomic Structure and the Periodic Table

The periodic trends among elements are the result of the regular pattern in electron configurations of atoms. By examining the early development of the periodic table and the Bohr theory we will see the historical background for our modern views of atomic structure. We will also explore how the results of the wave mechanical description provide the basis for later treatments of bonding and spectroscopy. Here we briefly consider atomic spectra, the derivation of spectroscopic term symbols, and periodic trends for electron configurations and the attraction for electrons (ionization energies and electron affinities).

1.1 HISTORICAL BACKGROUND

A systematic approach to inorganic chemistry is today almost synonymous with studying the periodic relationships of the elements and their compounds. This approach has an empirical foundation built during the last century and a theoretical justification of half a century.

Within a decade following the presentation of a consistent set of atomic weights by Stanislao Cannizzaro at the Karlsruhe Conference in 1860, various forms of the periodic table appeared in France, England, Germany, and Russia. The role played by Cannizzaro's list of atomic weights in the development of the periodic table can better be appreciated by recalling that an attempt in 1852 by Gladstone to find a relationship between the atomic weights and other properties of the elements failed, because of the lack of a consistent set of atomic weights. The greatest share of credit for the periodic table is usually given to Dimitri Mendeleyev, and properly so, for it was the realization of his bold prophecy of new elements and their properties that led to the almost immediate acceptance of the periodic law. In 1871, Mendeleyev made the following predictions about an element that was discovered by Boisbaudran in 1875.

The properties of ekaaluminum, according to the periodic law, should be the following: Its atomic weight will be 68. Its oxide will have the formula El_2O_3; its salts will present the formula ElX_3. Thus, for example, the chloride of ekaaluminum will be $ElCl_3$; it will give for analysis 39% metal and 61% chlorine and will be more volatile than $ZnCl_2$. The sulfide El_2S_3, or oxysulfide $El_2(S,O)_3$, will be precipitated by H_2S and will be insoluble in ammonium sulfide. The metal will be easily obtained by reduction; its density will be 5.9, accordingly its atomic volume will be 11.5; it will be soft, and fusible at a very low temperature. It will not be oxidized on contact with air; it will decompose water when heated to redness. The pure liquid metal will not be attacked by acids and only slowly by alkali. The oxide El_2O_3 will have a specific gravity of approximately 5.5; it should be soluble in strong acids, forming an amorphous hydrate insoluble in water, dissolving in acids and alkali. The oxide of ekaaluminum will form the neutral salts and basic $El_2(OH,X)_6$, but no acid salts; the alum $KEl(SO_4)_2 \cdot 12H_2O$ will be more soluble than the corresponding salt of aluminum and less crystallizable. The basic properties of El_2O_3 being more pronounced than those of Al_2O_3 and less than that of ZnO . . . it will be precipitated by barium carbonate. The volatility as well as the other properties of the salts of ekaaluminum will be a mean between those of Al and In. It is probable that the metal in question will be discovered by spectral analysis as have been In and Tl.

Examining the above predicted properties for gallium reveals some properties that vary systematically with the position of the element in the periodic table—physical properties of the element and its compounds (specific gravity, hardness, melting point, boiling point, etc.), spectrographic properties, and chemical properties (formulas of possible compounds, acidic and basic properties of compounds, etc.). In fact, properties of the elements or their compounds that cannot be correlated by means of the periodic table are somewhat exceptional.[1]

Much of this book is devoted to attempting to understand the underlying principles that bring about these periodic relationships: that is, properties that show greater than average similarity for elements which lie in the periodic table (a) in a vertical column (called a group), (b) in a horizontal row (called a period), (c) within a given area (bounded by elements of two or more groups and two or more periods), and (d) on diagonals. The following illustrate these types of relationships. Elements in a group have similar arc and spark spectra; often, similar valences; similar crystal structures both for the element and for particular series of compounds, etc. Elements in a given period have similar maximum coordination numbers in their compounds. The compounds of the second-period elements Li, Be, and B show many similarities to the compounds of the third-period elements Mg, Al, and Si, to which they are diagonally related. Finally, there are numerous properties, such as classification of the elements as metals, metalloids, and nonmetals, that have an area relationship to the periodic table. The area relationships are often the most difficult to explain, because of the wider possible variation of the factors involved. Thus it may be difficult to explain in an *a priori* fashion why the carbides

---

[1] One of the major classes of "exceptional" properties is nuclear properties—nuclear magnetic moments, isotopic abundance, etc. These properties, however, may be rationalized on the basis of a "shell theory of the nucleus," which is analogous to the theoretical justification of the chemical periodic table. See B. H. Flowers, *J. Chem. Educ.* 1960, *37*, 610. Maria Goeppert Mayer won the Nobel Prize in Physics in 1963 for this theory; for her Nobel address, see *Angew. Chem.* 1964, *76*, 729.

of a given area are explosive—but that is certainly worthwhile knowing if you contemplate making carbides or acetylides of elements lying in or near such a known area.

Before proceeding to the theoretical basis of the periodic table, let us note the following steps in its evolution.

By 1829, Döbereiner had pointed out that there were a number of cases in which three elements, or triads, have similar chemical properties; and further, that one member of a triad has properties very close to the mean value of the other two—this is particularly true of the atomic weights.

Between 1860 and 1870, Newland, Meyer, and Mendeleyev prepared periodic tables by listing the elements in the order of increasing atomic weights and then grouping them according to chemical properties. In Mendeleyev's table the triads of Döbereiner always fell within the same group. It may, at first, appear odd that for the group VIII elements more than one element is listed in a period, but the elements iron, cobalt, and nickel form one of Döbereiner's triads and hence have to appear in the same group. Mendeleyev reassigned atomic weights to a number of elements in order to obtain a fit with the chemical properties of the other elements in the group. Later evaluations confirmed the need for reordering the atomic weights of some of the elements but firmly established a reversal in atomic weights, as compared to the position in the periodic table, for several pairs of elements (Te and I, Co and Ni are the early known cases.)

As the rare earth elements were discovered, difficulty was encountered in fitting them in the table. This led Basset and later Thomsen to propose the extended form of the table generally accepted today (see Figure 1.1.) Further, from considering the change of group valence from -1 for the halogens to $+1$ for the alkali metals, Thomsen reasoned that one should expect a group of elements lying between groups VII and I and having either infinite or zero valence. Since a valence of infinity is unacceptable from a chemical viewpoint, he proposed that a group of elements of zero valence separated the highly electronegative halogens from the highly electropositive alkali metals. He proceeded to predict the atomic weights of these elements as 4, 20, 36, 84, 132, and 212. He felt that these elements should terminate each period. Unfortunately, Thomsen did not publish these remarkable predictions until after argon had been discovered.

The last stage in the empirical development of the periodic table came in 1913, when Moseley found the x-ray emission from different elements had characteristic frequencies (v), which varied in a regular fashion with the ordinal number of the elements as they appear in the table. The empirical relationship is

$$v = k(Z - \sigma)^2 \tag{1.1}$$

where Z is the ordinal or atomic number, v is the characteristic x-ray frequency, and k and σ are constants for a given series. No reversals in atomic number occur in the periodic table; hence, it is a more fundamental property of an element than the atomic weight.

The empirical evolution of the periodic table had reached its peak. It was now possible to make a strictly ordered list of the elements with definite indication of missing elements. Each period terminated with a noble gas and it was possible to tell how many elements belonged to each period.

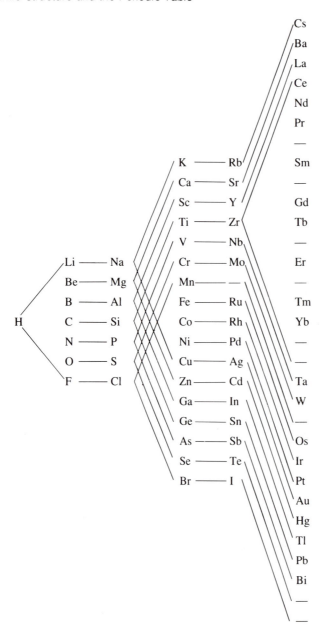

Figure 1.1 Long form of periodic table. (Proposed by J. Thomsen, *Z. Anorg. Chem.* 1895, *9*, 190.)

1.2 ATOMIC STRUCTURE AND THE THEORETICAL BASIS OF THE PERIODIC TABLE

The theoretical basis of the periodic table had to await the development of a clearer picture of the atom. The concept of atoms as fundamental or indivisible particles had to be abandoned at the beginning of this century. Studies of cathode rays and canal rays led to the recognition of the existence of negative and positive charges within the atom. Further complexity of the atom could be inferred from emission spectra of gaseous substances in magnetic fields (the Zeeman effect) and the discovery of radioactivity. Radioactivity not only indicated that the atom was not a fundamental unit, but also provided a probe with which to examine the atom. From the scattering of alpha particles by thin metal foils, Rutherford arrived at a nuclear model of the atom, with a nucleus carrying a number of unit charges equal to approximately one half the atomic weight of the element. Van den Broek pointed out that the use of the ordinal number of the element in the periodic table—that is, the atomic number—for the number of unit charges on the nucleus improved the fit of the alpha scattering data. Moseley also associated the atomic number with the nuclear charge.

1.2.1 Bohr Model of the Atom

A major advance in the understanding of the atom was Niels Bohr's development of a model of the atom that could account for the spectra of hydrogenlike atoms (i.e., one electron, one nucleus—H, He^+, Li^{2+}, etc.). In developing his model Bohr accepted some past notions, rejected others, and assumed some new ones.

1. The Rutherford nuclear model of the atom was accepted.
2. The theories of Planck and of Einstein that radiant energy is quantized in units of $h\nu$, where h is Planck's constant and ν is the frequency of the radiant energy, were accepted.
3. The classical electrodynamic theory that a charged particle undergoing acceleration must emit electromagnetic radiation was rejected for electrons within atoms.
4. The electron was assumed to travel in circular orbits.
5. Of all possible orbits, only those for which the electron had a specified angular momentum were acceptable (that is, the angular momentum was quantized).
6. It was postulated that radiation was emitted or absorbed only when the electron jumped from one orbit to another, the energy emitted or absorbed corresponding to the difference in the energies for the initial and final states of the system.
7. Except as noted above, classical physics was assumed to be applicable to the atom.

Before going further, we should note that Bohr's assumption of circular orbits has been shown to be much too restrictive. Assumptions **1, 2, 3,** and **6** are retained in wave mechanics, whereas **5** comes as a result of the one arbitrary assumption of wave mechanics. Accordingly, we will not pay too much attention to the geometry of the Bohr model, but rather shall be more concerned with the energy states of the atom based on Bohr's model.

From **1** and **4** above, the Bohr model for hydrogenlike atoms may be pictured as having a heavy nucleus bearing a charge of Ze (where Z is the atomic number and e is the magnitude of the charge on the electron) with an electron of charge e and mass m traveling with a velocity v in an orbit of radius r from the nucleus (See Figure 1.2.)

The following relationships result from the assumptions listed above.

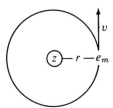

Figure 1.2 Bohr model of hydrogenlike atoms.

From classical physics **(7)** the centrifugal force may be equated with the coulombic attraction

$$\frac{mv^2}{r} = \frac{Ze^2}{r^2} \tag{1.2}$$

or

$$mv^2r = Ze^2 \tag{1.3}$$

The total energy, E, is the sum of the kinetic and potential energy

$$E = \frac{1}{2}mv^2 - \frac{Ze^2}{r} \tag{1.4}$$

substituting from (1.2) for $\frac{1}{2}mv^2$

$$E = -\frac{1}{2}\frac{Ze^2}{r} \tag{1.5}$$

Quantizing the angular momentum

$$mvr = n\left(\frac{h}{2\pi}\right) \tag{1.6}$$

where n (called the quantum number) must be an integer and h is Planck's constant.

From (1.2) ÷ (1.6)

$$v = Ze^2\frac{2\pi}{nh} \tag{1.7}$$

From (1.3) and (1.7)

$$r = \frac{Ze^2}{mv^2} = \frac{n^2h^2}{4\pi^2mZe^2} \tag{1.8}$$

From (1.5) and (1.8)

$$E = -\frac{2\pi^2mZ^2e^4}{n^2h^2} = \frac{E_{(n=1)}}{n^2} \tag{1.9}$$

This equation gives the energy of hydrogenlike atoms in various quantum states. For the hydrogen atom itself, the lowest energy state—that is, the quantum state for $n = 1$—has the value of -13.6 eV or -1312 kJ/mole. The lowest energy state for an atom (or ion or molecule) is called the *ground state*. The first higher energy state above the ground state is called the *first excited state;* the next higher state is called the *second excited state,* etc. The first excited state of the hydrogen atom would be that state having a quantum number of two. The amount of energy needed to promote an atom from the ground state to a given excited state is called the *excitation energy.* The amount of energy needed to remove an electron from an atom in its ground state is called the *ionization energy.* The *separation energy* is the amount of energy necessary to remove an electron from an atom in a particular excited state. These relationships are illustrated in Figure 1.3, an energy level diagram for the hydrogen atom that gives examples of some of the above. In such a diagram only the ordinate has meaning.

On the basis of the Bohr model, we may conclude about hydrogenlike atoms that (1) the ionization energy for the removal of the single electron is proportional to Z^2; (2) the radius of the hydrogen atom in the ground state is 52.9 pm, and for hydrogenlike atoms it is inversely proportional to Z; the radius of the atom in excited states is proportional to n^2; (3) in the ground state the electron is traveling with a velocity of 2.187×10^8 cm/sec. These values have become the standard units of atomic physics. The unit of distance a_0 is called the Bohr radius; the unit of energy is either the Rydberg (13.605 eV) or the Hartree which is e^2/a_0 and turns out to be 2×13.605 eV; the ratio of the ground-state velocity of the electron in the hydrogen atom to the velocity of light, $v/c = 1/137$, is called the fine structure constant. Finally, the Bohr hydrogen atom in the ground state should have a magnetic moment of $eh/4\pi mc$ or 9.18×10^{-21} erg/gauss. Although quantum mechanics indicates the ground state of the hydrogen atom has no orbital angular momentum and hence no orbital magnetic moment, the unit above, termed a Bohr magneton, is used as the measure of magnetic moment of atoms and ions.

As noted above, one of the major achievements of the Bohr model of the atom was its ability to account for the spectra of hydrogenlike atoms. By the time Bohr proposed his model of the atom, spectroscopists had formulated many empirical rules dealing with line spectra of atoms. Among these rules, the one having the greatest influence on Bohr was that frequency, v, or wave number, $\omega \equiv 1/\lambda$, of the numerous individual lines observed in a given spectrum can be reduced to the difference among a smaller number of terms. For atomic H the terms take the form R/n^2, where R is a constant, called the Rydberg constant (having a value of $109,677.581$ cm^{-1}) and n is an integer. Thus, empirically, all lines of the spectrum of atomic hydrogen have wave numbers given by the equation

$$\omega = \frac{R}{n_1^2} - \frac{R}{n_2^2} = T_1 - T_2 \tag{1.10}$$

According to Bohr's theory,

$$h\nu = E_2 - E_1 = hc\omega \tag{1.11}$$

The term values of the spectroscopist thus are virtually identical to energy levels within the atom, differing only in sign (as a result of defining the potential energy of the ionized atom as zero) and a constant factor of hc (which essentially takes care of the difference in units). From Bohr's theory the Rydberg constant[2] is given by $2\pi^2 me^4/h^3c$. The excellent agreement between the value calculated from these fundamental constants and the spectroscopically derived value gave strong support to Bohr's theory.

---

[2]More rigorously, the electron mass, m, in the Bohr equations should be replaced by the reduced mass, μ. The reduced mass is given by $1/\mu = 1/m + 1/M$, where M is the nuclear mass. If the nucleus is assumed to be stationary, that is, $M = \infty$—the value of R is $109,737.31$ cm^{-1}.

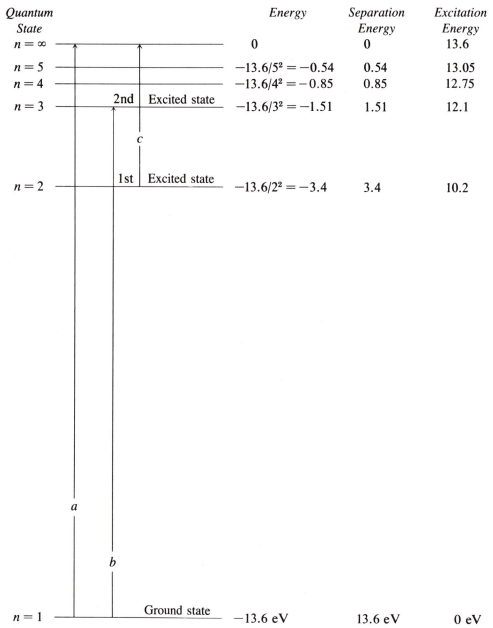

Quantum State				Energy	Separation Energy	Excitation Energy
$n = \infty$				0	0	13.6
$n = 5$				$-13.6/5^2 = -0.54$	0.54	13.05
$n = 4$				$-13.6/4^2 = -0.85$	0.85	12.75
$n = 3$		2nd	Excited state	$-13.6/3^2 = -1.51$	1.51	12.1
			c			
$n = 2$		1st	Excited state	$-13.6/2^2 = -3.4$	3.4	10.2
	a					
		b				
$n = 1$			Ground state	-13.6 eV	13.6 eV	0 eV

Figure 1.3   Energy-level diagram for the hydrogen atom. Line *a* corresponds to the ionization energy 13.6 eV. Line *b* corresponds to the excitation energy necessary to produce the second excited state: 12.1 eV. Line *c* corresponds to the separation energy of the first excited state: 3.4 eV.

The Bohr theory, with some modification, was found to be capable of explaining the Moseley relationship between characteristic x-ray spectra and atomic number. The excited state for emission of x rays was postulated as one in which a low-energy electron had been knocked out of a polyelectron atom. An electron from a higher energy state could then drop down to the lower empty orbit. In a polyelectron atom a given electron will be shielded partially from the positive charge of the nucleus by the electrons between it and the nucleus. The effective nuclear charge may be taken as $Z - \sigma$, where σ represents the shielding effect of underlying electrons. For an electron dropping from an $n = 2$ state to an $n = 1$ state in a polyelectron ion, the Bohr theory thus would predict

$$\nu = \frac{2\pi^2 m(Z - \sigma)^2 e^4}{h^3}\left(\frac{1}{n_2^2} - \frac{1}{n_1^2}\right) \tag{1.12}$$

or

$$\nu = \frac{3}{4}Rc(Z - \sigma)^2 \tag{1.13}$$

The values of the shielding constants for the K and L series support the conclusion of two and eight electrons, respectively, in these "shells."

1.2.2   Bohr-Sommerfeld Model of the Atom

Sommerfeld modified the Bohr treatment by specifying that electrons travel in elliptical orbits, with the nucleus at one of the foci—circular orbits being simply a special case of elliptical orbits. The electron traveling in an elliptical orbit would have, in addition to its angular momentum, a component of momentum along the radial direction. Both momenta were quantized: that is, taken as units of $h/2\pi$. The two resulting quantum numbers were called the azimuthal quantum number, designated as k, and the radial quantum number, n_r. The sum of these quantum numbers, $k + n_r$, corresponded to the principal quantum number of n of the Bohr theory. Except for relativity effects, which are not discussed here, the total energy still depended on n, as in the Bohr theory. Just as n determined the size of the Bohr orbits, the elliptical orbits in the Sommerfeld treatment have a major axis equal in length to the diameter calculated for a Bohr orbit of a given n state. The ratio of the major to the minor axis is equal to n/k. The possible orbits for $n = 3$ are shown in Figure 1.4. Although these levels, in the absence of relativity effects, would be of equal energy in a one-electron system, when lower orbits are filled, the most elliptical orbit (that is, 3_1) will have the lowest energy for a given n value, because of the penetration of the underlying filled orbits.

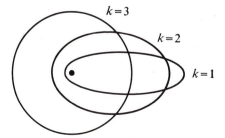

Figure 1.4   Bohr-Sommerfeld orbits for $n = 3$.

The quantum numbers n and k suffice to describe elliptical orbits in a plane, but another quantum number is necessary to describe the orientation of the plane of the ellipse in space. Not all orientations are possible—only those in which the component of momentum along the z axis is quantized. Since the energies of the states corresponding to different orientations differ in a directional magnetic field, the quantum number associated with these states is called the magnetic quantum number.

Finally, Uhlenbeck and Goudsmit postulated that the electron itself has a spin and a corresponding spin angular momentum, which is quantized, only a spin angular momentum of $\frac{1}{2}(h/2\pi)$ being allowed. In a directional magnetic field the momentum associated with the spin can only be parallel or antiparallel to a magnetic field, thus giving rise to a magnetic spin quantum number, m_s.

Thus, in the dozen years following Bohr's postulate of the quantization of orbital angular momentum of the electron in circular orbits, the postulate of quantization was extended to *all* momenta associated with a particle or system. For the electron in the atom at least four such quantized momenta must be specified to determine the system uniquely. Of these momenta, three may be associated with the three coordinates defining three-dimensional space and the fourth with the spin of the electron. Other momenta, derivable from these four or from which these four may be derived, serve equally well. The interaction of these momenta suggested a vector model of the atom, which we will discuss after briefly considering wave mechanics.

1.2.3 Wave Mechanics

Wave Properties of Matter

The Einstein equation for the energy of a photon, $E = h\nu$, may be combined with the Einstein equation relating mass and energy, $E = mc^2$, to give an expression for the momentum of a photon, $mc = h\nu/c$. Compton applied the laws of conservation of energy and conservation of momentum to the collision of a photon with a free electron and predicted the change in momentum or frequency of the photon for various angles of the scattered radiation relative to the incident radiation. Measurements of the scattering of x rays by the ''free'' electron of graphite verified the predictions, thus proving that photons have momenta.

De Broglie suggested that the particle-wave dualism phenomenon is not restricted to light, but that all particles must have an associated wave. For particles in general the wavelength of the associated waves is given by

$$\lambda = \frac{h}{mv} \text{ (de Broglie equation)} \tag{1.14}$$

where m is the mass of a particle traveling with a velocity v. The wave properties predicted for matter have been verified for free particles by experiments in which diffraction patterns have been observed for beams of electrons, neutrons, and atoms, respectively.

In a classical wave picture, the intensity of a light wave is proportional to the square of the amplitude of the wave. In a photon picture, the intensity of a light beam is proportional to the number of photons striking a given area. If both representations are to be valid, then the number of photons striking a given area in a given time period must be proportional to the square of the amplitude of the light wave striking that area. Since a given area may be subdivided into an infinite number of smaller areas, whereas a light beam would have a finite number of photons, the above statement may be modified to read that the *probability of finding a photon* in a given area is proportional to the square of the amplitude of the light wave falling on that area. Or, *for any particle,* the probability of finding a particle in a given volume is proportional to the square of the amplitude of the wave associated with the particle in the given volume. In terms of de Broglie's matter waves, the Bohr quantization condition specifies that the circumference of the orbit must be a whole multiple of the wavelength associated with the electron in the orbit—that is, $n\lambda = 2\pi r$—which would give rise to a standing wave. Replacing λ by h/mv gives Equation 1.6.

The Uncertainty Principle

We can see a reciprocal relationship between wave properties and properties of matter by rearranging the equations $E = h\nu$ and $\lambda = h/p$ (where p stands for momentum and replaces the mv of Equation 1.14) to give

$$h = ET = p\lambda \tag{1.15}$$

(where T is the period of vibration, $T = 1/\nu$).

Energy and momentum usually are associated with particle properties, whereas period and wavelength are associated with wave properties. When one of these *(E or p)* is large, the other *(T or λ)* will be small. Thus at long wavelengths, such as radiowaves, it is difficult to show particle behavior, whereas at very short wavelengths, such as γ rays, it is difficult to show wave behavior. For heavy particles, such as a ball, the wave treatment is not very useful.

An equation very similar in form to Equation 1.15 was proposed by Heisenberg and has become known as Heisenberg's uncertainty principle. This equation, $(\Delta P_x)(\Delta X) \geq h$, states that the product of the uncertainty of the momentum of particle with respect to a given coordinate and the uncertainty in position with respect to the same coordinate must be equal to or greater than Planck's constant. This uncertainty is caused not by experimental errors, but rather by the inherent indeterminacy in describing simultaneously both position and momentum. A corollary, from Bohr, states that a single experiment cannot show simultaneously particle and wave properties of radiation.

The Bohr-Sommerfeld model of the atom specified both the momentum and the position of the electron with respect to each coordinate. According to the uncertainty principle, it is not possible to have such precise knowledge of both momenta and position. On other grounds the quantum number k appeared to give an incorrect orbital angular momentum and had been replaced empirically by $k - 1$.

1.2.4 The Schrödinger Equation

In 1927, Schrödinger proposed an equation that specified no discrete orbits, but instead described the wave associated with the electron. This equation, given below, provided the basis for wave mechanics.[3]

---

[3]The equation given is known as the time-independent Schrödinger wave equation, and the solutions obtained are to be multiplied by a phase factor $e^{-i\omega t}$ (see below).

$$\frac{\partial^2 \psi}{\partial x^2} + \frac{\partial^2 \psi}{\partial y^2} + \frac{\partial^2 \psi}{\partial z^2} + \frac{8\pi^2 m}{h^2}(E - V)\psi = 0 \tag{1.16}$$

In this equation, ψ is the amplitude of the wave function associated with the electron, E is the total energy of the system, V is the potential energy of the system (equal to $-e^2/r$ for hydrogen), m is the mass of the electron, h is Planck's constant, and x, y, and z are the usual Cartesian coordinates. The frequency of the wave describing the electron is related to its energy by $E = h\nu$. Compared to Bohr's postulates given on p. 5, the last statement replaces **5** and the wave form of the equation replaces **4**, the other postulates being retained.

Since the probability of finding the electron in a given volume element is proportional to ψ^2, ψ itself must be a single valued function with respect to the spatial coordinates, must be a continuous function, and must become zero at infinity. These conditions are imposed on the wave equation as boundary conditions.

The transformation into polar coordinates of the wave equation for the hydrogen atom facilitates the separation of variables and solution of the equation. The position variables in polar coordinates are r, θ, and ϕ, where r is the radial distance of a point from the origin, θ is the inclination of the radial line to the z axis, and ϕ is the angle made with the x axis by the projection of the radial line in the xy plane. (See Figure 1.5.)

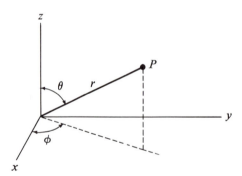

Figure 1.5   Variables of polar coordinates.

The solutions for ψ, called wave functions, may be expressed as the product of three functions, each of which depends on only one of the coordinates.

$$\psi(r, \theta, \phi) = R(r)\Theta(\theta)\Phi(\phi) \tag{1.17}$$

The boundary conditions require that certain constants that enter into the solution of the wave equation take on only integral values. These constants, called quantum numbers, are designated by n, l, and m_l. The principal quantum number, n, may take on the values 1, 2, 3, . . . , etc.; l may have the values of 0, 1, . . . , up to $n - 1$; m_l can have values ranging from $-l$ through 0 to $+l$. The wave functions, ψ, which are solutions of the Schrödinger equation, are commonly called orbitals. Orbitals for which $l = 0$, 1, 2, 3, and 4 are called, respectively, s, p, d, f, and g orbitals.

Radial Functions

The radial part of the wave function depends only on the n and l values and has an exponential term e^{-Zr/na_0} and a preexponential term involving a polynominal of the $n - 1$ degree. The preexponential consists of a normalization constant *(vide infra)*, a term $(2Zr/na_0)^l$, and a term known as an associated Laguerre polynominal, L, given in Table 1.1.

Table 1.1   Associated Laguerre polynomials[a,b]

Electron	n	l	$L_{n+l}^{2l+1}(x)$
$1s$	1	0	$-1!$
$2p$	2	1	$-3!$
$3d$	3	2	$-5!$
$4f$	4	3	$-7!$
$2s$	2	0	$2x - 4$
$3p$	3	1	$24x - 96$
$4d$	4	2	$720x - 5760$
$3s$	3	0	$-3x^2 + 18x - 18$
$4p$	4	1	$-60x^2 + 600x - 1200$
$4s$	4	0	$4x^3 - 48x^2 + 144x - 96$

[a]From H. E. White, *Introduction to Atomic Spectroscopy*
McGraw-Hill, New York, 1934, p. 67.
[b]$x = 2Zr/na_0$.

The exponential term has the effect of causing R to drop off more slowly with increasing values of the principal quantum number n: in other words, orbital size increases with increasing n. The $(2Zr/na_0)^l$ term causes the R value of all orbitals with $l > 0$ to go to zero as r goes to zero; that is, only s orbital electrons have any probability of being found in the vicinity of the atomic nucleus. (This has consequences in such areas as Mossbauer spectroscopy, in which only changes in electron density at the nucleus are detected.) The associated Laguerre polynomials produce sign changes in R as r varies for orbitals with $n - l \geqq 2$. This produces orthogonality in orbitals with equal l but different n: that is, $\int \psi_1 \psi_2 d\tau = 0$ or $\int_0^\infty R_1 R_2 dr = 0$.

Orthogonality guarantees the noninterference of different orbitals. Finally, the normalization constant is introduced to assure a unity probability of finding the electron in all of space. The normalization constant is the square root term in the formula for the radial functions given below (see also p. 134).

$$R_{n,l} = \sqrt{\frac{4(n - l - 1)! Z^3}{[(n + l)!]^3 n^4 a_0^3}} \cdot \left(\frac{2Zr}{na_0}\right)^l \cdot e^{-Zr/na_0} \cdot L_{n+l}^{2l+1}(x) \tag{1.18}$$

The probability of finding the electron in a volume element of fixed size centered about some given point at a distance r from the nucleus, is given by $R^2 d\tau$ (where $d\tau$ is the elemental volume unit). The probability that the electron will be found at a distance r from the nucleus, is given by $(4\pi r^2)R^2 dr$; this is the probability that the electron will be found in a spherical shell of thickness dr at a distance from the nucleus ranging from r to $r + dr$. Figure 1.6 gives

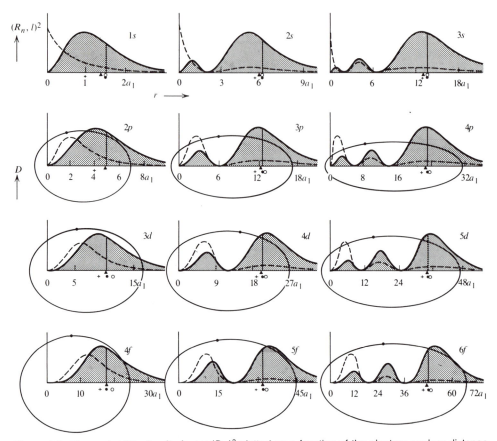

Figure 1.6   The probability-density factor $(R_{n,l})^2$ plotted as a function of the electron-nuclear distance r (r is given in units $a_1 = 0.53$ Å, the radius of the first Bohr circular orbit). The density distribution curves $D = 4\pi r^2 (R_{n,l})^2$, the shaded areas, are to be compared with the electron-nuclear distance of the classical electron orbits, where the orbital angular momentum is taken to be $\sqrt{l(l+1)} \cdot h/2\pi$. [From H. E. White, *Introduction to Atomic Spectra*, McGraw-Hill, New York, 1934, p. 68.]

curves showing R^2 and $(4\pi r^2)R^2$ versus r for various hydrogen atomic orbitals (note that the scale changes with each plot). The maximum in the shaded curve occurs at the distance where the probability of finding the electron is maximum. The vertical line is placed at the average distance of the electron from the nucleus, $\bar{r}$. This value is the same as that obtained from the Bohr-Sommerfeld orbits if k is replaced by $\sqrt{l(l+1)}$. These modified Bohr-Sommerfeld orbits also are shown in the figures; the *ns* orbits are depicted as vibrations along a line whose length just equals the diameter of the Bohr radii. Since this corresponds to elliptical motion with no angular momentum, the minor axis of the ellipse has no length.

Angular Functions

Figure 1.6 shows electron distribution curves for an atom in free space—that is, where there is no basis for assigning one or more Cartesian reference axes, and consequently, Θ^2 and Φ^2

are constant.[4] In general, we will be interested in atoms in molecules in which the symmetry is less than spherical and the electron distribution is dependent on the angular part of the wave function, as well as the radial part.

The angular functions may be generated from the equation

$$A_{(l)} = \frac{x^a y^b z^c}{r^l} \tag{1.19}$$

where l is the l quantum number and a, b, and c may take positive (or zero) integer values such that $a + b + c = l$. The variables x, y, and z are the Cartesian coordinates, and r is the distance from the origin (see Figure 1.5). In addition to the angular function of a given l value, all orbitals of lower l value of the same parity (even or odd l) will be generated. Linear combinations of the solutions may be made to achieve an orthogonal (linearly independent) set of solutions.

For $l = 0$, $A = \dfrac{x^0 y^0 z^0}{r^0} = 1$. There is no angular dependence of an s orbital, and the radial functions given in Figure 1.6 completely describe the s orbitals.

For $l = 1$, the possibilities for A are x/r, y/r, and z/r. To evaluate these, simply assign any value to x, y, and z and then calculate r from the Pythagorean relationship $r = \sqrt{x^2 + y^2 + z^2}$. Next, evaluate the desired A function and plot it by passing a line through the origin and the chosen x, y, z point and marking the value of A as a distance along the line from the origin. Any arbitrary choice of x, y, and z on a given line does not alter the value of A: that is, A is a function of θ and ϕ alone and, indeed, usually is expressed in trigonometric terms, as in Table 1.2. Note that in the equation for A the r value is always positive, whereas those of x, y, and z may be positive or negative. The three p orbitals having the angular functions x/r, y/r, and z/r are commonly denoted as p_x, p_y, and p_z.

Example   Make a plot of the angular part of the wave function for a p_x orbital.

Solution $A = x/r$. To make a plot in the xy-plane, we set $z = 0$ and choose values of x and y to evaluate A.

x	y	r	x/r	Comment
10	0	10	1	Note that the direction of the ray is the same and hence A is the same.
1	0	1	1	
4	3	5	0.8	
3	4	5	0.6	
1	1	$\sqrt{2}$	$\sqrt{2}/2 = 0.707$	
5	12	13	$5/13 = 0.384$	
-10	0	10	-1	
4	-3	5	0.8	
-4	-3	5	-0.8	
-4	3	5	-0.8	
	etc.			

[4]See the end of this section for the appropriate functions.

As might be guessed from the function x/r, the maximum occurs along the x axis, and increasing y with x constant simply reduces the value of the function. The same is true for increasing z, which results in the angular function for p_x being the surface of a set of tangent spheres—the value of the function in any direction being the distance from the origin to the surface. The sign of x/r is the same as the sign of x. The sign of the wave function is of particular importance in connection with the formation of molecular orbitals from atomic orbitals.

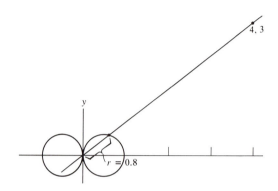

We may construct other orthogonal sets of angular p-orbital functions, but so long as we use real numbers as coefficients, the result will be simply to rotate the orbital set in space. One such set might be

$$p_{x+y+z} = (1/\sqrt{3})\,(p_x + p_y + p_z)$$

$$p_{x-y} = (1/\sqrt{2})\,(p_x - p_y) \tag{1.20}$$

$$p_{2z-x-y} = (1/\sqrt{6})\,(-p_x - p_y + 2p_z)$$

The p_{x+y+z} orbital would be directed along the $x = y = z$ axis with p_{x-y} perpendicular to it and directed along the $x = -y$ axis; the third p orbital would be perpendicular to these two.

If complex numbers are used, a new set of p orbitals may be found.

$$p_0 = p_z$$

$$p_+ = (1/\sqrt{2})\,(p_x + ip_y) \tag{1.21}$$

$$p_- = -(1/\sqrt{2})\,(p_x - ip_y)$$

where $i = \sqrt{-1}$. The subscripts denote the m_l values. Only in this set of orbitals, appropriate for linear molecules or atoms in a one-directional magnetic field, is m_l well defined. No simple drawing can represent angular functions containing complex numbers. When complex numbers occur in A, we replace A^2 by AA^*, where A^* is the complex conjugate of A—that is, whenever a term involving i appears in A, it is replaced in A^* by $-i$. Hence AA^* contains only real numbers. Sketches of p_0 and p_+ or p_- are shown in Figure 1.7.

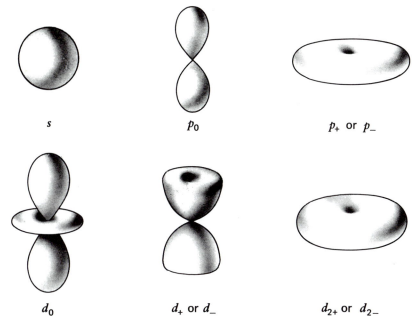

s 　　　　　 p_0 　　　　　 p_+ or p_-

d_0 　　　　　 d_+ or d_- 　　　　　 d_{2+} or d_{2-}

Figure 1.7   Angular dependence of $\psi\psi^*$ with a single designated Cartesian axis (z).

Table 1.2 Φ_{m_l} Functions for the hydrogen atom[a]

$$\Phi_0(\phi) = \frac{1}{\sqrt{2\pi}} \qquad \text{or} \qquad \Phi_0(\phi) = \frac{1}{\sqrt{2\pi}}$$

$$\Phi_1(\phi) = \frac{1}{\sqrt{2\pi}} e^{i\phi} \qquad \text{or} \qquad \Phi_{1\,\cos}(\phi) = \frac{1}{\sqrt{\pi}} \cos\phi$$

$$\Phi_{-1}(\phi) = \frac{1}{\sqrt{2\pi}} e^{-i\phi} \qquad \text{or} \qquad \Phi_{1\,\sin}(\phi) = \frac{1}{\sqrt{\pi}} \sin\phi$$

$$\Phi_2(\phi) = \frac{1}{\sqrt{2\pi}} e^{i2\phi} \qquad \text{or} \qquad \Phi_{2\,\cos}(\phi) = \frac{1}{\sqrt{\pi}} \cos 2\phi$$

$$\Phi_{-2}(\phi) = \frac{1}{\sqrt{2\pi}} e^{-i2\phi} \qquad \text{or} \qquad \Phi_{2\,\sin}(\phi) = \frac{1}{\sqrt{\pi}} \sin 2\phi$$

[a]From L. Pauling, *The Nature of the Chemical Bond*, 3rd ed. © 1960 by Cornell University. Used by permission of Cornell University Press.

Finally, we can construct a set of p orbitals using the hypercomplex numbers i, j, and k, where $i^2 = j^2 = k^2 = -1$ and $ij = -ji = k$, $jk = -kj = i$, and $ki = -ik = j$.

$$p_1 = (1/\sqrt{3})\,(ip_x + jp_y + kp_z)$$

$$p_2 = (1/\sqrt{2})\,(ip_x + jp_y) \tag{1.22}$$

$$p_3 = (1/\sqrt{6})\,(-ip_x - jp_y + 2kp_z)$$

AA^* for this set resembles the angular part of $\Psi\Psi^*$ obtained by Dirac's relativistic quantum mechanics. The p_1 orbital is spherically symmetric. If the coefficients $e^{i\alpha t}$, $e^{i\beta t}$ . . . are used to combine all the orbitals of a given type, where α, β . . . are random phases, a spherical orbital may be constructed for any value of l.

Turning to the angular functions for the d orbitals where $l = 2$, the possible solutions for A are x^2/r^2, y^2/r^2, z^2/r^2, xy/r^2, xz/r^2, and yz/r^2. Knowing that the s-orbital angular function ($l = 0$) is included in the above solutions we readily find it in the sum of the first three: $A_s = x^2/r^2 + y^2/r^2 + z^2/r^2 = 1$. The two other linear combinations we can construct from these three terms, which are orthogonal to the s orbital, are

$$d_{x^2 - y^2} = x^2/r^2 - y^2/r^2$$

$$d_{z^2} = -x^2/r^2 - y^2/r^2 + 2z^2/r^2 \tag{1.23}$$

These and other angular functions are shown in Figure 1.8.

Comparing x/r, xy/r^2, and xyz/r^3,

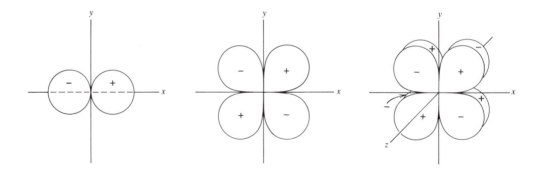

we note that nodal planes occur at $x = 0$ in the first, $x = 0$ and $y = 0$ in the second; and $x = 0$, $y = 0$, and $z = 0$ in the third. The sign of each lobe remains the same on the positive side of the "new coordinate" and inverts on the negative side of the new coordinate. In similar fashion, we may visualize other f orbitals from knowledge of the d orbitals: that is, $f_{z(x^2 - y^2)}$ and the similar $f_{y(x^2 - z^2)}$ and $f_{x(z^2 - y^2)}$. The "d_{z^2}" orbital is actually the $d_{2z^2 - x^2 - y^2}$; the $-x^2 - y^2$ terms give the torus of sign -1, which is maximized in the xy plane. The "f_{z^3}" is actually $(2z^3 - zx^2 - zy^2)/r^3$ and might be visualized as $z(2z^2 - x^2 - y^2)$: that is, a new nodal plane appears at $z = 0$, the part of the torus on the $+z$ side having a negative sign and the part on the $-z$ side having a positive sign.

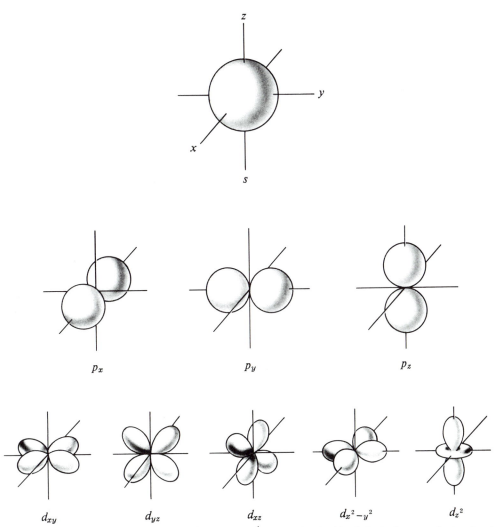

Figure 1.8 Total angular dependence of the wave function, $\psi(\theta, \phi)$ with all Cartesian coordinates fixed (*d*-functions after R. G. Pearson, *Chem. Eng. News* 1959, *37*, 72.)

There are several other linear combinations of the real *d*-orbital angular functions, including a rather interesting equivalent set of *d* orbitals that point along the slant edges of a pair of pentagonal bipyramids joined at the apex to give $\mathbf{D}_{5d}$ symmetry[5] (see Chapter 3 for symmetry notation).

[5]R. E. Powell, *J. Chem. Educ.* 1968, *45*, 45.

The following set of *d* orbitals of well-defined m_l values may be constructed from the "real" set.

$$d_{-2} = \sqrt{\frac{3}{8}} \frac{(x - iy)^2}{r^2}$$

$$d_{-1} = \sqrt{\frac{3}{2}} \frac{z(x - iy)}{r^2}$$

$$d_0 = \sqrt{\frac{1}{4}} \frac{3z^2 - r^2}{r^2} \qquad\qquad (1.24)$$

$$d_1 = -\sqrt{\frac{3}{2}} \frac{z(x + iy)}{r^2}$$

$$d_2 = \sqrt{\frac{3}{8}} \frac{(x + iy)^2}{r^2}$$

We must emphasize that the information needed to determine the angular functions of the orbitals is carried in the subscript of the orbital name ($d_{x^2 - y^2}$, etc.). In the few cases in which the subscript is an abbreviated expression (d_{z^2}, f_{z^3}, etc.), the orbital transforms under various symmetries as the polynomial given. Although several different sets of orbitals may exist for a particular *l* value, the appropriate set for any particular situation is determined by the symmetry of the environment of the atom. The *A* surfaces are often referred to as orbitals and, indeed, serve very satisfactorily for most arguments concerned with chemical bonding. Note, however, that the variation of ψ in space is the product of $\psi(R)$ and $\psi(\theta,\phi)$ at every point in space. A slice of the ψ function yields contours such as those shown in Figure 1.9.

Some consequences of the spatial orientation of the orbitals are discussed in Chapter 2, under the heading "Shapes of Molecules," and also in Chapter 7, under "Ligand Field Theory."

Interpretations of the Schrödinger Equation

Further interpretations of the Schrödinger equation and comparisons with the Bohr-Sommerfeld treatment may be in order. As explained earlier, if we performed an experiment to detect the electron as a particle, the probability of finding the electron in a given volume element would be given by $\psi\psi^* d\tau$. This leads to the polemical question, what is the nature of the electron in the unperturbed atom? One commonly held view is that the electron exists as a particle, which spends an amount of time in each volume unit proportional to $\psi\psi^*$ for that volume unit. Because of the very rapid motion of the electron, over a period of time as long as, say, a microsecond, the electron appears to be smeared out and can be said to form a cloud whose charge density in any given volume element would be $\psi\psi^* d\tau$. The path traveled by the electron would be unpredictable, according to the uncertainty principle. This view ignores not only the prediction of classical electrodynamics that an electron traveling around the nucleus would radiate energy and collapse, but also the question of how the electron passes through the surfaces of zero electron density corresponding to $\psi = 0$. Another view is that the electron in the atom *is* a wave. The electron distribution would be static and there would be no emission

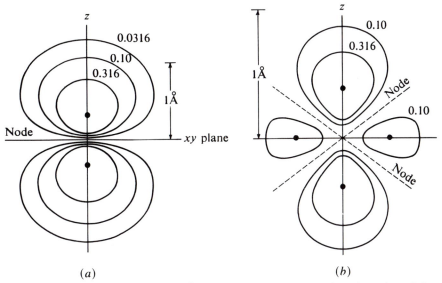

Figure 1.9   (a) Contours of constant ψ^2 at 0.0316, 0.10, and 0.316 of maximum for a C 2_{p_z} orbital. The *xy* plane is a nodal surface. (b) Contours of constant ψ^2 at 0.10 and 0.316 of maximum for a Ti(III) $3d_{z^2}$. (From E. A. Ogryzlo and G. Porter, *J. Chem. Educ.* 1963, *40*, 258.)

of radiation. Just as the total energy of a vibrating string may be withdrawn by a probe at any point except a node, so an interaction with the electron wave at any point other than a node would affect the entire electron wave.

We may say that the electron distributions with nodal surfaces resulting from the Schrödinger equation stem directly from *assuming* the wave character of the electron, just as elliptical orbits were assumed in the Bohr-Sommerfeld treatment. Since the wave equation assumes less about the position of the electron than does the assumption of elliptical orbits (just as the elliptical orbits are less restrictive than circular orbits), it has a greater chance of being universally applicable. For the one-electron case, many of the conclusions from the Bohr-Sommerfeld treatment quantitatively or qualitatively agree with the results from the Schrödinger equation, especially regarding energies, average distance of the electron from the nucleus, and penetration.

1.2.5   Electronic Configurations and the Periodic Table

According to the Schrödinger equation an orbital is determined uniquely by the three quantum numbers n, l, and m_l. Since l can have the values of zero up to $n - 1$ and m_l can have the values of $-l$ through zero to $+l$, for a given value of n there will be n^2 orbitals. Table 1.3 shows the quantum numbers, orbital designations, and shell descriptions corresponding to the orbitals through $n = 3$.

Table 1.3 Orbital and shell designations for different quantum states

Value of n	1	2	2	2	2	3	3	3	3	3	3	3	3	3
Value of l	0	0	1	1	1	0	1	1	1	2	2	2	2	2
Value of m_l	0	0	−1	0	1	0	−1	0	1	−2	−1	0	1	2
Orbital designation	1s	2s	2p	2p	2p	3s	3p	3p	3p	3d	3d	3d	3d	3d
Shell designation	K	L				M								

The shell designation stands for the orbitals of a given principal quantum number n; K, L, M, N . . . correspond to values of n of 1, 2, 3, 4 . . . , respectively. The letters s, p, d, f, g of the orbital designation indicate l values of 0, 1, 2, 3, 4 . . . , respectively, and the number preceeding the letter gives the principal quantum number. So for a $3p$ electron, $l = 1$ and $n = 3$.

Polyelectron atoms are formed by adding electrons successively to hydrogenlike orbitals. The number of electrons that may occupy each orbital is limited to two, and these must have the magnetic moment associated with their spins opposed.[6] The electrons enter the lowest available empty orbital, the order of increasing energy is that of increasing $(n + l)$ or, where $(n + l)$ are the same, of increasing n.

$$1s < 2s < 2p < 3s < 3p < 4s \gtrsim 3d < 4p < 5s \gtrsim 4d < 5p < 6s \sim 5d \sim 4f < 6p$$

Where approximate inequality is indicated, the order given is that for the lighter elements. At high atomic numbers the order of increasing energy is that of increasing n, and within a given shell, with increasing l. At very high atomic numbers, we can expect spin-orbit coupling to split the p-orbital energies to the extent that the $p_{1/2}$ orbital will be stabilized greatly compared to the $p_{3/2}$ orbitals (see below). Where more than one orbital has the same n and l value (such as the p orbitals), each individual orbital will be filled with a single electron before any of the equivalent orbitals are occupied by pairs of electrons. Table 1.4 gives the occupancy of the orbitals, usually called the *electron configuration*, for the known elements. The superscripts indicate the number of electrons in a given set of orbitals: for nitrogen, for example, the $1s^2 2s^2 2p^3$ indicates that there are two electrons in the $1s$ orbital, two in the $2s$ orbital, and one electron in each of the three p orbitals. Since more than one energy state usually will be associated with an electron configuration, more detailed microstates are sometimes shown in which m_s values are indicated by the direction of an arrow. Whereas more than one microstate may be associated with a given energy level, the ground-state energy level will always be associated with the microstate with the maximum number of electrons of similar spin. Thus for nitrogen, we have

[6]This is known as the Pauli exclusion principle. The nonelectrostatic repulsive force between two electrons of the same spin known as the Pauli force, comes into play very sharply at short range. The nonpenetrability of matter and the shapes of molecules are determined largely by Pauli forces. (See p. 65 and p. 189.)

The periodic chemical and physical behavior of the elements results from the periodic recurrence of the same outermost electron configuration. Each period of the table begins with an element having an s^1 configuration (hydrogen or the alkali metals) and ends with an element having a noble gas configuration ($1s^2$ or s^2p^6). Table 1.4 illustrates the makeup of the periodic table in terms of the orbitals being filled. Taking into account the number of elements involved, the periods are sometimes designated as the very short period (first), the short periods (second and third), the long periods (fourth and fifth), and the very long periods (sixth and seventh). Elements with an incompletely filled set of s and p orbitals (their other sets of orbitals being either completely filled or completely empty) are called representative elements. Within a given group of the periodic table the representative elements have the same outermost electron configuration, with the principal quantum numbers corresponding to the period number. The ionic states and covalent states formed by the representative elements are related directly to the electron configuration of the elements and are reasonably predictable (p. 43).

Elements having an incompletely filled set of d orbitals are called transition elements. This definition is not exact, since within the nickel group elements, the definition would exclude Pd, which has a d^{10} configuration. Despite this unique configuration, the chemistry of palladium belongs with that of nickel and platinum and Pd is properly termed a transition metal. Copper, silver, and gold, although unambiguously classified as representative elements by the definition given here, are often considered to be transition metals, because of the electronic configuration of their 2+ or 3+ ions. Although the ground-state configurations within a transition element group are not always the same (see V, Nb, Ta; Cr, Mo, W; Fe, Ru, Os; Co, Rh, Ir; Ni, Pd, Pt), the differences in energy between the ground-state configuration and the "group configuration" are small.

Filling the f orbitals gives rise to the inner transition series of 14 elements. There are two known series of inner transition elements, the lanthanides and the actinides. The elements from atomic number 106 up are not currently known, but since nuclear theory indicates a half-life on the order of years for some of them, both calculations and predictions based on periodic behavior are of current interest. The differences in energy levels between the $p_{1/2}$ and $p_{3/2}$ are sufficiently large here, according to calculations, that $8p_{3/2}$ will not be filled until the $9p_{1/2}$ has been filled (Table 1.4). The superactinides are thus predicted to end the eighth period without achieving the noble gas configuration.

1.3 ATOMIC SPECTRA

The spectrum of atomic hydrogen can be accounted for quite satisfactorily on the basis of the Bohr model. Each principal quantum number is associated with a single energy level.[7] We can observe in the emission spectrum various series of lines that are identified with their discoverers. They differ in the final state of the hydrogen atom as follows: Lyman series, $n = 1$; Balmer series, $n = 2$; Paschen series $n = 3$; Brackett series $n = 4$; Pfund series, $n = 5$.

The spectrum of atomic helium is much more complex than that of hydrogen. Initially, some observers even hypothesized that two types of helium existed, because of the discovery

---

[7]If the emission source is cooled sufficiently to reduce Doppler effects, at very high resolution a splitting of 0.36 cm^{-1} can be found for the $2p$ levels (*vide infra*).

Table 1.4 Electron configuration of the neutral atoms[a]

	1s
Z	H
	1
1s	1

	2s	2p
Z	Li **3**	Be **4**
2s	1	2
2p	—	—

Z	Na **11**	Mg **12**
3s	1	2
3p	—	—

Z	K **19**	Ca **20**	Sc **21**
4s	1	2	2
3d	—	—	1
4p	—	—	—

Z	Rb **37**	Sr **38**	Y **39**
5s	1	2	2
4d	—	—	1
5p	—	—	—

	Cs **55**	Ba **56**	La **57**		Ce **58**	Pr **59**
6s	1	2	2		2	2
5d	—	—	1		1	—
4f	—	—	—		1	3
6p	—	—	—		—	—

	Fr **87**	Ra **88**	Ac **89**		Th **90**	Pa **91**
7s	1	2	2		2	2
6d	—	—	1		2	1
5f	—	—	—		—	2
7p	—	—	—		—	—

Z	119	120	121	122	123	124	125	126	127	128	129	130	131	132	133	134	135	136	137	138	139	140	141	142	143
8s	1	2	2	2	2	2	2	2	2	2	2	2	2	2	2	2	2	2	2	2	2	2	2	2	2
$8p_{1/2}$	—	—	1	1	1	1	1	2	2	2	2	2	2	2	2	2	2	2	2	2	2	2	2	2	2
7d	—	—	—	1	1	—	—	1	2	2	2	2	2	2	3	4	4	3	1	1	2	1	2	2	2
6f	—	—	—	—	1	3	3	2	2	2	2	2	2	2	2	2	2	2	2	2	2	2	2	2	2
5g	—	—	—	—	—	—	—	1	2	3	4	5	6	7	8	8	9	10	11	12	13	14	15	16	17
9s	—	—	—	—	—	—	—	—	—	—	—	—	—	—	—	—	—	—	—	—	—	—	—	—	—

Z	165	166
9s	1	2
$9p_{1/2}$	—	—
$8p_{3/2}$	—	—

[a]After B. Fricke and J. McMinn, *Naturwissenschaften* 1976, *63*, 162.

Periodic table (electron configuration chart).

Period 1

He
2
2

Period 2 (p-block)

B	C	N	O	F	Ne
5	6	7	8	9	10
2	2	2	2	2	2
1	2	3	4	5	6

Period 3 (p-block)

Al	Si	P	S	Cl	Ar
13	14	15	16	17	18
2	2	2	2	2	2
1	2	3	4	5	6

Period 4

Ti	V	Cr	Mn	Fe	Co	Ni	Cu	Zn	Ga	Ge	As	Se	Br	Kr
22	23	24	25	26	27	28	29	30	31	32	33	34	35	36
2	2	1	2	2	2	2	1	2	2	2	2	2	2	2
2	3	5	5	6	7	8	10	10	10	10	10	10	10	10
—	—	—	—	—	—	—	—	—	1	2	3	4	5	6

Period 5

Zr	Nb	Mo	Tc	Ru	Rh	Pd	Ag	Cd	In	Sn	Sb	Te	I	Xe
40	41	42	43	44	45	46	47	48	49	50	51	52	53	54
2	1	1	2	1	1	—	1	2	2	2	2	2	2	2
2	4	5	5	7	8	10	10	10	10	10	10	10	10	10
—	—	—	—	—	—	—	—	—	1	2	3	4	5	6

Period 6

Nd	Pm	Sm	Eu	Gd	Tb	Dy	Ho	Er	Tm	Yb	Lu	Hf	Ta	W	Re	Os	Ir	Pt	Au	Hg	Tl	Pb	Bi	Po	At	Rn
60	61	62	63	64	65	66	67	68	69	70	71	72	73	74	75	76	77	78	79	80	81	82	83	84	85	86
2	2	2	2	2	2	2	2	2	2	2	2	2	2	2	2	2	2	1	1	2	2	2	2	2	2	2
—	—	—	—	1	—	—	—	—	—	—	1	2	3	4	5	6	7	9	10	10	10	10	10	10	10	10
4	5	6	7	7	9	10	11	12	13	14	14	14	14	14	14	14	14	14	14	14	14	14	14	14	14	14
—	—	—	—	—	—	—	—	—	—	—	—	—	—	—	—	—	—	—	—	—	1	2	3	4	5	6

Period 7 — (elements 104–118: NOT CURRENTLY KNOWN)

U	Np	Pu	Am	Cm	Bk	Cf	Es	Fm	Md	No	Lr	104	105	106	107	108	109	110	111	112	113	114	115	116	117	118
92	93	94	95	96	97	98	99	100	101	102	103	104	105	106	107	108	109	110	111	112	113	114	115	116	117	118
2	2	2	2	2	2	2	2	2	2	2	2	2	2	2	2	2	2	2	2	2	2	2	2	2	2	2
1	1	—	—	1	—	—	—	—	—	—	—	3	4	5	6	7	8	9	10	10	10	10	10	10	10	10
3	4	6	7	7	9	10	11	12	13	14	14	14	14	14	14	14	14	14	14	14	14	14	14	14	14	14
—	—	—	—	—	—	—	—	—	—	—	1	—	—	—	—	—	—	—	—	—	1	2	3	4	5	6

Hypothetical superheavy block (144–164)

144	145	146	147	148	149	150	151	152	153	154	155	156	157	158	159	160	161	162	163	164
2	2	2	2	2	2	2	2	2	2	2	2	2	2	2	2	2	2	2	2	2
2	2	2	2	2	2	2	2	2	2	2	2	2	2	2	2	2	2	2	2	2
3	3	3	3	3	3	3	3	3	3	3	3	3	3	3	4	5	6	8	9	10
1	3	4	5	6	6	6	8	9	11	12	13	14	14	14	14	14	14	14	14	14
18	18	18	18	18	18	18	18	18	18	18	18	18	18	18	18	18	18	18	18	18
—	—	—	—	—	—	—	—	—	—	—	—	—	—	—	1	1	1	—	—	—

Hypothetical block (167–172)

167	168	169	170	171	172
2	2	2	2	2	2
2	2	2	2	2	2
1	2	2	2	2	2
—	—	1	2	3	4

of two sets of spectral lines with no crossover between the terms of the two sets. The current interpretation is that these sets are caused by excitation which produces states in which electrons have either parallel spins ($\vec{s}_1 + \vec{s}_2 = 1 = S$) or antiparallel (paired) spins ($\vec{s}_1 + \vec{s}_2 = 0$). The selection rule that S remain unchanged during a transition (p. 283) accounts for the failure to observe crossover lines between the two sets of terms. Helium has a ground-state electron configuration of $1s^2$. Normally, the excited states are produced by the promotion of a single electron. If the remaining $1s$ electron were completely effective in screening the promoted electron from the nuclear charge, then the excited electron would see a nuclear charge of $+1$ and the terms would be the same as those for hydrogen. Although the $2p$, $3d$, $4f$, etc. levels are close to those of hydrogen, the $2s$ definitely is not. Interestingly, the lowest-lying excited state is a metastable one, a triplet state arising from the $1s^1 2s^1$ configuration with parallel spins.[8] The triplet state cannot return to the singlet ground state by emitting radiation, as this is a forbidden transition; instead, it returns by radiationless energy exchange during collisions. Accordingly, this state has a longer lifetime. These so-called metastable states of molecules and atoms play an important role in photochemistry.

Even within the set of singlet (or triplet) terms, the only combinations of transitions observed are those in which the l quantum number of the excited electron changes by one: that is, $1s^2 \leftarrow 1s^1 2p^1$, $1s^1 2p^1 \leftarrow 1s^1 3d^1$. . . .

The atomic spectra of the neutral alkaline earths resemble the spectrum of helium in consisting of singlets and triplets.

The atomic spectra of the neutral alkali metals can be accounted for by a set of spectroscopic terms arising from the occupancy of various atomic orbitals by the single valence electron available. Penetration of the core electrons causes differentiation of the s, p, d, etc. levels of a given principal quantum number—the lower the l value, the lower-lying the level. The selection rule of transitions occurring only for $\Delta l = \pm 1$ (p. 284) accounts for the observed series of lines. The series were named, in part, according to their spectral features: the series resulting from $2^2P \leftarrow n^2S$ is termed the *sharp* series; that from $2^2S \leftarrow 3^2P$, the *principal* series; from $2^2P \leftarrow n^2D$, the *diffuse* series; from $3^2D \leftarrow n^2F$, the *fundamental* series. The orbital labels s, p, d and f derive from the spectroscopic series and from the association of energy levels with the spectroscopic terms. Providing the prototypes for all other atomic spectra are the series observed in alkali metal spectra: that is, the series may be classified as sharp series, principal series, etc. Other spectra may show greater numbers of such series.

1.4 VECTOR MODEL OF THE ATOM AND SPECTROSCOPIC TERMS

The Bohr-Sommerfeld model for hydrogenlike atoms predicted that because of relativity effects, the energy state of the atom would be very slightly dependent on the angular momentum of the electron: that is, the $2s^1$ and $2p^1$ configurations were predicted to be slightly different.[9]

---

[8]We shall see later that it is a slight oversimplification to speak of the triplet state arising from a single configuration with designated spin.

[9]Present data indicate that for hydrogenlike atoms, terms of the same j value are identical in energy except for the $^2S_{1/2}$ ($2s^1$) term, which lies 0.0354 cm^{-1} *above* the $^2P_{1/2}$ term. C. E. Moore, *Atomic Energy Levels*, Vol. I, NSRDS-NBS 35, 1971.

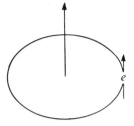
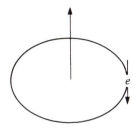

Figure 1.10 Possible alignments of orbit and spin angular momenta for a one electron atom. An "adjusted" angular momentum of $(k - 1) (h/2\pi)$ must be used.

Searching for these differences in hydrogenlike atoms by high-resolution spectroscopy revealed more energy levels than expected. These data led to the conclusion that the $2p^1$ configuration (and higher states beyond the s) could have two energy states, which differ in the interaction of the orbit angular momentum and the spin angular momentum of the electron (see Figure 1.10).

Although the wave equation abandons the picture of specific orbits for the electron, the property of angular momentum for an electron in an orbital is retained; angular momentum is associated with rotation of the electron cloud. The new angular momentum quantum number l is more satisfactory than k. The interaction of the quantized orbital angular momentum with the quantized spin angular momentum produces a resultant total angular momentum, which is also quantized. For a $2p$ electron, therefore, the vector sum of the orbital angular momentum, $l(h/2\pi)$, and the spin angular momentum, $s(h/2\pi)$, may be either $(l + \frac{1}{2})(h/2\pi)$ or $(l - \frac{1}{2})(h/2\pi)$. The resultant total angular momentum is $j(h/2\pi)$; the new quantum j sometimes is referred to as the "inner quantum number." For the hydrogen atom in the $2p^1$ configuration, the two possible j states have a difference in energy of 0.3651 cm^{-1}. (For hydrogenlike atoms the difference increases with Z^4.) These two states would be designated in spectroscopic notation as $2^2P_{1/2}$ and $2^2P_{3/2}$, the superprefix 2 indicating the term is one of a doublet, the subscript $\frac{1}{2}$ or $\frac{3}{2}$ indicating the j quantum number, the capital P indicating a term (or energy state) rather than an electron configuration, and the preceding 2 indicating the principal quantum number. The doublet terms of the hydrogenlike atoms (and the differences in s, p, and d orbitals of the same quantum number) give rise to fine structure in the spectrum. A further interaction of the total angular momentum of the electron with the spin angular momentum *of the nucleus* gives rise to a further, very slight splitting of the energy levels, which results in a hyperfine structure of the spectrum.

In a weak unidirectional magnetic field the two P terms are each split into $2j + 1$ terms differing in energy by an amount proportional to the applied strength. These states are characterized by a quantum number m_j which may have a value of $j, j - 1, \ldots -j$. The component of total angular momentum in the direction of the field is $m_j(h/2\pi)$. With a magnetic field present, the $^2P_{1/2}$ term thus yields two terms and the $^2P_{3/2}$ term yields four terms. The splitting into six terms for the ^2P state in a magnetic field is related to the fact that there are six ways of placing an electron in p orbitals that differ in the values of m_l and/or m_s. The number of different ways e electrons can be placed in a given set of n orbital sites in accordance with the Pauli exclusion principle is given by

$$\text{Number of microstates} = \frac{n!}{e!h!} \tag{1.25}$$

where h is the number of "holes" and is equal to $(n - e)$. Thus there are 252 different microstates for a d^5 state $\left(10 \times \dfrac{9}{2} \times \dfrac{8}{3} \times \dfrac{7}{4} \times \dfrac{6}{5} \right)$. This is equivalent to saying there are 10 different ways of placing the first electron in the set of orbitals, nine different ways of placing the second electron in the set of orbitals, eight for the third, and so on. Since these choices are not independent, however, one must divide by the number of electrons added at each stage: that is, $10 \times \dfrac{9}{2} \times \dfrac{8}{3} \cdots$. A few of these would be as follows:

m_l	2	1	0	-1	-2
	↑	↑	↑	↑	↑
	↓	↓	↓	↓	↓
etc.	↑	↓	↑	↑	↑

In a strong magnetic field each of these would have a different energy. If there are several types of partially occupied orbitals, the total number of electron configurations will be the product of those possible for each set of orbitals: for $d^5 s^1$, for example, the total would be 252×2, or 504.

In a polyelectron atom the individual l values combine vectorially to give a resultant L, and the individual s values combine to give a resultant S. Filled shells or subshells have a resultant $L = 0$ and $S = 0$. In general, L may have integer values ranging from the smallest difference of l values to the sum of the l values, whereas S may take on the values, differing by whole numbers from $\dfrac{1}{2}$ to $\dfrac{n}{2}$ (for odd n), or from 0 to $\dfrac{n}{2}$ (for even n), where n is the number of unpaired electrons. The M_L and M_S values are the sum of the respective m_l or m_s values for the system. For a given L value, possible M_L values are $-L, (-L + 1), \ldots 0, \ldots L - 1, L$; there are $2L + 1$ possible M_L states. Similarly, for a given S value there are $2S + 1$ possible M_S states. The L and S values, which represent the quantum-mechanically-allowed orbital angular momentum states and the spin angular momentum states, respectively, combine vectorially to produce J values, representing the total angular momentum states. J values may run from $(L - S)$ to $(L + S)$ in integer steps. Thus the number of J states for given L and S values is $2S + 1$, if $L \geq S$.

The different energy states of an atom or ion usually are given spectroscopic term designations of the form

$$n^a T_j$$

where n may be the principal quantum number, or simply a running number, or even the electron configuration (as, for example, $3d^4 \, {}^5D_2$, in which $n = 3d^4$); a is a number equal to $2S + 1$ and is called the multiplicity of the term; T is a capital letter S, P, D, F, G, H, I, K, L, M, . . . corresponding to the value of L; and j is the numerical J value. Note that J is omitted as a term symbol. Orbitals of odd l value are said to have odd parity, or to be ungerade. Determining the parity of the spectroscopic terms is the occupancy of the orbitals of odd parity (p, f, h, etc.). An even number of p and f electrons leads to an even parity of the term, whereas an odd number of p or f electrons leads to an odd parity. Odd parity is (sometimes) noted by a superscript $°$ following the term symbol, as in $P°$.

1.4.1 Derivation of Spectroscopic Terms from Electron Configurations

Spectroscopic terms may be derived for nonequivalent electrons (electrons differing in values of the quantum numbers n or l) by obtaining all possible values of L as the vector sums of the individual l values and all possible values of S as the vector sums of the individual s values. The terms for a $2p^1 3p^1$ electron configuration would have L values of 0, 1, and 2 and S values of 0 and 1, giving ^1S, ^1P, ^1D and ^3S, ^3P, and ^3D terms.

Fewer terms arise from equivalent electrons (electrons having the same values of n and l), because of the Pauli exclusion principle. So for a $2p^2$ configuration, a value of $L = 2$ results only in cases in which l_1 and l_2 are oriented similarly and hence m_{l_1} and m_{l_2} are identical. For $L = 2$, accordingly, the values of m_{s_1} and m_{s_2} must differ and S must equal 0. Hence only the singlet term is permitted for $L = 2$: that is, ^1D. The Pauli exclusion principle may be applied to help determine spectral terms for equivalent electrons by noting that the M_L and M_S values obtained from the L and S values must also correspond to those obtained from the sum of the individual m_l and m_s values, respectively. Obtaining all M_L and M_S values requires writing all possible individual microstates and obtaining M_L and M_S for each microstate. Thus for the $2p^2$ configuration we find the following Pauli-allowed microstates.

	$m_s = +1/2$			$m_s = -1/2$			$M_L = \Sigma m_l$	$M_S = \Sigma m_s$
$m_l =$	$+1$	0	-1	$+1$	0	-1		
1.	↑	↑					1	1
2.	↑				↑		0	1
3.		↑	↑				-1	1
4.				↓	↓		1	-1
5.				↓		↓	0	-1
6.					↓	↓	-1	-1
7.	↑			↓			2	0
8.	↑				↓		1	0
9.	↑					↓	0	0
10.		↑		↓			1	0
11.		↑			↓		0	0
12.		↑				↓	-1	0
13.			↑	↓			0	0
14.			↑		↓		-1	0
15.			↑			↓	-2	0

These M_L and M_S data are tabulated in the array shown in Figure 1.11.

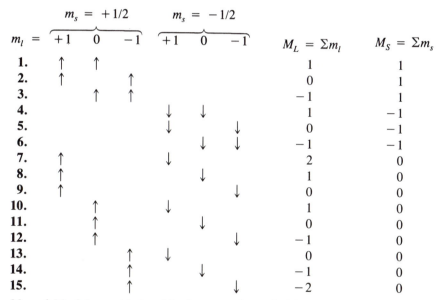

M_L	$M_S = -1$	$M_S = 0$	$M_S = +1$
2		1	
1	1	2	1
0	1	3	1
-1	1	2	1
-2		1	

$M_S \rightarrow$

Figure 1.11 An array for a p^2 configuration showing the number of individual microstates of a given M_L and M_S.

By inspecting the array of M_L and M_S values in Figure 1.11, we can see that it consists of the sum of the following arrays, which may be identified with their spectroscopic states through their maximum M_L and M_S values.

$$
\begin{array}{rr|r}
 & 2 & 1 \\
\uparrow & 1 & 1 \\
M_L & 0 & 1 \\
 & -1 & 1 \\
 & -2 & 1 \\
\hline
 & & 0 \\
 & & M_S
\end{array}
\quad
\begin{array}{l}
L = 2 \\
S = 0 \\
\text{Term: } {}^1D
\end{array}
\qquad\qquad
\begin{array}{rr|rrr}
\uparrow & 1 & 1 & 1 & 1 \\
M_L & 0 & 1 & 1 & 1 \\
 & -1 & 1 & 1 & 1 \\
\hline
 & & -1 & 0 & +1 \\
 & & & M_S\!\rightarrow
\end{array}
\quad
\begin{array}{l}
L = 1 \\
S = 1 \\
\text{Term: } {}^3P
\end{array}
$$

$$
\begin{array}{rr|r}
M_L & 0 & 1 \\
\hline
 & & 0 \\
 & & M_S
\end{array}
\quad
\begin{array}{l}
L = 0 \\
S = 0 \\
\text{Term: } {}^1S
\end{array}
$$

Thus a $2p^2$ configuration would give rise to the terms ${}^1D$, ${}^3P$, and ${}^1S$.[10]

The counting of microstates can be simplified by a procedure known as spin factoring,[11] which consists of obtaining "partial terms" for individual spin sets (all $m_s = +1/2$ or $-1/2$) and multiplying the partial terms to obtain entire columns of the array. Thus, when taken together microstates **1, 2,** and **3** have M_L values corresponding to a P partial term. An empty set obviously contributes nothing to the orbital angular momentum and corresponds to a partial S term. One electron of spin α (or β) in a set of p orbitals also yields a P partial term. Table 1.5 provides a complete listing of partial terms for various electron occupancies of the orbitals through g.

Table 1.5 Partial terms arising from the occupancy of a single spin set (either α or β)

Orbital Occupancy (Electrons or Holes)	0	1	2	3	4	5	6	7
Orbital Set								
s	S	S						
p	S	P	P	S				
d	S	D	PF	PF	D	S		
f	S	F	PFH	$SDFGI$	$SDFGI$	PFH	F	S
g	S	G	$PFHK$	$PF[2]G^a$ $HIKM$	$SD[2]FG[2]^a$ $HI[2]KLN$			

^{a}[2] means the preceding partial term occurs twice.

[10]For other examples of this approach, see K. E. Hyde, *J. Chem. Educ.* 1975, *52*, 87.

[11]D. H. McDaniel, *J. Chem. Educ.* 1977, *54*, 147.

The term symbol for any empty, filled, or half-filled subshell is S. A singly occupied subshell gives the corresponding term symbol, S for s^1, P for p^1, D for d^1, etc. One vacancy ("hole") in a subshell gives the same term symbol as for one electron—the *hole formalism*, or P for p^5, D for d^9, etc. The hole formalism applies to the partial terms (for one spin set), so p_α^2 (one "hole" in the half-filled set) gives P, d_α^4 gives D, and f_α^6 gives F. The derivations of partial terms for d^2, f^2, f^3, etc, are demonstrated by writing all of the microstates, as for Figure 1.11, in the reference by McDaniel. Note that in Table 1.5, the partial terms to the right of the solid line can be obtained from those to the left by the hole formalism. Thus the partial terms are the same for d_α^2 and d_α^3, f_α^3 and f_α^4, f_α^2 and f_α^5, g_α^4 and g_α^5, etc. Similarly, for complete configurations the hole formalism leads to the same terms for d^2 and d^8, d^3 and d^7, d^4 and d^6, f^2 and f^{12}, etc.

We will reexamine the p^2 case. The p^2 configuration gives rise to the spin configurations $p_\alpha^1 p_\beta^1$, $p_\alpha^2 p_\beta^0$, and $p_\alpha^0 p_\beta^2$, where α refers to the "spin up" and β to "spin down." Each spin configuration may be identified with one or more partial terms (as in Table 1.5), and the *product* of the partial terms gives the L values of a single column in an array of the type given in Figure 1.11. The product of partial terms with L values L_1 and L_2 includes all L values from $(L_1 + L_2)$, $(L_1 + L_2 - 1)$, . . . through $(L_1 - L_2)$. Thus for p^2 we have

$$p_\alpha^2 p_\beta^0 \qquad P \times S \rightarrow \quad P \ (M_L = +1, 0, -1) \quad M_S = +1 \text{ (since both}$$
$$\text{electrons}$$
$$p_\alpha^0 p_\beta^2 \qquad S \times P \rightarrow \quad P \ (M_L = +1, 0, -1) \quad M_S = -1 \text{ are of the}$$
$$\text{same spin)}$$

$$p_\alpha^1 p_\beta^1 \qquad P \times P \rightarrow \begin{cases} S \ (M_L = 0) \\ P \ (M_L = +1, 0, -1) \\ D \ (M_L = 2, 1, 0, -1, -2) \end{cases} \quad M_S = 0$$

We need not construct the array, since we already have the information we would extract from it. The P term is a triplet, since the maximum M_S is 1. The $M_S = 0$ of this ^3P are contained in the p^1p^1 product, so the new terms arising there are ^1S and ^1D. Note that for all terms of high spin-multiplicity, the lower M_S microstates will be replicated, along with terms of lower spin-multiplicity. The so-called Russell-Saunders terms derived in this fashion are the more important ones for light atoms. Here, the individual l values couple to give a total L value, and the individual s values of each electron couple to give a total S value. This coupling scheme is also called L and S coupling, since the resultant L and S couple to give a resultant J. Important for the heavier elements is another coupling scheme: the coupling of individual s and l values to give an individual j value. These j values for each electron then couple to give the J value (the quantum number for the total angular momentum of the atom). This coupling scheme is known as jj coupling. Most transition metals have terms that are intermediate rather than pure L and S or jj.

Example Determine the term symbols for the d^3 configuration using spin factoring.

Solution The possible configurations are $d_\alpha^3 d_\beta^0$, $d_\alpha^0 d_\beta^3$, $d_\alpha^2 d_\beta^1$, and $d_\alpha^1 d_\beta^2$. Taking the partial terms for each spin set from Table 1.5, the products are

$d_\alpha^3 d_\beta^0$
$d_\alpha^0 d_\beta^3$

$(P + F)(S) = {}^4P + {}^4F$ These are quartet terms; since there are three unpaired electrons, $S = 3/2$. The M_S values for $S = 3/2$ are $\pm 3/2, \pm 1/2$, so P and F terms with $M_S = \pm 1/2$ will appear from other configurations as ${}^2P$ and ${}^2F$, and we must subtract these terms below.

$d_\alpha^2 d_\beta^1$
$d_\alpha^1 d_\beta^2$

$(P + F)(D)$ $(P)(D) = {}^2F + {}^2D + {}^2P$
$(F)(D) = {}^2H + {}^2G + {}^2F + {}^2D + {}^2P$

Collecting the doublet terms and eliminating one ${}^2P$ and one ${}^2F$ gives ${}^2H \; {}^2G \; {}^2F \; 2{}^2D \; {}^2P$.

The number of microstates is $10 \times \dfrac{9}{2} \times \dfrac{8}{3} = 120$. The degeneracy of a spectral term is the product of the spin-multiplicity and the orbital degeneracy $(2L + 1)$ of the term, or $4 \times 7 = 28$ for ${}^4F$.

All terms $= {}^4F \; {}^4P \; {}^2H \; {}^2G \; {}^2F \; {}^2D[2] \; {}^2P$

Total degeneracy $= 28 + 12 + 22 + 18 + 14 + 10 \times 2 + 6 = 120$ microstates

1.4.2 Energy Levels in Polyelectron Atoms

Penetration Effects

Since filled shells have $L = 0$ and $S = 0$, the alkali metals should exhibit hydrogenlike spectroscopic terms. They do, but the energy differences between n^2S, n^2P, n^2D, etc. terms are much greater. As mentioned earlier, these differences arise from the greater penetration of the underlying filled shells by ns electrons than by np electrons, np electrons, in turn, show greater penetration than nd electrons, and so on.

An alternative way of discussing penetration effects is to speak of the effective nuclear charge acting on a particular electron. This effective nuclear charge is the actual nuclear charge less the screening effect of other electrons in the atom. Slater has developed the following set of empirical rules for calculating the screening effect of other electrons on a particular electron.

1. The electrons are grouped in the order $1s$; $2s$ and $2p$; $3s$ and $3p$; $3d$; $4s$ and $4p$; $4d$; $4f$; etc.—ns and np being considered as a single group.

2. Electrons in groups lying above that of a particular electron do not shield it at all.

3. A shielding of 0.35 is contributed by each other electron in the same group (except for a $1s$ electron, which contributes 0.30 to the shielding of the other $1s$ electron).

4. For d and f electrons the shielding from underlying groups is 1.00 for each electron in the underlying group. For s, p electrons the shielding from the immediately underlying shell (that is, $n - 1$) is 0.85 for each electron; the shielding from groups further in is 1.00 for each electron.

Ideally, Slater's rules should permit the calculation of the energy level of any electron in any atom or ion by the formula

$$E = \frac{-(Z - S)^2}{n^2} 13.6 \text{eV} \tag{1.26}$$

where Z is the nuclear charge, S is the shielding constant, and n is the principal quantum number of the electron. Unfortunately, values calculated in this fashion may differ from those obtained from spectroscopic studies by as much as a factor of five. Nevertheless, the effective nuclear charges $(Z - S)$ calculated from Slater's rules have been used for such purposes as calculating ionic radii (see Section 6.2.1) and electronegativities of the elements (see problem 2.22).

Spin-Orbit Interactions

The separation of the doublet terms, because of the spin orientation of the one outermost electron, falls between that expected for a hydrogenlike atom of atomic number Z and that for a hydrogenlike atom with an effective atomic number of one. The doublet separation is greatest for P terms and decreases with increasing L of the term, as we would expect on the basis of penetration of underlying filled shells. Except for cases of high atomic numbers (e.g., Rb, Cs) the splitting of doublet terms is relatively small. For example, for the $4p^1$ ^2P terms of K, the splitting is 57.72 cm^{-1}.

Spin-Spin Interactions

In a partially filled subshell with more than one electron, the effect of the electrons' spin interactions may be fairly large. Thus the energy difference between the $3d^4$ $^5D_2$ term and the $3d^4$ $^1D_2$ term of V$^+$ ion is 20944.87 cm^{-1}. The low energy term arises from a configuration involving *four unpaired* electrons, whereas the high energy term has two *pairs* of electrons. The primary reason for the differing energy of these terms is the differing exchange energy between pairs of electrons having the same m_s quantum number—that is, the same spin. Four unpaired electrons, ↑ ↑ ↑ ↑, have six distinct pairs that may have similar spin orientation, whereas two pairs, ↑↓ ↑↓, have only two distinct pairs that may have similar spin orientations.

Hund's Rules and Ground-State Terms

Using Hund's rules, we may select from the possible terms the lowest energy term arising from a given configuration. It will be the one with maximum S (and accordingly, maximum multiplicity) and, within this restriction, maximum L; the J value will be the minimum one for subshells less than half-filled and the maximum one for shells more than half-filled. The ground-state term for a given configuration may be found easily in the following manner, without deriving all the possible terms. An electron configuration is written with the orbitals filled in the order of decreasing m_l; each orbital is singly occupied before any orbital is doubly occupied. The sum of the m_l values for this configuration gives the L value of the term; the S value is one half the number of singly occupied orbitals. For a d^2 case, this would give

$$m_l = \quad +2 \qquad +1 \qquad 0 \qquad -1 \qquad -2$$
$$\odot \qquad \odot \qquad \bigcirc \qquad \bigcirc \qquad \bigcirc$$
$$L = 2 + 1 = 3 \qquad S = 1/2 \times 2 = 1$$

The L of 3 gives an F term; the S of 1 gives a multiplicity of 3. Combination of L and S gives

J values of 2, 3, and 4; the lowest is selected in this case, since the d orbitals are less than half-filled. The ground-state term for a d^2 configuration is accordingly a $^3F_2$ term. A half-filled subshell gives an S term, and since $L = 0$, $J = S$. Accordingly, the ground-state term for N is $^4S_{3/2}$.

Example   Determine the ground-state terms for d^3 and d^8 configurations, and for C.

Solution

$m_l =$	2	1	0	-1	-2	
d^3	↑	↑	↑			Max. $M_L = 3$, so $L = 3$ and $S = 3/2$
d^8	↑↓	↑↓	↑↓	↑	↑	Max. $M_L = 3$, so $L = 3$ and $S = 1$

For d^3 the ground state is $^4F_{3/2}$ (lowest J or $L - S$).
For d^8 the ground state is $^3F_4$ (highest J or $L + S$).
Although both of these are F terms, they are not hole counterparts like d^3 and d^7, both of which have ^4F as the ground state but with $J = 9/2$ for d^7.
For C $(2s^2 2p^2)$ we get

$m_l =$	1	0	-1		
	↑↓	↑	↑		Max. $M_L = 1$, so $L = 1$ and $S = 1$
	2s	2p			Ground state $= {}^3P_0$ (lowest J)

Note that the configuration written gives the maximum S and L values, this is only one microstate of the array for the ground-state term, however. There are $4 \times 7 = 28$ microstates for ^4F.

1.5   IONIZATION ENERGIES

The ionization energy of an atom has been defined as the energy needed to remove an electron from an atom in its ground state. The second ionization energy is the additional energy needed to remove the second electron, and so forth. Table 1.6 gives the ionization energies as derived from the analysis of atomic spectra.

Figure 1.12 plots the first ionization energies against the atomic numbers of the elements. *In general, the ionization energy increases on crossing a period (e.g., Li through Ne) and decreases on descending in a group (He through Rn).* These trends, and many of the variations within them, may be rationalized in terms of the electronic configurations of elements involved. On crossing a period, the principal quantum number for the electron being removed remains constant but the *effective nuclear charge* increases, because of incomplete nuclear shielding by electrons of the same quantum number. On descending in a given group, the principal quantum number increases regularly and relatively more rapidly than the effective nuclear charge. Variations from or within these trends result primarily from changes in the type of orbital from which the electron is taken, or from changes in the multiplicity of the ground-state term. Elements with a ground-state $^1S_0$ term have higher ionization energies than their neighboring elements (see exceptions below). This will be the ground-state term for any element that has an outermost completely-filled shell or subshell. The exceptions to this rule are the alkaline earth metals that have a transition metal as a neighbor. In this case the orbit of the s electron of the transition metal may be thought of in terms of the Bohr-Sommerfeld model as penetrating

Table 1.6   Ionization energies of the elements[a] (in eV; 1 eV/atom = 96.4869 kJ/mole)

Z	Element	I	II	III	IV	V	VI	VII	VIII
1	H	13.598							
2	He	24.587	54.416						
3	Li	5.392	75.638	122.451					
4	Be	9.322	18.211	153.893	217.713				
5	B	8.298	25.154	37.930	259.368	340.217			
6	C	11.260	24.383	47.887	64.492	392.077	489.981		
7	N	14.534	29.601	47.448	77.472	97.888	552.057	667.029	
8	O	13.618	35.116	54.934	77.412	113.896	138.116	739.315	871.387
9	F	17.422	34.970	62.707	87.138	114.240	157.161	185.182	953.886
10	Ne	21.564	40.962	63.45	97.11	126.21	157.93	207.27	239.09
11	Na	5.139	47.286	71.64	98.91	138.39	172.15	208.47	264.18
12	Mg	7.646	15.035	80.143	109.24	141.26	186.50	224.94	265.90
13	Al	5.986	18.828	28.447	119.99	153.71	190.47	241.43	284.59
14	Si	8.151	16.345	33.492	45.141	166.77	205.05	246.52	303.17
15	P	10.486	19.725	30.18	51.37	65.023	220.43	263.22	309.41
16	S	10.360	23.33	34.83	47.30	72.68	88.049	280.93	328.23
17	Cl	12.967	23.81	39.61	53.46	67.8	97.03	114.193	348.28
18	Ar	15.759	27.629	40.74	59.81	75.02	91.007	124.319	143.456
19	K	4.341	31.625	45.72	60.91	82.66	100.0	117.56	154.86
20	Ca	6.113	11.871	50.908	67.10	84.41	108.78	127.7	147.24
21	Sc	6.54	12.80	24.76	73.47	91.66	111.1	138.0	158.7
22	Ti	6.82	13.58	27.491	43.266	99.22	119.36	140.8	168.5
23	V	6.74	14.65	29.310	46.707	65.23	128.12	150.17	173.7
24	Cr	6.766	16.50	30.96	49.1	69.3	90.56	161.1	184.7
25	Mn	7.435	15.640	33.667	51.2	72.4	95	119.27	196.46
26	Fe	7.870	16.18	30.651	54.8	75.0	99	125	151.06
27	Co	7.86	17.06	33.50	51.3	79.5	102	129	157
28	Ni	7.635	18.168	35.17	54.9	75.5	108	133	162
29	Cu	7.726	20.292	36.83	55.2	79.9	103	139	166
30	Zn	9.394	17.964	39.722	59.4	82.6	108	134	174
31	Ga	5.999	20.51	30.71	64				
32	Ge	7.899	15.934	34.22	45.71	93.5			
33	As	9.81	18.633	28.351	50.13	62.63	127.6		
34	Se	9.752	21.19	30.820	42.944	68.3	81.70	155.4	
35	Br	11.814	21.8	36	47.3	59.7	88.6	103.0	192.8
36	Kr	13.999	24.359	36.95	52.5	64.7	78.5	111.0	126
37	Rb	4.177	27.28	40	52.6	71.0	84.4	99.2	136
38	Sr	5.695	11.030	43.6	57	71.6	90.8	106	122.3
39	Y	6.38	12.24	20.52	61.8	77.0	93.0	116	129

(Continued)

Table 1.6 Ionization energies of the elements[a] (in eV; 1 eV/atom = 96.4869 kJ/mole)
(Continued)

Z	Element	I	II	III	IV	V	VI	VII	VIII
40	Zr	6.84	13.13	22.99	34.34	81.5			
41	Nb	6.88	14.32	25.04	38.3	50.55	102.6	125	
42	Mo	7.099	16.15	27.16	46.4	61.2	68	126.8	153
43	Tc	7.28	15.26	29.54					
44	Ru	7.37	16.76	28.47					
45	Rh	7.46	18.08	31.06					
46	Pd	8.34	19.43	32.93					
47	Ag	7.576	21.49	34.83					
48	Cd	8.993	16.908	37.48					
49	In	5.786	18.869	28.03	54				
50	Sn	7.344	14.632	30.502	40.734	72.28			
51	Sb	8.641	16.53	25.3	44.2	56	108		
52	Te	9.009	18.6	27.96	37.41	58.75	70.7	137	
53	I	10.451	19.131	33					
54	Xe	12.130	21.21	32.1					
55	Cs	3.894	25.1						
56	Ba	5.212	10.004						
57	La	5.577	11.06	19.175					
58	Ce	5.47	10.85	20.20	36.72				
59	Pr	5.42	10.55	21.62	38.95	57.45			
60	Nd	5.49	10.72						
61	Pm	5.55	10.90						
62	Sm	5.63	11.07						
63	Eu	5.67	11.25						
64	Gd	6.14	12.1						
65	Tb	5.85	11.52						
66	Dy	5.93	11.67						
67	Ho	6.02	11.80						
68	Er	6.10	11.93						
69	Tm	6.18	12.05	23.71					
70	Yb	6.254	12.17	25.2					
71	Lu	5.426	13.9						
72	Hf	7.0	14.9	23.3	33.3				
73	Ta	7.89							
74	W	7.98							
75	Re	7.88							
76	Os	8.7							
77	Ir	9.1							

(Continued)

Table 1.6 Ionization energies of the elements[a] (in eV; 1 eV/atom = 96.4869 kJ/mole) *(Continued)*

Z	Element	I	II	III	IV	V	VI	VII	VIII
78	Pt	9.0	18.563						
79	Au	9.225	20.5						
80	Hg	10.437	18.756	34.2					
81	Tl	6.108	20.428	29.83					
82	Pb	7.416	15.032	31.937	42.32	68.8			
83	Bi	7.289	16.69	25.56	45.3	56.0	88.3		
84	Po	8.42							
85	At								
86	Rn	10.748							
87	Fr								
88	Ra	5.279	10.147						
89	Ac	6.9	12.1						
90	Th		11.5	20.0	28.8				
91	Pa								
92	U								
93	Np								
94	Pu	5.8							
95	Am	6.0							

[a]From C. E. Moore, "Ionization Potentials and Ionization Limits from Atomic Spectra," NSRDS-NBS 34, 1970.

the *d* orbits; hence, as the *d* electrons are filled in to build up the transition metals, the effective nuclear charge increases. Accordingly, the ionization energy increases in a fairly regular fashion from the alkali metal through zinc or its congeners. In the case of mercury, not only the *d* orbitals but also the *f* orbitals have been filled in, and the penetration effect is much larger. The ionization energy of mercury is almost as great as that of radon. Following the zinc family elements, the ionization energy drops sharply for Ga, In, and Tl, since the electron involved is a *p* electron that occupies a less penetrating orbit. A similar, but smaller, drop in ionization energy following Be and Mg may also be accounted for by the change in the ionizing electron from an *s* to a *p* electron. The higher ionization energies of nitrogen and its congeners, as compared with those of their neighbors, may be attributed to the stabilizing influence of the spin-spin interactions, which are highest in filled and half-filled subshells.

Successive ionization energies increase in magnitude because of the smaller number of electrons left to shield the ionizing electron from the nuclear charge. Variation with atomic number in the second ionization energy is very similar to that in the first ionization energy, except that it is displaced by one atomic number. Some similarity persists in the third and higher ionization energies, again provided that isoelectronic ions are compared. Thus, the highest ionization energies always occur for ions of the noble gas configurations.

Ahrens has pointed out that the oxidation states (see p. 44) of the elements in their compounds may be correlated by the difference in ionization energy for successive states, Δ_I.

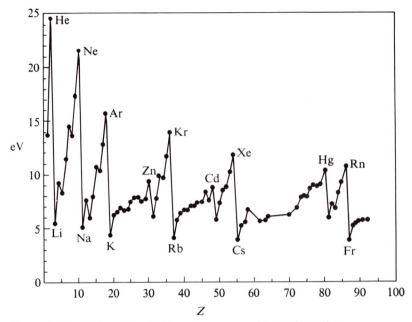

Figure 1.12   Variation of the first ionization energy with atomic number.

If Δ_I is around 10 or 11 eV or less, the lower state is not stable. Consider the values for Al.

$$\Delta_{1,2} = 12.8 \text{ eV}$$
$$\Delta_{2,3} = 9.6$$
$$\Delta_{3,4} = 91.5$$

Although compounds of Al(I) are stable as gaseous species, AlCl(g), they are unknown as solids. No compounds of Al(II) are known. Al(III) is the only stable state for Al compounds under usual conditions. Δ_I values of around 16 or above usually lead to stable states. Since Δ is always high at the noble gas configuration, these configurations frequently are found for the elements in their compounds. The $s^2p^6d^{10}$ configuration is also one in which Δ is high, and it, too, is a common configuration for positive ions.

1.6   ELECTRON AFFINITIES

The electron affinity of an atom is the energy involved in adding an electron to a gaseous atom in its ground state, to form a gaseous anion.

$$X(g) + e \rightarrow X(g)^{-} \tag{1.27}$$

Electron affinity sometimes is expressed as the ionization energy of $X(g)^{-}$. Since for most atoms, energy is released when an electron is added, the electron affinity in such cases is

Table 1.7 Electron affinities of the elements[a]

IA	IIA	III	IV	V	VI	VII	VIII	VIII	VIII	IB	IIB	IIIB	IVB	VB	VIB	VIIB
1 H 72.77 0.7542																
3 Li 59.8 0.620	4 Be <0 <0											5 B 26.8 0.278	6 C 122.3 1.268	7 N ≤0 ≤0	8 O 141.1 1.462	9 F 328.0 3.399
11 Na 52.7 0.546	12 Mg −14 −0.15											13 Al 42.6 0.442	14 Si 133.6 1.385	15 P 71.7 0.743	16 S 200.43 2.0772	17 Cl 348.8 3.615
19 K 48.36 0.5012	20 Ca <0 <0	21 Sc <0 <0	22 Ti 7.7 0.080	23 V 50.8 0.526	24 Cr 64.4 0.667	25 Mn <0 <0	26 Fe 24 0.25	27 Co 67 0.7	28 Ni 111 1.15	29 Cu 118.3 1.226	30 Zn −47 −0.49	31 Ga 29 0.3	32 Ge 116 1.2	33 As 77 0.80	34 Se 194.97 2.0206	35 Br 324.6 3.364
37 Rb 46.89 0.4860	38 Sr <0 <0	39 Y ~0 ~0	40 Zr 41.2 0.427	41 Nb 86.3 0.894	42 Mo 72.1 0.747	43 Tc 67 0.7	44 Ru 106 1.1	45 Rh 109.8 1.138	46 Pd 53.8 0.558	47 Ag 125.7 1.303	48 Cd −32 −0.33	49 In 29 0.3	50 Sn 121 1.25	51 Sb 101 1.05	52 Te 190.16 1.9708	53 I 295.4 3.061
55 Cs 45.50 0.4715	56 Ba <0 <0	57 La 48 0.5	72 Hf <0 <0	73 Ta 31.2 0.323	74 W 78.7 0.816	75 Re 14 0.15	76 Os 106 1.1	77 Ir 151.1 1.566	78 Pt 205.3 2.128	79 Au 222.76 2.3086	80 Hg −61 −0.63	81 Tl 29 0.3	82 Pb 35.2 0.365	83 Bi 91.4 0.947	84 Po 183 1.9	85 At 270 2.8

Legend: kJ/g-atom / eV/atom

[a]Data are from H. Hotop and W. C. Lineberger, *J. Phys. Chem. Ref. Data* 1975, 4, 539; C. S. Feigerle, R. R. Corderman, and W. C. Lineberger, *J. Chem. Phys.* 1981, 74, 1513 (B, Al, Bi, and Pb); P. D. Burrow, J. A. Michejda, and J. Comer, *J. Phys. B: Atom. Molec. Phys.* 1976, 9, 3225 (Zn, Cd, and Hg); C. S. Feigerle, R. R. Corderman, S. V. Bobashev, and W. C. Lineberger, *J. Chem. Phys.* 1981, 74, 1580 (Ti, V, Cr, Zr, Nb, Mo, Rh, Pd, Ta, W, and Ir).

expressed as a positive number (the amount of energy released). The factors affecting electron affinities are the same as those affecting ionization energies, so the general trend is for electron affinities to increase (more energy *released*) from left to right in the periodic table, because of the increase in nuclear charge *and* decrease in atomic radius. Table 1.7 displays the electron affinities of most elements. The trend within a group is less regular than for ionization energies, but we must remember that the increase in size would be expected to decrease the attraction for an electron, whereas the increase in nuclear charge would be expected to increase the attraction. For the alkali metals, as for ionization energies, the size effect is more important and the electron affinities decrease from Li → Cs. This is also the general trend for most of the nonmetals, with some notable exceptions.

The electron affinities of N, O, and F are *much* lower than expected from the group trends, which are regular for S → Po and Cl → At. The ionization energies are regular within each of these groups and highest for N, O, and F. Proceeding across the second period, the sizes decrease and the nuclear charges increase, so that the negative charge density increases, just about reaching the limit for F. Thus, N, O, and F have very high ionization energies, but the addition of another electron is not so favorable as for P, S, and Cl, respectively. The electronegativities (Section 2.4) of N, O, and F are very high, but electronegativity reflects the attraction for electrons *without* adding an electron to form a negative ion. The amazingly low electron affinity of N and the trends in the VB and VIB groups are considered later (p. 556). The factors causing low electron affinities for N, O, and F lead to unusually low bond energies when these elements are bonded to themselves or to one another by single bonds (p. 572).

The noble gases have stable ns^2np^6 configurations ($1s^2$ for He), and the electron affinities are zero or slightly negative (corresponding to endothermic processes), indicating that there is no bound state for the added electron. The IIA metals with the ns^2 configuration (beyond a noble gas configuration) and the IIB metals with the $(n-1)d^{10}ns^2$ configuration have negative electron affinities. The next higher energy state (^2P) is too high in energy for the favorable addition of an electron. The electron affinities are very high for the neighboring elements Pt and Au.

GENERAL REFERENCES

R. L. DeKock and H. B. Gray, *Chemical Structure and Bonding*, Benjamin/Cummings, Menlo Park, Cal., 1980. One of the better recent treatments.

G. Herzberg, *Atomic Spectra and Atomic Structure*, Dover, New York, 1944. A classic book that is still useful.

R. M. Hochstrasser, *Behavior of Electrons in Atoms*, Benjamin, New York, 1964. Good treatment at an appropriate level.

L. Pauling, *The Nature of the Chemical Bond*, 3rd ed., Cornell University Press, Ithaca, N.Y., 1960. This classic book gives much detail of interest to the chemist.

PROBLEMS

1.1   Bohr postulated that lines in the emission spectrum from hydrogen in highly excited states, such as $n = 20$, would not be observed under ordinary laboratory conditions, because of the large size of such atoms and the much greater probability of atom collision deactivation as compared with radiation deactivation. Using the Bohr model, calculate the ratio of the cross-sectional area of a hydrogen atom in the $n = 20$ state to that of one in the $n = 1$ state.

1.2 According to the Bohr model of the atom, what would be the size of a Ne^{9+} ion? What would the ionization energy be for this ion? What would the excitation energy be for the first excited state?

1.3 Calculate the energy (eV) released in the transition of a hydrogen atom from the state $n = 3$ to $n = 1$. The wavelength of the radiation emitted in this transition may be found from the relation λ (in Å) $= 12398/E(eV)$ (often remembered as the approximate $\lambda = 12345/eV$). Calculate the wavelengths of all spectral lines that could be observed from a collection of hydrogen atoms excited by a potential of 12 V.

1.4 Assuming a screening of ½ for an s electron, calculate the wavelength expected for the K_α x-ray line of Tc.

1.5 The Balmer series in the hydrogen spectrum originates from transitions between $n = 2$ states and higher states. Compare the wavelengths for the first three lines in the Balmer series with those expected for similar transitions in Li^{2+}.

1.6 Obtain expressions for $(\partial E/\partial Z)_n$ and $(\partial^2 E/\partial Z^2)_n$ using the Bohr model (Equation 1.9). Using the ionization energies for isoelectronic sequences of atoms and ions from Table 1.6, and finite differences to obtain $(\Delta E/\Delta Z)_n$ and $(\Delta^2 E/\Delta Z^2)_n$, show that the $n = 1$ level is filled when the electron occupancy is 2. How would you estimate unknown ionization energies using a finite-difference approximation? Estimate the electron affinities of F and O by this procedure.

1.7 Explain briefly the observation that the energy difference between the $1s^2 2s^1 {}^2S_{1/2}$ state and the $1s^2 2p^1 {}^2P_{1/2}$ state for Li is 14904 cm^{-1}, whereas for Li^{2+} the $2s^1 {}^2S_{1/2}$ and the $2p^1 {}^2P_{1/2}$ states differ by only 2.4 cm^{-1}.

1.8 Using the p orbitals for an example, distinguish between the angular part of the probability function, the radial part of the probability function, and a probability contour. Draw simple sketches to illustrate. How would each of these be affected by a change in the principal quantum number, n?

1.9 Give the characteristic valence shell configuration for the following periodic groups (e.g., ns^1 for the alkali metals).

    a. Noble gases.   **d.** Ti family.

    b. Halogens.   **e.** N family.

    c. Coinage metals (Group IIB).

1.10 Write the electron configuration beyond a noble gas core for [for example, F, $(He)2s^2 2p^5$] Rb, La, Cr, Fe, Cu, Tl, Po, Gd, and Lu.

1.11 Write the electron configuration beyond a noble gas core and give the number of unpaired electrons for (for example, F^-, $[He]2s^2 2p^6$) K^+, Ti^{3+}, Cr^{3+}, Fe^{2+}, Cu^{2+}, Sb^{3+}, Se^{2-}, Sn^{4+}, Ce^{4+}, Eu^{2+}, and Lu^{3+}.

1.12 If an atom is in an energy state in which $L = 3$ and $S = 2$, how would the state be described in spectroscopic notation?

1.13 What is the number of microstates for an f^3 configuration? Which of these is unique to the ground-state term arising from this configuration?

1.14 Derive the spectral terms for the d^7 configuration. Identify the ground-state term. Give all J values for the ground-state term and indicate which is lowest in energy.

1.15 Derive the spectral terms for the f^2 configuration. Identify the ground-state term. Give all J values for the ground-state term and indicate which is lowest in energy.

1.16 What spectroscopic terms arise from the configuration $1s^2 2s^2 2p^6 3s^2 3p^5 3d^1$?

1.17 What nodal plane(s) might be expected for the $f_{x^3 - (3/5)xr^2}$ orbital?

1.18 The angular function of an orbital is described by $\dfrac{1}{r^6}\{x^4(y^2 - z^2) + y^4(z^2 - x^2) + z^4(x^2 - y^2)\}$.

Is the orbital of g or u symmetry? What is the basis of your assignment?

1.19 The trend in electron affinities for Group VB (N family) is opposite that for VIB (O family). Explain the trend for each family.

1.20 Rearrange equation 1.24 to express the "real" set of d orbitals in terms of the set of d orbitals of well defined m_l values.

II

Localized Bonding

You have been introduced to theories of chemical bonding in earlier courses. Here we review briefly and then extend the principles needed for understanding and predicting molecular geometry. The more sophisticated molecular orbital theory is useful for molecules involving delocalized bonding (Chapter 4), and ligand field theory (Chapter 7) is useful for transition metal complexes. Applying molecular orbital and ligand field theories usually depends upon prior knowledge of molecular shapes based upon the more qualitative treatments discussed here.

2.1 REVIEW OF COVALENT BONDING

2.1.1 Molecules That Obey the Octet Rule

Those metals that have one, two, or three electrons in the outer shell and a noble-gas configuration in the next-to-outermost shell commonly combine with nonmetals by losing the outermost electron(s) to form simple cations. This leaves an ion with an s^2p^6 configuration, such as K^+, Mg^{2+}, and Al^{3+}. Other common electronic configurations for metal ions are $s^2p^6d^{10}$, the pseudo noble-gas configuration, found in such ions as Ag^+, Zn^{2+}, Cd^{2+}, Ga^{3+}; and $(n-1) s^2p^6d^{10}ns^2$, as in Sn^{2+}, Tl^+, and Sb^{3+} (n is the principal quantum number of the outermost shell). Most of the lanthanide metals form tripositive ions, leaving partially filled $4f$ orbitals, whereas many transition elements form ions having partially filled d orbitals (Au^{3+}, Fe^{2+}, Fe^{3+}, etc.). The nonmetals that require one or two electrons to complete a noble-gas configuration commonly accept the required number of electrons from metals to form simple anions. In these cases, valence refers to an ionic state and has a sign associated with it (for example, oxygen most commonly has an ionic valence of -2). For a monoatomic ion the *oxidation number* (or *oxidation state*) of the element is the same as the ionic valence—that is, the charge on the ion. Ionic compounds have structures and compositions determined almost entirely by the coulombic interactions among the ions.

Just as ionic valence states may be predicted from the electronic configuration of the atom, the number of electrons an atom may share is related to the atomic configuration. The Pauli exclusion principle allows only two electrons with opposed spins to occupy an atomic orbital, and when a pair of electrons is shared by two atoms, the electron pair is assumed to occupy an atomic orbital on each atom.

As a first approximation, the number of covalent bonds formed is the same as the number of unpaired s or p electrons initially present in the isolated atom. The halogen atoms thus have an electronic configuration of s^2p^5. We expect one bond here, since each atom has only one unpaired electron to share. An electron dot picture would represent the X_2 molecule as

$$:\overset{..}{\underset{..}{X}}:\overset{..}{\underset{..}{X}}:$$

In writing electron dot formulas (Lewis structures) we place dots representing the shared pair between the symbols for the two bonded atoms. Singly occupied orbitals are indicated by the electron dot picture, as in $:\overset{.}{C}\cdot$ for the carbon atom. The valence bond representation replaces the pair of dots by a bar and represents the simultaneous occupancy of two atomic orbitals by a pair of electrons. In cases in which the occupancy of nonbonding orbitals appears obvious, we may show only the bonding pairs, as in

$$X—X$$

A pair of electrons occupying a nonbonding atomic orbital commonly is referred to as a lone pair.

For oxygen the electron configuration of the atom may be represented by

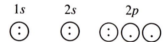

Normally, we expect oxygen to share two electrons, as it does in almost all of its covalent compounds. The Lewis structures below are typical.

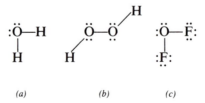

Although oxygen forms two bonds in each compound shown, the oxidation number of oxygen is $-$II in water, $-$I in hydrogen peroxide, and II in oxygen difluoride. The *oxidation number for atoms in covalent compounds* is obtained by assigning shared electrons to the more electronegative atom (p. 72) and then counting the charge on the quasi ion (see Figure 2.1 for additional examples). Lewis structures are limited to cases obeying the octet rule. Molecules in which the octet rule is exceeded for one or more atoms are represented similarly by valence bond structures.

2.1.2 Expanded Octet

Sulfur should be chemically similar to oxygen, since they have the same outer electronic configuration—and indeed, sulfur forms H_2S, H_2S_2, and SF_2 with structures similar to *(a)*, *(b)*, and *(c)* above. However, sulfur also has empty *d* orbitals of the same principal quantum number as the valence electrons; that is,

The expenditure of promotional energy will produce states having a greater number of unpaired electrons; that is,

and

If the formation of electron pair bonds by another element and by sulfur in one of these excited states releases enough energy to pay for the promotion of electrons, then the sulfur can form four or even six covalent bonds. In general, such promotion seems to be important only when sulfur combines with the more electronegative elements, F, O, Cl, or Br. Examples are given in Figure 2.1. The assignment of the oxidation number of $-II$ to S in *(c)* follows the convention of naming CS_2 as a sulfide, even though the electronegativities of C and S are the same (see Table 2.7).

The availability of empty *d* orbitals constitutes one of the major differences between the second-period elements and elements of higher periods, as noted for oxygen and the sulfur family. Similar differences may be pointed out in the halogen family and the nitrogen group of elements. Chlorine, bromine, iodine, and astatine have outermost electron configurations that may be represented by

and electron promotion accordingly could result in the states represented by

$$\begin{array}{ccc} ns & np & nd \end{array}$$

Sharing of the unpaired electrons would lead to compounds containing one, three, five, and seven covalent bonds. These are found, respectively, in the elements Cl_2, etc. and in the

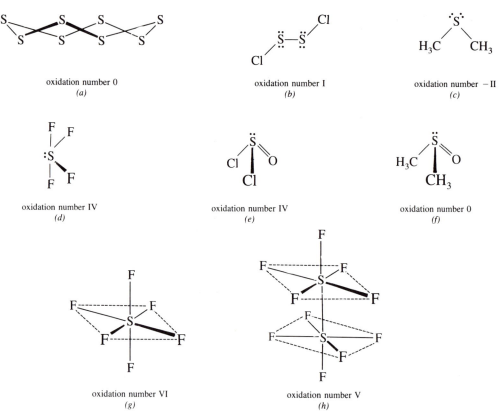

Figure 2.1   Valence bond structures for S compounds.

highest fluorides of chlorine (ClF_3), bromine (BrF_5), and iodine (IF_7). The effect of the increasing difference in electronegativity in this series in forming more bonds is apparent. Valence bond structures of these are given in Figure 2.2.

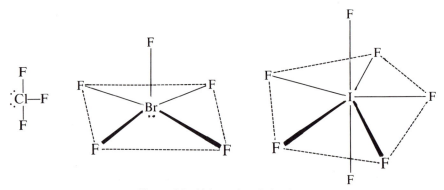

Figure 2.2   Valence bond structures.

In the group V elements, nitrogen, $1s^2 2s^2 2p^3$, forms three bonds, whereas phosphorus, arsenic, and antimony commonly form three or five bonds (Figure 2.3).

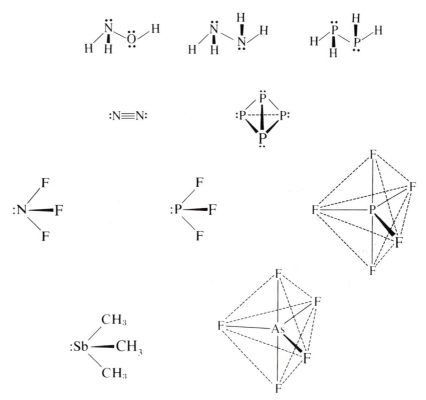

Figure 2.3 Valence bond structures.

In the carbon family the outermost configuration of the isolated atom is

ns   *np*

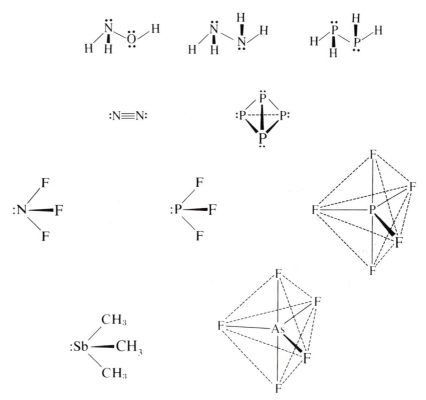

The promotional energy necessary to produce the $ns^1 np^3$ state is, however, more than compensated for by the formation of the two additional bonds.

2.1.3 Coordinate Covalence or Dative Bonding

Carbon monoxide provides a common case of molecule formation in which an *s* electron of carbon is *not* promoted. Here we would expect carbon to have achieved the configuration

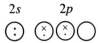

and oxygen to have an s^2p^6 configuration. However, the high dissociation energy of CO and the very short C—O bond distance suggest that the Lewis structure is better represented by

$$^{(-)}:C{\equiv}O:^{(+)} \qquad \text{or simply} \qquad :C{\equiv}O:$$

The arrow in the right-hand formula is a convention sometimes used to indicate that oxygen donates *both* of the electrons in the bond. This so-called *coordinate covalent bond,* or *dative bond,* cannot be distinguished from an ordinary covalent bond once it is formed. The charges shown, called *formal charges,* always arise in dative bonding. Formal charges may be calculated by assigning half of the electrons shared between two atoms to each atom and calculating the resulting charge on the quasi ion. The sum of the formal charges in a molecule will be zero; the sum of the formal charges in an ion will be the same as the charge on the ion. Formal charges, although useful in evaluating possible alternative valence bond structures, should not be interpreted too literally as representing the actual charge distribution within molecules.

Formal charges are assigned assuming 100% covalent character, whereas oxidation number assignments assume no electron sharing. The actual charge distribution lies somewhere between the extremes. Pauling's electroneutrality principle (p. 60) is a good guide to reasonable charge distributions.

Nitrogen can form dative bonds as follows.

An arrow may be used for the N → O bond, since it can be identified as the dative bond. Arrows are not appropriate for NH_4^+ or $N(CH_3)_4^+$, because the four bonds are equivalent. Phosphorus can donate its unshared pair in dative bonding, as in

$$\overset{(+)}{Cl_3 P} \rightarrow \overset{(-)}{B Br_3}$$

or it can accept a pair of electrons into a *d* orbital, as in

$$Cl_3 \overset{(-)}{\ddot{P}} \leftarrow \overset{(+)}{N} (CH_3)_3$$

Phosphorus in PF_5 can add a fluoride ion by accepting a pair of electrons into one of its empty *d* orbitals, as shown below.

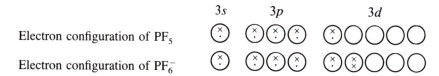

Through dative bonding, phosphorus can form four and six bonds, as well as the usual compounds such as PX_3 and PX_5. This series may be illustrated by the following species.

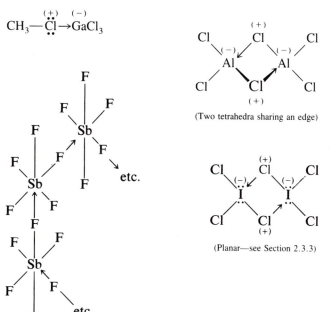

In addition to the compounds XX', XX_3', XX_5', and XX_7', mentioned earlier for the heavier halogens, dative bonding allows the formation of ICl_2^-, ICl_4^-, and IF_6^-. Halogens also can form two bridging bonds, as shown in Figure 2.4.

Figure 2.4 Dative bonds (formal charges shown).

In the case of boron and other group III elements, accepting electrons *via* dative bonds is the rule rather than the exception. Thus the following series of ions or compounds is known for boron.

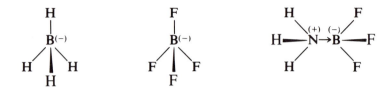

2.1.4 Resonance

Some molecules or ions cannot be represented adequately by a single valence bond formula. The rules of combination would lead to the representation $\begin{smallmatrix}(-)O\\[2pt](-)O\end{smallmatrix}C{=}O$ for the carbonate ion, but the oxygen atoms are known to be equivalent and the bond angles are all 120° in the planar ion. The difficulty arises simply because electrons cannot be represented as dots or lines restricted to particular positions. The "fourth" electron pair, which forms a double bond wherever it is written, is not localized in any one of the three bonds, but rather, is somewhat "smeared out" over all three bonds. There are three equivalent bonds, each of which is something between a single and a double bond. Here the "fourth" bond can be written in any one of the three positions, so that each bond is described as having a *bond order* of $1\frac{1}{3}$. The bond order of a single bond is 1; of a double bond, 2; of a triple bond, 3.

Pauling introduced the concept of *resonance* to adapt the simple valence bond notation to situations in which electrons are delocalized.[1] Three structures are written for the carbonate ion.

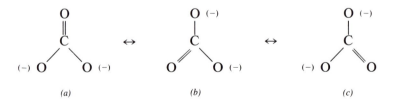

Do not interpret the three *contributing structures* as having any independent existence. The carbonate ion does *not* consist of a mixture of the three structures, nor is there an equilibrium among the three structures. The simple valence bond notation is inadequate to represent the

[1]Resonance is needed because of the inadequacy of our notation, which represents electrons by dots or lines. Although the use of a circle inside the benzene ring emphasizes the delocalization of π electrons, it does not take care of the electron bookkeeping.

structure of the carbonate ion, which is neither one nor all of the above structures, but something in between, a *resonance hybrid*. The fact that several structures are written, rather than one, does not in any way increase the physical reality of any of the structures. The three bonds and the three oxygen atoms are equivalent. The contributing structures are sometimes separated by double-headed arrows ($\leftrightarrow$), which should not be confused with the reversible arrows ($\rightleftharpoons$) used in chemical equilibrium.

The assignment of a bond order of $1\frac{1}{3}$ to the C—O bond in carbonates is straightforward, since the three most reasonable contributing structures are equivalent. The problem is more complex in the oxoanions of the halogens, the sulfur family, and the phosphorus family. For example, the sulfate ion, may be viewed as arising from dative bonding between a sulfide ion and four oxygen atoms (a), or from the sharing of two pairs of electrons between each oxygen and sulfur (b). In addition to these extremes, we may write the intermediate structures (c), (d), and (e).

		Four equivalent structures	Six equivalent structures	Four equivalent structures
(a)	(b)	(c)	(d)	(e)

The actual structure of the sulfate ion would be an average of all the structures shown (and even, perhaps, some additional ones) weighted according to the number of equivalent structures and their relative energies. It has been estimated that the bond order in sulfates is approximately 1.5, corresponding most closely to (d).

The familiar application of resonance to benzene includes, the usual Kekulé structures (a and b); the less important Dewar structures (c–e); and unimportant structures such as (f), in which there is charge separation and one carbon has only six electrons.

(a)	(b)	(c)	(d)	(e)	(f)

The elimination of unlikely structures and the selection of the more important contributing structures, that is, those closer to the true resonance hybrid, is facilitated by the following rules.

1. The contributing structures should have the same or nearly the same atomic positions. The most important contributing structures normally differ only in the positions of electrons. The closer the bond angles and lengths of a contributing structure are to those of the actual molecule, the greater the "contribution" of that structure to the hybrid. The Dewar structures are less important than the Kekulé structures, because of the very long bond across the ring. The shape of a molecule corresponding to the Dewar structures would be considerably different from that of benzene.

2. All contributing structures of a particular molecule must have the same number of un-paired electrons. In structure (*f*) for benzene, the charge separation could be avoided by placing one unpaired electron on the top carbon atom and another on the bottom carbon atom, but this possibility is ruled out.

3. Contributing structures should not differ too widely in the positions of electrons—in other words, the contributing structures should not differ greatly in energy. We would expect structure (*f*) for benzene to be a high-energy (or less stable) structure compared with (*a*) and (*b*), and to be much less important: that is, the resonance hybrid is much closer to the representations of (*a*) and (*b*) than to (*f*). This rule also eliminates structures in which there are more than eight electrons around an atom with no additional low energy orbitals.

4. Like charges should not reside on atoms close together in a contributing structure, but unlike charges should not be greatly separated. This can be interpreted as a special case of rule 3, since the positions of electrons in the unlikely structures will be much different from those in other structures. In HN_3, structure (*c*) is relatively unimportant, because of the (+) charges on adjacent N atoms and also because of the double negative charge on the remaining N atom.

<table>
<tr><td align="center">HN_3</td><td></td><td align="center">N_3^-</td></tr>
<tr><td align="center">(+)   (−)
H—N̈=N=N̈:</td><td align="center">(*a*)</td><td align="center">(−)   (+)   (−)
:N̈=N=N̈:</td></tr>
<tr><td align="center">(−)   (+)
H—N̈—N≡N:</td><td align="center">(*b*)</td><td align="center">(2−)   (+)
:N̈—N≡N:</td></tr>
<tr><td align="center">(+)   (+)   (2−)
H—N≡N—N̈:</td><td align="center">(*c*)</td><td align="center">(+)   (2−)
:N≡N—N̈:</td></tr>
</table>

Since the similar structure for the N_3^- ion does not give (+) charges on adjacent N atoms, it should be more important than for the molecular acid. The change in the relative importance of the last contributing structure should result in the same bond order (about 2) for each bond in N_3^-, but in a bond order less than 2 for the N—N bond nearer the H atom and a bond order greater than 2 for the terminal N—N bond in HN_3. The fact that there is an additional important contributing structure for the N_3^- ion, as compared with HN_3, results in resonance stabilization of the anion. This might be related to the rather great strength of HN_3 as an acid, and possibly to the instability of covalent azides generally.

5. Contributing structures in which negative charge resides on an electronegative element and positive charge resides on an electropositive element are more important than those in which the reverse is true. For HF the contributing structure $H^{(+)}F^{(-)}$ is an important one; the ionic structure $H^{(-)}F^{(+)}$ is not. When acetone loses a proton an anion is obtained.

$$:\ddot{O}\!:^{(-)}$$

The contributing structure in which the negative charge is on the oxygen, $CH_3C{=}CH_2$, is more important than the one with negative charge on carbon.

6. The greater the number of covalent bonds, the greater the importance of a contributing

structure. The doubly bonded structures of BF_3,

$$\underset{F}{\overset{F}{\diagdown}} B \overset{(-)\ (+)}{=\!=} F\ ,$$

are important because of

the formation of an additional covalent bond, even though a fluorine atom acquires a positive formal charge (see p. 48). Here the tendency to remove the electron deficiency of boron, or to utilize all low-energy orbitals in the molecule, is more important than the charge distribution. In the case of pyridine-*N*-oxide, structures (*c–e*) have a favorable distribution of charges from the standpoint of the electronegativity of the atoms involved, but the fact that these structures contain one less bond than the (*a* and *b*) structures diminishes their importance. Rule **5** indicates that structures (*f–h*) should be even less important than (*a*) and (*b*), because the (+) charge is on the more electronegative nitrogen and the (−) charge is on a carbon atom.

The experimental heats of formation of molecules such as CH_4 agree with values obtained by summing the C—H bond energies. A molecule for which resonance is important is more stable than could be predicted from any one of the contributing structures. The discrepancy between calculated and observed heats of formation is called the *resonance energy;* the greater the number of significant contributing structures, the greater the resonance energy. Typical resonance energies (kJ/mole) are benzene, 155; naphthalene, 314; C≡O, 439, and O=C=O, 151. The increased stability of molecules stabilized by resonance is reflected in the shortening of bond lengths compared to the average expected from the contributing structures.

Planarity is common for molecules of second-period elements in which resonance stabilization is important. Contributing structures frequently differ in the positions of double bonds. In order that a double bond may be written in any one of several positions with good *p-p* overlap (see p. 58), the molecule must be planar with sp^2 or sp hybridization.

The greater the number of contributing structures, the greater the delocalization of the electronic charge, and hence the more stable the molecule. Localization of charge on an atom is usually an unstable situation, occurring only under special circumstances. Since ions can be stabilized very greatly if the charge can be delocalized, resonance, in general, will be more important in charged species than in neutral ones. The increasing acidity in the series HClO, $HClO_2$, $HClO_3$, $HClO_4$ is determined largely by the greater stability of the anions containing several oxygen atoms, because of the greater delocalization of the negative charge. For the most part, the negative charge in ClO^- must reside on the oxygen atom. In ClO_4^- the negative charge is spread over four equivalent oxygen atoms.

2.2 THEORETICAL TREATMENT OF THE COVALENT BOND

2.2.1 Heitler-London Theory

The covalent bond was first developed theoretically by Heitler and London, whose treatment of the hydrogen molecule provides a good introduction to the discussion of the covalent bond that follows. When two hydrogen atoms are brought together without exchanging electrons (structure I), there is a weak attraction at large distances that becomes a strong repulsion at short distances (dashed curve Figure 2.5). (The hydrogen nuclei are represented as H_a and H_b and the electrons as ·1 and ·2.) Since at short distances the alternative structure, in which the electrons exchange positions (structure II), is just as likely, the potential energy curve (lower solid curve) corresponds to the normal H—H distance by the exchange between two equivalent structures,

$$\text{I} \quad H_a^{·1 ·2} H_b$$
$$\text{II} \quad H_a^{·2 1·} H_b$$

as in resonance stabilization. The wave functions for the individual structures can be written

$$\psi_I = \psi_a(1)\psi_b(2) \tag{2.1}$$
$$\psi_{II} = \psi_a(2)\psi_b(1) \tag{2.2}$$

The structures being equivalent, their wave functions represent the same energy; so Equation

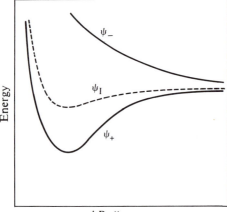

Energy

ψ_-

ψ_I

ψ_+

AB distance

Figure 2.5 Potential-energy curves for H_2.

2.3, which can be written as Equation 2.4, will also be a solution to the wave equation (c_1 and c_2 are mixing coefficients).

$$\psi_+ = c_1\psi_{\mathrm{I}} + c_2\psi_{\mathrm{II}} \qquad (2.3)$$

or

$$\psi_+ = c_1\psi_a(1)\psi_b(2) + c_2\psi_a(2)\psi_b(1) \qquad (2.4)$$

The minimum in the potential energy curve indicates that the ψ_+ wave function leads to a stable molecule. The wave equation $\psi_- = c_1\psi_{\mathrm{I}} - c_2\psi_{\mathrm{II}}$ represents the situation in which no attraction is observed at any distance, because the two electrons have parallel spins (see the upper solid curve in Figure 2.5). Electrons must have opposed spins for favorable interaction.

Pauling has shown that the results may be improved by a consideration of ionic structures for the hydrogen molecule. This results in the addition of two more terms to the wave equation 2.4.

Ionic structures	New terms to be added to Equation 2.4
III $\quad$ H$_a^{(+)}$ $\quad$:H$_b^{(-)}$	$c_3\psi_b(1)\psi_b(2)$
IV $\quad$ H$_a$:$^{(-)}$ $\quad$ H$_b^{(+)}$	$c_4\psi_a(1)\psi_a(2)$

The Heitler-London treatment can be extended to other molecules with the covalent bond resulting from the interaction of electrons with opposed spins on two atoms. The valence electrons of a combining atom must be unpaired. If they are not unpaired in the ground state of the atom, vacant orbitals of low energy must be available to accommodate in separate orbitals the previously spin-paired electrons. For second-period elements the atomic orbitals must be of the same major quantum number, n, because the orbitals must not differ greatly in energy. An atom can then form a covalent bond for each of its stable valence orbitals. Thus nitrogen with the outer configuration $2s^2 2p_x^1 2p_y^1 2p_z^1$ can form three bonds, as in NF$_3$, but cannot form NF$_5$, because the second shell can accommodate only eight electrons. Phosphorus with the configuration $3s^2 3p_x^1 3p_y^1 3p_z^1$ can form PF$_3$; and by promoting one $3s$ electron to a vacant $3d$ orbital to give the configuration $3s^1 3p_x^1 3p_y^1 3p_z^1 3d^1$, it can form five bonds in PF$_5$, as discussed earlier (p. 47).

The Heitler-London treatment features the formation of a chemical bond as the result of the pairing of two electrons, one from each bonded atom. The electron density of the shared pair is a maximum between the two bonded atoms. Atoms tend to combine to form closed shells, but the octet is exceeded in PF$_5$ and SF$_6$ and probably is not attained in BI$_3$. As presented, the Heitler-London treatment does not consider the formation of odd-electron molecules such as H$_2^+$, where no pairing occurs. Nor does it include the formation of the coordinate covalent bond where one of the bonded atoms furnishes both of the electron pairs, although extensions could cover these bonding situations.

2.2.2 Valence Bond (Pauling-Slater) Theory

Overlap of Atomic Orbitals

Pauling and Slater extended the Heitler-London theory, making it more general and accounting for the directional character of covalent bonds. The main consideration is that stable compounds result from the tendency to fill all stable orbitals with electron pairs, shared or unshared.

The shape of a molecule is determined primarily by the directional charcteristics of the orbitals involved. However, molecular shape and the directional character of the orbitals are interrelated. For a given coordination number, only a few molecular shapes are reasonable. The atomic or hybrid (see below) orbitals used for bonding are those with directional characteristics compatible with the symmetry of the molecule. A covalent bond may be described as resulting from the overlap of orbitals on two atoms so that the two orbitals can be occupied by an electron pair. In HF the spherically symmetrical *s* orbital of H overlaps with the singly occupied $2p$ orbital of F.

The H_2O molecule results from the overlap of the two singly-occupied *p* orbitals of oxygen with the *s* orbital of each of two hydrogen atoms, to form an angular molecule (see Figure 2.6*b*). The *p* orbitals are mutually perpendicular, but the bond angle in water is 104.45°. Repulsion between the hydrogen atoms accounts for the fact that the angle is greater than 90°. (A better approach to the correlation of bond angles is presented on p. 61). The ammonia molecule is pyramidal with a bond angle of 107.3°, as a result of bonding using the three *p* orbitals. One might have expected the bond angle to be larger than the angle in the water molecule, because of the greater repulsion among the three hydrogen atoms.

Hybridization

Carbon has the outer configuration $2s^2 2p_x^1 2p_y^1$, but unpairing the $2s$ electrons and promoting one to the completely vacant $2p_z$ orbital results in $2s^1 2p_x^1 2p_y^1 2p_z^1$, as required for the formation of four bonds as in CH_4. The molecular shape is not immediately apparent from the characteristics of the atomic orbitals: the *s* orbital is spherically symmetrical and the three *p* orbitals are mutually perpendicular. Nevertheless, physical and chemical evidence show that the methane molecule is tetrahedral, with four equivalent bonds. In an atom surrounded by four electron pairs, the mutual repulsion among the electron pairs will orient them toward the apices of a tetrahedron. The wave functions, ψ, for the tetrahedrally oriented orbitals are the linear combinations of the atomic wave functions. There are four combinations, differing in the weighting coefficients a, b, c, and d:

$$\psi = a\psi_s + b\psi_{p_x} + c\psi_{p_y} + d\psi_{p_z} \tag{2.5}$$

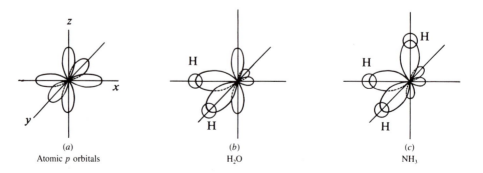

(*a*)	(*b*)	(*c*)
Atomic *p* orbitals	H_2O	NH_3

Figure 2.6   Angular characteristics of *p* orbitals.

This process of combining atomic orbitals to give mixed orbitals is known as *hybridization*. The combination of the *s* and three *p* orbitals gives *tetrahedral hybrid sp³ orbitals*. One of the hybrid *sp³* orbitals is represented in Figure 2.7. The *p* lobe with the same sign of the ψ function as that of the *s* orbital is enlarged, and the one of opposite sign is diminished. The enlarged lobe of the resulting orbital can give more favorable overlap with the orbital of another atom and thus form a stronger bond than can either the *p* or *s* orbitals alone. The orbital represented is one of the four equivalent *sp³* orbitals.

There are three equivalent *sp²* hybrid orbitals directed toward the corners of an equilateral triangle and two equivalent *sp* hybrid orbitals pointing in opposite directions. The shapes of the major lobes for *sp³*, *sp²*, and *sp* hybrid orbitals differ very little, mainly in the relative sizes of the minor and the major lobe. The minor lobe is largest for *sp³* and smallest for *sp*, so the greatest fraction of electron density would be in the bonding lobe for the *sp* hybrid orbital. Overlap integrals increase in the order *sp³* < *sp²* < *sp*, indicating the greatest overlap for *sp* bonds. The increasing C—H bond energies in the series CH_4, C_2H_4, and C_2H_2 support increasing bond strengths with increasing *s* character for *s-p* hybrids[2]—that is, greatest for *sp*. Pauling used the extension of the orbitals, rather than the overlap integrals, to arrive at a different order of relative bond strengths: 1.93 for *sp*, 1.99 for *sp²*, and 2.00 for *sp³*, compared with 1.00 for pure *s*. The overlap integrals provide better estimates of bond strengths. Figure 2.8 displays some of the important hybrids—including cases involving *d* orbital participation—along with the corresponding geometrical configurations.

Multiple Bonds for Second-Period Elements

The bonding requirements of each carbon in ethylene, C_2H_4, are satisfied by single bonds to two hydrogen atoms and a double bond between the carbon atoms. A double bond is represented by two bent or "banana" bonds formed by the sharing of an edge of two tetrahedra. Two electrons can be accommodated in each of the orbitals, one above and one below the plane of the molecule (see Figure 2.9*a*).

The usual description of multiple bonds is adapted from molecular orbital theory (see p. 136). In the ethylene molecule, the carbon orbitals can be considered to be hybridized to give

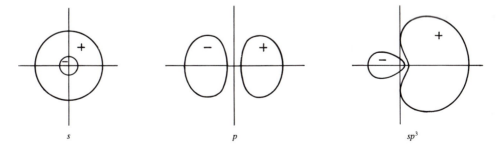

Figure 2.7 Combination of orbitals to give an sp³ hybrid.

[2]R. McWeeney, *Coulson's Valence*, 3rd ed., Oxford, Oxford, 1979, p. 204.

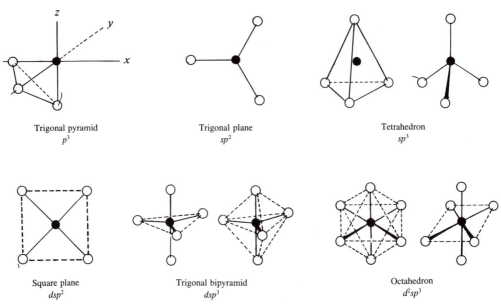

Figure 2.8   Some bonding configurations.

three equivalent sp^2 orbitals (at 120° to each other). Each carbon uses two of the sp^2 hybrid orbitals for overlap with the $1s$ orbitals of two hydrogens, and the remaining sp^2 orbital for the formation of the carbon-carbon bond. These five bonds are known as *sigma(σ) bonds,* covalent bonds in which the electron density reaches a maximum along the line joining the bonded atoms. The remaining two electrons from the unmixed p atomic orbitals of the carbon atoms pair to form a *pi(π) bond.* The electron pair of the π bond occupies the π orbital, which is obtained by the overlap of the two atomic p orbitals. As shown in Figure 2.9b, the combination of two atomic p wave functions gives the molecular π wave function. The π orbital picture

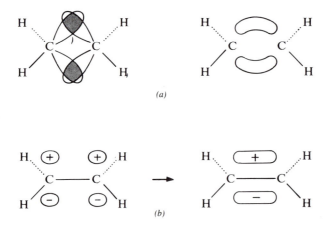

Figure 2.9   Representations of a double bond.

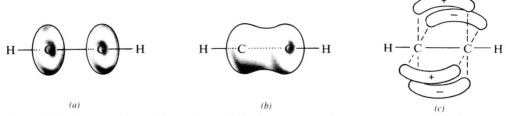

Figure 2.10 Representation of the orbitals in C_2H_2. Contours of maximum electron density are sketched.

shows two regions of high electron density, one above and one below the plane of the molecule, but only one electron pair occupies the pair of lobes. In contrast, the banana bond picture shows two distinct bonds, each of which may accommodate an electron pair for a total of four electrons.

Acetylene, C_2H_2, is a linear molecule, which can be represented by two tetrahedra sharing a face to produce three badly bent bonds joining the carbon atoms. However, the π bond representation seems much more satisfactory. The σ bonds form because overlap of sp hybrids leaves each carbon with two singly occupied p orbitals. Since only one direction is defined (Figure 2.10a), the unhybridized p orbitals of each carbon atom will be doughnut-shaped and will overlap to give a π orbital, which has cylindrical symmetry and extends over both carbon atoms (Figure 2.10b). This can also be viewed as resulting from the combination of two π orbitals, each having two lobes whose nodal planes intersect at right angles along the inter-nuclear line (Figure 2.10c). With four π electrons, the concentration of more electronic charge between the nuclei shortens the bond distance and increases the bond strength from single to double to triple bonds (Table 2.1).

Multiple Bonds for Elements beyond the Second Period

The double bonding encountered in second period elements differs in extent and type from that encountered in other elements. Double bonding produced by the formation of p-p-π bonds (overlap of p orbitals on bonded atoms) is common for the second-period elements and diminishes in importance down through a given family. Fluorine and oxygen often form double bonds even at the expense of acquiring a formal positive charge. This serves to delocalize the electrons, diminishing the very high charge density of these small atoms and decreasing the

Table 2.1 Bond lengths and bond energies[a] for carbon and nitrogen

	C—C	C=C	C≡C
Bond length, pm	154	134	120
Bond energy, kJ/mole	356	682	962

	N—N	N=N	N≡N
Bond length, pm	145	125	110
Bond energy, kJ/mole	240	450	942

[a]Data from S. W. Benson, *J. Chem. Educ.* 1965, *42*, 502, or Table 2.9.

repulsion among the unshared electrons on F or O. The need for such delocalization, or for decreasing the repulsion among lone pairs, is not so great for larger atoms. For larger atoms the p-p-π orbital overlap is less favorable, because the p orbitals are larger and more diffuse. The increase in atomic radius also favors an increase in coordination number, thereby allowing electrons that might have formed π bonds to form σ bonds. Table 2.2 displays the significant differences in structure and degree of aggregation between formally similar combinations involving elements of the second and third periods. The elements of the second period tend to form discrete molecules involving p-p-π bonding (BCl_3, CO_2, N_2, N_2O_3, and O_2). Third-period elements tend to form larger or polymeric molecules. Thus Al_2Cl_6 exists as a dimer, P_4 is a tetrahedral molecule, in P_4O_6 the P_4 tetrahedron is expanded to incorporate an O into each tetrahedral edge, S_8 forms rings, and SiO_2 is a three-dimensional network of SiO_4 tetrahedra. In each of these cases there are only single bonds. Carbon disulfide exists as discrete CS_2 molecules (CO_2 structure), whereas SiS_2 forms infinite chains of SiS_4 tetrahedra. In $(HPO_3)_x$ and P_4O_{10} the coordination number (C.N.) of P is four, whereas the C.N. for N is three in the corresponding nitrogen compounds (p. 561). At room temperature the multiple-bonded molecules for Si and P analogous to C_2H_4 and C_2H_2 and to N_2, respectively, do not exist in appreciable amounts.*

Although p-p-π bonding is of little importance in elements of the third and higher periods, multiple bonding may occur in other ways. A coordination number of four leaves no empty p orbitals on the central atom for the formation of p-p-π bonds, but the bond length in an ion such as SO_4^{2-} is shorter than would be expected for single bonds. Sulfur atoms, unlike those of the second-period elements, have available vacant d orbitals of low energy for the formation of d-p-π bonds, which result from the overlap of a filled p orbital on the oxygen with an empty d orbital on sulfur (see Figure 2.11). The d-p overlap may be more favorable than p-p overlap for large atoms, because the d orbitals project out in the general direction of the bond.

Pauling proposed an *electroneutrality principle* stating that electrons are distributed in a molecule in such a way as to make the residual charge on each atom zero or very nearly zero, except that hydrogen and the most electropositive metals can acquire partial positive charge and the most electronegative atoms can acquire partial negative charge. The charge on an atom

Table 2.2 Comparison of melting points and boiling points of some elements and compounds from the second and third periods

$BCl_3(g)$ b. 12.5°C	CO_2 (g) subl. − 78.5°	$N_2(g)$ b. − 195.8°	$N_2O_3(g)$ dec.	$O_2(g)$ b. − 182.96°
$Al_2Cl_6(s)$ subl. 177.8°	$SiO_2(s)$ m. 2230°	$P_4(s)$ m. 590°	$P_4O_6(s)$ m. 23.8°	$S_8(s)$ m. 95.5°
	$CS_2(l)$ b. 46.3°	$HNO_3(l)$ b. 83°	$N_2O_5(l)$ m. 30°	
	$(SiS_2)_x(s)$ subl. 109°	$(HPO_3)_x(s)$ dec.	$P_4O_{10}(s)$ subl. 300°	

*An $\backslash$Si$=$Si$/$ derivative has been reported recently by R. West, M. J. Fink, and J. Michl (*Science* 1981, *214*,

1343). It is stabilized with respect to polymerization, which would convert π- into σ-bonds, by bulky mesityl substituents.

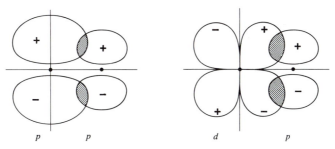

Figure 2.11 Sketches of *p-p-π* and *d-p-π* bonding.

can be decreased by a change in the polarity, or amount of ionic character, of a bond (see p. 76) or by a change in the amount of multiple bond character of a bond. The formal charge of $+3$ on Cl in the hypothetical

structure $\left[\begin{array}{c} O \\ | \\ Cl \\ O \diagup \quad \diagdown O \\ | \\ O \end{array}\right]^{-}$ is reduced to $+2$ in $\left[\begin{array}{c} O \\ \| \\ Cl \\ O \diagup \quad \diagdown O \\ | \\ O \end{array}\right]^{-}$ and to zero in $\left[\begin{array}{c} O \\ | \\ Cl \\ O \diagup \quad \diagdown O \\ \| \\ O \end{array}\right]^{-}$

As expected, the double-bond character (*d-p-π*) increases in the series PO_4^{3-}, SO_4^{2-}, ClO_4^{-}, as the formal charge on the central atom increases for the single-bonded structures. Bond polarity makes the central atom positive and thus tends to offset the decrease in positive charge accompanying double bonding. The bond order is about 1.5 for ClO_4^{-}. An extended Hückel molecular orbital treatment of S—O and P—O bonds gives reasonable bond lengths even without using *d* orbitals, but some participation of *d* orbitals provides better results overall. For SO_4^{2-} a bond order of about 1.25, giving S a formal charge of $+1$, and a bond order only slightly greater than 1 for PO_4^{3-} seem reasonable. The electroneutrality principle, considering polarization as well as formal charge, is a good guide. Oxoanions of fluorine are not stable (but see p. 594 for a discussion of HOF), partly because the high electronegativity of fluorine prohibits a formal positive charge. The positive charge thus cannot be eliminated by double-bond formation, because the fluorine does not have vacant low-energy orbitals available.

2.3 SHAPES OF MOLECULES

2.3.1 Molecular Geometry and Hybridization

The shape of a molecule, to a first approximation, is determined by its σ bond structure. Since they often occupy hybridized orbitals, unshared electron pairs can be treated as bonded groups in determining the hybridization: accordingly, the hybridization in NH_3 and H_2O can be considered to be sp^3. The importance of unshared electron pairs in determining molecular shapes is illustrated in Table 2.3. In writing Lewis structures for AX_n compounds other than those of H, A is generally more electropositive than X. Orbitals on X are filled with unshared electrons except as needed for bonding to complete the octet of A.

Table 2.3   Shapes of some simple molecules

General Formula	Number of Unshared Electrons on A	σ Bonding Orbital Hybridization	Shape	Examples
AX_2	None	sp	Linear	$O{=}C{=}O$, $S{=}C{=}S$, $O{=}\overset{(+)}{N}{=}O$ $H{-}C{\equiv}N$ $H{-}C{\equiv}C{-}H^a$ ZnX_2, CdX_2, HgX_2, $AgCl_2^-$
	One or more	sp^2 or sp^3	Angular	$X{-}\ddot{O}{:}$, NH_2^-, $:\overset{..}{S}{(+)}$, O_3, $:N$ ..., $:\overset{..}{Cl}$ ...; $(+){\cdot}N$, $:N$, $:\overset{..}{S}$, $SnCl_2$, PbX_2
AX_3	None	sp^2	Planar	BX_3 type, NO_3^-, CO_3^{2-}, $H_2C{=}CH^a$, COX_2, SO_3
	One or two	sp^3	Pyramidal	NX_3 type, H_3O^+, PX_3, ClO_3^-
AX_4	None	sp^3	Tetrahedral	CX_4, NH_4^+, BX_4^-, SiX_4, TiX_4, ZnX_4^{2-} SO_4^{2-}, SO_2Cl_2, POX_3, PO_4^{3-}, ClO_4^-

^{a}Each carbon in C_2H_2 and C_2H_4 corresponds to A in AX_2 and AX_3, respectively.

In many molecules it is difficult to assess *a priori* the relative importance of various contributing structures or the contribution of a particular type of hybridization. Atoms with more low-lying orbitals than valence electrons will have reasonably unambiguous σ bond configurations and hence predictable structures (at least for simple molecules). Atoms with more valence electrons than low-lying orbitals may contribute these electrons in varying degrees in diverse situations (that is, in different molecules) and accordingly, may have several distinct and not easily predictable σ bond structures. In the trimethoxoboron, $B(OCH_3)_3$, molecule the boron hybridization must be sp^2 with bond angles of 120°, but the orbital hybridization of the oxygen atoms is apparently intermediate between sp^3 and sp^2 (Figure 2.12). Any contribution of resonance structures involving double bonding between boron and oxygen would require a lone pair in an atomic *p* orbital on oxygen for the formation of a π bond. Hence double

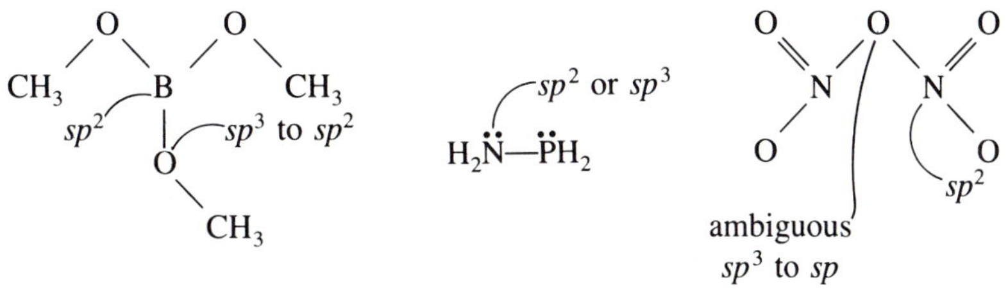

Figure 2.12 Possible assignments of hybridization.

bonding would favor sp^2 hybridization here. Similarly, the P—N bond in H_2N—PH_2 might be sp^2, sp^3, or some intermediate combination of N orbitals. The hybridization of the σ bonding orbitals of each nitrogen in N_2O_5 must be sp^2, but the bridging oxygen might be sp^3, sp^2, or even sp.

The geometry of a molecule in an excited electronic state likely will differ drastically from that of the ground state, as the promotion of an electron can change the hybridization of the orbitals. The first excited state of acetylene has C—C—H bond angles of 120° and a C—C distance similar to that found in benzene. The hybridization of the C orbitals changes from sp to sp^2 in order to accommodate nonbonding electrons.

2.3.2 Isoelectronic Molecules

The species CH_4, NH_4^+, and BH_4^- are isoelectronic, since they contain the same number of electrons and the same number of atoms. A molecule (or ion) with a central atom, four hydrogen atoms and a total of 10 electrons (all shells included) should have the tetrahedral methane structure, that is, be isostructural with methane. The SiH_4 molecule is also isostructural with CH_4, since the added inner shell of Si does not change the bonding. Molecules with the same number of *valence electrons*—for example, CH_4 and SiH_4, CO_2 and CS_2, and O_3 and SO_2—are commonly described as being isoelectronic. Isoelectronic molecules containing central atoms of different periods often do not have the same structures: examples include CO_2 and SiO_2, and N_2O_5 and P_4O_{10} (see p. 60).

The BN unit is isoelectronic with C_2. Borazine, $B_3N_3H_6$, has been described as "inorganic benzene" because its physical properties so closely resemble those of benzene (see p. 682 for the benzenelike structure). Likewise, boron nitride, BN, forms both layer (graphite-type) and diamond structures. We encounter the diamond structure in the elements Si, Ge, and Sn, as well as in such compounds as the Group IIIB phosphides, arsenides, and antimonides—excepting SbB and those of Tl.

The molecules CH_4 and CF_4 and the ions BH_4^- and BF_4^- differ in the number of valence electrons, but the additional electrons on F are unshared pairs that are nonbonding. The molecules have the same structure. Replacing H by another atom that forms only a single bond should not change the structure. The BF_3 molecule is isoelectronic and isostructural with CO_3^{2-}, but BH_3 is the unstable monomer of B_2H_6 (diborane). The molecules BF_3 and BH_3 differ because of the π bonding in BF_3 (see pp. 159 and 161), which cannot occur for a B—H bond.

It is sometimes convenient to view the Ne atom as having four electron pairs arranged tetrahedrally. The molecules HF, H_2O, NH_3, and CH_4 might be considered to differ from Ne in the number of electron pairs that are protonated. Bent, extending the isoelectronic concept to include these cases,[3] regards one, two, three, and four nuclear protons, respectively, as being removed to share electron pairs. Thus all of these species can be related by the tetrahedral arrangement of the electron pairs. Of course, the tetrahedra are distorted when there are both bonding and nonbonding electron pairs, as will be discussed in the next section.

If nonbonding electron pairs are localized, protonation does not cause much change. The series of molecules presented in Figure 2.13 are related structurally, although this fact is not immediately evident from the simplest formulas. Basically, there is a linear grouping of three atoms involving *sp* hybridization of the central atom and two π bonds. The number of protonated electron pairs on the terminal atoms does not alter the bonding greatly. Bent's generalized definition states that molecules are isoelectronic with each other when they have the same number of heavy atoms (Z > 3) and the same number of valence electrons—a very useful approach in relating the structures of simple inorganic and organic molecules. The series

$$\overset{\displaystyle O}{\underset{\displaystyle \|}{}}CH_3CCH_3 \text{ (acetone), } \overset{\displaystyle O}{\underset{\displaystyle \|}{}}H_2NCNH_2 \text{ (urea), } \overset{\displaystyle O}{\underset{\displaystyle \|}{}}HOCOH \text{ (carbonic acid), and } \overset{\displaystyle O}{\underset{\displaystyle \|}{}}FCF \text{ (carbonyl fluoride)}$$

is related by the replacement of —CH_3 by the "isoelectronic" —NH_2, —OH, and —F groups. (Figure 2.14 shows other "isoelectronic" groups.) Do not apply these generalizations blindly, however. Be alert for changes in bond order: for example, possible *p-d-*π bonding for S—F where corresponding *p-p-*π bonding is not possible for O—F or S—H. Also, O_2 is not isoelectronic with C_2H_4, as the O_2 requires a completely different bonding description (p. 138).

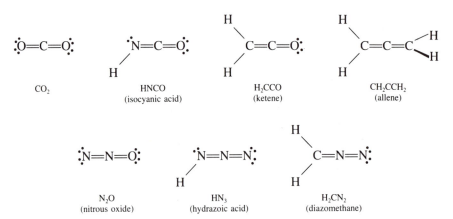

Figure 2.13 Structurally related molecules with three heavy atoms and 16 electrons.

[3]H. A. Bent, *J. Chem. Educ.* 1966, *43*, 170.

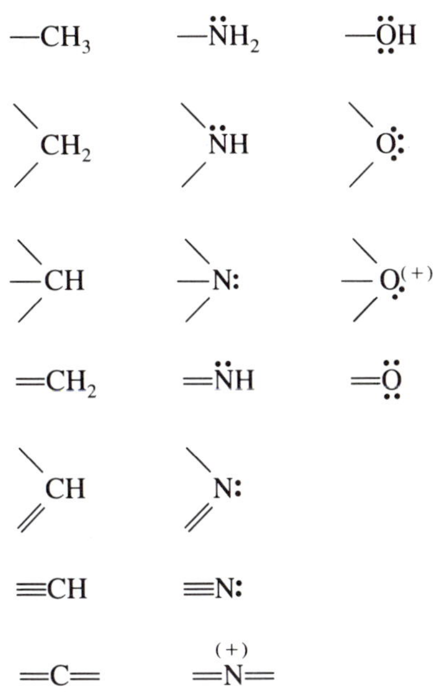

$$-CH_3 \qquad -\ddot{N}H_2 \qquad -\ddot{O}H \qquad -\ddot{\ddot{F}}:$$

$$\diagdown_{/}CH_2 \qquad \diagdown_{/}\ddot{N}H \qquad \diagdown_{/}\ddot{O}:$$

$$-\diagdown_{/}CH \qquad -\diagdown_{/}N: \qquad -\diagdown_{/}\ddot{O}^{(+)}$$

$$=CH_2 \qquad =\ddot{N}H \qquad =\ddot{\ddot{O}}$$

$$\diagdown_{/\!/}CH \qquad \diagdown_{/\!/}N:$$

$$\equiv CH \qquad \equiv N:$$

$$=C= \qquad =\overset{(+)}{N}=$$

Figure 2.14 Isoelectronic groupings (read horizontally).

2.3.3 The Pauli Exclusion Principle and the Prediction of the Shapes of Molecules (Valence Shell Electron-Pair Repulsion Theory)

Molecules Covered by the Octet Rule

You can predict the approximate shapes of simple molecules by considering unshared electron pairs (lone pairs) to be equivalent to bonded groups. To obtain more exact descriptions of the molecular shapes and variations of bond angles, you must take into account repulsion among electron pairs in the valence shell of the central atom. Repulsion between bonding pairs is not as great as repulsion between lone pairs. This approach was introduced by Gillespie and Nyholm and developed extensively by Gillespie[4] as the *valence shell electron pair repulsion* (VSEPR) *theory*. Repulsion among the electrons can be considered to arise from the operation of the Pauli Exclusion Principle. Although this principle usually is stated in terms of the energy or the quantum numbers of the electrons, it effectively rules out the possibility that two electrons might be at the same point at the same time. Two electrons can be confined to the same orbital only if they have opposed spins. Since repulsion is greater for electrons of the same spin, the total repulsion is minimized by arranging the electrons of a complete octet in four localized pairs of electrons of opposed spin, each pair directed toward the apex of a tetrahedron. This is the configuration expected for Ne, F^-, and O^{2-}.

[4]R. J. Gillespie and R. S. Nyholm, *Quart. Rev.* 1957, *11*, 339; R. J. Gillespie, *J. Am. Chem. Soc.* 1960, *82*, 5978; *J. Chem. Educ.* 1970, *47*, 18; *Angew. Chem. Internat. Edit.* 1967, *6*, 819.

If one of the electron pairs is used for bonding, as in HF, the four pairs are no longer equivalent. The bonding pair now becomes more localized, since its electrons are restricted in their motion by two positive nuclei. Repulsion among the electrons in the nonbonding pairs will cause them to spread out to occupy the space made available by the localization of the bonding pair. If two of the electron pairs are used for bonding, as in H_2O, the bond angle will be determined by the repulsions among the four electron pairs. Instead of the normal tetrahedral angle (109.47°) for four equivalent pairs, as found in CH_4 and as expected for O^{2-}, the bond angle in H_2O is 104.45°. Because of the differences in localization, the repulsion between electron pairs decreases in the order lone pair-lone pair > lone pair-bonding pair > bonding pair-bonding pair. (See Figure 2.15 for a representation of the distribution of electron pairs confined to a plane.) The localization of the bonding pairs permits the bond angle to close, relative to the tetrahedral angle, in order to decrease repulsion between the lone pairs. In NH_3 the HNH bond angles are 107.3°—an angle greater than that in H_2O because there are no lone pair-lone pair repulsions. In addition to the restriction of two nuclei, each bonding pair is localized by the repulsion of two other bonding pairs and one lone pair, whereas in water each bonding pair is localized by the repulsion of one other bonding pair and two lone pairs. We can approach the variation in bond angles by starting with the tetrahedral CH_4 molecule as a reference. The bond angles in NH_3 decrease because of the presence of the lone pair, and in H_2O the bond angle decreases further because of the presence of two lone pairs.

Here we are considering four equivalent hybrid orbitals about O and N, with the same amount of *s* character in each. Hall (*J. Am. Chem. Soc.*, 1978, *100*, 6333; *Inorg. Chem.*, 1978, *17*, 2261) has shown that the bent shape of H_2O can be explained on the basis of the preferential occupancy of the lower energy *s* orbital by a lone pair, which gives more *p* character to the bonding orbitals. Although his treatment of H_2O and other cases raises questions about the theoretical basis for the VSEPR theory, in practice, the VSEPR theory has an exceptionally good record for predicting molecular shapes.

Comparing the bond angles in NO_2 (132°) and in NO_2^- (115°) demonstrates that a single unshared electron causes less repulsion than an unshared electron pair. The nitryl ion, NO_2^+, with no unshared electron on nitrogen, is linear (see Figure 2.16).

The bond angles decrease in the series NH_3, PH_3, AsH_3, and SbH_3 and in the series H_2O, H_2S, H_2Se, and H_2Te (see Table 2.4), presumably because the increasing size and lower electronegativity of the central atom permit the bonding electrons to be drawn out further, thus decreasing the repulsion between bonding pairs.[5] The variation in bond angles in a series such

(a)   (b)

Figure 2.15 (a) Most-probable spatial distribution and corresponding trigonal orbitals for three electron pairs confined to a plane. (b) Orbitals for two bonding electron pairs and one lone pair; ● nuclei of bonded atoms. (Reprinted with permission from R. J. Gillespie, *J. Am. Chem. Soc.* 1960, *82*, 5978. Copyright 1960, American Chemical Society.)

[5]In line with the approach presented earlier (assuming p^3 bonding—see p. 56), one might consider that the angles approach 90° for the larger central atoms, because of the increasing effectiveness of the protons' screening as the size of the central atom increases.

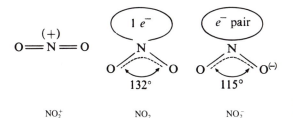

Figure 2.16 Bond angles in NO_2^+, NO_2, and NO_2^-.

as NH_3, PH_3, AsH_3, and SbH_3 is in the direction expected for an increase in the bonds' p character: that is, the bond angles approach the values expected for p^3 bonds, rather than those for sp^3 hybrid bonds. This is just the opposite of what might be expected, since the energy difference between the s and p orbitals is greatest for nitrogen.

The bond angles decrease by large amounts from NH_3 to PH_3 or from H_2O to H_2S, then by very small amounts for the following members of each series. Gillespie suggests that the difference arises because N and O are members of the second period, for which the octet represents a closed shell. Strong repulsions between the electron pairs result from the fact that

Table 2.4[a] Comparison of bond angles

CH_4		109.47°	H₂C=CH₂	119.9°	H₂CO	115.8°
CH_3F	⟨HCH	110°				
CH_3Cl	⟨HCH	110.3°				
CHF_3	⟨FCF	108.8°				
$CHCl_3$	⟨ClCCl	110.4°				
			$F_2C=CF_2$	114°	Cl_2CO	111.3°
			$Cl_2C=CCl_2$	113.5°	F_2CO	108.0°

NH_3	107.3°	NF_3	102.4°	NCl_3	106.8°	
PH_3	93.3°	PF_3	97.8°	PCl_3	100.1°	
AsH_3	91.8°	AsF_3	96.2°	$AsCl_3$	98.5°	
SbH_3	91.3°	SbF_3	95.0°	$SbCl_3$	97.2°	
				PBr_3	101.0°	
				PI_3	102°	

H_2O	104.45°	F_2O	101.5°	F_2SO	⟨FSO 106.8°
H_2S	92.2°				⟨FSF 92.8°
H_2Se	91°				
H_2Te	89.5°				

[a]Data from R. J. Gillespie, *Molecular Geometry*, Van Nostrand Reinhold, New York, 1972; or M. C. Favas and D. L. Kepert, *Prog. Inorg. Chem.* 1980, **27**, 325.

the space around the central atom is filled completely. For compounds of elements of the second period, the bond angles rarely fall below the tetrahedral angle by more than a few degrees. Members of the later periods can accommodate more than four electron pairs in the valence shell; the maximum usually is six pairs. Six electron pairs in the valence shell of an atom should produce an octahedral arrangement with angles of 90° between electron pairs, and the bond angles of the hydrides of the heavier group V and VI elements do seem to approach 90° (see Table 2.4).[6] Thus the repulsion caused by the lone pair in a molecule such as PH_3 would tend to force the bonding pairs closer together until the repulsion among them became great at an angle close to 90°. The differences between the bond angles for the hydrides of each successive member of a family are small because the maximum number of electrons (six) in the valence shell remains constant after the first member of each family, even though the radius continues to increase.

The most striking argument for the great importance of repulsion among electron pairs, in contrast to steric repulsion between atoms arises from a comparison of bond angles between hydrogen and fluorine compounds of nitrogen and oxygen. The decrease in the bond angles for NF_3 and F_2O as compared with NH_3 and H_2O may be explained in terms of the decrease in repulsion between bonding pairs resulting from the electrons being drawn out further by the fluorine. The FCF bond angles in the C—F compounds shown in Table 2.4 are also smaller than the HCH bond angles in the corresponding C—H compounds.

The bond angles of the fluorine compounds are smaller than those of the corresponding hydrogen compounds only for second-period elements. Multiple bonding is not possible in NH_3, NF_3, or PH_3 but is expected to occur in PF_3, using a filled *p* orbital on F and an empty *d* orbital on P.

Effects of *p-d-π* Bonding

Earlier values of bond angles for PF_3 and AsF_3 were significantly larger than those of the corresponding chlorides. Recent values indicate that the bond angles increase regularly as the size of the halogen increases in each series for the phosphorous and arsenic halides (Table 2.4). The tendency to form multiple bonds is generally greater for small atoms (O, N, F) than for larger atoms (p. 59). Since the bond angles increase in the order $PF_3 < PCl_3 < PBr_3 < PI_3$, the effect of any decrease in bond order for the larger halogens might be offset by the increasing size of the bonding orbitals.

The F_2SO molecule is isoelectronic with ClO_3^- and, as expected, is pyramidal, with S at the apex. The formal charges are 0 for each F, +1 for S, and −1 for O, which forms a dative bond. There is the possibility of *p-d-π* bonding to S for F and O, but the negative formal charge on O and its lower electronegativity, compared with F, should favor greater π donation for O than for F. Because of the lone-pair repulsion all bond angles should be less than 109°, and the expected greater bond order for S—O compared with S—F should result in a larger-angle FSO than the angle FSF. (The angles can be compared in Table 2.4.) The bond lengths (S—O, 141.2 pm; S—F, 158.5 pm) are consistent with the predicted bond orders.

---

[6]The angle reported for H_2Te (89.5°) falls below 90° by such a small amount that it cannot be taken to signify a trend.

Table 2.5 Configurations from the number of electron pairs

Number of σ Pairs Plus Lone Pairs	Configuration (Considering Lone Pairs Directed as Any Other Group)	Hybridization
2	Linear	sp
3	Trigonal planar	sp^2
4	Tetrahedral	sp^3
5	Trigonal bipyramidal	$d_{z^2}sp^3$
5	Square pyramidal	$d_{x^2-y^2}sp^3$
6	Octahedral	d^2sp^3
7	Pentagonal bipyramidal	d^3sp^3

Molecules with More Than Four Sigma Bonding Plus Nonbonding Electron Pairs

Table 2.5 gives the overall geometrical configurations usually encountered for different total numbers of σ and lone pairs. Where the total number of σ bonding pairs and lone pairs is five or greater, several arrangements often are possible. The most significant characteristic of lone pairs seems to be a tendency to occupy as much space as available. In structures where axial and equatorial positions are not equivalent, lone pairs preferentially occupy those positions that provide the most space. Thus lone pairs prefer equatorial positions in a trigonal bipyramid, where the bond angles are 120° in the equatorial plane, compared with 90° for the axial positions. The reverse is true for a pentagonal bipyramid, where the equatorial positions are more crowded. If all positions are equivalent, as in an octahedron, two lone pairs will occupy opposite or *trans* positions.

Applying these guidelines to examples shown in Figure 2.17 is simple. Since there are five pairs about I in ICl_2^-, we begin with the trigonal bipyramid. The lone pairs occupy the more spacious equatorial positions, giving structure *(a)*, a linear arrangement of atoms. Given the preference of lone pairs for equatorial positions in a trigonal bipyramid, we would correctly predict *(a)* structures for $TeCl_4$ and ClF_3. However, the greater repulsions caused by lone pairs distort the regular geometry shown in Figure 2.17: the lone pair in $TeCl_4$ forces the chlorines closer together to produce a flattened and distorted tetragonal pyramid, and the axial fluorines in ClF_3 bend away from the lone pair to give a bent "T". The positions are equivalent for an octahedral arrangement of six pairs, so the two lone pairs of ICl_4^- occupy opposite positions and thus produce a square planar structure *(a)*. Obviously, only one structure is possible for BrF_5, since the lone pair can occupy any one of the equivalent positions of the octahedron. The influence of the lone pair is evident from the fact that the four equivalent fluorine atoms lie *above* the plane of the bromine, giving an F(axial)—Br—F(eq) bond angle of 84.5°. In the isoelectronic series TeF_5^-, IF_5, and XeF_5^+, the bond lengths are remarkably constant and the F(apical)—M—F(basal) bond angles are 79°, 81°, and 79°, respectively. You can also predict the pentagonal bipyramidal structure of IF_7.

As we have seen, repulsion is small between bonding pairs directed toward very electronegative atoms—for example, compare OF_2 and OH_2 (p. 67). The same considerations apply

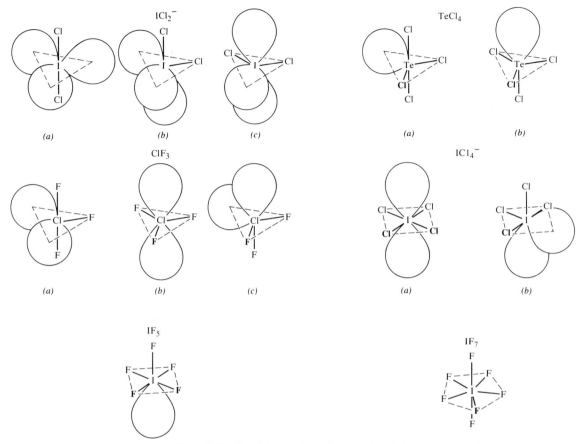

Figure 2.17 Possible structures for some interhalogen compounds.

in choosing between axial and equatorial positions in a trigonal bipyramid. The *least electro-negative atoms preferentially occupy equatorial positions*. Thus the methyl groups occupy equatorial positions in CH_3PF_4 and $(CH_3)_2PF_3$ (Figure 2.18). The fluorines bend *away* from the methyl groups. Similarly, double bonds are larger and much more delocalized than single bonds. Double bonds also prefer equatorial positions—as observed for SOF_4, where the oxygen occupies an equatorial position because the double-bond character of the S—O bond should be greater than that of the S—F bond. The structure of XeO_2F_2 indicates that the two F atoms are in axial positions and both oxygens and the lone pair in equatorial positions of a trigonal bipyramid. The bond angles deviate as expected from a regular trigonal bipyramid, because of differing repulsions among the various groups.

The Roles of Unshared Electrons in Metal Complexes

Unshared electrons in the valence shell are not always stereochemically active: that is, they do not behave as equivalent to substituents in determining molecular shapes. This is the case for

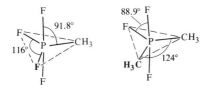

Figure 2.18 Structures of $PF_4(CH_3)$ and $PF_3(CH_3)_2$.

transition metal complexes that have unshared electrons in d orbitals. The Cr^{3+} ion has three unpaired d electrons; its six-coordinate complexes, such as $[Cr(NH_3)_6]^{3+}$, still have three unpaired electrons. Six electron pairs for the NH_3 molecules are accommodated in the d^2sp^3 hybrid orbitals. The d orbitals involved ($d_{x^2-y^2}$ and d_{z^2}) are those with lobes along the x, y, and z axes. Cobalt(III) complexes, such as $[Co(NH_3)_6]^{3+}$, generally are diamagnetic; the use of the $d_{x^2-y^2}$ and d_{z^2} orbitals for bonding forces the six electrons to occupy the other three d orbitals (Figure 2.19). For Cr(III) and Co(III) complexes, the unshared electrons are accommodated in d orbitals that are directed between the ligands. Since these nonbonding orbitals are not involved in the hybridization, the complexes are octahedral whether these orbitals are empty ($[Sc(H_2O)_6]^{3+}$), half-filled (Cr(III) complexes), or filled (Co(III) complexes). In general, the metal d electrons are not the equivalent of bonded groups for metal complexes.

Where nonbonding orbitals, or particularly, antibonding orbitals (p. 152) are occupied unequally, distortion of the regular polyhedron is common. This distortion results from the Jahn-Teller effect (p. 272), and not because lone pairs behave as equivalent to bonded groups.

The complex ions $TeCl_6^{2-}$, $TeBr_6^{2-}$, and $SbBr_6^{3-}$ have no partially filled d orbitals. Since each case has one lone pair on the metal, VSEPR leads us to expect a distorted octhedron, but that is not the case. In these exceptions, ligand-ligand repulsion is considered to be of overriding importance. Possibly, the stereochemically inactive lone pair occupies a spherical s orbital. (The case of XeF_6, which also contains one lone pair on Xe, is discussed on p. 602.) However, the complex ion $[Sb(C_2O_4)_3]^{3-}$ has the pentagonal pyramidal structure expected for a lone pair in an axial position of a pentagonal bipyramid. The Sb is 35 pm below the pentagonal plane and the Sb—O axial bond is about 20 pm shorter than the average Sb—O equatorial bond length. It has been pointed out that for elements to the right of the periodic table with coordination number six plus an unshared electron pair, the electron pair is active if the donor

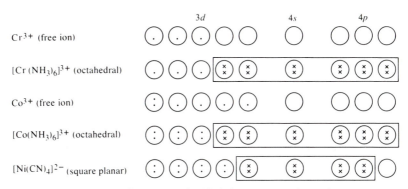

Figure 2.19 Population of orbitals for some metal complexes.

atoms are O^{2-} or F^- and inactive for Cl^- or Br^- donors. As we shall see later (p. 541), the electron clouds of O^{2-} and F^- are not easily deformed (polarized) in an electrostatic field and so are considered to be *hard* donors (bases). Chloride ion and bromide ion are larger and more polarizable (*soft* donors or bases). In these terms, small hard donors cause the unshared pair to be stereochemically active. Of course, repulsion among the large donors would favor the smaller effective coordination number (that is, for the unshared pair not to act as a seventh substituent). Also, where polarization is important, accommodating the lone pair in the spherically symmetrical *s* orbital would be most favorable.

2.4 ELECTRONEGATIVITY

2.4.1 Pauling's Electronegativity Scale

Electronegativity, which usually is introduced in modern elementary chemistry courses and has already been used here in the discussion of oxidation numbers and formal charges in resonance structures, can be defined as the attraction of an atom in a molecule for electrons. Students soon acquire a feeling for the relative electropositive character of metals and electronegative character of nonmetals. Within a periodic family the electronegative character generally decreases with increasing atomic radius, although there are some exceptions. A more detailed consideration of the basis for an electronegativity scale follows.

Pauling established a scale of electronegativities based on "excess" bond energies. The energy of the bond, $D(A—B)$, is the sum of a nonpolar contribution, $D_{np}(A—B)$, and a polar contribution, $D_p(A—B)$. The nonpolar contribution is the energy expected from the mean of the nonpolar bond energies $D(A—A)$ and $D(B—B)$. The polar contribution, designated Δ, is a measure of the bond's expected polarity, which results from a displacement of the electronic charge toward one atom. Ordinarily, the polarity of the bond increases with increasing differences in electronegativities of A and B.

$$D(A—B) = D_{np}(A—B) + D_p(A—B) \tag{2.6}$$

$$D_{np}(A—B) = \frac{D_{np}(A—A) + D_{np}(B—B)}{2} \tag{2.7}$$

$$D_p(A—B) = \Delta = D(A—B) - \frac{D_{np}(A—A) + D_{np}(B—B)}{2} \tag{2.8}$$

The Δ values are negative for some of the active metal hydrides, but the negative values can be avoided by using the geometric mean of the bond energies ($\sqrt{D(A—A) \times D(B—B)}$) to give Δ' values. From Table 2.6, which provides data for a few important elements, it can be seen that the sum of the Δ' values for Si—O and O—F does not give the Δ' value for Si—F that might be expected if Δ' were the actual difference in the electronegativities of the respective elements. The $\sqrt{\Delta'}$ values are more nearly additive, so they are used instead. A table of $\sqrt{\Delta'}$ values is a table of differences in electronegativities. We must make an arbitrary assignment to one element in order to convert the $\sqrt{\Delta'}$ values to electronegativity values for individual elements.

Table 2.6 Electronegativities from bond polarities

Atom A	Atom B	$\sqrt{D(A-A) \times D(B-B)}$	D(A—B) (observed)	Difference Δ'	$\sqrt{\Delta'}$
Si	O	172 kJ	372 kJ	200	14.1
O	F	146	184	38	6.2
Si	F	180	536	356	18.9
H	Cl	326	431	105	10.2
Cl	O	180	205	25	5.0
H	O	247	452	205	14.3

Applying the postulate of the geometric mean is more difficult than applying the postulate of additivity, since values of Δ can be obtained directly from heats of reaction. We need individual bond energies to calculate Δ'. Because of the paucity of single bond energies, the Δ values usually are used; for most bonds there is not much difference between the two mean values. Pauling used the relationship

$$\Delta = 96.49(X_A - X_B)^2 \tag{2.9}$$

or

$$X_A - X_B = 0.102\sqrt{\Delta} \tag{2.10}$$

Combining Equations 2.6–2.9 gives the relationship

$$D(A-B) = \tfrac{1}{2}\{D(A-A) + D(B-B)\} + 96.49(X_A - X_B)^2 \tag{2.11}$$

which implies that the contribution of the new bond to the heat of formation of the substance is equal to Δ, or $96.49(X_A - X_B)^2$. So if the substance consists of diatomic molecules, AB, its heat of formation (ΔH_f) is equal to Δ. In the more general case of the compound AB_n, the heat of formation is obtained by summing the expression over all of the bonds in the molecule.

$$-\Delta H_f = n \times 96.49(X_A - X_B)^2 \tag{2.12}$$

or

$$(X_A - X_B) = 0.102\sqrt{\frac{-\Delta H_f}{n}} \tag{2.13}$$

These relationships assume that the reactants and products have the same number of covalent bonds.

We must apply a correction for compounds containing N and O, because of the extra stability of N_2 and O_2 in their standard states. The triply bonded N_2 molecule is more stable than a hypothetical molecule for which the bond energy is taken as three times the N—N single-bond energy (160 kJ/mole). This extra stability is $942 - 3 \times 160 = 462$ kJ/mole N_2 or 231 kJ/g-atom N. The quantity 231 kJ/g-atom N serves as a correction for the calculation of the heats of formation of nitrogen compounds. The extra stability of the O_2 molecule is 494 kJ/mole O_2 − 2 × 142 kJ/mole O—O = 210 kJ/mole O_2 or 105 kJ/g-atom O.

Applying the corrections to Equation 2.12 gives

$$-\Delta H_f = n \times 96.49(\chi_A - \chi_B)^2 - 231 n_N - 105 n_O \qquad (2.14)$$

where n_N and n_O are the numbers of N and O atoms, respectively, in the compound. In this way Pauling was able to extend the table of electronegativities to include most of the elements.

2.4.2 Allred's Electronegativity Values

Allred used the most recent thermochemical data to calculate the electronegativities of 69 elements. Where the necessary data were available, he calculated electronegativities from single-bond energies as described by Pauling. For the many metals whose M—M bond energies are not known, electronegativities were obtained from the heats of formation of the compounds, using Equation 2.14. Electronegativities were calculated for several compounds of M in most cases, then averaged to give the values in Table 2.7. Electronegativities calculated using Equations 2.9 and 2.14 agreed to a significant degree.

Most of Allred's electronegativity values are essentially the same as Pauling's. In several respects however, Allred's values agree better with the chemical behavior of the elements than those of Pauling. The alternation in electronegativities of the main group III and IV elements should be expected from the chemistry of these elements and can be explained by the transition metal contraction and the lanthanide contraction.

As the oxidation number or formal charge on an atom increases, the attraction for electrons increases, and hence the electronegativity must increase. The electronegativity values in Table 2.7 are for the oxidation numbers indicated, and Table 2.8 displays Allred's values for a few

Table 2.7 The electronegativity scale of the elements[a]

I	II	III	II	II	II	II	II	II	II	I	II	III	IV	III	II	I
H 2.2																
Li 1.0	Be 1.6											B 2.0	C 2.6	N 3.0	O 3.4	F 4.0
Na 0.9	Mg 1.3											Al 1.6	Si 1.9	P 2.2	S 2.6	Cl 3.2
K 0.8	Ca 1.0	Sc 1.4	Ti 1.5	V 1.6	Cr 1.7	Mn 1.6	Fe 1.8	Co 1.9	Ni 1.9	Cu 1.9	Zn 1.7	Ga 1.8	Ge 2.0	As 2.2	Se 2.6	Br 3.0
Rb 0.8	Sr 1.0	Y 1.2	Zr 1.3	Nb *1.6*	Mo 2.2	Tc *1.9*	Ru *2.2*	Rh 2.3	Pd 2.2	Ag 1.9	Cd 1.7	In 1.8	Sn 2.0	Sb 2.1	Te *2.1*	I 2.7
Cs 0.8	Ba 0.9	La–Lu 1.1–1.3	Hf *1.3*	Ta *1.5*	W 2.4	Re *1.9*	Os *2.2*	Ir 2.2	Pt 2.3	Au 2.5	Hg 2.0	Tl 2.0	Pb 2.3	Bi 2.0	Po *2.0*	At *2.2*
		Ac *1.1*	Th *1.3*	Pa *1.5*	U *1.4*	Np–No *1.3*	Pu *1.3*									

[a]The oxidation numbers are specified at the top of each group. Values are from A. L. Allred, *J. Inorg. Nucl. Chem.* 1961, *17*, 215—except those underlined, which are from Pauling.

Table 2.8 Electronegativities of some elements in different oxidation states

Mo(II)	2.18	Fe(II)	1.83	Sn(II)	1.80
Mo(III)	2.19	Fe(III)	1.96	Sn(IV)	1.96
Mo(IV)	2.24	Tl(I)	1.62	Pb(II)	1.87
Mo(V)	2.27	Tl(III)	2.04	Pb(IV)	2.33
Mo(VI)	2.35				

elements in different oxidation states. Since the variation is not large, a single value to the nearest 0.1 unit can be used for most purposes. The concept of electronegativity is not sufficiently quantitative to make small differences significant.

2.4.3 Mulliken's Electronegativity Scale

Soon after Pauling introduced his scale of electronegativities, Mulliken proposed another basis for an electronegativity scale. Seeking a more exact relationship between the properties of an atom and its electronegativity, Mulliken used the *ionization energy* (IE), the energy required for the removal of an electron from a gaseous atom, and the *electron affinity* (EA), the energy involved in the addition of an electron to the gaseous atom. An element's electronegativity is the average of the ionization energy and the electron affinity (both in electron volts, 1 eV/molecule = 96.49 kJ/mole).

$$\text{Electronegativity} = \frac{\text{IE} + \text{EA}}{2} \tag{2.15}$$

The resulting values are about 2.8 times as large as Pauling's values. The advantage of Mulliken's electronegativity scale is that different values for elements can be obtained by taking into account different ionic states and changes in hybridization.

2.4.4 Group Electronegativities

Some observers have criticized the electronegativity concept for its weak theoretical foundation. A single electronegativity value for an element is obviously inadequate, since the value varies with oxidation number, substituents (compare the hydrogen bonding ability of $HCCl_3$ with that of CH_4), and hybridization (compare the acidity and hydrogen bonding ability of HCCH with those of CH_4). Using an extension of Mulliken's approach, Jaffé and others have calculated group electronegativities allowing for the charge of a group, effects of substituents, and the hybridization of a bonding orbital. As you might expect, positive groups have higher electronegativities than neutral groups, CF_3 is much more electronegative than CH_3, and electronegativity increases with an increase in the *s* character of a hybrid orbital—that is, $sp > sp^2 > sp^3$. These refinements extend the usefulness of the electronegativity concept and make it more generally applicable, but like those modifications designed to remove some shortcomings of valence bond theory, they destroy much of the value of the simple concept. The average electronegativity value of an element is still useful for qualitative predictions. The group electronegativities have greater significance, but for exacting work you might prefer to use directly the information needed to calculate the group electronegativities.

> Simple concepts often play a very important role in the development of science, stimulating ideas and research that ultimately reveal the inadequacy of the qualitative concept. When this stage is reached, we should still recognize the value of the concept in the progress made and either discard it if it is no longer useful or use it with an eye to its limitations.

2.4.5 Partial Ionic Character in Covalent Compounds

It has been pointed out that the electron distribution in covalent bonds joining dissimilar atoms is usually unsymmetrical, with the electron density greater near the more electronegative atom. Polar covalent bonds can be described either in terms of shared electrons distributed unsymmetrically or as the resonance hybrid of a covalent structure (electrons shared equally) and an ionic structure (electrons transferred).

Some confusion exists in the use of the terms *polar* and *nonpolar,* which have been used synonymously with "ionic" and "covalent," respectively. We shall describe substances as ionic if they give ionic lattices, conducting melts and solutions, etc., reserving *polar* and *nonpolar* for different types of covalent bonds. A *nonpolar covalent bond* involves equal or nearly equal sharing of the bonding electrons; the difference in the electronegativities should be less than 0.5. A *polar covalent bond* is one with an appreciable amount of ionic character.

The greater the difference in the electronegativities of bonded atoms, the greater the polarity or ionic character of the bond. Thus we should be able to determine the ionic character of a bond from an empirical relationship between ionic character and electronegativity differences. The difficulty has been the lack of a method of giving reliable ionic characters for a wide range of compounds in order to establish the empirical relationship. Pauling has used the *dipole moment* as a measure of the polarity and hence of the ionic character of a bond. A diatomic molecule, such as HF, in which the electrons are not shared equally, will have a positive and a negative end. Such a molecule acts as a dipole and tends to become aligned in an electrical field. The force that acts on a dipole to align it in an electrical field is a measure of the dipole moment. The greater the ionic character of a bond, the greater the charge separation and the larger the dipole moment.

The electric dipole moment, μ, is the product of the charge at one pole, q, and the distance between poles, d, as in

$$\mu = qd \qquad (2.16)$$

Dipole moments are expressed commonly in *Debye units*. One Debye unit is 10^{-8} pm-esu (10^{-18} cm-esu). Two charges equal in magnitude to the charge of an electron (4.8×10^{-10} esu) and separated by a distance of 91.7 pm (the interatomic distance for the HF molecule) would give a dipole moment of 4.4 Debye units. A value of 4.4 Debye units represents the expected dipole moment for 100% ionic HF. The percentage of ionic character is obtained from the ratio of the observed dipole moment to the dipole moment for complete electron transfer. The observed dipole moment for HF is 1.98 Debye units, corresponding to 1.98/4.4 or 45% ionic character. This treatment assumes that the total dipole moment results from the unsymmetrical distribution of charge in the bond.

Pauling derived the relationship

$$\text{Amount of ionic character} = 1 - e^{-(1/4)(\chi_A - \chi_B)^2} \tag{2.17}$$

from a plot of electronegativity difference vs. percentage of ionic character. Percentages of ionic character originally were obtained from the measured dipole moments of HCl, HBr, and HI, which gave 19%, 11% and 4% ionic character, respectively, and an estimated 60% ionic character for HF. The relationship has been revised by others to take into account the corrected value for the percentage of ionic character of HF from dipole moment measurements. However, Pauling argues that the deviation of HF from the equation is justified and that his relationship is probably as useful as the revised ones. These relationships can be used as qualitative guides only. Rough and easy-to-remember values for electronegativity differences and percentage of ionic character are 1.0, 20%; 1.5, 40%; 2.0, 60%; and 2.5, 80%.

In discussing the use of dipole moments as a guide to the ionic character of bonds, we must emphasize that it is the bond moment, not the total dipole moment of a molecule, that must be considered in all cases. The C—F bond is distinctly polar, but the tetrahedral arrangement of the four bonds in CF_4 results in a molecule with zero dipole moment. The hybridization of a bonding orbital also affects the nonbonding orbitals, causing the lone pairs of electrons to be distributed dissymmetrically. Coulson reported that the bond moments of the water molecule contribute only about one quarter of the total dipole moment. The lone pair moment provides the major contribution.

Our approach to describing polar bonding started with the nonpolar covalent model as the limiting case—an approach particularly suited to discrete molecules. You also can start with an ionic model as the limiting case and consider varying degrees of distortion or polarization of the electron cloud of the anion, or even of both ions. Extreme polarization is equivalent to electron sharing. The latter approach is most useful in understanding cases in which the ionic model is a reasonable approximation (p. 220).

2.5 RADII OF ATOMS IN COVALENT COMPOUNDS

2.5.1 Covalent Radii

The bond length—the distance between two bonded atoms—in a covalent molecule such as Cl_2 is the sum of the covalent radii of the atoms. Thus the covalent radius of the chlorine atom is one half of the chlorine-chlorine distance (198.8 pm). The covalent radius for nonmetals is often called the atomic radius. The carbon-carbon bond distance in diamond (154.4 pm) is close to that in ethane (153.4 pm) and other saturated hydrocarbons. Carbon has a single-bond covalent radius of 77 pm, so combining the covalent radii of carbon and chlorine gives 176 pm as the expected C—Cl bond distance—in close agreement with that observed for CCl_4.

The single-bond radii of nitrogen and oxygen are not obtained from the bond lengths in N_2 (110 pm) and O_2 (120.8 pm), since these molecules contain multiple bonds. Instead, we

Hollis F. Price Library
LeMoyne-Owen College
Memphis, Tennessee

take the single-bond radii from the bond lengths in compounds such as hydrazine, H_2NNH_2, and hydrogen peroxide, HOOH. The radius of nitrogen obtained from N_2 (55 pm) is the triple-bonded radius. A triple-bonded radius for carbon can be obtained from the carbon-carbon bond length in acetylene (120 pm). Combining the triple-bonded radii of carbon and nitrogen gives the expected bond length ($55 + 60 = 115$ pm) for a $C \equiv N$ bond, as compared with the observed bond length of 116 pm for CH_3CN. A double-bonded radius of carbon can be obtained from the carbon-carbon bond length (134 pm) in ethylene (see Table 2.1). The presence of a multiple bond changes the hybridization (sp^2 for C in $H_2C = CH_2$ and sp for C in $HC \equiv CH$) and the C—H single-bond distance. The C—H bond length decreases with increasing s character: 109.6 pm for CH_4, 108.4 pm for $H_2C = CH_2$, and 105.8 pm for $HC \equiv CH$ (see p. 57). Table 2.9 presents bond dissociation energies and bond lengths for calculating covalent radii for molecules of interest.

We usually can assume that covalent radii can be added to give bond lengths, provided that the bond for which the length is calculated is similar to the bonds used for the evaluation of the covalent radii in bond order, bond hybridization, and bond strength. Schomaker and Stevenson have attributed the discrepancy between calculated and observed bond lengths to variations in bond polarity for most cases where multiple bonding is thought to be unimportant. Thus the bond length, r_{AB} in pm, can be calculated from the nonpolar radii of A and B and the difference in electronegativities of A and B.

$$r_{AB} = r_A + r_B - 9|\chi_A - \chi_B| \qquad (2.18)$$

Table 2.9   Bond energies and bond lengths

Bond	Molecule or Crystal	Bond Dissociation Energy[a] (kJ/mole)	Bond Distance[b] (pm)
H—H	H_2	432.08	74.14
H—C	CH_4	**425.1 ± 8**	109.6
H—N	NH_3	**431 ± 8**	101.2
H—O	H_2O	**493.7 ± 0.8**	95.7
H—F	HF	566.3	91.7
H—Si	SiH_4	**398.3**	148.0
H—P	PH_3	~322	143.7
H—S	H_2S	**377 ± 4**	133.5
H—Cl	HCl	427.78	127.4
H—Ge	GeH_4	**365**	153
H—As	AsH_3	~247	151.9
H—Se	H_2Se	276(?)	146
H—Br	HBr	362.6	141.4
H—Sn	SnH_4	<320	170.1
H—Sb	SbH_3	—	170.7
H—Te	H_2Te	238(?)	170
H—I	HI	294.68	160.9
B—F	BF_3	**665**	129
C—C	C_2	599	124.3
	Diamond	—	154.45
	C_2H_6	350	153.4

(*Continued*)

Table 2.9 Bond energies and bond lengths (*Continued*)

Bond	Molecule or Crystal	Bond Dissociation Energy[a] (kJ/mole)	Bond Distance[b] (pm)
C—N	CH_3NH_2	**331 ± 13** (298 K)	147.4
C—O	CO	1070.2	112.8
	CO^+	804.5	111.5
	CO^-	784	—
C—F	CH_3F	**452 ± 21** (298 K)	138.5
C—Si	SiC	320	—
	$Si(CH_3)_4$	301	187.0
C—P	$P(CH_3)_3$	260	184.1
C—S	CH_3SH	**297 ± 13**	181.9
	$S(CH_3)_2$	270	181.7
C—Cl	CH_3Cl	**335.1**	178.1
	CH_2Cl_2	**314 ± 75**	177.2
	$CHCl_3$	—	176.2
	CCl_4	280	176.6
C—Br	CBr_4	**205 ± 13**	194
C—I	CH_3I	**226 ± 13**	213.9
N—N	N_2	941.66	109.77
	N_2^+	840.67	111.6
	N_2^-	765	119
	N_2H_4	**247 ± 13** (298 K)	145.1
	N_2F_4	88	
N—O	NO	626.86	115.1
	NO^+	1046.9	106.3
	NO^-	487.8	125.8
N—F	NF_3	**238 ± 8**	136
N—Cl	NCl_3	**381** (298 K)	—
O—O	O_2	493.59	120.75
	O_2^+	642.9	111.6
	O_2^-	395.0	(135)
	H_2O_2	**207.1 ± 2.1**	148
O—F	OF_2	**268 ± 13**	142
O—Cl	Cl_2O	**139.3 ± 4**	170
F—F	F_2	154.6	141.2
Si—Si	Si_2	310	224.6
	Si (diamond str.)	—	235.2
Si—F	SiF_4	564	156.1
Si—Cl	$SiCl_4$	380	201.9
Si—Br	$SiBr_4$	309	216
Si—I	SiI_4	234	243.5
P—P	P_2	485.6	189.3
	P (black)	—	222.4
P—F	PF_3	—	153.5
P—Cl	PCl_3	326	203
P—Br	PBr_3	263	223
P—I	PI_3	184	246
S—S	S_2	421.58	188.9
	S_8	226	205

(*Continued*)

Table 2.9   Bond energies and bond lengths (*Continued*)

Bond	Molecule or Crystal	Bond Dissociation Energy[a] (kJ/mole)	Bond Distance[b] (pm)
S—F	SF_6	284	156
S—Cl	S_2Cl_2	255	207
S—Br	S_2Br_2	217(?)	224
Cl—Cl	Cl_2	239.23	198.8
Cl—F	ClF	252.53	162.8
Cl—Br	BrCl	215.46	213.6
Cl—I	ICl	207.75	232.1
Ge—Ge	Ge_2	272	—
	Ge (diamond str.)	188	245
Ge—F	GeF_4	—	168
Ge—Cl	$GeCl_4$	332.5	208
Ge—Br	$GeBr_4$	276	229.7
As—As	As_2	382	210.3
	As_4	146	243
As—O	As_4O_6	—	178
As—F	AsF_3	464	171.2
As—Cl	$AsCl_3$	**444**	216.1
As—Br	$AsBr_3$	242	233
As—I	AsI_3	213	254
Se—Se	Se_2	320	216.6
	α-Se	—	237.4
Br—Br	Br_2	190.15	228.1
Br—F	BrF	245.8	175.9
Br—I	IBr	175.4	246.9
Sn—Sn	Sn_2	190	—
	α-Sn (gray)	—	281.0
Sn—Cl	$SnCl_4$	318	233
Sn—Br	$SnBr_4$	272	246
Sn—I	SnI_4	—	269
Sb—Sb	Sb_2	298	234.2
	Sb	—	290
Sb—Cl	$SbCl_3$	309	232.5
Sb—Br	$SbBr_3$	—	251
Te—Te	Te_2	258.3	255.6
	α-Te	—	283.4
Te—Cl	$TeCl_4$	—	233
Te—Br	$TeBr_2$	—	251
I—I	I_2	148.82	266.6
I—F	IF	277.8	191.0

[a]Bond energies and bond lengths for diatomic molecules and ions are for 0 K, from K. P. Huber and G. Herzberg, *Molecular Spectra and Molecular Structure,* Vol. IV, *Constants of Diatomic Molecules,* Van Nostrand Reinhold, New York, 1979. Values for molecules given in boldface type are from B. deB. Darwent, *Bond Dissociation Energies in Simple Molecules,* NSRDS-NBS31, 1970. The values from Darwent are for 0 K except as specified for 298 K. In many cases the differences between the values at 0 K and 298 K are smaller than the uncertainty. Other values are from T. L. Cottrell, *The Strengths of Chemical Bonds,* 2nd ed., Butterworths, London, 1958. The uncertainties in these values likely will be great.
[b]Bond lengths for diatomic molecules and ions are from Huber and Herzberg (see footnote a); for most other elements, from J. Donohue, *The Structures of the Elements,* Wiley, New York, 1974. Most of the data for other compounds are from L. E. Sutton, Ed., "Tables of Interatomic Distances and Configuration in Molecules and Ions," Special Publication No. 18, The Chemical Society, London, 1965. In a few cases data are from Sutton's original compilation, Special Publication No. 11, The Chemical Society, London, 1958.

GENERAL REFERENCES

J. K. Burdett, *Molecular Shapes*, Wiley-Interscience, New York, 1980. Excellent coverage of inorganic stereochemistry, including VSEPR, Walsh diagrams, main-group and transition-metal complexes, and cluster compounds.

E. Cartmell and G. W. A. Fowles, *Valency and Molecular Structure*, 4th ed., Butterworths, London, 1977. Good treatment of bonding at about the level of this text.

T. L. Cottrell, *The Strengths of Chemical Bonds*, 2nd ed., Butterworths, London, 1958. Good source of data, even though it is dated.

R. L. DeKock and H. B. Gray, *Chemical Structure and Bonding*, Benjamin/Cummings, Menlo Park, Cal., 1980. Good supplement to this text.

R. McWeeny, *Coulson's Valence*, 3rd ed., Oxford, Oxford, 1979. A classic book brought up to date.

L. Pauling, *The Nature of the Chemical Bond*, 3rd ed., Cornell University Press, Ithaca, N.Y., 1960. A classic.

PROBLEMS

2.1 Indicate the electronic configuration expected for the possible covalent compounds, MX_n (where $X^- = Cl^-$ or Br^-), of Sn, At, and Ra. Assuming only σ bonding, predict the geometry of each molecule.

2.2 What ionic states are expected for the elements given in Problem 2.1? What ionic states are expected for Ti, Pr, and Se?

2.3 Assign formal charges and oxidation numbers, and evaluate the relative importance of three Lewis structures of OCN^-, cyanate ion. Compare these to the Lewis structures of the fulminate ion, CNO^-.

2.4 Write a reasonable electron dot structure and assign formal charges and oxidation numbers for each of the following: ClF, ClF_3, ICl_4^-, $HClO_3$ ($HOClO_2$).

2.5 Give the oxidation number, formal charge, and hybridization of the central atom in each of the following: NO_3^-, BF_4^-, $S_2O_3^{2-}$, ICl_2^+, ClO_3^-. What are the molecular shapes?

2.6 Select the reasonable electron-dot structures for each of the following compounds. Indicate what is wrong with each incorrect or unlikely structure.

2.7   Give the expected hybridization of P, O, and Sb in Cl_3P—O—$SbCl_5$. The P—O—Sb bond angle is 165°.

2.8   Predict both the gross geometry (from the σ orbital hybridization) and the fine geometry (from bond-electron pair repulsion, etc.) of the following species: F_2SeO, $SnCl_2$, I_3^-, and $IO_2F_2^-$.

2.9   Which of the following in each pair will have the larger bond angle? Why? CH_4, NH_3; OF_2, OCl_2; NH_3, NF_3; PH_3, NH_3.

2.10   Trimethylphosphine has been reported by Holmes to react with $SbCl_3$ and $SbCl_5$ to form, respectively, $(Me_3P)(SbCl_3)$ or $(Me_3P)_2(SbCl_3)$ and $(Me_3P)(SbCl_5)$ and $(Me_3P)_2(SbCl_5)$. Suggest valence bond structures for each of these and indicate approximate bond angles around the Sb atom.

2.11   Indicate by a sketch the following hybrid orbitals (indicate the sign of the amplitude of the wave function on your sketch): *a.* an *sd* hybrid   *b.* a *pd* hybrid.

2.12   The Ru—O—Ru bond angle in $(Cl_5Ru)_2O$ is 180°. What is the state of hybridization of the oxygen? Explain the reasons for the large bond angle. (See R. J. Gillespie, *J. Am. Chem. Soc.* 1960, *82*, 5978.)

2.13   Using H. A. Bent's isoelectronic groupings, identify the species of line **(2)** that are ''isoelectronic'' with each of those of line **(1)**. Each entry might have more than one ''isoelectronic'' partner in the other line.
   1. $(H_2N)_2CO$ $HONO_2$ OCO   CO $(CN)_2$ ClO_2^+ $Si_3O_9^{6-}$ $TeCl_2$ Bi_9^{5+}
   2. H_2CCO $(HO)_2CO$ ONN H_2CNN H_3CNO_2 F_2CO BF B_2O_2 $(CH_3)_2CO$ CH_2CN^-
   SO_2 cyclic $(SO_3)_3$ ICl_2^+ Pb_9^{4-}

2.14   Consult Pauling's electroneutrality principle (p. 60) and describe the reasonable charge distribution in NH_4^+ and in SO_4^{2-}.

2.15   The bromine atom in BrF_5 is below the plane of the base of the tetragonal pyramid. Explain.

2.16 H_2C=SF_4 gives the expected isomer with the double-bonded methylene group in an equatorial position. From the orientation of the π bond, would you expect the H atoms to be in the axial plane or the equatorial plane? (See K. O. Christe and H. Oberhammer, *Inorg. Chem.* 1981, *20*, 296.)

2.17   Compare the expected bond orders of ClO_4^- and IO_4^-, and of IO_3^-, BrO_3^-, and ClO_3^-. (See R. Nightingale, *J. Phys. Chem.* 1960, *64*, 162.)

2.18   The C—C bond distance in methyl acetylene is unusually short (146 pm) for a single bond. Show how this short bond can be rationalized in more than one way.

2.19   Calculate the dipole moment to be expected for the ionic structure H^+Cl^- using the same internuclear separation as for the HCl molecule. Calculate the dipole moment for HCl assuming 19% ionic character.

2.20   Calculate the heats of formation (from electronegativities) for: H_2S, H_2O, SCl_2, NF_3, and NCl_3. Calculate the heats of formation of these compounds from the bond energies for comparison.

2.21   Calculate the electronegativity differences from the bond energies for H—Cl, H—S, and S—Cl. Using Allred's electronegativity of H, compare the electronegativity values that can be obtained for S and Cl with those tabulated.

2.22   Allred and Rochow (*J. Inorg. Nucl. Chem.* 1958, *5*, 264–8, 269–88) have proposed the following empirical equation for the calculation of electronegativities

$$\chi = 0.359 \frac{Z_{eff}}{r^2} + 0.744$$

where Z_{eff} is calculated using Slater's rules (p. 32) and r is the covalent radius (in Å). Using radii (in Å) from data in Table 2.9, calculate the electronegativities of As and Br. Little and Jones used this equation to calculate a complete set of electronegativities (*J. Chem. Educ.* 1960, *37*, 231). What are the advantages and disadvantages of this method of obtaining electronegativities?

III
Symmetry

Most of this chapter deals with symmetry groups. To make full use of symmetry, we need the group theoretical treatment, but symmetry and symmetry groups will be adequate for the entire text except for brief sections at the ends of Chapters 4 and 7. So the development of group theory (Section 3.5) can be omitted, along with the group theoretical approach to molecular orbital theory (Sections 4.3, 4.4, and 7.10). The terminology and notation used here are very important in modern inorganic chemistry. By introducing the terminology and notation of group theory (Section 3.6), we gain an appreciation of the results of the molecular orbital treatments using group theory, without working through all of the detail. Matrix notation is used in introducing symmetry operations, and matrix multiplication is treated in Appendix 3.1. However, the use of matrices is not essential for understanding symmetry.

3.1 INTRODUCTION

Symmetry greatly enhances the beauty of nature, as is very evident in leaves, flowers, and crystals. Molecular structure reveals remarkable ramifications of symmetry, which are developed further in the orderly packing of molecules in crystals. From Chapter I, you are familiar with the descriptions of atomic orbitals for the hydrogen atom. In a chemical compound the orbital shapes are dictated by symmetry, which can even be used to describe their energy states. Animated films may make molecular vibrations seem random, but they can be resolved into characteristic vibrational modes that are determined by and described in terms of the molecular symmetry. Spectroscopic selection rules can be expressed in terms of the symmetry of energy states, whether these are electronic or vibrational levels. Utilizing symmetry greatly diminishes the work necessary in determining a crystal structure or in carrying out molecular orbital calculations. Thus familiarity with symmetry simplifies our task and aids our understanding, even as it greatly enhances our appreciation of the awesome beauty of chemistry, and of nature in general.

3.2 SYMMETRY ELEMENTS

Symmetry is defined here as an invariance to transformation. The nature of the transformation defines the type of symmetry. Thus the equation $xy + yz + xz = xyz$ shows *permutational* symmetry, as does the palindrome "Able was I ere I saw Elba." The equation is unchanged by any of the possible permutations of x, y, and z among themselves, whereas the palindrome is unchanged by the specific permutation involving exchange of the first and last letters, the second and penultimate letters, . . . x and $n - x$ letters, and thus is invariant to being read forwards or backwards.

3.2.1 Mirror Plane (σ)

Mirror symmetry denotes invariance to reflection, as illustrated in Figure 3.1. It may be possessed by a single object, or by a *set* of objects that need not individually exhibit the particular symmetry. Thus a unit cell of a racemate (a 50–50 mixture of optical isomers) may have a mirror plane because of the presence of both d and l molecules in the unit cell (more commonly, a center of symmetry is present—*vide infra*).

To be invariant under the reflection operation, the reflection must be carried out in the plane that causes the image to be coincident with the object or set of objects. Such a plane is called a *symmetry plane* (or *mirror plane*) and is designated by the Greek letter σ (or by the letter m). The *reflection operation* that carries out this reflection also is given the designation σ (or m).

Reflection in the yz plane takes a point from the position (x,y,z) to a new position $(-x,y,z)$—only the out-of-plane coordinate changes sign. Alternatively, we can say that the transformation equations for σ_{yz} are

$$x' = (-1)x + (0)y + (0)z$$
$$y' = (0)x + (1)y + (0)z \tag{3.1}$$
$$z' = (0)x + (0)y + (1)z$$

We can write these equations more compactly in matrix notation as

$$\begin{bmatrix} x' \\ y' \\ z' \end{bmatrix} = \begin{bmatrix} -1 & 0 & 0 \\ 0 & 1 & 0 \\ 0 & 0 & 1 \end{bmatrix} \begin{bmatrix} x \\ y \\ z \end{bmatrix} \tag{3.2}$$

or state simply that the σ_{yz} transformation matrix is $\begin{bmatrix} -1 & 0 & 0 \\ 0 & 1 & 0 \\ 0 & 0 & 1 \end{bmatrix}$.

Example 3.1   Write the transformation matrix for the reflection of a point with coordinates (x,y,z) through the σ_{xy} plane.

Solution   Since σ_{xy} leaves the x and y coordinates unchanged and changes the sign of z,

$$\begin{bmatrix} x' \\ y' \\ z' \end{bmatrix} = \begin{bmatrix} 1 & 0 & 0 \\ 0 & 1 & 0 \\ 0 & 0 & -1 \end{bmatrix} \begin{bmatrix} x \\ y \\ z \end{bmatrix}.$$

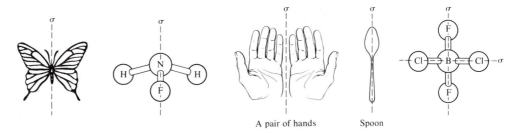

Figure 3.1 Mirror symmetry—the planes of symmetry are the solid lines labeled σ perpendicular to the plane of the page.

Example 3.2 Write the transformation matrix for the reflection of a point through the plane containing the z axis and at $45°$ to the x and y axes.

Solution The z coordinate will be unchanged, so you can check your result in the xy plane. The point $(x,0)$ would become $(0,x)$ and the point $(0,y)$ would become $(y,0)$, *or* $(x,y) \rightarrow (x',y')$, where x' $= y$ and $y' = x$. So the matrix interchanges x and y, $\begin{bmatrix} x' \\ y' \\ z' \end{bmatrix} = \begin{bmatrix} 0 & 1 & 0 \\ 1 & 0 & 0 \\ 0 & 0 & 1 \end{bmatrix} \begin{bmatrix} x \\ y \\ z \end{bmatrix}$. See Appendix 3.1 at the end of the chapter for a discussion of matrix multiplication. You can also draw the points and symmetry plane in the xy plane to verify the result.

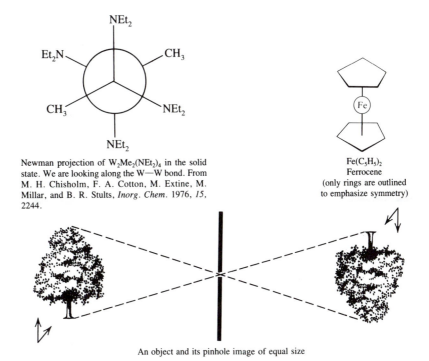

Newman projection of $W_2Me_2(NEt_2)_4$ in the solid state. We are looking along the W—W bond. From M. H. Chisholm, F. A. Cotton, M. Extine, M. Millar, and B. R. Stults, *Inorg. Chem.* 1976, *15*, 2244.

$Fe(C_5H_5)_2$
Ferrocene
(only rings are outlined to emphasize symmetry)

An object and its pinhole image of equal size

Figure 3.2 Inversion symmetry.

3.2.2 Center of Symmetry *(i)*

The *inversion operation* takes a point on a line through the origin, the *inversion center,* to an equal distance on the other side, thus transforming a point with the coordinates (x,y,z) to one with the coordinates $(-x, -y, -z)$. The transformation matrix for inversion is, therefore

$$\begin{bmatrix} -1 & 0 & 0 \\ 0 & -1 & 0 \\ 0 & 0 & -1 \end{bmatrix}$$, the matrix giving the coefficients for Equation 3.1. An object invariant under

the inversion operation is said to possess a *center of symmetry* or *inversion center* (see Figure 3.2). Both the center of symmetry and the inversion operation are designated by the symbol *i.*

Example 3.3   Which of the following molecules have a center of symmetry? Which have mirror planes?

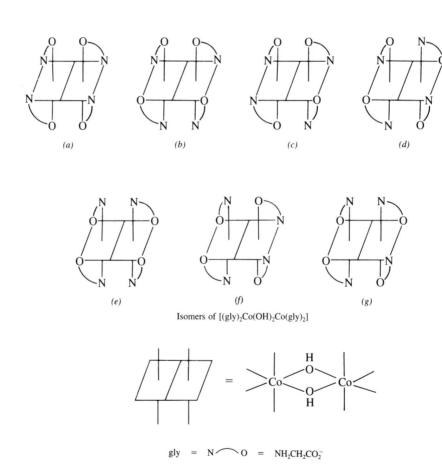

Isomers of $[(gly)_2Co(OH)_2Co(gly)_2]$

$gly = N\frown O = NH_2CH_2CO_2^-$

Solution   Isomers *(a), (d),* and *(e)* have *i.* Isomers *(a), (b)* and *(e)* have σ through the shared edge. The chelate rings eliminate possible σ planes perpendicular to the shared edge.

3.2.3 Axis of Rotation (C_n)

An object has *axial symmetry* (see Figure 3.3) when it is invariant to rotation by some fraction of 360° (or 2π radians). It is said to have an *n-fold axis of symmetry*, or a C_n axis, if it is invariant to rotation by $2\pi/n$. In the rotation operation (also designated by C_n) the object conventionally is rotated *clockwise*. The transformation matrix for a clockwise C_n operation about the z axis is

$$\begin{bmatrix} \cos 2\pi/n & \sin 2\pi/n & 0 \\ -\sin 2\pi/n & \cos 2\pi/n & 0 \\ 0 & 0 & 1 \end{bmatrix} \qquad (3.3)$$

For a $(C_2)_z$ operation, the transformation matrix is thus $\begin{bmatrix} -1 & 0 & 0 \\ 0 & -1 & 0 \\ 0 & 0 & 1 \end{bmatrix}$.

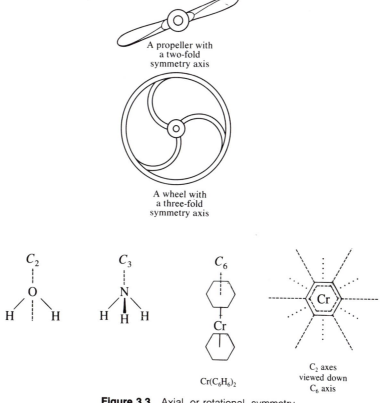

A propeller with
a two-fold
symmetry axis

A wheel with
a three-fold
symmetry axis

C_2

C_3

C_6

$Cr(C_6H_6)_2$

C_2 axes
viewed down
C_6 axis

Figure 3.3 Axial, or rotational, symmetry.

Example 3.4 What is the transformation matrix for a point (x,y,z) for a C_3 operation about the axis $x=y=z$? See Appendix 3.1 for discussion of matrix multiplication.

Solution   Rotation by C_3 about this axis takes x into y, y into z, and z into x, so

$$\begin{bmatrix} 0 & 1 & 0 \\ 0 & 0 & 1 \\ 1 & 0 & 0 \end{bmatrix} \begin{bmatrix} x \\ y \\ z \end{bmatrix} = \begin{bmatrix} y \\ z \\ x \end{bmatrix}.$$

Example 3.5   Find the highest n-fold rotation axis for each of the following species. Which of the species has other symmetry axes not coincident with the n-fold axis already found?

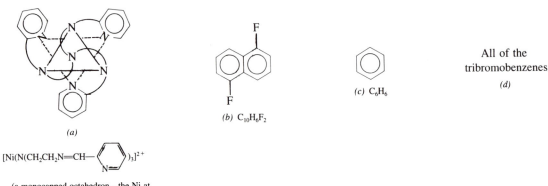

(a)

$[\text{Ni(N(CH}_2\text{CH}_2\text{N}{=}\text{CH}{-} \langle \rangle\text{)}_3]^{2+}$

(a monocapped octahedron—the Ni at the center of the octahedron is not shown. From L. J. Wilson and N. J. Rose, *J. Am. Chem. Soc.* 1968, *90*, 6041.

(b) $C_{10}H_6F_2$

(c) C_6H_6

All of the tribromobenzenes

(d)

Solution   (a) C_3, (b) C_2, (c) C_6,
C_6H_6 has $6C_2 \perp C_6$.

C_2 C_1 C_3, $3C_2 \perp C_3$ Lines represent C_2 and σ_v.

3.2.4   Identity (E)

The C_1 operation, consisting of a rotation by 360° or an integral multiple of 360°, leaves any object unchanged. It is given the special symbol E (German *Einheit*) and is called the *identity* operation. The transformation matrix for the identity operation is $\begin{bmatrix} 1 & 0 & 0 \\ 0 & 1 & 0 \\ 0 & 0 & 1 \end{bmatrix}$, the *unit matrix*.

Example 3.6   Show that the product of the identity transformation matrix and any other transformation matrix leaves the other matrix unchanged.

Solution

$$\overset{E}{\begin{bmatrix} 1 & 0 & 0 \\ 0 & 1 & 0 \\ 0 & 0 & 1 \end{bmatrix}} \begin{bmatrix} a & b & c \\ e & f & g \\ h & i & j \end{bmatrix} = \begin{bmatrix} (a+0+0) & (b+0+0) & (c+0+0) \\ (0+e+0) & (0+f+0) & (0+g+0) \\ (0+0+h) & (0+0+i) & (0+0+j) \end{bmatrix}$$

3.2.5 Improper Rotation (S_n)

The rotation operations discussed above are sometimes termed *proper rotations,* to distinguish them from improper rotations. An *improper rotation* (rotation-reflection) consists of a rotation by C_n followed by reflection in the plane perpendicular to the C_n axis. Both the symmetry element and the operation are given the symbol S_n for a rotation-reflection of $2\pi/n$ radians.

Since the transformation matrix for σ_{xy} is $\begin{bmatrix} 1 & 0 & 0 \\ 0 & 1 & 0 \\ 0 & 0 & -1 \end{bmatrix}$, the transformation matrix for

$\sigma_{xy}\,(C_n)_z = (S_n)_z$ is $\begin{bmatrix} \cos 2\pi/n & \sin 2\pi/n & 0 \\ -\sin 2\pi/n & \cos 2\pi/n & 0 \\ 0 & 0 & -1 \end{bmatrix}$, where the rotation is clockwise. To obtain

the product of two symmetry operations, multiply the transformation matrices or perform the operations in sequence from right (C_n first, as written above) to left.

Figure 3.4 gives some examples of molecules with S_n axes. Example *(a)* has a mirror plane through the line joining Sn, Mn, and CO, through the upper phenyl group and between a pair of phenyls and two pairs of CO. Since σ is unchanged by multiplication by C_1, $\sigma C_1 = \sigma = S_1$. Example *(b)* has a center of symmetry at the center of the P—P bond. Rotating

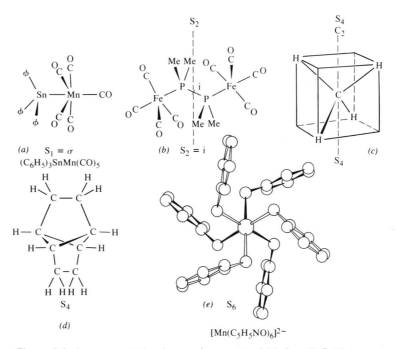

Figure 3.4 Improper rotational axes of symmetry. *(a)* is from H. P. Weber and R. F. Bryan, *Acta Cryst.* 1967, *22,* 822. *(d)* after J. F. Chiang and S. H. Bauer, *Trans. Faraday Soc.* 1968, *64,* 2248. *(e)* is reprinted with permission from T. J. Bergendahl and J. S. Wood, *Inorg. Chem.* 1975, *14,* 338; copyright 1975, American Chemical Society.

about this point by 180° (C_2) and then reflecting in a plane perpendicular to C_2 reproduces the original positions, so $i = \sigma C_2 = S_2$. Molecules with an S_n axis have a plane of symmetry where n is odd, so the S_n operation simply results from having both C_n and σ operations. The $B(OCH_3)_3$ molecule in Figure 3.8 has an S_3 axis. Rotating the CH_4 molecule by 90° about a C_2 axis shifts the H atoms to the formerly vacant sites of a cube. Reflection through a plane perpendicular to C_2 restores the original arrangement. A tetrahedron has an S_4 axis, but no C_4 axis or σ perpendicular to S_4; the S_4 axis coincides with a C_2 axis. A second example having an S_4 axis is shown (Figure 3.4*d*); the $[Mn(C_5H_5NO)_6]^{2+}$ complex ion has an S_6 axis coincident with the C_3 axis perpendicular to the plane of the paper.

Example 3.7 Using transformation matrices, show that the S_1 operation is identical to the σ operation, and that the S_2 operation is identical to the i operation.

Solution

$$S_1 = \sigma_{xy}C_1 = \begin{bmatrix} 1 & 0 & 0 \\ 0 & 1 & 0 \\ 0 & 0 & -1 \end{bmatrix}\begin{bmatrix} 1 & 0 & 0 \\ 0 & 1 & 0 \\ 0 & 0 & 1 \end{bmatrix} = \sigma_{xy}$$

$$S_2 = \sigma_{xy}C_2 = \begin{bmatrix} 1 & 0 & 0 \\ 0 & 1 & 0 \\ 0 & 0 & -1 \end{bmatrix}\begin{bmatrix} -1 & 0 & 0 \\ 0 & -1 & 0 \\ 0 & 0 & 1 \end{bmatrix} = \begin{bmatrix} -1 & 0 & 0 \\ 0 & -1 & 0 \\ 0 & 0 & -1 \end{bmatrix} = i$$

It is often said that a molecule is optically active if it has no center of symmetry or plane of symmetry. These are necessary, but not sufficient conditions. The example using matrices and the molecules in Figure 3.4 demonstrate that $S_1 = \sigma$ and $S_2 = i$. The criterion for a molecule to be optically active is that it must not have *any* S_n symmetry axis. A valid test is that it should not be possible to superimpose a model of the molecule on its mirror image.

3.3 INTRODUCTION TO GROUPS

3.3.1 Kaleidoscopes

A kaleidoscope has two mirrors arranged at an angle (usually 60°) to each other to produce multiple reflections of bits of colored glass. Consider the reflections of the letter P by the two mirrors σ_1 and σ_2 having a 45° dihedral angle, as illustrated in Figure 3.5.

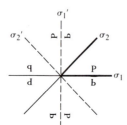

Figure 3.5 Reflections in a kaleidoscope.

Reflection of P through σ_1 produces the image b; reflection of P through σ_2 produces the image ᑫ. Each of these images may be reflected in turn, with the following results.

$$\sigma_1 \text{ on } ᑫ \rightarrow ᆮ \equiv \sigma_1\sigma_2 \text{ on P} \equiv C_4 \text{ on P}$$

$$\sigma_2 \text{ on } b \rightarrow ᑫ \equiv \sigma_2\sigma_1 \text{ on P} \equiv C_4^3 \text{ on P}$$

Note that the result of reflection through σ_1 and σ_2 *depends on the order* ($\sigma_1\sigma_2$ means σ_2 followed by σ_1) in which the reflection operations are performed: that is, the reflection operations *do not commute*. The successive performance of symmetry operations is the *product* of the operations.

In arithmetic and ordinary algebra, multiplication commutes: that is, $2 \times 3 = 3 \times 2$ and $x \times y = y \times x$. Matrix multiplication and the multiplication of symmetry operations do not commute, except in special cases. Thus $E \times \sigma = \sigma \times E$ and the unit matrix $\left(\begin{bmatrix} 1 & 0 & 0 \\ 0 & 1 & 0 \\ 0 & 0 & 1 \end{bmatrix} \text{ for } E \right)$ commutes with other matrices, but $\sigma_1\sigma_2 \neq \sigma_2\sigma_1$. Note also that the result of successive reflection through two different mirror planes is the same as a single rotation through an angle which is twice that between the mirror planes [for $\sigma_1\sigma_2 \equiv C_4$, $2 \times 45° = 90°$; and for $\sigma_2\sigma_1$, $2 \times 315° = 630° = 270°$ (C_4^3)]. The rotational axis coincides with the intersection of the two mirror planes.

Another look at the 45° kaleidoscope shows that the images ᑫ and ᆮ may be reflected further.

$$\sigma_1 \text{ on } ᑫ \rightarrow ᆮ \equiv \sigma_1\sigma_2\sigma_1 \text{ on P} \quad \equiv C_4\sigma_1 \quad \equiv \sigma_2' \text{ on P}$$

$$\sigma_2 \text{ on } ᆮ \rightarrow ᑫ \equiv \sigma_2\sigma_1\sigma_2 \text{ on P} \quad \equiv C_4^3\sigma_2 \quad \equiv \sigma_1' \text{ on P}$$

$$\text{and } \sigma_1 \text{ on } ᑫ \rightarrow d \equiv \sigma_1\sigma_2\sigma_1\sigma_2 \text{ on P} \equiv C_4C_4 \quad \equiv C_2 \text{ on P}$$

$$\sigma_2 \text{ on } ᆮ \rightarrow d \equiv \sigma_2\sigma_1\sigma_2\sigma_1 \text{ on } \quad \text{P} \equiv C_4^3 \, C_4^3 \equiv C_2 \text{ on P}$$

σ_1' and σ_2' are mirror planes perpendicular to σ_1 and σ_2, respectively, and are indicated by dashed lines in Figure 3.5. In the sequence of operations $\sigma_1\sigma_2\sigma_1\sigma_2$, to be performed in order from *right to left*, we may add parentheses in any fashion we choose; for example $(\sigma_1\sigma_2)(\sigma_1\sigma_2)$ or $\sigma_1(\sigma_2\sigma_1\sigma_2)$ or $(\sigma_1\sigma_2\sigma_1)\sigma_2$, the operations within the parentheses being performed first (again, from right to left). From the products evaluated above, note that the groupings can be written as $(\sigma_1\sigma_2)(\sigma_1\sigma_2) = (C_4)(C_4)$, $\sigma_1(\sigma_2\sigma_1\sigma_2) = \sigma_1(\sigma_1')$, and $(\sigma_1\sigma_2\sigma_1)(\sigma_2) = (\sigma_2')\sigma_2$—each of which is equivalent to a C_2 operation. When the result does not depend on the choice of parentheses or the grouping of operations, the operations are said to be *associative* (capable of being joined).

The two mirror operations σ_1 and σ_2 performed in exhaustive combination generate a *group*. In the mathematical sense, a group is a collection of operations having the following properties.

1. The product of any two operations must be an operation of the group (a group is said to be closed under multiplication).

2. Every operation must have an *inverse:* that is, for every operation there must be an operation that will undo the effect of the first operation.

3. Every group must have an identity operation, E, that has the effect of leaving every group member, or element of the group, unchanged. The identity operation commutes with all other operations of the group: that is, $EA = AE$.

4. All operations of the group are associative: $ABC = (AB)C = A(BC)$.

5. The product of any two operations, or elements, is defined.

A group is realized by any object (or set of objects, equations, etc.) that appears to be unchanged after carrying out *any* of the operations of the group. Such an object, or set, is said to belong to the group. The properties of a group can be understood most readily by examining a group multiplication table.

3.3.2 Group Multiplication Tables

In the kaleidoscope group constructed on the letter P, the set of letters is invariant under each of the operations of the group. However, for convenience in constructing a group multiplication table we arbitrarily may label one letter with the identity operation and each of the other letters with the single operation that produces it from the letter labeled E. The product rule, or law of multiplication, is to perform the operations sequentially (from right to left) or, in table form, to carry out the operation at the top of the table (labeling the columns) first and the operation at the left of the table (labeling the rows) second. The product is the single operation that produces the same result. Table 3.1 is our example multiplication table.

Table 3.1 Multiplication table for the kaleidoscope group (Fig. 3.6)

	E	C_4	C_4^2	C_4^3	σ_1	σ_2	σ_1'	σ_2'
E	E	C_4	C_4^2	C_4^3	σ_1	σ_2	σ_1'	σ_2'
C_4	C_4	C_4^2	C_4^3	E	σ_2'	σ_1	σ_2	σ_1'
C_4^2	C_4^2	C_4^3	E	C_4	σ_1'	σ_2'	σ_1	σ_2
C_4^3	C_4^3	E	C_4	C_4^2	σ_2	σ_1'	σ_2'	σ_1
σ_1	σ_1	σ_2	σ_1'	σ_2'	E	C_4	C_4^2	C_4^3
σ_2	σ_2	σ_1'	σ_2'	σ_1	C_4^3	E	C_4	C_4^2
σ_1'	σ_1'	σ_2'	σ_1	σ_2	C_4^2	C_4^3	E	C_4
σ_2'	σ_2'	σ_1	σ_2	σ_1'	C_4	C_4^2	C_4^3	E

The entries were obtained by taking from Figure 3.6 a letter whose label gives the operation at the top of the table, performing one of the operations indicated at the left of the table, and entering the label of the product letter in the table. Thus, performing a C_4 operation on σ (labeled σ_2') gives ꟼ (labeled σ_1'). Two operations that have E as their product are each other's *inverse;* an operation and its inverse always commute. Note that each σ is its own inverse—or, $\sigma^2 = E$.

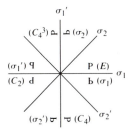

Figure 3.6 Kaleidoscope group with symmetry labels.

3.4 SYMMETRY POINT GROUPS

The symmetry of a molecule located on symmetry axes, cut by planes of symmetry, or centered at an inversion center is known as *point symmetry*. Crystallographers (along with designers of wallpaper and textiles) are concerned with translational symmetry (space groups), which might involve reorientation of a molecule from one lattice site to another.

3.4.1 C_n, C_{nh}, and C_{nv} Groups

The groups generated by repetition of a C_n operation are termed $\mathbf{C}_n$ *point groups*. Since $C_n^n = E$, after n repetitions we will be back where we started; hence the $\mathbf{C}_n$ group has n operations, or the *order* of the group is n. Some examples of molecules exhibiting $\mathbf{C}_n$ point group symmetry are shown in Figure 3.7. Molecules belonging to the $\mathbf{C}_n$ point group will be optically active and will have a dipole moment (except where accidental cancellation of individual bond moments occurs).

Adding to the C_n operation as a new generating operation a horizontal mirror plane, σ_h, forms new groups termed $\mathbf{C}_{nh}$, examples of which are shown in Figure 3.8. The additional operations (besides rotations and reflection in a horizontal plane) generated in a $\mathbf{C}_{nh}$ group, as products of C_n and σ_h, include an inversion operation, where n is even, and various S_n operations. The $\mathbf{C}_{1h}$ group (only a mirror plane) is called $\mathbf{C}_s$.

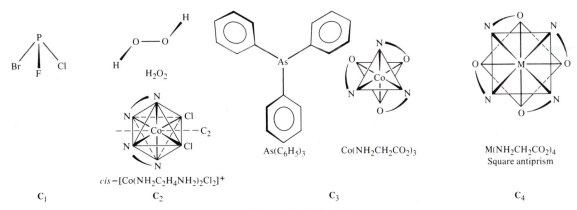

Figure 3.7 Examples of C_n point groups.

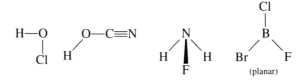

$$\mathbf{C}_{1h} \equiv \mathbf{C}_s$$

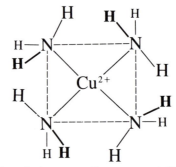

$\mathbf{C}_{2h}$ $\mathbf{C}_{3h}$

(NH$_3$ are fixed to preserve $\mathbf{C}_4$ symmetry; actually, they rotate freely)

$\mathbf{C}_{4h}$

Figure 3.8   Examples of $\mathbf{C}_{nh}$ point groups.

If a mirror plane contains the rotational axis, the group is called a $\mathbf{C}_{nv}$ group—the v indicating a vertical mirror plane, σ_v. The mirror plane will be reproduced n times in a $\mathbf{C}_{nv}$ group, examples of which are shown in Figure 3.9. The $\mathbf{C}_{4v}$ group can be recognized as our 45° kaleidoscope group, which was generated by multiple σ operations. This illustrates that the choice of generators is arbitrary. Both the $\mathbf{C}_{nv}$ and $\mathbf{C}_{nh}$ groups have $2n$ elements or operations.

3.4.2   Dihedral Groups ($\mathbf{D}_n$, $\mathbf{D}_{nh}$, and $\mathbf{D}_{nd}$)

Adding a C_2 axis perpendicular to a C_n axis generates one of the *dihedral groups,* $\mathbf{D}_n$, as shown in Figure 3.10. Because of the C_n axis, there must be n C_2 axes, which, together with

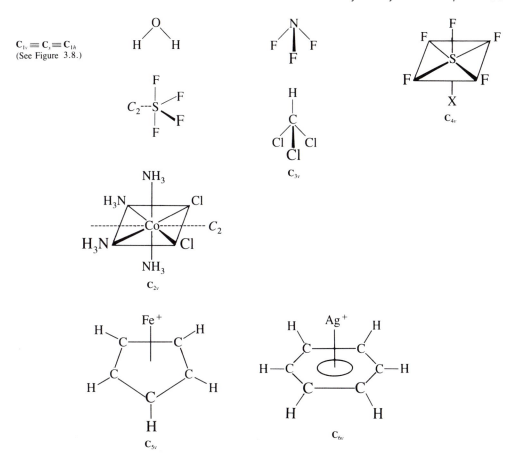

Figure 3.9 Examples of C_{nv} point groups.

the C_n axis, give a total of $2n$ symmetry elements for a $\mathbf{D}_n$ group. Molecules in this group must have a zero dipole moment and will be optically active.

Adding a σ_h to a $\mathbf{D}_n$ group generates a $\mathbf{D}_{nh}$ group, with $4n$ symmetry elements. For even values of n, an inversion center will be found. Figure 3.11 provides examples of some $\mathbf{D}_{nh}$ groups.

Adding a vertical mirror plane to a $\mathbf{D}_n$ group in such a fashion as to bisect adjacent C_2 axes, generates a $\mathbf{D}_{nd}$ group (see Figure 3.12 for examples). The mirror planes are referred to as σ_d *(dihedral planes)*, n of which are present. For odd values of n, an inversion center will be found. There are $4n$ symmetry elements in a $\mathbf{D}_{nd}$ group.

3.4.3 $\mathbf{S}_n$ Groups

The S_n operation serves as a generator for the $\mathbf{S}_n$ groups. For odd n, $S_n^{2n} = E$ and the $\mathbf{S}_n$ group is the same as the $\mathbf{C}_{nh}$ group (the latter labels are used). For even n, the $\mathbf{S}_n$ group has no mirror planes, the absence of which may be used to distinguish $\mathbf{S}_n$ groups from $\mathbf{D}_{nd}$ groups, which

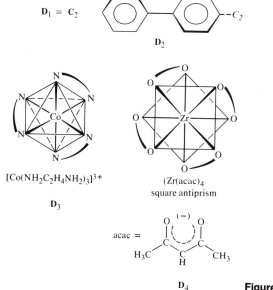

$\mathbf{D}_1 \equiv \mathbf{C}_2$

$\mathbf{D}_2$

$[Co(NH_2C_2H_4NH_2)_3]^{3+}$

$\mathbf{D}_3$

$(Zr(acac)_4)$
square antiprism

acac =

$\mathbf{D}_4$

Figure 3.10 Examples of $\mathbf{D}_n$ point groups.

also contain an S_n axis. For examples, see Figure 3.1 for $\mathbf{S}_1 \equiv \mathbf{C}_s$ and Figure 3.4 for $\mathbf{S}_2 \equiv \mathbf{C}_i$. The group having only an inversion center is called $\mathbf{C}_i$. The examples in Figure 3.4 illustrating the S_4 axis belong to the $\mathbf{D}_{2d}$ group. You should check the $\mathbf{D}_{2d}$ character table (Appendix 3.3) for all of the symmetry elements and identify them in Figure 3.4. Figure 8.28 illustrates a Zr complex with $\mathbf{S}_4$ symmetry—an example of a compound that is optically inactive, even though it lacks a center or plane of symmetry.

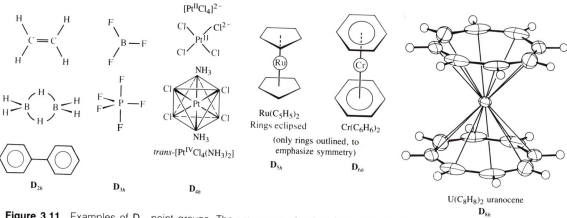

Figure 3.11 Examples of $\mathbf{D}_{nh}$ point groups. The uranocene structure is reprinted with permission from A. Zalkin and K. N. Raymond, *J. Am. Chem. Soc.* 1969, *91*, 5667. Copyright 1969, American Chemical Society.

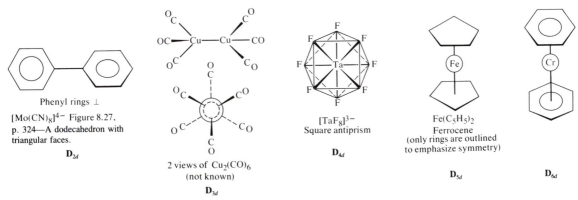

Figure 3.12 Examples of D_{nd} point groups.

3.4.4 Linear Groups ($C_{\infty v}$ and $D_{\infty h}$)

Linear molecules possess a C_∞ axis along a line of the nuclei, and there are an infinite number of reflection planes (σ_v) passing through the atoms. If the molecule has no C_2 axis perpendicular to the C_∞ axis, it belongs to the $C_{\infty v}$ group. Examples of molecules in the $C_{\infty v}$ group include CO, HCl, HCN, and SCO. If you find a C_2 axis perpendicular to the C_∞ axis (if one is found, an infinite number exist), the molecule or ion belongs to the $D_{\infty h}$ group. The subscript h indicates the presence of σ_h. These are the symmetrical linear molecules: N_2, O_2, HCCH, OCO, and [NCAgCN]$^-$, for example.

3.4.5 The Platonic Solids (Cubic and Icosahedral Groups)

The remaining class of symmetry groups take their names from the *Platonic solids,* which are examples of solid figures belonging to these groups. In a Platonic solid, all vertices, edges, and faces are equivalent. In three-dimensional space there are only five such polyhedra—the tetrahedron, the octahedron, the cube, the regular dodecahedron (with pentagonal faces), and the icosahedron. If the centers of the adjacent faces of a Platonic solid are connected, the solid figure outlined is also a Platonic solid, and the two figures are said to be *conjugates.* Conjugates belong to the same symmetry group. The octahedron and the cube are conjugates, as are the icosahedron and dodecahedron; the tetrahedron is its own conjugate. The Platonic solid groups thus include the icosahedral group along with the tetrahedral and octahedral groups, both of which are classified as belonging to the cubic groups. The *rotation* subgroup (eliminating all operations except C_n) of these groups is designated by a single-letter abbreviation (**I, T,** and **O**), and the full symmetry group of the Platonic solids is designated by a two-letter abbreviation (**I$_h$, T$_d$,** and **O$_h$**).

In generating a group, we will follow the position of a slash mark located near an edge of a cube (but not falling on any symmetry element) as we go through the operations of a group (Figure 3.13); each location of the slash mark indicates a unique operation of the group. The cube is used as an example to develop the **O** and **O$_h$** groups.

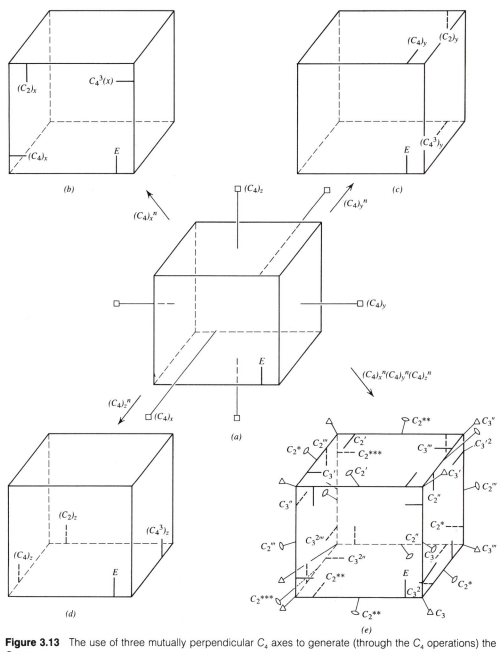

Figure 3.13 The use of three mutually perpendicular C_4 axes to generate (through the C_4 operations) the O group.

The most easily used generators of the rotations of **O** are the three mutually perpendicular C_4 axes passing through the centers of opposite faces of the cube, which are labeled $(C_4)_x$, $(C_4)_y$, and $(C_4)_z$. With the arbitrarily chosen slash marked E in Figure 3.13a, the operation of E, C_4, C_4^2, and C_4^3 produces Figure 3.13b, c, or d, depending on the rotational axis chosen. Any set of four slash marks on a face may be rotated about one of the other C_4 axes to another face, yielding, finally, Figure 3.13e. Each slash mark represents also a single rotational operation of the **O** group. The new rotational symmetry elements may be identified as three-fold axes through the body diagonals of the cube and two-fold axes through the center of an edge and the center of the cube (and accordingly, through the center of the opposite edge). The 24 slash marks correspond to the number of operations of the rotational group, **O**. Adding a mirror plane doubles the number of slash marks, bringing the total number of operations to 48 in the **O**$_h$ group. The added symmetry elements are a center of symmetry *(i)*, S_6 axes through the body diagonals (coincident with C_3), S_4 axes coincident with the C_4 axes, and mirror planes normal to all the C_2 axes (including the C_4^2 axes).

The number of operations belonging to a group is termed the *order* of the group and denoted by the letter h. The order of the *rotation group* of a Platonic solid is simply the product of the number of faces and the number of edges per face. For the full symmetry group of a Platonic solid, it is doubled. Thus a tetrahedron has four triangular faces, so $h = 12$ for **T** or 24 for **T**$_d$. An octahedron has eight triangular faces, so $h = 24$ for **O** or 48 for **O**$_h$, as we saw above. A regular dodecahedron has 12 pentagonal faces and a regular icosahedron has 20 triangular faces, so the order of the **I** group is 60. The rotational operations arise from the two-fold, three-fold, and five-fold axes lying along the center of the polyhedra and the centers of the edges, faces, and vertices. The **I**$_h$ group (Figure 3.14) has 120 operations, the added 60 improper rotations arising from a center of symmetry, S_{10} axes coincident with the C_5 axes, S_6 axes coincident with the C_3 axes, and reflection planes normal to the C_2 axes.

Example 3.8   Identify all symmetry operations associated with the **T**$_d$ group.

Solution   The rotations of the tetrahedron (**T** group) are $4C_3$ (through each apex), $4C_3^2$, and $3C_2$ (through opposite edges). The order is 12, including E. The orientation of the tetrahedron is chosen for coincidence of the symmetry elements in common with the cube and octahedron (see Figure 3.4 for orientation within a cube). The C_2 axes here coincide with the *x, y* and *z* axes. We can get the **T**$_d$ group by adding σ_d planes that bisect the edges. On adding a new symmetry element, we must take the product of it with each other symmetry element. The product of σ_d and C_2 merely interchanges σ_d planes. We can take the product of σ_d and C_3 by transforming models or using matrices. First, examine the models

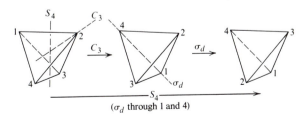

$(\sigma_d$ through 1 and 4$)$

The single operation that will accomplish $\sigma_d C_3$, as shown, is S_4 about the axis shown. Using matrices for C_3 about the axis $x = y = z$ (see Example 3.4) and σ_d containing the *y* axis, we have

$$
\overset{\sigma_d}{\begin{bmatrix} 1 & 0 & 0 \\ 0 & 0 & -1 \\ 0 & -1 & 0 \end{bmatrix}} \overset{C_3}{\begin{bmatrix} 0 & 1 & 0 \\ 0 & 0 & 1 \\ 1 & 0 & 0 \end{bmatrix}} = \overset{S_4}{\begin{bmatrix} 0 & 1 & 0 \\ -1 & 0 & 0 \\ 0 & 0 & -1 \end{bmatrix}}
$$

There is a σ_d through each of the six edges. There are $3C_2$ axes, because clockwise and anticlockwise rotation (or clockwise rotation from above and below) are equivalent; but since they are not equivalent for S_4 if we number positions, there are $6S_4$ axes. With the added $6\sigma_d$ and $6S_4$, the order of the $\mathbf{T}_d$ group is 24.

The $\mathbf{O}_h$ group is commonly encountered, but the $\mathbf{O}$ group is rare (see Figure 3.14). Most tetrahedral molecules belong to the $\mathbf{T}_d$ group. The rotational group ($\mathbf{T}$) would be achieved for

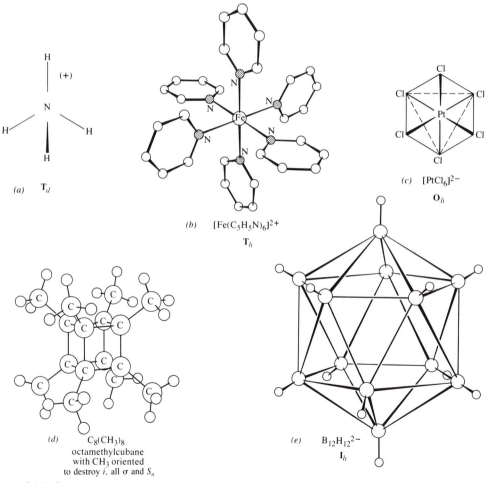

Figure 3.14 Examples of cubic and icosahedral groups. *(b)* is reprinted with permission from R. J. Doedens and L. E. Dahl, *J. Am. Chem. Soc.* 1966, *88*, 4847; copyright 1966, American Chemical Society. *(d)* is from *Symmetry: A Stereoscopic Guide for Chemists* by Ivan Bernal, Walter C. Hamilton and John S. Ricci. W. H. Freeman and Company. Copyright © 1972. *(e)* is from E. L. Muetterties, *The Chemistry of Boron and Its Compounds*, Wiley, New York, 1967, p. 233.

$Si(CH_3)_4$ with the CH_3 staggered to destroy σ_d (and S_4) but to retain the C_3 and C_2 axes. The tetrahedron does not have a center of symmetry, and if the inversion operation is used to generate a new set of operations from the **T** group, the improper rotations will be i, σ_h (reflections normal to the C_2 axes), and S_6 operations along the C_3 axes. The total of 24 proper and improper rotations comprise the $\mathbf{T}_h$ group. Molecules belonging to this group do not look like tetrahedra (see Figure 3.14 for a beautiful example), but they can be distinguished from $\mathbf{O}_h$ molecules, which have C_4 axes.

The sphere belongs to an infinite order group, $\mathbf{O}(3)$, containing all possible proper and improper rotations. The rotation subgroup is labeled $\mathbf{R}(3)$ for all possible rotations in three dimensions. All isolated atoms belong to $\mathbf{O}(3)$ symmetry.

3.4.6 Assigning Molecules to Symmetry Groups

The most direct route to assigning molecules to symmetry groups is to assign them first to a rotation group and then, for optically inactive species, to the full symmetry group.

Neglecting for the moment mirror planes and improper rotations (including i), all molecules belong either to one of the Platonic solid rotation groups (**T**, **O**, or **I**) or to a $\mathbf{C}_n$ or $\mathbf{D}_n$ rotation group (Figure 3.15). Those belonging to the Platonic solid groups usually are easy to recognize, because of the multiple C_n axes of high order—even when additional atoms are added along all edges or faces [$Mo_6Cl_8^{4+}$, $Ta_6Cl_{12}^{2+}$, $Be_4O(O_2CCH_3)_6$, etc.—see Example 3.9]. If only one n-fold axis is found, the species belongs to a $\mathbf{C}_n$ group. (Obviously, every C_n axis implies coincident C_n^m axes, where m takes on integer values up to $n-1$.) Finding a C_2 axis perpendicular to the C_n puts the molecule in a $\mathbf{D}_n$ rotation group. (There will be n C_2 axes perpendicular to the C_n axis, but finding just one is sufficient for identification.) Linear molecules belong to either $\mathbf{C}_\infty$ or $\mathbf{D}_\infty$ rotation groups, depending on whether or not they have a

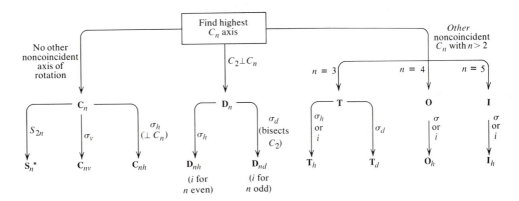

Figure 3.15 Flow chart for assignment of molecules to point groups.

center of symmetry (or σ_h). Thus linear species such as H_2, NCCN, FHF^-, and CO_2 belong to D_∞, whereas N_2O, HF, and BrF belong to C_∞. The C_1 group has no symmetry other than the identity operation.

If we cannot find a mirror plane, a center of symmetry, or any other S_n axis, the point group is the rotation group already identified. Actually, in addition to σ and i, we need only to look for an S_4 axis. Molecules having S_n axes with $n = 2, 3, 5$, or 6 have σ or i. Existence of only an S_4 axis is rare (see above), and no examples of the S_8 group (or those with higher n) are known. If we find an additional element of symmetry, equivalent to S_n, we must identify the full group. Molecules identified above as **T**, **O**, or **I** generally become T_d, O_h, or I_h; the assignment may be checked by finding any *one* of the improper rotations of the group, including σ (S_1) or i (S_2). The C_{nv} group has a set of vertical mirror planes containing the C_n axis— finding one of these is sufficient. The C_{nh} group has a "horizontal" mirror plane normal to the C_n axis, as does the D_{nh} group. The D_{nd} group has a set of "dihedral" mirror planes bisecting adjacent C_2 axes. A mirror plane, in addition to only a C_1 axis, identifies the C_s group. All of the full groups mentioned above can be identified by finding planes of symmetry. If there are no mirror planes, few possibilities remain. For C_1, finding i identifies the C_i group. If the rotation group is C_2 or C_4, we should look for S_4 and S_8 axes, respectively; if these are found, the groups are S_4 or S_8.

Example 3.9 Assign the following species to rotation groups and to full symmetry groups: $\tilde{M}o_6Cl_8^{4+}$ (Figure 15.38, p. 708), $Ta_6Cl_{12}^{2+}$ (Figure 15.38), $Be_4O(O_2CCH_3)_6$ (Figure 14.3, p. 621), P_4, P_4O_6, and P_4O_{10}.

Solution

	$Mo_6Cl_8^{4+}$	$Ta_6Cl_{12}^{2+}$	$Be_4O(O_2CCH_3)_6$	P_4, P_4O_6, and P_4O_{10}
Rotation Group	**O**	**O**	**T**	**T**
Full Group	O_h	O_h	T_d	T_d

Example 3.10 Assign the following species to rotation groups and to full symmetry groups.

Solution

(a)

(b)

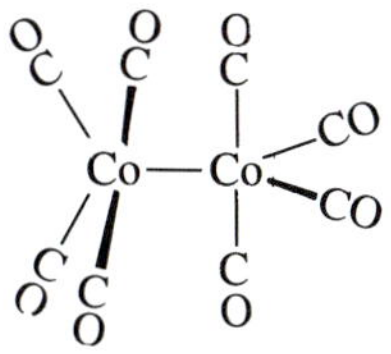

(c)

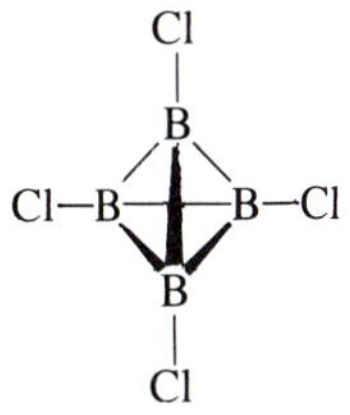

Structures for Co_2 $(CO)_8$. (Reprinted with permission from R. L. Sweany and T. L. Brown, *Inorg. Chem.* 1977, *16*, 415. Copyright 1977, American Chemical Society.)

	P_2Cl_4	B_2F_4	B_2Cl_4	$Co_2(CO)_8$ isomers			B_4Cl_4
				(a)	(b)	(c)	
Rotation Group	$\mathbf{C}_2$	$\mathbf{D}_2$	$\mathbf{D}_2$	$\mathbf{D}_3$	$\mathbf{C}_2$	$\mathbf{D}_2$	$\mathbf{T}$
Full Group	$\mathbf{C}_{2h}$	$\mathbf{D}_{2h}$	$\mathbf{D}_{2d}$	$\mathbf{D}_{3d}$	$\mathbf{C}_{2v}$	$\mathbf{D}_{2d}$	$\mathbf{T}_d$

3.5 CLASS STRUCTURE, REPRESENTATIONS, AND CHARACTER TABLES

3.5.1 Symmetry Classes and Character Tables

Symmetry elements are said to be in the same class if an operation of the group carries one into the other. Thus an examination of the NH_3 molecule ($\mathbf{C}_{3v}$) shows that all three σ_v mirror planes are in the same class, since each is carried into another by a C_3 rotation. In the $\mathbf{C}_{4v}$ group (SXF_5, Figure 3.9), however, there are two classes of σ_v. The two σ_v through the F—S—F bonds belong to one class, and the two σ_v that bisect the FSF angles belong to a separate class.

Rotations have a clockwise or counterclockwise sense that must be taken into account when determining class structure. Thus in a $\mathbf{C}_{nv}$ group, a σ_v operation shows that all C_n^+ (normal clockwise rotation) are in the same class as the inverse C_n^- [counterclockwise rotation, which is equivalent to clockwise rotation by $C_n^{(n-1)}$]. For example, C_4^+ (normally written simply as C_4) is in the same class as C_4^- (normally written as C_4^3). In $\mathbf{C}_4$, $\mathbf{C}_{4h}$, or $\mathbf{S}_8$ groups, C_4 and C_4^- (C_4^3) are unique classes.

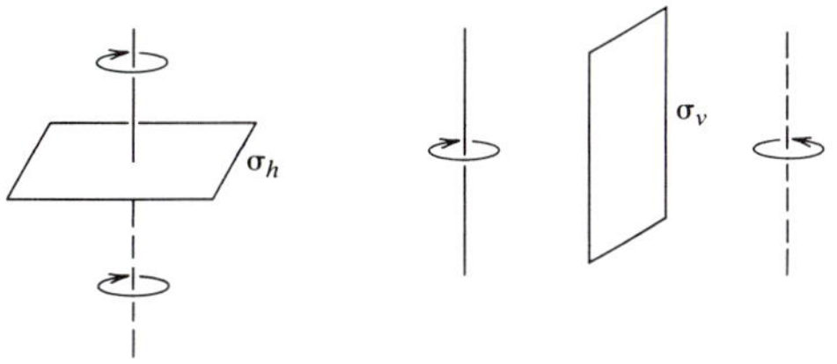

Rotational sense unchanged

Rotational sense reversed

As will be discussed below, the number of elements in each class is given at the top of the *character table* (Appendix 3.3) for each point group. Thus the character table identifies all symmetry operations by class and also describes the way various properties of a molecule behave with respect to each class of operations.

Example 3.11   Group the C_n, C_n^m axes, and the σ planes by classes for the following groups: $\mathbf{D_3}$, $\mathbf{D_4}$, $\mathbf{D_{4h}}$, $\mathbf{T}$, and $\mathbf{T_d}$.

Solution   For $\mathbf{D_3}$, a C_2 axis takes the C_3 axis from above into the one below; and since clockwise rotation as viewed from below is equivalent to counterclockwise rotation from above, C_3 and C_3^- (or C_3^2) are in the same class.

For $\mathbf{D_4}$, the same situation applies for C_4 from above and from below (equivalent to C_4^- or C_4^3), but $C_4^2 = C_2$ is unique. There are two classes of $C_2 \perp C_4$, one class of the two axes along the x and y axes, and one class of the two C_2 between the coordinate axes.

For $\mathbf{D_{4h}}$, the C_4 and C_2 axes are in classes as for $\mathbf{D_4}$. The σ_h is unique. There are two classes of vertical planes—the two through the coordinate axes (σ_v) and the two between the axes (σ_d). The two σ_v are interchanged by C_4, as are the two σ_d.

For $\mathbf{T}$, C_3 and C_3^2 are in different classes, since they are not interchanged by C_2, which is not perpendicular to C_3. All C_2 axes are in the same class, since they are interchanged by the C_3 operations. For $\mathbf{T_d}$, S_4 interchanges C_3 and C_3^2 (or C_3 from below), so they are in the same class.

C_n^+ and C_n^- ($C_n^{(n-1)}$) rotations are interchanged by σ_v and C_2 perpendicular to C_n, but not by σ_h. They belong to the same class for all groups except $\mathbf{C_n}$, $\mathbf{C_{nh}}$, $\mathbf{S_n}$, and $\mathbf{T}$.

3.5.2 Representations

A *representation* of a group is a set of matrices (one for each operation of the group) whose multiplication table is the same (under the rules of matrix multiplication—see Appendix 3.1) as that for the symmetry operations of the group (Section 3.3.2).

For the 45° kaleidoscope (or the $\mathbf{C_{4v}}$ group), the transformation matrices for each operation will form a representation. These may be written (letting the z axis be the vertical axis) as

$$
\underset{E}{\begin{bmatrix} 1 & 0 & 0 \\ 0 & 1 & 0 \\ 0 & 0 & 1 \end{bmatrix}} \quad
\underset{\sigma_1}{\begin{bmatrix} 1 & 0 & 0 \\ 0 & -1 & 0 \\ 0 & 0 & 1 \end{bmatrix}} \quad
\underset{\sigma_1'}{\begin{bmatrix} -1 & 0 & 0 \\ 0 & 1 & 0 \\ 0 & 0 & 1 \end{bmatrix}} \quad
\underset{\sigma_2}{\begin{bmatrix} 0 & 1 & 0 \\ 1 & 0 & 0 \\ 0 & 0 & 1 \end{bmatrix}} \quad
\underset{\sigma_2'}{\begin{bmatrix} 0 & -1 & 0 \\ -1 & 0 & 0 \\ 0 & 0 & 1 \end{bmatrix}} \quad
\underset{C_4}{\begin{bmatrix} 0 & 1 & 0 \\ -1 & 0 & 0 \\ 0 & 0 & 1 \end{bmatrix}} \quad
\underset{C_4^3(C_4^-)}{\begin{bmatrix} 0 & -1 & 0 \\ 1 & 0 & 0 \\ 0 & 0 & 1 \end{bmatrix}} \quad
\underset{C_2(C_4^2)}{\begin{bmatrix} -1 & 0 & 0 \\ 0 & -1 & 0 \\ 0 & 0 & 1 \end{bmatrix}}
$$

Note that transformations of the z coordinate are completely independent of the x and y coordinates, but that x and y are interchanged by some operations. Accordingly, each of the above representations can be broken into two independent representations consisting of the transformation matrices for x and y (two-dimensional matrices) and the transformation matrices for z (one-dimensional matrices).

$$
x,y \quad
\underset{E}{\begin{bmatrix} 1 & 0 \\ 0 & 1 \end{bmatrix}} \quad
\underset{\sigma_1}{\begin{bmatrix} 1 & 0 \\ 0 & -1 \end{bmatrix}} \quad
\underset{\sigma_1'}{\begin{bmatrix} -1 & 0 \\ 0 & 1 \end{bmatrix}} \quad
\underset{\sigma_2}{\begin{bmatrix} 0 & 1 \\ 1 & 0 \end{bmatrix}} \quad
\underset{\sigma_2'}{\begin{bmatrix} 0 & -1 \\ -1 & 0 \end{bmatrix}} \quad
\underset{C_4}{\begin{bmatrix} 0 & 1 \\ -1 & 0 \end{bmatrix}} \quad
\underset{C_4^3}{\begin{bmatrix} 0 & -1 \\ 1 & 0 \end{bmatrix}} \quad
\underset{C_2}{\begin{bmatrix} -1 & 0 \\ 0 & -1 \end{bmatrix}}
$$

$$
z \quad [1] \qquad [1] \qquad [1] \qquad [1] \qquad [1] \qquad [1] \qquad [1] \qquad [1]
$$

As will be seen later, these matrices cannot be reduced further and hence are termed *irreducible representations*.

In general, spotting irreducible representations is not as easy as in the example above. We will not discuss here the techniques for bringing a reducible matrix into block diagonal form $\begin{bmatrix} \Box & 0 & \\ & 0 & \\ 0 & 0 & \Box \end{bmatrix}$ where only zero elements appear outside the blocks; however, when each representation appears in the same block diagonal form each set of blocks with similar row and column indices forms a representation.

Example 3.12 Show that the transformation matrices for x and y given above have the same multiplication table as Table 3.1 (p. 92).

Solution C_2^2 and σ^2 give E.

$$\sigma_1^2 = \begin{bmatrix} 1 & 0 \\ 0 & -1 \end{bmatrix} \begin{bmatrix} 1 & 0 \\ 0 & -1 \end{bmatrix} = \begin{bmatrix} 1 & 0 \\ 0 & 1 \end{bmatrix} = E$$

$$\sigma_2^2 = \begin{bmatrix} 0 & 1 \\ 1 & 0 \end{bmatrix} \begin{bmatrix} 0 & 1 \\ 1 & 0 \end{bmatrix} = \begin{bmatrix} 1 & 0 \\ 0 & 1 \end{bmatrix} = E$$

$$C_2^2 = \begin{bmatrix} -1 & 0 \\ 0 & -1 \end{bmatrix} \begin{bmatrix} -1 & 0 \\ 0 & -1 \end{bmatrix} = \begin{bmatrix} 1 & 0 \\ 0 & 1 \end{bmatrix} = E$$

The products RC_4 (giving the column under C_4) are

$$EC_4 = C_4E = C_4$$

$$C_4^2 = \begin{bmatrix} 0 & 1 \\ -1 & 0 \end{bmatrix} \begin{bmatrix} 0 & 1 \\ -1 & 0 \end{bmatrix} = \begin{bmatrix} -1 & 0 \\ 0 & -1 \end{bmatrix} = C_2$$

$$C_4^2C_4 = \begin{bmatrix} -1 & 0 \\ 0 & -1 \end{bmatrix} \begin{bmatrix} 0 & 1 \\ -1 & 0 \end{bmatrix} = \begin{bmatrix} 0 & -1 \\ 1 & 0 \end{bmatrix} = C_4^3 \ (C_4^-) \;\Bigg\rbrace$$

$$C_4^3C_4 = \begin{bmatrix} 0 & -1 \\ 1 & 0 \end{bmatrix} \begin{bmatrix} 0 & 1 \\ -1 & 0 \end{bmatrix} = \begin{bmatrix} 1 & 0 \\ 0 & 1 \end{bmatrix} = E \;\Bigg\rbrace \text{These commute.}$$

$$\sigma_1C_4 = \begin{bmatrix} 1 & 0 \\ 0 & -1 \end{bmatrix} \begin{bmatrix} 0 & 1 \\ -1 & 0 \end{bmatrix} = \begin{bmatrix} 0 & 1 \\ 1 & 0 \end{bmatrix} = \sigma_2 \;\Bigg\rbrace$$

$$\sigma_2C_4 = \begin{bmatrix} 0 & 1 \\ 1 & 0 \end{bmatrix} \begin{bmatrix} 0 & 1 \\ -1 & 0 \end{bmatrix} = \begin{bmatrix} -1 & 0 \\ 0 & 1 \end{bmatrix} = \sigma_1'$$

$$\sigma_1'C_4 = \begin{bmatrix} -1 & 0 \\ 0 & 1 \end{bmatrix} \begin{bmatrix} 0 & 1 \\ -1 & 0 \end{bmatrix} = \begin{bmatrix} 0 & -1 \\ -1 & 0 \end{bmatrix} = \sigma_2'$$

$$\sigma_2'C_4 = \begin{bmatrix} 0 & -1 \\ -1 & 0 \end{bmatrix} \begin{bmatrix} 0 & 1 \\ -1 & 0 \end{bmatrix} = \begin{bmatrix} 1 & 0 \\ 0 & -1 \end{bmatrix} = \sigma_1$$

Take the other products and check the results against Table 3.1, as practice in multiplying matrices and in reading multiplication tables.

The sum of the diagonal terms of a matrix is called the *trace* (or *spur*) of the matrix. For a representation, the trace of each matrix is called a *character*. Reducible representations furnish *compound characters,* which are the sum of the *simple characters* of the irreducible representations contained in the reducible representation. An irreducible representation giving a unique set of characters is said to belong to a particular *symmetry species*. The number of symmetry species, or irreducible representations, for any group is the same as the number of classes of operations in the group. For a given symmetry species, the character is the same for all operations in the same class. The character under the identity operation is equal to the dimension of the matrix representation: 1 for [1], 2 for $\begin{bmatrix} 1 & 0 \\ 0 & 1 \end{bmatrix}$, etc. If the character under the identity operation is squared, and the sum of the squares is taken over all symmetry species, Γ, the result is equal to the order of the group

$$\sum_{\Gamma} [\chi(E)]^2 = h$$

More generally

$$n_g \sum_{\Gamma} \chi^2 = h \tag{3.4}$$

where χ is the character under a given operation, n_g is the number of operations in the class, $\sum_{\Gamma}$ indicates the sum is taken over all symmetry species in the group, and h is the order of the group. The criterion that a set of characters be simple (that is, that the representation be irreducible) is that

$$\sum_g n_g \chi^2 = h \tag{3.5}$$

For each irreducible representation, the sum of each character squared and multiplied by the number of operations in the particular class must be equal to the order of the group. Here the summation is over the g classes of the group.

Returning to the 45° kaleidoscope group, we find that since there are five classes of operations (E, $2C_4$, C_2, $2\sigma_1$, and $2\sigma_2$), there must be five symmetry species. To satisfy Equation 3.4, four of these must be one-dimensional representations and the remaining one must be two-dimensional.

$$1^2 + 1^2 + 1^2 + 1^2 + 2^2 = h = 8$$

By applying Equation 3.5 we see that the characters obtained from the x,y transformation matrices satisfy the criterion for a set of simple characters,

	E	$2C_4$	C_2	$2\sigma_1$	$2\sigma_2$
(x,y)	2	0	-2	0	0

$$2^2 + 2(0)^2 + (-2)^2 + 2(0)^2 + 2(0)^2 = h = 8$$

and thus the representation is irreducible. This symmetry species is given the label "E" (not the identity) which denotes a two-dimensional representation. Since the x,y transformation matrices yield the E symmetry species, x and y are said to form a basis for this symmetry species or simply to transform as the E species (or E representation). The characters for z are all $+1$. This is the totally symmetric symmetry species, which is labeled A_1.

3.5.3 Orthogonality of Representations

Another important requirement is that the irreducible representations must be orthogonal. The corresponding elements of the matrices for all symmetry operations comprise a vector, as do the characters for all symmetry operations of a representation. Two vectors **A** and **B** are orthogonal if the projection of one on the other is zero, or if **A** · **B** (the *scalar* or *dot* product) is zero. The scalar product for the vectors **A** and **B** in two dimensions is

$$\mathbf{A} \cdot \mathbf{B} = AB \cos \theta$$

where θ is the angle between the vectors and A and B are their lengths. Since $\cos 90° = 0$, the vectors are orthogonal if $\mathbf{A} \cdot \mathbf{B} = 0$. The irreducible representations of a group comprise a set of mutually orthogonal vectors that must satisfy the requirement that $\Gamma_i \cdot \Gamma_j = 0$ (where Γ_i and Γ_j are two different representations). If we take as the components of each vector (representation) the character (χ) for each symmetry operation (R), orthogonality requires that

$$\sum_R \chi_i(R)\chi_j(R) = 0 \text{ if } i \neq j \qquad (3.6)$$

Let us check for the orthogonality of A_1 and E

$\mathbf{C}_{4v}$	E	$2C_4$	C_2	$2\sigma_1$	$2\sigma_2$
A_1	1	1	1	1	1
E	2	0	-2	0	0

$$A_1 \cdot E = 2 \quad + 0 \quad -2 \quad + 0 \quad + 0 = 0 \quad \therefore \quad \text{orthogonal}$$

We have two of the five representations. We know that a mirror operation carried out twice must give the identity—$\sigma^2 = E$, and $C_2^2 = E$. Since the matrix of a one-dimensional representation for the identity is 1, the matrix (and the character) for each σ and for C_2 must be $\sqrt{1}$, or ± 1. The matrices (and the characters) for one-dimensional representations for C_4 also must be ± 1 and since $C_4^2 = C_2$, the matrices (and the characters) for C_2 are actually $+1$ $(\pm 1)^2$. The representations must be different, and we know that the characters for E and C_2 are all $+1$, so the others will have different combinations of ± 1 for C_4, σ_1, and σ_2. Because of the requirement for orthogonality (Equation 3.6), only one of these can be $+1$, so one of the three additional representations has $+1$ for C_4, one has $+1$ for σ_1, and the third has $+1$ for σ_2. Checking the orthogonality of one of these with A_1:

	E	$2C_4$	C_2	$2\sigma_1$	$2\sigma_2$
A_1	1	1	1	1	1
Γ_i	1	1	1	-1	-1

$$A_1 \cdot \Gamma_i = 1 \quad + 2(1) \quad + (1) \quad + 2(-1) + 2(-1) = 0 \quad \therefore \quad \text{orthogonal}$$

The full character table for the kaleidoscope group, known as the $\mathbf{C}_{4v}$ group, is given in Table 3.2. Looking at all the representations, we see that the characters differ for σ_1 and σ_2 (different classes) but that both C_4 $(C_4^1$ and $C_4^3)$ are in the same class and have the same characters.

Table 3.2   Character table for $\mathbf{C}_{4v}$ (also the 45° kaleidoscope group)

$\mathbf{C}_{4v}$	E	$2C_4$	C_2	$2\sigma_1$	$2\sigma_2$
A_1	1	1	1	1	1
A_2	1	1	1	-1	-1
B_1	1	-1	1	1	-1
B_2	1	-1	1	-1	1
E	2	0	-2	0	0

3.5.4 Mulliken Labels for Representations

The labels of the symmetry species, which follow a set of rules devised by Mulliken, tell us about the dimensionality of the representation, as well as some of its symmetry behavior. One-dimensional representations are denoted A or B, two-dimensional by E, three-dimensional by T, four by G, and five by H. The totally symmetric representation is denoted always by A (or by A_1, A_{1g}, or A'). If more than one symmetry species is symmetric with respect to the highest-order rotational axis, the additional species are labeled A_2, A_3, etc. A one-dimensional symmetry species that is antisymmetric under the highest-order rotational axis is labeled B, or B_1, B_2, etc., if there is more than one. For groups with a center of inversion, a subscript g is used to indicate symmetric (gerade) behavior and a subscript u is used to indicate antisymmetric (ungerade) behavior under the inversion operation. For $\mathbf{C}_{nh}$ and $\mathbf{D}_{nh}$ groups that lack an inversion center (that is, for odd values of n), the horizontal mirror plane is used to assign $'$ or $''$ superscripts—the $'$ signifying symmetric behavior and the $''$ signifying antisymmetric behavior under the σ_h operation. (In the above discussion, symmetric behavior means the character will be positive and antisymmetric behavior means the character will be negative under the specified operation.)

Example 3.13   Verify the orthogonality of A_2 and B_1 for $\mathbf{C}_{4v}$ and show that B_1 obeys Equation 3.5.

Solution

$\mathbf{C}_{4v}$	E	$2C_4$	C_2	$2\sigma_1$	$2\sigma_2$
A_2	1	1	1	-1	-1
B_1	1	-1	1	1	-1

$$A_2 \cdot B_1 = (1)(1) + 2(1)(-1) + (1)(1) + 2(-1)(1) + 2(-1)(-1) = 0$$

Now verify the orthogonality of the other representations for yourself.
Applying Equation 3.5 to B_1,

$$(1)^2 + 2(-1)^2 + (1)^2 + 2(1)^2 + 2(-1)^2 = 8 = h$$

Example 3.14   Derive the matrices for the transformation of x,y,z for each class of operations of $\mathbf{D}_3$. Determine the characters and derive the character table.

Solution   There are three classes of operations for $\mathbf{D}_3$—E, $2C_3$, and $3C_2$. The matrices for the transformation of x,y,z are obtained using Matrix 3.3.

$$
E \qquad\qquad C_3 \qquad\qquad (C_2)_x
$$

$$
\begin{bmatrix} 1 & 0 & 0 \\ 0 & 1 & 0 \\ 0 & 0 & 1 \end{bmatrix} \qquad
\begin{bmatrix} -1/2 & \sqrt{3}/2 & 0 \\ -\sqrt{3}/2 & -1/2 & 0 \\ 0 & 0 & 1 \end{bmatrix} \qquad
\begin{bmatrix} 1 & 0 & 0 \\ 0 & -1 & 0 \\ 0 & 0 & -1 \end{bmatrix}
$$

We choose one of the C_2 axes along x for convenience.

$$
3 \qquad\qquad 0 \qquad\qquad -1 \qquad\qquad \text{Characters}
$$

So the C_3 operation on x or y gives a function of x and y—or, x and y are "mixed" by the C_3 operation—but z is independent. This reducible representation can be reduced using the blocks as for $\mathbf{C}_{4v}$.

$$
E \qquad\qquad\qquad C_3 \qquad\qquad\qquad C_2
$$

$$
x,y \quad
\begin{bmatrix} 1 & 0 \\ 0 & 1 \end{bmatrix} \qquad
\begin{bmatrix} -1/2 & \sqrt{3}/2 \\ -\sqrt{3}/2 & -1/2 \end{bmatrix} \qquad
\begin{bmatrix} 1 & 0 \\ 0 & -1 \end{bmatrix}
$$

$$
2 \qquad\qquad\qquad -1 \qquad\qquad\qquad 0 \qquad\qquad \text{Characters}
$$

$$
z \qquad [1] \qquad\qquad\qquad [1] \qquad\qquad\qquad [-1]
$$

$$
1 \qquad\qquad\qquad 1 \qquad\qquad\qquad -1 \qquad\qquad \text{Characters}
$$

Since there are three classes of operations, there are three symmetry species. We have two of them (A_2 and E) and know that there must be a totally symmetric representation, A_1. The character table is

$\mathbf{D_3}$	E	$2C_3$	$3C_2$
A_1	1	1	1
A_2	1	1	-1
E	2	-1	0

Check that these are irreducible representations (Equation 3.5) and that they are orthogonal (Equation 3.6).

3.5.5 Symmetry Species for Functions and Orbitals

We already have seen how the transformation matrices for x and y give an E symmetry species, and z transforms as the A_1 species in $\mathbf{C}_{4v}$ (and $\mathbf{D}_3$). To see how the xy or x^2-y^2 functions transform, we must first find the behavior of x and y under the group operations and then construct the behavior of the desired functions, as illustrated below.

	E	C_4	C_4^3	C_2	σ_1	σ_1'	σ_2	σ_2'
$x \rightarrow$	x	y	$-y$	$-x$	x	$-x$	y	$-y$
$y \rightarrow$	y	$-x$	x	$-y$	$-y$	y	x	$-x$
$xy \rightarrow$	xy	$-xy$	$-xy$	xy	$-xy$	$-xy$	xy	xy
$x^2-y^2 \rightarrow$	x^2-y^2	y^2-x^2	y^2-x^2	x^2-y^2	x^2-y^2	x^2-y^2	y^2-x^2	y^2-x^2

We find the way xy transforms by taking the product of the transformation of x and y under each operation. For x^2-y^2, we square the element for the transformation of x under each operation, do the same for y, and take the difference. For xy and x^2-y^2, the result of performing each operation is equivalent to multiplying by 1 or -1. Tabulating these characters for each operation gives

E	$2C_4$	C_2	$2\sigma_1$	$2\sigma_2$	
1	-1	1	-1	1	for xy
1	-1	1	1	-1	for $x^2 - y^2$

Examining the $\mathbf{C}_{4v}$ character table shows us that xy transforms as the B_2 symmetry species and $x^2 - y^2$ as the B_1 species. Most character tables used by chemists, including those given in Appendix 3.3, list polynomials, or combinations of polynomials, that transform as the various symmetry species. The polynomials listed are also those used for the subscripts of various atomic orbitals (p_x, p_y, d_{xy}, $d_{x^2-y^2}$, etc.), and these orbitals also transform as the various symmetry species indicated. Thus we can identify all of the p and d orbitals, and s is totally symmetric (A_1, A_{1g}, etc.).

The p orbitals transform as x, y, and z, and the representations of the d_{xy}, d_{xz}, and d_{yz} orbitals are identified by those labels (xy, etc.). Orbitals belonging to a two- or three-dimensional representation are grouped in parentheses—for example, (xy, yz) for E and (x,y,z) for T. The d_{z^2} orbital is identified as z^2, or occasionally as the full polynomial $2z^2 - x^2 - y^2$. The $d_{x^2-y^2}$ orbital is identified as $x^2 - y^2$, except when x^2, y^2, and z^2 independently belong to A_1 or A_g—as for $\mathbf{C}_{2v}$, $\mathbf{D}_2$, etc.—in which case both d_{z^2} and $d_{x^2-y^2}$ belong to A_1 or A_g. The symbols R_x, R_y, and R_z refer to rotations about x, y, and z, respectively. The corresponding representations are used in applying selection rules for magnetic dipole transitions and optical activity (see p. 286).

Example 3.15   From the character tables in Appendix 3.3, identify the symmetry species for p_x, p_z, d_{z^2}, $d_{x^2-y^2}$, and d_{xy} for $\mathbf{D}_3$, $\mathbf{C}_{4v}$, $\mathbf{D}_{3h}$, $\mathbf{D}_{4h}$, and $\mathbf{O}_h$.

Solution $\mathbf{D}_3$ p_z (A_2), (p_x,p_y) (E), d_{z^2} (A_1), $(d_{x^2-y^2}, d_{xy})$ (E).

$\mathbf{C}_{4v}$ p_z (A_1), (p_x,p_y) (E), d_{z^2} (A_1), $d_{x^2-y^2}$ (B_1), d_{xy} (B_2)

$\mathbf{D}_{3h}$ p_z (A_2''), (p_x,p_y) (E'), d_{z^2} (A_1'), $(d_{x^2-y^2}, d_{xy})$ (E')

$\mathbf{D}_{4h}$ p_z (A_{2u}), (p_x,p_y) (E_u), d_{z^2} (A_{1g}), $d_{x^2-y^2}$ (B_{1g}), d_{xy} (B_{2g})

$\mathbf{O}_h$ (p_x,p_y,p_z) (T_{1u}), $(d_{z^2},d_{x^2-y^2})$ (E_g), (d_{xz},d_{yz},d_{xy}) (T_{2g})

We note that p_x does not itself transform as a symmetry species in any of the groups listed, but must be combined with one or more of the remaining p orbitals to give a function that does.

3.5.6   Character Tables for Cyclic Groups

A $\mathbf{C}_n$ (cyclic) group has n elements—C_n^1, C_n^2, etc.—each of which comprises a class by itself. Accordingly, a $\mathbf{C}_n$ group has n one-dimensional representations, for which the matrices, because they are one-dimensional, are the same as the characters. All $\mathbf{C}_n$ groups have the totally symmetric representation, A, with all characters $= 1$. For $\mathbf{C}_n$ groups with n even, there is a B representation with characters of $+1$ or -1, which alternate for E, C_n^1, C_n^2, etc. All other $\mathbf{C}_n$ groups (except $\mathbf{C}_1$, which has only the A representation; and $\mathbf{C}_2$, which has only A and B representations) have one-dimensional representations that occur in pairs and carry the Mulliken label E. Their characters, except for the identity operation, are imaginary or complex numbers.

They can be used as such, but for applications to physical problems they can be combined to give real numbers. The complex characters are expressed in terms of ϵ (defined for the particular group beside each character table), where

$$\epsilon = \exp(2\pi i/n) \tag{3.7}$$

The characters for rotations are evaluated by the relationships

$$\exp(\theta i) = \cos \theta + i \sin \theta \tag{3.8}$$

giving for rotations by C_n,

$$\epsilon_n = \cos 2\pi/n + i \sin 2\pi/n \tag{3.9}$$

and for rotations by C_n^m,

$$\epsilon_n^m = \cos 2\pi m/n + i \sin 2\pi m/n \tag{3.10}$$

For C_4, $\epsilon = \cos 90° + i \sin 90° = i$. This is the character for one of the pair of one-dimensional representations; since the paired representations are complex conjugates, the character for C_4 is $-i$ for the other representation. The character table is given in Table 3.3.

Table 3.3 Character table for $\mathbf{C}_4$

$\mathbf{C}_4$	E	C_4	C_4^2	C_4^3
A	1	1	1	1
B	1	-1	1	-1
E	$\begin{cases} 1 \\ 1 \end{cases}$	$\begin{matrix} i \\ -i \end{matrix}$	$\begin{matrix} -1 \\ -1 \end{matrix}$	$\left. \begin{matrix} -i \\ i \end{matrix} \right\}$

Example 3.16 The E representation for C_4 occurs as a pair of one-dimensional representations, with characters 1 and 1 for the identity and i and $-i$ for C_4. From these values, derive the characters for C_4^2 and C_4^3.

Solution Since the characters are the one-dimensional matrices, you can get the characters for new operations as products of other operations.

$$C_4^2 = (i)^2 = -1 \qquad\qquad C_4^3 = C_4 C_4^2 = i(-1) = -i$$
$$\text{and} \quad (-i)^2 = -1 \qquad \text{and} \qquad -i(-1) = i$$

For all cyclic groups, if the characters are evaluated for one rotation, the others can be obtained as products. For C_n groups, all operations commute. Groups for which all operations commute are called *Abelian* groups.

3.5.7 Reducing Reducible Representations

In the example of the C_{4v} group (or the 45° kaleidoscope group—Section 3.3.1), a reducible representation was reduced, by inspection, into its irreducible representations. Such reductions are carried out more readily in the general case by using the set of characters for the reducible representation and a character table for the group. To find the number of times, n_Γ, a particular irreducible representation is contained in a reducible representation, we use the formula

$$n_\Gamma = \frac{1}{h} \sum_g n_g \chi_R \chi_\Gamma \tag{3.11}$$

where h is the order of the group, n_g the number of operations in class g, χ_R is the character of the reducible representation and χ_Γ that of the irreducible representation under the operations of class g, and the summation is taken over all classes.

Example 3.17   What symmetry species (irreducible representations) are contained in the representation having the following characters in the $\mathbf{T}_d$ group?

E	C_3	C_2	S_4	σ_d
9	0	1	1	1

Solution   Appendix 3.3 supplies the character table for the $\mathbf{T}_d$ group. For each irreducible representation, multiply the number of operations in each class by the corresponding characters for that irreducible representation *and* the reducible representation. These products are summed and divided by 24 (h, the total number of operations), as follows.

$$
\begin{aligned}
&\quad\quad\quad E \quad\quad 8C_3 \quad\quad 3C_2 \quad\quad 6S_4 \quad\quad 6\sigma_d \\
n_{A_1} &= 1/24[1 \cdot 1 \cdot 9 + 8 \cdot 0 \cdot 1 \; + 3 \cdot 1 \cdot 1 \; + 6 \cdot 1 \cdot 1 \; + 6 \cdot 1 \cdot 1] \; = 1\,A_1 \\
n_{A_2} &= 1/24[1 \cdot 1 \cdot 9 + 8 \cdot 0 \cdot 1 \; + 3 \cdot 1 \cdot 1 \; + 6 \cdot 1 \cdot -1 + 6 \cdot 1 \cdot -1] = 0\,A_2 \\
n_{E} &= 1/24[1 \cdot 2 \cdot 9 + 8 \cdot 0 \cdot -1 + 3 \cdot 1 \cdot 2 \; + 6 \cdot 1 \cdot 0 \; + 6 \cdot 1 \cdot 0] \; = 1\,E \\
n_{T_1} &= 1/24[1 \cdot 3 \cdot 9 + 8 \cdot 0 \cdot 0 \; + 3 \cdot 1 \cdot -1 + 6 \cdot 1 \cdot 1 \; + 6 \cdot 1 \cdot -1] = 1\,T_1 \\
n_{T_2} &= 1/24[1 \cdot 3 \cdot 9 + 8 \cdot 0 \cdot 0 \; + 3 \cdot 1 \cdot -1 + 6 \cdot 1 \cdot -1 + 6 \cdot 1 \cdot 1] \; = 1\,T_2
\end{aligned}
$$

The reduction must give either zero or a whole number for each irreducible representation. As a check, the dimension of the reducible representation (character for E) must be the sum of the dimensions of the irreducible representations found. Here,

$$
\begin{aligned}
A_1 + E + T_1 + T_2 \\
1 + 2 + 3 + 3 = 9
\end{aligned}
$$

Because large or small characters for some operations of the reducible representation indicate the irreducible representations contained, you might be able to reduce it by inspection, summing the characters of the irreducible representations for each class of operations as a check. In applying the formula, first examine those representations you think are contained, stopping when you can check the sum of the dimensions of representations involved.

3.6 SUMMARY OF TERMINOLOGY AND NOTATION FROM SYMMETRY AND GROUP THEORY

3.6.1 Symmetry Elements and Operations

Identity (E): Leaving an object unchanged—the equivalent of rotation by 360°. This operation is required by group theory, since the product of two operations (performing them in succession) must be equivalent to another operation of the group.

Rotation (C_n): Rotation (clockwise) by $360°/n$ about an (n-fold) axis. Also called a *proper rotation*. Performing a C_n operation n times is equivalent to E. (See Figure 3.3.)

$$C_n^n = E \qquad C_4^2 = C_2$$

Mirror Plane (σ): Reflection through a plane (a plane of symmetry). (See Figure 3.1.) A mirror plane perpendicular to C_n is called a *horizontal plane*, σ_h. Planes containing C_n are called *vertical planes*, σ_v, or, if they bisect angles between symmetry axes, *dihedral planes*, σ_d.

$$\sigma^2 = E$$

Improper Rotation (S_n): Rotation by $360°/n$ followed by reflection in a plane perpendicular to the axis of rotation. Also called a *rotation-reflection operation*. (See Figure 3.4.)

$$S_n = \sigma C_n$$
$$S_1 = \sigma C_1 = \sigma, \text{ since } C_1 = E$$
$$S_n^2 = C_n^2, \text{ since } \sigma^2 = E$$

Inversion (i): Reflection through a center of symmetry. (See Figure 3.2.)

$$i^2 = E \qquad iC_2 = \sigma$$
$$i = \sigma C_2 = S_2$$

Symmetry elements and symmetry operations are set in *italics* in this book.

3.6.2 Symmetry Point Groups

The presence of some combinations of symmetry elements requires the existence of others. If a molecule has a C_3 axis and σ_v, there must be three σ_v. The natural groupings of symmetry elements are called symmetry point groups, of which only the chemically important concern us:

C$_1$—only E, equivalent to a one-fold axis.
C$_s$—only σ (all groups contain E). (See Figure 3.9.)
C$_i$—only i.

Axial groups

C$_n$—C_n axis. (See Figure 3.7.)
C$_{nv}$—C_n and n vertical planes (σ_v) through C_n. (See Figure 3.9.)
C$_{nh}$—C_n and a horizontal plane (σ_h) $\perp$ to C_n. (See Figure 3.8.)
D$_n$—C_n axis and nC_2 axes $\perp$ to C_n. (See Figure 3.10.)
D$_{nh}$—**D$_n$** elements plus σ_h. For even n, there is an inversion center. (See Figure 3.11.)
D$_{nd}$—**D$_n$** elements plus n dihedral planes (σ_d) bisecting the angles between C_2 axes. For odd n, there is an inversion center. (See Figure 3.12.)
S$_n$—S_n axis (used for even n).

$$\left. \begin{array}{l} \text{for odd } n, \ \mathbf{S}_n = \mathbf{C}_{nh} \\ \mathbf{S}_2 = \mathbf{C}_i \\ \mathbf{S}_1 = \mathbf{C}_s \end{array} \right\} \text{ The latter symbols are used.}$$

Cubic groups. (See Figure 3.14)

T—$4C_3$, and $3C_2$ (mutually $\perp$).

T$_d$—**T** elements plus 6σ and $3S_4$ and $3S_4^3$.

O—$3C_4$ (mutually $\perp$), $4C_3$, and $6C_2$.

O$_h$—**O** elements, i, $3\sigma_h$, $6\sigma_d$.

Icosahedral group

I$_h$—$12C_5$, $20C_3$, $15C_2$, i, 15σ.

Linear groups

C$_{\infty v}$—C_∞ axis and an infinite number of σ_v.

D$_{\infty h}$—**C**$_{\infty v}$ elements plus σ_h.

Point group labels are set in **bold face** in this book.

3.6.3 Some Terminology of Group Theory (Arranged Alphabetically)

Characters: Performing a symmetry operation on a point in space described by the coordinates x, y, and z yields new coordinates x', y', and z'—a transformation that can be accomplished mathematically by multiplying by a matrix. The sum of the diagonal elements of a matrix is its *character*. Characters can be used instead of matrices for many purposes.

Character Table: A character table contains the characters for all of the irreducible representations of a group for all the classes of symmetry operations; in other words, it summarizes the symmetry properties for all of the representations. Each row in a character table is an irreducible representation of the group. To the right of the character table, x, y, and z identify the representations of the p_x, p_y, and p_z orbitals, respectively. The representations of the d_{xz}, d_{yz}, and d_{xy} orbitals correspond to those of the respective products xz, yz, and xy. The d_{z^2} orbital belongs to the same representation as z^2. If $x^2 - y^2$ does not appear, then x^2, y^2, and z^2 all belong to the totally symmetric representation (first row, A, A_1, A_{1g}, etc.), as do the d_{z^2} and $d_{x^2 - y^2}$ orbitals. The s orbital is not listed, but it always belongs to the totally symmetric representation.

Classes of Symmetry Elements: Symmetry elements (or operations) belong to the same class if one can be transformed into another by some operation of the group. Symmetry elements of the same class have the same characters and form a single column in a character table.

Matrix: A matrix is a rectangular array of numbers. Here we are concerned primarily with the use of matrices for transforming the coordinates of a point in space. In general, matrix multiplication (Appendix 3.1) does not *commute*, that is $\mathscr{AB} \neq \mathscr{BA}$.

Mulliken Symbols: One-dimensional representations are labeled A if they are symmetric with respect to the rotational axis of highest order; they are labeled B if there is a change in sign after this C_n operation. Two-dimensional representations (such as for doubly degenerate orbitals) are E and three-dimensional representations (such as for triply degenerate orbitals) are T. For centrosymmetric point groups, the subscript g (gerade) indicates no change in sign on reflection through the center of symmetry, and the subscript u (ungerade) indicates that the signs change for this operation.

Orthogonality: All irreducible representations of a point group are mutually orthogonal. A test for orthogonality is that the scalar product of the representations must equal zero.

Products of Representations: The *scalar* (or dot) *product* $\mathcal{A} \cdot \mathcal{B}$ gives a number (a scalar). The scalar product is used as a test for orthogonality of representations, since $\mathcal{A} \cdot \mathcal{B} = 0$ if $\mathcal{A}$ and $\mathcal{B}$ are orthogonal. The *direct product* (Appendix 3.2) of two representations gives a representation of the group.

Representation: A set of matrices (or their characters), one for each class of symmetry operations, comprises a *representation*. Properties of molecules, such as wave functions and vibrational modes, have symmetry characteristics that are described by one of the representations of the point group for the molecule. The dimension of a representation gives the dimensions of the matrices for the representation. Thus a one-dimensional representation *(A or B)* has 1×1 matrices—that is, a simple number. A two-dimensional representation *(E)* has 2×2 matrices such as $\begin{bmatrix} 1 & 0 \\ 0 & 1 \end{bmatrix}$ for which the *character* is 2. A three-dimensional representation *(T)* has 3×3 matrices with a maximum character of 3. The representations that have matrices of lowest order (or smallest values for their characters) are *irreducible representations* or *symmetry species*. Representations for which the matrices can be reduced to others of lower order are *reducible representations*. For each symmetry operation, the character of the reducible representation is the sum of the characters of the irreducible representations contained.

GENERAL REFERENCES

I. Bernal, W. C. Hamilton, and J. S. Ricci, *Symmetry*, W. H. Freeman, San Francisco, 1972. Beautiful stereoscopic illustrations of molecules belonging to the important point groups.

F. A. Cotton, *Chemical Applications of Group Theory*, 2nd ed., Wiley-Interscience, New York, 1971. The best general treatment for chemists.

D. C. Harris and M. D. Bertolucci, *Symmetry and Spectroscopy*, Oxford, New York, 1978. Application to spectroscopy and molecular orbital theory.

M. Orchin and H. H. Jaffé, *Symmetry, Orbitals, and Spectra*, Wiley-Interscience, New York, 1971. A good supplement to Cotton, with more discussion of applications.

M. Orchin and H. H. Jaffé, "Symmetry, Point Groups, and Character Tables," *J. Chem. Educ.*, 1970, *47*, 246, 372, 510. Good treatment of topics covered here, at a suitable level.

J. A. Salthouse and M. J. Ware, *Point Group Character Tables*, Cambridge University Press, Cambridge, 1972. Very useful collection of character tables and other tables of interest to chemists.

APPENDIX 3.1 MATRIX REPRESENTATION OF TRANSFORMATION AND MATRIX MULTIPLICATION

Equations and Matrix Multiplication

If a point with Cartesian coordinates x, y, and z is transformed to a new point x', y', and z' by the equations

$$\begin{aligned} x' &= ax + by + cz \\ y' &= dx + ey + fz \\ z' &= gx + hy + iz \end{aligned} \tag{1}$$

and subsequently x', y', and z' is transformed to the point x'', y'', and z'' by the equations

$$
\begin{aligned}
x'' &= kx' + ly' + mz' \\
y'' &= nx' + oy' + pz' \\
z'' &= qx' + ry' + sz'
\end{aligned}
\tag{2}
$$

then algebraic substitution of (1) into (2) yields x'' in terms of x as

$$
\begin{aligned}
x'' &= k(ax + by + cz) + l(dx + ey + fz) + m(gx + hy + iz) \\
\text{similarily, } y'' &= n(ax + by + cz) + o(dx + ey + fz) + p(gx + hy + iz) \\
\text{and } z'' &= q(ax + by + cz) + r(dx + ey + fz) + s(gx + hy + iz)
\end{aligned}
\tag{3}
$$

The equations grouped under (3) may be factored to give

$$
\begin{aligned}
x'' &= (ka + ld + mg)x + (kb + le + mh)y + (kc + lf + mi)z \\
y'' &= (na + od + pg)x + (nb + oe + ph)y + (nc + of + pi)z \\
z'' &= (qa + rd + sg)x + (qb + re + sh)y + (qc + rf + si)z
\end{aligned}
$$

In matrix notation we have

$$
\begin{bmatrix} x'' \\ y'' \\ z'' \end{bmatrix}
= \begin{bmatrix} k & l & m \\ n & o & p \\ q & r & s \end{bmatrix}
\begin{bmatrix} a & b & c \\ d & e & f \\ g & h & i \end{bmatrix}
\begin{bmatrix} x \\ y \\ z \end{bmatrix}
$$

$$
= \begin{bmatrix} (ka + ld + mg) & (kb + le + mh) & (kc + lf + mi) \\ (na + od + pg) & (nb + oe + ph) & (nc + of + pi) \\ (qa + rd + sg) & (qb + re + sh) & (qc + rf + si) \end{bmatrix}
\begin{bmatrix} x \\ y \\ z \end{bmatrix}
$$

Each element of the product matrix is made up of the sum of n terms, where n is the number of columns of the first matrix that must be equal to the number of rows of the second matrix. A mnemonic aid for carrying out the multiplication is ⌐ or "row into column." The row of the first matrix and the column of the second matrix give the row and column of the element in the product matrix: that is,

$$
p_{ik} = \sum_j a_{ij} b_{jk}
$$

where the first subscript denotes the row and the second the column.

APPENDIX 3.2   THE DIRECT PRODUCT OF REPRESENTATIONS

The *direct product* of two representations gives a representation of the group. For each class of operations, the product of the characters of two representations gives the character of the resulting representation. Using the $\mathbf{C}_{4v}$ group,

C_{4v}	E	$2C_4$	C_2	$2\sigma_1$	$2\sigma_2$
A_1	1	1	1	1	1
A_2	1	1	1	-1	-1
B_1	1	-1	1	1	-1
A_2B_1 gives	1	-1	1	-1	1

From the character table, we see that this is B_2 or $A_2B_1 = B_2$. We see that $A_1\Gamma_i = \Gamma_i$, or the multiplication (direct product) of any representation (Γ_i) by A_1 leaves the representation unchanged. Here E times any other representation gives E. Any one-dimensional representation times itself gives A_1, $\Gamma_i^2 = A_1$, whereas E^2 gives a four-dimensional (reducible) representation. Although we shall not try to prove it here, it is important for spectroscopic selection rules (p. 282) that the direct product of any representation by itself gives the totally symmetric representation (here, A_1)—or, if the product is reducible, that it contains A_1.

APPENDIX 3.3   CHARACTER TABLES[a]

1. The nonaxial groups

C_1	E
A	1

C_s	E	σ_h		
A'	1	1	x, y, R_z	$x^2, y^2,$ z^2, xy
A''	1	-1	z, R_x, R_y	yz, xz

C_i	E	i		
A_g	1	1	R_x, R_y, R_z	$x^2, y^2, z^2,$ xy, xz, yz
A_u	1	-1	x, y, z	

2. The C_n groups

C_2	E	C_2		
A	1	1	z, R_z	x^2, y^2, z^2, xy
B	1	-1	x, y, R_x, R_y	yz, xz

C_3	E	C_3	C_3^2			$\epsilon = \exp(2\pi i/3)$
A	1	1	1	z, R_z	$x^2 + y^2, z^2$	
E	$\begin{cases}1 \\ 1\end{cases}$	$\begin{matrix}\epsilon \\ \epsilon^*\end{matrix}$	$\begin{matrix}\epsilon^* \\ \epsilon\end{matrix}$	$(x, y); (R_x, R_y)$	$(x^2 - y^2, xy); (yz, xz)$	

C_4	E	C_4	C_2	C_4^3		
A	1	1	1	1	z, R_z	$x^2 + y^2, z^2$
B	1	-1	1	-1		$x^2 - y^2, xy$
E	$\begin{cases}1 \\ 1\end{cases}$	$\begin{matrix}i \\ -i\end{matrix}$	$\begin{matrix}-1 \\ -1\end{matrix}$	$\begin{matrix}-i \\ i\end{matrix}$	$(x, y); (R_x, R_y)$	(xz, yz)

[a]Tables as compiled in F. A. Cotton, *Chemical Applications of Group Theory*, 2nd ed., Wiley-Interscience, New York, 1971.

The **C**$_n$ groups (*Continued*)

C$_5$	E	C_5	C_5^2	C_5^3	C_5^4		$\epsilon = \exp(2\pi i/5)$
A	1	1	1	1	1	z, R_z	$x^2 + y^2, z^2$
E_1	$\left\{\begin{matrix}1\\1\end{matrix}\right.$	$\begin{matrix}\epsilon\\\epsilon^*\end{matrix}$	$\begin{matrix}\epsilon^2\\\epsilon^{2*}\end{matrix}$	$\begin{matrix}\epsilon^{2*}\\\epsilon^2\end{matrix}$	$\left.\begin{matrix}\epsilon^*\\\epsilon\end{matrix}\right\}$	$(x, y); (R_x, R_y)$	(yz, xz)
E_2	$\left\{\begin{matrix}1\\1\end{matrix}\right.$	$\begin{matrix}\epsilon^2\\\epsilon^{2*}\end{matrix}$	$\begin{matrix}\epsilon^*\\\epsilon\end{matrix}$	$\begin{matrix}\epsilon\\\epsilon^*\end{matrix}$	$\left.\begin{matrix}\epsilon^{2*}\\\epsilon^2\end{matrix}\right\}$		$(x^2 - y^2, xy)$

C$_6$	E	C_6	C_3	C_2	C_3^2	C_6^5		$\epsilon = \exp(2\pi i/6)$
A	1	1	1	1	1	1	z, R_z	$x^2 + y^2, z^2$
B	1	-1	1	-1	1	-1		
E_1	$\left\{\begin{matrix}1\\1\end{matrix}\right.$	$\begin{matrix}\epsilon\\\epsilon^*\end{matrix}$	$\begin{matrix}-\epsilon^*\\-\epsilon\end{matrix}$	$\begin{matrix}-1\\-1\end{matrix}$	$\begin{matrix}-\epsilon\\-\epsilon^*\end{matrix}$	$\left.\begin{matrix}\epsilon^*\\\epsilon\end{matrix}\right\}$	$\begin{matrix}(x, y);\\(R_x, R_y)\end{matrix}$	(xz, yz)
E_2	$\left\{\begin{matrix}1\\1\end{matrix}\right.$	$\begin{matrix}-\epsilon^*\\-\epsilon\end{matrix}$	$\begin{matrix}-\epsilon\\-\epsilon^*\end{matrix}$	$\begin{matrix}1\\1\end{matrix}$	$\begin{matrix}-\epsilon^*\\-\epsilon\end{matrix}$	$\left.\begin{matrix}-\epsilon\\-\epsilon^*\end{matrix}\right\}$		$(x^2 - y^2, xy)$

C$_7$	E	C_7	C_7^2	C_7^3	C_7^4	C_7^5	C_7^6		$\epsilon = \exp(2\pi i/7)$
A	1	1	1	1	1	1	1	z, R_z	$x^2 + y^2, z^2$
E_1	$\left\{\begin{matrix}1\\1\end{matrix}\right.$	$\begin{matrix}\epsilon\\\epsilon^*\end{matrix}$	$\begin{matrix}\epsilon^2\\\epsilon^{2*}\end{matrix}$	$\begin{matrix}\epsilon^3\\\epsilon^{3*}\end{matrix}$	$\begin{matrix}\epsilon^{3*}\\\epsilon^3\end{matrix}$	$\begin{matrix}\epsilon^{2*}\\\epsilon^2\end{matrix}$	$\left.\begin{matrix}\epsilon^*\\\epsilon\end{matrix}\right\}$	$\begin{matrix}(x, y);\\(R_x, R_y)\end{matrix}$	(xz, yz)
E_2	$\left\{\begin{matrix}1\\1\end{matrix}\right.$	$\begin{matrix}\epsilon^2\\\epsilon^{2*}\end{matrix}$	$\begin{matrix}\epsilon^{3*}\\\epsilon^3\end{matrix}$	$\begin{matrix}\epsilon^*\\\epsilon\end{matrix}$	$\begin{matrix}\epsilon\\\epsilon^*\end{matrix}$	$\begin{matrix}\epsilon^3\\\epsilon^{3*}\end{matrix}$	$\left.\begin{matrix}\epsilon^{2*}\\\epsilon^2\end{matrix}\right\}$		$(x^2 - y^2, xy)$
E_3	$\left\{\begin{matrix}1\\1\end{matrix}\right.$	$\begin{matrix}\epsilon^3\\\epsilon^{3*}\end{matrix}$	$\begin{matrix}\epsilon^*\\\epsilon\end{matrix}$	$\begin{matrix}\epsilon^2\\\epsilon^{2*}\end{matrix}$	$\begin{matrix}\epsilon^{2*}\\\epsilon^2\end{matrix}$	$\begin{matrix}\epsilon\\\epsilon^*\end{matrix}$	$\left.\begin{matrix}\epsilon^{3*}\\\epsilon^3\end{matrix}\right\}$		

C$_8$	E	C_8	C_4	C_2	C_4^3	C_8^3	C_8^5	C_8^7		$\epsilon = \exp(2\pi i/8)$
A	1	1	1	1	1	1	1	1	z, R_z	$x^2 + y^2, z^2$
B	1	-1	1	1	1	-1	-1	-1		
E_1	$\left\{\begin{matrix}1\\1\end{matrix}\right.$	$\begin{matrix}\epsilon\\\epsilon^*\end{matrix}$	$\begin{matrix}i\\-i\end{matrix}$	$\begin{matrix}-1\\-1\end{matrix}$	$\begin{matrix}-i\\i\end{matrix}$	$\begin{matrix}-\epsilon^*\\-\epsilon\end{matrix}$	$\begin{matrix}-\epsilon\\-\epsilon^*\end{matrix}$	$\left.\begin{matrix}\epsilon^*\\\epsilon\end{matrix}\right\}$	$\begin{matrix}(x, y);\\(R_x, R_y)\end{matrix}$	(xz, yz)
E_2	$\left\{\begin{matrix}1\\1\end{matrix}\right.$	$\begin{matrix}i\\-i\end{matrix}$	$\begin{matrix}-1\\-1\end{matrix}$	$\begin{matrix}1\\1\end{matrix}$	$\begin{matrix}-1\\-1\end{matrix}$	$\begin{matrix}-i\\i\end{matrix}$	$\begin{matrix}i\\-i\end{matrix}$	$\left.\begin{matrix}-i\\i\end{matrix}\right\}$		$(x^2 - y^2, xy)$
E_3	$\left\{\begin{matrix}1\\1\end{matrix}\right.$	$\begin{matrix}-\epsilon\\-\epsilon^*\end{matrix}$	$\begin{matrix}i\\-i\end{matrix}$	$\begin{matrix}-1\\-1\end{matrix}$	$\begin{matrix}-i\\i\end{matrix}$	$\begin{matrix}\epsilon^*\\\epsilon\end{matrix}$	$\begin{matrix}\epsilon\\\epsilon^*\end{matrix}$	$\left.\begin{matrix}-\epsilon^*\\-\epsilon\end{matrix}\right\}$		

3. The **D**$_n$ groups

D$_2$	E	$C_2(z)$	$C_2(y)$	$C_2(x)$		
A	1	1	1	1		x^2, y^2, z^2
B_1	1	1	-1	-1	z, R_z	xy
B_2	1	-1	1	-1	y, R_y	xz
B_3	1	-1	-1	1	x, R_x	yz

The **D**$_n$ groups *(Continued)*

D$_3$	E	$2C_3$	$3C_2$		
A_1	1	1	1		$x^2 + y^2, z^2$
A_2	1	1	-1	z, R_z	
E	2	-1	0	$(x, y); (R_x, R_y)$	$(x^2 - y^2, xy); (xz, yz)$

D$_4$	E	$2C_4$	$C_2(=C_4^2)$	$2C_2'$	$2C_2''$		
A_1	1	1	1	1	1		$x^2 + y^2, z^2$
A_2	1	1	1	-1	-1	z, R_z	
B_1	1	-1	1	1	-1		$x^2 - y^2$
B_2	1	-1	1	-1	1		xy
E	2	0	-2	0	0	$(x, y); (R_x, R_y)$	(xz, yz)

D$_5$	E	$2C_5$	$2C_5^2$	$5C_2$		
A_1	1	1	1	1		$x^2 + y^2, z^2$
A_2	1	1	1	-1	z, R_z	
E_1	2	$2 \cos 72°$	$2 \cos 144°$	0	$(x, y); (R_x, R_y)$	(xz, yz)
E_2	2	$2 \cos 144°$	$2 \cos 72°$	0		$(x^2 - y^2, xy)$

D$_6$	E	$2C_6$	$2C_3$	C_2	$3C_2'$	$3C_2''$		
A_1	1	1	1	1	1	1		$x^2 + y^2, z^2$
A_2	1	1	1	1	-1	-1	z, R_z	
B_1	1	-1	1	-1	1	-1		
B_2	1	-1	1	-1	-1	1		
E_1	2	1	-1	-2	0	0	$(x, y); (R_x, R_y)$	(xz, yz)
E_2	2	-1	-1	2	0	0		$(x^2 - y^2, xy)$

4. The **C**$_{nv}$ groups

C$_{2v}$	E	C_2	$\sigma_v(xz)$	$\sigma_v(yz)$		
A_1	1	1	1	1	z	x^2, y^2, z^2
A_2	1	1	-1	-1	R_z	xy
B_1	1	-1	1	-1	x, R_y	xz
B_2	1	-1	-1	1	y, R_x	yz

C$_{3v}$	E	$2C_3$	$3\sigma_v$		
A_1	1	1	1	z	$x^2 + y^2, z^2$
A_2	1	1	-1	R_z	
E	2	-1	0	$(x, y); (R_x, R_y)$	$(x^2 - y^2, xy); (xz, yz)$

The **C**$_{nv}$ groups *(Continued)*

C_{4v}	E	$2C_4$	C_2	$2\sigma_v$	$2\sigma_d$		
A_1	1	1	1	1	1	z	$x^2 + y^2,\ z^2$
A_2	1	1	1	-1	-1	R_z	
B_1	1	-1	1	1	-1		$x^2 - y^2$
B_2	1	-1	1	-1	1		xy
E	2	0	-2	0	0	$(x, y);\ (R_x, R_y)$	(xz, yz)

C_{5v}	E	$2C_5$	$2C_5^2$	$5\sigma_v$		
A_1	1	1	1	1	z	$x^2 + y^2,\ z^2$
A_2	1	1	1	-1	R_z	
E_1	2	$2\cos 72°$	$2\cos 144°$	0	$(x, y);\ (R_x, R_y)$	(xz, yz)
E_2	2	$2\cos 144°$	$2\cos 72°$	0		$(x^2 - y^2,\ xy)$

C_{6v}	E	$2C_6$	$2C_3$	C_2	$3\sigma_v$	$3\sigma_d$		
A_1	1	1	1	1	1	1	z	$x^2 + y^2,\ z^2$
A_2	1	1	1	1	-1	-1	R_z	
B_1	1	-1	1	-1	1	-1		
B_2	1	-1	1	-1	-1	1		
E_1	2	1	-1	-2	0	0	$(x, y);\ (R_x, R_y)$	(xz, yz)
E_2	2	-1	-1	2	0	0		$(x^2 - y^2,\ xy)$

5. The **C**$_{nh}$ groups

C_{2h}	E	C_2	i	σ_h		
A_g	1	1	1	1	R_z	$x^2,\ y^2,\ z^2,\ xy$
B_g	1	-1	1	-1	R_x, R_y	$xz,\ yz$
A_u	1	1	-1	-1	z	
B_u	1	-1	-1	1	x, y	

C_{3h}	E	C_3	C_3^2	σ_h	S_3	S_3^5		$\epsilon = \exp(2\pi i/3)$
A'	1	1	1	1	1	1	R_z	$x^2 + y^2,\ z^2$
E'	$\begin{cases}1\\1\end{cases}$	$\begin{matrix}\epsilon\\ \epsilon^*\end{matrix}$	$\begin{matrix}\epsilon^*\\ \epsilon\end{matrix}$	$\begin{matrix}1\\1\end{matrix}$	$\begin{matrix}\epsilon\\ \epsilon^*\end{matrix}$	$\left.\begin{matrix}\epsilon^*\\ \epsilon\end{matrix}\right\}$	(x, y)	$(x^2 - y^2,\ xy)$
A''	1	1	1	-1	-1	-1	z	
E''	$\begin{cases}1\\1\end{cases}$	$\begin{matrix}\epsilon\\ \epsilon^*\end{matrix}$	$\begin{matrix}\epsilon^*\\ \epsilon\end{matrix}$	$\begin{matrix}-1\\-1\end{matrix}$	$\begin{matrix}-\epsilon\\ -\epsilon^*\end{matrix}$	$\left.\begin{matrix}-\epsilon^*\\ -\epsilon\end{matrix}\right\}$	(R_x, R_y)	(xz, yz)

The C_{nh} groups (Continued)

C_{4h}	E	C_4	C_2	C_4^3	i	S_4^3	σ_h	S_4		
A_g	1	1	1	1	1	1	1	1	R_z	$x^2+y^2,\ z^2$
B_g	1	-1	1	-1	1	-1	1	-1		$x^2-y^2,\ xy$
E_g	1	i	-1	$-i$	1	i	-1	$-i$	(R_x, R_y)	(xz, yz)
	1	$-i$	-1	i	1	$-i$	-1	i		
A_u	1	1	1	1	-1	-1	-1	-1	z	
B_u	1	-1	1	-1	-1	1	-1	1		
E_u	1	i	-1	$-i$	-1	$-i$	1	i	(x, y)	
	1	$-i$	-1	i	-1	i	1	$-i$		

$\epsilon = \exp(2\pi i/5)$

C_{5h}	E	C_5	C_5^2	C_5^3	C_5^4	σ_h	S_5	S_5^7	S_5^3	S_5^9		
A'	1	1	1	1	1	1	1	1	1	1	R_z	$x^2+y^2,\ z^2$
E_1'	1	ϵ	ϵ^2	ϵ^{2*}	ϵ^*	1	ϵ	ϵ^2	ϵ^{2*}	ϵ^*	(x, y)	
	1	ϵ^*	ϵ^{2*}	ϵ^2	ϵ	1	ϵ^*	ϵ^{2*}	ϵ^2	ϵ		
E_2'	1	ϵ^2	ϵ^*	ϵ	ϵ^{2*}	1	ϵ^2	ϵ^*	ϵ	ϵ^{2*}		$(x^2-y^2,\ xy)$
	1	ϵ^{2*}	ϵ	ϵ^*	ϵ^2	1	ϵ^{2*}	ϵ	ϵ^*	ϵ^2		
A''	1	1	1	1	1	-1	-1	-1	-1	-1	z	
E_1''	1	ϵ	ϵ^2	ϵ^{2*}	ϵ^*	-1	$-\epsilon$	$-\epsilon^2$	$-\epsilon^{2*}$	$-\epsilon^*$	(R_x, R_y)	(xz, yz)
	1	ϵ^*	ϵ^{2*}	ϵ^2	ϵ	-1	$-\epsilon^*$	$-\epsilon^{2*}$	$-\epsilon^2$	$-\epsilon$		
E_2''	1	ϵ^2	ϵ^*	ϵ	ϵ^{2*}	-1	$-\epsilon^2$	$-\epsilon^*$	$-\epsilon$	$-\epsilon^{2*}$		
	1	ϵ^{2*}	ϵ	ϵ^*	ϵ^2	-1	$-\epsilon^{2*}$	$-\epsilon$	$-\epsilon^*$	$-\epsilon^2$		

$\epsilon = \exp(2\pi i/6)$

C_{6h}	E	C_6	C_3	C_2	C_3^2	C_6^5	i	S_3^5	S_6^5	σ_h	S_6	S_3		
A_g	1	1	1	1	1	1	1	1	1	1	1	1	R_z	$x^2+y^2,\ z^2$
B_g	1	-1	1	-1	1	-1	1	-1	1	-1	1	-1		
E_{1g}	1	ϵ	$-\epsilon^*$	-1	$-\epsilon$	ϵ^*	1	ϵ	$-\epsilon^*$	-1	$-\epsilon$	ϵ^*	(R_x, R_y)	(xz, yz)
	1	ϵ^*	$-\epsilon$	-1	$-\epsilon^*$	ϵ	1	ϵ^*	$-\epsilon$	-1	$-\epsilon^*$	ϵ		
E_{2g}	1	$-\epsilon^*$	$-\epsilon$	1	$-\epsilon^*$	$-\epsilon$	1	$-\epsilon^*$	$-\epsilon$	1	$-\epsilon^*$	$-\epsilon$		$(x^2-y^2,\ xy)$
	1	$-\epsilon$	$-\epsilon^*$	1	$-\epsilon$	$-\epsilon^*$	1	$-\epsilon$	$-\epsilon^*$	1	$-\epsilon$	$-\epsilon^*$		
A_u	1	1	1	1	1	1	-1	-1	-1	-1	-1	-1	z	
B_u	1	-1	1	-1	1	-1	-1	1	-1	1	-1	1		
E_{1u}	1	ϵ	$-\epsilon^*$	-1	$-\epsilon$	ϵ^*	-1	$-\epsilon$	ϵ^*	1	ϵ	$-\epsilon^*$	(x, y)	
	1	ϵ^*	$-\epsilon$	-1	$-\epsilon^*$	ϵ	-1	$-\epsilon^*$	ϵ	1	ϵ^*	$-\epsilon$		
E_{2u}	1	$-\epsilon^*$	$-\epsilon$	1	$-\epsilon^*$	$-\epsilon$	-1	ϵ^*	ϵ	-1	ϵ^*	ϵ		
	1	$-\epsilon$	$-\epsilon^*$	1	$-\epsilon$	$-\epsilon^*$	-1	ϵ	ϵ^*	-1	ϵ	ϵ^*		

6. The **D**$_{nh}$ groups

$\mathbf{D}_{2h}$	E	$C_2(z)$	$C_2(y)$	$C_2(x)$	i	$\sigma(xy)$	$\sigma(xz)$	$\sigma(yz)$		
A_g	1	1	1	1	1	1	1	1		x^2, y^2, z^2
B_{1g}	1	1	-1	-1	1	1	-1	-1	R_z	xy
B_{2g}	1	-1	1	-1	1	-1	1	-1	R_y	xz
B_{3g}	1	-1	-1	1	1	-1	-1	1	R_x	yz
A_u	1	1	1	1	-1	-1	-1	-1		
B_{1u}	1	1	-1	-1	-1	-1	1	1	z	
B_{2u}	1	-1	1	-1	-1	1	-1	1	y	
B_{3u}	1	-1	-1	1	-1	1	1	-1	x	

$\mathbf{D}_{3h}$	E	$2C_3$	$3C_2$	σ_h	$2S_3$	$3\sigma_v$		
A_1'	1	1	1	1	1	1		$x^2 + y^2, z^2$
A_2'	1	1	-1	1	1	-1	R_z	
E'	2	-1	0	2	-1	0	(x, y)	$(x^2 - y^2, xy)$
A_1''	1	1	1	-1	-1	-1		
A_2''	1	1	-1	-1	-1	1	z	
E''	2	-1	0	-2	1	0	(R_x, R_y)	(xz, yz)

$\mathbf{D}_{4h}$	E	$2C_4$	C_2	$2C_2'$	$2C_2''$	i	$2S_4$	σ_h	$2\sigma_v$	$2\sigma_d$		
A_{1g}	1	1	1	1	1	1	1	1	1	1		$x^2 + y^2, z^2$
A_{2g}	1	1	1	-1	-1	1	1	1	-1	-1	R_z	
B_{1g}	1	-1	1	1	-1	1	-1	1	1	-1		$x^2 - y^2$
B_{2g}	1	-1	1	-1	1	1	-1	1	-1	1		xy
E_g	2	0	-2	0	0	2	0	-2	0	0	(R_x, R_y)	(xz, yz)
A_{1u}	1	1	1	1	1	-1	-1	-1	-1	-1		
A_{2u}	1	1	1	-1	-1	-1	-1	-1	1	1	z	
B_{1u}	1	-1	1	1	-1	-1	1	-1	-1	1		
B_{2u}	1	-1	1	-1	1	-1	1	-1	1	-1		
E_u	2	0	-2	0	0	-2	0	2	0	0	(x, y)	

D_{5h}

D_{5h}	E	$2C_5$	$2C_5^2$	$5C_2$	σ_h	$2S_5$	$2S_5^3$	$5\sigma_v$		
A_1'	1	1	1	1	1	1	1	1		$x^2+y^2,\ z^2$
A_2'	1	1	1	-1	1	1	1	-1	R_z	
E_1'	2	$2\cos 72°$	$2\cos 144°$	0	2	$2\cos 72°$	$2\cos 144°$	0	(x,y)	
E_2'	2	$2\cos 144°$	$2\cos 72°$	0	2	$2\cos 144°$	$2\cos 72°$	0		$(x^2-y^2,\ xy)$
A_1''	1	1	1	1	-1	-1	-1	-1		
A_2''	1	1	1	-1	-1	-1	-1	1	z	
E_1''	2	$2\cos 72°$	$2\cos 144°$	0	-2	$-2\cos 72°$	$-2\cos 144°$	0	(R_x, R_y)	(xz, yz)
E_2''	2	$2\cos 144°$	$2\cos 72°$	0	-2	$-2\cos 144°$	$-2\cos 72°$	0		

D_{6h}

D_{6h}	E	$2C_6$	$2C_3$	C_2	$3C_2'$	$3C_2''$	i	$2S_3$	$2S_6$	σ_h	$3\sigma_d$	$3\sigma_v$		
A_{1g}	1	1	1	1	1	1	1	1	1	1	1	1		$x^2+y^2,\ z^2$
A_{2g}	1	1	1	1	-1	-1	1	1	1	1	-1	-1	R_z	
B_{1g}	1	-1	1	-1	1	-1	1	-1	1	-1	1	-1		
B_{2g}	1	-1	1	-1	-1	1	1	-1	1	-1	-1	1		
E_{1g}	2	1	-1	-2	0	0	2	1	-1	-2	0	0	(R_x, R_y)	(xz, yz)
E_{2g}	2	-1	-1	2	0	0	2	-1	-1	2	0	0		$(x^2-y^2,\ xy)$
A_{1u}	1	1	1	1	1	1	-1	-1	-1	-1	-1	-1		
A_{2u}	1	1	1	1	-1	-1	-1	-1	-1	-1	1	1	z	
B_{1u}	1	-1	1	-1	1	-1	-1	1	-1	1	-1	1		
B_{2u}	1	-1	1	-1	-1	1	-1	1	-1	1	1	-1		
E_{1u}	2	1	-1	-2	0	0	-2	-1	1	2	0	0	(x, y)	
E_{2u}	2	-1	-1	2	0	0	-2	1	1	-2	0	0		

D_{8h}

D_{8h}	E	$2C_8$	$2C_8^3$	$2C_4$	C_2	$4C_2'$	$4C_2''$	i	$2S_8$	$2S_8^3$	$2S_4$	σ_h	$4\sigma_d$	$4\sigma_v$		
A_{1g}	1	1	1	1	1	1	1	1	1	1	1	1	1	1		$x^2+y^2,\ z^2$
A_{2g}	1	1	1	1	1	-1	-1	1	1	1	1	1	-1	-1	R_z	
B_{1g}	1	-1	-1	1	1	1	-1	1	-1	-1	1	1	1	-1		
B_{2g}	1	-1	-1	1	1	-1	1	1	-1	-1	1	1	-1	1		
E_{1g}	2	$\sqrt{2}$	$-\sqrt{2}$	0	-2	0	0	2	$\sqrt{2}$	$-\sqrt{2}$	0	-2	0	0	(R_x, R_y)	(xz, yz)
E_{2g}	2	0	0	-2	2	0	0	2	0	0	-2	2	0	0		$(x^2-y^2,\ xy)$
E_{3g}	2	$-\sqrt{2}$	$\sqrt{2}$	0	-2	0	0	2	$-\sqrt{2}$	$\sqrt{2}$	0	-2	0	0		
A_{1u}	1	1	1	1	1	1	1	-1	-1	-1	-1	-1	-1	-1		
A_{2u}	1	1	1	1	1	-1	-1	-1	-1	-1	-1	-1	1	1	z	
B_{1u}	1	-1	-1	1	1	1	-1	-1	1	1	-1	-1	-1	1		
B_{2u}	1	-1	-1	1	1	-1	1	-1	1	1	-1	-1	1	-1		
E_{1u}	2	$\sqrt{2}$	$-\sqrt{2}$	0	-2	0	0	-2	$-\sqrt{2}$	$\sqrt{2}$	0	2	0	0	(x, y)	
E_{2u}	2	0	0	-2	2	0	0	-2	0	0	2	-2	0	0		
E_{3u}	2	$-\sqrt{2}$	$\sqrt{2}$	0	-2	0	0	-2	$\sqrt{2}$	$-\sqrt{2}$	0	2	0	0		

7. The D_{nd} groups

D_{2d}	E	$2S_4$	C_2	$2C'_2$	$2\sigma_d$		
A_1	1	1	1	1	1		x^2+y^2, z^2
A_2	1	1	1	-1	-1	R_z	
B_1	1	-1	1	1	-1		x^2-y^2
B_2	1	-1	1	-1	1	z	xy
E	2	0	-2	0	0	(x, y); (R_x, R_y)	(xz, yz)

D_{3d}	E	$2C_3$	$3C_2$	i	$2S_6$	$3\sigma_d$		
A_{1g}	1	1	1	1	1	1		x^2+y^2, z^2
A_{2g}	1	1	-1	1	1	-1	R_z	
E_g	2	-1	0	2	-1	0	(R_x, R_y)	(x^2-y^2, xy); (xz, yz)
A_{1u}	1	1	1	-1	-1	-1		
A_{2u}	1	1	-1	-1	-1	1	z	
E_u	2	-1	0	-2	1	0	(x, y)	

D_{4d}	E	$2S_8$	$2C_4$	$2S_8^3$	C_2	$4C'_2$	$4\sigma_d$		
A_1	1	1	1	1	1	1	1		x^2+y^2, z^2
A_2	1	1	1	1	1	-1	-1	R_z	
B_1	1	-1	1	-1	1	1	-1		
B_2	1	-1	1	-1	1	-1	1	z	
E_1	2	$\sqrt{2}$	0	$-\sqrt{2}$	-2	0	0	(x, y)	
E_2	2	0	-2	0	2	0	0		(x^2-y^2, xy)
E_3	2	$-\sqrt{2}$	0	$\sqrt{2}$	-2	0	0	(R_x, R_y)	(xz, yz)

D_{5d}	E	$2C_5$	$2C_5^2$	$5C_2$	i	$2S_{10}^3$	$2S_{10}$	$5\sigma_d$		
A_{1g}	1	1	1	1	1	1	1	1		x^2+y^2, z^2
A_{2g}	1	1	1	-1	1	1	1	-1	R_z	
E_{1g}	2	$2\cos 72°$	$2\cos 144°$	0	2	$2\cos 144°$	$2\cos 72°$	0	(R_x, R_y)	(xz, yz)
E_{2g}	2	$2\cos 144°$	$2\cos 72°$	0	2	$2\cos 72°$	$2\cos 144°$	0		(x^2-y^2, xy)
A_{1u}	1	1	1	1	-1	-1	-1	-1		
A_{2u}	1	1	1	-1	-1	-1	-1	1	z	
E_{1u}	2	$2\cos 72°$	$2\cos 144°$	0	-2	$-2\cos 144°$	$-2\cos 72°$	0	(x, y)	
E_{2u}	2	$2\cos 144°$	$2\cos 72°$	0	-2	$-2\cos 72°$	$-2\cos 144°$	0		

$\mathbf{D}_{6d}$	E	$2S_{12}$	$2C_6$	$2S_4$	$2C_3$	$2S_{12}^5$	C_2	$6C'_2$	$6\sigma_d$		
A_1	1	1	1	1	1	1	1	1	1		$x^2 + y^2,\ z^2$
A_2	1	1	1	1	1	1	1	-1	-1	R_z	
B_1	1	-1	1	-1	1	-1	1	1	-1		
B_2	1	-1	1	-1	1	-1	1	-1	1	z	
E_1	2	$\sqrt{3}$	1	0	-1	$-\sqrt{3}$	-2	0	0	(x, y)	
E_2	2	1	-1	-2	-1	1	2	0	0		$(x^2 - y^2,\ xy)$
E_3	2	0	-2	0	2	0	-2	0	0		
E_4	2	-1	-1	2	-1	-1	2	0	0		
E_5	2	$-\sqrt{3}$	1	0	-1	$\sqrt{3}$	-2	0	0	(R_x, R_y)	(xz, yz)

8. The $\mathbf{S}_n$ group.

S_4	E	S_4	C_2	S_4^3		
A	1	1	1	1	R_z	$x^2 + y^2,\ z^2$
B	1	-1	1	-1	z	$x^2 - y^2,\ xy$
E	$\begin{cases} 1 \\ 1 \end{cases}$	$\begin{matrix} i \\ -i \end{matrix}$	$\begin{matrix} -1 \\ -1 \end{matrix}$	$\begin{matrix} -i \\ i \end{matrix}$	$(x, y);\ (R_x, R_y)$	(xz, yz)

S_6	E	C_3	C_3^2	i	S_6^5	S_6		$\epsilon = \exp(2\pi i/3)$
A_g	1	1	1	1	1	1	R_z	$x^2 + y^2,\ z^2$
E_g	$\begin{cases} 1 \\ 1 \end{cases}$	$\begin{matrix} \epsilon \\ \epsilon^* \end{matrix}$	$\begin{matrix} \epsilon^* \\ \epsilon \end{matrix}$	$\begin{matrix} 1 \\ 1 \end{matrix}$	$\begin{matrix} \epsilon \\ \epsilon^* \end{matrix}$	$\begin{matrix} \epsilon^* \\ \epsilon \end{matrix}$	(R_x, R_y)	$(x^2 - y^2,\ xy);$ (xz, yz)
A_u	1	1	1	-1	-1	-1	z	
E_u	$\begin{cases} 1 \\ 1 \end{cases}$	$\begin{matrix} \epsilon \\ \epsilon^* \end{matrix}$	$\begin{matrix} \epsilon^* \\ \epsilon \end{matrix}$	$\begin{matrix} -1 \\ -1 \end{matrix}$	$\begin{matrix} -\epsilon \\ -\epsilon^* \end{matrix}$	$\begin{matrix} -\epsilon^* \\ -\epsilon \end{matrix}$	(x, y)	

S_8	E	S_8	C_4	S_8^3	C_2	S_8^5	C_4^3	S_8^7		$\epsilon = \exp(2\pi i/8)$
A	1	1	1	1	1	1	1	1	R_z	$x^2 + y^2,\ z^2$
B	1	-1	1	-1	1	-1	1	-1	z	
E_1	$\begin{cases} 1 \\ 1 \end{cases}$	$\begin{matrix} \epsilon \\ \epsilon^* \end{matrix}$	$\begin{matrix} i \\ -i \end{matrix}$	$\begin{matrix} -\epsilon^* \\ -\epsilon \end{matrix}$	$\begin{matrix} -1 \\ -1 \end{matrix}$	$\begin{matrix} -\epsilon \\ -\epsilon^* \end{matrix}$	$\begin{matrix} -i \\ i \end{matrix}$	$\begin{matrix} \epsilon^* \\ \epsilon \end{matrix}$	$(x, y);$ (R_x, R_y)	
E_2	$\begin{cases} 1 \\ 1 \end{cases}$	$\begin{matrix} i \\ -i \end{matrix}$	$\begin{matrix} -1 \\ -1 \end{matrix}$	$\begin{matrix} -i \\ i \end{matrix}$	$\begin{matrix} 1 \\ 1 \end{matrix}$	$\begin{matrix} i \\ -i \end{matrix}$	$\begin{matrix} -1 \\ -1 \end{matrix}$	$\begin{matrix} -i \\ i \end{matrix}$		$(x^2 - y^2,\ xy)$
E_3	$\begin{cases} 1 \\ 1 \end{cases}$	$\begin{matrix} -\epsilon^* \\ -\epsilon \end{matrix}$	$\begin{matrix} i \\ -i \end{matrix}$	$\begin{matrix} \epsilon \\ \epsilon^* \end{matrix}$	$\begin{matrix} -1 \\ -1 \end{matrix}$	$\begin{matrix} \epsilon^* \\ \epsilon \end{matrix}$	$\begin{matrix} -i \\ i \end{matrix}$	$\begin{matrix} -\epsilon \\ -\epsilon^* \end{matrix}$		(xz, yz)

9. The cubic groups

T ($\epsilon = \exp(2\pi i/3)$)

T	E	$4C_3$	$4C_3^2$	$3C_2$		
A	1	1	1	1		$x^2+y^2+z^2$
E	$\begin{cases}1\\1\end{cases}$	$\begin{matrix}\epsilon\\\epsilon^*\end{matrix}$	$\begin{matrix}\epsilon^*\\\epsilon\end{matrix}$	$\begin{matrix}1\\1\end{matrix}$		$\begin{matrix}(2z^2-x^2-y^2,\\x^2-y^2)\end{matrix}$
T	3	0	0	-1	$(R_x,R_y,R_z);\,(x,y,z)$	(xy,xz,yz)

T$_h$ ($\epsilon = \exp(2\pi i/3)$)

T$_h$	E	$4C_3$	$4C_3^2$	$3C_2$	i	$4S_6$	$4S_6^5$	$3\sigma_h$		
A_g	1	1	1	1	1	1	1	1		$x^2+y^2+z^2$
A_u	1	1	1	1	-1	-1	-1	-1		
E_g	$\begin{cases}1\\1\end{cases}$	$\begin{matrix}\epsilon\\\epsilon^*\end{matrix}$	$\begin{matrix}\epsilon^*\\\epsilon\end{matrix}$	$\begin{matrix}1\\1\end{matrix}$	$\begin{matrix}1\\1\end{matrix}$	$\begin{matrix}\epsilon\\\epsilon^*\end{matrix}$	$\begin{matrix}\epsilon^*\\\epsilon\end{matrix}$	$\begin{matrix}1\\1\end{matrix}$		$\begin{matrix}(2z^2-x^2-y^2,\\x^2-y^2)\end{matrix}$
E_u	$\begin{cases}1\\1\end{cases}$	$\begin{matrix}\epsilon\\\epsilon^*\end{matrix}$	$\begin{matrix}\epsilon^*\\\epsilon\end{matrix}$	$\begin{matrix}1\\1\end{matrix}$	$\begin{matrix}-1\\-1\end{matrix}$	$\begin{matrix}-\epsilon\\-\epsilon^*\end{matrix}$	$\begin{matrix}-\epsilon^*\\-\epsilon\end{matrix}$	$\begin{matrix}-1\\-1\end{matrix}$		
T_g	3	0	0	-1	3	0	0	-1	(R_x,R_y,R_z)	(xz,yz,xy)
T_u	3	0	0	-1	-3	0	0	1	(x,y,z)	

T$_d$

T$_d$	E	$8C_3$	$3C_2$	$6S_4$	$6\sigma_d$		
A_1	1	1	1	1	1		$x^2+y^2+z^2$
A_2	1	1	1	-1	-1		
E	2	-1	2	0	0		$(2z^2-x^2-y^2,\,x^2-y^2)$
T_1	3	0	-1	1	-1	(R_x,R_y,R_z)	
T_2	3	0	-1	-1	1	(x,y,z)	(xy,xz,yz)

O

O	E	$6C_4$	$3C_2(=C_4^2)$	$8C_3$	$6C_2$		
A_1	1	1	1	1	1		$x^2+y^2+z^2$
A_2	1	-1	1	1	-1		
E	2	0	2	-1	0		$(2z^2-x^2-y^2,\,x^2-y^2)$
T_1	3	1	-1	0	-1	$(R_x,R_y,R_z);(x,y,z)$	
T_2	3	-1	-1	0	1		(xy,xz,yz)

O_h	E	$8C_3$	$6C_2$	$6C_4$	$3C_2(=C_4^2)$	i	$6S_4$	$8S_6$	$3\sigma_h$	$6\sigma_d$		
A_{1g}	1	1	1	1	1	1	1	1	1	1		$x^2 + y^2 + z^2$
A_{2g}	1	1	-1	-1	1	1	-1	1	1	-1		
E_g	2	-1	0	0	2	2	0	-1	2	0		$(2z^2 - x^2 - y^2,$ $x^2 - y^2)$
T_{1g}	3	0	-1	1	-1	3	1	0	-1	-1	(R_x, R_y, R_z)	
T_{2g}	3	0	1	-1	-1	3	-1	0	-1	1		(xz, yz, xy)
A_{1u}	1	1	1	1	1	-1	-1	-1	-1	-1		
A_{2u}	1	1	-1	-1	1	-1	1	-1	-1	1		
E_u	2	-1	0	0	2	-2	0	1	-2	0		
T_{1u}	3	0	-1	1	-1	-3	-1	0	1	1	(x, y, z)	
T_{2u}	3	0	1	-1	-1	-3	1	0	1	-1		

10. The groups $C_{\infty v}$ and $D_{\infty h}$ for linear molecules

$C_{\infty v}$	E	$2C_\infty^\Phi$	$\cdots$	$\infty\sigma_v$		
$A_1 \equiv \Sigma^+$	1	1	$\cdots$	1	z	$x^2 + y^2, z^2$
$A_2 \equiv \Sigma^-$	1	1	$\cdots$	-1	R_z	
$E_1 \equiv \Pi$	2	$2\cos\Phi$	$\cdots$	0	(x, y); (R_x, R_y)	(xz, yz)
$E_2 \equiv \Delta$	2	$2\cos 2\Phi$	$\cdots$	0		$(x^2 - y^2, xy)$
$E_3 \equiv \Phi$	2	$2\cos 3\Phi$	$\cdots$	0		
$\cdots$		$\cdots$		$\cdots$		

$D_{\infty h}$	E	$2C_\infty^\Phi$	$\cdots$	$\infty\sigma_v$	i	$2S_\infty^\Phi$	$\cdots$	∞C_2		
$A_{1g} \equiv \Sigma_g^+$	1	1	$\cdots$	1	1	1	$\cdots$	1		$x^2 + y^2, z^2$
$A_{2g} \equiv \Sigma_g^-$	1	1	$\cdots$	-1	1	1	$\cdots$	-1	R_z	
$E_{1g} \equiv \Pi_g$	2	$2\cos\Phi$	$\cdots$	0	2	$-2\cos\Phi$	$\cdots$	0	(R_x, R_y)	(xz, yz)
$E_{2g} \equiv \Delta_g$	2	$2\cos 2\Phi$	$\cdots$	0	2	$2\cos 2\Phi$	$\cdots$	0		$(x^2 - y^2, xy)$
$\cdots$										
$A_{1u} \equiv \Sigma_u^+$	1	1	$\cdots$	1	-1	-1	$\cdots$	-1	z	
$A_{2u} \equiv \Sigma_u^-$	1	1	$\cdots$	-1	-1	-1	$\cdots$	1		
$E_{1u} \equiv \Pi_u$	2	$2\cos\Phi$	$\cdots$	0	-2	$2\cos\Phi$	$\cdots$	0	(x, y)	
$E_{2u} \equiv \Delta_u$	2	$2\cos 2\Phi$	$\cdots$	0	-2	$-2\cos 2\Phi$	$\cdots$	0		
$\cdots$										

11. The icosahedral groups[b]

I_h	E	$12C_5$	$12C_5^2$	$20C_3$	$15C_2$	i	$12S_{10}$	$12S_{10}^3$	$20S_6$	15σ		
A_g	1	1	1	1	1	1	1	1	1	1		$x^2 + y^2 + z^2$
T_{1g}	3	$\frac{1}{2}(1+\sqrt{5})$	$\frac{1}{2}(1-\sqrt{5})$	0	-1	3	$\frac{1}{2}(1-\sqrt{5})$	$\frac{1}{2}(1+\sqrt{5})$	0	-1	(R_x, R_y, R_z)	
T_{2g}	3	$\frac{1}{2}(1-\sqrt{5})$	$\frac{1}{2}(1+\sqrt{5})$	0	-1	3	$\frac{1}{2}(1+\sqrt{5})$	$\frac{1}{2}(1-\sqrt{5})$	0	-1		
G_g	4	-1	-1	1	0	4	-1	-1	1	0		$(2z^2 - x^2 - y^2,$
H_g	5	0	0	-1	1	5	0	0	-1	1		$x^2 - y^2,$ $xy, yz, zx)$
A_u	1	1	1	1	1	-1	-1	-1	-1	-1		
T_{1u}	3	$\frac{1}{2}(1+\sqrt{5})$	$\frac{1}{2}(1-\sqrt{5})$	0	-1	-3	$-\frac{1}{2}(1-\sqrt{5})$	$-\frac{1}{2}(1+\sqrt{5})$	0	1	(x, y, z)	
T_{2u}	3	$\frac{1}{2}(1-\sqrt{5})$	$\frac{1}{2}(1+\sqrt{5})$	0	-1	-3	$-\frac{1}{2}(1+\sqrt{5})$	$-\frac{1}{2}(1-\sqrt{5})$	0	1		
G_u	4	-1	-1	1	0	-4	1	1	-1	0		
H_u	5	0	0	-1	1	-5	0	0	1	-1		

[b]For the pure rotation group I, the outlined section in the upper left is the character table; the g subscripts should, of course, be dropped and (x, y, z) assigned to the T_1 representation.

PROBLEMS

3.1 Use matrices to show that $\sigma_1\sigma_2 = C_4$ and $\sigma_2\sigma_1 = C_4^3$ for the kaleidoscope group in Figure 3.5. Take the intersection of σ_1 and σ_2 as the z axis (that is, the axis perpendicular to the plane of the paper) and let the x axis be the intersection of σ_1 and the plane of the paper.

3.2 Select the point group to which each of the species in Example 3.5 belongs.

3.3 Assign the molecules in Example 3.3 to the appropriate point groups.

3.4 Assign the species in Figure 3.4 to the appropriate point groups.

3.5 A scalene triangle has three unequal sides. A regular scalene tetrahedron may be constructed by folding the pattern obtained by joining the midpoints of the sides of a scalene triangle having acute angles. To what point group does the regular scalene tetrahedron belong?

3.6 Assign point groups to the following figures and their conjugates. What are the shapes of the conjugate figures?

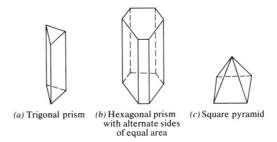

(a) Trigonal prism   *(b)* Hexagonal prism   *(c)* Square pyramid
with alternate sides
of equal area

3.7 Page 130 contains figures that can be copied and used for the construction of a tetrahedron, an octahedron, and a cube. Identify the point groups for each, considering the shading of the faces.

3.8 Pages 130 and 131 contain figures that can be copied and used for the construction of a dodeca-hedron with trigonal faces, a dodecahedron with pentagonal faces, and an icosahedron. Determine the point groups for the polyhedra, taking into account any shading on the faces. Ignoring the shading, which of the polyhedra are conjugate Platonic solids?

3.9 Determine the order of the $\mathbf{D}_{3h}$ group, the number of classes of operations, and the dimensions of the irreducible representations.

3.10 Indicate how the slash mark on the "vee-bar" shown below would be shifted under each of the operations of the $\mathbf{C}_{2v}$ group. Use your results as a guide to construct the multiplication table for the $\mathbf{C}_{2v}$ group.

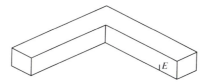

3.11 Determine the independent operations of the $\mathbf{C}_{3h}$ group and construct the group multiplication table. Determine whether C_3 and C_3^2 are in the same class.

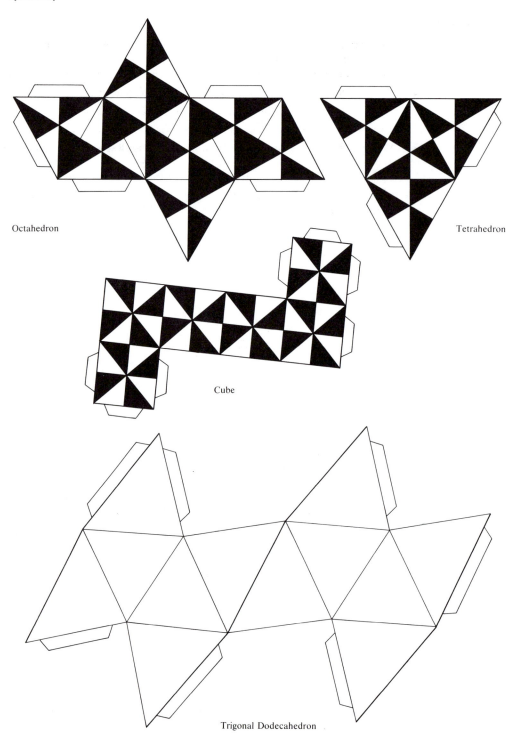

Octahedron

Tetrahedron

Cube

Trigonal Dodecahedron

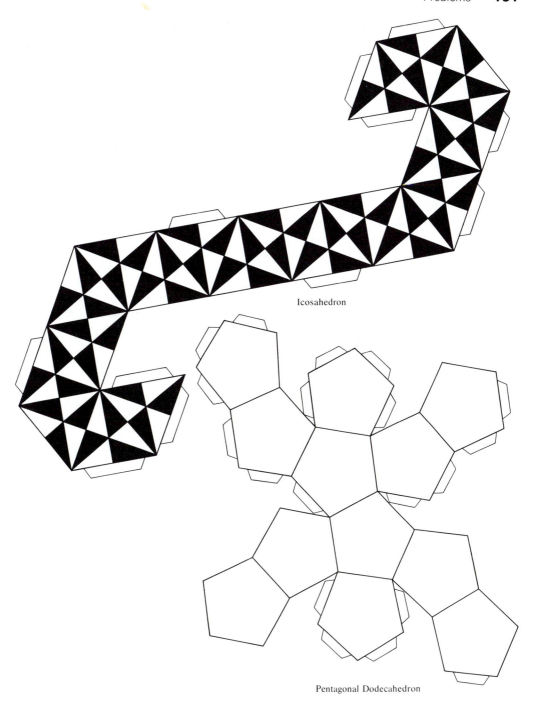

Icosahedron

Pentagonal Dodecahedron

3.12   Test the mutual orthogonality of the representations of the $\mathbf{D}_{3h}$ group.

3.13   Derive the $\mathbf{C}_{3v}$ character table.

3.14   Derive the character table for the $\mathbf{C}_3$ group using Equations 3.9 and 3.10, expressing the characters as simple and complex numbers.

3.15   Take the following direct products and find the irreducible representations contained.
  a. $E_g \times E_u$ in $\mathbf{D}_{3d}$ **b.** $E_g \times T_{1g}$ in $\mathbf{O}_h$ **c.** $E_g \times T_{2g}$ in $\mathbf{O}_h$
  d. $E_u \times E_u$ in $\mathbf{O}_h$

3.16   Give the symmetry labels for the p and d orbitals for $[Co(NH_3)_6]^{3+}$, $[Co(en)_3]^{3+}$, $[Co(edta)]^-$, and *trans*-$[Cr(Cl)_2(NH_3)_4]^+$.

3.17   The representations for the same orbitals in centrosymmetric groups and closely related subgroups usually differ only in the dropping of the g and u subscripts. Explain why the p orbitals are T_{1u} in $\mathbf{O}_h$ and T_2 in $\mathbf{T}_d$.

3.18   Instead of using Cartesian coordinates x and y, the position of a point in the xy plane can be specified by the polar coordinates r and ϕ.
  a. What is the mathematical relation between x, y, and r, ϕ?
  b. Rotation around the z axis through an angle θ takes the point (x,y) into (x',y'). The new polar coordinates are $(r, (\phi - \theta))$. Express mathematically the relation between (x',y') and the polar coordinates $(r, (\phi - \theta))$. Convert the relationship into one between (x',y') and (x,y).
  c. Express the results of **(b)** in matrix notation, thereby obtaining the rotation matrix (3.3).

IV
Molecular Orbital Theory

Molecular orbital theory has emerged as the most important approach to chemical bonding. Although molecular orbital calculations even for fairly simple molecules require computers, we can obtain much of the bonding description from symmetry and group theory. These results can be appreciated even if we do not work through all of the steps to get the results. The notation and terminology should be familiar to any chemist who uses the current literature of inorganic chemistry. Group theory is used only for the latter part of the chapter. After studying Section 3.5, you should find it easy to follow the detailed steps involved in getting the descriptions of the molecular orbitals. If the discussion of group theory in Section 3.5 was omitted, coverage after the treatment of triatomic molecules can be limited to describing the molecular orbitals and energy-level diagrams. The results obtained should seem rather obvious from the symmetry of the molecules, and they will become part of your intuitive feel for molecules, even without working through the detailed group-theoretical treatment.

4.1 DIATOMIC MOLECULES

4.1.1 The H_2 and He_2^+ Molecules

According to molecular orbital (MO) theory, valence electrons are influenced by all nuclei and other electrons of the molecule.[1] Inner electrons, of course, are localized on a particular atom, and some valence shell electrons are rather highly localized on one atom or a small group of atoms. We need be concerned only with the valence shell electrons. In order to obtain wave

---

[1] R. S. Mulliken, "Spectroscopy, Molecular Orbitals and Chemical Bonding," *Science* 1967, *157*, 13. The Nobel Prize address—for a broad audience.

functions for the molecular orbitals, we assume that these are *linear combinations of atomic orbitals (LCAO)*. Atomic orbitals that can combine are limited by the symmetry of the molecule and the symmetries of the orbitals.

Consider first the diatomic molecule H_2. The $1s$ orbitals give two linear combinations. *The number of molecular orbitals is the same as the number of atomic orbitals combined*—we conserve orbitals. The sum of the atomic orbitals (actually the wave functions are added, $1s_A + 1s_B$) gives a *bonding molecular orbital* (Figure 4.1), because the maximum electron density (the square of the sum of the wave functions) occurs in the region between the atoms. The other linear combination, $1s_A - 1s_B$, gives cancellation in the region of overlap, resulting in a nodal plane separating the atoms (Figure 4.1). Since the electron density is depleted between the atoms (zero in the nodal plane), the nuclei are poorly screened from one another and a repulsive interaction results; so this is an *antibonding orbital*. Molecular orbitals with rotational symmetry about the bond axis (C_∞) are called *sigma* (σ) orbitals. The bonding orbital is designated σ and the antibonding orbital is designated σ^*. An occupied σ orbital forms a σ bond.

Molecular orbitals in a linear (including diatomic) molecule are labeled by the Greek letters σ, π, δ, ϕ, etc., corresponding to the Roman letters s, p, d, f, etc. for atomic orbitals. The molecular orbitals have λ angular momentum quantum numbers 0, 1, 2, 3 . . . , corresponding to the l angular momentum quantum numbers of the atomic orbitals. The MO label may be obtained pictorially from the appearance of the cross-section of the MO perpendicular to the line of the bond and the label of the atomic orbital to which the cross-section corresponds. These labels are not strictly appropriate for non-linear molecules, where the labels must correspond to the symmetry species of the point group to which the molecule belongs. Nevertheless, the general labels (σ, π, etc.) are used to describe the local bonding symmetry between any pair of atoms.

The wave functions for molecular orbitals are normalized by multiplying by a normalization constant N such that $\int (N\Psi)^2 d\tau = 1$—or the probability of finding the electron somewhere outside of the nucleus is unity. For our bonding orbital, the normalized wave equation is

$$\int [N(\psi_{1s_A} + \psi_{1s_B})]^2 d\tau = N^2 \int (\psi_{1s_A}^2 + 2\psi_{1s_A}\psi_{1s_B} + \psi_{1s_B}^2)d\tau$$

$$= N^2 \left(\int \psi_{1s_A}^2 d\tau + 2 \int \psi_{1s_A}\psi_{1s_B} d\tau + \int \psi_{1s_B}^2 d\tau \right) = 1 \quad (4.1)$$

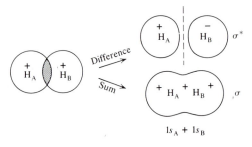

$$1s_A + 1s_B$$

Figure 4.1 The linear combination of atomic orbitals for H_2

Since we are using normalized atomic wave functions, $\int \psi_{1s_A}^2 d\tau = \int \psi_{1s_B}^2 d\tau = 1$; so our expression reduces to

$$1 = N^2(2 + 2\int \psi_{1s_A}\psi_{1s_B}d\tau)$$

or   (4.2)

$$N = \pm \frac{1}{\sqrt{2(1 + \int \psi_{1s_A}\psi_{1s_B}d\tau)}}$$

We choose the positive sign to give a positive wave function. The integral $\int \psi_{1s_A}\psi_{1s_B}d\tau$, called the *overlap integral*, is neglected in the *LCAO* approximation, giving

$$N = 1/\sqrt{2} \tag{4.3}$$

Our two molecular orbitals (Figure 4.2) are

$$\Psi_\sigma = \frac{1}{\sqrt{2}}(\psi_{1s_A} + \psi_{1s_B}) \text{ and } \Psi_{\sigma*} = \frac{1}{\sqrt{2}}(\psi_{1s_A} - \psi_{1s_B}) \tag{4.4}$$

In the more general case the normalization constant N equals 1 divided by the square root of the sum of the squares of the coefficients in the *LCAO* for all orbitals.

$$\Psi = a\psi_1 + b\psi_2 - c\psi_3 \cdots$$

$$N = \frac{1}{\sqrt{a^2 + b^2 + c^2 \cdots}} \tag{4.5}$$

> In calculating the normalization constant, it seems strange to disregard the overlap integral, which describes the extent of overlap of the bonding orbitals. Neglect of the overlap intergral does not affect the symmetry properties of the MO. With this approximation, the antibonding orbitals for an X_2 molecule are raised in energy by the same amount by which the bonding orbitals are lowered. But if we include the overlap integral, the antibonding orbitals will be more antibonding than the corresponding bonding orbitals are bonding. Consequently, the He_2 molecule discussed below would be unstable; it is not just that there is no net gain in its formation.

The H_2^+ ion is stable in gaseous discharge tubes. Its single electron occupies the lower-energy molecular orbital, σ^1. The resulting one-electron bond, bond order 1/2, is about half as strong as the electron pair bond encountered for H_2—σ^2, bond order 1. Adding an electron to H_2 produces H_2^-, whose additional electron can be accommodated in the σ^* orbital. Pop-

Figure 4.2 The relative energies of the molecular and atomic orbitals for H_2.

ulation of the antibonding orbital weakens the bond, since the net number of bonding electrons (bonding electrons minus antibonding electrons) is reduced to one, bond order 1/2. The bond order would be zero for the unstable H_2^{2-}. These same orbitals apply for possible combinations of helium atoms. The bond order is zero for He_2, an unstable species. However, if an electron is ejected from an He atom in a gaseous discharge tube, the He^+ ion formed would be isoelectronic with the H atom and can form bonds to other species. Two He^+ ions can combine to form He_2^{2+} with bond order 1, isoelectronic with H_2. The He^+ ion can combine with the He atom to form He_2^+. Since there are three electrons, one must occupy the σ^* orbital, giving a bond order of 1/2. Note that the higher bond order of He_2^{2+} does not mean that this is a more stable species; since it has only two electrons, its positive nuclei are more poorly screened from one another and it would be expected to have only transitory existence.

4.1.2 Diatomic Molecules of Second-Period Elements

Sigma Bonding

Sigma bonds also can be formed using p orbitals ($s + p_z$ or $p_z + p_z$) directed along the molecular axis (Figure 4.3) or d orbitals ($s + d_{z^2}$, $p_z + d_{z^2}$, or $d_{z^2} + d_{z^2}$). A hybrid orbital (e.g., sp, sp^2, sp^3) of one atom can form a sigma bond to another atom, as long as the bond formed has rotational symmetry. An s orbital (or p_z) on one atom cannot form a sigma bond by overlap with a p_x or p_y orbital on another atom (Figure 4.3), because net bonding is precluded by the cancellation caused by the opposite signs of the amplitudes of the wave functions in one region of overlap. This combination does not have the rotational symmetry of a σ orbital, nor does it meet the requirements for other types of bonds (π or δ).

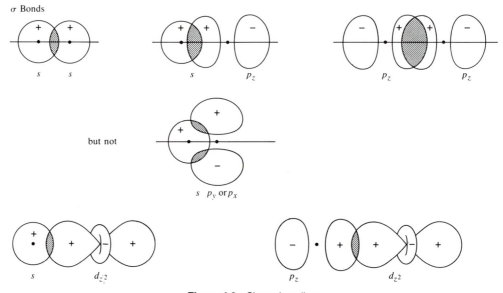

Figure 4.3 Sigma bonding.

Pi Bonding

With the z axis chosen as the bond direction (the unique axis is conventionally labeled z), the p_x (or p_y) orbitals on the two atoms can combine to form *pi* (π) orbitals (Figure 4.4). Each of the two equivalent π orbitals ($p_x + p_x$ or $p_y + p_y$) has a nodal plane through the nuclei bonded. The parallel p orbitals overlap, to cause an increase in electron density on each side of this plane. Pi orbitals formed from combinations of d orbitals also result in molecular orbitals with a single nodal plane (e.g., $p_x + d_{xz}$, or $d_{xz} + d_{xz}$). Combinations of d orbitals to produce molecular orbitals with two nodal planes through the nuclei bonded are called *delta* (δ) orbitals (e.g., $d_{x^2-y^2} + d_{x^2-y^2}$ or $d_{xy} + d_{xy}$; see Figure 4.4). There are now many examples of compounds, such as $Re_2Cl_8^{2-}$, involving M—M quadruple bonds (1 σ, 2 π, and 1 δ); see p. 648.

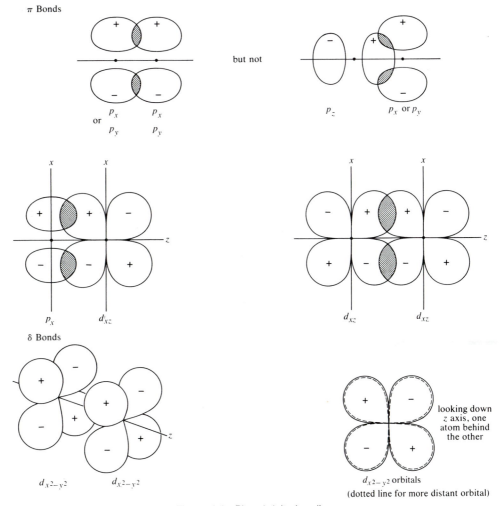

Figure 4.4 Pi and delta bonding.

Figures 4.3 and 4.4 show orbitals that combine to give bonding molecular orbitals. In the corresponding antibonding orbitals, the signs of the wave functions[2] are opposed in the region' of overlap. The molecular orbitals resulting from combinations of atomic *s* and *p* orbitals for an X_2 molecule are sketched in Figure 4.5 as electron density plots,[3] but the signs of the *wave functions* for individual lobes are shown because they are important in determining the proper combinations. The orbitals are designated as *g* (gerade) if they are centrosymmetric and *u* (ungerade) if there is a change of sign of the wave function on reflection through a center of symmetry.

Molecular Orbital Energy Levels

The combinations of atomic orbitals are obtained unambiguously from symmetry considerations. For an X_2 molecule where X is a second-period element, the order of some of the energy levels depends on the number of electrons and the relative energies, and hence on the extent of mixing of the $2s$ and $2p$ orbitals on the same atom (hybridization). For Li_2 there is no problem. The $2s$ orbitals combine in the same way as the $1s$ orbitals, giving a bonding and an antibonding orbital. The two valence-shell electrons enter the lower energy σ (bonding) orbital to produce a bond order of 1. Inner shells are nonbonding, since there is little interaction and as many antibonding electrons as bonding electrons. The bond order of Be_2 is zero, since there are as many antibonding electrons as bonding electrons (see Table 4.1); it is not a stable molecule.

The B_2 molecule has two unpaired electrons, so the energy of the doubly degenerate π orbitals must be lower than that for σ_{2p_z}, the order shown in Figure 4.6 and Table 4.1. The same order of energy levels applies for C_2, because it is diamagnetic—$\sigma_{2s}^2\ \sigma_{2s}^{*2}\ \pi_x^2\ \pi_y^2$. The N_2 molecule would be diamagnetic with a triple bond regardless of the relative energies of the π and σ_{p_z} orbitals. However, since it is known that N_2^+ has a single σ_{p_z} electron, the σ_{p_z} orbital must be higher in energy, giving $\sigma_{2s}^2\ \sigma_{2s}^{*2}\ \pi_x^2\ \pi_y^2\ \sigma_{p_z}^2$ for N_2 (Figure 4.7).

The oxygen molecule is paramagnetic with two unpaired electrons. Any simple valence-bond description would predict a diamagnetic molecule, $\overset{..}{O}{=}\overset{..}{O}$—which must be wrong. The molecular orbital description has as the highest filled level the doubly degenerate π^* orbitals, $\pi_x^{*1}\ \pi_y^{*1}$, giving two unpaired electrons and a bond order of 2 (see Table 4.1). As we shall see, the relative order of energies of the $\pi_{x,y}$ and σ_{p_z} orbitals is reversed for O_2 and F_2 in comparison with that of N_2. The significance of the π^* orbitals as the highest-energy occupied orbitals is shown by the *shortening* of the bond length on removal of an (antibonding) electron to give O_2^+ (Table 4.1). Adding an electron to form the superoxide ion, O_2^-, weakens the bond (Table 4.1), and adding another electron to form the peroxide ion, O_2^{2-}, gives a diamagnetic species with bond order 1. The configuration of the F_2 molecule is identical to that of O_2^{2-}. The bond energy of F_2 is very low compared with that of Cl_2 (239 kJ/mole), because of the strong repulsion between lone pairs for F_2 (see p. 226).

---

[2]The signs of the wave functions have only mathmetical significance—*not* physical significance. These correspond to the signs of the amplitudes of light waves. In combining light waves for diffraction experiments, the amplitudes are combined with the proper signs. The resulting intensity of light is given by the square of the sum of the amplitudes, eliminating negative signs for intensities. Electron density is given by the square of the wave function.

[3]A. C. Wahl, "Molecular Orbital Densities: Pictorial Studies," *Science* 1966, *151*, 961. Useful orbital contour plots for diatomic molecules.

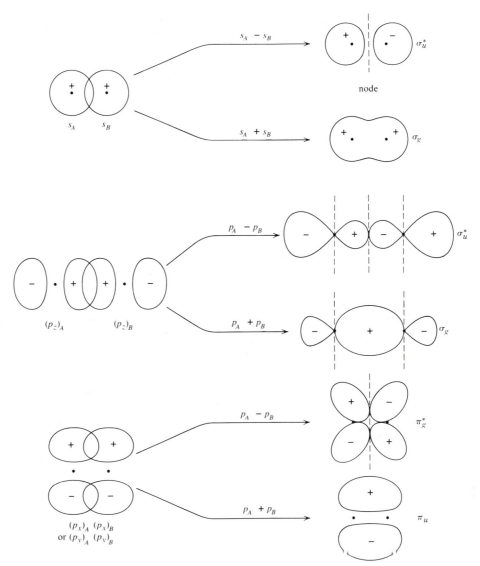

Figure 4.5 Molecular orbitals from combinations of s and p atomic orbitals. The signs of the wave functions (Ψ) are shown even though these are electron density (Ψ^2) contour plots.

The magnetic properties of B_2 and C_2 and the spectroscopic study of N_2^+ make certain the order of the highest filled energy levels for these molecules: that is, $\pi_{x,y}$ is lower in energy than σ_{p_z}. This is the order expected when the energies of the $2s$ and $2p$ orbitals differ slightly. These circumstances permit mixing of the s and p orbitals (hybridization), so that σ_s has some p character and σ_{p_z} some s character. As a result, the lower-energy σ orbital (which we have labeled σ_s) is lowered still further in energy, and the other σ orbital (which we labeled σ_{p_z}) is raised in energy such that its energy exceeds that of the $\pi_{x,y}$ orbitals. The effects can be

Table 4.1 Molecular orbital configurations for some diatomic species[a]

	Configuration	Bond Order	Number of Unpaired Electrons	Bond Length (pm)	Bond Dissociation Energy (kJ/mole)
H_2	σ_{1s}^2	1	0	74.14	432.1
Li_2	σ_{2s}^2	1	0	267.3	101
Be_2	$\sigma_{2s}^2\,\sigma_{2s}^{*2}$	0	0	—	—
B_2	$\sigma_{2s}^2\,\sigma_{2s}^{*2}\,\pi_x^1\,\pi_y^1$	1	2	159	291
C_2	$\sigma_{2s}^2\,\sigma_{2s}^{*2}\,\pi_x^2\,\pi_y^2$	2	0	124.3	599
N_2	$\sigma_{2s}^2\,\sigma_{2s}^{*2}\,\pi_x^2\,\pi_y^2\,\sigma_{p_z}^2$	3	0	109.77	942
O_2	$\sigma_{2s}^2\,\sigma_{2s}^{*2}\,\sigma_{p_z}^2\,\pi_{x,y}^4\,\pi_x^{*1}\,\pi_y^{*1}$ [b]	2	2	120.75	494
O_2^+	$\sigma_{2s}^2\,\sigma_{2s}^{*2}\,\sigma_{p_z}^2\,\pi_{x,y}^4\,\pi^{*1}$	2.5	1	111.6	643
O_2^-	$\sigma_{2s}^2\,\sigma_{2s}^{*2}\,\sigma_{p_z}^2\,\pi_x^4\,\pi_x^{*2}\,\pi_y^{*1}$	1.5	1	135	395
O_2^{2-}	$\sigma_{2s}^2\,\sigma_{2s}^{*2}\,\sigma_{p_z}^2\,\pi_x^4\,\pi_x^{*2}\,\pi_y^{*2}$	1	0	149	126
F_2	$\sigma_{2s}^2\,\sigma_{2s}^{*2}\,\sigma_{p_z}^2\,\pi_{x,y}^4\,\pi_x^{*2}\,\pi_y^{*2}$	1	0	141	155
Ne_2	$\sigma_{2s}^2\,\sigma_{2s}^{*2}\,\sigma_{p_z}^2\,\pi_{x,y}^4\,\pi_{x,y}^{*4}\,\sigma_{p_z}^{*2}$	0	0	(315)	0.2

[a]Data (except for O_2^{2-}) from K. P. Huber and G. Herzberg, *Molecular Spectra and Molecular Structure*, Vol. IV, *Constants of Diatomic Molecules*, Van Nostrand–Reinhold, New York, 1979.
[b]See text for discussion of the reversal of the order of energies for σ_{p_z} and $\pi_{x,y}$.

seen more easily if we consider first the formation of *sp* hybrid orbitals on each N and then combine these to form molecular orbitals (Figure 4.7). Here, in the linear combinations each hybrid contributes to each σ molecular orbital, but the closer in energy a hybrid orbital is to the resulting molecular orbital, the greater the contribution. The energy of the more stable bonding orbital is lowered greatly and the energy of the least stable antibonding orbital is increased greatly.

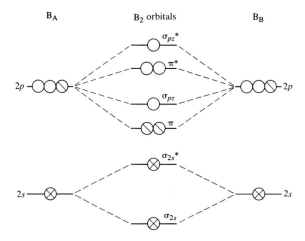

Figure 4.6 Molecular orbital energy-level diagram for B_2.

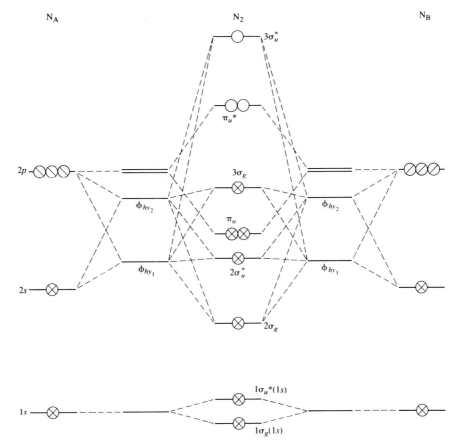

Figure 4.7 Molecular orbital energy-level for N_2 using *sp* hybrid orbitals. Adapted from M. Orchin and H. H. Jaffé, *Symmetry, Orbitals, and Spectra*, Wiley-Interscience, New York, 1971, p. 41.

The two *sp* hybrid orbitals considered earlier (p. 57) are equivalent and have equal energy ($\lambda = 1$).

$$\varphi_{h_1} = \frac{1}{\sqrt{1 + \lambda^2}} (2s + \lambda 2p)$$

$$\varphi_{h_2} = \frac{1}{\sqrt{1 + \lambda^2}} (\lambda 2s - 2p)$$

In the case of an X_2 molecule, one hybrid orbital is used for bonding to the other X atom (the combination with greater *p* character, φ_{h_2} with $\lambda < 1$), whereas the lone pair occupies the *sp* hybrid with greater *s* character. The difference in the energies of the two hybrids thus depends on the relative energies of the *s* and *p* orbitals. As the nuclear charge increases across a period, the energy separation between $2s$ and $2p$ increases. The more penetrating $2s$ orbital (p. 32) decreases in energy with increasing atomic number, as *p* electrons are added. The greater the energy separation between $2s$ and $2p$, the less mixing occurs and the greater the energy separation between φ_{h_1} and φ_{h_2}.

Where the energy separation between the 2s and 2p orbitals is great enough, there will be little mixing and the energy of σ_{p_z} will fall below that of $\pi_{x,y}$ (Figure 4.8). This appears to be the order of levels for O_2 and F_2 (and was the only order considered in early MO treatments). The order of these two filled levels is not very important for O_2 and F_2, because the bond order and magnetic properties are not affected.

Heteronuclear Molecules

Heteronuclear diatomic molecules can be described similarly. We can get the bond order and number of unpaired electrons from the population of energy levels in Figure 4.6 for an XY molecule, where X and Y are second-period elements. However, where mixing of orbitals (on the same atom) is important and the relative energies of the orbitals of X and Y differ significantly, a hybridization approach is preferable. In the MO energy-level diagram for CO illustrated in Figure 4.9, the orbital energies are lower for the more electronegative oxygen, so that the lower-energy hybrid of carbon is comparable in energy with the higher-energy hybrid of oxygen. These orbitals of comparable energy combine to give bonding and antibonding molecular orbitals. The widely separated (in energy) very low-energy hybrid on oxygen and the high-energy hybrid on carbon are *nonbonding;* rather than combining to give molecular orbitals, they remain as localized orbitals on O and C, respectively. These are the orbitals occupied by the two lone pairs directed away from the CO bond. Since the lone-pair orbital on oxygen is close in energy to the oxygen atomic 2s orbital, it is said to be mostly s in character: that is, it is much like an oxygen 2s orbital. The lone-pair orbital on carbon, on the other hand, is close in energy to the carbon p orbitals, so it is said to have a great deal of carbon p character.

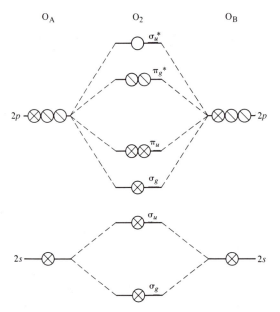

Figure 4.8 Molecular orbital energy-level diagram for O_2.

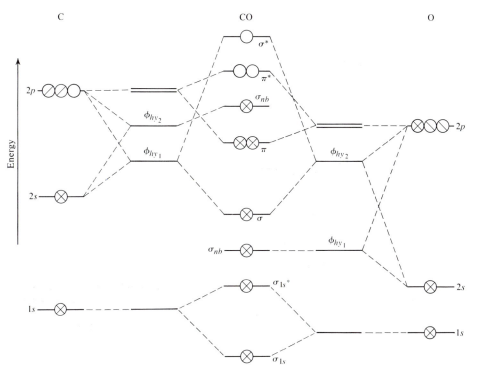

Figure 4.9 The molecular orbital energy-level diagram for CO. Adapted from M. Orchin and H. H. Jaffé, *Symmetry, Orbitals, and Spectra*, Wiley-Interscience, New York, 1971, p. 47.

With its higher energy, the carbon lone pair is more loosely bound and thus is the pair donated to metals in forming carbonyl complexes (p. 405). The electron cloud for the π orbitals will be polarized toward the more electronegative oxygen atom (Figure 4.10). Actually, since only one direction *(z)* is defined for the CO molecule, x and y are indistinguishable for the atomic and molecular orbitals. Consequently, π_x and π_y together form a sheath having cylindrical symmetry. Although these bonds have axial symmetry, they are π, not σ, since there is no maximum in the electron density along the internuclear axis—it is a node. The empty π^* orbitals would be more localized on carbon.

Consider a molecule containing H bonded to a second-period element (F). First we must judge the availability and symmetry of the orbitals on the two atoms. Hydrogen has only a $1s$ orbital, so symmetry permits σ-bond formation only by overlap with the $2s$ and $2p_z$ orbitals of F. The energies of the F orbitals are much lower than those of H (the energy of the highest filled orbital can be obtained from the ionization energy: -1.312 MJ/g-atom for H and -1.681 MJ/g-atom for F). The $2s$ orbital of F is nonbonding, since it is so much lower in energy than the $1s$ orbital of H (Figure 4.11). The σ bond involves overlap of the H $1s$ orbital and F $2p_z$ orbital. With no orbitals of proper symmetry to combine with them, the other p orbitals on F are nonbonding.

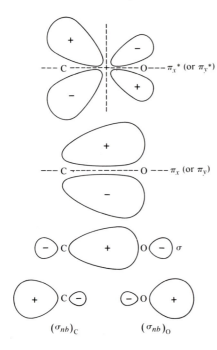

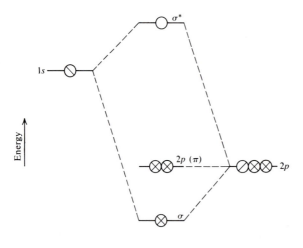

Figure 4.10 Sketches of molecular orbitals for CO. The π and π^* orbitals are shown in cross-section since the π_x and π_y or π_x^* and π_y^* combine, giving cylindrical symmetry along the bond axis.

Figure 4.11 Molecular orbital energy-level diagram for HF.

4.2 TRIATOMIC MOLECULES

The formation of stable bonding molecular orbitals requires similar energies and good overlap of the orbitals combined. In the case of diatomic molecules, the symmetry restrictions for good overlap are obvious. In general the possibility of good overlap is determined by the symmetry of the molecule and the symmetry of the atomic orbitals. Symmetry and group theory are valuable in deducing the combinations permitted for larger molecules, where the possible combinations might not be obvious. For triatomic molecules, we will use the terminology of and results from group theory, without having to provide a mathematical treatment.

4.2.1 Linear Molecules

The CO_2 molecule is linear and belongs to the $\mathbf{D}_{\infty h}$ point group, whose character table[4] is given in Chapter 3, Appendix 3. The rotational axis (C_∞, taken as the z direction) must be the internuclear axis for σ bonding. The σ bonding orbitals available for carbon and each oxygen are $2s$ and $2p_z$. Since the oxygen $2s$ orbitals are much lower in energy than the carbon orbitals, they are essentially nonbonding. Both oxygen $2p_z$ orbitals combine with the carbon orbitals in the same way, so we may conveniently consider them together using the symmetry of the molecule and the symmetry characteristics of the oxygen orbitals. The z axis on each O is defined as $+$ in the direction pointing toward C.

The totally symmetric carbon $2s$ orbital is unchanged after carrying out any of the symmetry operations of the group: that is, it belongs to the A_{1g} representation.[5] Only those combinations of oxygen orbitals that also are totally symmetric (belong to the A_{1g} representation) can combine with the carbon $2s$ orbital. The oxygen combination or *group orbitals*—the two linear combinations of the $2p_z$ orbitals—are

$$a_{1g} \qquad\qquad\qquad \text{and} \qquad\qquad a_{1u}$$

The a_{1g} oxygen group orbital combines with the carbon $2s$ orbital to give a bonding molecular orbital; with the signs reversed for both oxygen orbitals (still a_{1g}), it combines with the carbon $2s$ orbital to give an antibonding molecular orbital.

$$\sigma_{a_{1g}} = (2s)_C + (2p_{zO_1} + 2p_{zO_2})$$

and

$$\sigma^*_{a_{1g}} = (2s)_C - (2p_{zO_1} + 2p_{zO_2})$$

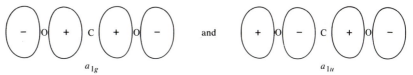

---

[4] M. Orchin and H. H. Jaffé, "Character Tables and Their Significance," *J. Chem. Educ.* 1970, *47*, 510. This is the third in a series of three articles on symmetry and the use of character tables. See also p. 104.

[5] The representations themselves and electronic states will be designated by uppercase letters (A_1, B_1, etc.) while lowercase letters (a_1, b_1, etc.) will be used for atomic or molecular orbitals. The special symbols often used to represent linear point groups are included in the character table, along with the symbols used for other point groups.

The carbon $2p_z$ orbital has the symmetry of the a_{1u} representation, so the oxygen group orbital to be used for bonding also must have this symmetry. The molecular orbitals (C plus O group orbitals) are

$$\sigma_{a_{1u}} = (2p_z)_C + (-2p_{zO_1} + 2p_{zO_2})$$

and

$$\sigma^*_{a_{1u}} = (2p_z)_C - (-2p_{zO_1} + 2p_{zO_2})$$

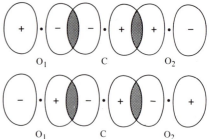

These a_{1u} and a^*_{1u} orbitals exceed in energy the respective a_{1g} and a^*_{1g} orbitals, since for carbon the energy of $2p$ is higher than that of $2s$.

The π bonds are formed similarly, but there are two equivalent C p_π orbitals (the degenerate $2p_x$ and $2p_y$ orbitals) and corresponding pairs of O group orbitals. The C p_π orbitals belong to the two-dimensional representation E_{1u}. The four O p_π orbitals can combine to give four group orbitals. One of the pair of degenerate e_{1u} O group orbitals $(p_x + p_x)$ is shown below with the corresponding C e_{1u} orbital.

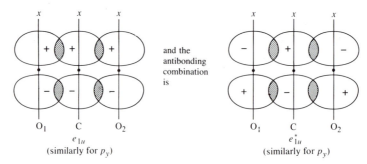

and the antibonding combination is

e_{1u}
(similarly for p_y)

e^*_{1u}
(similarly for p_y)

The other pair of O group orbitals $(p_x - p_x$ and $p_y - p_y)$ belonging to a representation of the point group are

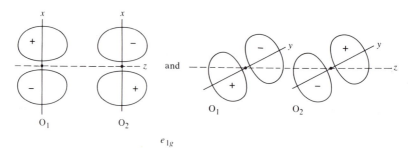

and

e_{1g}

These group orbitals are centrosymmetric and together belong to the E_{1g} representation. Since no C orbitals belong to this representation—that is, have this symmetry—these orbitals are *nonbonding*. Since the e_g orbitals do not combine with C orbitals at all, electrons in these orbitals will be localized on the oxygens.

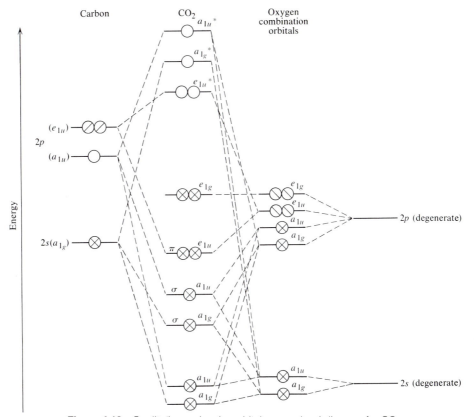

Figure 4.12 Qualitative molecular orbital energy-level diagram for CO_2.

Figure 4.12 shows a qualitative molecular orbital energy level diagram. In general, the splitting between the π and π^* energy levels is smaller than that between σ and σ^* levels from the same set, because of less overlap for the π and π^* orbitals. The nonbonding "lone" pairs on the oxygens are in the lowest-energy a_{1g} and a_{1u} and in the e_{1g} orbitals. There are two σ bonding orbitals (a_{1g} and a_{1u}) and two π bonding orbitals (e_{1u}). We should verify that the number of molecular orbitals (12) is the same as the number of atomic valence orbitals (four orbitals for each of three atoms). A similar bonding scheme applies to other isoelectronic molecules or ions—CS_2, N_2O, N_3^-, OCN^-, NO_2^+, etc. (see p. 63).

4.2.2 Angular Molecules

The NO_2 molecule is angular because of the unshared electron on N. The symmetry of the molecule is C_{2v} (see the character table in Chapter 3, Appendix 3). We choose the z axis as the C_2 axis and the x axis perpendicular to the molecular plane. Although primarily nonbonding, the O $2s$ orbitals are involved to a somewhat greater extent than is true for CO_2, because the energy separation between the O and N orbitals is not as great. Thus, relative to the atomic O orbitals, the energies of these molecular orbitals are lowered more than for CO_2. The symmetry species (irreducible representations) for the O $2s$ group orbitals in the C_{2v} point group are a_1 ($s + s$) and b_2 ($s - s$). (See Figure 4.13).

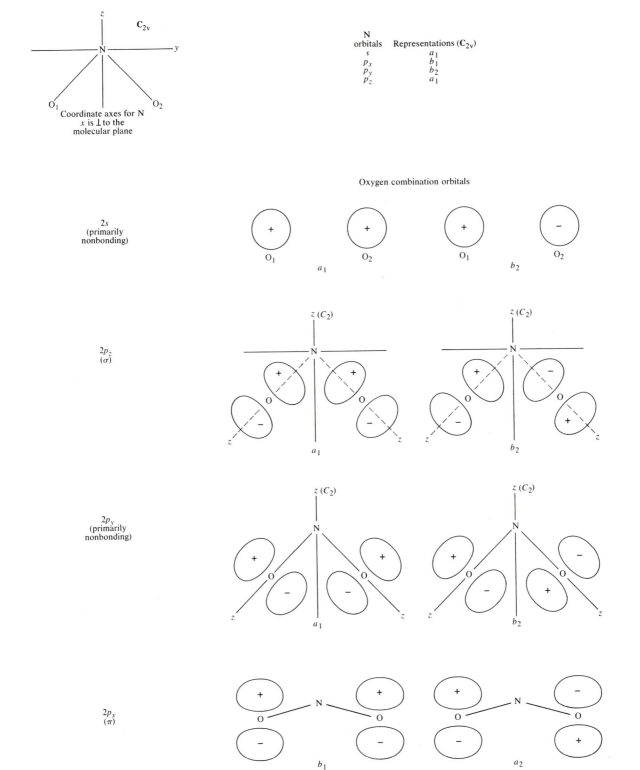

Figure 4.13 Nitrogen and oxygen combination orbitals for NO_2.

The N $2s$ and $2p_z$ orbitals belong to the same representation (a_1), so they combine in the same way with O group orbitals that also have a_1 symmetry. This includes the O $2s$ orbitals, to the limited extent to which they participate. For convenience, we choose to orient the axes for the oxygens so that the z axis for each O is directed toward the N and the x axis is perpendicular to the molecular plane (Figure 4.13). Thus the O $2p_z$ orbitals are oriented ideally for σ bonding. The group orbitals from these $2p_z$ orbitals have a_1 and b_2 symmetry, as shown. The a_1 ($2p_z$) O group orbital can combine with the N a_1 orbitals (s and p_z). Since two N orbitals and one O group orbital are combining, three molecular orbitals must result from the linear combination. We can visualize these best by looking at the combinations of N $2s$ and $2p_z$ orbitals (sp hybrids) and at the way these two hybrids could combine with the a_1 group orbital (Figure 4.14). The best combination (the one with overlapping lobes having the same signs) is strongly bonding, and the worst combination (overlapping lobes having opposite signs) is strongly antibonding. The third combination is mostly nonbonding, because of poor overlap. The result will be the same whether the N orbitals are combined first (hybrids) or linear combinations of the three orbitals are taken directly, as shown at the top of Figure 4.14. The b_2 group orbital can combine with the $2p_y$ orbital of N to give a bonding orbital (Figure 4.14) and an antibonding orbital.

Example   Sketch the b_2^* antibonding orbital. Why is there no b_2 nonbonding orbital?

Solution   The b_2^* orbital is

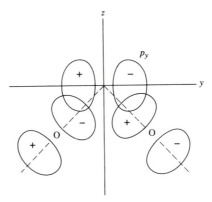

There is one b_2 group orbital and one b_2 orbital on N ($2p_y$). Since we must conserve orbitals, only two MO are obtained from the combination—the bonding and antibonding orbitals.

The O $2p_y$ orbitals give combinations of a_1 and b_2 symmetry (Figure 4.13), but these do not give good overlap with the corresponding N orbitals. The $2p_x$ orbitals, which are perpendicular to the molecular plane, are available for all three atoms for π bonding. The O group orbitals are b_1 ($p_x + p_x$) and a_2 ($p_x - p_x$). The N p_x orbital gives bonding and antibonding molecular orbitals (Figure 4.14) with the b_1 group orbital. The a_2 group orbital is nonbonding, since there is no N orbital with this symmetry.

The σ bonding orbitals are σ_{a_1} and σ_{b_2}, and the π bonding orbital, involving all three atoms, is π_{b_1} (Figure 4.15). The a_1 and b_2 orbitals from combining the O $2s$ orbitals are

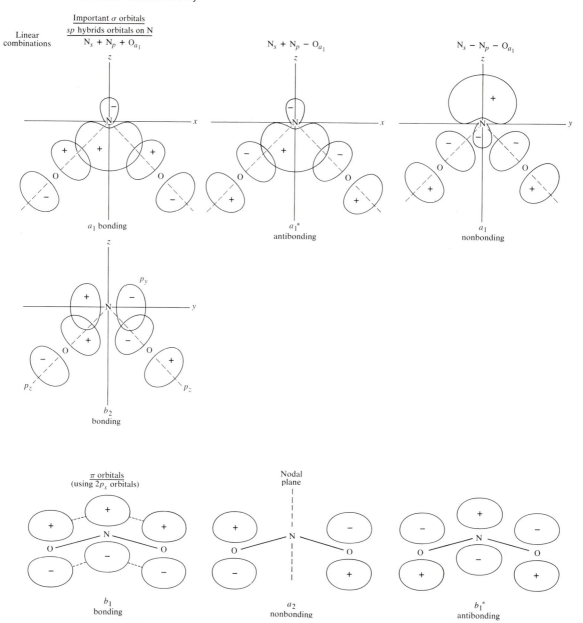

Figure 4.14 Molecular orbitals for NO_2.

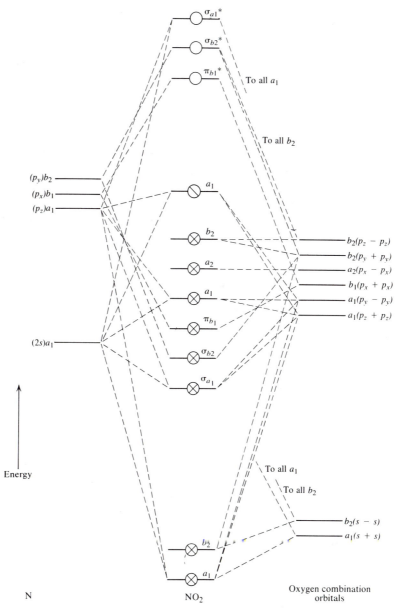

Figure 4.15 Qualitative molecular orbital energy-level diagram for NO_2.

nonbonding and accommodate O lone pairs. The a_2 orbital is strictly nonbonding, because of symmetry, and accommodates another lone pair localized on the oxygens. The other two O lone pairs of the five lone pairs in the preferred Lewis structure $\cdot N \begin{smallmatrix} \ddot{O} \cdot \\ \parallel \\ \ddot{O} \colon \end{smallmatrix}$ are accommodated in the a_1 $(p_y - p_y)$ and b_2 $(p_y + p_y)$ orbitals. The highest-energy a_1 orbital (close in energy to N $2p_z$) accommodates the odd electron localized primarily on N. Another electron can be added to this orbital to form NO_2^-, for which the bonding description is similar. The bonding descriptions of SO_2 and O_3 are similar to those for NO_2^-, with allowance for the relative energies of the orbitals of the bonded atoms. In the case of O_3, the $2s$ orbitals on all three atoms have the same energy, so they are strongly bonding.

Removing an electron from NO_2 to form NO_2^+ produces a linear ion with the CO_2 structure—another p orbital becomes available for π bonding. An NO_2 electron can be promoted from the b_2 level to the a_1 level in the electronic excited state without significant change in bonding or geometry. Promoting an electron in CO_2 to the π^* level can lead to removal of the double degeneracy of this level, with accompanying bending of the molecule. These levels (p_x, p_y) have different energies for a bent molecule, because the x and y directions are nonequivalent. The electron would go into the lower-energy orbital in the bent molecule, lowering the energy of the molecule by β, the splitting parameter.

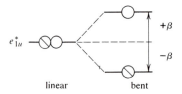

This is an application of the Jahn-Teller theorem (see p. 272).

The actual situation is not so simple, since the orbitals from e_{1u}^* that result from the bending can mix with other orbitals belonging to the same representations, raising or lowering their energies. Because of this, they do not split symmetrically. It is important to remember that electronically excited states need not have the same geometry as the ground state. One of the excited states of CO_2 would be much like NO_2, but with a vacancy in a lower nonbonding orbital.

4.2.3 Walsh Diagrams

The bonding description of an XH_2 molecule resembles that for CO_2 or NO_2, depending on the shape, but without π bonding; also, the energy of the H $1s$ orbital is high compared with the energies of the O orbitals. The molecular orbital description of the bonding in H_2O can treat the O $2s$ and $2p$ orbitals separately, as we have done, to obtain different representations for the two σ bonds, each of which involves both O orbitals. Such orbitals often are called delocalized molecular orbitals—although they are not delocalized in the same sense as π electrons in an aromatic molecule. Alternatively, we can start with sp^3 hybrid orbitals on O and use two of these equivalent orbitals to form σ bonds to the hydrogens. Such orbitals often are called localized molecular orbitals. The latter description, although very similar to the valence bond description, provides additional information about excited states and permits correlation with other molecular shapes.

Consider an XH_2 molecule, where X is a second-period element with a maximum of eight electrons in four bonding or nonbonding orbitals. The relative energies of these orbitals and their occupancy will determine the molecular shape. For the angular molecule, with the limiting bond angle of 90°, the axes are chosen as for NO_2. For the linear case, the y axis is chosen as the unique σ bond direction, so that when the angle is decreased the axes agree with the choice for the angular molecule.

Figure 4.16 depicts a *Walsh diagram* showing the correlation of the energies of the orbitals for linear and bent XH_2 molecules. Since the hydrogens have only $1s$ orbitals, the only bonding orbitals available are those with the symmetry of the H group orbitals (a_{1g} and a_{1u} for the linear case). In the bent molecule the a_1 orbital involves both s and p_z orbitals on X, but a_{1g} involves only the s orbital of X in the linear molecule. Since p orbitals have higher energy than s orbitals, mixing in of p_z must raise the energy of the MO for the bent molecule relative to the linear case. The b_2-a_{1u} pair involves opposite signs of the wave functions for the two H atoms. Since it is antibonding with respect to the H atoms themselves, this orbital must become less stable as the H atoms come closer together with decreasing bond angle. The nonbonding e_{1u} orbitals for the linear molecule give two orbitals (a_1 and b_1) for the bent molecule. The p_x orbital, perpendicular to the plane of the molecule for any bond angle, is affected relatively little by bending. This is the b_1 orbital for the bent molecule. The e_{1u} orbitals (linear case) are pure p orbitals, but with increased bending, p_z mixes more and more with the s orbital of X, lowering the energy of the a_1 orbital. One of the a_1 orbitals acquires more s character as the other loses it. In fact, $2a_{1g}$ (linear) uses a pure s orbital, but $2a_1$ (90°) uses the p_z orbital, with

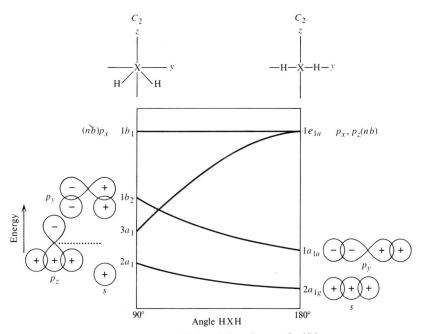

Figure 4.16 Walsh correlation diagram for XH_2.

sp hybrids involved at intermediate angles. Also, the e_{1u} orbitals (linear case) are pure *p* orbitals, but one of these correlates with $3a_1$ (bent), which at 90° is a nonbonding *s* orbital. At intermediate angles this a_1 orbital also involves *sp* hybrids.

The BeH_2, BH_2^+, and HgH_2 molecules with four valence-shell electrons prefer the linear arrangement, filling the lower-energy bonding orbitals. We cannot get two bonds for the bent molecule by filling both a_1 orbitals, since one is nonbonding because of symmetry. Molecules of the XH_2 type with five to eight valence electrons should be bent. The first excited state of BeH_2 and HgH_2 should be bent. The BH_2 molecule is bent (131°), but has a linear excited state.

In a series of detailed papers, Walsh[6] treated a range of simple molecules in an empirical and rather qualitative manner. In the past, MO treatments usually started with an assumed geometry. Now, at least for simple molecules, we can use computers to determine the optimum geometry, as well as the ordering of the energy levels. The more complex the molecule, the greater the importance of symmetry in reducing the problem to manageable proportions.

4.3   BORON TRIFLUORIDE—A GROUP THEORETICAL TREATMENT

The more complex the molecule, the more important it is to use the group theoretical approach. The following bonding treatments use detailed applications of group theory as presented in Chapter 3. Even if Section 3.5 was omitted or treated lightly, however, you should be able to understand the qualitative descriptions obtained. The terminology used is summarized in Section 3.6.

The BF_3 molecule is planar with 120° bond angles (**D**$_{3h}$ symmetry). The bonding description applies also to the isoelectronic CO_3^{2-} and NO_3^- ions. We choose the C_3 axis along *z* for B, with the *z* axes of each F parallel to C_3 and the *x* axes directed toward B. The coordinate axes of the F atoms are brought into coincidence by the C_3 operation (Figure 4.17).

> The choice of axes on B is not really arbitrary. If we want to use the character tables available, we must assign axes in the same way as for the construction of the character table. The major C_n axis is, conventionally, the *z* axis.

4.3.1   Sigma Bonding

Irreducible Representations

We first must find the irreducible representations to which the σ bonds belong. The matrices for the transformations of the σ bonds can be written as follows.

---

[6]A. D. Walsh, *J. Chem. Soc. 1953*, 2260–2331. See also R. M. Gavin Jr., "Simplified Molecular Orbital Approach to Inorganic Stereochemistry," *J. Chem. Educ.* 1969, *46*, 413; N. C. Baird, "Molecular Geometry Predictions Using Simple MO Theory," *J. Chem. Educ.* 1978, *55*, 412; B. M. Gimarc, "Applications of Qualitative Molecular Orbital Theory," *Acc. Chem. Res.* 1974, *7*, 384; and J. K. Burdett, *Molecular Shapes*, Wiley-Interscience, New York, 1980.

$$E \begin{pmatrix} \sigma_1 \\ \sigma_2 \\ \sigma_3 \end{pmatrix} = \begin{pmatrix} 1 & 0 & 0 \\ 0 & 1 & 0 \\ 0 & 0 & 1 \end{pmatrix} \begin{pmatrix} \sigma_1 \\ \sigma_2 \\ \sigma_3 \end{pmatrix} = \begin{pmatrix} \sigma_1 \\ \sigma_2 \\ \sigma_3 \end{pmatrix}$$
$$x = 3$$

$$\sigma_h \begin{pmatrix} \sigma_1 \\ \sigma_2 \\ \sigma_3 \end{pmatrix} = \begin{pmatrix} 1 & 0 & 0 \\ 0 & 1 & 0 \\ 0 & 0 & 1 \end{pmatrix} \begin{pmatrix} \sigma_1 \\ \sigma_2 \\ \sigma_3 \end{pmatrix} = \begin{pmatrix} \sigma_1 \\ \sigma_2 \\ \sigma_3 \end{pmatrix}$$
$$x = 3$$

$$C_3 \begin{pmatrix} \sigma_1 \\ \sigma_2 \\ \sigma_3 \end{pmatrix} = \begin{pmatrix} 0 & 1 & 0 \\ 0 & 0 & 1 \\ 1 & 0 & 0 \end{pmatrix} \begin{pmatrix} \sigma_1 \\ \sigma_2 \\ \sigma_3 \end{pmatrix} = \begin{pmatrix} \sigma_2 \\ \sigma_3 \\ \sigma_1 \end{pmatrix}$$
$$x = 0$$

$$S_3 \begin{pmatrix} \sigma_1 \\ \sigma_2 \\ \sigma_3 \end{pmatrix} = \begin{pmatrix} 0 & 1 & 0 \\ 0 & 0 & 1 \\ 1 & 0 & 0 \end{pmatrix} \begin{pmatrix} \sigma_1 \\ \sigma_2 \\ \sigma_3 \end{pmatrix} = \begin{pmatrix} \sigma_2 \\ \sigma_3 \\ \sigma_1 \end{pmatrix}$$
$$x = 0$$

$$C_2 \begin{pmatrix} \sigma_1 \\ \sigma_2 \\ \sigma_3 \end{pmatrix} = \begin{pmatrix} 1 & 0 & 0 \\ 0 & 0 & 1 \\ 0 & 1 & 0 \end{pmatrix} \begin{pmatrix} \sigma_1 \\ \sigma_2 \\ \sigma_3 \end{pmatrix} = \begin{pmatrix} \sigma_1 \\ \sigma_3 \\ \sigma_2 \end{pmatrix}$$
$$x = 1$$

$$\sigma_v \begin{pmatrix} \sigma_1 \\ \sigma_2 \\ \sigma_3 \end{pmatrix} = \begin{pmatrix} 1 & 0 & 0 \\ 0 & 0 & 1 \\ 0 & 1 & 0 \end{pmatrix} \begin{pmatrix} \sigma_1 \\ \sigma_2 \\ \sigma_3 \end{pmatrix} = \begin{pmatrix} \sigma_1 \\ \sigma_3 \\ \sigma_2 \end{pmatrix}$$
$$x = 1$$

Character table

D_{3h}	E	$2C_3$	$3C_2$	σ_h	$2S_3$	$3\sigma_v$	
A_1'	1	1	1	1	1	1	
A_2'	1	1	-1	1	1	-1	
E'	2	-1	0	2	-1	0	(x,y)
A_1''	1	1	1	-1	-1	-1	
A_2''	1	1	-1	-1	-1	1	z
E''	2	-1	0	-2	1	0	

Orientation of axes for BF_3:
z is $\perp$ to molecular plane.

Labels for σ orbitals,
C_2 axes and σ_{v1}.

Transformation properties of σ orbitals

D_{3h}	E	C_3	C_2	σ_h	S_3	σ_v
σ_1	1	0	1	1	0	1
σ_2	1	0	0	1	0	0
σ_3	1	0	0	1	0	0
Γ_σ	3	0	1	3	0	1

Figure 4.17 Boron trifluoride: character table, orientation of axes, and σ bonds.

The set of characters obtained comprises a representation for the σ bonds in the group. The characters can be obtained without writing out the matrices by examining the effect of one symmetry operation of each class (character table) on each σ bond treated as a vector. If the vector is unchanged by an operation, we write $+1$—as is the case for σ_1 for E, C_2, σ_h, and σ_v, choosing the particular C_2 axis and σ_v plane shown. The choice is arbitrary, but we use the *same* elements chosen for σ_1 for examining σ_2 and σ_3. Operations C_3 and S_3 exchange σ_1 with σ_2, so σ_1 "goes into something else." We write 0 unless the vector (or an orbital) is unchanged or merely has its direction (or sign) changed. None of the operations reverses the direction of σ_1, so no -1 values are recorded. Using the *same set* of symmetry elements as before, only E and σ_h leave σ_2 and σ_3 unchanged. The sum of the numbers in each column is the character of the representation (Γ_σ) for the σ bonds of BF_3; it is a reducible representation, since it does not appear in the character table. By inspection (or by using the formula in Section 3.5.7), we can see that it is the sum of A_1' and E'. [Since the character 3 for σ_h requires that only single-prime representations are involved, one must be E'. The character $+1$ for C_2 (or σ_v) indicates that the other representation must be A_1'.] The boron orbitals belonging to these representations are s (a_1') and p_x, p_y (e').

Group Orbitals

Next we find the combination of orbitals for the group orbitals belonging to a_1' and e'—if such combinations exist—by using the *projection operator method*. Choosing one F σ orbital, we perform *every* symmetry operation (not just one of each class) on it and write down the result. The E operation leaves σ_1 unchanged, C_3 gives σ_2, C_3^2 (C_3^{-1}, the counterclockwise rotation) gives σ_3, etc.

E	C_3	C_3^2	$(C_2)_1$	$(C_2)_2$	$(C_2)_3$	σ_h	S_3	S_3^2	σ_v	σ_v'	σ_v''
σ_1	σ_2	σ_3	σ_1	σ_3	σ_2	σ_1	σ_2	σ_3	σ_1	σ_3	σ_2

To find the linear combinations of F σ orbitals belonging to the A_1' representation, we multiply each orbital generated by the character of A_1' for the operation used. Since these are all $+1$, we add and then divide the result by 4 to simplify the coefficients.

$$\Psi_\sigma \, a_1' = 4\sigma_1 + 4\sigma_2 + 4\sigma_3 \quad \text{or} \quad \sigma_1 + \sigma_2 + \sigma_3$$

It is always permissible to simplify coefficients at this stage, since next we normalize to obtain the proper coefficients for the normalization condition (p. 134).

The wave function for this F group orbital is the normalized *LCAO* (see p. 135).

$$\Psi_{a_1'} = \frac{1}{\sqrt{3}} (\sigma_1 + \sigma_2 + \sigma_3)$$

(This orbital is sketched in Figure 4.18.) What if we try to get a group orbital for A_2', which was not included in Γ_σ? Multiplying the orbitals generated above by the respective characters for A_2' demonstrates that all of the orbitals cancel. There is no $\sigma_{a_2'}$ group orbital for BF_3.

The other group orbital is e', so we multiply the σ orbitals generated above by the respective characters for E', giving

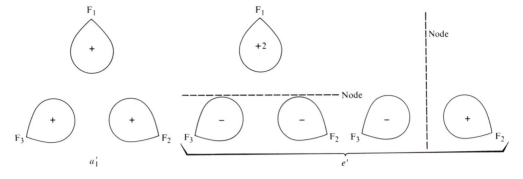

σ group orbitals

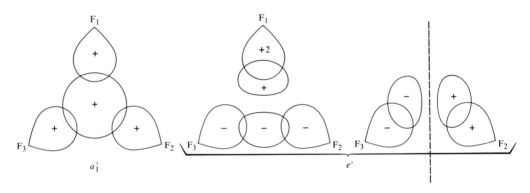

Bonding σ molecular orbitals

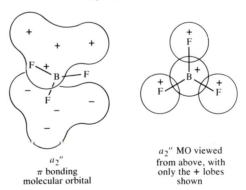

Figure 4.18 Sigma group orbitals and sigma and pi bonding orbitals for BF_3.

$$4\sigma_1 - 2\sigma_2 - 2\sigma_3 \quad \text{or} \quad 2\sigma_1 - \sigma_2 - \sigma_3$$

and

$$\Psi(\sigma_{e'})_a = \frac{1}{\sqrt{6}} (2\sigma_1 - \sigma_2 - \sigma_3)$$

Note that this orbital (Figure 4.18) has a nodal plane. It is one of the *LCAO* for e', and we must find the other one of the degenerate pair.[7] Performing a C_3 operation on the *LCAO* above gives $2\sigma_2 - \sigma_3 - \sigma_1$; a C_3^2 operation gives $2\sigma_3 - \sigma_1 - \sigma_2$. These are not independent orbitals—we have merely interchanged numbers—but subtracting the second from the first gives

$$
\begin{array}{r}
2\sigma_2 - \sigma_3 - \sigma_1 \\
-(-\sigma_2 + 2\sigma_3 - \sigma_1) \\
\hline
3\sigma_2 - 3\sigma_3
\end{array}
\quad \text{or} \quad \sigma_2 - \sigma_3
$$

and

$$
\Psi(\sigma_{e'})_b = \frac{1}{\sqrt{2}} (\sigma_2 - \sigma_3)
$$

It helps to know what we are looking for in trying to arrive at the second *LCAO*. The a_1' orbital has no nodal plane and $(\sigma_{e'})_{a_1}$ has one nodal plane. The other e' orbital must have a nodal plane also, and since it will be perpendicular to that for e_a', it will slice through F_1. This means that the coefficient for σ_1 will be zero and the other σ orbitals will have opposite signs. The group orbitals and the σ-bonding MO are shown in Figure 4.18.

Orthonormal Wave Functions

How do we know that these two orbitals are a basis for the E' representation? A basis set of functions must be *orthogonal*, and, if also normalized, they are *orthonormal*. Normalization requires that

$$
\int \Psi^2 \, d\tau = 1
$$

which is taken care of by the normalization factors (p. 134). Orthogonality of two wave functions requires that the scalar product of the functions must equal zero, as for representations of groups (see Section 3.5.3).

$$
\int \Psi_i \cdot \Psi_j \, d\tau = 0
$$

Checking our pair of *LCAO*,

$$
\int \frac{1}{\sqrt{6}} (2\sigma_1 - \sigma_2 - \sigma_3) \cdot \frac{1}{\sqrt{2}} (\sigma_2 - \sigma_3) \, d\tau
$$

$$
= \frac{1}{\sqrt{12}} \int (2\sigma_1\sigma_2 - 2\sigma_1\sigma_3 - \sigma_2^2 + \sigma_2\sigma_3 - \sigma_3\sigma_2 + \sigma_3^2) \, d\tau
$$

Since σ_1, σ_2, and σ_3 are orthonormal atomic wave functions—that is, since we start with atomic wave functions that are orthonormal—

---

[7]The systematic way to find the other *LCAO* of a degenerate set is the *Gram-Schmidt* method for orthogonalization. See M. Orchin and H. H. Jaffé, *Symmetry, Orbitals, and Spectra*, Wiley-Interscience, New York, 1971, p. 153.

$$\int \sigma_i \cdot \sigma_j \, d\tau = 0 \quad \text{and} \quad \int \sigma_i^2 d\tau = 1$$

so

$$\int \Psi_i \Psi_j d\tau = \frac{1}{\sqrt{12}} [0 - 0 - 1 + 0 - 0 + 1] = 0$$

This is an orthonormal pair of functions fulfilling the requirements for the E' representation.

4.3.2 Pi Bonding

The π bonding is handled in the same way. The p_z orbitals on the three F atoms, treated as vectors perpendicular to the molecular plane, give the following characters.

	E	C_3	C_2	σ_h	S_3	σ_v
Γ_π	3	0	-1	-3	0	1

and $\Gamma_\pi = A_2'' + E''$. The boron p_z orbital belongs to A_2'', so it is available for π bonding. The e'' group orbital must be nonbonding, since there is no e'' orbital for boron. We can get the *LCAO* for the a_2'' group orbital as before. Using all operations on the p_z orbital of one F (π_1), we get

E	C_3	C_3^2	$(C_2)_1$	$(C_2)_2$	$(C_2)_3$	σ_h	S_3	S_3^2	σ_v	σ_v'	σ_v''
$+\pi_1$	$+\pi_2$	$+\pi_3$	$-\pi_1$	$-\pi_3$	$-\pi_2$	$-\pi_1$	$-\pi_2$	$-\pi_3$	$+\pi_1$	$+\pi_3$	$+\pi_2$

and multiplying by the respective characters for A_2'', we get

$$4\pi_1 + 4\pi_2 + 4\pi_3 \quad \text{or} \quad \pi_1 + \pi_2 + \pi_3$$

and

$$\Psi_{a_2''} = \frac{1}{\sqrt{3}} (\pi_1 + \pi_2 + \pi_3)$$

Figure 4.18 shows the combination of this π group orbital with the p_z orbital of boron. Since there is only one π bonding group orbital for BF_3, we could sketch the obvious combination and check to see that it has a_2'' symmetry without using the projection operator method.

We found that the doubly degenerate e'' group orbitals are nonbonding. These two orbitals resemble the e' orbitals in sign patterns, except that the sketches would apply to the lobes of the p_z orbitals on one side of the molecule and the corresponding lobes on the opposite side would have opposite signs.

There are four bonding orbitals and four corresponding antibonding orbitals. Adding the nonbonding e'' orbitals gives a total of 10 orbitals. We used all four boron orbitals and two for each fluorine. The σ orbital for B is primarily $2p_x$. The much lower-energy $2s$ orbitals (a_1' and e' groups) are primarily nonbonding, and the p_y orbitals (a_2' and e' groups) are nonbonding because they are not properly oriented for σ or π bonding. A qualitative energy-level diagram is depicted in Figure 4.19, in which two or more orbitals of a given representation are numbered in order of increasing energy ($1a_1'$, $2a_1'$, etc.).

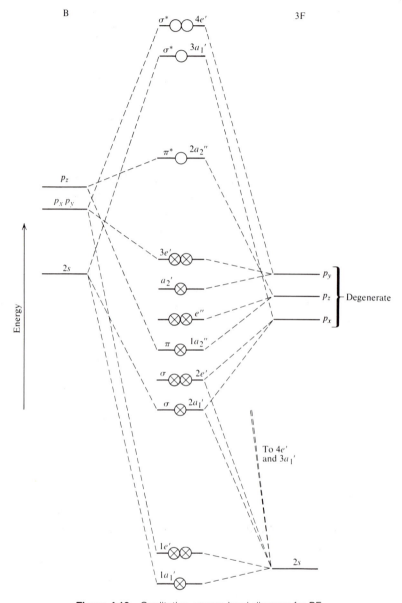

Figure 4.19   Qualitative energy-level diagram for BF$_3$.

4.4   DIBORANE

Diborane, B$_2$H$_6$, presented a challenge in early attempts to describe its bonding (for the chemistry and history of the boranes, see p. 656). Diborane often is described as electron-deficient, because there are only six electron pairs available for a structure that seems to require eight

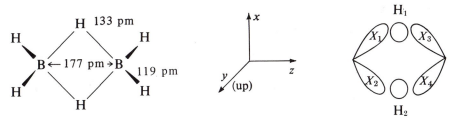

Figure 4.20 Structure and orientation of the orbitals involved in the bridges for B_2H_6.

bonds.[8] The borons and four terminal hydrogens are in a planar ethylene arrangement. The other two hydrogens form bridges between borons in a plane perpendicular to the ethylenelike framework (Figure 4.20). Since this point group ($\mathbf{D}_{2h}$; see Appendix 3.3 for the character table) has no axis of highest order, the choice of axes is arbitrary—no choice will be advantageous with respect to bond directions. We will orient the framework in the xz plane with the z axis through the borons.

Each boron has about it a distorted tetrahedral arrangement of four hydrogens, so for a simple treatment we can use sp^3 hybrid orbitals on each boron. The B—H terminal bonds are ordinary two-center bonds, so we will focus on the bridge bonding. We are left with two sp^3 orbitals per B, one s orbital per H, and four electrons.

4.4.1 Representations for the Boron and Hydrogen Orbitals

First we must determine the representations for the B and H orbitals as shown in Figure 4.20. We carry out the symmetry operations on each orbital, recording whether it is left unchanged ($+1$) or changes into another orbital (0). No operation changes the direction (sign) of any of the orbitals. We can write down the totals for the representations for the B orbitals directly as the number of orbitals left unchanged by an operation.

D_{2h}	E	$C_2(z)$	$C_2(y)$	$C_2(x)$	i	σ_{xy}	σ_{xz}	σ_{yz}
Γ_{boron}	4	0	0	0	0	0	4	0
Γ_H	2	0	0	2	0	2	2	0

$$\Gamma_{boron} = A_g + B_{2g} + B_{1u} + B_{3u} \qquad \Gamma_H = A_g + B_{3u}$$

Only A_g, B_{2g}, B_{1u}, and B_{3u} have $+1$ characters for σ_{xz}, so these must be the irreducible representations contained. The other operations can be used to check this result to ensure that one of these does not occur more than once. (See Section 3.5.7 for the systematic reduction of reducible representations.) The results for the H orbitals are obtained similarly, as shown above. The A_g and B_{3u} representations for boron and hydrogen can combine, and the boron B_{2g} and B_{1u} representations must be nonbonding.

[8]The MO correlation between the C_2H_6 and B_2H_6 structures is instructive. See B. M. Gimarc, *Acc. Chem. Res.* 1974, 7, 384.

4.4.2 Group Orbitals

We obtain the *LCAO* for the group orbitals by using the projection operator, as before.

D_{4h}	E	$C_2(z)$	$C_2(y)$	$C_2(x)$	i	σ_{xy}	σ_{xz}	σ_{yz}
X_1	X_1	X_2	X_4	X_3	X_4	X_3	X_1	X_2
H_1	H_1	H_2	H_2	H_1	H_2	H_1	H_1	H_2

Multiplying by the respective characters of the representations involved produces the following *LCAO* and, after normalizing, the wave functions:

$$a_g \quad 2X_1 + 2X_2 + 2X_3 + 2X_4 \qquad \Psi_{a_g} = \frac{1}{2}(X_1 + X_2 + X_3 + X_4)$$

$$b_{2g} \quad 2X_1 - 2X_2 - 2X_3 + 2X_4 \qquad \Psi_{b_{2g}} = \frac{1}{2}(X_1 - X_2 - X_3 + X_4)$$

$$b_{1u} \quad 2X_1 + 2X_2 - 2X_3 - 2X_4 \qquad \Psi_{b_{1u}} = \frac{1}{2}(X_1 + X_2 - X_3 - X_4)$$

$$b_{3u} \quad 2X_1 - 2X_2 + 2X_3 - 2X_4 \qquad \Psi_{b_{3u}} = \frac{1}{2}(X_1 - X_2 + X_3 - X_4)$$

$$a_g \quad 4H_1 + 4H_2 \qquad \Psi_{a_g} = \frac{1}{\sqrt{2}}(H_1 + H_2)$$

$$b_{3u} \quad 4H_1 - 4H_2 \qquad \Psi_{b_{3u}} = \frac{1}{\sqrt{2}}(H_1 - H_2)$$

Applying the signs obtained to the sketch of the orbitals gives us the bridge molecular orbitals shown in Figure 4.21. A qualitative energy level diagram (Figure 4.22) can be drawn for the bridge bonding. The a_g bonding orbital should be lower in energy than the b_{3u} orbital, because the latter has a nodal plane, which necessarily causes more localization of the electron cloud. Similarly, the nonbonding b_{2g} orbital (two nodes) should be higher in energy than b_{1u} (one node). The four electrons can be accommodated in the two bonding orbitals, leaving the nonbonding and antibonding orbitals empty. We must be aware that a quantitative treatment might lead to inversions in the order of some of the higher-energy orbitals, because of differences in the energy separation between B and H orbitals.

The six atomic or hybrid orbitals are all involved in the bonding a_g and b_{3u} molecular orbitals. We usually view these as two three-center bridge bonds, each involving one electron pair. In these so-called electron-deficient molecules, only the lowest (bonding) orbitals are filled. Many of them can accept electrons to form anions, but only with accompanying structural changes. In most stable molecules, all bonding and nonbonding orbitals are filled.

4.5 XENON DIFLUORIDE

4.5.1 Participation of *d* Orbitals

The electron dot structure for XeF_2 places 10 electrons around Xe, implying the use of Xe $5d$ orbitals.

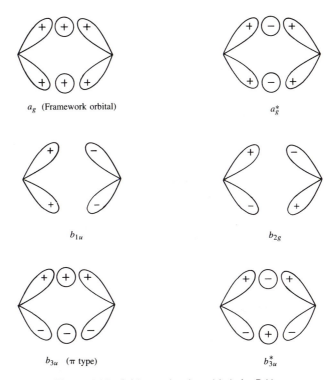

Figure 4.21   Bridge molecular orbitals for B$_2$H$_6$.

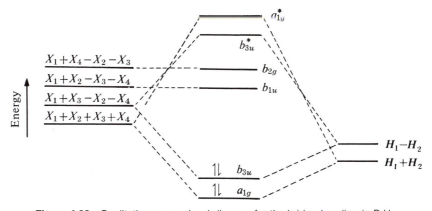

Figure 4.22   Qualitative energy-level diagram for the bridge bonding in B$_2$H$_6$.

$$:\ddot{F}—\ddot{X}e—\ddot{F}:$$

We shall see later that this is not the only bonding description, but first we will examine the possible bonding interactions using the d orbitals. The symmetry is $\mathbf{D}_{\infty h}$ and the possible combinations of s and p orbitals are the same as for CO_2. The F group and Xe orbitals are

	s	a_{1g}		$s + s$	a_{1g}
Xenon	p_z	a_{1u}	*F group*	$s - s$	a_{1u}
orbitals	p_x, p_y	e_{1u}	*orbitals*	$p_z + p_z$	a_{1g}
	d_{z^2}	a_{1g}		$p_z - p_z$	a_{1u}
	d_{xz}, d_{yz}	e_{1g}		$(p_x, p_y) + (p_x, p_y)$	e_{1u}
	$d_{x^2 - y^2}, d_{xy}$	e_{2g}		$(p_x, p_y) - (p_x, p_y)$	e_{1g}

The seven molecular orbitals shown in Figure 4.23 can accommodate 14 electrons, but one MO is nonbonding. The F $2s$ orbitals are nonbonding because of their low energy. Since there are no e_{2g} F group orbitals, the Xe $d_{x^2 - y^2}$ and d_{xy} (e_{2g}) orbitals are nonbonding. Ten electrons occupy the nonbonding orbitals. Accordingly, with 12 bonding electrons for two Xe—F bonds, the bond order for each would be 3! The bonds just are not nearly as strong as this implies, since the average bond energy is 130 kJ/mole and the Xe—F distance is 200 pm. For comparison, the bond energy for IF, which involves an ordinary two-center single bond, is 278 kJ/mole with a bond distance of 191 pm. A bond order of about 0.5 for each Xe-F bond is more appropriate. The Xe d orbitals are too high in energy to be important in bonding unless they mix with lower-energy orbitals of the same symmetry. Thus, only the d_{z^2} orbital, which can hybridize with s, might be expected to contribute to the bonding. This could give a bond order of one for each Xe—F bond.

4.5.2 No *d* Orbital Participation

If we neglect the d orbital participation, we might examine the CO_2 description (Figure 4.12). The 22 electrons of XeF_2 would fill all except the a_{1u}^* orbital; such a bonding description, requiring the use of most of the antibonding orbitals, is not very realistic—some of these orbitals must be nonbonding.

Rundle proposed a half-bond description, without using d orbitals, that is appealing. Assume the three electron pairs on Xe in the electron dot structure to be nonbonding—they can be accommodated in Xe sp^2 hybrid orbitals (localized MO). The F $2s$, $2p_x$, and $2p_y$ orbitals are also nonbonding, accommodating six lone pairs. The F p_z group orbitals are a_{1g} and a_{1u} (as for CO_2). The a_{1g} orbital is nonbonding, since the Xe s orbital is energetically unavailable. The combinations are shown in Figure 4.24. One pair of electrons is delocalized over the three-center bond—or, each Xe—F bond has a bond order of 1/2. No antibonding orbitals are populated.

4.6 CYCLIC PLANAR π MOLECULES

The planar molecules, such as C_6H_6 and C_5H_5 (a radical that forms the stable $C_5H_5^-$ ion), are of interest because of their delocalized π bonding. They form "sandwich"-type metal com-

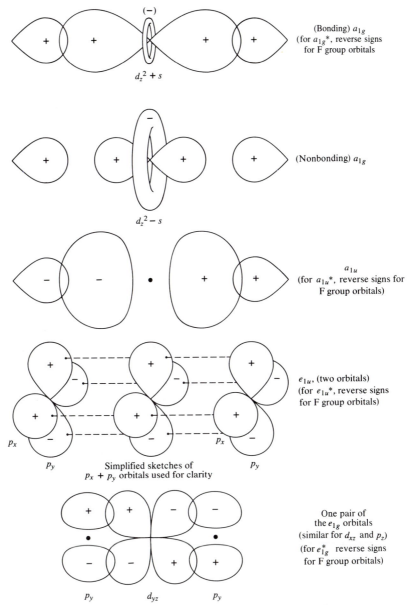

Figure 4.23 Sketches of molecular orbitals for XeF_2.

plexes such as ferrocene, $Fe(C_5H_5)_2$, and dibenzenechromium, $Cr(C_6H_6)_2$. The bonding description of ferrocene (p. 442) utilizes the π bonding scheme we shall now examine for C_5H_5.

The point group for C_5H_5 is $\mathbf{D}_{5h}$, but it is more convenient to use the cyclic subgroup $\mathbf{C}_5$. If desired, the labels for the representations in the $\mathbf{D}_{5h}$ group can be found by examining the symmetry of the orbitals obtained and adding appropriate subscripts or primes. For a planar

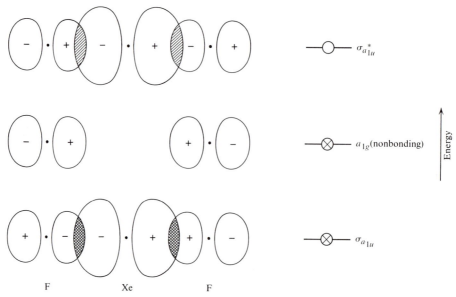

Figure 4.24 The three-center bonding of XeF$_2$.

C_nH_n molecule, each carbon can be regarded as using sp^2 hybrid orbitals for σ bonding. There are n π orbitals, and in the C_n point group one of the orbitals will belong to each of the n representations. We do not have to determine the representations of the π orbitals as before. For C_5H_5 these representations are A, E_1, and E_2. In the cyclic groups each E representation consists of 2 one-dimensional representations—which obviates the problem of finding the right linear combinations for the E pair. Since the operations of the group merely involve the successive application of C_5 rotations, operating on π_1 gives

C_5	E	C_5	C_5^2	C_5^3	C_5^4
π_1	π_1	π_2	π_3	π_4	π_5

Using the projection operator method, we obtain

$$a \qquad\qquad \pi_1 + \pi_2 + \pi_3 + \pi_4 + \pi_5$$

$$e_1 \qquad \begin{cases} e_1(1)\ \pi_1 + \epsilon\pi_2 + \epsilon^2\pi_3 + \epsilon^{2*}\pi_4 + \epsilon^*\pi_5 \\ e_1(2)\ \pi_1 + \epsilon^*\pi_2 + \epsilon^{2*}\pi_3 + \epsilon^2\pi_4 + \epsilon\pi_5 \end{cases} \qquad \epsilon = e^{-2\pi i/5}$$

$$e_2 \qquad \begin{cases} e_2(1)\ \pi_1 + \epsilon^2\pi_2 + \epsilon^*\pi_3 + \epsilon\pi_4 + \epsilon^{2*}\pi_5 \\ e_2(2)\ \pi_1 + \epsilon^{2*}\pi_2 + \epsilon\pi_3 + \epsilon^*\pi_4 + \epsilon^2\pi_5 \end{cases}$$

The complex numbers can be eliminated by taking linear combinations. Adding $e_1(1)$ and $e_1(2)$,

$$e_{1a} = 2\pi_1 + (\epsilon + \epsilon^*)\pi_2 + (\epsilon^2 + \epsilon^{2*})\pi_3 + (\epsilon^{2*} + \epsilon^2)\pi_4 + (\epsilon^* + \epsilon)\pi_5$$

and, from trigonometric formulas (see Section 3.5.6),

$$\epsilon = \cos\frac{2\pi}{5} + i\sin\frac{2\pi}{5} \qquad \epsilon^* = \cos\frac{2\pi}{5} - i\sin\frac{2\pi}{5}$$

$$\epsilon + \epsilon^* = 2 \cos 2\pi/5 = 0.618$$

$$\epsilon^2 = \cos \frac{4\pi}{5} + i \sin \frac{4\pi}{5} \qquad \epsilon^{2*} = \cos \frac{4\pi}{5} - i \sin \frac{4\pi}{5}$$

$$\epsilon^2 + \epsilon^{2*} = 2 \cos 4\pi/5 = -1.618$$

giving

$$e_{1a} = 2\pi + 0.618\pi_2 - 1.618\pi_3 - 1.618\pi_4 + 0.618\pi_5$$

Dividing by two and normalizing gives

$$\Psi(e_{1a}) = \frac{1}{\sqrt{2.5}}(\pi_1 + 0.314\pi_2 - 0.809\pi_3 - 0.809\pi_4 + 0.314\pi_5)$$

Subtracting $e_1(2)$ from $e_1(1)$ and normalizing gives

$$\Psi(e_{1b}) = \frac{1}{\sqrt{7.24}}(1.62\pi_2 + \pi_3 - \pi_4 - 1.62\pi_5)$$

Similarly, for e_2 we get

$$\Psi(e_{2a}) = \frac{1}{\sqrt{2.5}}(\pi_1 - 0.809\pi_2 + 0.314\pi_3 + 0.314\pi_4 - 0.809\pi_5)$$

$$\Psi(e_{2b}) = \frac{1}{\sqrt{7.24}}(\pi_2 - 1.62\pi_3 + 1.62\pi_4 - \pi_5)$$

These orbitals are sketched in Figure 4.25, which shows the relative coefficients for the wave functions of the individual p orbitals. The relative order of the energy levels can be determined easily. The completely delocalized a orbital (a five-center orbital) is most stable, and the others become less stable with increasing numbers of nodes.[9]

A simple molecular orbital description, without the coefficients for the *LCAO* but with the ordering of the energies of the π orbitals, can be obtained for any of the planar cyclic C_nH_n systems without going through all the steps. The labels for the levels can be written from the character tables: C_3, a and e; C_4, a, e, and b; C_5, a, e_1, and e_2; C_6, a, e_1, e_2, and b; C_7, a, e_1, e_2, and e_3; etc. The completely delocalized a orbital is always lowest in energy, with the e's increasing in energy in the order $e_1 < e_2 < e_3$.[9] When n is even there is a b orbital that is highest in energy and has an alternating sign pattern around the ring. From Figure 4.26, which shows the relative energies of the orbitals for C_nH_n molecules, we readily see why C_5H_5 gives a stable anion $C_5H_5^-$ and why C_7H_7 forms the cation $C_7H_7^+$. The empirical rule for aromaticity ($4n + 2$ electrons) is also evident, since the orbitals being filled after the first come in pairs *(e)*.

Sketching the π orbitals for C_nH_n molecules is also easy (Figure 4.27). Each e_1 orbital has a single nodal plane. We superimpose one of these orbitals on the molecule in a manner consistent with its symmetry and arrange the other e_1 orbital so that the nodal plane is perpendicular to the first choice. For e_2 there are two perpendicular nodal planes. After placing one e_2 orbital on the framework of the molecule, the second e_2 is placed so that the nodal

---

[9]Just as the energies of *ns* (or *np*, *nd*, etc.) orbitals increase with the number of nodes, so do molecular orbitals. The greater the number of nodes, the greater the *localization* of charge.

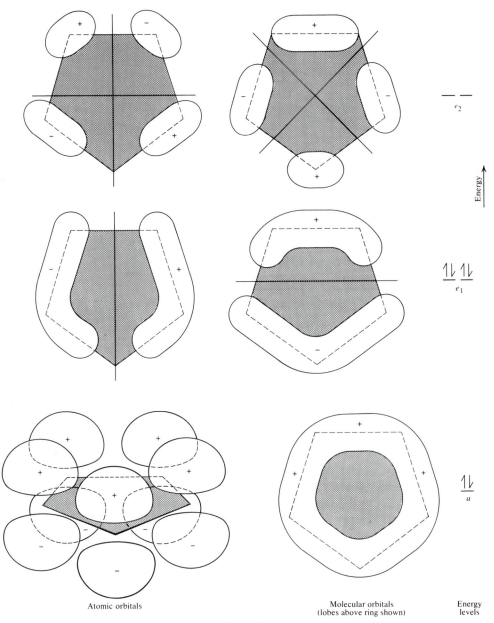

Atomic orbitals

Molecular orbitals
(lobes above ring shown)

Energy
levels

Figure 4.25 Pi molecular orbitals for C_5H_5. The orbitals are designated for the C_5 point group.

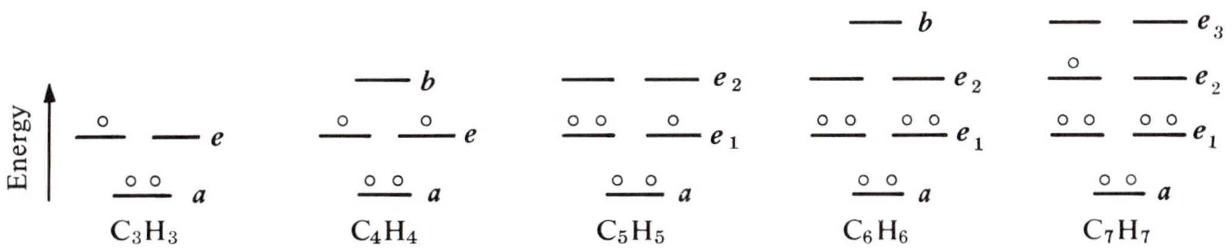

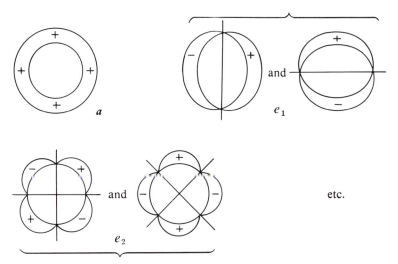

Figure 4.26 Relative energies of the π orbitals for planar cyclic C_nH_n molecules.

planes bisect the angles formed in the first e_2 sketch. For C_4H_4 or C_6H_6, superimposing the nodal planes for the highest orbital *(b)* on the framework of carbons such that the nodes—two and three planes, respectively—slice between the carbons, produces a strongly antibonding orbital with alternating signs around the ring. If we try the other orientation, with the nodal planes bisecting the angles for the first case, the planes go through all of the carbon atoms. This corresponds to coefficients of zero for each atomic orbital, and the wave function vanishes. There is only one such orbital.

An interesting series of cyclic planar anions $C_nO_n^{2-}$ $(4-$, in some cases)[10] forms complexes with several metals. The $H_2C_4O_4$ has been called "squaric acid." The treatment of C_nH_n molecules can be extended to $C_nO_n^{2-}$ ions, considering two concentric shells that can or cannot match in sign. See Figure 4.28 for sketches of the π orbitals for $C_4O_4^{2-}$ using $\mathbf{D}_{4h}$ labels for the representations. The energy-level diagrams (twice as many π orbitals as for C_nH_n) involve repeating the same patterns as in Figure 4.26 at higher energy.

Figure 4.27 Molecular-orbital wave patterns for cyclic π systems, showing the lobes on one side of the molecular plane.

---

[10]R. West and J. Niu, "Oxocarbons and Their Reactions," in Vol. 2 of *The Chemistry of the Carbonyl Group*, J. Zabicky, Ed., Wiley, New York, 1970.

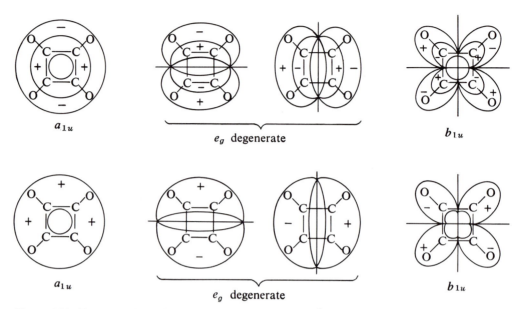

Figure 4.28 Representations of the π molecular orbitals of $C_4O_4^{2-}$, showing the lobes on one side of the molecular plane.

4.7 MOLECULAR ORBITAL DESCRIPTIONS OF OTHER MOLECULES

We can regard the bonding in the tetrahedral P_4 molecule as similar to that of MnO_4^- with the Mn removed, and that in the square planar Te_4^{2+} ion as analogous to that of $Ni(CN)_4^{2-}$ with the Ni removed. In the cases of P_4 and Te_4^{2+}, however, the simplest choices of axes for bonding interaction would differ from those chosen for MnO_4^- and $Ni(CN)_4^{2-}$, respectively. Bonding descriptions involving the participation of d orbitals are given for octahedral complexes (p. 289) and for ferrocene (p. 442).

GENERAL REFERENCES

H. Bock and P. D. Mollere, "Photoelectron Spectra," *J. Chem. Educ.* 1974, *51*, 506. An experimental approach to teaching molecular orbital theory.

J. K. Burdett, *Molecular Shapes*, Wiley-Interscience, New York, 1980. Basically a molecular orbital treatment of inorganic stereochemistry, but other approaches are treated well.

F. A. Cotton, *Chemical Applications of Group Theory*, 2nd ed., Wiley-Interscience, New York, 1971. A chemist's approach to group theory.

R. L. DeKock and H. B. Gray, *Chemical Structure and Bonding*, Benjamin/Cummings, Menlo Park, Cal., 1980. Good treatment.

W. E. Hatfield and R. A. Palmer, *Problems in Structural Inorganic Chemistry*, Benjamin, New York, 1971. Good problems of current interest, with solutions. There are chapters on symmetry and group theory, and on chemical bonding.

W. E. Hatfield and W. E. Parker, *Symmetry in Chemical Bonding and Structures,* Merrill, Columbus, Ohio, 1974. The treatment is close to the level of this text, but uses more quantum mechanics.

M. Orchin and H. H. Jaffé, *Symmetry, Orbitals and Spectra,* Wiley-Interscience, New York, 1971. Molecular orbital theory and spectra using symmetry and group theory.

Many articles of interest are included as reprints from *J. Chem. Educ.* in *Collected Readings in Inorganic Chemistry,* Vol. I, G. W. Watt and W. F. Kieffer, Eds., 1962; and Vol. II, G. Galloway, Ed., Chemical Education Publishing, Easton, Pa., 1972.

PROBLEMS

4.1 Give the bond order and the number of unpaired electrons for Be_2^+, B_2^+, C_2^+, O_2, O_2^+, O_2^-, O_2^{2-}.

4.2 Sketch sigma bonding orbitals that result from combination of the following orbitals on separate atoms: p_z and d_{z^2}, s and p_z, $d_{x^2-y^2}$ and $d_{x^2-y^2}$ (let the σ bond be along the x axis in the latter case).

4.3 Sketch π bonding orbitals that result from combination of the following orbitals on separate atoms: p_x and p_x, p_x and d_{xz}, d_{xz} and d_{xz}.

4.4 Sketch a delta bond for an X—Y molecule, identifying the atomic orbitals and showing the signs of the amplitudes of the wave functions (signs of the lobes).

4.5 Explain why the CO ligand donates the lone pair on carbon in bonding to metals, rather than the pair on oxygen.

4.6 Write the molecular orbital configurations and give the bond orders of NO^+, NO, and NO^-. Which of these species should be paramagnetic?

4.7 *a.* Draw a molecular-orbital energy diagram for the molecular orbitals that would arise in a diatomic molecule from the combination of unhybridized *d* orbitals. Label the atomic orbitals being combined and the resulting molecular orbitals. Let the z axis lie along the bond. *b.* Sketch the shape of these orbitals.

4.8 Account for the differences in dissociation energies and bond lengths in Table 2.9 on addition and removal of an electron from *a.* O_2, *b.* N_2, *c.* CO, *d.* NO.

4.9 The bond dissociation energy of C_2 (599 kJ/mole) decreases slightly on forming C_2^+ (513 kJ/mole) and increases greatly on forming C_2^- (818 kJ/mole). Why is the change much greater for the addition of an electron?

4.10 The Lewis structure for CO_2 shows four lone pairs of electrons on the oxygens. There are two nonbonding pairs in the MO description. How can you reconcile the two descriptions?

4.11 The preferred electron dot structure for NO_2 shows five lone pairs on the oxygens and one unshared electron on N. Show how this description is consistent with the MO description.

4.12 In forming NO_2^+ from NO_2, from which molecular orbital is the electron lost? On which atom is this electron localized? To which orbital is an electron added in forming NO_2^-? What are the consequences, in terms of molecular shape, of the loss and gain of an electron ?

4.13 Why is the first excited state of BeH_2 bent, whereas that of BH_2 is linear?

4.14 Apply the group theoretical treatment to obtain the bonding description for NO_2^-.

4.15 Apply the group theoretical treatment to obtain the MO bonding description for σ bonding in $PF_5(\mathbf{D}_{3h})$. Note that the two axial sigma bonds are treated as one set and the three sigma bonds in the equatorial plane are in a separate set.

4.16 Apply the group theoretical treatment to obtain the MO bonding description for sigma bonding for $SF_6(\mathbf{O}_h)$.

V
Hydrogen Bonding and Weak Interactions

Covalent bonding and ionic "bonding," are strong interactions that are very important in determining the stability of chemical compounds. Other interactions usually are weaker, although the strongest hydrogen bonds are stronger than some weak covalent bonds. These "other" interactions can affect profoundly the properties and structures of substances and often account for differences among compounds of similar formula types. Here we consider briefly the nature of these forces, their relative strengths, and when they are important.

5.1 THE HYDROGEN BOND

5.1.1 Historical Background

By 1900 much data had accumulated on molecular association. Interpretation of vapor pressures, osmotic pressures, surface tension, freezing point depressions and boiling point elevations, Trouton constants, and similar data pointed to the existence of a class of associated molecular substances such as water, formic acid, formamide, and hydrogen cyanide. However, one of the earliest proposals for the hydrogen bond came from data on dissociation, not association. In 1907, T. S. Moore published a paper on the equilibrium constants in the system

$$NH_3 \overset{K_h}{\rightleftharpoons} H_3NHOH \overset{K_b}{\rightleftharpoons} NH_4^+ + OH^-$$

and the related system of the ethylamines. K_h was obtained from experiments on the distribution of ammonia (or the amine) between water and an organic solvent. The product of K_h and K_b was obtained by conductivity studies. Moore's comments in 1907:

The introduction of three ethyl groups into ammonia increases the constant (K_b) only twelve-fold. Now the introduction of a fourth ethyl group gives tetraethylammonium hydroxide, the degree of ionization of which is comparable with that of sodium hydroxide. It is difficult to imagine that this large change of ionization is produced solely by the introduction of the fourth ethyl group when the first three have produced a comparatively small effect. These results point rather to a difference in constitution between the quaternary ammonium hydroxides and the hydroxides of primary, secondary, and tertiary amines.

Five years later he published a clear picture of this difference:

The following formulae, where thick strokes mean strong unions and thin strokes weak unions, show roughly the difference between trimethylammonium hydroxide and tetramethylammonium hydroxide.

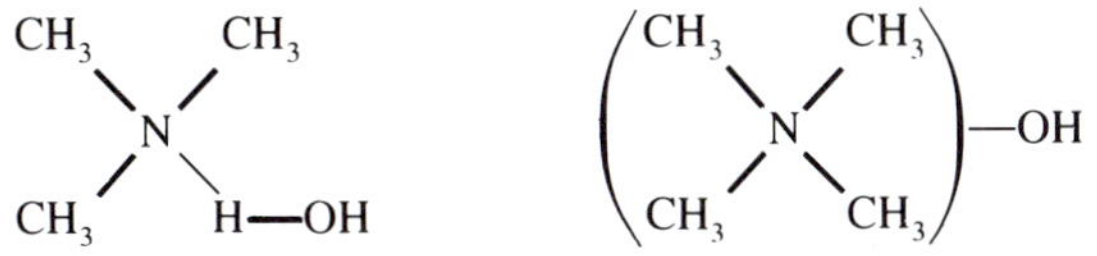

But Moore did not generalize his idea to cover other phenomena. Latimer and Rodebush in 1920 introduced the term "hydrogen bond" to describe the nature of association in the liquid state of water, hydrogen fluoride, etc. Furthermore, they pointed out the necessity for a slightly acidic hydrogen and a nonbonding electron pair in order to form a hydrogen bond, which they formulated as A:H:B. On the basis of this picture they were able to account for the high dielectric constant of water, the mobility of the hydrogen ion in water, the weakness of ammonium hydroxide as a base (apparently without knowledge of Moore's publications), as well as melting points and boiling points of liquids capable of hydrogen bonding. Within the next decade or so, the importance of hydrogen bonding in determining crystal structures, solubilities, and spectra had been pointed out. Some of these effects, as well as some theoretical aspects of the hydrogen bond, are discussed in the following sections.

5.1.2 Evidence for Hydrogen Bonding— Influence on the Vaporization Process for Pure Liquids

We find some of the most striking effects of hydrogen bonding on the physical properties of substances by contrasting the properties of water with those of dimethyl ether (Table 5.1). As Figure 5.1 indicates, the dimethyl derivatives of the Group VI elements usually boil about 80 to 100°C lower than the hydrides. Accordingly, from Figure 5.1 it appears that water boils about 200 degrees higher than it would in the absence of hydrogen bonding. Replacing a hydroxylic hydrogen with a methyl group in other compounds produces a similar decrease in boiling point. Thus dimethylsulfate boils at 188°C, trimethylphosphate at 193°C, and methylborate at 65°C; the corresponding hydrogen-bonded compounds—sulfuric acid, phosphoric acid, and boric acid—do not boil but decompose with the loss of water at considerably higher temperatures (340, 213, and 185°C, respectively). Boric acid does not melt before decomposing.

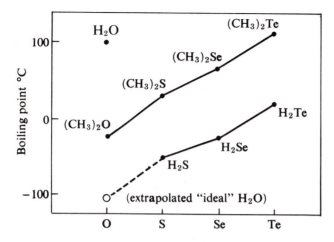

Figure 5.1 Boiling points of hydrogen- and methyl-derivatives of the Group VI elements.

Hydrogen bonding is not restricted to hydroxylic compounds, although it is most common in this class of compound. Hydrogen fluoride boils at 19.5°C, whereas methyl fluoride boils at −78°C. Hydrazoic acid, HN_3 boils at 37°C, whereas methylazide boils at 20 to 21°C. The heat of vaporization of hydrogen-bonded substances is also greater than that of their methyl derivatives.

Table 5.1 Some physical properties of water, methanol, and dimethyl ether

	H_2O	CH_3OH	$(CH_3)_2O$
Critical temperature (°C)	374.2	240.6	126.7
Boiling point (°C)	100	64.7	− 23.7
Melting point (°C)	0	− 97.8	− 138.5
Heat of vaporization (kJ/mole)	40.6	35.2	21.5
Heat of fusion (kJ/mole)	6.0	2.2	4.9
Trouton constant (ΔH_v/b.p. K) (kJ/mole-degree)	109	102	86
Dielectric constant	78.54	32.63	5.02
Dipole moment (Debye units)	1.84	1.68	1.30
Viscosity at 20°C (centipoise)	1.005	0.597	(0.2332 for diethyl ether)

Intermolecular hydrogen bonding in pure substances increases the heat of vaporization in two ways. (1) By adding to the attraction between molecules. Either this attraction must be overcome on vaporization (as in the case of H_2O) and/or small polymeric molecules of the liquid must be vaporized—for example, $(HF)_x$. In the latter event the heat of vaporization is higher than normal because the dispersion forces between the polymeric units are greater than between the monomeric units. (2) By restricting rotation of the molecules in the liquid. Such rotation is possible in the gas, and the energy absorbed in exciting this rotation adds to the

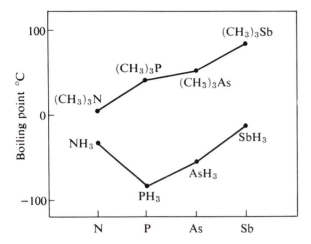

Figure 5.2 Boiling points of hydrogen- and methyl-derivatives of the Group V elements.

heat of vaporization. However, the additional degees of freedom such rotation allows the molecules are reflected in an increase in the entropy of vaporization. So, the extra heat of vaporization from the rotational contribution does not produce a further increase in the boiling point of the substance. Hydrogen bonding thus (1) increases the boiling point of a liquid beyond that expected from a consideration of dispersion forces alone (that is, expected from comparison with comparable non–hydrogen-bonded liquids) and (2) gives rise to a greater entropy of vaporization (that is, Trouton constant, when the temperature is the boiling point) than that of unassociated liquids.

Heats of vaporization and heats of sublimation sometimes are used to estimate the energy of the hydrogen bond involved. For water, the hydrogen bonds have an energy of about -25kJ/mole H-bond. For substances such as HF and acetic acid, which are partially associated in the gas phase, the hydrogen bond energy is obtained not from ΔH_v, but directly from the variation of the degree of association of a gaseous sample with temperature. This in turn is obtained from PVT data with appropriate assumptions about the dissociating species.

The boiling point of ammonia, although higher than might be expected from a comparison with phosphine, etc., unlike the oxygen compounds, is lower than the completely methylated compound trimethylamine (see Figure 5.2). X-ray studies of solid ammonia indicate that three hydrogen atoms from different nitrogen atoms point toward the lone pair of electrons. Thus a normal hydrogen bond is not formed. From heats of sublimation the energy of interaction with the lone pair is about -5.4 kJ/g-atom of hydrogen.

5.1.3 Energies of Hydrogen Bonds Formed between Molecules

Table 5.2 lists the range of energies observed for a number of different hydrogen bonds formed between neutral molecules. For the same hydrogen-bonded molecules, the energies determined from measurements of gas-phase association equiiibria generally tend to give values somewhat higher than those obtained from solution equilibria. For neutral molecules, hydrogen bonds are strongest when formed between the more electronegative atoms.

Table 5.2   Energy range of some hydrogen bonds formed between neutral molecules[a]

Bond	Energy Range (kJ/mole)	Formed by
F—H···O	46	HF with ketones, alcohols, ethers
F—H···F	29	HF(g)
Cl—H···O	25–34	HCl with ketones, alcohols
O—H···X	6–15	C_6H_5OH with alkyl halides
O—H···O	8–34	H_2O, ROH, RCO_2H, C_6H_5OH with themselves and with ketones, ethers
O—H···N	6–38	H_2O, ROH, C_6H_5OH with amines, ammonia
O—H···S	17–21	ROH, C_6H_5OH, R_FOH with thioethers, thioketones
O—H···Se	13–17	ROH, C_6H_5OH, R_FOH with selenoethers
O—H···π	4–8	ROH, C_6H_5OH with aromatic hydrocarbons
S—H···O	4–8	C_6H_5SH with $(BuO)_3PO$, ketones
S—H···N	8–12	C_6H_5SH with C_5H_5N
S—H···π	1–4	C_6H_5SH with aromatic hydrocarbons
C—H···O	6–15	$HCCl_3$ with ketones and ethers
C—H···N	6–17	HCN with itself, $HCCl_3$ with amines
N—H···N	6–17	NH_3, RNH_2
N—H···O	17–21	RNH_2 with ROH, $CH_3CONHCH_3$
N—H···S	21–25	*N*-methylaniline with thiocamphor
N—H···π	4–8	RNH_2 with aromatic hydrocarbons

[a]Data largely from S. Singh, A. S. N. Murthy, and C. N. R. Rao, *Trans. Faraday Soc.* 1966, *62*, 1056.

5.1.4 Some Weak Hydrogen Bonds Involving Methyl Hydrogens

We have assumed up to this point that a hydrogen atom attached to a methyl group does not participate in hydrogen bonding. Yet CH_3CN (b.p. 82°C), acetonitrile, has an even higher boiling point than HCN (b.p. 25°). It seems likely that the polar nature of the methyl C—H bond in this case is enhanced greatly by the strong electron-withdrawing inductive effect of the CN group, and also by hyperconjugative interactions with the CN group. The moderately high dielectric constant (38.8) of CH_3CN supports the notion of hydrogen bonding here. A *weak* hydrogen bond may exist also in acetone, CH_3CCH_3, dimethylcarbonate, ethylene car-
$$\overset{\|}{O}$$
bonate, and other places where a basic center exists in the molecule and the methyl group is attached to a relatively positive center.

5.1.5 Consequences of Hydrogen Bonding

Azeotropic Behavior

When two substances are mixed, an ideal solution results if the molecules cannot distinguish between solute and solvent molecules (that is, if the intermolecular forces between solute and solvent are the same as between solvent and solvent or solute and solute, and the molecules are of the same size). In such cases the partial pressure of each component, P_i, is given by the product of the mole fraction of the component, N_i, and its vapor pressure in the pure state, P_i^0—in other words, the solution obeys Raoult's law $P_i = N_i P_i^0$. For an ideal two-component system the total pressure, P_T, is a linear function of the mole fraction composition; that is, $P_T = N_i P_i^0 + (1 - N_i)P_j^0 = P_j^0 + N_i(P_i^0 - P_j^0)$. Solutions whose total pressure is greater than the ideal pressure are said to show positive deviation from Raoult's law, and *vice versa*.

We usually interpret a negative deviation from Raoult's law as caused by a greater interaction between unlike molecules than between like molecules. If the negative deviation is very large, there will be a minimum in the total vapor pressure at some particular composition for a given temperature. Such solutions give rise to maximum boiling mixtures (see Figure 5.3). Mixing liquids that themselves show no strong hydrogen bonds but that together may form strong hydrogen bonds, produces negative deviation from Raoult's law and often results in maximum boiling mixtures. Examples include mixtures containing highly halogenated hydrocarbons (thus making the remaining C—H bonds more polar) and ketones or esters—chloroform + acetone, cyclohexanone + bromoform, butylacetate + 1,2,3-trichloropropane, and pyridine + chloroform.

Positive deviation from Raoult's law indicates that the like molecules have greater attraction for each other than the unlike molecules. This often occurs when solutions are made with liquids of which only one contains strong hydrogen bonds in the pure state. Thus mixtures of carbon disulfide or hydrocarbons (which have no hydrogen bonding capacity) with alcohols or primary or secondary amines will produce positive deviations from Raoult's law and often, minimum boiling mixtures. If the deviation is extreme, the system will separate into two phases, as in the case of water and hydrocarbons. Codistillation processes utilize the fact that immiscible liquids produce minimum boiling mixtures (steam distillation is the most common of these codistillations).

Solubility

When an uncharged organic compound dissolves to any appreciable extent in water, the solubility may be attributed to hydrogen bonding. Thus dimethylether is completely miscible with water, whereas dimethylsulfide is only slightly soluble in water.

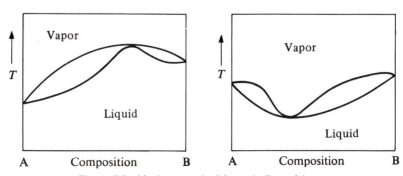

Figure 5.3 Maximum and minimum boiling mixtures.

Benzene is only slightly soluble in water, whereas pyridine is completely miscible with water. The number of carbon atoms in a molecule that is still soluble in water depends greatly on the number of atoms per molecule capable of forming hydrogen bonds with water. Usually, about three carbon atoms per oxygen atom will give a highly water-soluble compound. Thus dioxane, $C_4H_8O_2$, is completely miscible with water, whereas diethyl ether, $(C_2H_5)_2O$, is only moderately soluble (7.5 g/100). Sugars are water soluble because of hydrogen bonding with water. The strong hydrogen bonds formed between the solvent and hydrated ions and oxoanions contribute greatly to the solubility of salts in water. (See Chapter 6, p. 234.)

Hydrogen bonding also plays a role in many nonaqueous solvents. Chloroform is a good solvent for fatty acids, since its polar C—H bond may engage in hydrogen bonding. Ethers can serve as solvents for hydrogen chloride because of hydrogen-bonding ability.

Acetylene is very soluble in acetone because of hydrogen bond formation. Because of its sensitivity to shock, acetylene itself under pressure is dangerous to handle and is handled commercially by dissolution in acetone under pressure. It does not dissolve appreciably in water, since the hydrogen-bonding interaction between water molecules is much stronger than that between acetylene and water molecules.

Intramolecular Hydrogen Bonding

Many organic compounds permit intramolecular (within the molecule) hydrogen bonding, which will be favored over intermolecular hydrogen bonding, because of the less favorable entropy change associated with the latter. Changes from *inter* to *intra*molecular hydrogen bonding in going from *meta* and *para* to *ortho* aromatic compounds may cause large differences in the melting points, boiling points, and solubilities of these isomers. Intramolecularly hydrogen-bonded substances are more soluble in organic solvents and have lower melting and boiling points than their intermolecularly hydrogen-bonded isomers.

Dielectric Constants

Note in Table 5.1 the decrease in dielectric constant on successively substituting methyl groups for hydrogen in H_2O. Other cases of a decrease in dielectric constant on replacing hydrogen by methyl include *N*-methylacetamide, dielectric constant of 175.7 (25°C); *N,N*-dimethylacetamide, 37.8; HCN, 120; CH_3CN, 38.8.

The unusually high dielectric constants of the hydrogen-bonded materials result from the existence of polymeric molecules having an enhanced electric dipole moment, which is greater than the sum of the dipole moments of the monomers of which it is composed. The polymeric species have structures similar to the solids but, of course, with fewer molecules per unit.

An apparent anomaly occurs with acetamide, which has a dielectric constant of 74—much less than its *N*-methyl derivative. This would be expected if acetamide forms a randomly oriented polymer rather than a linear polymer, as in the case of *N*-methylacetamide.

Infrared Spectra

Various absorption bands in the infrared spectrum of a compound often may be associated with particular vibrational modes of the molecule or of functional groups in the molecule. Hydrogen bonding of a group such as the —O—H group affects the frequency, band width, and intensity of the infrared absorption bands assigned to the —O—H stretching vibration and —O—H bending vibration. Dilute solutions of an alcohol (or other substance capable of association through hydrogen bonding) in an inert solvent such as carbon tetrachloride show an absorption band characteristic of the stretching vibration of a free —O—H group. More concentrated solutions, on the other hand, display a new absorption band, at a lower frequency and with a greater absorption, resulting from a hydrogen-bonded species (see Figure 5.4). Information on the equilibria between monomeric and polymeric species is obtained from such spectral data. Studies of the effect of temperature on these equilibria allow us to evaluate the enthalpy, entropy, and free energy changes associated with the hydrogen bond formation in these cases. We find that the enthalpy change for hydrogen bond formation often parallels the shift in the —O—H (or —N—H) absorption frequency that takes place on going from the free O—H to the hydrogen-bonded O—H.

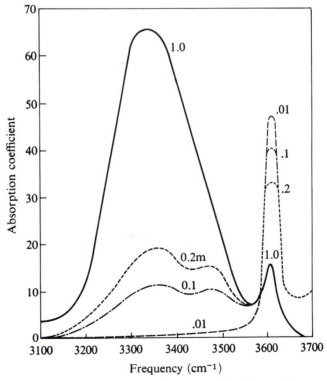

Figure 5.4 Absorption spectra of *tert*-butanol in CCl$_4$ at various concentrations at 25°C. (From U. Liddel and E. D. Becker, *Spectrochim. Acta*, 1957, *10*, 70.)

Crystal Structure

When hydrogen bonding is possible in molecular crystals, the structure is one that yields the maximum number of hydrogen bonds. This gives rise to structures in which the individual molecules are connected through hydrogen bonds to form chains, sheets, or three-dimensional networks. In some cases the crystals contain units consisting of dimeric or trimeric hydrogen-bonded species.

Hydrogen-bonded chains are found in the linear HCN polymer. Zigzag chains are more common, occurring in solid HF, CH_3OH, formic acid, acetanilide, and probably *N*-methylamides.

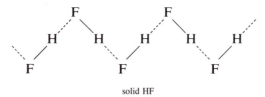

solid HF

Oxalic acid forms a maximum number of hydrogen bonds in either chain or sheet structures. Both modifications actually are known (see Figure 5.5).

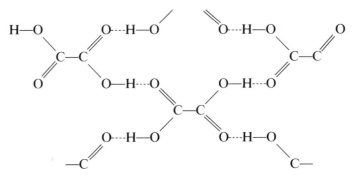

Figure 5.5 α-Oxalic acid.

The sheet and three-dimensional structures are typical in cases where each molecule has several hydrogens for hydrogen bonding and also several lone electron pairs. Boric acid provides another example of a compound with a sheet structure. Three-dimensional networks occur in such compounds as water and telluric acid.

In structures such as ice (p. 206), each oxygen is attached tetrahedrally to four H, as this is required in the solid for the maximum number of hydrogen bonds to be formed. The recent molecular beam study of the water dimer[1] examined the orientation of a single hydrogen bond, which is linear within 1° and oriented tetrahedrally with respect to the electron donor molecule. This result is expected for sp^3 hybridization of the oxygen electron donor and a significant covalent contribution to the hydrogen bond. The hydrogen bond in the $(HF)_2$ dimer[2] also is

[1] T. R. Dyke, K. M. Mack, and J. S. Muenter, *J. Chem. Phys.* 1977, *66*, 498.

[2] T. R. Dyke, B. J. Howard, and W. Klemperer, *J. Chem. Phys. 1972, 56*, 2442.

oriented at essentially a tetrahedral angle relative to the covalent bond of the HF electron donor. The isolated HF molecule has axial (C_∞) symmetry without localized lone pairs, so purely electrostatic interaction should lead to a linear dimer. In the $H_2S\cdots H$—F dimer, the plane of the H_2S molecule is at 91° to the $S\cdots H$—F bond.[3] The bond angles for covalent molecules of third-period elements (e.g., H_2S and PH_3) are close to 90° (see p. 67). Observation for $H_2S\cdots H$—F suggests that the lone pair of H_2S used for hydrogen bonding is essentially p in character.

Hydrogen bonding is very important in polypeptides and cellulose structures, providing the primary bonding that results in the cross-linking responsible for the α-helix of proteins.

In ionic substances, hydrogen bonding may also play a major role in determining the structure. Discrete hydrogen-bonded ions exist in acid salts such as KHF_2 (discussed below). Other acid salts, such as sodium hydrogen carbonate, form polyanion chains or three-dimensional networks, as in KH_2PO_4.

The difference in structure between NH_4F (wurtzite structure) and NH_4Cl (CsCl structure) stems from the greater strength of the N—HF bond than the N—HCL bond.

5.1.6 Theory

Coulson[4] estimated the various contributions to the energy of the hydrogen bond in ice as follows:

Type of Energy	
Electrostatic	−25kJ/mole
Delocalization	−33
Repulsive	+35
Dispersion	−12
Total	Theoretical −35 Experimental −25.5

These factors may be of greater or lesser importance in other hydrogen-bonded systems. Let us examine each of these factors.

Repulsion and Dispersion

In most molecular compounds the closest interatomic distance between atoms of neighboring molecules is set by the balance of dispersive forces and repulsive forces arising from interpenetration of the electron clouds. Where a hydrogen bond is formed, the forces specifically resulting from hydrogen bonding—that is, the electrostatic force and the force caused by delocalization—will cause the two atoms sharing the hydrogen atom to have an internuclear separation less than the sum of their van der Waals radii. Indeed, this is one of the most reliable experimental criteria for hydrogen bond formation.

If two groups that may hydrogen-bond are forced into close proximity because of other steric factors present in the molecule, then not all of the increase in repulsion, caused by the close approach of atoms, will be at the expense of the H-bond energy, and a very strong

[3]T. R. Dyke, to be published.
[4]C. A. Coulson, *Research* (London) 1957, *10*, 149.

H-bond may be formed. This occurs in highly alkylated malonic acids, *gem*-dimethylcaronic acid, nickel *bis*(dimethyglyoximate) and some other compounds.

Electrostatic Term

In neutral molecules the electrostatic term may be thought of as arising from a dipole-dipole interaction unique in that the hydrogen has no underlying shell of electrons; this permits an unusually close approach of the positive end of the dipole to the negative end of a neighboring dipole. Also, a major component of the negative end of the dipole may reside in a free pair of electrons, relatively near the surface of the molecule, and with directional orientation in space. These factors could prevent even diatomic molecules that possess H-bonds from behaving like simple point dipoles. The structure of HF (p. 180) thus is not inconsistent with an electrostatic model of the hydrogen bond. The electrostatic model has been rather successful because the energies of the other contributions cancel to a great extent.

Since the interaction of an *ion* with a dipole is much greater than dipole-dipole interaction (see p. 189), you should expect that the electrostatic contribution to the energy of the hydrogen bond will be much greater in ion-molecule interactions, as in FHF^-, $ClHCl^-$, $HOHOH^-$, etc. Such interactions would decrease, as the ion increases in size, and also as the charge density is drained from the surface of a polyatomic ion by π bonding between the outer and central atoms.

Delocalization

In terms of resonance, the following will be contributing structures of the resonance hybrid of the hydrogen bond AHB.

$$A\text{---}H \quad B \qquad\qquad A^-H^+\cdots B \qquad\qquad A^-H\text{---}B^+$$
$$(1) \qquad\qquad\qquad (2) \qquad\qquad\qquad (3)$$

$$A^+H^- \quad B \qquad\qquad \overline{A \quad H^- \quad B^+}$$
$$(4) \qquad\qquad\qquad (5)$$

Structure (1) represents no interaction; (2) and (4) represent electrostatic interaction; (3) and (5) represent charge delocalization and might be said to represent the covalent character of the hydrogen bond. To the extent that A—H has partial ionic character, charge delocalization becomes more important. When A—H is a positive ion, or B a negative ion—that is, when there is a full unit charge—charge delocalization may become quite important. You can see that the importance of charge delocalization should decrease with increasing size of the ions involved in hydrogen bonding.

Other Viewpoints and Further Considerations

The possibility of covalence contributing appreciably to the stability of the hydrogen bond was rejected by Pauling on the basis that "the hydrogen atom, with only one stable orbital (the $1s$ orbital), can form only one covalent bond . . . ." But Pimentel[5] has given a molecular orbital

[5]G. C. Pimentel, *J. Chem. Phys.* 1951, *19*, 446.

description of the hydrogen bond that surmounts this objection to covalency. From the three atomic orbitals, the $1s$ of hydrogen and the two p orbitals of A and B, three molecular orbitals are constructed—a bonding orbital, a nonbonding orbital, and an antibonding orbital. The four electrons then occupy the bonding and nonbonding orbitals to produce a three-center, four-electron bond. On this basis, we may consider the bridge bond in diborane (p. 161) as a hydrogen bond in which the nonbonding orbitals are not occupied. This is a three-center, two-electron bond as described for XeF_2 (p. 164). Pimentel's picture indicates that atoms A and B must have nearly the same electronegativity; otherwise, the nonbonding orbital would be virtually the p orbital of the highly electronegative atom and the bonding orbital would be virtually a two-center covalent bond A—H, and the hydrogen bond would lose its covalent character.

The availability of high-speed computers has encouraged many theoretical studies of hydrogen bonding using *ab initio* approaches for small molecules and semiempirical approaches for larger ones. Still, the best water dimer calculations give total energies that are in error by 100 times the hydrogen bond energy,[6] because the latter is only a small fraction of the calculated total energy.

Allen[7] has presented a model, based on theoretical studies, that provides a good description of the hydrogen bond and of the factors contributing to its strength. The model depends on three variables. (1) The bond dipole moment, $\mu_{A—H}$, for the A—H bond that provides the bonding H. This is an indication of the effectiveness of the screening of the proton. (2) The ability of the electron donor to donate electrons. To provide a reference, this ability is expressed as ΔI, the difference between the ionization energy of the electron donor atom and the noble gas atom in its row. In this way atoms of different rows can be compared. (3) The extension of the hydrogen-bonding lone pair, taken as the internuclear separation, R, between the hydrogen-bonded atoms A and B. The first two variables are inherent properties of A—H and B, whereas the third depends on the specific bonding situation. The energy required to break the hydrogen-bonded dimer, A—H$\cdots$B, to give A—H and B, is given by

$$E = K(\mu_{A—H})\Delta I/R \tag{5.1}$$

where K is an energy scale factor. Hydrogen bond dissociation energies obtained from Equation 5.1 are in good agreement with those calculated from molecular orbital wave functions.

5.1.7 Anion-Molecule Interactions

When hydrogen fluoride comes into contact with an alkali metal fluoride at room temperature, an exothermic reaction occurs and an acid salt is formed. This reaction is used to remove HF from gas streams—particularly in the commercial production of F_2, where HF is almost always a contaminant.

$$NaF(s) + (HF)_x(g) \rightarrow NaHF_2(s) + 69 \text{ kJ/mole}$$

Reversal of the reaction—that is, thermal decomposition of $NaHF_2$—serves as a convenient source of anhydrous HF. The heat released in reaction with HF is least with LiF (43 kJ/mole)

[6]M. D. Joesten and L. J. Schaad, *Hydrogen Bonding,* Dekker, New York, 1974, p. 109.
[7]L. C. Allen, *J. Am. Chem. Soc.* 1975, *97,* 6921.

and increases with the size of the cation: 88 kJ with KF, 93 with RbF, 98 with CsF, and 155 with tetramethylammonium fluoride. X-ray diffraction studies indicate that the alkali metal hydrogendifluorides are salts with a positive metal cation and $(FHF)^-$ as an anion. The F—F distance in KHF_2 is only 225 pm, with the hydrogen nucleus equidistant from the fluorine nuclei.

As the HF is absorbed, the crystal lattice of KF must expand to accommodate the $(FHF)^-$ ion. The energy necessary for this expansion is greatest for $LiHF_2$ and least for $CsHF_2$. The loss of lattice energy (see Chapter 6, p. 220) on going from the halide to the HF adduct is reflected in a decrease in m.p.—the m.p. of KF is 846°C, whereas that of KHF_2 is 239°C. The larger alkali metal ions permit the formation of even higher adducts—KH_2F_3 and KH_3F_4, CsH_2F_3, etc.—with still lower melting points. The low melting points of these compounds facilitate the production of fluorine by the electrolysis of fused KHF_2, KH_3F_4, etc. Liquid HF (b.p. 19°C) also is used for this purpose, but the volatility of HF and its corrosive nature introduce technical difficulties.

The solubility of many fluorides in liquid HF results from the high degree of solvation of the fluoride ion through hydrogen bonding—that is, $F(HF)_n^-$. The strong acid behavior of liquid HF, in contrast to aqueous solutions of HF, indicates that the solvation energy of the fluoride ion is much higher in HF than in water.

Water in its reactions with alkali metal hydroxides is quite similar to HF with MF. Thus solid KOH occasionally is used to remove water vapor from gas streams (especially where it is also desirable simultaneously to remove CO_2).

$$KOH(s) + H_2O(g) \rightarrow KOH \cdot H_2O(s) + 84 \text{ kJ}$$

As noted earlier, strong hydrogen bonds should be expected in ion-molecule interactions where there is equal competition for the proton, a charge to delocalize, and a dipole to interact with the ion. Table 5.3 gives the experimental energies of some hydrogen bonds for ion-molecule systems, measured either in the gas phase or in systems specifically designed to minimize lattice energy effects. The relative strengths of the hydrogen bonds in the H_2S—X^- systems and in the NH_3—X^- systems cannot be explained by either an electrostatic approach or an acid-base approach. The molecular orbital approach accounts for these bond strengths in terms of the better match of orbital electronegativity of Br and S, and of Cl and N, resulting in a more stable three-center bonding orbital.

Hydrogen bonds are much stronger in ion–molecule than in molecule-molecule interactions (see Tables 5.2 and 5.3). The wide variety of known oxoacids furnish numerous possibilities of similar molecule-ion interactions. Some of the compounds whose formulas have puzzled chemists until recently are $RbNO_3 \cdot HNO_3$; $RbNO_3 \cdot 2HNO_3$; $KHCO_2 \cdot HCO_2H$; $KHCO_2 \cdot 2HCO_2H$; $KHCO_2 \cdot 3HCO_2H$; and $NaCH_3CO_2 \cdot CH_3CO_2H$. That hydrogen bonds in these compounds are strong is shown by the short O—H—O distance of 241_7 pm found in sodium hydrogen diacetate, as compared with the value 276 pm found in ice.

5.1.8 Cation-Molecule Interactions

The nature of the hydrogen ion in water has been speculated about for the last 60 years. Largely on the basis of the existence of isomorphous crystals of perchloric acid hydrate ($H_3O^+ClO_4^-$) and ammonium perchlorate ($NH_4^+ClO_4^-$), it was assumed that the hydrogen ion in water existed

Table 5.3 Some energies of hydrogen bonds in ion-molecule systems ($-\Delta H$ in kJ/mole H-bond)

Gas phase[a]	$H_3NHNH_3^+$ 105	$H_2OHOH_2^+$ 134	$HOHOH^-$ 100	$HOH \cdots F^-$ 96	FHF^- 188
	$H_3N \cdots HN{<}$ 55–105	$H_2O \cdots HO{<}$ 75–135		$HOH \cdots Cl^-$ 55	$ClHCl^-$ 100
		$H_2O \cdots HN{<}$ 33–71		$HOH \cdots Br^-$ 53	$BrHBr^-$ 75
				$HOH \cdots I^-$ 43	
Solid systems[b]	$Cl(HCF_3)_2^-$ 32	$H_2NH \cdots F^-$ 37	$HSHSH^-$ 59	$CH_3OHOC_4H_9^-$ 81	FHF^- 155
	$Cl \cdots HCCl_3^-$ 59	$H_2NH \cdots Cl^-$ 43	$HSH \cdots Cl^-$ 41	$HOH \cdots Cl^-$ 44	$ClHCl^-$ 59
	$Br \cdots HCCl_3^-$ 53	$H_2NH \cdots Br^-$ 26	$HSH \cdots Br^-$ 43		$BrHBr^-$ 55
	$I(HCCl_3)_3^-$ 32	$H_2NH \cdots I^-$ 23	$I(H_2S)_2^-$ 39		IHI^- 52

[a]Data from P. Kebarle, *Ann. Rev. Phys. Chem.* 1977, *28*, 445; W. R. Davidson, J. Sunner, and P. Kebarle, *J. Am. Chem. Soc.* 1979, *101*, 1675; P. Yamdagni and P. Kebarle, *J. Am. Chem. Soc.* 1971, *93*, 7139; S. A. Sullivan and J. L. Beauchamp, *J. Am. Chem. Soc.* 1975, *97*, 1160.
[b]Data from D. H. McDaniel *Ann. Reports Inorg. and Gen. Synthesis* 1972, 293; D. H. McDaniel and R. E. Vallee, *Inorg. Chem.* 1963, *2*, 996.

as H_3O^+, termed the oxonium ion. In 1936, Huggins suggested that in water the hydrogen ion would exist as

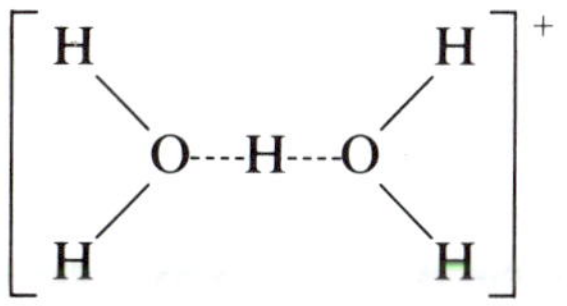

in which, unlike in ice, the hydrogen in the hydrogen bond would be equidistant from both oxygen atoms. In 1954 Eigen proposed that the hydrogen ion in water exists as $H_9O_4^+$ with the following structure.

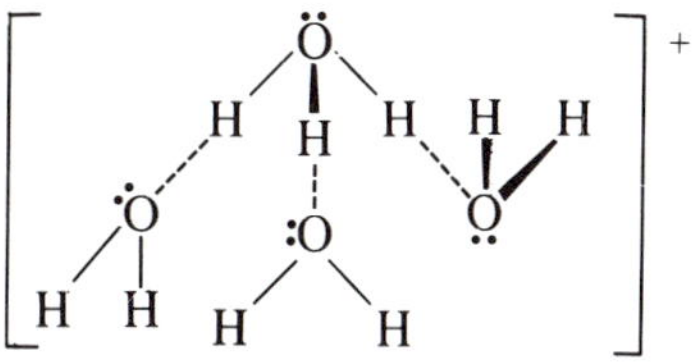

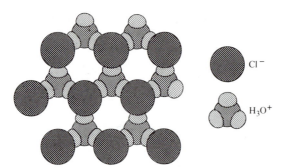

Figure 5.6   Layer structure of oxonium chloride. (From Yoon and Carpenter *Acta Cryst.* 1959, *17*, 12.)

Support for a hydration number of four for the hydrogen ion comes from specific heat studies of aqueous hydrochloric acid solutions, from extraction studies on HCl, HBr, and $HClO_4$, from activity coefficients of aqueous acids, and more recently, from data obtained by field emission mass spectrometry.

The exceptionally high mobility of the hydrogen ion in water is unexpected for a species as large as $H_9O_4^+$ and can be explained by the ease with which hydrogen bonds realign to form new $H_9O_4^+$ species incorporating water previously in a secondary hydration sphere.

Solids may contain the hydrated hydrogen ion in species varying from H_3O^+ up to $H_9O_4^+$ as illustrated by the hydrates of hydrogen bromide given below with their melting points.

	m.p. (°C)	
$HBr \cdot 4H_2O$	-56.8	
$HBr \cdot 3H_2O$	-48	
$HBr \cdot 2H_2O$	-11	
$HBr \cdot H_2O$	-4	(incongruent melting point)

The increase in melting point on going from HBr (m.p. $-86°C$) to $HBr \cdot H_2O$ (oxonium bromide) (m.p. $-4°C$) results from a change from a covalent to an ionic lattice (with hydrogen bonding). The decrease in melting points on going to higher hydrates results from expansion of the crystal lattice to accommodate the larger hydrated species accompanied by a loss in lattice energy. Similar hydrates exist for HI, whereas HCl forms only $HCl \cdot H_2O$, $HCl \cdot 2H_2O$, and $HCl \cdot 3H_2O$, but not a higher hydrate. A layer of the structure of oxonium chloride, $HCl \cdot H_2O$, is shown in Figure 5.6.

5.2 STRONG BONDING INTERACTIONS

Thus far we have considered covalent bonding (Chapters 2 and 4) and hydrogen bonding (Section 5.1). Covalent bonds generally are very strong: for example, H—H, 432.08 kJ/mole; C—C (C_2H_6), 350 kJ/mole; Cl—Cl, 239.23 kJ/mole. Strong polar covalent bonds are formed between elements differing greatly in electronegativity: for example, H—F, 566.3 kJ/mole; C—F (CH_3F), 452 kJ/mole. Weak single bonds are formed between the most electronegative elements (see p. 40), as in the case of F—F, 154.6 kJ/mole. Metals form weak (single) covalent bonds in dimeric molecules: for example, Li_2, 100.9 kJ/mole; Cs_2, 38.0 kJ/mole. Multiple bonding can enhance the bond strength greatly—N_2, 941.66 kJ/mole, being the most notable example. Covalent bonds are highly directional. The study of the stereochemistry of simple molecules and coordination compounds (Chapter 8) merely suggests the incredibly intricate stereochemistry of biologically active compounds and the key roles of metal ions, which are often at the active sites.

Ions of opposite charge attract each other very strongly. Ionic solids are treated in Chapter 6. The dissociation energy for an ion pair varies with the product of the ionic charges and inversely with the internuclear distance, as shown in Equation 5.2, where Z_1 and Z_2 are the ionic charges and e is the charge on the electron.

$$\text{Potential energy for } M^+X^- = \frac{Z_1 Z_2 e^2}{d} \tag{5.2}$$

Thus the ionic "bond" energy is strongly dependent on ionic radius, decreasing from LiF (895 kJ/mole, 155 pm) to CsI (418 kJ/mole, 332 pm).[8] The great increase in "bond" energy with increasing ionic charge—2140 kJ/mole, d = 130 pm (est.) for BeF^+ and 4180 kJ/mole, d = 133 pm for BeO, both calculated for formation of the ion pairs from separate ions—is limited, since polarization effects become important for high ionic charges (see p. 196), leading to covalency. $TiCl_4$ is a volatile liquid. The attraction between ions in ion pairs or formula units provides a reasonable comparison with covalent bond energies. However, the stabilities of ionic compounds in their standard states (solids) are enhanced further by the lattice energies—the energy released in the formation of the crystal (1 mole) from separate ions (see Section 6.3). The lattice energy for LiF is 1041 kJ/mole, whereas the formation of ion pairs releases 754 kJ/mole. The LiF distance is shorter in the ion pair (155 pm) than in the crystal (209 pm) where there is repulsion among the 6 F^- around each Li^+.

Along with covalent and ionic bonding, bonding in the metallic state (Section 6.8) is unique. In this type of bonding, positive metal ions interact with electron bands delocalized throughout the lattice. This might be regarded as the limiting case of multicenter covalent bonding. Except for Hg, the standard states of metals are solids. The dissociation energy for a mole of a metal is the sublimation energy, which varies greatly from 770 kJ/mole for W to 68 kJ/mole for Cs (values at the boiling points). See Figure 6.32 for a plot of the sublimation (cohesive) energies of the elements.

5.3 WEAK INTERACTIONS IN COVALENT SUBSTANCES

The atoms within a covalent molecule are bound firmly together by the covalent bonds, but the molecules usually are attracted very weakly to other molecules by *van der Waals* forces. This weak attraction is responsible for the low boiling points and melting points of molecular substances. The van der Waals force can be pictured as the attraction exerted by a positive nucleus on electrons beyond its own radius. There is a finite probability of finding the electrons of an atom at any point out to infinity; the electron density does not fall abruptly to zero at the

---

[8]These values are calculated as the energy for the dissociation to form ions from Equation 5.2. The experimental value for LiF(g) is 754 kJ/mole. The dissociation of LiF(g) to neutral atoms requires 575 kJ/mole.

distance used as the radius of the atom. Since a nucleus must hold its own electrons at distances even greater than the atomic radius, it will attract electrons of other atoms that approach closely. The attractive force increases with the nuclear charge and the number of electrons the two atoms possess. Table 5.4 indicates the magnitude of the energy of various types of molecular interactions.

5.3.1   Instantaneous Dipole-Induced Dipole Interaction

F. London has presented a quantum mechanical treatment of the attractive forces between neutral molecules in terms of dispersion effects. The attraction between the spherical atoms of the noble gas elements is greater than we might expect for rigidly spherical atoms. Although the distribution of electronic charge is spherically symmetrical on a time average, momentary dipoles can exist. These short-lived dipoles polarize adjacent atoms, causing induced dipoles and increasing the force of attraction, which is called the *London force* or *dispersion force*. The dispersion force increases with increasing size and polarizability of the atoms of molecules

Table 5.4   Comparison of intermolecular and intramolecular interactions

Type of Interaction	*Example*	*Energy of Interaction between Molecules or Units*
Van der Waals (instantaneous dipole–induced dipole)	H_2 (b.p. 20 K) CH_4 (b.p. 112 K) CCl_4 (b.p. 350 K) CF_4 (b.p. 112 K) n-$C_{28}H_{58}$ (m.p. 336 K)	~0.1–5 kJ/mole or ~10 (T_{bp} K) J/mole
Dipole–induced dipole	$Xe(H_2O)_x$ solvation of noble gases or hydrocarbons (see text)	
Ion–induced dipole	Ions in a molecular matrix (see text)	
Dipole–Dipole	NF_3—NF_3 (b.p. 144 K) BrF—BrF (b.p. 293 K)	5–20 kJ/mole
Ion–Dipole	$K(OH_2)_6^+$ Ions in aqueous solution and solid hydrates	67 kJ/mole (energy per bond)
Hydrogen bond	$(H_2O)_x$, $(HF)_x$, $(NH_3)_x$ alcohols, amines HF_2^-	4–50 kJ/mole for neutral molecules
Cation–Anion	NaCl, CaO	400–500 kJ/mole of MX ''molecules''
Covalent bond	H_2 F_2 Cl_2 Li_2	432.08 kJ/mole 154.6 kJ/mole 239.32 kJ/mole 100.9 kJ/mole

and varies inversely with the sixth power of the distance between molecules. A repulsive term, caused in the last analysis by the Pauli exclusion principle, drops off more rapidly with distance. The twelfth and ninth powers of intermolecular distance have been used.

For substances such as N_2 and hexane, interaction between molecules arises primarily from the dispersion forces and usually increases with increasing molecular weight. Note, however, that increasing the molecular weight *per se,* without a corresponding increase in the dispersion forces, has little or no effect on the boiling point. Thus CD_4 has a slightly lower boiling point than CH_4. Similarly, many fluorocarbons' electrons are held so tightly by the highly electronegative fluorine atoms that the fluorocarbons boil at temperatures close to, or even less than, the corresponding hydrocarbons. A high degree of symmetry or compactness also favors a low boiling point, by shielding the electrons deep within the molecule and lessening their interaction with neighboring molecules. Thus branching generally lowers the boiling points of isomeric organic compounds.

5.3.2 Ion-Induced Dipole or Dipole-Induced Dipole Interactions

The attraction between a symmetrical neutral molecule and an ion or a molecule with a permanent dipole moment is enhanced greatly because of an induced dipole in the symmetrical molecule. The noble gases form hydrates that could be the result of dipole-induced dipole interactions, since the more stable hydrates are formed by the larger and more polarizable noble gas atoms. These hydrates probably are more properly regarded as clathrate compounds (see below).

There are few clear-cut cases of ion-induced dipole interactions, because more important forces are present in situations where such attractions might occur. Thus the interactions between ions and a nonpolar molecule in an ionic medium would be much less important than the ionic interactions.

5.3.3 Ion-Dipole and Dipole-Dipole Interactions

Ions attract the polar water molecules very strongly to form hydrated ions. The hydration energy released provides the energy required for the separation of the ions from the ionic crystal. We shall see later (Section 7.4) that treating metal complexes of water and ammonia in terms of ion-dipole attraction has serious limitations, especially for transition metal ions. The greater stability of ammonia complexes of some metals as compared to the aqua (water) complexes can be explained by the greater polarizability of ammonia, which causes the total dipole moment—permanent plus induced—of ammonia in such complexes to be greater than that of water in similar complexes.

We can see the effect of the attraction between dipoles by comparing the boiling points of NF_3 ($-129°C$) and OF_2 ($-144.8°C$) with that of CF_4 ($-161°C$), which has a higher molecular weight but zero dipole moment. Hydrogen bonding might be considered a special case of dipole–dipole interaction; it is much stronger than ordinary dipole–dipole interactions, because the positive center (H^+) is screened poorly.

5.3.4 Van der Waals Radii

The internuclear separation between nonbonded atoms that are in contact is determined by the sum of their *van der Waals radii*. Partitioning of the separation distance between the two atoms may be achieved by using data from systems where both atoms are identical—then, the van der Waals radius is simply half the nonbonding separation. From a somewhat oversimplified viewpoint, the van der Waals radius for an atom of a nonmetallic element will be approximately the same as the radius of the anion formed by the element, since both present to the outside world a completed octet of electrons (see p. 221 for a table of ionic radii). Table 5.5 lists some van der Waals radii for the nonmetals.

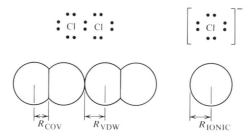

Table 5.5 Van der Waals radii of nonmetals (in pm)[a]

	H			
	120			
C	N	O	F	Ne
170	155	152	147	154
Si	P	S	Cl	Ar
210	180	180	175	188
	As	Se	Br	Kr
	185	190	185	202
		Te	I	Xe
		206	198	216

[a]From A. Bondi, *J. Phys. Chem.* 1964, *68*, 441. Values are mean van der Waals radii for single-bonded forms of the elements.

Although the van der Waals radius of an atom depends on the hybridization of the atom and the nature of the forces in the crystal lattice, it usually will vary by less than 10 pm in different nonbonding situations. When a hydrogen bond is formed, the approach of the hydrogen atom to the acceptor atom is considerably less than the sum of their van der Waals radii—the difference between the calculated and observed distances being as much as 100 pm for strong hydrogen bonds or as little as 20 pm for weak hydrogen bonds formed by donors such as chloroform. When in a crystal structure, two atoms are found to be closer than the sum of their van der Waals radii, a hydrogen bond between them usually supplies the explanation. The shortening of the A···B distance in an A—H···B bond over that calculated for van der Waals contact of A and B varies from no observable deviation to as much as 70 pm in FHF$^-$ and H$_2$OHOH$_2^+$. Typical values for the A···B shortening in oxygen-to-oxygen or oxygen-to-nitrogen hydrogen bonded systems are around 30 pm.

5.4 CLATHRATE COMPOUNDS

In a number of compounds, the composition appears very strange and unaccountable in terms of the usual bonding forces. The compound $Ni(CN)_2(NH_3)C_6H_6$ originally was thought to be a coordination compound in which there were four groups bonded to the nickel, including the benzene molecule. We now know that the compound has a very interesting structure in which there are two different Ni^{2+} coordination sites. Each Ni^{2+} is coordinated to four bridging CN^- to form sheets. Half of the Ni^{2+} ions are six-coordinate, with ammonia molecules coordinated above and below the plane. These Ni^{2+} ions are bonded to the N of CN^-. The benzene molecules are trapped in the "holes" in the lattice, where they just fit (Figure 5.7). Compounds in which a molecule is trapped in the "cage" of the crystal lattice are called *clathrate compounds*. Hydroquinone forms a series of compounds that approach the composition $(C_6H_6O_2)_3 \cdot Y$, where Y can be HCl, HBr, H_2O, SO_2, etc.

Clathrate compounds have become important in the separation of isomeric hydrocarbons. Urea and thiourea have been used widely as the solid phase and inorganic complexes have been used for separations such as the recovery of *p*-xylene either from gasolines or from aromatic hydrocarbons of comparable molecular weights by single-step operations.

The tendency for water molecules to form hydrogen bonds produces open structures in the solid state. The holes permit the formation of clathrate compounds of the type $Y \cdot xH_2O$ (where Y may be Xe, Cl_2, Br_2, SO_2, $CHCl_3$, etc., and *x* varies from about 6 to 17). Several tetraalkyl ammonium salts have been found to give hydrates in which the ions occupy cavities in some type of hydrogen-bonded ice structure.[9] The compounds are of the type $[(n\text{-}C_4H_9)_4N^+]_m X^{m-} \cdot myH_2O$ (where X is F^-, Cl^-, Br^-, CrO_4^{2-}, etc.; and *y* is approximately 32) and $[(i\text{-}C_5H_{11})_4N^+]_m X^{m-} \cdot my'H_2O$ (where X is F^-, Cl^-, CrO_4^{2-}, or WO_4^{2-} and y' is approximately 40). Each type of compound shown represents an isomorphous series. It is unusual to find an isomorphous series of compounds in which the same cation is combined with anions of different charge.

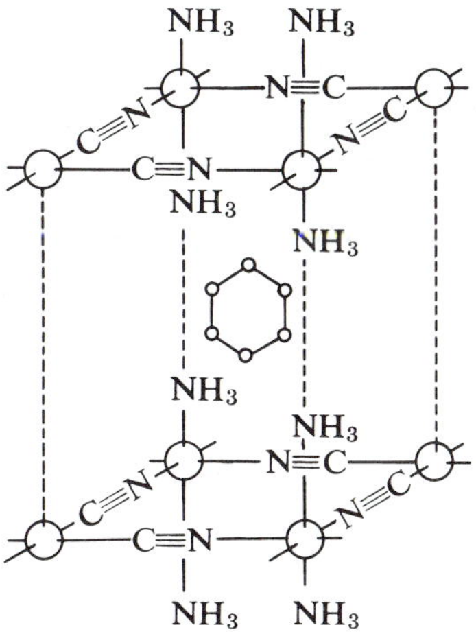

Figure 5.7 The $Ni(CN)_2(NH_3)C_6H_6$ clathrate compound. (From J. H. Rayner and H. M. Powell, *J. Chem. Soc.*, 1952, 319.)

[9]G. A. Jeffrey and R. K. McMullan, *Prog. Inorg. Chem.* 1967, *8*, 43.

GENERAL REFERENCES

R. L. DeKock and H. B. Gray, *Chemical Structure and Bonding,* Benjamin/Cummings, Menlo Park, Cal., 1980.

W. C. Hamilton and J. A. Ibers, *Hydrogen Bonding in Solids,* Benjamin, New York, 1968.

M. D. Joesten and L. J. Schaad, *Hydrogen Bonding,* Dekker, New York, 1974. Good coverage in a single volume.

G. C. Pimentel and A. L. McClellan, *The Hydrogen Bond,* W. H. Freeman, San Francisco, 1959. Dated but thorough.

P. Shuster, P. G. Zundel, and C. Sandorfy, Eds., *The Hydrogen Bond,* 3 vols., North Holland, Amsterdam, 1976.

S. N. Vinogradov and R. H. Linnell, *Hydrogen Bonding,* Van Nostrand-Reinhold, New York, 1971. General treatment at the undergraduate level.

PROBLEMS

5.1 List the substances in each of the following groups in order of increasing boiling points. (*Hint:* first group the substances according to the type of interaction involved.)
  a. LiF, LiBr, CCl_4, NH_3, CH_4, SiC, CsI.
  b. Xe, NaCl, NO, CaO, BrF, Al_2O_3, SiF_4.

5.2 Which of the following mixtures would be expected to have maximum boiling points and which to have minimum boiling points?
  a. Methyl acetate and chloroform.
  b. C_6H_{12} and C_2H_5OH.

5.3 Although ΔH_{vap} for HF is lower than that for H_2O, HF forms stronger H-bonds. Explain.

5.4 When no chemical reaction occurs, the solubility of a gas in a liquid is proportional to the magnitude of the van der Waals interaction energy of the gas molecules (see Table 5.4). Indicate the relative solubility of O_2, N_2, Ar, and He in water. Why do deep-sea divers use a mixture of He and O_2 instead of air?

5.5 Estimate the dispersion energy for H_2O from its "ideal" boiling point given in Figure 5.1, and compare your estimate to Coulson's estimate on page 181.

5.6 Use appropriate radii from pages 90 and 190 to calculate the expected S $\cdots$ S distance for SF_6 molecules in contact in the solid.

5.7 Calculate the expected O $\cdots$ O separation in a system containing O—H O if no hydrogen bonding were to occur and the O—H O atoms are linearly arranged. Compare your results with the nearest O—O distance, 275 pm, in ice I_h.

5.8 The intensity of an infrared absorption is proportional to the change in the dipole moment occurring during the vibration. The asymmetric stretching vibration in IHI^- is far more intense than that found for the stretching vibration of HI. Offer a reasonable explanation. Does your explanation also hold for the intensities shown in Figure 5.4.?

5.9 Indicate whether positive, negative, or no deviation from Raoult's law might be expected for the following binary systems:
  a. HCl—$(CH_3)_2O$ ***c.*** $HCCl_3$—CCl_4
  b. H_2O—C_8H_{18} ***d.*** $HCCl_3$—$(C_2H_5)_3N$.

5.10 The heat released in the reaction $Br^-(g) + HCl(g) \rightarrow BrHCl^-(g)$ has been found to be 38 kJ. By devising a suitable thermochemical cycle, and using available data, evaluate the expected heat released in the reaction $Cl^-(g) + HBr(g) \rightarrow ClHBr^-(g)$.

5.11 Explain why comparing a normal melting point with a melting point taken with the compound under water sometimes helps distinguish between intra- and intermolecular hydrogen bonding. (See E. D. Amstutz, J. J. Chessick, and I. M. Hunsberger, *Science* 1950, *111*, 305.

5.12 Why does CsCl(s) react with HCl(g) at low temperatures whereas NaCl(s) does not?

5.13 It has been suggested that the higher m.p. of *p*-methylpyridine-*N*-oxide compared with the *o*- and *m*- isomer may stem from hydrogen bonding. How might this be tested experimentally? Be specific. Mention the data to be gathered, with what the data might be compared, and how this would serve to establish hydrogen bonding or its absence.

5.14 Explain why the anhydrous acid $HICl_4$ cannot be isolated, but the crystalline hydrate $HICl_4 \cdot 4H_2O$ may be obtained from ICl_3 in aqueous HCl.

VI
Inorganic Solids

Perhaps your first impressions led you to regard metal salts, particularly those forming ionic crystals, as typical inorganic compounds. As we shall see, inorganic compounds include gases (such as NH_3, SO_2, and SF_6), liquids (such as $TiCl_4$ and $SOCl_2$), and solids covering a melting point range of almost 2000 degrees. The idealized model for ionic crystals treats the lattice as an array of ions, regarded as hard spheres for monoatomic ions, arranged according to simple principles of geometry and electrostatics. We will begin our study with idealized descriptions and then consider cases that deviate from the model. In examining structures, we will emphasize those features of primary interest to the chemist, not minor distortions that are of greater concern to the crystallographer. In this way we can relate larger numbers of compounds and gain a feeling for periodic and structural trends. Most of the elements are metals and share to a remarkable extent properties considered to be characteristic of the "metallic state." The structures and bonding of metals also are discussed.

Some individuals are better than others at visualizing a drawing as a three-dimensional structure. We all improve with practice in handling models and relating these to drawings. Building styrofoam-ball models of close-packed arrangements and other structures discussed in detail helps in this process. Perhaps some models are available for individual use.

6.1 SIMPLE INORGANIC SOLIDS

6.1.1 Properties

Crystals of simple inorganic salts consist of positive and negative ions packed in such a way as to allow a minimum distance between cation and anion and to provide maximum shielding from one another of ions of like charge. This arrangement produces the maximum attractive forces and the minimum repulsive forces. The number of ions of opposite charge around an ion in the crystal is the *coordination number* (C.N.) of that ion. Spherical ions have no directed bonding forces. The arrangement in the crystal is determined by the relative sizes of the ions and coulombic forces between ions. Many of the inorganic structures commonly encountered can be described most easily in terms of close-packed arrangements of one or more of the ions or atoms. The degree of covalent character varies greatly within a series of compounds of the same formula type, and even the same structure. Except when the distinction is clear, we will use the terms ''atom'' and ''ion'' loosely in describing structures, to avoid lengthy qualifications.

The ions of a salt are held rigidly in place by strong coulombic forces. Because the ions are not free to move about, most ionic crystals are very poor conductors of electricity. If the crystal structure is broken down by melting the substance or by dissolving it in a polar solvent, the ions can migrate freely in an electrical field and hence carry current which flows with mass transfer as the ions travel through the melt or solution. The charge transfer processes at the electrodes require oxidation-reduction reactions involving the electrodes and/or the ions in a fused salt. In solution the electrode reactions also might involve oxidation and/or reduction of the solvent. The conductivity of the fused salt or of a solution of the salt is determined by the ions' mobility, which depends on their size and charge. Ions of high charge decrease in mobility when associated with ions of opposite charge (ion pairing) and with solvent molecules (solvation).

6.1.2 Hardness

The strong electrostatic forces between ions cause ionic crystals to be relatively hard and to have high melting and boiling points. Melting points, boiling points, and equivalent conductance vary with polarization effects. Similarly, the hardness of crystals generally increases with decreasing ionic radius or increasing ionic charge. Salts of cations with noble-gas type configurations generally are harder (for the same ionic charges and comparable ionic distances) than those with pseudo-noble-gas (18 e) or similar (18 + ns^2) configurations (see Table 6.1).

Most ionic substances have low solubilities in all but the most polar solvents. Water is a good solvent for ionic substances because of its large dipole moment, which contributes to the high hydration energies of ions, and its high dielectric constant, which reduces the attractive forces between cations and anions in solution. The solubility of ionic substances and the specific case of water as a solvent will be discussed in greater detail below (p. 234).

Table 6.1 Variation of hardness of some crystals with similar structures

NaCl structure	LiF	MgO	MgS	CaO	SrO
M–X distance (pm)	202	210	259	240	257
Hardness (Mohs' scale)	3.3	6.5	4.5	4.5	3.5
CaF_2 structure	CaF_2	CdF_2	SrF_2	PbF_2	
M–X distance (pm)	236	234	250	257	
Hardness (Mohs' scale)	6	4	3.5	3.2	

6.1.3   Fajans' Rules for the Prediction of Relative Nonpolar Character

Typical ionic compounds, such as NaCl and CaF_2, differ strikingly from covalent compounds such as H_2, CH_4, and H_2O in such properties as melting points, boiling points, and electrical conductivity. These properties roughly indicate the degree of nonpolar character of similar compounds. During melting, solid and liquid phases are in equilibrium, and thus no free energy change accompanies the process. Accordingly, $\Delta H_m = T\Delta S_m$. If the C.N. of the atoms/ions does not change on melting (as for CH_4, CO_2, Al_2Cl_6, and $SnCl_4$) or vaporizing, the solid is termed a molecular solid. For molecular solids the ΔH_m is generally small, since primary bonding interactions are unchanged. Because most binary compounds of nonmetals are molecular compounds and the bonding is primarily covalent, low melting points have become associated with covalent bonding character. In view of such high melting compounds as BN, where the bonding is primarily covalent, such an association of melting points and covalency can be misleading. The differences between the melting points of AlF_3 (m.p. 1040°C) and SiF_4 (sublimes at -77°C) does not indicate a great difference in the ionic character of the bonds. The silicon in the SiF_4 molecule is shielded effectively from the fluorides of other SiF_4 molecules and the attractive forces are weak. Since the aluminum is not shielded on all sides, Al^{3+} interacts very strongly with neighboring F^- ions to give a three-dimensional network and C.N. 6.

Electrostatic forces in a crystal are essentially isotropic (lacking in directional character). Polarization interactions, on the other hand, are anisotropic (directional). As the C.N. decreases during the melting of nonmolecular solids, the contributions of polarization increase, compensating to some extent for the energy lost through the decrease in C.N. This compensation depends on the factors examined below (Fajans' rules).

Whereas Pauling dealt with variations in polarity in terms of the degree of ionic character in covalent bonds (p. 76), Fajans discussed the variation in the degree of nonpolar character in ionic compounds in terms of polarization effects. It should be noted that neither treatment need be limited to only one class of compounds. Fajans' rules, which focus on size and charge relationships and on the electronic configuration of the cation, may be summarized as follows:

1. Nonpolar character increases with decreasing cation size or increasing cation charge. Very small cations or cations with high charge have high charge density and consequently tend to distort or polarize the electron cloud around the anion. The greater the polarization of the anion, the more nonpolar the bond between the atoms (see Figure 6.1). Table 6.2 compares NaCl and $CaCl_2$, and $MgCl_2$ and $AlCl_3$ or $ScCl_3$ for charge effects. $AlCl_3$ and $ScCl_3$, $BeCl_2$, $MgCl_2$, and $CaCl_2$ are compared for size effects. The result of large changes in size and charge can be seen by comparing $SnCl_2$ and $SnCl_4$.

$SnCl_4$ is best described as a covalent substance, not unlike CCl_4. The Sn^{4+} represents the formal oxidation state of the hypothetical cation. If you start with Sn^{4+} and four Cl^-, the electron clouds of the Cl^- ions are strongly polarized to envelope the Sn^{IV}. The electrons are shared, with each Cl retaining a partial negative charge and Sn retaining a partial positive charge. The neutral $SnCl_4$ molecules are attracted to one another by weak van der Waals forces.

In the initial printings of the first edition of this book, we discussed Fajans' rules in terms of the degree of *covalent* character of ionic compounds. Since Professor Fajans did not believe in covalent bonding at all, *nonpolar* was substituted for *covalent* (same number of letters) in later printings, so as to conform with his views.

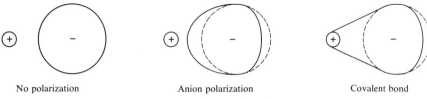

No polarization   Anion polarization   Covalent bond

Figure 6.1 Polarization effects.

2. Nonpolar character increases with an increase in anion size or anion charge. The larger the anion or the higher its negative charge, the more easily it is polarized by cations, because the electrons are held more loosely.

3. Nonpolar character is greater for cations with a non-noble-gas configuration than for noble-gas type cations. Cations with eight-electron structures cause less distortion, and undergo less distortion themselves, than those with 18-electron structures (pseudo–noble gas structures). If we compare cations of about the same size (Na^+, Cu^+), the pseudo–noble gas cation has a much higher nuclear charge and the d electrons in the 18-electron outer shell do not shield the nuclear charge effectively. An anion near a Cu^+ ion, compared with a Na^+ ion, behaves as though it were under the influence of a greater ionic charge and is distorted to a greater extent. The spherical filled d electron cloud is also more easily polarized by anions than is an eight-electron shell. This results in some polarization of a psuedo–noble gas cation by the anion. Such polarization of the cation results in even greater "exposure" of the cationic nuclear charge, because the electron cloud of the cation is "squeezed" back toward the far side of the cation. The increased effective nuclear charge further enhances the polarization of the anion (see Figure 6.2 and Table 6.3).

Table 6.2 Effect of cation charge and size upon nonpolar character of anhydrous chlorides

Cation	Cation Radius (pm, C.N. 6)	Melting Point (°C)	Boiling Point (°C)	Equivalent Conductance at m.p.
Na^+	116	800	1470	133
Ca^{2+}	114	772	>1600	51.9
Mg^{2+}	86	712	1412	28.8
Al^{3+}	67.5	Subl.	Subl. 178	1.5×10^{-5}
Sc^{3+}	88.5	Subl.	Subl. ~800	15
Be^{2+}	41 (C.N. 4)	405	~550	0.066
Mg^{2+}	86	712	1412	28.8
Ca^{2+}	114	772	>1600	51.9
Sn^{2+}	112 (C.N. 8)	246	606	21.9
Sn(IV)	83	−33	114	0

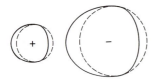

Figure 6.2 Cation and anion polarization.

Table 6.3 Effect of electronic structure of the cation upon nonpolar character of anhydrous chlorides

Cation	Cation Radius (pm, C.N.6)	Melting Point (°C)
Na^+	116	800
Cu^+	91	430
K^+	152	776
Ag^+	129	455
Ca^{2+}	114	772
Cd^{2+}	109	568

Colored Compounds of "Colorless" Ions

One of the manifestations of an increase in nonpolar character of inorganic salts is the appearance or enhancement of color. Although oxides of colorless cations usually are white, the corresponding sulfides likely will be deeply colored if the cation is one that tends to polarize anions. With few exceptions (including ZnS) the only white metal sulfides are those of the alkali and alkaline earth metals. In a series of halides of ions such as Ag^+, Hg^{2+}, and Pb^{2+}, the fluorides and chlorides are colorless and the bromides are colorless ($PbBr_2$) or faintly colored. The appearance of color in an inorganic salt containing only normally colorless ions usually indicates an appreciable amount of nonpolar character or some unusual structural feature. An appreciable amount of polarization leads to intense absorption bands known as charge transfer bands. The absorption corresponds to the transfer of an electron from one atom to another in the excited state (p. 288). Compounds in which the same metal is present in two oxidation states (e.g., Fe_3O_4, Mn_3O_4, Pb_3O_4, $KFe[Fe(CN)_6]$) usually are very intensely colored whenever the metal ions are bridged in such a way as to permit electronic interaction. Defect structures such as encountered in the tungsten bronzes, Na_xWO_3 ($x < 1$), also are likely to produce colored substances. In tungsten bronze some of the Na^+ sites are vacant. A W^V ion is oxidized to W^{VI} for each omission of Na^+ to maintain charge balance (p. 230).

6.1.4 Close Packing

The most efficient two-dimensional packing of uniform spheres leaves a close-packed layer in which each sphere has six neighbors arranged hexagonally (Figure 6.3). An identical second layer can be placed over the first, with each sphere of the second layer in an indentation formed by three spheres of the first layer. The positions of the spheres of the first layer are designated

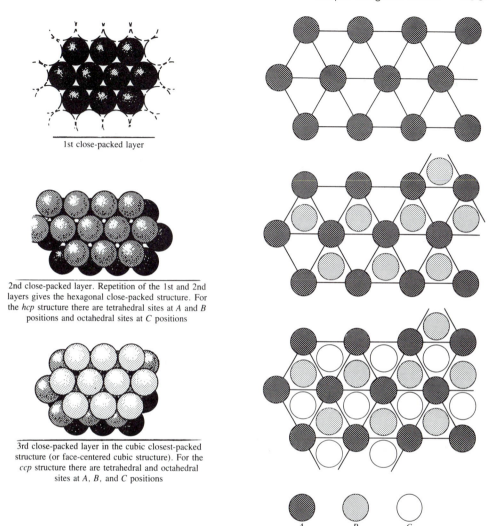

1st close-packed layer

2nd close-packed layer. Repetition of the 1st and 2nd layers gives the hexagonal close-packed structure. For the *hcp* structure there are tetrahedral sites at *A* and *B* positions and octahedral sites at *C* positions

3rd close-packed layer in the cubic closest-packed structure (or face-centered cubic structure). For the *ccp* structure there are tetrahedral and octahedral sites at *A, B,* and *C* positions

A B C

Figure 6.3 Close packing layers.

as *A* positions and those of the second layer as *B* positions. Two arrangements are possible for a third layer in a close-packed structure. The spheres in the third layer might line up with those in the first layer to give an *ABA* arrangement (Figure 6.3). If continued, *ABABAB* . . . , this sequence is described as a hexagonal close-packed *(hcp)* arrangement. The other possibility is to have the spheres of the third layer in the identations *not* directly over the spheres of the first layer (Figure 6.3): these are *C* positions. *A, B,* and *C* are the only packing positions in any close-packed structure. The sequence *ABCABC* . . . describes a cubic close-packed *(ccp)* arrangement. Any sequence—e.g., *ABAC* . . . *ABCB* . . .—is possible, but the *hcp* and *ccp* sequences are most common. For a close-packed structure each sphere has 12 nearest neighbors,

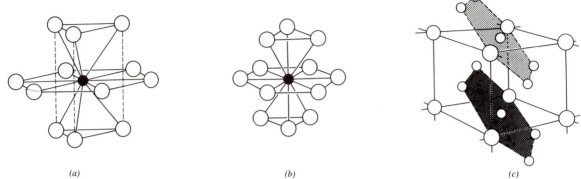

(a)   *(b)*   *(c)*

Figure 6.4 Hexagonal (*a*) and (*b*) cubic close-packed structures. (*c*) The *ccp* arrangement along the body-diagonal of a face-centered cubic structure.

with six arranged hexagonally in the same layer, three above and three below (Figure 6.4). The local symmetry of the packing unit consisting of an atom and its 12 neighbors is $\mathbf{D}_{3d}$ for *ccp* and $\mathbf{D}_{3h}$ for *hcp*. The face-centered cubic and *ccp* structures are the same—the names emphasize different features. The packing direction—the line perpendicular to the close-packed layers—is along a body diagonal of a face-centered cubic structure, as shown in Figure 6.4(*c*); but it coincides with the crystallographic axis of the *hcp* structure.

Between two close-packed layers (packing layers, *P*) there are interstitial sites or holes with octahedral and tetrahedral symmetry (Figure 6.5). The octahedral holes, which are formed by staggered triangular groups of spheres in each of the two layers, themselves form a layer (*O*) that is centered exactly between the two packing layers, at 1/2 *d,* where *d* is the distance between *P* layers (Figure 6.6). There are two layers of tetrahedral holes (*T*). One *T* layer is formed by tetrahedra, each of which has a sphere in the upper *P* layer at the tetrahedral apex and three spheres in the lower *P* layer forming the base. The other *T* layer is formed by tetrahedra, each of which has a sphere in the lower *P* layer at the apex and three spheres in the upper *P* layer forming the base. These layers can be distinguished as T_+ for those tetrahedra pointing upward and T_- for those pointing downward. The T_+ layer is nearer the lower *P* layer (1/4 *d*), and the T_- layer is nearer the upper *P* layer (3/4 *d*). The arrangements of the *T* and *O* sites are exactly the same between any two adjacent *P* layers for any close-packed structure.

In both *hcp* and *ccp* structures, each *P* sphere has one tetrahedron above and one below, so there are two *T* sites/*P* position. Each *P* sphere is shared by six octahedra for a close-packed structure, so there is one *P*/octahedron or one *P*/octahedral site (1/6 × 6 *P* spheres/*O* site). The octahedra share faces within each *P* layer in *hcp:* that is, all octahedra are stacked with all *O* sites at *C* positions. In *ccp* the octahedra share edges only.

The general sequence for any close-packed structure is PT_+OT_- (the distinction between the T_+ and T_- layers usually is not made). For *hcp* the sequence is $P_A TOTP_B TOT$ . . . (Figure 6.7), and for *ccp* it is $P_A TOTP_B TOTP_C TOT$ . . . (Figure 6.8). Since *T* sites must line up above or below *P* spheres, they are located only at *A* or *B* positions for *hcp*, and at *A*, *B*, and *C* positions for *ccp*. The *O* sites are staggered relative to the *P* spheres, so they occur only at *C* positions for *hcp*, but at *A*, *B*, and *C* positions for *ccp*.

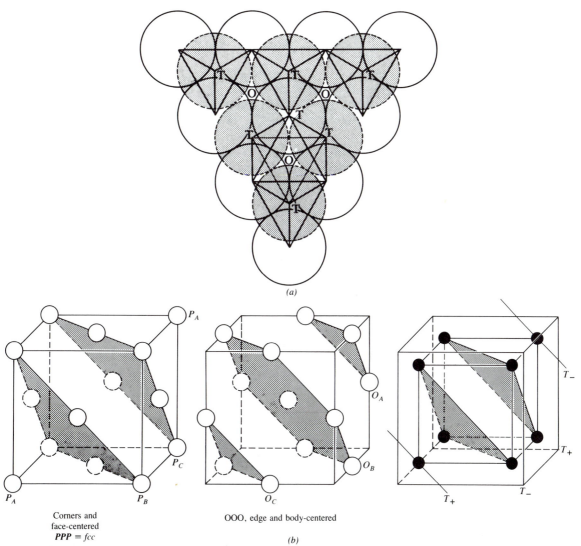

(a)

Corners and
face-centered
PPP ≡ *fcc*

OOO, edge and body-centered

(b)

Figure 6.5 (*a*) Tetrahedral (*T*) and octahedral (*O*) holes in a close-packed arrangement of spheres. (*b*) Positions of *P*, *O*, and *T* sites in the usual view of a face-centered cubic *(ccp)* structure.

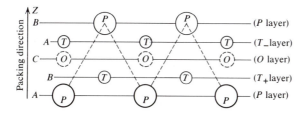

Figure 6.6 Construction of the basic close-packed unit (two packing layers for *any* close-packed structure). *P*, packing atom; *O*, octahedral site; *T*, tetrahedral site; *A*, *B*, *C*, the three relative packing positions. (From S.-M. Ho and B. E. Douglas, *J. Chem. Educ.* 1969, *46*, 208.)

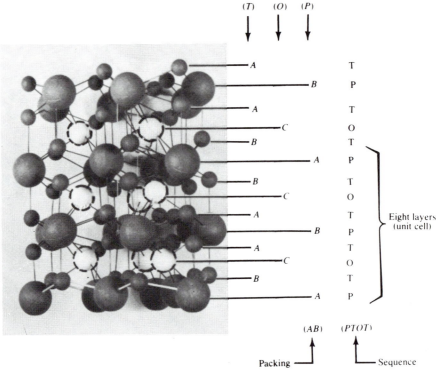

Figure 6.7 The basic hexagonal close-packed (*hcp*) structure. (From S.-M. Ho and B. E. Douglas, J. Chem. Educ. 1969, *46*, 208.)

6.1.5 Notation for Close-Packed Structures

Most inorganic structures can be regarded as close-packed arrays of one or more kinds of atoms or ions, with or without other atoms or ions in octahedral and/or tetrahedral sites. In the ideal case where the packing spheres touch, the spheres in O sites must be smaller than the P spheres ($r_O = 0.414\ r_P$) and those in T sites must be still smaller ($r_T = 0.225\ r_P$). (See the radius ratio rules, p. 217). Larger spheres can be accommodated in T and O sites by pushing the P spheres apart, but without changing the relationships described for close-packed structures. The close-packed sequence provides the basis for describing most inorganic structures using simple notation.

PO and *PO*$_{1/2}$ Structures

The repeating unit for *hcp* is $P_A P_B \ldots$, which is designated as 2P; that for *ccp* is designated 3P for $P_A P_B P_C$. In these cases the T and O layers are empty. The NaCl structure can be regarded as a *ccp* array of Cl$^-$ with Na$^+$ in octahedral sites. The notation is 6PO, indicating that there are six layers ($P_A O P_B O P_C O$) in the repeating unit. Since half of these are P layers, the three P layers (6/2) must be in a *ccp* arrangement. In the unit cell for NaCl shown in Figure 6.9, the packing direction is along a body diagonal. Both Na$^+$ and Cl$^-$ have C.N. 6. The

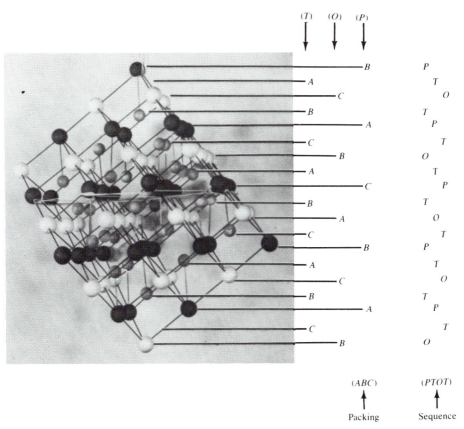

Figure 6.8 The basic cubic close-packed (*ccp*) structure. (From S.-M. Ho and B. E. Douglas, *J. Chem. Educ.* 1969, *46*, 208.)

NiAs structure is similar, in that all octahedral sites are filled by Ni atoms; but unlike the Cl^- in NaCl, the As atoms are in an *hcp* arrangement (Figure 6.9; see also p. 208). The stacking unit (P_AOP_BO) is described as 4*PO*. Rutile, one of the modifications of TiO_2, has Ti atoms (these are not simple Ti atoms, but there are no Ti^{4+} *ions* in such a structure) in only half of the *O* sites formed by *hcp* layers of oxygens. The notation is $4PO_{1/2}$. However, the unit cell is orthorhombic with $a = b$, so the structure deviates from the close-packed model. Ti has C.N. 6 and oxygen has C.N. 3 in rutile (Figure 6.10). Half of the *O* sites are filled, giving alternate layers of TiO_6 octahedra (but not corresponding to *A* or *B* layers). In order to achieve effective screening of the ions and the three-fold symmetry about oxygen, slight folding or puckering of the packing layers must occur.

PT Structures

There are two modifications of ZnS involving close-packed arrangements of S with Zn in half of the *T* sites: that is, all of the *T* sites in one of the two *T* layers. The S are *hcp* in wurtzite and *ccp* in zinc blende (Figure 6.9). Zn and S have C.N. 4 for both structures. Wurtzite is

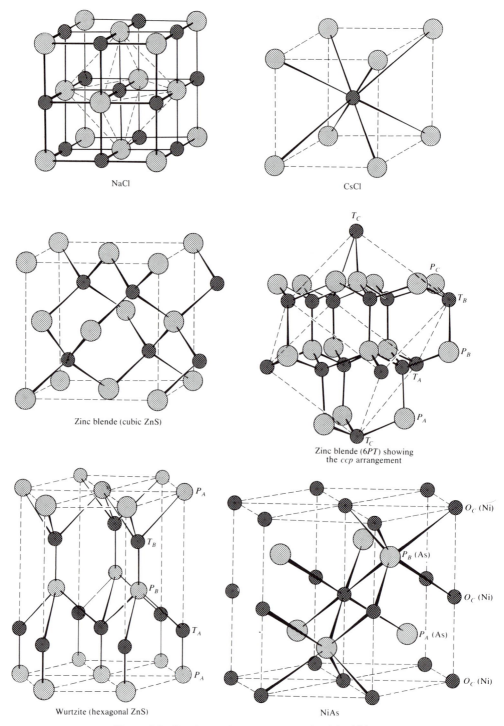

NaCl

CsCl

Zinc blende (cubic ZnS)

Zinc blende (6*PT*) showing
the *ccp* arrangement

Wurtzite (hexagonal ZnS)

NiAs

Figure 6.9 Structures of some compounds of the MX type.

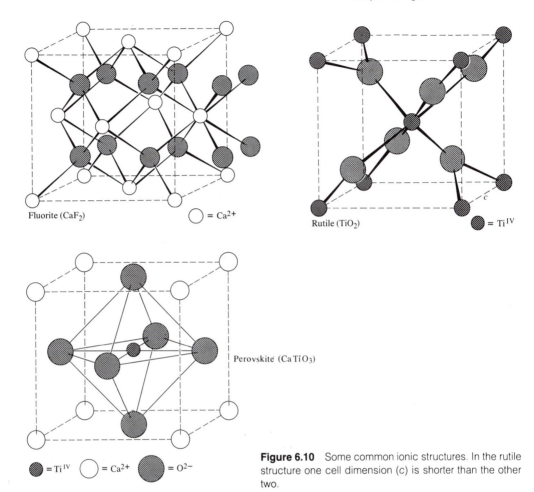

Fluorite (CaF$_2$)　　○ = Ca^{2+}

Rutile (TiO$_2$)　　● = Ti^{IV}

Perovskite (CaTiO$_3$)

● = Ti^{IV}　○ = Ca^{2+}　◉ = O^{2-}

Figure 6.10 Some common ionic structures. In the rutile structure one cell dimension (*c*) is shorter than the other two.

designated *4PT*, to indicate the *hcp* arrangement (P_ATP_BT), and zinc blende is designated *6PT*, since S are *ccp* ($P_ATP_BTP_CT$). The *C* sites are empty in wurtzite and form large hexagonal channels along the crystallographic *c* axis. This can be seen for ice I_h (Figure 6.11), where the oxygens replace all Zn and S in wurtzite. Only two thirds of the *T* sites in one layer are occupied in the *ccp* γ-Ga$_2$S$_3$, as described by $6PT_{2/3}$. Many *giant molecules* (C, CSi, SiO$_2$, etc.—p. 237) commonly have tetrahedral structures characterized by *PT* packing.

PTT Structures

All *T* sites are filled by F$^-$ in a *ccp* array of Ca^{2+} (in *P* sites) in fluorite, CaF$_2$ (Figure 6.10). The notation *9PTT* indicates the *ccp* arrangement, since one third of the nine layers are *P* layers and $1/3 \times 9 = 3$ for an *ABC* sequence. The stoichiometry of a *PTT* structure is 1:2— MX$_2$ (fluorite) or M$_2$X (antifluorite). MX (1:1) stoichiometry is achieved for *PO, PT* or $PT_{1/2}T_{1/2}$, and *PTOT* (with M in *T* sites and X in *P* and *O* sites—see CsCl). In CaF$_2$ the C.N.

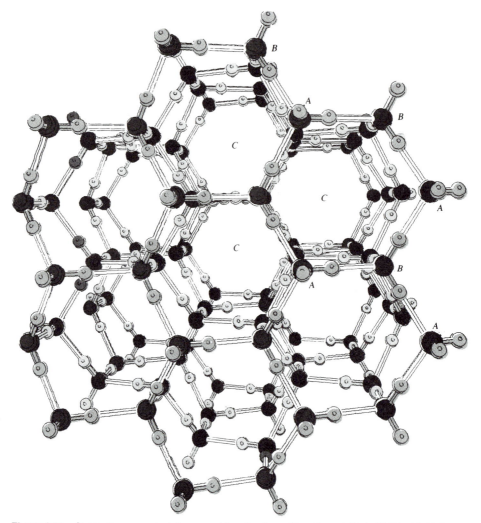

Figure 6.11 Computer-generated diagram of the structure of ice. Copyright © 1976 by W. G. Davies and J. W. Moore, used by permission; reprinted from *Chemistry*, J. W. Moore, W. G. Davies, and R. W. Collins, McGraw-Hill, New York, 1978. All rights reserved.

is eight for Ca^{2+} and four for F^-. PtS has a *ccp* array of *PT*, with S occupying half of the *T* sites in each *T* layer, as designated by $9PT_{1/2}T_{1/2}$. Pt has C.N. 4, but unlike zinc blende with tetrahedral ZnS_4 units, the PtS_4 units are planar (Figure 6.12). Here we can see the effect of covalent bonding between Pt and S. From the lattice, we would expect the bond angles to be 90° for Pt and for S, instead of the normal tetrahedral angle of 109.5° for S. Both give a little so that the Pt-S-Pt angle is about 97.5° and the S-Pt-S angle is about 82.5°. Because *P, O,* and *T* sites are equivalent for a *ccp* arrangement, we could also describe the PtS structure as $9P_{1/2}O_{1/2}T$, with Pt in all of one set of *T* sites and S in half of the *P* and *O* sites. For a *ccp* structure each *T* site is at the center of a cube formed by four *P* sites and four *O* sites, each

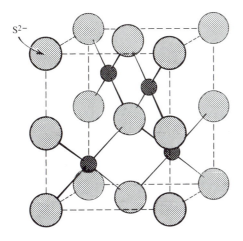

S^{2-}

Figure 6.12 The PtS structure showing the planar PtS$_4$ units. The larger ions are S^{2-}

set describing a tetrahedron. For PtS the P and O sites filled are along parallel face diagonals of the cube, giving planar PtS$_4$ units.

In some structures, one layer, such as a P layer, is occupied by more than one kind of atom. The perovskite structure, CaTiO$_3$, is designated $6PO_{1/4}$, with calcium and oxygen *together* forming the P layers (*ccp*). Ti fills only one fourth of the O sites—those surrounded only by oxygens (Figure 6.10).

The structures encountered for metals are $2P$ (*hcp*), $3P$ (*ccp*), and body-centered cubic (*bcc*; see p. 242). If all O sites of a *ccp* arrangement are occupied by atoms identical to those in the P sites (T sites empty), a simple cubic structure results. Although the full sequence is $P_AO_CP_BO_AP_CO_B$, since the P and O atoms are the same and the sites are identical, the stacking unit is really $3PO$ ($P = O$), with only three layers in the unit cell. This is the structure of the element polonium. If all T **and** O sites are occupied by atoms identical to those in the P sites, a *bcc* structure results. The full sequence is $P_AT_BO_CT_AP_BT_C(O_AT_BP_CT_AO_BT_C)$, but since P and O sites are identical here, the repeating unit is $P_AT_BO_CT_AP_BT_C$ or $P_AT_BP_CT_AP_BT_C$. To emphasize that the structure is derived from a *ccp* arrangement (ABC), the designation is $3 \cdot 2PTOT$ rather than $6PTOT$. Here the index number, 6, is not a multiple of the number of $PTOT$ layers (4) in the basic sequence, because of the equivalence of the P and O layers.

In a space-filling model (spheres touching) for a close-packed structure we see that the spheres filling the octahedral sites are smaller than those in the packing sites, and that those filling the tetrahedral sites are smaller still. If we use spheres of the same size for the P and O sites ($P = O$), the P layers are forced apart and the structure is no longer closest-packed. The relationships, however, are still as indicated in Figure 6.6, except for the radius ratio ($r_P:r_O$). The structure is the same as for NaCl, but with all spheres identical (paint all the same color). Similarly forcing the same atoms into all P, O, and T positions in the *bcc* structure forces the P layers apart, so that the structure is not closest-packed. We obtain the model by taking a NaCl model and adding a sphere at the center of each cube (formed by 4Na$^+$ and 4Cl$^-$) to fill the T sites (the P spheres form the tetrahedron, but the O spheres form another tetrahedron completing the cube). If all of these spheres are the same size and color, it is a $PTOT$ model for *bcc*.

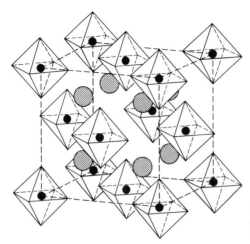

Figure 6.13 The K_2PtCl_6 structure showing the $PtCl_6^{2-}$ octahedra in a *ccp* arrangement, with K^+ in tetrahedral sites.

The CsCl structure can be regarded as having Cl^- in all P (*ccp*) *and* O positions, with Cs^+ in all T sites. The designation is $3 \cdot 2PTOT$. Both ions have C.N. 8 (Figure 6.9). We can view the BiF_3 structure as a *ccp* array of Bi (P positions), with F in all O and T sites, or $12PTOT$. This is also the structure of the intermetallic compound $BiLi_3$ and of cryolite, Na_3AlF_6. The cryolite structure is closely related to that of K_2PtCl_6 (Figure 6.13). The AlF_6^{3-} octahedra are in P positions with 2 Na^+ in T positions (corresponding to K_2PtCl_6) and the other set of Na^+ ions in O positions. Figure 6.13 shows these positions at the centers of the cube edges and at the body center. Structures based upon *hcp* with all T sites filled are not encountered and are not expected to be stable. The pairs of T sites nearest the P layers are too close in *hcp* for both to be occupied *fully*. *Hcp* structures involving *partial*, but staggered, occupancy of both T layers are known. The mineral forsterite, Mg_2SiO_4, is designated $8PT_{1/8}O_{1/2}T_{1/8}$ with P layers of oxygens, Mg in half of the O sites, and Si in one eighth of the T sites in each T layer (see p. 238). The structure of Al_2Br_6 is described as $2 \cdot 6PT_{1/6}T_{1/6}$, with O sites empty and Al in one sixth of the T sites in each T layer.

6.1.6 Common Structures of Inorganic Crystals

MX Compounds

Over 700 compounds display the structures represented in Figures 6.9 and 6.10. Most of the alkali halides, alkaline earth oxides, sulfides, etc.—as well as many other 1:1 compounds—crystallize in the $6PO$ (NaCl) structure. In the less common NiAs structure, $4PO$, the As is at the center of a trigonal prismatic arrangement of 6Ni and the Ni fills all octahedral holes. Since all O sites, and only O sites, occupy C positions for a *hcp* structure, the Ni are lined up along the C direction. The short Ni—Ni distances account for the semimetallic character of NiAs. Such a structure should occur only in appreciably covalent MX compounds—as in salts of highly polarizable anions such as sulfides and selenides of Ti, V, Fe, Co, and Ni, and in the tellurides of Ti, V, Cr, Mn, Co, and Ni. The corresponding oxides have the NaCl ($6PO$)

Table 6.4 Structures[a] of some compounds of the type MX

		Halides					*Oxygen Family*		
Ions	*F*	*Cl*	*Br*	*I*	*Ions*	*O*	*S*	*Se*	*Te*
Li	------------6PO(NaCl)----------------				Be	4PT	6PT	6PT	6PT
Na	------------6PO(NaCl)----------------				Mg	6PO	6PO	6PO	6PT
K	------------6PO(NaCl)----------------				Ca	6PO	6PO	6PO	6PO
Rb	------------6PO(NaCl)----------------				Sr	6PO	6PO	6PO	6PO
Cs	6PO	-----------3·2PTOT(CsCl)----------			Ba	6PO	6PO	6PO	6PO
Cu	—	6PT (4PT)[b]	6PT (4PT)[b]	6PT (4PT)[b]	Zn	4PT	4PT 6PT	6PT	4PT 6PT
Ag	6PO	6PO	6PO	4PT	Cd	6PO	4PT 6PT	4PT 6PT	6PT
					Hg	other[c]	6PT	6PT	6PT

[a]The structural notation corresponds to 6PO, NaCl; 3·2PTOT, CsCl; 4PT, wurtzite (*hcp* X); and 6PT, zinc blende (*ccp* X).
[b]High-temperature structure.
[c]See p. 629.

structure. Some of the halides of the larger alkali metal ions have the $3 \cdot 2PTOT$ (CsCl) structure with C.N. 8. Table 6.4 summarizes the structures encountered for some common inorganic compounds.

The arrangements of ions in a crystal depend to a great extent on the relative sizes of the ions. Table 6.5 provides the geometrical arrangements for the packing of spheres. The consequences of radius ratio effects will be considered later (p. 217). The expected radius ratios should be used only as rough guidelines. Of the alkali metal halides, only LiF (0.76), LiCl, LiBr, LiI, NaCl, NaBr, NaI, and KI (0.74) fall within the expected range for the $r_M:r_X$ for C.N. 6. All other alkali metal halides have higher ratios (0.8–1.5). An important limitation for MX (1:1) compounds is that the C.N. of the cation is the *same* as that of the anion. Consequently, CsF $\left(\dfrac{r_M}{r_X} = 1.52, \text{ using } r_M \text{ for C.N. 6—Table 6.8} \right)$ has C.N. 6, even though

Table 6.5 Stable arrangements of rigid spheres X about sphere M

Coordination Number of M	Arrangement of X	Radius ratio $r_M:r_X$
2	Linear	To 0.15
3	Triangular	0.15–0.22
4	Tetrahedral	0.22–0.41
4	Planar	0.41–0.73
6	Octahedral	0.41–0.73
8	Cubic	Greater than 0.73

this is the *highest* ratio for any of the cesium halides. Here, the C.N. 8 for F^- is unfavorable. These effects limit the CsCl structure to MX compounds containing *only* univalent ions. In fact, the CsCl structure is uncommon and found only for CsX ($X = Cl^-$, Br^-, I^-, CN^-, SH^-, SeH^-, and NH_2^-), TlX ($X = Cl^-$, Br^-, and CN^-), NH_4X ($X = Cl^-$, Br^-, and I^-), and intermetallic compounds where high coordination numbers for both metals are tolerable.

The MX compounds CsCN, TlCN, and CsSH have cubic structures even though the anions are not spherical. Cubic symmetry results from staggered orientations of the ions, whether regular or disordered. Some other salts, containing complex anions, give structures closely related to the 6*PO* structure. Thus the CaC_2 (calcium carbide) structure is face-centered, with the linear C_2^{2-} ions all oriented in the same direction to extend the unit cell along one axis. The calcite ($CaCO_3$) structure occurs for several carbonates of divalent cations; for some nitrates, such as $LiNO_3$ and $NaNO_3$; and for some borates, such as $ScBO_3$, YBO_3, and $InBO_3$. The calcite structure resembles the NaCl (6*PO*) structure, with the lattice expanded as required by the replacement of the Cl^- by the plane triangular carbonate ions.

Some substances of the MX type, where M is a small cation or MX is appreciably covalent, give the zinc blende, 6*PT* (the diamond structure differs only in that *all* Zn and S sites are occupied by carbon atoms), or the wurtzite, 4*PT*, structure. These structures are found for the copper(I) halides, and the oxides, sulfides, etc., of Be and the Group IIB metals. The radius ratio expected for C.N. 4 (tetrahedral) alone is a poor criterion in determining structure. Of the silver halides, only AgI has C.N. 4. This is the silver halide with the lowest value of r_M/r_X (0.55, using r_M for C.N. 4), but that still is larger than 0.41. Polarization of the large I^- by Ag^+ transfers negative charge to Ag^+ and favors the lower C.N. Plotting an ionic size parameter (actually, the average of the values of the valence shell quantum number $\bar{n}$ for the ions—a quantity interpreted by Pearson[1] as a measure of the directional character of the bonds formed) against the difference in electronegativities ($\Delta\chi$) of the ions rather cleanly separates those MX compounds with the NaCl structure from those with one of the ZnS structures. We can improve the separation by plotting $\bar{n}$ against $\Delta\chi \dfrac{r_-}{r_+}$. Plotting a covalence parameter against an ionic parameter[2] improves the separation even more, and in this plot we see that the wurtzite structure (4*PT*) is favored over the zinc blende (6*PT*) structure for large $\Delta\chi$ values.

We usually treat structures of MX-type solids in terms of the packing of ions, except where covalent bonding is obviously of primary importance (as in the diamond). Such is the case for giant molecules (for example, SiC; see Section 6.6). Using a molecular orbital approach, Burdett[3] described the zinc blende (6*PT*) and wurtzite (4*PT*) structures as the stacking of puckered sheets (the *P* and nearest *T* layers). Since all close-packed layers are identical (differing only in the way they are stacked), both the 6*PT* and 4*PT* structures consist of puckered sheets of linked hexagons in chair conformations. You can readily see the chair in the center of the zinc blende structure shown in Figure 6.9 by focusing your attention on the hexagon consisting of three light spheres (*T* sites) above the plane of three dark spheres (*P* sites). Hexagons also can be traced in the vertical stacking, revealing boat conformations for 4*PT* but chair conformations for 6*PT*. As we have seen (p. 200) there are three equivalent

[1]W. B. Pearson, *J. Phys. Chem. Solids*, 1962, *23*, 103.

[2]J. C. Phillips, "Chemical Bonds in Solids" in *Treatise on Solid State Chemistry*, N. B. Hannay, Ed. Vol. 1, Plenum, New York, 1973, p. 36.

[3]J. K. Burdett, *J. Am. Chem. Soc.* 1980, *102*, 450; *Nature* 1979, *279*, 121. See also *Gen. Ref.*

packing directions for the cubic 6PT, so hexagons traced in any direction have the chair conformation. There is only one packing direction for 4PT, and the chair conformation occurs only for the puckered sheets consisting of the $P + T$ layers. MO calculations reveal that the 6PT structure is favored for small $\Delta\chi$ values, whereas the 4PT becomes relatively more favorable for large $\Delta\chi$ values. Burdett shows how changing the number of valence electrons for MX (eight for ZnS) leads to structural changes for GaS (nine electrons), elemental As (10 electrons for As_2, M = X = As), and other cases. Layer structures such as those of $CdCl_2$, CdI_2, and MoS_2 (p. 213) yield to the same approach.

Exercise Build ball-and-stick models of the cage units for 6PT and 4PT structures to see the conformations of the hexagonal rings. For 6PT the cage consists of three S in one layer (P_A sites), three Zn above these in T_B sites, three S in the next P_B layer, and a Zn in the T_C site to complete the cage. The 4PT cage has three S in P_A sites, three Zn in T_B sites, three S in P_B sites, and three Zn in T_A sites. If ball-and-stick models are not available, examine models for zinc blende and wurtzite and outline the positions of the structures described above with masking tape.

MX_2, M_2X, and MX_3 Compounds

Compounds of the type MX_2 which are predominantly ionic commonly give the 9PTT (fluorite, CaF_2) or $4PO_{1/2}$ (rutile, TiO_2) structures (Figure 6.10). Compounds of large cations likely will display the 9PTT structure with the cations in the P positions (C.N. 8). Since the cations occupy the O sites in $4PO_{1/2}$, the C.N. is 6. These typical ionic structures are commonly encountered for fluorides and oxides, but not for the more covalent chlorides, bromides, iodides, sulfides, selenides, and tellurides. The antifluorite structure, in which the positions of the cations and anions are reversed as compared with fluorite, is encountered for alkali metal oxides, etc., of the type M_2X. Table 6.6 lists some compounds with these structures.

The radius ratio rules (Table 6.5) are more useful in predicting structures for MX_2 and M_2X compounds than for MX compounds, probably because the C.N. of one ion is twice that of the other. Large cations that form MX_2 compounds with a high degree of covalency often have the rather complex $PbCl_2$ structure (see references such as Wells). Smaller cations, or those that form even more covalent MX_2 compounds, commonly display one of the layer structures ($CdCl_2$, CdI_2, or MoS_2—see p. 214). The very small Si^{IV} gives one of the SiO_2 structures (p. 241). GeS_2 shows a related three-dimensional network of tetrahedra sharing all edges; the Ge is larger than Si, but the compound is less ionic than SiO_2. Figure 6.14 shows the grouping of compounds with these structures in a plot of $\bar{n}$ (the average quantum number

Table 6.6 Structures of some compounds of the type MX_2 and M_2X

Fluorite (CaF_2, 9PTT)			Rutile (TiO_2, $4PO_{1/2}$)			Antifluorite (9PTT)			
CaF_2	CdF_2	ZrO_2	MgF_2	NiF_2	TiO_2	Li_2O	Li_2S	Li_2Se	Li_2Te
SrF_2	HgF_2	ThO_2	MnF_2	ZnF_2	MnO_2	Na_2O	Na_2S	Na_2Se	Na_2Te
BaF_2	PbF_2	CeO_2	FeF_2	PdF_2	MoO_2	K_2O	K_2S	K_2Se	K_2Te
	$SrCl_2$	UO_2	CoF_2		GeO_2		Rb_2S		
					SnO_2				

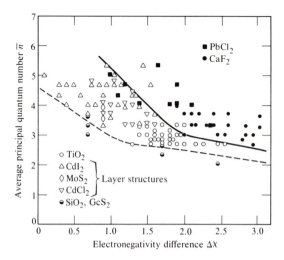

Figure 6.14 The structures of AX_2 compounds. (From E. Mooser and W. B. Pearson, *Acta Cryst.* 1959, *12*, 1015.)

n for the ions, a measure of size) against electronegativity difference (ionic character increases with an increase in $\Delta\chi$). "Ionic" MX_3 compounds of large cations commonly have LaF_3 or BiF_3 structures. Smaller cations and/or those forming more covalent MX_3 compounds will give ReO_3, AlF_3, or layer structures [$CrCl_3$ (p. 215) or BiI_3 type]. Figure 6.15 illustrates the grouping of MX_3 structure types.

We encounter the $6PO_{1/4}$ (perovskite, $CaTiO_3$) structure (Figure 6.10) in over 250 compounds of the types $M^{II}M^{IV}O_3$ and $M^IM^{II}F_3$. Researchers have prepared many compounds with the idealized or slightly distorted perovskite structure, because of interest in their ferroelectric properties. Another common structure for compounds of the type $M^{II}M^{IV}O_3$ is ilmenite, $FeTiO_3$, which is related structurally to α-Al_2O_3, $2\cdot 6PO_{2/3}$. The cations fill two-thirds of the O sites with the oxygens *hcp*. The perovskite structure is encountered for large M^{II} cations, which occupy the P sites; the ilmenite structure is favored if both cations have comparable sizes. The spinel ($MgAl_2O_4$) or $18PO_{3/4}PT_{1/4}O_{1/4}T_{1/4}$ structure is found in complex salts of the

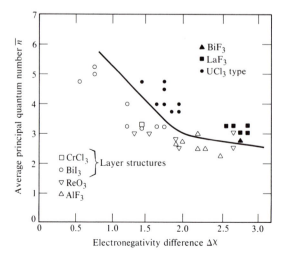

Figure 6.15 The structures of AX_3 compounds. (From E. Mooser and W. B. Pearson, *Acta Cryst.* 1959, *12*, 1015.)

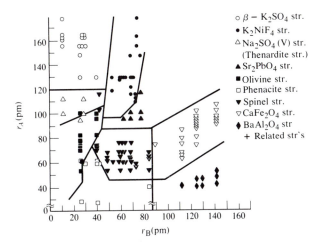

Figure 6.16 Structures of AB_2O_4 compounds. From J. C. Phillips, "Chemical Bonds in Solids" in *Treatise on Solid State Chemistry*, N. B. Hannay, Ed., Plenum, New York, 1973. Copyright © 1973 Bell Telephone Laboratories.

type $M^{II}M_2^{III}O_4$, $M^{IV}M_2^{II}O_4$, or $M^{VI}M_2^{I}O_4$, as well as in some sulfides, selenides, and fluorides of the same formula types. The P atoms (oxygen) are in a *ccp* pattern, with the Mg^{2+} ions in T sites and Al^{3+} ions in O sites for $MgAl_2O_4$, which is regarded as a *normal* spinel. In an *inverse* spinel, such as $MgFe_2O_4$ or $Fe(Mg\ Fe)O_4$, half of the Fe^{III} ions are in T sites, and the Mg^{II} and the remaining Fe^{III} ions are in the O sites. The formula $Fe(Mg\ Fe)O_4$ emphasizes the inverse roles of the cations. We often can predict normal- or inverse-type spinels on the basis of ionic radii (larger ions prefer the octahedral sites) and the preference of some cations for octahedral sites, because of ligand field effects (p. 270). The preference of Mn^{III} for octahedral sites results in a normal spinel for Mn_3O_4. Because of ligand field effects, Fe^{II} has a greater preference than Fe^{III} for octahedral sites and Fe_3O_4 is an inverse spinel $[Fe^{III}(Fe^{II}Fe^{III})O_4]$. The strong preference of Ni^{II} for octahedral sites leads to inverse spinels such as $NiFe_2O_4$, but Cr^{III} has a stronger preference for octahedral sites and gives normal spinels, even in competition with Ni^{II}.

In addition to the spinel structure, AB_2O_4-type compounds display several other structures.[4] The remarkable clustering in domains is shown by plotting the size of one cation against the other (Figure 6.16).

Often we can visualize complex structures more readily by recognizing their relationship to close-packed arrangements. Thus the K_2PtCl_6 structure (Figure 6.13) can be viewed as a *ccp* array of $PtCl_6^{2-}$ octahedra (each octahedron treated as a unit occupying a lattice site) with K^+ ions in *all* of the T sites, the antifluorite structure, $9PTT$. This is a common structure for compounds of the type $M_2M'X_6$. The K_2PtCl_4 structure is similar, replacing the octahedral $PtCl_6^{2-}$ by planar $PtCl_4^{2-}$ units, but the cell length perpendicular to the $PtCl_4^{2-}$ planes is shortened.

Layer Structures

Many salts of highly polarizing cations and easily polarizable anions (p. 197) have structures consisting of layers of cations sandwiched between layers of anions, but without additional

---

[4]For details of structures not covered here, see A. F. Wells, *Structural Inorganic Chemistry*, 4th ed., Oxford, Oxford, 1975.

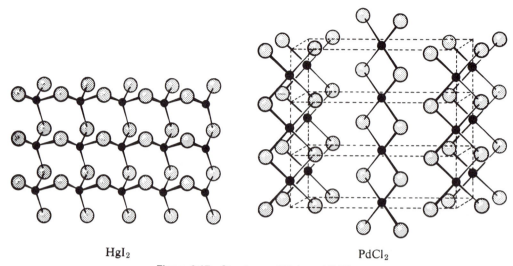

HgI₂   PdCl₂

Figure 6.17   Structures of HgI₂ and PdCl₂.

cations between the sandwiches. Such salts have relatively low melting points and low solubilities in polar solvents. The effects of polarization are carried further in $CuCl_2$ and $PdCl_2$, which form infinite chains by sharing Cl^- between planar MCl_4 groups (Figure 6.17).

We can view the $CdCl_2$ structure as *ccp* layers of Cl^- with Cd^{2+} filling O sites in alternate O layers, or $9PPO$. The triple Cl-Cd-Cl layers are stacked without cations between them. In effect, the negative charge of the Cl^- layers is polarized toward the Cd^{2+} layer to such an extent that the interaction between the Cl layer of one sandwich and the Cl layer of the next sandwich is comparable to the van der Waals interaction between CCl_4 molecules. Other salts with this structure are $MnCl_2$, $FeCl_2$, $CoCl_2$, $NiCl_2$, NiI_2, $MgCl_2$, and $ZnBr_2$. The CdI_2 structure (Figure 6.18) differs in that the I^- are *hcp*, so the structure is described as $2(3/2)PPO$. The index of the notation is written $2(3/2)$ to make it clear that while the packing arrangement is *hcp* (2), the total number of layers in the packing unit is three, not six $[(P_AP_BO_C)P_AP_BO_C \cdots]$. Other compounds with the CdI_2 structure are $CdBr_2$, $FeBr_2$, $CoBr_2$, $NiBr_2$, MgI_2, CaI_2, ZnI_2,

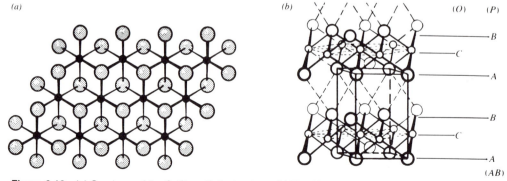

Figure 6.18   (a) One layer of the CdCl₂ or CdI₂ structure. (b) The 2(3/2)PPO structure (CdI₂ layer structure). The hexagonal unit cell is outlined by dark lines. The P layers are *hcp*.

Table 6.7 Some examples illustrating the filling of *P*, *T*, and *O* layers

Layers Occupied	Cubic Close Packing	Hexagonal Close Packing
P	Cu,*ccp*(3*P*)	Mg,*hcp*(2*P*)
PO	NaCl(6*PO*)	NiAs(4*PO*)
PPO	$CdCl_2(9PPO)$	$CdI_2(2 \cdot \frac{3}{2}PPO)$
	$CrCl_3(9PPO_{2/3})$	
PT	ZnS(6*PT*)	ZnS(4*PT*)
PTT	$CaF_2(9PTT)$	None
	$PtS(9PT_{1/2}T_{1/2})$	
	$SnI_4(3 \cdot 6PT_{1/8}T_{1/8})$	$Al_2Br_6(2 \cdot 6PT_{1/6}T_{1/6})$
PTOT	Na,*bcc*(3 · 2*PTOT*, *P* = *T* = *O*)	None with all *T* filled
	CsCl(3 · 2*PTOT*, *P* = *O*)	$Mg_2SiO_4(8PT_{1/8}O_{1/2}T_{1/8})$
	$BiF_3(12PTOT, T = O)$	—

PbI_2, MnI_2, FeI_2, CoI_2; hydroxides of divalent Mg, Ca, Cd, Mn, Fe, Co, and Ni; and sulfides of quadrivalent Zr, Sn, Ti, and Pt. The $CrCl_3$ layer structure resembles that of $CdCl_2$ with only one third of the octahedral sites filled: that is, with two thirds of the sites in alternate layers filled, $9PPO_{2/3}$.

The mercury(II) halides display a unique structural variation. Mercury(II) fluoride has the typically ionic fluorite structure. The $HgCl_2$ structure consists of discrete $HgCl_2$ molecules. Mercury(II) bromide gives a layer-type structure similar to that of $CdCl_2$, but the octahedral arrangement is distorted, with two Br atoms much closer than the other four. In the layer structure of HgI_2, each Hg atom is surrounded by four equivalent iodide ions (Figure 6.17). The HgI_2 structure can be regarded as a *ccp* arrangement of I^-, with Hg^{2+} ions in one half of the *T* sites in alternate layers (the other layer of *T* sites is empty).

Molybdenum disulfide and tungsten disulfide have layer structures in which each metal atom has six S neighbors at the apices of a trigonal prism, rather than the more common octahedral arrangement seen in $CdCl_2$ and CdI_2. The layer structure of MoS_2 accounts for the properties that make it a good solid lubricant. The commercial product (Moly-S) is similar to graphite in properties and appearance. As might be expected, the replacement of S by the more polarizable Se or Te improves the properties that make layer-structured sulfides good lubricants.

Looking at inorganic crystal structures in relation to close packed arrangements aids in recognizing the coordination number and local symmetry of each atom or ion and in seeing relationships among structures. Table 6.7 summarizes the sequence of filling *O* and *T* sites for some simple inorganic substances in cubic and hexagonal close-packed arrangements.

6.2   IONIC RADII

6.2.1   Methods of Evaluating Ionic Radii

The radius of an isolated atom or ion has little meaning. It might be taken as infinite. Since "ionic radius" refers to the distance of closest approach by another ion, the radius commonly is evaluated from the observed distance between centers of nearest neighbors.

Several approaches have been used to evaluate the individual ionic radii. Lande's method for obtaining the radius of the anion assumes that relatively large anions will be in contact with one another when packed around a small cation such as Li^+ in LiI (see Figure 6.19c). The radius of the iodide ion is taken as half of the I—I internuclear distance. In crystals such as KI, where the I—I distance is greater than that found in LiI, it is assumed that the anion and cation are in contact with each other. From the known radius of I^- (that is, from LiI) and the internuclear distance in KI, we may calculate the radius of K^+. Similarly, an estimate of the ionic radius of the oxide ion is obtained from the O—O distance in silicates.

Another approach by Pauling gives a set of *univalent radii* based on four alkali halides— NaF, KCl, RbBr, and CsI. In order to partition the internuclear distance between the individual ions, (1) the cation and anion are assumed to be in contact, that is, $r_+ + r_- = d$ ($d =$ internuclear separation); (2) for a given noble-gas configuration, the radius is assumed to be inversely proportional to the effective nuclear charge, $r = C/Z_{eff}$, where C is the proportionality constant. The effective nuclear charges are evaluated by subtracting a screening constant from the actual nuclear charge. For the lighter elements the screening constants used by Pauling are similar to those obtained from Slater's rules (see p. 32). Once the proportionality constant has been evaluated, we can calculate the univalent radius for any isoelectronic ion.

Example Calculate the univalent radii for Na^+, F^-, and oxide ion.

Solution The internuclear distance in NaF is 231 pm, the sum of the ionic radii. Both Na^+ and F^- have the Ne configuration, for which the Slater screening constant is $(2 \times 0.85) + (8 \times 0.35)$ $= 4.5$, giving an effective atomic number for Na^+ of $11 - 4.5 = 6.5$ and for F^- of $9 - 4.5 = 4.5$. For ions with the Ne configuration, the proportionality constant $C = 614$, from

$$\frac{C}{6.5} + \frac{C}{4.5} = 231$$

The ionic radii are

$$r_{Na^+} = \frac{C}{Z_{eff}} = \frac{614}{6.5} = 95 \text{ pm} \qquad r_{F^-} = \frac{C}{Z_{eff}} = \frac{614}{4.5} = 136 \text{ pm}$$

A *univalent* radius for oxide ion (Ne configuration) is

$$r_0 = \frac{C}{Z_{eff}} = \frac{614}{8 - 4.5} = 176 \text{ pm}$$

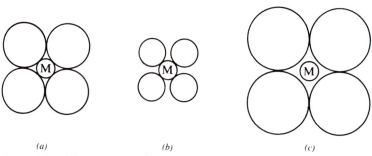

(a)   (b)   (c)

Figure 6.19 Effect of variation of the ratio of radius of cation to radius of anion.

Pauling's univalent radii agree quite well with observed interionic distances for salts containing univalent ions. However, if we attempt to calculate internuclear separation of O and Mg in MgO using the sum of the univalent radii, the value obtained (258 pm) does not agree with the observed Mg—O distance (205 pm). Although all the ions involved in the calculation are isoelectronic, the Mg—O distance is much shorter than expected, because of the higher charge on the ions. The calculated radii are *univalent radii*—the radii the ions would have if they had unit charge *without any change* in electronic configuration. We can convert the univalent radii into crystal radii, which will then agree with the observed interionic distances, by taking into account the charge on the ions. The ratio of the crystal radius, r_c, to the univalent radius, r_1, is given by the equation

$$\frac{r_c}{r_1} = Z^{-2/(n-1)} \tag{6.1}$$

where Z is the ionic charge and n is a constant, called the Born exponent, that has a particular integral value for each type of electronic configuration (see p. 221).

Just as the Mg—O distance is shorter than that calculated from univalent radii, the radius of a univalent ion is shorter when its neighbors have a charge greater than one. The correction for the charge of *neighboring ions* can be made by the relationship

$$r_{11} = r_{z_1 z_2}(Z_1 Z_2)^{1/(n-1)} \tag{6.2}$$

where r_{11} is the univalent radius (or the sum of univalent radii), $r_{z_1 z_2}$ is the radius (or the internuclear distance) in the salt with ions of charges Z_1 and Z_2, and n is the Born exponent. This equation reduces to Equation 6.1 for the special case in which $Z_1 = Z_2$ and $r_{z_1 z_2}$ is the crystal radius.

Ionic radii have been obtained also from electron density plots obtained from x-ray analysis. Taking the radius of each of two adjacent ions as the distance from the nuclear center (electron density maximum) to the electron density minimum between the ions yields radii that are self-consistent, but the radii obtained are larger for cations and smaller for anions than those from other sources. A scholarly review of methods for the evaluation of ionic radii is given by Waddington.[5]

6.2.2 Radius Ratio Effects

The ratios of (cation radius)/(anion radius) expected for various C.N. are given in Table 6.5. Ratios outside the ranges given are encountered without a change in structure, but in those cases the interionic distances no longer agree with the sum of the ionic radii. We can illustrate this phenomenon by considering a structure with C.N. 6. The spatial relationship can be seen most clearly in a sectional view through any plane of four anions around the central cation. The stable arrangement, where there is cation-anion contact and the anions are almost in contact, is shown in Figure 6.19a. If the radius ratio is larger (Figure 6.19b, larger cation or smaller anion), the decreased anion repulsion would produce an M—X distance shorter than the sum of the ionic radii. Actually, for an exaggerated situation such as that shown in Figure 6.19b, we would expect an increase in coordination number. If the radius ratio is smaller than

[5]T. C. Waddington, *Trans. Faraday Soc.* 1966, *62*, 1482.

in Figure 6.19a, the anion repulsion prevents anion-cation contact and the apparent M—X distance is increased beyond the sum of the ionic radii (Figure 6.19c). Where anion-anion repulsion is important, as for many lithium salts, we must apply a radius ratio correction to relate the ionic radii to the internuclear distance.

The alkali halides represent a series in which properties dependent on the stability of the ionic lattice should vary in a consistent way. The heats of fusion, melting points, boiling points, etc. might be expected to vary regularly with the M—X distance. These properties do follow a consistent pattern for salts of K^+, Rb^+, and Cs^+, but Li^+ salts and some Na^+ salts show significant deviations, as illustrated by the melting points in Figure 6.20. We can eliminate these irregularities by applying radius ratio corrections, to allow for the lower stability of the salts for which anion-anion repulsion is important. Similar corrections are needed to remove irregularities in the boiling points.

Some salts crystallize in two or more modifications differing in coordination number. Anion-anion repulsion is always greater for the higher coordination number, resulting in an increase in the cation-anion distance. Rubidium halides (RbCl, RbBr, and RbI) normally show C.N. 6 (NaCl structure) but adopt C.N. 8 (CsCl structure) at high pressures. We can calculate the radius for C.N. 8 (r_8) from the radius for C.N. 6 (r_6, as given in Table 6.8) by the equation

$$\frac{r_8}{r_6} = \left(\frac{8A_6}{6A_8}\right)^{1/(n-1)} \tag{6.3}$$

where n is the Born exponent and A_6 and A_8 are the Madelung constants (see p. 223) for the NaCl structure (C.N. 6) and CsCl structure (C.N. 8), respectively. The Madelung constants for these structures are almost the same, so that the ratio r_8/r_6 is approximately $(\frac{4}{3})^{1/(n-1)}$; for $n = 10$, $r_8/r_6 = (\frac{4}{3})^{1/9} = 1.032$.

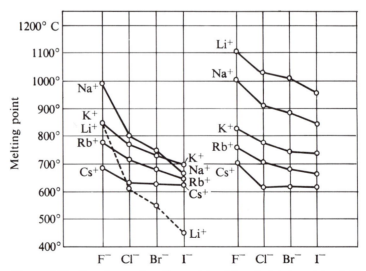

Figure 6.20 The observed melting points of the alkali halogenides (left) and values corrected for the radius-ratio effect (right). From Linus Pauling, *The Nature of the Chemical Bond*, 3rd ed., Cornell University, Ithaca, N.Y., Copyright © 1960 by Cornell University. Used by permission of Cornell University Press.

6.2.3 Crystal Radii

Several sets of crystal radii have been used. Each set is self-consistent, but because of the different approaches employed, the sets differ sufficiently from each other to prevent combining radii from different sets. The most comprehensive compilation of ionic radii is that of Shannon and Prewitt, as revised and extended by Shannon.

The crystal radii for ions with different coordination numbers are given in Table 6.8, and those for ions of variable oxidation number are given in Table 6.9. The paper by Shannon includes some other oxidation states and coordination numbers—in some cases, for C.N. as

Table 6.8 Crystal radii (pm)[a]
(the positive oxidation number is the same as the group number except as noted for actinides)

Coordination Number	I	II	III	IV	V	VI	VII		I	II	III	IV	V	VI	VII
	Li	Be									B	C	N		
3	—	30									15	6	4.4		
4	73	41									25	29	—		
6	90	59									41	30	27		
	Na	Mg									Al	Si	P	S	Cl
4	113	71									53	40	31	26	22
6	116	86.0									67.5	54.0	52	43	41
8	132	103									—	—	—	—	—
	K	Ca	Sc	Ti	V	Cr	Mn		Cu	Zn	Ga	Ge	As	Se	Br
4	151	—	—	56	49.5	40	39		74	74	61	53.0	47.5	42	39
6	152	114	88.5	74.5	68	58	60		91	88.0	76.0	67.0	60	56	53
8	165	126	101.0	88	—	—	—		—	104.0	—	—	—	—	—
	Rb	Sr	Y	Zr	Nb	Mo	Tc		Ag	Cd	In	Sn	Sb	Te	I
2	—	—	—	—	—	—	—		81	—	—	—	—	—	—
4	—	—	—	73	62	55	51		114 / 116^{Sq}	92	76	69	—	57	56
6	166	132	104.0	86	78	73	70		129	109	94.0	83.0	74	70	67
8	175	140	115.9	98	88	—	—		142	124	106	95	—	—	—
	Cs	Ba	La	Hf	Ta	W	Re		Au	Hg	Tl	Pb	Bi	Po	At
4	—	—	—	72	—	56	52		—	110	89	—	—	—	—
6	181	149	117.2	85	78	74	67		151	116	102.5	79	90	81	76
8	188	156	130.0	97	88	—	—		—	128	112	91.5	—	—	—

	La	Ce	Pr	Nd	Pm	Sm	Eu	Gd	Tb	Dy	Ho	Er	Tm	Yb	Lu
6	117.2	115	113	112.3	111	109.8	108.7	107.8	106.3	105.2	104.1	103.0	102.0	100.8	100.1
7	124	121	—	—	—	116	115	114	112	111	—	108.5	—	106.5	—
8	130.0	128.3	126.6	124.9	123.3	121.9	120.6	119.3	118.0	116.7	115.5	114.4	113.4	112.5	111.7

	Ac^{III}	Th^{IV}	Pa^{V}	U^{VI}	Np^{IV}	Pu^{IV}	Am^{IV}	Cm^{IV}	Bk^{IV}	Cf^{IV}				No^{II}	
6	126	108	92	87	101	100	99	99	97	96.1				124	
8	—	119	105	100	112	110	109	109	107	106					

[a]Values from R. D. Shannon, *Acta Cryst.* 1976, *A32*, 751. The reference includes other oxidation states and coordination numbers.

Table 6.9   Crystal radii (pm)[a] for positive oxidation numbers, C.N. 6 (except as designated in *italics*)

Oxida-
tion
Number

	Ti	V	Cr	Mn	Fe	Co	Ni	Cu	Zn	Ga	Ge	As	Se
II	100	93	94 87LS	97.0 81LS	77(4) 92 75LS	72(4) 88.5 79LS	69(4) 63(4)Sq 83	71(4) 71(4)Sq 87	74(4) 88.0		87		
III	81.0	78.0	75.5	78.5 72LS	78.5 69LS 63(4)	75	74 70LS	68LS		76.0		72	
IV	74.5	72	69	67	72.5	67	62LS				67.0		64

	Zr	Nb	Mo	Tc	Ru	Rh	Pd	Ag	Cd	In	Sn	Sb	Te
II	—	—	—	—	—	—	78(4)Sq	93(4)Sq	109	—	122[b](8)	—	—
III	—	86	83	—	82	80.5	90	81(4)Sq	—	94.0	—	94(5)	—
IV	86	82	79	78.5	76.0	74	75.5	—	—	—	83.0	—	111

	Hf	Ta	W	Re	Os	Ir	Pt	Au	Hg	Tl	Pb	Bi	Po
I	—	—	—	—	—	—	—	151	133 111(3)	164	—	—	—

	Hf	Ta	W	Re	Os	Ir	Pt	Au	Hg	Tl	Pb	Bi	Po
II	—	—	—	—	—	—	74(4)Sq	—	116	—	133	—	—
III	—	86	—	—	—	82	—	82(4)Sq	—	102.5	—	117	—
IV	85	82	80	77	77.0	76.5	76.5	—	—	—	91.5	—	108

Ce^{IV}	Pr^{IV}	Sm^{II}	Eu^{II}	Tb^{IV}	Tm^{II}	Yb^{II}
101	99	141(8)	131	90	117	116

Pa^{IV}	U^{IV}	Np^{VI}	Pu^{VI}	Am^{III}	Cm^{III}	Bk^{III}	Cf^{III}
104	103	86	85	111.5	111	110	109

[a]Values from R. D. Shannon, *Acta Cryst.* 1976, *A32*, 751. Low-spin values (LS) and values for square planar (Sq) coordination are designated by superscripts.
[b]Value for C.N. 8 from R. D. Shannon and C. T. Prewitt, *Acta Cryst.*, 1969, *B25*, 925. The value is probably doubtful, since it was not included in the revised tabulation (footnote a).

high as 12 or 14. Table 6.10 lists crystal radii for anions, with van der Waals radii of the noble gases for comparison. Table 6.11 lists radii for some complex ions.

6.3   LATTICE ENERGY

6.3.1   Born Equation

We can determine the stability of an ionic lattice from the coulombic interactions among the ions. A pair of ions of opposite charge attract one another, and the potential energy varies inversely with the first power of d, the internuclear separation. (The force between the ions is $Z_1 Z_2/d^2$.) As the ions come very close together, they repel one another because of interpene-

Table 6.10 Crystal radii (pm for C.N. 6)[a]

	OH⁻	H⁻	He
	123	153[b]	(93)[c]
N^{3-}	O^{2-}	F^-	Ne
132	126	119	(112)
(C.N. 4)	S^{2-}	Cl^-	Ar
	170	167	(154)
	Se^{2-}	Br^-	Kr
	184	182	(169)
	Te^{2-}	I^-	Xe
	207	206	(190)

[a]R. D. Shannon, *Acta Cryst.* 1976, *A32*, 751.
[b]D. F. C. Morris and G. L. Reed, *J. Inorg. Nucl. Chem.* 1965, *27*, 1715.
[c]Van der Waals radii of the noble gases are given for comparison.

Table 6.11 Radii of some complex anions[a]

BeF_4^{2-}	CO_3^{2-}	NO_3^-				
245	185	189				
BF_4^-						
228						
		PO_4^{3-}	SO_4^{2-}	ClO_4^-	CrO_4^{2-}	MnO_4^-
		238	230	236	230	240
		AsO_4^{3-}	SeO_4^{2-}		MoO_4^{2-}	
		248	243		254	
		SbO_4^{3-}	TeO_4^{2-}	IO_4^-		
		260	254	249		
$TiCl_6^{2-}$	$IrCl_6^{2-}$	SiF_6^{2-}	$GeCl_6^{2-}$			
248	254	194	243			
$TiBr_6^{2-}$	$PtCl_6^{2-}$	GeF_6^{2-}	$SnCl_6^{2-}$			
261	259	201	247			
$ZrCl_6^{2-}$			$PbCl_6^{2-}$			
247			248			

[a]Radii for oxo anions, BeF_4^{2-}, and BF_4^- are "thermochemical" radii from T. O. Waddington, *Adv. Inorg. Chem. Radiochem.* 1959 *1*, 157; and A. F. Kapustinskii, *Quart. Rev.* 1956 *10*, 283. Radii for MX_6^{2-} in compounds with the antifluorite structure are from R. H. Prince, *Adv. Inorg. Chem. Radiochem.* 1979 *22*, 349.

tration of electron clouds. The repulsion energy is inversely proportional to the *n*th power of *d*. The Born exponent, *n*, increases with an increase in electron density around the ions (*n* = 5 for the He configuration, 7 for Ne, 9 for Ar or Cu^+, 10 for Kr or Ag^+, and 12 for Xe or Au^+; an average value for *n* is used if the cation and anion have different configurations).

The Born equation gives the potential energy (PE) for a pair of ions where Z_1 and Z_2 are

$$\text{PE} = \frac{Z_1 Z_2 e^2}{d} + \frac{be^2}{d^n} \qquad (6.4)$$

integral charges (with appropriate signs), e is the charge on the electron, d is the internuclear separation, n is the Born exponent, and b is a repulsion coefficient. The potential energy is negative (corresponding to the release of energy) when d is greater than the internuclear separation, d_0, since the first term is negative for ions of opposite charge. The repulsion term increases to become dominant for very small values of d. The potential energy is at a minimum at d_0 and becomes positive when d is very small.

Equation 6.4 gives the energy released when a cation and anion, separated by an infinite distance, are brought together until they are separated by the distance d. In a crystal of NaCl, however, each Na^+ is surrounded by six Cl^- at a distance d, not by one. Twelve other Na^+ are located at a distance of $\sqrt{2}d$, eight other Cl^- at $\sqrt{3}d$, six more Na^+ at $2d$, 24 Cl^- at $\sqrt{5}d$, 24 Na^+ at $\sqrt{6}d$, etc. (see Figure 6.21). Summing all of the interactions gives the first term in the potential-energy expression for the energy released in bringing a sodium ion from infinity to its stable position in the NaCl lattice.

$$PE_{(1)} = -\frac{6e^2}{d} + \frac{12e^2}{\sqrt{2}d} - \frac{8e^2}{\sqrt{3}d} + \frac{6e^2}{2d} - \frac{24e^2}{\sqrt{5}d} + \frac{24e^2}{\sqrt{6}d} \cdots$$

or

$$PE_{(1)} = -\frac{e^2}{d}\left(6 - \frac{12}{\sqrt{2}} + \frac{8}{\sqrt{3}} - \frac{6}{2} + \frac{24}{\sqrt{5}} - \frac{24}{\sqrt{6}} \cdots\right) \qquad (6.5)$$

The product Z_1Z_2 is omitted for simplicity, since it is unity in this case; only the appropriate signs are carried. Equation 6.5 is the sum of an infinite series converging toward 1.747558, which is the *Madelung constant*[6] for the NaCl (6PO) structure. This Madelung constant is used for any salt with the 6PO structure, since it depends only on the geometrical arrangement of the ions. Values of the Madelung constant for other structures are evaluated similarly. Table 6.12 provides the values for some common structures.

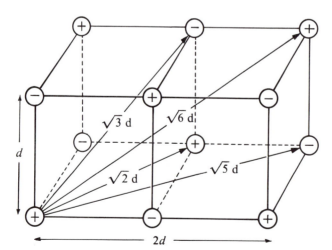

Figure 6.21 Distances to neighboring ions in the NaCl lattice.

[6]E. L. Burrows and S. F. Kettle, *J. Chem. Educ.* 1975, *52*, 58.

Table 6.12 Values of Madelung constants[a]

Structure	Madelung Constant	Structure	Madelung Constant
Sodium chloride	1.74756	Rutile (TiO_2)	4.816
Cesium chloride	1.76267	Anatase (TiO_2)	4.800
Zinc blende (ZnS)	1.63806	Cadmium iodide	4.71
Wurtzite (ZnS)	1.64132	β-Quartz (SiO_2)	4.402
Fluorite (CaF_2)	5.03878	Corundum (Al_2O_3)	24.242

[a]Values given are for Z defined as the highest common factor of the ionic charges. Values differ for compounds that have variations of the rutile, β-quartz, and corundum structures.

The Madelung constant values given here are those usually available in the literature—"conventional" values. If Z^2 in Equation 6.6 is replaced by Z_+Z_- (actual charges on the ions), the purely geometric Madelung constant is the same as the conventional value for NaCl and for ZnS (MX type), but is 2.51939 for CaF_2 and 4.040 for Al_2O_3. See D. Quane, *J. Chem. Educ.* 1970, 47, 396.

The second term in the potential-energy expression for ions in a real crystal allows for the repulsive forces resulting from the interpenetration of electron clouds. It is simpler to handle than the first term, since the repulsion varies inversely with d^n and hence only nearest neighbors need be considered. Each sodium ion is surrounded by six chloride ions, so the repulsion energy term becomes $6be^2/d^8$. The number of nearest neighbors times b is designated by B, the repulsion coefficient.

The potential energy of an ion in a crystal, considering the forces of all neighboring ions, becomes

$$\text{PE} = -\frac{Ae^2Z^2}{d} + \frac{Be^2}{d^n} \tag{6.6}$$

where A is the Madelung constant, B is the repulsion coefficient, and Z is the highest common factor of the ionic charges (1 for NaCl, Na_2O, $MgCl_2$, Al_2O_3, etc.; 2 for MgO, TiO_2, etc.). An additional repulsion term must be added where anion-anion repulsion is unusually great, as with many lithium salts.

An ion achieves its most stable equilibrium position when attractive and repulsive forces are balanced; then, the PE is at a minimum and $d = d_0$. Differentiating the PE with respect to d and equating to zero results in

$$d_0 = \left(\frac{nB}{AZ^2}\right)^{\frac{1}{(n-1)}} \tag{6.7}$$

and, solving for B,

$$B = \frac{d_0^{n-1}AZ^2}{n} \tag{6.8}$$

By substituting the value of B in Equation 6.6, the potential energy becomes

$$(PE)_0 = \frac{Ae^2Z^2}{d_0}\left(\frac{1}{n} - 1\right) \tag{6.9}$$

The lattice energy, U_0, is defined as the energy released in the formation of a mole of MX (crystal) from the gaseous ions separated from each other by infinite distances ($U_0 = -(PE)_0 N$, where N is Avogadro's number).

$$U_0 = \frac{NAe^2Z^2}{d_0}\left(1 - \frac{1}{n}\right) \tag{6.10}$$

U_0 is defined as positive; ΔH for the process is negative. The lattice energy is very useful in correlating properties of ionic substances, because the formation or the destruction of the crystal is frequently the most important step in reactions involving ionic substances.

The Born treatment assumes that the ions are hard spheres undistorted by the neighboring ions. Since increasing distortion corresponds to increasing covalent character, the lattice energy calculated from Equation 6.10 should agree poorly with the experimental value where the covalent character of the bonds is appreciable (see p. 229).

6.3.2 The Born-Haber Cycle

Lattice energies, which until recently could not be obtained by direct measurement, can be evaluated using the Born-Haber cycle, which relates the lattice energy to other thermochemical quantities. The formation of a solid salt (MX) from the elements by two different paths is formulated as

$$
\begin{array}{ccc}
M(g) + X(g) & \xrightarrow{\; I + E(-) \;} & M^+(g) + X^-(g) \\
{\scriptstyle S}\uparrow {\scriptstyle \frac{1}{2}D} & & \downarrow {\scriptstyle U_0(-)} \\
M(c) + \frac{1}{2}X_2(g) & \xrightarrow{\; \Delta H_f(-) \;} & MX(c)
\end{array}
$$

where U_0 is the lattice energy, I the ionization energy of $M(g)$, E the electron affinity of $X(g)$, D the heat of dissociation of $X_2(g)$, S the sublimation energy of the metal, and ΔH_f the heat of formation of MX(c) from the elements. The negative signs after ΔH_f, E, and U indicate that these processes correspond to the release of energy (as written). Since the change in energy is independent of the path, ΔH_f can be equated to the algebraic sum of the other thermochemical quantities with the appropriate thermodynamic signs.

We thus can determine the lattice energy from thermochemical quantities that can be evaluated experimentally. In early applications, the lattice energy was calculated from Equation 6.10 and used with the experimental thermochemical quantities to evaluate electron affinities, for which experimental values were not available. Where electron affinities are available, an experimental U_0 can be obtained from a Born-Haber cycle and compared with the theoretical value from Equation 6.10. Agreement usually is very good, except for salts containing ions of high charge or pseudo–noble gas type cations, for which the bonding has appreciable covalent character. The experimental value of U_0 usually exceeds appreciably the theoretical value for salts such as HgS, HgSe, and PbO_2.

By comparing the heats of formation of the alkali halides (Table 6.13), we can see that we must consider all of the thermochemical quantities involved in order to explain the variations. The heats of formation (ΔH_f) decrease in magnitude through the series MF, MCl, MBr, and MI, as anion repulsion increases and the coulombic energy decreases because of the increasing interionic distances. The relative contributions of the lattice energy (U_0), dissociation energy *(D)*, and electron affinity *(E)* vary through the series. As M increases in size, the ΔH_f values increase in magnitude for the chlorides, bromides, and iodides—corresponding to a decrease in anion repulsion—but decrease for the fluorides. Anion repulsion is at a minimum for F⁻ because of its small size. The lattice energy of fluorides is greater with small cations, because of the short M—X distance. The sublimation energy *(S)* and ionization energy *(I)* of the metal and the lattice energy are the major factors determining the trends.

Table 6.13 Thermochemical data for the alkali metal halides (kJ/mole)

	$-\Delta H_f{}^a$ 298 K (MX)	S 298 K (M)	$\frac{1}{2}D$ 298 K	I (M)	E (X)	U_0 298 K	$\Delta H_1{}^b$	U_0 0 K	$U_0{}^c$ (Theo.)
LiF	616.9	160.7	78.9	520.5	328.0	1049.0	−6	1043	966
NaF	573.6	107.8	78.9	495.4	328.0	927.7	−5	923	885
KF	567.4	89.2	78.9	418.4	328.0	825.9	−3	823	786
RbF	553.1	82.0	78.9	402.9	328.0	788.9	−2	787	730
CsF	554.7	77.6	78.9	375.3	328.0	758.5	−1	757	723
LiCl	408.3	160.7	121.3	520.5	348.8	862.0	−5	857	809
NaCl	411.1	107.8	121.3	495.4	348.8	786.8	−2	785	752
KCl	436.7	89.2	121.3	418.4	348.8	716.8	−1	716	677
RbCl	430.5	82.0	121.3	402.9	348.8	687.9	0	688	651
CsCl	442.8	77.6	121.3	375.3	348.8	668.2	+1	669	622
LiBr	350.2	160.7	111.8	520.5	324.6	818.6	−4	815	772
NaBr	361.4	107.8	111.8	495.4	324.6	751.8	−1	751	718
KBr	393.8	89.2	111.8	418.4	324.6	688.6	0	689	650
RbBr	389	82.0	111.8	402.9	324.6	661	+1	662	629
CsBr	395	77.6	111.8	375.3	324.6	635	+2	637	600
LiI	270.1	160.7	106.8	520.5	295.4	762.7	−2	761	723
NaI	288	107.8	106.8	495.4	295.4	703	0	703	674
KI	327.9	89.2	106.8 ·	418.4	295.4	646.9	+1	648	615
RbI	328	82.0	106.8	402.9	295.4	625	+1	626	594
CsI	337	77.6	106.8	375.3	295.4	602	+2	604	569

[a] I. Barin and O. Knacke, Eds., *Thermochemical Properties of Inorganic Substances,* Springer-Verlag, West Berlin, 1973.
[b] ΔH for changing M(g) and X(g) at 298 K to 0 K (−12 kJ) plus ΔH for changing MX(c) at 298 K to 0 K (6 to 15 kJ) See D. Cubicciotti, *J. Chem. Phys.* 1959, *31*, 1646, and 1961, *34*, 2189; as well as D. A. Johnson, *Some Thermodynamic Aspects of Inorganic Chemistry,* Cambridge Press, Cambridge, 1968, App. 2, and J. L. Holm, *J. Chem. Educ.* 1974, *51*, 460.
[c] Lattice energy calculated using Equation 6.10.

Considering the individual steps for the formation of MX(c) from the elements, the only quantities corresponding to the release of energy are the lattice energy and the electron affinity. The electron affinity (the energy involved in adding an electron to a gaseous atom in its lowest energy state to form a gaseous anion) for the addition of a single electron is greater for each of the halogens than for the member of the oxygen family in the same period, because of the greater nuclear charge, the smaller size, and the formation of a closed electronic configuration (see Table 1.8 for electron affinities). The very small size of the F atom, with its high electron density, makes adding an electron to it slightly less favorable energetically than for Cl. Fluorine is a better oxidizing agent than chlorine partly because of the much lower dissociation energy of F_2, which more than compensates for the difference in electron affinity. In addition, the very high hydration energy of F^- and the high lattice energy of fluorides contribute greatly to the energy released in the formation of fluorides in solution or as solids.

The fact that the bond energy of F_2 is much lower than that of Cl_2 is not surprising in view of the low single-bond energies (in kJ/mole) for O (142) and N (167) as compared with S (268) and P (239). The low values for N, O, and F may result from the greater repulsion among the nonbonding electrons of the two small atoms with such high electron densities. The other members of the families, moreover, have vacant d orbitals of the same quantum number, which might be hybridized with the p orbitals to some extent to contribute to the bonding and reduce repulsion. The increased polarizability of the larger atoms also helps decrease repulsion.

The formation of the oxide ion and other anions with a charge greater than one is an endothermic process. Energy is released as the first electron is gained by a gaseous oxygen atom, but adding the second electron requires the expenditure of a greater amount of energy, because it is repelled by the negative charge on the O^- ion. Because of the sulfur atom's greater size and lower charge density, the electron affinity for the formation of S^{2-} from S is less endothermic than that for the formation of O^{2-} from O. For oxides and salts of other simple $2-$ or $3-$ ions, the lattice energy is the only quantity involved in the formation of the solid that releases energy. The only simple $3-$ ion encountered in ionic crystals is the nitride ion—and then only when combined with the most electropositive metal ions. Forming the N^{3-} ion from N would require 2200 kJ/mole.

6.3.3 Oxygen Compounds of the Alkali Metals

The oxygen compounds of the alkali metals demonstrate the delicate balance of the energy terms involved in the formation of the compounds. Burning the metals in air produces the compounds Li_2O, Na_2O_2, KO_2, RbO_2, and CsO_2, which contain O^{2-}, O_2^{2-} (peroxide ion), and O_2^- (superoxide ion), respectively (Table 6.14).

The heats of formation and ΔG^0 values decrease from Li to Cs for each type of compound except the superoxides, for which the values vary only within the limits of uncertainty. A major factor in establishing the order observed for M_2O and M_2O_2 is the decrease in lattice energy with increasing size of M^+.

The lattice energies decrease for each metal as the size of the anion increases. From the peroxides to the superoxides, an even greater decrease accompanies the decrease in charge of the anion. In spite of a decrease in lattice energy, the ΔH^0 values increase in the order oxide–peroxide–superoxide for each metal. The only other thermochemical quantities involved

Table 6.14 Thermodynamic data for oxocompounds of the alkali metals (kJ for two moles of M, 25°C)

		M_2O	M_2O_2	$2MO_2$
Li[a]	$-\Delta H^0$	598.73	632.62	—
	$-\Delta G^0$	562.12	571.0	—
	Lattice type	9PTT (antifluorite)	Hexagonal	—
	$U_0{}^b$	2830	2460	—
Na[a]	$-\Delta H^0$	418.0	513.2	521.33
	$-\Delta G^0$	379.1	449.7	437.4
	Lattice type	9PTT	Hexagonal	
	$U_0{}^b$	2650	2190	1690
K[a]	$-\Delta H^0$	363.2	495.80	565.68
	$-\Delta G^0$	322.1	429.78	496.98
	Lattice type	9PTT	Rhombic	Tetragonal CaC_2
	$U_0{}^b$	2250	1980	1540
Rb	$-\Delta H^0$	330[c]	435[d]	569[d]
	$-\Delta G^0$	290	360	439
	Lattice type	9PTT	Rhombic	Tetragonal CaC_2
	$U_0{}^b$	2170	1880	1500
Cs	$-\Delta H^0$	318[c]	405[d]	635 ± 30[c]
	$-\Delta G^0$	280	330	594
	Lattice type	anti-$CdCl_2$	Rhombic	Tetragonal CaC_2
	$U_0{}^b$	2090	1780	1490

[a]JANAF Thermochemical Tables.
[b]Born-Haber cycle.
[c]I. Barin and O. Knacke, Eds., *Thermochemical Properties of Inorganic Substances,* Springer-Verlag, West Berlin, 1973; Supplement, 1977.
[d]M. K. Karapet'yants and M. L. Karapet'yants, *Thermodynamic Constants of Inorganic and Organic Compounds,* Ann Arbor-Humphrey, Ann Arbor, Mich., 1970.

in the Born-Haber cycle that change in such a series (for the same metal) are those involved in the formation of the anion. Adding two electrons to O_2 to form the O_2^{2-} ion requires less energy than adding two electrons to O to form O^{2-}, because in the former case the charge can be distributed over the two atoms. The addition of a single electron to O_2 is an exothermic process. In addition, the formation of O_2^{2-} or O_2^- does not require the dissociation of O_2. The formation of the O^{2-} ion (from O) is endothermic to the extent of 623 kJ/mole, whereas forming the O_2^{2-} ion from O_2 requires *ca.* 470 kJ/mole and forming the O_2^- ion releases 42.5 kJ. The energies of formation of the anion differ most significantly for the larger cations, where the changes in the lattice energy (which are in the opposite direction) are relatively small.

Metals (K, Rb, and Cs) that form superoxides upon burning in air have significantly greater ΔH° and ΔG° values for MO_2 (per mole of M) than for M_2O or M_2O_2. The differences between ΔH° and ΔG° values for Na_2O_2 and NaO_2 are small, the small difference between the ΔG° values being consistent with the observed formation of Na_2O_2 by combustion. Thermodynamic values for Li_2O and Li_2O_2 do not differ greatly (particularly ΔG°), but the order does not agree with the observation that burning Li in air produces Li_2O, with little Li_2O_2 being formed. The conditions prevailing at the site of the reaction during the combustion of a metal are likely to be very much different from those conditions to which the thermodynamic data refer. If equilibrium conditions do not prevail, we cannot reliably predict the products when the free-energy values are nearly the same.

6.3.4   Which MX_n Compounds Should Be Stable?

The steps required to form a gaseous metal ion involve sublimation energy, to convert the solid metal in its standard state to the separate gaseous atoms, and ionization energy *(I)*—the energy required to remove the outermost electron from the gaseous atom (p. 34). The energy needed to remove the second, third, etc. electrons corresponds to the second, third, etc., ionization energies, respectively, and the energy needed to form M^{2+} is the sum of the first and second ionization energies. The periodic variation in *I* indicates that the elements most likely to form simple cations (those with low *I* values) in ionic compounds will be found at the beginning of each period, and for the members of a family with higher atomic number.

The very great increase in successive ionization energies is the major factor limiting the charge of cations in simple ionic substances to $1+$ or $2+$, with $3+$ uncommon and $4+$ encountered only for very large ions, such as Th^{4+}. The other factor involves the increased covalent character of compounds containing ions of high charge. As Equation 6.10 makes clear, the lattice energy increases greatly with increasing ionic charge. We might expect the lattice energy of a salt such as NaF_2 to be very large (the value for MgF_2 can be taken as an estimate), but if we calculate the energy for the reaction

$$NaF_2 \rightarrow NaF + \tfrac{1}{2}F_2$$

we find it to be highly exothermic, primarily because of the very high second ionization energy of Na. Similarly, although the formation of MgF should be an exothermic process (the lattice energy of MgF should be about the same as that of NaF), the much greater lattice energy of MgF_2 makes the formation of solid MgF_2 much more favorable, and the disproportion of the solid MgF to give MgF_2 and Mg is exothermic.[7]

Example   Calculate the lattice energy, the heat of formation, and the heat of disproportionation of CaCl.

Solution   Assume that CaCl has the NaCl structure. The crystal radius of Ca^{2+} is 114 pm and its univalent radius is 118 pm. Use 120 pm as the approximate radius of Ca^+. The value of *e* is 4.80 $\times 10^{-10}$ esu, or $\sqrt{dyne}$ cm. Here the cgs units offer an advantage, and we can convert ergs to kJ easily.

$$U_0 = \frac{(4.80 \times 10^{-10}\ \sqrt{dyne}\ cm)^2\ 1^2(6.02 \times 10^{23})1.75}{2.9 \times 10^{-8}\ cm}\left(1 - \frac{1}{9}\right)$$

$$= \frac{7.44 \times 10^{12}\ erg}{mole} \times \frac{1\ kJ}{10^{10}\ erg} = 744\ kJ/mole$$

Using $S = 201$ kJ, $\tfrac{1}{2}D = 121$ kJ, $I = 589$ kJ, and $E = 349$ kJ

$$\Delta H_f = 201 + 121 + 589 - 349 - 744 = -182\ kJ/mole\ CaCl.$$

[7]For other examples, see J. L. Holm, *J. Chem. Educ.* 1974, *51*, 460.

For the disproportionation reaction,

$$2\ CaCl \rightarrow CaCl_2 + Ca$$

$$\Delta H = -799 + 0 + 2(182) = -435 \text{ kJ/mole } CaCl_2 \text{ formed.}$$

The rate of evaporation of Al from liquid Al can be greatly increased by passing a stream of $AlCl_3$ over the surface. Upon cooling, the AlCl formed and carried along in the gas stream disproportionates to deposit Al. This process has been patented (Gross, 1946) for the purification of Al.

The low heat of formation of some noble metal halides, as compared with alkali metal halides, results largely from high ionization energies, but frequently the differences in sublimation energy also are great. Consider the steps of the Born-Haber cycle that differ for NaCl and AgCl (energies in kJ/mole).

$$Ag(c) \xrightarrow[S]{284} Ag(g) \xrightarrow[I]{727} \begin{matrix} Ag^+(g) \\ Cl^-(g) \end{matrix} \xrightarrow[-U_0]{-910} AgCl(c) \quad \Delta H_f = -127 \text{ kJ/mole}$$

$$Na(c) \xrightarrow[S]{108} Na(g) \xrightarrow[I]{495} \begin{matrix} Na^+(g) \\ Cl^-(g) \end{matrix} \xrightarrow[-U_0]{-787} NaCl(c) \quad \Delta H_f = -411 \text{ kJ/mole}$$

Although the lattice energy is greater for AgCl, the heat of formation is much less. Both the ionization energy and sublimation energy contribute significantly to the "nobility" of Ag as compared with Na. (See p. 244 for a discussion of periodic variations in sublimation energies.) AgCl's heat of formation would be even lower were it not for the importance of polarization effects. Silver chloride is one of the compounds (p. 224) for which the polarization effects (or increased covalent character) cause the lattice energy obtained from the Born-Haber cycle (910 kJ/mole) to be appreciably larger than that calculated using the Born equation and the ionic radii (735).

6.3.5 Stabilization of Complex Ions by Large Counter Ions[8]

To isolate a compound such as MIF_4 (p. 589), you would not try to make $LiIF_4$, because the reaction

$$LiIF_4 \rightarrow LiF + IF_3$$

is strongly favored by the high lattice energy of LiF and the low lattice energy of $Li^+(IF_4)^-$. The IF_4^- ion is more stable in a solid with a large cation such as Cs^+ or $N(C_2H_5)_4^+$. The nitrate ion is a weak bidentate ligand (p. 325) for which complexes such as $Co(NO_3)_4^{2-}$ and $Zr(NO_3)_6^{2-}$ can be isolated only with large cations. The value of n in polyhalide complexes MX_n^{x-} commonly decreases as the size of the halide ion increases. Hexafluorides are much more common than hexachlorides or hexabromides, with the latter more likely to be isolated as salts of large cations.

The stabilizing effect of the large counter ion can take several forms.

Lattice Energy Limiting The decomposition of LIF_4 is favored because of the great gain in lattice energy in the formation of LiF. With a cation large enough for cation–cation contact in the MIF_4 structure, the lattice energy would favor MIF_4 over MF.

[8]D. H. McDaniel, *Ann. Reports Inorg. Gen. Syntheses* 1972, 293; F. Basolo, *Coord. Chem. Rev.* 1968, *3*, 213.

Polarization Effects   Small cations may polarize some complex ions so much that they are pulled apart.

Insulating Effects   Ions, such as SbF_4^-, that tend to form polyanions can be isolated and stabilized by large cations in the solid.

6.4 DEFECT STRUCTURES

Thus far, we have treated crystals as perfectly repeating arrays of atoms. But crystals are far less perfect than has been assumed. Imperfections in crystals arise from dislocation of ions, ion vacancies in the lattice, or nonstoichiometric proportion of the ions present—or simply from foreign ions or ''impurities'' in the lattice.[9]

6.4.1 Schottky Defects

Stoichiometric crystals may display two types of lattice defects. Vacancies of anions and cations, which are equal in number, are termed Schottky defects. Figure 6.22 shows a single-layer sketch representation of a sodium chloride crystal containing Schottky defects, along with a schematic representation.

6.4.2 Frenkel Defects

Frenkel defects occur when an ion occupies an interstitial site, leaving its normal site vacant. These defects are most likely in crystals in which the anion and cation differ greatly in size, so that the smaller ion can fit into the interstitial sites. Frenkel defects in AgBr are illustrated in Figure 6.23.

The large amount of heat released in the formation of a crystal lattice from gaseous ions is accompanied by a large unfavorable entropy change, because of the formation of the rigid well ordered crystal lattice. The introduction of vacancies or dislocations in the crystal lattice certainly will decrease a crystal's heat of formation, but the increase in entropy accompanying the decrease in order favors the process. Obviously, a balance will be achieved between entropy

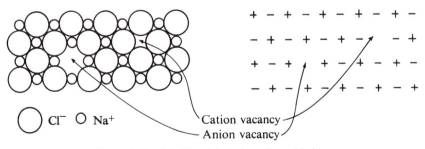

Figure 6.22   Schottky defects in sodium chloride.

[9]M. E. Fine, ''Introduction to Chemical and Structural Defects in Crystalline Solids,'' Chapter 5 in *Treatise on Solid State Chemistry*, Vol. 1, N. B. Hannay, Ed., Plenum Press, New York, 1973; R. Rohmer, *Bull. Soc. Chim. France* **1955**, 159; H. W. Etzel, *J. Chem. Educ.* 1961 **38**, 225.

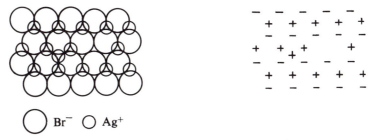

Figure 6.23 Frenkel defects in silver bromide.

and enthalpy, and defects will be more abundant in crystals where the enthalpy needed to create a defect is low. Defects are less likely for crystals of the extreme ionic type, where the balance of coulombic forces is critical. Frenkel defects, in particular, are more likely to occur in partially covalent substances, where polarization helps decrease the effects of charge dislocations. Pure stoichiometric crystals exhibit both Frenkel and Schottky defects, because of a balance in enthalpy and entropy terms. Stoichiometric CrO has 8% vacancies in the anion and cation sites.

6.4.3 Solid Electrolytes

The low-temperature modification (β) of AgI has the wurtzite (4*PT*) structure and is a poor conductor of electricity—which is generally true for solid salts. At 145.8°C there is a sharp transformation to α-AgI, which has a body-centered cubic arrangement of I^-, accompanied by a tremendous increase in ionic conductivity. Such substances are referred to as *solid electrolytes,* since their conductivities are closer to those of liquid electrolytes than those of ordinary solid salts. The Ag^+ ions in α-AgI (which melts at 555°C) can move from their normal sites through positions of lower coordination number. Other good solid electrolytes, such as $RbAg_4I_5$ and $[(CH_3)_4N]_2Ag_{13}I_{15}$, have passageways of face-sharing tetrahedra, fewer than half of which are occupied by Ag^+. At 50.7°C a phase transition forms the solid electrolyte α-Ag_2HgI_4, which has a *ccp* arrangement of anions and passageways formed by face-sharing tetrahedra and octahedra. The passageways generally are zigzag—except for $(C_5H_5NH)Ag_5I_6$ (pyridinium salt), which has an essentially *hcp* arrangement of I^- (at *A* and *B* positions). The pyridinium ions are located at sites centered about *C* positions, and all octahedral holes for Ag^+ are lined up along the *C* directions (see p. 205), which form straight channels for motion of Ag^+. This motion is limited by the number of vacancies, but vacancies can be created by the movement of Ag^+ ions from octahedral holes into empty tetrahedral holes. The structure is ordered at −30°, with Ag^+ occupying octahedral sites and only one of the two sets of tetrahedral sites; but the Ag^+ positions become disordered at room temperature and above.[10] Solid electrolytes generally display Frenkel defects and can be expected for large, highly polarizable anions and highly polarizing cations of low charge.

Some new high-energy batteries (p. 568) use nonaqueous systems requiring *(a)* high temperature, in the case of liquid Na-liquid S or liquid Li-metal sulfide with a molten electro-

[10]S. Geller, *Science* 1972, *176*, 1016.

lyte; *(b)* solid electrolytes; or *(c)* nonaqueous solvents for the electrolyte for ambient temperature batteries. One of the solid electrolytes used is β-alumina,[11] which is not actually an isomorph of Al_2O_3. Based on x-ray diffraction studies, the empirical formula of β-alumina is $Na_2O \cdot 11Al_2O_3$. Usually, the β-alumina used contains ~25% more Na_2O than indicated by the formula. Beta-alumina has a layer structure featuring four close-packed layers of oxide ions and Al^{3+} in octahedral and tetrahedral sites. The Na^+ occupy planes between the quadruple oxide layers; these planes contain equal numbers of loosely packed Na^+ and O^{2-} ions. An excess of this ratio of Na^+ ions is accompanied by Al^{3+} vacancies, to maintain charge balance. Adding MgO or Li_2O creates a stabilized form referred to as β″-alumina. Both β- and β″-alumina have high ionic conductivity for Na^+ and can be used as solid electrolytes for the liquid Na-liquid S battery.

6.4.4 Nonstoichiometric Compounds

Nonstoichiometric compounds often result when a cation can have different oxidation states, as in FeS, FeO, Cu_2O, NiO, CuO, CuI, etc. These might better be represented by formulas such as $Fe_{1-\delta}S$, where δ represents the fraction of vacant cation sites per formula weight. Electrical neutrality can be maintained by oxidizing the equivalent number of Fe^{2+} to Fe^{3+}, giving the composition $Fe^{2+}_{1-\delta}Fe^{3+}_{2\delta/3}S^{2-}$. The lattice of $Fe_{1-\delta}S$ may be represented as shown in Figure 6.24. The structures of FeO, γ-Fe_2O_3, and Fe_3O_4 all display the *ccp* array of oxide ions. In the case of FeO, which has the $6PO$ or NaCl structure, the octahedral sites are up to 95% filled, giving the limiting stoichiometric ratio $Fe_{0.95}O$. The "mixed" oxide Fe_3O_4 is an inverse spinel ($18PO_{3/4}PT_{1/4}O_{1/4}T_{1/4}$), with 32 oxide ions in the unit cell, eight Fe^{2+} in O sites, and 16 Fe^{3+} in T and O sites. On average, $21\frac{1}{3}$ Fe^{3+} for the same unit cell (32 oxide ions) are distributed randomly among the O and T sites. Interconversion between γ-Fe_2O_3 and Fe_3O_4 is accomplished easily. The structure of α-Fe_2O_3 is the same as that of α-Al_2O_3 (corundum, $2 \cdot 6PO_{2/3}$, based on an *hcp* scheme), with only O sites occupied by Fe^{3+}.

John Dalton interpreted stoichiometric relationships as resulting from the combination of discrete atoms. Berthollet believed that compounds had variable composition. Unfortunately, the beautiful crystals of minerals, considered to represent perfection, were used to test the opposing views. Minerals commonly have variable composition, because of isomorphous substitution of ions. The existence of nonstoichiometric compounds was an obstacle for the acceptance of atomic theory. The term *daltonide* is now used for stoichiometric compounds, and *berthollide* for nonstoichiometric compounds.

Nonstoichiometric compounds may occur even if the metal is not present in two or more oxidation states—as in the case of zinc oxide, which loses oxygen on heating to give $Zn_{1+\delta}O$. Sodium chloride reacts with sodium vapor to give $Na_{1+\delta}Cl$. In these cases an electron or an electron pair may occupy the vacant anion site (Figure 6.25).

Square-planar metal complexes can stack in the solid so as to set up direct M–M interaction along a chain.[12] Partially oxidized Pt(II) complexes such as $K_2Pt(CN)_4X_{0.3} \cdot 3H_2O$ (X = Cl⁻

---

[11]G. C. Farrington and J. L. Briant, *Science* 1979, *204*, 1371; J. T. Kumer, *Prog. Solid State Chem.* 1972, *7*, 141, and *Inorg. Synth.* 1979, *19*, 51.

[12]J. S. Miller and A. J. Epstein, *Prog. Inorg. Chem.* 1976, *20*, 1; *Inorg. Synth.* 1979, *19*, 1.

Fe^{3+}	S^{2-}	Fe^{2+}	S^{2-}		Na^+	Cl^-	Na^+	Cl^-	Na^+
S^{2-}	☐	S^{2-}	Fe^{3+}		Cl^-	☐	Cl^-	Na^+	Cl^- Schottky defects
Fe^{2+}	S^{2-}	Fe^{2+}	S^{2-}		Na^+	e	Na^+	Cl^-	Na^+ "trapped electron" or "F" center

Figure 6.24 Defects in FeS.   **Figure 6.25** Defects in NaCl.

or Br^-) stacked in this way provide metallic conduction along the direction of M–M interaction—a one-dimensional metal. Cation-deficient complexes such as $K_{1.75}[Pt(CN)_4] \cdot 1.5H_2O$ behave similarly. The compounds $Hg_{2.86}AsF_6$ and $Hg_{2.91}SbF_6$ are anisotropic metallic conductors that contain infinite linear chains of mercury atoms in two mutually perpendicular directions.

The dichalcogenides, MX_2, of the transition-series groups IV, V, and VI have layer structures (p. 213). Since the bonding is weak between the layers of X^{2-} where the cation sites are vacant, other species can fit between these layers to form *intercalation compounds*.[13] Layer structures involving octahedral coordination of M (TiS_2) and trigonal prismatic coordination of M (Nb, Ta, Mo, and W disulfides) are important for intercalation. Intercalation compounds are formed with NH_3, pyridine, various amines and amides, alkali metal hydroxides, and many metals. The metal sulfides capable of forming intercalation compounds reversibly with Na or Li are of interest as solid-state cathodes capable of replacing S in the Na-S or Li-S batteries. The reversible reactions are

$$x\,Na^+ + x\,e + TaS_2 \rightleftarrows Na_x TaS_2$$

$$x\,Li^+ + x\,e + TiS_2 \rightleftarrows Li_x TiS_2$$

Among other materials of interest as solid-state electrodes are metal oxides with cavities such as that occupied by Ca^{2+} in the perovskite structure (p. 205); these include the vanadium oxides V_2O_5, V_6O_{13}, and VO_2. Bronzes with the composition $Li_x V_2O_5$ have been obtained from the reaction of V_2O_5 and LiI in acetonitrile.[14] The acetonitrile serves as a solvent and a mild reducing agent: for each Li^+ incorporated in the lattice, one vanadium(V) is reduced. Such oxides are of interest as cathodes for nonaqueous lithium batteries.

6.4.5 Semiconductors

In addition to chemical defects, impurities can cause imperfections in the lattice. In germanium containing a trace amount of Ga, the Ga goes into the Ge lattice (diamond structure) but lacks one electron, thus creating an electron hole. Movement of electrons into this hole is the equivalent of movement of the hole about the lattice. When placed in an electric field, such a hole behaves as if it were a positive charge. A sample of Ge "doped" with Ga is called a

[13]F. R. Gamble and T. H. Geballe, "Inclusion Compounds," Chapter 2 in *Treatise on Solid State Chemistry*, Vol. 3, N. B. Hannay, Ed., Plenum, New York, 1976; G. C. Farrington and J. L. Briant, *Science* 1979, *204*, 1371.

[14]D. W. Murphy *et al.*, *Inorg. Chem.* 1979, *18*, 2800.

p-type semiconductor, because it behaves as if there were mobile positive centers in it. In contrast, a sample of Ge "doped" with As adds electrons to the conduction band of Ge (see the section on metals) and is called an *n*-type semiconductor, since it behaves as if there were negative charge carriers in it (as indeed there are) (see p. 246). $Fe_{1-\delta}S$ and $Fe_{1-\delta}O$ are *p*-type semiconductors, since electron transfer from Fe^{2+} to Fe^{3+} makes the Fe^{3+} appear to move through the lattice. $Na_{1+\delta}Cl$ and $Zn_{1+\delta}O$, on the other hand, are *n*-type semiconductors, because of the mobility of the "trapped electrons."

Occasionally, impurities are added to reduce rather than promote semiconduction—as in the case of TiO_2 ceramic insulators, where pentavalent impurities are added to reduce the effect of Ti^{3+} ions.

In addition to applications to semiconductors (transistors, etc.), lattice defect compounds are important in catalysis, luminescence (especially color TV), and photography, and they are being studied with regard to radiation damage in materials and corrosion phenomena.

6.5 SOLUBILITIES OF IONIC SUBSTANCES

Ionic substances are only very slightly soluble in most common solvents, except for those that are quite polar. The strong attractive forces between ions in the crystal must be overcome, and this can be accomplished only if the attractive forces between the ions and solvent molecules are at least comparable to the lattice energy. The energy necessary for the separation of the ions from the crystal comes from the solvation of the ions. The attractive forces between nonbonded neutral molecules are usually very weak, and the forces between an ion and a neutral molecule are not much stronger unless the molecule has a fairly high dipole moment and/or high polarizability. The greater the dipole moment, the stronger the attraction by an ion and, usually, the greater the solvation energy. Solvation energies also increase with the polarizability of the solvent molecules. The energy required to separate ions or keep them apart is diminished by a decrease in the forces between ions, which are dependent on the *dielectric constant*[15] of the medium. Since the dielectric constant of water is 78.54 (25°C), the attractive forces between two ions in water is 1/78.54 of the force between the ions separated by the same distance in a vacuum. Good solvents for ionic substances usually have high dipole moments and high dielectric constants—although few solvents have dielectric constants as high as that of water.

Water is a much better solvent for ionic substances than other solvents with high dielectric constants (e.g., HF). We could view water molecules as donating electron pairs to the metal ions, but for many metal ions the interaction can be interpreted quite satisfactorily in terms of electrostatic forces (see the Ligand Field Theory, p. 268). According to either interpretation, a metal ion interacts strongly with a layer of water molecules referred to as the primary hydration sphere. Since the water molecules in the primary hydration sphere form stronger hydrogen bonds to other water molecules than they would otherwise, a secondary hydration sphere arises. The water molecules in the secondary hydration sphere, in turn, are hydrogen-bonded to other water molecules with hydrogen bonds weaker than those formed by water in the primary hydration sphere but stronger than in water itself. Anions—particularly oxoan-

---

[15]The dielectric constant of a medium is defined as the ratio of a condenser's (capacitor's) capacity with the medium between the plates to its capacity with the space between the plates evacuated.

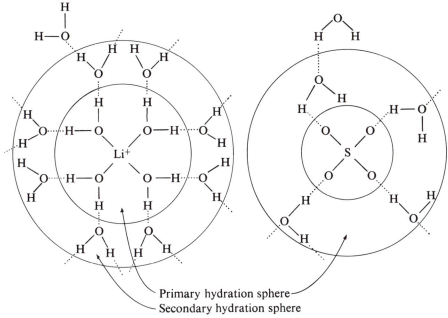

Figure 6.26 Hydrated Li^+ and SO_4^{2-} ions.

ions—also can be strongly hydrogen-bonded to water (see Figure 6.26). HF is a poorer solvent for salts, in spite of its high dipole moment and ability to form strong hydrogen bonds, because it is a much weaker base than water and hence a poorer electron donor.

The energy changes involved in the dissolution of a salt can be handled conveniently by a Born-Haber type cycle.

$$MX(c) \xrightarrow{-U_0} M^+(g) + X^-(g)$$

where U_0 is the lattice energy of the crystal MX (positive thermodynamic sign as shown), H_+ and H_- are the energies released (negative sign) as a result of the solvation of the gaseous positive and negative ions, and L is the observed heat of solution at infinite dilution. Since the total energy change in going from MX(c) to the solvated ions is independent of the path, the heat of solution is given by

$$L = H_+ + H_- - U_0$$

The heat of solution can be positive or negative, depending on the relative magnitude of the lattice energy and the heats of solvation of the ions. The heat of solution is often, but not always, negative (exothermic) for very soluble substances. The other factor that can account for high solubility—even though the process is endothermic—is the entropy change that accompanies dissolution. Accompanying the destruction of the well-ordered crystal lattice is a large favorable entropy change, but the ions orient the solvent molecules, causing a decrease in entropy. Using the thermodynamic relationships $\Delta G^0 = \Delta H^0 - T\Delta S^0$ and $\Delta G^0 = -RT \ln K$, the heat of solution, L, can be related to the solubility product constant, K, and the change in entropy, ΔS^0, when one mole of solute dissolves to give an ideal one molar solution.

$$RT \ln K = -L + T\Delta S^0 \tag{6.11}$$

where R is the gas constant and T the absolute temperature. Care must be exercised in obtaining thermodynamic quantities from solubility products, because of complications caused by non-ideal behavior and competing equilibria—complex formation, for example.

Equation 6.10 gives the effects of size and charge on lattice energy. The solvation energy, H, of an ion is given by the Born equation

$$H = -\frac{z^2}{2r}\left(1 - \frac{1}{\epsilon}\right) \tag{6.12}$$

where z is the charge on the ion, r the radius,[16] and ϵ the dielectric constant of the solvent. The solvation energies and solubilities of salts usually increase as the dielectric constant of the solvent increases. Both solvation energy and lattice energy are inversely proportional to the ionic radii. The solubilities of salts usually increase as the size of cations or anions increases, presumably because of more favorable entropy changes on solvation—and also, possibly, because of polarization effects in some cases. The charge effect usually is considerably greater than the size effect. With increasing ionic charge, the lattice energy increases more than the solvation energy whenever there is an accompanying increase in the Madelung constant. The entropy change accompanying the dissolution of a salt is usually less favorable for small ions, and particularly for ions of high charge. Ions of high charge density cause a great deal of ordering of the solvent molecules, as a result of solvation.

Equation 6.12 does not take into account polarization effects in evaluating solvation energies, and the greater polarization of cations with pseudo–noble-gas type configurations is allowed for only in the value of the Born exponent used in Equation 6.10 for the calculation of lattice energy. Solubility depends on the relative polarizability of the anion and of the solvent molecules. Unless the solvent molecules are easily polarizable, the solubility of the salt usually is low if the cation is strongly polarizing and the anion is easily polarizable. The solubilities of salts containing pseudo-noble-gas type cations—$AgCl$, $PbCl_2$, $HgCl_2$, for example—usually are lower in water than those containing noble-gas type cations—alkali and alkaline earth halides, for example—because of the greater anion polarization. Ammonia has a lower dipole moment than water but is more polarizable; hence, it is a poorer solvent than water for substances of the extreme ionic type, but better than water for salts of strongly polarizing cations and for salts of easily polarized anions. Salts of pseudo-noble-gas type cations are commonly more soluble in ammonia than in water, because of the greater polarizability and basicity of ammonia, giving stable complexes. The solubilities of the silver halides in liquid ammonia increase with increasing anion size (and polarizability), whereas the reverse order is observed in water.

---

[16]Latimer, Pitzer, and Slansky (*J. Chem. Phys.* 1939, *7*, 108) showed that the Born equation gives poor results unless an empirical adjustment of the radius is made. Reasonable results are obtained if the effective cationic radius is taken to be the crystal radius plus 70 pm and the effective anionic radius is 25 pm larger than the crystal radius. Probably the major factor requiring some adjustment is the expected decrease in ϵ in the immediate vicinity of an ion.

Equation 6.12 also neglects specific solvent interactions such as hydrogen bonding. The hydration energy of F^- is much higher than that of K^+, even though the ionic radii are about the same. The unusually high hydration energy of F^- (and OH^-) results from the formation of strong hydrogen bonds (see Chapter 5).

6.6 GIANT MOLECULES

Compounds of elements in the same family and of the same formula type sometimes differ strikingly in properties (see p. 60). Thus CO_2 is a gas at room temperature and pressure and the solid (Dry Ice) is easily broken into smaller pieces, yet SiO_2 is a very hard, dense, high-melting solid. Within a multiple-bonded CO_2 molecule the valence forces are satisfied, so only weak van der Waals forces hold the molecules together in the liquid or solid state. There are no discrete SiO_2 molecules, even at quite high temperatures. The larger size of Si (compared with C) dictates a higher coordination number, which, together with the decreased tendency of Si to form multiple bonds, leads to the formation of Si—O single bonds in tetrahedral SiO_4 units. The tetrahedra are linked together in a three-dimensional network by the sharing of each O between two Si. In a single crystal of quartz, SiO_2, all SiO_4 units are cross-linked into a single unit—one "molecule." Such substances are called *giant molecules* or *covalent crystals.*

Diamond provides a good example of a giant molecule. Here, each C atom is bonded tetrahedrally to four others. The structure is the same as that described for zinc blende (see Figure 6.9), except that all sites are occupied by carbon atoms. Diamond's extreme hardness stems from this three-dimensional network of covalent bonds. Carborundum, SiC, has the zinc-blende structure, and its great hardness makes it useful as an abrasive. Graphite features carbon atoms arranged in two-dimensional layers of hexagonal rings; with conjugated double bonds through each layer; each carbon has four bonds.

6.7 STRUCTURES OF SILICATES

6.7.1 Common Silicates

Silicates provide an interesting array of structural types that show greater variety than usually encountered for compounds of other elements. The silicates are also of great technical importance. The crust of the earth is made up primarily of metal silicates. Silicon and oxygen account for almost 75% by weight of the earth's crust, and on a volume basis, oxygen alone accounts for 92% of the earth's crust. Moreover, the eight most abundant elements in the earth's crust[17] usually occur in silicates. The ubiquitous silicates occupy a unique position in minerology.

The composition of the silicate minerals used to be given in terms of mole ratios of oxides: for example, forsterite, $2MgO \cdot SiO_2$ for Mg_2SiO_4; and orthoclase, $K_2O \cdot Al_2O_3 \cdot 6SiO_2$ for $KAlSi_3O_8$. The mixed oxide formulations were used partly because in many cases, the structures or manner of chemical combination were not known. Also, most minerals are not pure chemical compounds that can be represented accurately by simple chemical formulas. The formulas given are idealized; the actual composition might vary considerably.

---

[17]These eight elements account for 98.5% by weight of the crust: O, 46.6%; Si, 27.7%; Al, 8.1%; Fe, 5.0%; Ca, 3.6%; Mg, 2.1%; Na, 2.8%; and K, 2.6%.

Variations in the composition of minerals principally result from the isomorphous replacement of one ion by another. The extent of isomorphous replacement of ions of the same charge is determined by the relative sizes of the ions. If the sizes are very similar (within about 15%), complete isomorphous replacement can occur. Mg^{2+} (86 pm) and Fe^{2+} (92 pm) can combine in any proportion in a mineral of the type $M_2^{II}SiO_4$. If little iron is present the mineral is called forsterite, represented as Mg_2SiO_4. Less commonly, where little magnesium is present the mineral is called fayalite, Fe_2SiO_4. Most often the mineral encountered contains varying proportions of Mg and Fe and is known as olivine, represented by $(Mg,Fe)_2SiO_4$.

Ions of comparable size can replace one another even if the charges differ. The feldspars can be represented by the general formula $M(Al,Si)_4O_8$, where M can be Na, K, Ca, or Ba and the ratio of Si to Al varies from 3:1 to 2:2. For each Ca^{2+}, which substitutes for Na^+ in albite, $NaAlSi_3O_8$, an additional Al^{3+} ion substitutes for Si^{4+}. In the mineral anorthite, $CaAl_2Si_2O_8$, the substitution of Ca^{2+} for Na^+ is complete. Actually, the formulas for albite and anorthite are idealized, and the presence of small amounts of the other cation (Ca^{2+} or Na^+) does not necessitate changing the name of the mineral. Nevertheless, many minerals of intermediate composition ranges are characterized and given their own names.

The silicate minerals have a number of essential features that should be kept in mind. All silicates contain tetrahedral SiO_4 units, which may be linked together by sharing corners, but never by sharing edges or faces. When other cations (alkali or alkaline earth metal ions, Fe^{2+}, etc.) are present in the structure, they usually share oxygens of the SiO_4 groups, to give an octahedral configuration around the cation. Aluminum can replace Si in the SiO_4 tetrahedra, requiring the addition of another cation or the replacement of one by another of higher charge to maintain charge balance. Aluminum also can occupy octahedral sites.

6.7.2   Silicates Containing "Discrete" Anions

In discussing silicates the term "discrete ion" is used when the oxygen atoms are not shared between silicate anions. Silicate ions are not discrete in the sense of independent rotation, as are the perchlorate ions in alkali metal perchlorates at temperatures above *ca.* 300°C. The simplest discrete silicate anion is the orthosilicate ion, SiO_4^{4-}. In the orthosilicates forsterite, $Mg_2[SiO_4]$, and olivine, $(Mg,Fe)_2[SiO_4]$, stacking of the SiO_4 tetrahedra around the divalent cation produces an octahedral configuration. The structure is designated $8PT_{1/8}O_{1/2}T_{1/8}$, with oxide ions *hcp* (P layers), one-eighth of the T sites occupied by Si, and half of the O sites occupied by Mg. Orthosilicates (discrete SiO_4^{4-}) are the most compact of the silicate structures. Sharing oxygens between silicons actually opens up the structure.

The uncommon mineral phenacite, $Be_2[SiO_4]$, contains the very small Be^{2+} ions in tetrahedral sites. In zircon, $Zr[SiO_4]$, the large Zr^{4+} ion has C.N. 8. The garnets, $M_3^{II}M_2^{III}[SiO_4]_3$ where M^{II} is Ca, Mg, or Fe and M^{III} is Al, Cr, or Fe, have a more complex structure, in which the M^{II} ions have C.N.8 and the M^{III} ions have C.N. 6. Each oxygen is shared by one Si and three Mg in forsterite, by one Si and two Be in phenacite, by one Si and two Zr in zircon, and by one Si, one Al, and two Ca in garnet.

We might expect discrete anions containing short chains of SiO_4 tetrahedra, but these are rare. The $Si_2O_7^{6-}$ anion is encountered in thortveitite, $Sc_2[Si_2O_7]$, and in a few other minerals of greater complexity. There are very few examples of short chains containing more than two

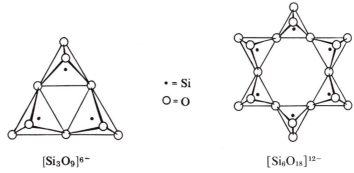

$[Si_3O_9]^{6-}$ • = Si   O = O $[Si_6O_{18}]^{12-}$

Figure 6.27 Cyclic silicate anions.

SiO_4 groups. Discrete anions consisting of rings of SiO_4 groups are encountered more commonly. Rings of three tetrahedra (to give six-membered rings) containing the anion $[Si_3O_9]^{6-}$ are encountered in wollastonite, $Ca_3[Si_3O_9]$, and benitoite, $BaTi[Si_3O_9]$. The anion $[Si_6O_{18}]^{12-}$, a ring of six tetrahedra, is found in the emerald, which is the mineral beryl, $Al_2Be_3[Si_6O_{18}]$ (Figure 6.27). In beryl an oxygen is shared by one Si, one Al (C.N. 6) and one Be (C.N. 4). One Al is substituted for an Si in the ring in cordierite, $Mg_2Al_3[AlSi_5O_{18}]$.

6.7.3 Silicates Containing Infinite Chains

Single Chains

Each SiO_4 tetrahedron can share two oxygens to form single chains of indefinite length (Figure 6.28). We can represent the anion by the formula of the repeating unit, SiO_3^{2-}. Minerals of this type are called the *pyroxenes,* which include enstatite, $Mg[SiO_3]$, diopside, $CaMg[SiO_3]_2$, and spodumene, $LiAl[Si_2O_6]$. Spodumene is an important lithium ore. The nonbridging oxygens are shared with Mg^{2+} (C.N. 6) in enstatite. In diopside the Mg has C.N. 6 and the Ca has C.N. 8. Both Li and Al are six-coordinate in spodumene.

Double Chains

A class of minerals known as the amphiboles contain double chains of SiO_4 tetrahedra (Figure 6.28) joined to form rings of six tetrahedra. The repeating unit is $Si_4O_{11}^{6-}$, with half of the silicon atoms sharing three oxygens with other Si atoms and half sharing only two oxygens with other Si atoms. The amphiboles always contain some OH^- groups associated with the metal ion. Tremolite, $Ca_2Mg_5[(OH)_2|(Si_4O_{11})_2]$, is a typical amphibole. The vertical bar used in the formula indicates that the OH is not a part of the silicate framework.

The asbestos minerals are related to the amphiboles, but their structures are more complex.

6.7.4 Silicates Containing Sheets

Each silicon atom can share three oxygen atoms with other Si atoms, to give large sheets. Cross-linking similar to that in the amphiboles, but extending indefinitely in two dimensions, produces rings of six Si in talc, $Mg_3[(OH)_2|Si_4O_{10}]$, and biotite, $K(Mg,Fe)_3[(OH)_2|AlSi_3O_{10}]$.

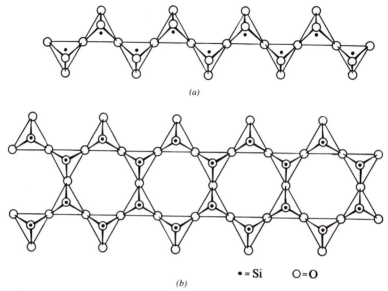

$\bullet$ = Si   O = O

Figure 6.28   Chains of SiO_4 tetrahedra in (*a*) pyroxenes and (*b*) amphiboles.

Biotite is one of the mica minerals. The SiO_4 (and AlO_4) tetrahedra form sheets of interlocking rings with the unshared oxygens pointed in the same direction. Two of these sheets are parallel with the unshared oxygens pointed inward. These oxygens and the OH^- ions are bonded to Mg^{2+} and Fe^{2+} between the sheets. Double sheets weakly bonded together by the K^+ ions account for the characteristic cleavage into thin sheets. The flashes of light reflected off granite often are from flat surfaces of biotite crystals.

Apophyllite, $Ca_4K[F|(Si_4O_{10})_2] \cdot 8H_2O$, contains sheets made up of SiO_4 tetrahedra linked to form alternating four- and eight-membered rings. In this case the sheets are not doubled because the oxygens not shared between Si atoms do not all point in the same direction. The K^+ and Ca^{2+} ions lie between the sheets associated with the oxygens uninvolved in the interlocking network of the sheets.

6.7.5   Framework Silicates

The partial replacement of Si by Al requires adding another cation in the framework silicates. The SiO_4, along with randomly distributed AlO_4, are linked by the sharing of all four oxygens, as in SiO_2. We encounter three groups of framework silicates: the feldspars; the zeolites (characterized by a very open structure that makes the minerals effective cation exchangers); and the ultramarines which have basketlike frameworks and are used as pigments because they commonly are colored.

The feldspars, comprising about two-thirds of the igneous rocks, are the most important of the rock-forming minerals. The general formula for the feldspars was given above (p. 238) as $M(Al,Si)_4O_8$. When the ratio of Si:Al is 3:1, M is Na or K; and when it is 2:2, M is Ca or Ba. The two classes of feldspars are based not on the charge on the cation, but on the crystal symmetry. The *monoclinic* feldspars, including orthoclase, $K[AlSi_3O_8]$, and celsian,

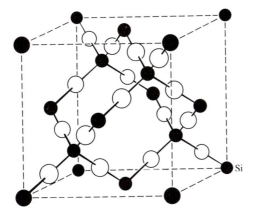

Figure 6.29 The β-cristobalite (SiO$_2$) structure.

Ba[Al$_2$Si$_2$O$_8$], contain the larger cations. Substitution of Na$^+$ for K$^+$ or Ca^{2+} for Ba^{2+} does not occur extensively. *Triclinic* feldspars are referred to as the plagioclase feldspars. We know of a number of minerals in this series varying in composition from albite, Na[AlSi$_3$O$_8$], to anorthite, Ca[Al$_2$Si$_2$O$_8$], corresponding to the isomorphous substitution of Ca^{2+} for Na$^+$ and Al^{3+} for Si^{4+}.

Silica

All crystalline forms of SiO$_2$ have three-dimensional framework structures with sharing of all four oxygens of the SiO$_4$ tetrahedra. Of the forms, quartz is stable to 870°, tridymite is stable from 870° to 1470°, and cristobalite is stable from 1470° to 1710° (m.p.). Each of these forms has a low (α) and high (β) temperature modification. Since transitions among the three are very sluggish, each of these polymorphic forms can be studied at temperatures outside its range of stability.

The structure of β-cristobalite (Figure 6.29) is related to diamond or zinc blende (6*PT*), with Si in all the C, or Zn and S, positions. The cubic close-packed oxygens lie midway between the Si atoms but slightly shifted away from the line joining the silicons. In the β-tridymite structure, which is related similarly to wurtzite (4*PT*; Figure 6.9), the α modifications involve slight rotations of the SiO$_4$ tetrahedra relative to one another. There are helices of linked tetrahedra in a crystal of β-quartz. The crystal is optically active, since all of the helices will be either right- or left-handed in a particular crystal. Although slight distortions occur among the linked tetrahedra in α-quartz, the crystals are optically active.

6.8 CRYSTAL STRUCTURES OF METALS[18]

Metal atoms in the solid state display three common arrangements: cubic close-packed (also called face-centered cubic), hexagonal close-packed, and body-centered cubic. As their names imply, the first two correspond to the closest packing of similar spheres. These are related to

[18]J. Donahue, *The Structures of the Elements,* Wiley, New York, 1974; S. M. Ho and B. E. Douglas, ''The Structures of the Elements in the *PTOT* System,'' *J. Chem. Educ.* 1972, *49,* 74.

the close-packed layers as shown in the drawings in Figure 6.30 (see also Figures 6.3 and 6.4). In both close-packed structures, each metal atom has 12 nearest neighbors. In the body-centered cubic structure (see Figure 6.30) each metal atom has eight nearest neighbors and six next-nearest neighbors only slightly (~15%) farther away. Although the *bcc* structure is only about 90% as dense as the close-packed structures, we can view it as a *ccp* structure expanded to accommodate like spheres in all of the tetrahedral and octahedral sites, 3·2*PTOT* (see p. 207).

The structures encountered for metals are shown in Table 6.15. Many metals adopt more than one structure, depending on temperature and pressure. The relatively small differences in the stabilities of the structures is illustrated by iron—the *bcc* structure is stable at low temperature, the *ccp* structure is more stable in the range 906–1400°C, but the *bcc* structure is again more stable from 1400°C to the melting point (1535°C).

The softness of a metal, as noted by its malleability (ability to be beaten or rolled into sheets) and ductility (ability to be drawn into wire), depends on the crystal structure, as well as on other factors such as bond strength. These metallic characteristics are shaped by the ability of layers to slide over one another to give equivalent arrangement of the identical spheres. The highly symmetrical *ccp* structure has slip planes of close-packed layers along four directions (corresponding to the three-fold axes—the body diagonals), as compared with only one direction for the *hcp* structure. Metallic characteristics are important in applications of the *ccp* metals Al, Cu, Ag, Au, Ni, Pd, Pt, Rh, Ir, and Pb. Impurities can create dislocations (local bonding), causing an increase in hardness. Soft metals become "work-hardened" because of dislocations created in the deformation of the metal and disruptions of slip planes, as can be seen in contrasting the ease with which you can bend new copper tubing to the greater

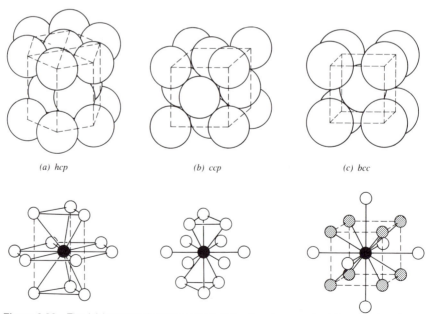

(a) hcp   *(b) ccp*   *(c) bcc*

Figure 6.30   The (a) hexagonal close-packed, (b) cubic close-packed (face-centered cubic), and (c) body-centered cubic structures. The nearest neighbors for each structure are shown below the crystal models.

Table 6.15 Crystal structures of metals[a]

3 Li **PTOT**[b] *2P* *3P*	4 Be **2P** *PTOT*	2P *hcp* 3P *ccp* PTOT *bcc* 3P′ *distorted ccp* 4P *double hexagonal, length of c axis is doubled* 9P *has ABABCBCA . . . sequence*											
11 Na **PTOT** *2P*	12 Mg **2P**											13 Al **3P**	14 Si **6PT**
19 K **PTOT**	20 Ca **3P** *PTOT*	21 Sc **2P** *PTOT*	22 Ti **2P** *PTOT*	23 V **PTOT**	24 Cr **PTOT**	25 Mn **PTOT** *3P*	26 Fe **PTOT** *3P*	27 Co **2P** *3P*	28 Ni **3P**	29 Cu **3P**	30 Zn **2P**	31 Ga *other*	32 Ge **6PT**
37 Rb **PTOT**	38 Sr **3P** *PTOT*	39 Y **2P** *PTOT*	40 Zr **2P** *PTOT*	41 Nb **PTOT**	42 Mo **PTOT**	43 Tc **2P**	44 Ru **2P**	45 Rh **3P**	46 Pd **3P**	47 Ag **3P**	48 Cd **2P**	49 In **3P′**	50 Sn **6PT**
55 Cs **PTOT**	56 Ba **PTOT**	71 Lu **2P** *PTOT*	72 Hf **2P** *PTOT*	73 Ta **PTOT**	74 W **PTOT**	75 Re **2P**	76 Os **2P**	77 Ir **3P**	78 Pt **3P**	79 Au **3P**	80 Hg *other*	81 Tl **2P** *PTOT*	82 Pb **3P**

57 La **4P** *2P* *3P* *PTOT*	58 Ce **2P** *3P* *4P* *PTOT*	59 Pr **4P** *2P* *3P* *PTOT*	60 Nd **4P** *3P* *PTOT*	61 Pm	62 Sm **9P** *PTOT*	63 Eu **PTOT**	64 Gd **2P** *PTOT*	65 Tb **2P** *PTOT*	66 Dy **2P** *PTOT*	67 Ho **2P** *PTOT*	68 Er **2P** *PTOT*	69 Tm **2P** *PTOT*	70 Yb **3P** *PTOT* *2P*
89 Ac **3P**	90 Th **3P** *PTOT*	91 Pa **PTOT** *3P*	92 U **other** *PTOT*	93 Np **other** *PTOT*	94 Pu **other** *PTOT*	95 Am **4P** *3P*	96 Cm **4P** *3P*	97 Bk **4P** *3P*	98 Cf	99 Es	100 Fm	101 Md	102 No

[a] Adapted from S.-M. Ho and B. E. Douglas, *J. Chem. Educ.* 1972, *49*, 74.
[b] Room-temperature structures in boldface.

difficulty encountered in bending it a second time. A few metals—for example, Sb, Bi, and α-Mn—are brittle. Directional bonding in Sb and Bi forms puckered layers, destroying any slip planes.

6.9 BONDING IN METALS

6.9.1 Valence Bond Approach

The large coordination numbers of metal atoms in the solid state vis-á-vis the number of valence electrons per atom indicates that the two-center valence bond cannot be of importance here. Nevertheless, the valence bond approach, using the resonance concept to bring in electron delocalization, provides a good qualitative explanation of many of the properties of metals and interstitial compounds.

Figure 6.31 illustrates the variation in cohesive energies displayed by the elements of the three long periods. The related properties, such as boiling points, melting points, heats of fusion, and hardness, show similar variation. The third long period (the rare earth elements are omitted) shows the greatest regularity in cohesive energies and may be interpreted as showing an increase in the number of metallic bonds per atom, from one in cesium to six in tungsten. Any further increase in atomic number and corresponding number of electrons does not increase the bonding but indeed decreases it, until a minimum is reached at mercury. Griffith based his rationalization of the cohesive energies of Figure 6.31 on the use of d and s orbitals for bonding until the d and s orbitals are filled completely in the isolated atom, with the s and p orbitals used in subsequent bonding. Forming a bond requires a half-filled orbital, which then pairs its electron through bond formation. According to this picture, the number of covalent bonds should reach a maximum with six electrons occupying the d and s orbitals. Further occupancy of these orbitals decreases the number of bonds formed. When the d orbitals are filled completely, s and p orbitals are used, and the number of covalent bonds reaches a maximum with four electrons in the s and p orbitals.

This picture also can be used to explain some of the trends within the groups of the periodic table. Thus the d and s orbitals have more closely matched energies in the later periods, and the reverse is true of the s and p orbitals. Accordingly, $d^n s$ hybridization gives stronger bonds in the later periods, whereas sp^n hybridization gives weaker bonds in the later periods.

The above view does not account for the magnetic properties of the metals. A picture presented by Pauling accomplishes this by assuming that of the nine orbitals in the valence level [the $(n - 1)d^5$, ns, and np^3], six may be used for bonding, 0.72 are kept empty for

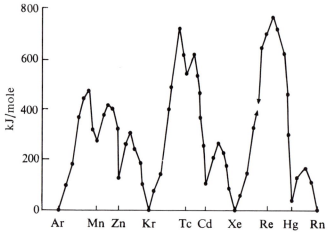

Figure 6.31 Cohesive energies in the three long periods (from J. S. Griffith, *J. Inorg. Nucl. Chem.* 1956, *3*, 15.) The energies are those required to produce a gaseous atom from the elements in their standard states.

metallic conduction, and the remaining 2.28 are nonbonding. The empirical value 0.72 orbitals per atom merely indicates that at a given instant not all atoms have equivalent electronic configurations. Nevertheless, the Pauling picture does not predict the cohesive energies as satisfactorily as the simpler picture: that is, it does not account for the bimodal shape of the cohesive energy curve in a period.

Structures of metals do not occur in perfectly regular periodic patterns—a situation complicated by the polymorphism of many metals. Nevertheless, a search for broad patterns (Table 6.15) reveals that the *ccp* structure occurs for filled or nearly filled electron subshells: Cu, Ni, and Co families; Lu; and Ca and Sr of group IIA. Also, we note that *bcc* is the primary structure for half-filled or nearly half-filled subshells: s^1 for IA, s^2d^5 or s^1d^5 near the center of each transition series, and f^7 for Eu. The *bcc* arrangement is the most common exclusive structure (in cases of no polymorphism), and also the most common secondary structure. The most common primary structure for intermediate electron configurations is *hcp*.

Of the likely *s-d-p* hybrid combinations, the most suitable for the symmetry of each of the structures have been examined. The relative importance of each of these is determined by the symmetry of equivalent sets of neighbors, the population of the orbitals, and the relative energies of the *s, d,* and *p* orbitals. There are two distinct sets of neighbors for the *bcc* structure—eight closest and six at about 15% greater distance. The specific bonding interaction should be most important in this case, where this is the preferred structure. Because of its lower density, it is a likely prospect for a high-temperature structure. In the ideal *hcp* arrangement, the 12 neighbors are equidistant but only the in-plane neighbors are arranged centrosymmetrically about a particular sphere. Hybrids of different symmetry might be used for bonding to the in-plane and out-of-plane neighbors, which are clearly distinguished along the unique packing direction. The minor distortions common in *hcp* metals cause the distances to the two sets of neighbors to differ. Since the centrosymmetric *ccp* arrangement has four equivalent "in-plane sets" of neighbors, distortions occur less commonly.

6.9.2 Band Theory of Metals[19]

The covalent bond approach used thus far has given us some understanding of the strength of bonding in metals and, to a lesser extent, of the lattice types found for different metals. This approach, however, has not yet accounted for the properties that allow us to distinguish metals from nonmetals: electrical conductivity, metallic luster, etc. These properties can be explained by the band theory of metals, which we shall now consider.

When atomic orbitals combine to form molecular orbitals, the energy levels of the original atomic orbitals split. Since the magnitude of this splitting depends upon the extent of overlap of the original atomic orbitals, it is strongly dependent upon the internuclear separations involved and the orientation of the atomic orbitals. In a metal, the combination of the *n* atomic orbitals of a given type gives rise to a series of molecular orbitals of closely spaced energies, called energy bands. Because of the low degree of orbital overlap, the band width for inner shell electrons is very small and the energy of the electrons in these bands is virtually the same

---

[19]It should be pointed out that although the approach used here starts with the individual atomic orbitals from which the molecular orbitals of the metal may be constructed, a more useful approach is to start with the electrons in the force field of an array of nuclei. Energy bands appear naturally, then, from particle in a box type calculations, with the inclusion of multiple barrier potentials. Although we do not deal with this theory here, the qualitative conclusions from a band picture are independent of the starting point from which the bands may be derived.

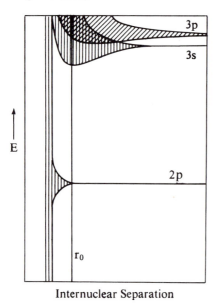

Internuclear Separation

Figure 6.32 Energy bands of sodium. (From J. C. Slater, *Introduction to Chemical Physics,* McGraw-Hill, New York, 1939, p. 494.)

as in the isolated atoms. For the outer or valence electrons, however, the energy range of electrons in a given band may be relatively large. The width of different bands may cause some overlap. These features are illustrated in Figure 6.32.

We can construct an energy-level diagram for a particular internuclear separation from Figure 6.32 by taking the band width and position corresponding to the internuclear separation, as shown in Figure 6.33. The energy-level diagram does not tell us the number of electrons of given energy for the system. Figure 6.34 shows the distribution function of electrons with energy in a given band. As electrons are added to a given metal lattice, the lowest energy levels are filled first. The spacing and filling of the bands determines whether a substance is a conductor, an insulator, or a semiconductor (metal, nonmetal, or semimetal). If the bands are completely filled or completely empty and the energy spacing between bands is large (Figure 6.34c), the substance will be an insulator. Only a substance with a partially filled band will serve as a conductor.

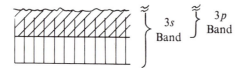

Figure 6.33 Energy bands of sodium at equilibrium distance between atoms.

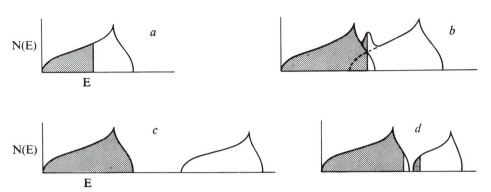

N(E)

E

N(E)

E

Figure 6.34 Distribution of energy states. The shading indicates occupied states.

Thus in Figure 6.34, *a* and *b* would show metallic conduction, *c* would be an insulator, and *d* would be a semiconductor. Figure 6.34*a* is typical of one-electron metals that would have exactly half the number of electrons the band could hold. Two-electron metals, such as Mg and Zn, would have enough electrons to fill completely the *s* band, were it not for overlap with the *p* band, which causes the *p* band to begin to fill before the *s* band is filled completely. For a semiconductor with an electron distribution similar to that shown in Figure 6.34*d*, some electrons may be excited thermally from the nearly filled band to the nearly empty band. Since the number of electrons so excited increases as the temperature rises, the conductivity of a semiconductor, unlike that of a metal, increases with increasing temperature. If the energy gap between the filled band and the empty band corresponds to the energy of the optical spectrum, electrons may be promoted from the filled band to the empty band by the absorption of light— a phenomenon called photoconductivity. Conductivity of a semiconductor may be controlled by the introduction of atoms from a neighboring periodic group (see p. 233).

6.10 ALLOYS

6.10.1 Solid Solutions and Intermetallic Compounds

If we add Zn to Cu to form brass, we will find different structure types, or phases, at different compositions. The temperature-composition phase diagram shown in Figure 6.35 illustrates the boundaries of existence of the phases; here, the new phases are designated as they appear from left to right by letters of the Greek alphabet.

The α phase of brass, being essentially a solid solution of Zn in Cu, has a face-centered cubic structure that goes over into a body-centered cubic structure for the β phase, a more complex cubic structure for the γ phase, and finally an *hcp* structure for the ϵ phase. Although there is a range of composition for these various phases, the formulas $CuZn$, Cu_5Zn_8, and $CuZn_3$ may be taken as the approximate composition of the β, γ, and ϵ phases, respectively. Hume-Rothery has pointed out that many intermetallic compounds have structures similar to β, γ, and ϵ brass at the same electron-to-atom ratio as the corresponding brass compounds. Some examples of these compounds, which are often referred to as *electron* compounds, are given in Table 6.16.

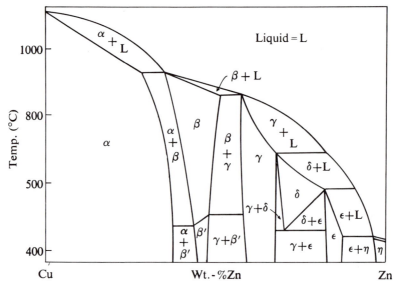

Figure 6.35 Phase diagram for the Cu-Zn system (From J. C. Slater, *Introduction to Chemical Physics*, McGraw-Hill, New York, 1939, p. 287.)

In counting the electron-to-atom ratio, the metals' valence electrons are considered to equal in number the group number in the periodic table—with the exception of the Group VIII metals, which fit into the scheme if they are thought to have no valence electrons for metallic bonding. Thus NiAl is considered to have an electron-to-atom ratio of 3:2, similar to β brass, CuZn. The existence of these similar binary metallic phases at similar electron-to-atom ratios can be best understood on the basis of the filling of electronic bands in such a fashion as to

Table 6.16 Hume-Rothery phases

β-Brass $\left(\dfrac{electron}{atom} = \dfrac{3}{2} \text{ or } \dfrac{21}{14}\right)$ (bcc)	$\gamma\left(\dfrac{21}{13}\right)$ (complex cubic)	$\epsilon\left(\dfrac{21}{12} \text{ or } \dfrac{7}{4}\right)$ (hcp)
CuZn	Cu_5Zn_8	$CuZn_3$
AgZn	Ag_5Zn_8	$AgZn_3$
AuZn	Cu_9Al_4	Ag_5Al_3
AgCd	$Cu_{31}Sn_8$	Cu_3Sn
Cu_3Al	$Na_{31}Pb_8$	Cu_3Si
Cu_5Sn	Rh_5Zn_{21}	
CoAl	Pt_5Zn_{21}	
FeAl	etc.	
NiAl		

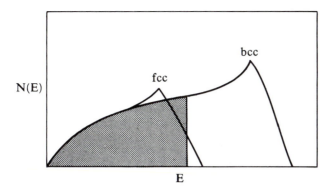

Figure 6.36 Distribution of energy states for an electron/atom ratio of 3/2 favoring the *bcc* structure.

give the lowest possible energy. Figure 6.36 illustrates the electron energy distributions corresponding to face-centered cubic and body-centered cubic lattices.

Although these electron compounds still have incompletely filled electron bands, a number of other intermetallic compounds with electron-to-atom ratios of 8:3 have completely filled bands. Depending on the energy separation of the next-higher empty band, the properties of these vary from insulators to conductors. Thus Mg_2Si is almost an insulator, whereas Mg_2Pb is a good metallic conductor. These compounds have the CaF_2 structure (see p. 205).

In many cases the radius ratio of the metal atoms determines the stoichiometry of the compound. Because numerous examples have been investigated by Laves, these compounds are often referred to as Laves phases. Thus the compounds Cu_2Mg, W_2Zr, KBi_2, and Au_2Bi have in common a relatively constant radius ratio of 1.25. Likewise, the series of compounds $CeFe_5$, $SmFe_5$, $DyFe_5$, $HoFe_5$, $CeCo_5$, $CeNi_5 - HoNi_5$ shares a constant radius ratio.

Since metals differ greatly in electronegativity, electrons may be transferred to the more electronegative species, permitting them to fill, or nearly fill, the bonding orbitals of a particular framework. Such compounds are called Zintl phases. The electropositive metal involved is usually an alkali metal or an alkaline earth metal, and the electronegative metal is usually a metal with an underlying d^{10} configuration. In NaTl, LiCd, and LiZn the atoms of the more electronegative metal display an arrangement similar to a diamond structure. NaTl has sufficient electrons to fill completely the electron bands for a diamond structure (that is, four per Tl), and the compound is colorless. With LiCd and LiZn, there are not enough electrons to fill completely the diamond structure and the compounds are colored. The compound $CaZn_2$ has a structure, which places Zn atoms in a graphitelike structure, with Ca atoms lying in the planes between sheets of the Zn network. Here it appears that Ca has transferred its electrons to Zn. There are some more complex Zintl phases with the formula AM_{13}, where A may be Na, K, Ca, Sr, or Ba with Zn, or K, Rb, or Cs with Cd.

Distinguishing between metals and nonmetals in compounds is often difficult. The description of MB_2 borides, where M = Al, Ti, Zr, V, Nb, or Ta, is similar to that of $CaZn_2$ (discussed above). The graphitelike layer apparently is made up of B_2^{2-} units. It is interesting that the electrical conductance of TiB_2 and ZrB_2 is greater than that of the metals themselves. The borides MB_4 (where M = Ce, Th, or U) are intermediate between the MB_2 structure and the MB_6 structure. In the MB_4 compounds, B_6^{2-} octahedra are joined through B_2^{2-} ethylenic-type units. The metal in MB_4 may thus be assumed to be M^{2+}.

6.10.2 Interstitial Compounds of Metals

Examining models of either *ccp* or *hcp* structures reveals the presence of voids or interstices of several different types (p. 203). Considering only two adjacent close-packed layers, these voids may be classified as either tetrahedral sites with C.N. 4 or octahedral sites with C.N. 6. For N atoms in a close-packed structure, there are $2N$ tetrahedral sites and N octahedral sites. Filling of the octahedral sites of a metal having a *ccp* structure with nonmetallic atoms leads to a NaCl-type structure (6*PO*). Although a *bcc* lattice also has octahedral and tetrahedral sites, in contrast to the close-packed structures the polyhedra describing these sites are slightly distorted, not regular. This distortion causes the tetrahedral sites for the *bcc* case to be slightly larger than the octahedral sites, again in contrast to the close-packed structures.

The geometry and number of the sites are particularly important in determining the stoichiometry and nature of many compounds formed by the transition metals and the second-period elements B, C, N, and O. Many compounds of the transition metals in Groups III through VI have the stoichiometry MX and an NaCl-type structure (see Table 6.17).

Table 6.17 Some binary transition-metal compounds having the NaCl structure (6*PO*)

ScN	TiC	TiN	TiO	VC	VN	VO	CrC[a]		CrN[a]
	ZrC	ZrN	ZrO	NbC	NbN	NbO	MoC[a]		MoN[a]
LaN	HfC	—	—	TaC	—	—	WC[a]		WN[a]
	ThC	ThN	ThO				UC	UN	UO

[a]*Hcp* structures for metal atoms in the compound (4*PO*).

These compounds retain several properties of metals—such as electrical conductivity, which decreases with increasing temperature, high thermal conductivity, and metallic luster—but unlike metals, they are brittle. Many of these compounds have very high melting points and are extremely hard. Rundle interpreted both the preference for the NaCl structure and the brittleness in terms of metal-nonmetal covalent bonding, which has a directional nature. Metal-nonmetal covalent bonding would also increase the overall bonding in the solid, despite the loss of some metal-metal bonding. Since each nonmetal atom is bonded to six nearest-neighbor metal atoms and has only four low-energy orbitals, a localized two-center valence bond description cannot be used. Instead, Rundle proposed p^3 bonding in which each lobe of the nonmetal p orbital overlaps with an orbital of the metal (see p. 164 for a MO treatment). Since a single p orbital can accommodate only two electrons and a single nonmetal orbital is being used to bond with two metal atoms, we call these bonds "half bonds."

You may notice that not all parent metals in Table 6.17 have a close-packed structure; presumably, these go over to such a structure in their interstitial compounds, at some slight expense to the metal-metal bonding that is more than compensated for by the metal-nonmetal bonding. Some of the metals have an *hcp* structure in the interstitial compounds. With these substances the metal orbitals are used efficiently in the metal itself, for bonding, and there is less net gain in forming an interstial compound. Beginning with the Cr group, a number of interstitial compounds is formed in which a lower number of possible sites are occupied—as in Mo_2C and W_2C. The series of compounds Fe_3C, Co_3C, Ni_3C, Mn_3C, and Fe_2MnC represents an occupancy of trigonal prismatic voids with a low occupancy of the voids. These compounds react with water to produce methane, hydrogen, and trace amounts of other hydrocarbons. Still lower occupancy is shown in Fe_4C, Cr_4C, Fe_4N, and Mn_4N, in which one fourth of the octahedral sites of a *ccp* structure are occupied.

Transition-metal hydrides tend to have the limiting formulas LaH_3, HfH_2, and TaH; experimental values such as $LaH_{2.76}$ and $TaH_{0.76}$ are typical. The limiting stoichiometry here suggests that the electron from hydrogen is going, at least in part, to the metal bonds—thus providing the particular metal with a net of six electrons for bonding. Accordingly, the hydrides of Groups VI, VII, and VIII transition metals are either of lower stability or completely unknown. With compounds such as HfH_2 or Zr_4H the hydrogen apparently occupies tetrahedral sites. These transition metal hydrides decompose at high temperatures and use has been made of this fact in using the hydrides for powder metallurgical fabrications. Palladium forms the compound Pd_2H. Hydrogen will readily diffuse through a hot Pd plug. The rate of diffusion is proportional to the square root of the hydrogen pressure indicating the diffusing species is atomic hydrogen. This is used in obtaining extremely pure H_2.

GENERAL REFERENCES

D. M. Adams, *Inorganic Solids,* Wiley, New York, 1974. Good treatment from the chemical viewpoint.

J. K. Burdett, "New Ways to Look at Solids," *Acc. Chem. Res.* 1982, *15,* 34.

J. K. Burdett, *Molecular Shapes,* Wiley-Interscience, New York, 1980. Chapter 15 deals with a molecular orbital treatment of some solids.

F. S. Galasso, *Structure and Properties of Inorganic Solids,* Pergamon, Oxford, 1970. Extensive tables of compounds as examples of the common structures—good figures showing relationships among structures, and a good index.

N. B. Hannay, Ed., *Treatise on Solid State Chemistry,* Vol. I: *The Chemical Structure of Solids,* Plenum, New York, 1973. Authoritative treatment of selected topics.

A. F. Wells, *Structural Inorganic Chemistry,* 4th ed., Oxford, Oxford, 1975. The single most important source for inorganic structures and structural insight for chemists.

R. W. G. Wyckoff, *Crystal Structures,* Vols. I and II, 2nd. ed., Wiley-Interscience, New York, 1963 and 1964 (respectively).

PROBLEMS

6.1 Variations in hardness for ionic substances correlate well with what one thermodynamic property?

6.2 Which of each of the following pairs might be expected to be more ionic?
  a. $CaCl_2$ or $MgCl_2$.   *c.* NaCl or CuCl (similar radii).
  b. NaCl or $CaCl_2$ (similar radii).   *d.* $TiCl_3$ or $TiCl_4$.

6.3 For each of the following pairs indicate which substance is expected to be:
  a. More covalent (Fajans' rules):   *b.* Harder:
    $MgCl_2$ or $BeCl_2$ $SnCl_2$ or $SnCl_4$ NaF or NaBr MgF_2 or TiO_2
    $CaCl_2$ or $ZnCl_2$ $CdCl_2$ or CdI_2 Al_2O_3 or Ga_2O_3
    $CaCl_2$ or $CdCl_2$ ZnO or ZnS
    $TiCl_3$ or $TiCl_4$ NaF or CaO

6.4 Which of the following are *not* possible close-packing schemes?
  a. ABCABC . . .   *d.* ABCBC . . .
  b. ABAC . . .   *e.* ABBA . . .
  c. ABABC . . .   *f.* ABCCAB . . .

6.5 The wurtzite (*4PT*) structure of ZnS has open channels along the packing direction, but the zinc blende (*6PT*) structure does not. Explain.

6.6 Many MX type compounds with C.N. 6 have the NaCl (*6PO*) structure, whereas few have the NiAs (*4PO*) structure. What features of the NiAs structure limit its occurrence? What characteristics of the compound are necessary for the NiAs structure? What unusual physical property of NiAs results from its structure?

6.7 Give the *PTOT* notation for
  NaCl   ZnS (both structures)
  CaF$_2$ TiO$_2$

6.8 What do the following pairs or groups of structures have in common, and how do they differ?
  a. 6PO, 6PT, and 4PT.   *c.* 9PTT and 4PO$_{1/2}$.
  b. 9PT$_{1/2}$T$_{1/2}$ and 6PT.   *d.* 2($\frac{2}{3}$)PPO and 4PO$_{1/2}$.

6.9 Why are layer structures such as those of CdCl$_2$ and CdI$_2$ usually not encountered for metal fluorides or compounds of the most active metals?

6.10 Rubidium choride assumes the CsCl structure at high pressures. Calculate the Rb—Cl distance in the CsCl structure from that for the NaCl structure (from ionic radii for C.N. 6). Compare with the Rb—Cl distance from radii for C.N. 8.

6.11 Calculate the cation/anion radius ratio (by using plane geometry) for a triangular arrangement of anions in which the cation is in contact with the anions but does not push them apart.

6.12 Estimate the density of MgO (*6PO*) and ZnS (*6PT*) using radii to determine the cell dimensions and the number of formula units per unit cell.

6.13 NaSbF$_6$ has the NaCl (*6PO*) structure. The density is 4.37 g/cm^3. Calculate the radius of SbF$_6^-$ using the radius of Na$^+$, the density, the formula weight, and the number of formula units per unit cell.

6.14 Compare the thermochemical data for the alkali halides. What factors are important in establishing the order of increasing heats of formation within each group of halides (fluorides, chlorides, etc.)?

6.15 Calculate the heat of formation of NaF$_2$ and the heat of reaction to produce NaF + 1/2F$_2$. (Assume rutile structure.)

6.16 Calculate the heats of formation of Ne$^+$F$^-$ and Na$^+$Ne$^-$, estimating radii of Ne$^+$ and Ne$^-$. What factors prohibit the formation of these compounds in spite of favorable lattice energies?

6.17 From spectral data the dissociation energy of ClF has been determined to be 253 kJ/mole. The ΔH_f^0 of ClF(g) is 50.6 kJ/mole. The dissociation energy of Cl$_2$ is 239 kJ/mole. Calculate the dissociation energy of F$_2$.

6.18 Although the electron affinity of F is lower than that of Cl, F$_2$ is much more reactive than Cl$_2$. Account for the higher reactivity of F$_2$ **a.** with respect to the formation of solid halides MX or MX$_2$ and **b.** with respect to the formation of aqueous solutions of MX or MX$_2$.

6.19 Compare the ionic radii of Na$^+$ and Mg^{2+}, and also those of S^{2-} and Cl$^-$. Explain the differences within each pair.

6.20 If Ge is added to GaAs, the Ge is about equally distributed between the Ga and As sites. Which sites would the Ge prefer if Se is added also? Would GaAs doped with Se be an *n*-type or a *p*-type semiconductor?

6.21   *a.* Under what circumstances with regard to relative sizes of ions and degree of nonpolar character are Frenkel and Schottky defects likely?

b. The phenomenon of "half-melting" of solid electrolytes is related to which type of defect?

6.22   What is the significance of the term "molecular weight" with respect to diamond or SiO_2 (considering a perfect crystal of each)?

6.23   The triclinic feldspars form an isomorphous series involving replacement of Ca^{2+} for Na^+ from albite, $Na[AlSi_3O_8]$ to anorthite, $Ca[Al_2Si_2O_8]$. How can the series be isomorphous with changing ratio of Si/Al? Why do not K^+ and Ba^{2+} occur in this series?

6.24   Metals that are very malleable (can be beaten or rolled into sheets) and ductile (can be drawn into wire) have the *ccp* structure. Why are these characteristics favored for *ccp* rather than *hcp*?

6.25   Discuss the possible effects of extreme pressure on a metal (assume *bcc* at low *P*) and on a solid nonmetal.

6.26   Two of the three structures encountered for metals are *ccp* and *hcp*. How can the body-centered cubic structure be described in terms of a close-packed structure?

6.27   Sketch the curves for the distribution of energy states and their electron populations for a metallic conductor, an insulator, and a semiconductor.

6.28   Soft metals become "work-hardened." Explain this and how the softness is restored by heating the metal.

6.29   Consider the sliding of one layer of a crystal over another until an equivalent arrangement is achieved. What are the consequences with respect to hardness, brittleness, and malleability for (1) a metal, (2) an ionic crystal, (3) a covalent crystal, and (4) a molecular crystal. In a crude way, how do these considerations relate to melting point?

VII

Coordination Compounds— Bonding and Spectra

A *coordination compound* or metal *complex* might be defined as a central atom or ion attached to a sheath of ions or molecules. We can think of coordination chemistry either as a branch of inorganic chemistry or as an approach to—an interpretation of—many aspects of chemistry. Certainly the chemistry of metal compounds in solution, in the melt, or in the solid is best understood in terms of coordination chemistry. To a great extent, coordination interactions dictate crystal structures and site preferences of metal ions. We are treating coordination chemistry early so that its pervasive influence can be considered in acid-base chemistry, most aspects of the chemistry of metals and nonmetals, organometallic compounds, compounds of biological significance, and mechanisms of reactions—in fact, in much of the remainder of this book! It is somewhat artificial to treat the bonding of coordination compounds apart from other species—$TiCl_4$, RuF_6, and SF_6 could be treated as "simple" MX_n compounds or as coordination compounds with C.N. 4 or 6. Bonding in metal complexes was omitted from earlier treatments to avoid complications arising from special features of these compounds.

7.1 INTRODUCTION

Perhaps your first obvious encounter with a coordination compound or complex ion was seeing the deep blue $[Cu(NH_3)_4]^{2+}$ ion formed by adding aqueous ammonia solution to a solution of a copper(II) salt. Although this reaction is usually described by Equation 7.1, Equation 7.2

furnishes a more complete representation.[1] The copper ion in solution is already a complex ion—an aqua or water complex.

$$Cu^{2+} + 4NH_3 \rightleftharpoons [Cu(NH_3)_4]^{2+} \tag{7.1}$$

$$[Cu(H_2O)_4]^{2+} + 4NH_3 \rightleftharpoons [Cu(NH_3)_4]^{2+} + 4H_2O. \tag{7.2}$$

The reaction is a substitution, rather than an addition, reaction in which ammonia displaces water from the *coordination sphere,* which includes those molecules or ions *(ligands)* bonded directly to the metal ion.

If acid is added to the solution of $[Cu(NH_3)_4]^{2+}$, the protons combine with NH_3 to give NH_4^+ and the ammonia complex is destroyed: that is, the ammonia molecules are removed by the stronger acid H^+ and the less basic water molecules take their place to give $[Cu(H_2O)_4]^{2+}$. When a minimum amount of hydrochloric acid is added to neutralize the ammonia, the solution takes on the pale blue color of the $[Cu(H_2O)_4]^{2+}$ ion. Adding an excess of concentrated hydrochloric acid turns the solution green, because some of the yellow $[CuCl_4]^{2-}$ ion is formed. The intermediate species $[Cu(H_2O)Cl_3]^-$, $[Cu(H_2O)_2Cl_2]$, and $[Cu(H_2O)_3Cl]^+$ will also be present in varying amounts, depending on the concentration of Cl^-. The $[Cu(H_2O)_4]^{2+}$ ion is never converted completely to $[CuCl_4]^{2-}$, even in concentrated hydrochloric acid solution, because the concentration of water is still greater than that of Cl^-. Upon dilution, the green solution again becomes blue, because the Cl^- loses out in the competition when the chloride ion concentration is not very great. Adding HBr or NaBr to a solution of $[Cu(H_2O)_4]^{2+}$ turns the solution green and finally, in the presence of a high Br^- concentration, dark brownish-green, because of the formation of brown $[CuBr_4]^{2-}$. This complex ion also is destroyed by dilution.

The chemistry of metals in solution is essentially the chemistry of their complexes. A metal ion in solution is coordinated to water molecules or to other ligands. The transition metal ions are fairly good Lewis acids and their complexes are quite stable. The cations of the more electropositive metals such as the alkali metals and alkaline earth metals are weaker Lewis acids that form fewer complexes. Although they are hydrated in solution, the interaction with water is much weaker than in the case of transition-metal ions. Reactions of ions generally require that the water layers (hydration spheres) be stripped away so that the reactants can come together.

Metal ions are almost never encountered without effective shielding from one another. In solution the shielding is provided by the solvent or some other ligand. In the solid state, cations are surrounded by anions. For an ionic substance such as NaCl, the interactions between the Na^+ and the six surrounding Cl^- ions is primarily electrostatic. As mentioned previously (see p. 213), the more acidic metal ions, such as Cd^{2+} and Pd^{2+}, tend to form lattices in which the interactions are not the same in all directions. $CuCl_2$ crystallizes as the dihydrate, which

$$\begin{array}{c} Cl \\ | \\ H_2O—Cu—OH_2 \\ | \\ Cl \end{array}$$

contains planar $H_2O—Cu—OH_2$ units. In many cases the complexes found in well-ordered crystals do not persist in solution, because of competition with the solvent.

---

[1]Actually, these representations for both the aqua and the ammonia complexes are also inaccurate. The Cu^{2+} ion usually has four molecules held rather strongly and two more held more loosely. Attached to the ammonia complex might be two molecules of water or one or two additional ammonia molecules, depending on the concentrations.

The role of coordination compounds in nature is discussed in Chapter 16. Coordinated metal ions are at the active centers of many enzymes.

7.2 ANALYTICAL APPLICATIONS

Equilibria involving complex formation are encountered frequently in analytical chemistry. Ag^+ is separated from Hg_2Cl_2 and $PbCl_2$ in the qualitative analysis scheme by dissolving AgCl as $[Ag(NH_3)_2]^+$. We separate antimony, arsenic, and tin sulfides from the remaining Group II sulfides by dissolving them in the presence of an excess of sulfide ion as the complexes SbS_4^{3-}, AsS_4^{3-}, and SnS_3^{2-}. Organic reagents that give insoluble complexes are used widely for separating and determining metals. Two of the most familiar examples are dimethylglyoxime, which is a fairly specific reagent for nickel, and 8-hydroxyquinoline (oxine), which precipitates many metals. These compounds usually are insoluble, because the metal is incorporated into a large organic molecule and the resulting complex has no net charge. Some uncharged complexes are fairly volatile and can be sublimed: for example, Fe (acetylacetonate)$_3$ can be purified by sublimation at reduced pressure. Since many uncharged complexes are

Bis(8-hydroxyquinolinato)zinc(II)

insoluble in water but soluble in organic solvents, they also are used for separations based on solvent extraction procedures. At an appropriate pH we can get reasonable separations of a metal such as Ga(III) from others by extracting the 8-hydroxyquinoline complex with chloroform. The equilibria involved are

$$HQ \rightleftharpoons H^+ + Q^- \qquad K_a = \frac{[H^+][Q^-]}{[HQ]} \tag{7.3}$$

$$3Q^- + M^{3+} \rightleftharpoons MQ_3(aq) \qquad K_f = \frac{[MQ_3(aq)]}{[M^{3+}][Q^-]^3} \tag{7.4}$$

$$MQ_3(aq) \rightleftharpoons MQ_3(org) \qquad K_d = \frac{[MQ_3(org)]}{[MQ_3(aq)]} \tag{7.5}$$

K_a depends only on the complexing agent. K_f depends on the specific metal ion and the complexing agent. K_d, the partition coefficient, depends on the solvation of the complex in the two solvents; it is particularly dependent on the choice of the organic solvent. At low pH (low concentration of Q^-) only the more stable complexes (high K_f) are formed to be extracted. Complexes of the same ligand with different metal ions but the same stoichiometry usually have partition coefficients of the same magnitude, because of the similar external appearance to the solvent.

Acetylacetone also forms complexes used in solvent extraction procedures. Acetylacetone exists in the keto and enol forms, which are in equilibrium. The enol form coordinates through both oxygen atoms with the loss of a proton to form a *chelate ring* (see Figure 7.1). The term "chelate" (Greek "crab's claw") was first used by Morgan to describe the formation of similar complexes. The term metal chelate is now used to refer to the compounds formed as a result of chelation. Chelating ligands must have two or more points of attachment. Ammonia, with only one point of attachment, is a unidentate ligand. Ligands such as acetylacetone and ethylenediamine ($NH_2C_2H_4NH_2$) are bidentate. Table 7.1 lists some common multidentate ligands. For the rules of nomenclature as applied to ligands and coordination compounds, see Appendix B.

Table 7.1 Some common multidentate ligands

Name	Formula	Abbreviation	Classification
Carbonato	CO_3^{2-}		Bidentate
Oxalato	$C_2O_4^{2-}$	ox	Bidentate
Ethylenediamine	$NH_2C_2H_4NH_2$	en	Bidentate
1,2-Propanediamine	$NH_2CH(CH_3)CH_2NH_2$	pn	Bidentate
Acetylacetonato	$CH_3{-}C{=}CHC{-}CH_3$ (with O^- and O)	acac	Bidentate
8-Hydroxyquinolinato		oxine	Bidentate
2,2′-Bipyridine		bipy	Bidentate
1,10-Phenanthroline		phen	Bidentate
Glycinato	$NH_2CH_2CO_2^-$	gly	Bidentate
Diethylenetriamine	$NH(C_2H_4NH_2)_2$	dien	Tridentate
Triethylenetetraamine	$(-CH_2NHC_2H_4NH_2)_2$	trien	Quadridentate
Nitrilotriacetato	$N(CH_2CO_2)_3^{3-}$	nta	Quadridentate
Tetraethylenepentaamine	$NH(C_2H_4NHC_2H_4NH_2)_2$	tetraen	Quinquedentate
Ethylenediamine-tetraacetato	$[-CH_2N(CH_2CO_2)_2]_2^{4-}$	edta[a]	Sexidentate

[a]Capitals usually have been used for edta and other amino polyacid anions. The IUPAC rules recommend lowercase letters for abbreviations of all ligands.

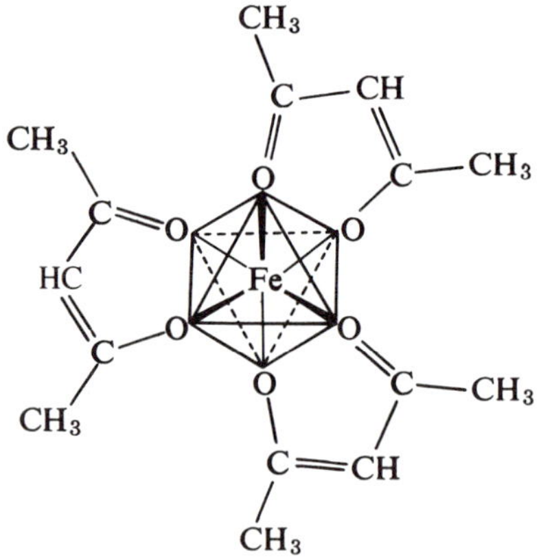

Figure 7.1 Chelate rings in the tris(acetylacetonato)iron(III) compound.

Many titration procedures now use edta in the determination of metals. These procedures usually involve competition for the metal between edta and a dye that also can serve as a ligand. One of the common oxidation-reduction indicators, ferroin, is a complex containing the $[Fe(phen)_3]^{2+}$ ion, which is deep red, whereas the Fe(III) compound is pale blue. In a suitable oxidation-reduction titration the removal of the color of the ferroin indicates that the end point has been reached. The ferroin is oxidized only after the reducing agent being titrated is oxidized completely.

7.3 VALENCE BOND THEORY

7.3.1 Werner's Views

Compounds now considered to be coordination compounds have been known for more than 150 years, but no satisfactory explanation for their formation was available until the imaginative work of Alfred Werner.[2] In 1893, Werner formulated his coordination theory, which provided the basis for modern theories. In order to appreciate Werner's insight, you must remember that the electron, the basis for all modern theories of chemical bonding was unknown at the time. Werner suggested that each metal has two kinds of valence: primary or ionizable valence, which can be satisfied only by negative ions, as in simple salts such as $CrCl_3$; and secondary valence, which can be satisfied by negative ions or neutral molecules. The secondary valences are responsible for the addition of ammonia to produce compounds such as $CrCl_3 \cdot 6NH_3$. Whereas each metal has a characteristic number of secondary valences directed in space to give a definite geometrical arrangement, the primary valences are nondirectional. Werner assumed that the six secondary valences were arranged octahedrally (see Figure 7.2). When a uninegative ion is in the inner coordination sphere, it satisfies one of the primary, as well as one of the secondary, valences of the metal (see Figure 7.3). When two anions are in the

---

[2]Translations of early papers are available. See G. B. Kauffman, Ed., *Classics in Coordination Chemistry,* Dover, New York, Part 1: *The Selected Papers of Alfred Werner,* 1968; Part 2: *Selected Papers* (1798–1899), 1976; Part 3: *Twentieth Century Papers* (1904–1935), 1978. Kauffman reviewed the early work in *J. Chem. Educ.* 1959, *36*, 521.

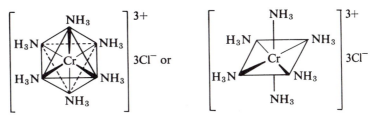

Figure 7.2 Octahedral structure of $[Cr(NH_3)_6]Cl_3$.

coordination sphere, the directional character of the secondary valences gives rise to *cis* and *trans* isomers—for example, $CrCl_3 \cdot 4NH_3$ (see Figure 7.3). Only the negative ions outside of of the coordination sphere are readily ionizable. Thus, the molecule $CrCl_3 \cdot 3NH_3$ should be a nonelectrolyte and the chloride ion should not be precipitated by silver nitrate, except as it is displaced from the coordination sphere. The species bonded to the metal ion in the coordination sphere are said to be in the *inner sphere*. The next sheath of ions or solvent molecules, held more loosely, make up the *outer sphere*.

Werner spent his productive life placing his intuitive coordination theory on a firm experimental basis. He compiled such an impressive amount of evidence for the octahedral configuration from the isomerism and reactions of six-coordinate complexes that his configuration was accepted generally long before it had been confirmed by modern structural investigations.

Some critics were skeptical of the optical activity attributed to the octahedral arrangement around a central ion in a complex such as $[Co(en)_3]Cl_3$. Figure 7.4 shows the optical isomers looking down the three-fold axis, in order to illustrate that the Δ and Λ isomers are really right- and left-handed spirals and are not superimposable (see p. 310 for notation). X-ray methods have proven that the Λ isomer (left-handed spiral) is the isomer formerly designated as *D* (for *dextro*, since it exhibits a positive rotation at the Na—D line). Werner prepared and resolved

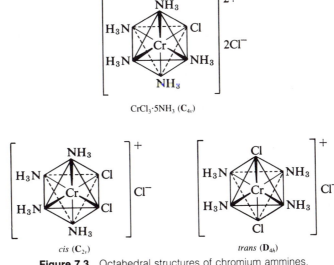

$CrCl_3 \cdot 5NH_3$ ($\mathbf{C_{4v}}$)

cis ($\mathbf{C_{2v}}$)   *trans* ($\mathbf{D_{4h}}$)

Figure 7.3 Octahedral structures of chromium ammines.

$$\left[Co \left\{ \begin{array}{c} OH \\ \\ OH \end{array} Co(NH_3)_4 \right\}_3 \right]^{6+}$$

which contains no carbon (see Figure 7.5), to answer the criticism that the optical activity of [Co(en)$_3$]Cl$_3$ in some way might be caused by carbon. He was awarded the Nobel Prize in 1913.

7.3.2 Valence Bond Interpretation

Effective Atomic Number

The nature of secondary valence follows from G. N. Lewis' treatment of the coordinate covalent bond in which the ligand atom—for example, N in NH$_3$—furnishes an electron pair that is shared with the central metal ion. Sidgwick argued that the coordination process provided the opportunity for a transition metal ion to reach a noble gas configuration, or the effective atomic number of a noble gas. The *effective atomic number* (EAN) of a metal ion is calculated by adding the electrons of the metal ion to those shared with it through coordination. In [Co(NH$_3$)$_6$]$^{3+}$ the cobalt(III) ion has 24 electrons plus the six electron pairs from the ammonia molecules, for a total of 36 electrons—the configuration of krypton. Quite a few, but not all, metals, achieve the EAN of a noble gas through coordination (see Table 7.2). The EAN concept has been particularly successful for the metal carbonyls, which are thought to contain the metal with the zero oxidation number and for some π complexes such as ferrocene (see p. 439). In general, EAN can be taken only as a rough guide. Most of the complexes of Ni(II) give EAN values which are 34 (C.N. 4) or 38 (C.N. 6) rather than 36.

This electron-counting scheme is also called the *rule of eighteen,* in contrast to the usual rule of eight for "simple" compounds. The rule of eight applies to main-group elements for which *s* and *p* are the only low-energy orbitals to be filled. For transition metals the *d* orbitals are included in the valence shell, since they are comparable in energy to the filled *s* and *p* orbitals of the same quantum number. The energy of the *d* orbitals is lowered with increasing nuclear charge through a transition series, so that after several *d* electrons have been added, the addition of ligand electrons can complete a shell of 18 electrons (a pseudo-noble-gas configuration—see p. 43).

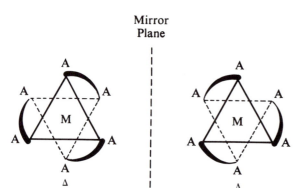

Figure 7.4   Representation of the mirror images of a complex of the type M(AA)$_3$ as right- and left-handed spirals (**D**$_3$).

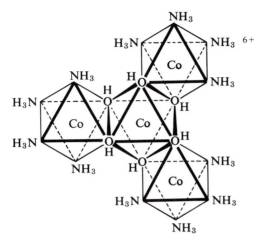

Figure 7.5 A completely inorganic optically-active complex ion.

Hybridization and Orbital Occupancy

Much of the modern valence bond treatment of coordination compounds was developed by Pauling. Although the ligand-field and molecular orbital treatments (see Sections 7.5, 7.9, and 7.10) are more generally useful, you should still be familar with the valence bond treatment and its terminology. We should recognize the importance of the qualitative valence bond model in fostering the development of this field to its present status.

According to this interpretation, coordination compounds result from the use of available bonding orbitals on the metal for the formation of coordinate covalent bonds (see p. 48). The coordination number and geometry are determined in part by size and charge effects, but also to a great extent by the orbitals available for bonding. The common hybridized orbitals en-

Table 7.2 Valence electron count for metals

Complex	Number of Electrons on M^{n+}	Number of Electrons from Ligands	EAN
$Co(NH_3)_6^{3+}$	$27 - 3 = 24$	$6 \times 2 = 12$	36 (Kr)
$Pt(NH_3)_6^{4+}$	$78 - 4 = 74$	$6 \times 2 = 12$	86 (Rn)
$Fe(CN)_6^{4-}$	$26 - 2 = 24$	$6 \times 2 = 12$	36
$Fe(CO)_5$	$26 - 0 = 26$	$5 \times 2 = 10$	36
$Cr(CO)_6$	$24 - 0 = 24$	$6 \times 2 = 12$	36
$Ni(CO)_4$	$28 - 0 = 28$	$4 \times 2 = 8$	36
$Ni(NH_3)_6^{2+}$	$28 - 2 = 26$	$6 \times 2 = 12$	38
$Ni(CN)_4^{2-}$	$28 - 2 = 26$	$4 \times 2 = 8$	34
$Cr(NH_3)_6^{3+}$	$24 - 3 = 21$	$6 \times 2 = 12$	33

countered in coordination compounds are shown in Table 7.3, which also lists the orbitals suitable for forming strong π bonds.

Magnetic susceptibility measurements have been used widely to determine the number of unpaired electrons in complexes, and from this information the number of d orbitals used for bond formation usually can be inferred for ions containing four to eight d electrons. The complexes in which the d electrons are forced to pair off to allow bond formation were referred to as "covalent" complexes, by Pauling. Those complexes in which the inner d orbitals are not used for bond formation as in $[Ga(C_2O_4)_3]^{3-}$, $[Ni(NH_3)_6]^{2+}$, or $[Fe(C_2O_4)_3]^{3-}$ (five unpaired electrons), were referred to as "ionic" complexes. The terms *low-spin* (or spin-paired) and *high-spin* (or spin-free) are preferred, since they describe the population of the d orbitals, as determined from magnetic properties, without any assumptions concerning the nature of the bonding.

Nickel(II) forms octahedral complexes with ligands such as H_2O and NH_3. However, with ligands such as CN^-, which display a greater tendency to form strong covalent bonds, nickel(II) forms diamagnetic planar complexes. Planar complexes are also common for Pd^{2+}, Pt^{2+}, Cu^{2+}, Au^{3+}, Rh^+, and Ir^+. The necessary d orbital either is available or can be made available by electron pairing in all these ions except Cu^{2+}. The interpretation applied above to the $[Co(NH_3)_6]^{2+}$ leads to the prediction that the 9th d electron in $[Cu(NH_3)_4]^{2+}$ would be promoted and might be lost easily, giving a Cu^{3+} complex. But in fact, the Cu^{2+} complexes are not oxidized easily. The interpretation of the planar Cu^{2+} complexes as "ionic" or using outer $4d$ orbitals is also unsatisfactory, because of the great stability of the complexes

Table 7.3   Hybrid orbitals for metal complexes[a]

Coord. Number	Hybridized Orbitals (σ)	Configuration	Strong π Orbitals	Examples
2	sp	Linear	p^2, d^2	$Ag(NH_3)_2^+$
3	sp^2	Trigonal plane	p, d^2	$BF_3, NO_3^-, Ag(R_3P)_3^+$
4	sp^3	Tetrahedral	—	$Ni(CO)_4, MnO_4^-, Zn(NH_3)_4^{2+}$
4	dsp^2	Planar	d^3, p	$Ni(CN)_4^{2-}, Pt(NH_3)_4^{2+}$
5	$d_{z^2}sp^3$ or d^3sp	Trigonal bipyramid	d^2	$TaF_5, CuCl_5^{3-}, [Ni(PEt_3)_2Br_3]$
5	$d_{x^2-y^2}sp^3$, d^2sp^2, d^4s, or d^4p	Tetragonal pyramid	d	$IF_5, [VO(acac)_2]$
6	d^2sp^3	Octahedral	d^3	$Co(NH_3)_6^{3+}, PtCl_6^{2-}$
7	d^5sp or d^3sp^3	Pentagonal bipyramid	—	ZrF_7^{3-}
7	d^4sp^2 or d^5p^2	Trigonal prism with an extra atom in one tetragonal face	—	TaF_7^{2-}, NbF_7^{2-}
8	d^4sp^3	Dodecahedral	d	$Mo(CN)_8^{4-}, Zr(C_2O_4)_4^{4-}$
8	d^5p^3	Antiprismatic	—	$TaF_8^{3-}, Zr(acac)_4$
8	d^3fsp^3 or d^3f^4s	Cubic	—	$U(NCS)_8^{4-}$

[a]G. E. Kimball, *J. Chem. Phys.* 1940, *8*, 188.

and because the bonding does not seem to be particularly different from that in planar nickel complexes. The lack of a really satisfactory account of the bonding in the very stable "ionic" complexes is a major weakness of the valence bond treatment. Also, absorption spectra and other properties involving excited states are not treated adequately by valence bond theory.

Electroneutrality Principle

The coordinate bond treatment of coordination compounds might seem to imply the accumulation of negative charge on a metal atom. The formal charge on Fe in $[Fe(H_2O)_6]^{3+}$ is -3, and the formal charge on Fe in $[Fe(H_2O)_6]^{2+}$ or $[Fe(CN)_6]^{4-}$ is -4. Pauling has shown that these unlikely situations need not arise. His electroneutrality principle (see p. 60) leads us to expect that the metal atom will be neutral or will bear only a partial positive charge. In the cationic aqua or ammonia complexes most of the positive charge can be spread over the H atoms so that no atom has more than a small partial positive charge. In an anionic complex the negative charge should be spread over the electronegative atoms present. The charge distribution is much different from that implied by the formal charges, because the bonds are polar: that is, they have partial ionic character. The formal charge on Fe in $[Fe(H_2O)_6]^{3+}$ would be -3, assuming 100% covalent character. If the Fe—O bonds were only 50% covalent the charge on the Fe would be zero. In $[Fe(H_2O)_6]^{2+}$ the Fe would have zero charge if the Fe—O bonds were 33.3% covalent, a reasonable approximation. The Fe—O bond in $[Fe(H_2O)_6]^{3+}$ should be slightly more covalent than that in $[Fe(H_2O)_6]^{2+}$, because the higher oxidation state of Fe increases its electronegativity.

Application of the electroneutrality principle leads to useful correlations and predictions about the stabilization of various oxidation states of transition metals with different ligands. The oxidation number common to all of the metals of the first transition series is II, as encountered in the aqua complexes. The approximately one-third covalent character of the M—O bond for divalent transition metals leads to zero formal charge on the metal. Because of the smaller difference in electronegativity of M and N, the M—N bond should be more covalent (closer to 50%) than the M—O bond. The more covalent M—N bond results in more charge transfer to the metal. This would favor a higher oxidation state in ammonia complexes in order to prevent a negative formal charge on the metal. It is difficult to oxidize $[Co(H_2O)_6]^{2+}$, but $[Co(NH_3)_6]^{2+}$ is oxidized to diamagnetic $[Co(NH_3)_6]^{3+}$ by O_2 in air. The difference is not so great for other first transition series ions, but in general, coordination with ligand atoms less electronegative than oxygen stabilizes the oxidation state next higher than that of the most stable aqua ion, unless other factors, such as π bonding, are involved. Compounds of the highest oxidation states are strongly oxidizing and are obtained only with the most electronegative ligands, F and O, as in fluoride complexes and oxoanions.

The general electron configuration for first transition series metals is $3d^x4s^2$, with Cr ($3d^54s^1$) and Cu ($3d^{10}4s^1$) as exceptions. All transition elements give M^{2+} ions, corresponding to the removal of the $4s$ electrons (or one s and one d electron for Cr and Cu). For early members of the transition series, however, higher oxidation states are more stable. Sc(III) is the only important oxidation state for Sc and for Ti, Ti(IV) is most important, with Ti(III) and especially Ti(II) obtainable only under strongly reducing conditions. V(II) and Cr(II), like Ti(III) and Ti(II), are good reducing agents. Manganese is the first member of the series for which II is the most important oxidation number.

Table 7.4   Stabilization of oxidation states through coordination

Complex	Oxidation Number of M	Comment
$Ni(CO)_4$, $Fe(CO)_5$, $Cr(CO)_6$	0	Stable
$Ni(CN)_4^{4-}$, $Pd(CN)_4^{4-}$	0	Easily oxidized
$Ni(CN)_4^{3-}$, $Mn(CN)_6^{5-}$	I	Easily oxidized
$Fe(phen)_3^{2+}$	II	Fe(II) more stable than Fe(III)

Pi Bonding

There are notable exceptions to the generalization that higher oxidation states are stabilized in compounds containing ligand atoms of low electronegativity. The most striking common examples are compounds of CO, CN^-, and 1,10-phenanthroline; some examples are given in Table 7.4. A ligand atom such as C (CO and CN^- coordinate through C normally) could give a predominantly covalent bond with a transition metal atom, leading to the accumulation of negative charge on the metal atom. The Ni—C bond distance in $Ni(CO)_4$, however, is shorter than expected for a single bond, suggesting that the bond has an appreciable amount of double-bond character. The valence bond representation of a contributing structure, containing a metal–carbon double bond (Figure 7.6) requires *that* particular contributing structure to have a carbon–oxygen double bond (instead of the triple bond in the other main contributing structure). The sketch of the orbitals available for metal–ligand π bonding in an MO description (Figure 7.6) reveals that the formation of an M—C π bond does not require eliminating one of the two C—O π bonds. The occupancy of the π^* orbital (resulting from donation from a filled metal d orbital) decreases the multiple-bond character of CO. This weakens (and lengthens) the C—O bond, while the accompanying M—C π bonding strengthens the M—C bond. Bonding descriptions are similar for cyanide complexes. Both CO and CN^- stabilize low oxidation numbers (see Table 7.4).

Several transition-metal complexes of bipyridine or 1,10 phenanthroline can be reduced to unusually low—even zero and negative—oxidation numbers. Thus $Fe(bipy)_3^{2+}$ can be reduced polarographically in 3 one-electron steps to $Fe(bipy)_3^-$, and the $Ni(bipy)_2^{2+}$ ion can be reduced in two steps to $Ni(bipy)_2^0$. The tris(bipy) complexes of Ti(0), Ti(−I), Cr(I), Cr(0), and V(0) can be obtained by chemical reduction using the lithium salt of $bipy^{2-}$ as the reducing agent in tetrahydrofuran.

$$MX_y + y\, Li_2(bipy) + n\, bipy \xrightarrow{\text{THF}} M(bipy)_n + y\, LiX + y\, Li(bipy)$$

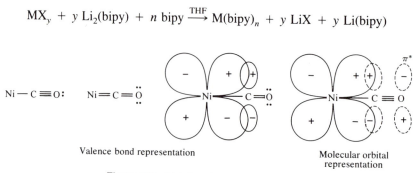

Valence bond representation   Molecular orbital representation

Figure 7.6   Metal-carbon double bonding.

The compounds M(bipy)$_2$ of Be, Mg, Ca, and Sr appear as formally M(0) compounds, but various experimental techniques lead to conflicting conclusions as to whether they should be represented as $M^0(bipy)_2$ or $M^{2+}(bipy^-)_2$. Because of extensive electron delocalization, the assignment of oxidation states is not clear for the reduced complexes.

7.4 SIMPLE ELECTROSTATIC APPROACH TO BONDING

7.4.1 Ion-Dipole Interaction

The electrostatic interpretation of the interaction between metal ions and ligands in complexes developed concurrently with the covalent bond approach. In the early 1920's the formation of complexes such as AlF_6^{3-} was discussed in terms of the attraction between oppositely charged ions. Water (aqua) and ammonia complexes were considered in terms of ion-dipole interaction. This approach leads to the correct prediction that large cations of low charge, such as Na^+ and K^+, form few complexes. In Group IIA the larger cations show little tendency to form stable complexes, but Be^{2+} does form complexes (e.g., BeF_4^{2-}), some of which have great stability. The *ionic potential* $\left(\dfrac{q}{r}\right)$ permits ions of different charges to be compared. We expect ions with high ionic potentials, such as Be^{2+}, Al^{3+}, and many of the transition metal ions, to form stable complexes. The simple electrostatic interpretation, however, does not account for the great stability of complexes of some cations, such as Hg^{2+}, whose ionic potential is small compared with Mg^{2+}, which has a high ionic potential, but forms few stable complexes.

Since the permanent dipole moment of the water molecule is greater than that of ammonia, aqua complexes should be considerably more stable than those of ammonia. Fulfilling this expectation, ammonia complexes of the alkaline earth and alkali metals are unstable in aqueous solution. Nevertheless, the ammonia complexes are much more stable than the aqua complexes for many of the transition metal ions, such as Ni^{2+}. Ni^{2+} and Mg^{2+} have about the same ionic potentials, so the difference in the relative stabilities of the aqua and ammonia complexes of these ions cannot be explained by simple electrostatic theory.

The simple electrostatic treatment views metal ions and ligands as rigid and undistorted. The picture can be improved by taking into account polarization of the ligands to explain the reversal in the stabilities of water and ammonia complexes. Further refinement, to include polarization of the cation as well as that of the ligand, is required to explain the relative stabilities of some of the halide complexes. The observed stabilities of many complexes agree reasonably well with "bond" energies obtained by taking into account polarization effects. The differences among the transition metals require that ligand-field splitting of the d orbitals be taken into account.

7.4.2 Ligand Polarization

The stability of complex ions containing neutral ligands depends on the ionic potential of the cation and the total dipole moment of the ligand—the sum of the permanent (μ) and induced (μ_i) dipole moments.

$$\text{Total dipole moment} = \mu + \mu_i$$

The induced dipole moment is given by the relationship $\mu_i = \alpha q/r$, where q/r (cation charge/distance between cation and ligand) represents the strength of the electrical field and α is the electronic polarizability of the molecule. The energy of interaction, E (polarization of the cation is neglected), is given by

$$E = \frac{q}{r}(\mu + \mu_i)$$

$$= \frac{q}{r}\mu + \frac{q^2}{r^2}\alpha$$

(7.6)

The permanent dipole moment of the water molecule is greater than that of ammonia, but the ammonia molecule is more polarizable and the induced dipole moment in a strong field can be quite large. Cations with small ionic potential form stable complexes with water because of its high permanent dipole moment. Cations with high ionic potential form more stable ammonia complexes because both α and q/r are large and make considerable contributions to the total dipole moment of NH_3. Thus the total dipole moment of the NH_3 can be greater than that of H_2O in a strong electric field.

7.4.3 Mutual Polarization of Cation and Ligand

Polarization of the ligand molecule does not fully explain the differences in the stabilities of the complexes of the metals in Groups IA and IIA compared with those of Groups IB and IIB. Even for metals with about the same ionic potential (e.g., Na^+ and Cu^+, Ca^{2+} and Cd^{2+}), the B subgroup metals form much more stable complexes than the A group metals. The polarization of the B subgroup cations themselves must be considered in order to account for the differences in the coordinating ability of ions such as Na^+ and Cu^+. Noble-gas type cations with an outer shell of eight electrons are not easily deformed, but polarization effects are quite important for cations that have the pseudo-noble-gas type configuration (18 electrons in the outer shell) or those with similar configurations [$(n-1)s^2p^6d^{10}ns^2$, or an almost complete group of d orbitals in the outer shell—see p. 197]. Cation polarization is particularly important, because it decreases the screening of the nuclear charge of the cation and decreases the metal–ligand distance. Both factors favor increased polarization of the ligand. Mutual polarization of the cation and of the ligand accounts for the greater stability of the complexes of ammonia, compared with aqua complexes, for the metals near the end of and just following the transition series.

Halide complexes of cations with noble-gas type configurations decrease in stability in the order

$$F^- > Cl^- > Br^- > I^-$$

This is the order expected on the basis of the anion sizes: the smaller anions release more energy (q_1q_2/r) because of the closer approach to the cation. The reverse order of stability is observed for cations such as Hg^{2+}, for which mutual polarization of the cation and anion favors combination with the larger, more polarizable anions.

Alkyl substitution in H_2O and NH_3 (an alkyl group is represented by R) decreases the tendency toward complex formation in the order

$$H_2O > ROH > R_2O$$
$$NH_3 > RNH_2 > R_2NH > R_3N$$

The order corresponds to a decrease in the dipole moments of the molecules, but hydrogen-bonding stabilization of aqua and ammine complexes also will play a great—perhaps predominant—role in aqueous solution. Alkyl substitution *increases* the tendency toward complex formation for H_2S and PH_3, corresponding to the order of *increasing* dipole moment in the series

$$H_2S < RSH < R_2S$$
$$PH_3 < RPH_2 < R_2PH < R_3P$$

The electrostatic approach, including polarization effects, permits us to predict the "bond" energies for many complexes. Table 7.5 compares the calculated energies per bond and those obtained from thermochemical data for the reaction

$$M^{m+}(g) + nL(g) \rightarrow ML_n^{m+}(g) \tag{7.7}$$

Since the metals of the first transition series have nearly the same radii, the electrostatic treatment predicts that the complexes they form in the same oxidation state will have approximately the same stability. Failure to account for the striking differences among the complexes of the transition metals is a great weakness of the simple electrostatic approach. The $[Cr(NH_3)_6]^{3+}$ and $[Co(NH_3)_6]^{3+}$ complexes are very stable, but manganese and iron do not form stable ammonia complexes in aqueous solution. Such differences among the transition metal complexes can be explained by applying the ligand field theory. The corrected values agree more closely with the experimental values than the uncorrected values for those complexes in Table 7.5 to which ligand field corrections apply. Even though the ligand field corrections are important in explaining differences among the transition metal complexes, note that simple electrostatic attraction accounts for most of the bond energy.

Table 7.5 Some metal-ligand bond energies[a]

Complex	Energy per Bond (Calc.)	Energy per Bond ("Exp.")	Energy per Bond Ligand Field Corrected
$[Fe(H_2O)_6]^{2+}$	209 kJ	243 kJ	218 kJ
$[Fe(H_2O)_6]^{3+}$	456	485	456
$[K(H_2O)_6]^{+}$	54	67	54
$[Cr(H_2O)_6]^{3+}$	464	510	502
$[Co(NH_3)_6]^{3+}$	490	560	523
$[AlF_6]^{3-}$	887	975	887
$[Zn(NH_3)_4]^{2+}$	360	372	360

[a]From F. Basolo and R. Pearson, *Mechanisms of Inorganic Reactions*, 2nd ed., Wiley, New York, 1967, p. 63.

7.5 THE LIGAND FIELD THEORY

The five *d* orbitals are degenerate (of equal energy) in gaseous metal ions. The simple electrostatic approach failed to predict differences among complexes of different transition metals because it ignored the fact that the degeneracy of the *d* orbitals may be removed in the electrostatic field created by the presence of the ligands. The symmetrical field caused by the ligands resembles the electric field around ions in ionic crystals. This similarity accounts for the name *crystal field theory*. Although crystal field theory was first proposed by Bethe in 1929, its application to coordination compounds is more recent. The name *ligand field theory* is used to refer to the approach in present use; it is essentially the same as the pure crystal field approach, except that covalent character is taken into account as necessary.

7.5.1 Octahedral Complexes

Ligand Field Stabilization Energy

In an octahedral complex oriented as shown in Figure 7.7, two of the *d* orbitals, d_{z^2} and $d_{x^2-y^2}$, are directed toward the ligands. The repulsion caused by the ligands raises the energy of these orbitals[3] (e_g^*) more than that of the other three orbitals, d_{xy}, d_{xz}, and d_{yz} (t_{2g}), which are directed at 45° to the axes (see Figure 7.8). The stronger the field caused by the ligands, the greater the splitting of energy levels ($10Dq$). The splitting causes no change in the energy of the system if all five orbitals are occupied equally. The t_{2g} level is lowered by $4Dq$ and the e_g^* level is raised by $6Dq$ relative to the average energy of the *d* orbitals. Hence the net change is zero for a d^{10} ion with $6t_{2g}$ and $4e_g$ electrons. Similarly, no stabilization results from the

$$6(-4Dq) + 4(+6Dq) = -24Dq + 24Dq = 0$$

splitting for a d^5 configuration if each orbital is singly occupied. For all configurations other than d^0, d^5 (high-spin), and d^{10}, splitting lowers the total energy of the system. The decrease in energy caused by the splitting of the energy levels is the *ligand field stabilization energy* (LFSE).

Electrons enter the t_{2g} orbitals in accordance with Hund's rule for d^1, d^2, and d^3 configurations. Thus $[Cr(NH_3)_6]^{3+}$ has three unpaired electrons with LFSE of $3(-4Dq) = -12Dq$. The LFSE for any octahedral chromium(III) complex is $-12Dq$, but the value of Dq varies with the ligand. There are two possible configurations for a d^4 ion in the ground state, $t_{2g}^3 e_g^{*1}$ and t_{2g}^4 (see Figure 7.9). In the first case the fourth electron occupies the e_g^* level, giving four unpaired electrons. The LFSE is $3(-4Dq) + 1(6Dq) = 6Dq$. This is referred to as the *weak-field* case, since the splitting is not great, or the *high-spin* case, since there is the maximum number of unpaired electrons. Here the splitting is small enough (implying a weak field or small value of Dq) so that the possible gain in LFSE that would result if the fourth electron were placed in the t_{2g} level is not great enough to provide the pairing energy required. If Dq were large (strong field), the energy required to pair two electrons would be less than the gain

---

[3]See Section 3.5.4 for an explanation of the MO notation. The e_g^* orbital carries the asterisk to indicate that this is really an antibonding MO. The asterisk usually is omitted in the literature, because the "pure" crystal field theory treatment neglects covalent bonding.

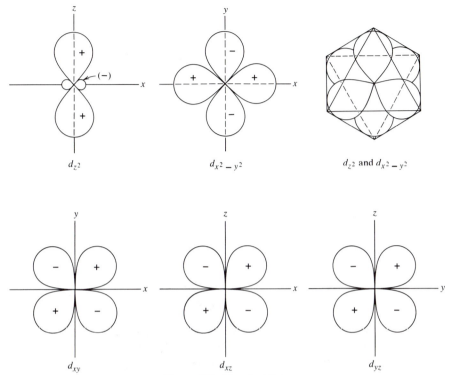

Figure 7.7 The *d* orbitals.

in LFSE that results. The t_{2g}^4 configuration, with two unpaired electrons, is referred to as the *strong-field* case (or the *low-spin* or spin-paired case). The LFSE for this case is $4(-4Dq)$ $= -16Dq$, and Dq is larger than in the weak-field case. The net energy gain is less than $-16Dq$ by the amount of energy required for pairing the electrons. Maximum LFSE is obtained for the d^6 configuration in a strong field, since the lower t_{2g} levels are occupied completely and the high energy levels are empty.

Table 7.6 lists the LFSE and the numbers of unpaired electrons for strong and weak fields in all d^n configurations. Of course, an LFSE of $-4Dq$ for the high-spin case is lower than $-4Dq$ for the low-spin case, because of the smaller Dq for the weak field.

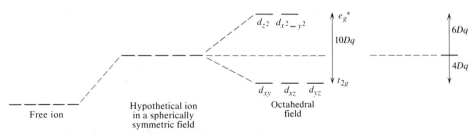

Figure 7.8 Splitting of the *d* energy levels in an octahedral complex.

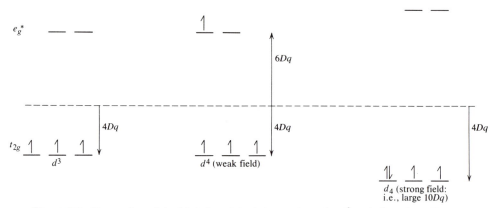

Figure 7.9 Occupation of *d* orbitals in octahedral complexes for d^3 and d^4 configurations.

Spectrochemical Series

The order of the ligand field strength for common ligands is approximately

$$CN^- > phen \sim NO_2^- > en > NH_3 \sim py > H_2O > C_2O_4^{2-} > OH^- > F^- > S^{2-} > Cl^- > Br^- > I^-$$

This is called the *spectrochemical series,* since the original order was determined from spectral shifts for complexes such as $[Co^{III}(NH_3)_5X]^{n+}$. The halides, at least, are in the order expected from electrostatic effects. In many other cases, we must consider covalent bonding to explain the order. The high ligand-field effect of CN^- and 1,10-phenanthroline is attributed to π bonding in which the metal donates electrons from one of the filled t_{2g} orbitals to a vacant

Table 7.6 Ligand field stabilization energy of octahedral complexes

Configu-ration	Examples	Strong Field				Weak Field			
		t_{2g}	e_g	No. of Un-paired e^-	LFSE $(-Dq)$	t_{2g}	e_g	No. of Un-paired e^-	LFSE $(-Dq)$
d^0	Ca^{2+}, Sc^{3+}	0	0	0	0	0	0	0	0
d^1	Ti^{3+}	1	0	1	4	1	0	1	4
d^2	V^{3+}	2	0	2	8	2	0	2	8^a
d^3	Cr^{3+}, V^{2+}	3	0	3	12	3	0	3	12^a
d^4	Cr^{2+}, Mn^{3+}	4	0	2	16	3	1	4	6
d^5	Mn^{2+}, Fe^{3+}	5	0	1	20	3	2	5	0
d^6	Fe^{2+}, Co^{3+}	6	0	0	24	4	2	4	4
d^7	Co^{2+}	6	1	1	18	5	2	3	8^a
d^8	Ni^{2+}	6	2	2	12	6	2	2	12^a
d^9	Cu^{2+}	6	3	1	6	6	3	1	6
d^{10}	Cu^+, Zn^{2+}	6	4	0	0	6	4	0	0

[a]See text.

orbital of the ligand. This type of π bonding further stabilizes the t_{2g} orbitals and shortens the M—X distance. The donation of electrons from the metal increases the effective positive charge on the metal ion. All three factors increase the splitting. Perhaps the greatest surprise in the order is the fact that the ligand field strength is greater for H_2O than for OH^-. Generally, charged ligands produce greater splitting than neutral ligands, because of the stronger interaction between a positive metal ion and a negative ligand. The observed order for OH^- and H_2O has been explained as the result of ligand-to-metal π bonding in OH^- complexes. The oxygen atom of the negative OH^- ion can share a pair of electrons in a filled p orbital with an empty t_{2g} metal orbital, if available for bond formation. Such bonding raises the energy level of the metal t_{2g} orbitals and lowers the effective charge on the metal. Both effects decrease the ligand field splitting (see p. 297). If the t_{2g} orbitals are filled, the interaction with the filled p orbitals is antibonding. Such interaction would be more important for the negative OH^- ion than for H_2O.

Cyanide and 1,10-phenanthroline complexes usually have low spin, corresponding to a large splitting of the energy levels and maximum occupancy of the t_{2g} orbitals. Hydrated ions and halide complexes usually have high spin. As might be expected, the splitting ($10Dq$) increases as the charge on the metal ion increases, because the cation radius decreases with increasing charge and the ligands would be more strongly attracted by the higher charge (see Table 7.7). The splitting, for the same ligands, increases and the pairing energy decreases for the second and third transition series metals compared with the first transition series. Both trends, which favor low-spin complexes, stem from the larger size of the $4d$ and $5d$ orbitals compared with the $3d$ orbitals. The larger orbitals, which extend farther from the central atom, are affected to a greater extent by the ligands. Pairing energies (Table 7.8) for the same number of d electrons increase with increasing oxidation number, because of increased electron repulsion. The pairing energies are higher for the d^5 configuration than for the d^6 configuration, because there is a greater loss in exchange energy (see p. 33) for the d^5 case. Consequently, there are more low-spin d^6 complexes than d^5 complexes. The pairing energies in complexes generally are 10–30% lower than those in free ions, because of lowered interelectron repulsion resulting from delocalization of the d electrons. Thus $[Co(NH_3)_6]^{3+}$ is a low-spin complex

Table 7.7 Ligand field splitting for some octahedral complexes ($10Dq$ values in italics indicate low-spin complex)[a]

Ions		6 Cl^-	6H_2O	6NH_3	3 en	6CN^-
d^3	Cr^{3+}	13.7 kK	17.4 kK	21.5 kK	21.9 kK	26.6 kK
d^5	Mn^{2+}	7.5	8.5	—	10.1	~30
	Fe^{3+}	11.0	14.3	—	—	35
d^6	Fe^{2+}	—	10.4	—	—	32.8
	Co^{3+}	—	20.7	22.9	23.2	34.8
	Rh^{3+}	20.4	27.0	34.0	34.6	45.5
d^8	Ni^{2+}	7.5	8.5	10.8	11.5	—

[a]The values of $10Dq$ are in kilokaysers (1 kK = 1000 cm^{-1}); 10 kK, or 1000 nm, corresponds to 119.7 kJ/mole (28.6 kcal/mole); 20 kK, or 500 nm, corresponds to 239.4 kJ/mole (57.2 kcal/mole).

Table 7.8 Pairing energies[a]

	Ion	P		Ion	P
d^4	Mn^{3+}	26.0 kK	d^6	Fe^{2+}	18.0 kK
d^5	Mn^{2+}	24.0	d^6	Co^{3+}	23.0
d^5	Fe^{3+}	30.0	d^7	Co^{2+}	22.5

[a]Pairing energies in kK (1 kK = 1000 cm^{-1}) for free ions; the values in complexes
are likely to be 10–30% lower because of decreased interelectron repulsion.

even though the free-ion pairing energy (23.0 kK) slightly exceeds the value of $10Dq$ (22.9 kK). This lowered interelectron repulsion in complexes is called the *nephelauxetic effect* (see Section 7.7.3). This effect is greatest for highly polarizable ligands—for example, I^- and those containing S ligand atoms—and for ligands that favor π bonding to metal.

We can evaluate the magnitude of the ligand field splitting from spectra. For a d^1 ion, such as $[Ti(H_2O)_6]^{3+}$, only one transition, from t_{2g} to e_g^*, is expected. The difference in energy between t_{2g} and e_g^* is $10Dq$. There is one absorption band with a maximum at 20.4 kK (490 nm), which we take as the value of $10Dq$. Although theoretical (calculated) values of $10Dq$ can be used as approximations, the spectral values are more reliable. The empirical $10Dq$ values from spectra reflect the actual splitting of energy levels resulting from electrostatic and covalent bonding effects. Spectra of complexes are discussed in Section 7.7

7.5.2 Distorted Octahedral Complexes

The discussion of ligand field splitting of the d orbitals into two levels applies to a regular octahedral arrangement with six identical ligands. In many cases some distortion of the octahedron can increase the stability of the complexes. Just as removing the degeneracy of the five d orbitals can result in ligand field stabilization, further removal of the degeneracy of the d orbitals can result in additional stabilization. The Jahn-Teller theorem predicts that distortion will occur whenever the resulting splitting of energy levels yields additional stabilization. If the octahedron is elongated along the z axis by increasing the metal-ligand distances along this axis, the degeneracy of the t_{2g} and of the e_g^* levels is removed. Because the ligand field decreases rapidly as the distance is increased, the d_{z^2} orbital is lowered in energy, while the $d_{x^2-y^2}$ orbital is raised in energy (see Figure 7.10). Withdrawing the ligands along the z axis decreases the ligand-ligand repulsion and tends to shorten the metal-ligand distances along the x and y axes. The splitting is such that no stabilization results if the two e_g^* orbitals are occupied equally. The t_{2g} orbitals are affected to a much smaller extent by the distortion, because they are not directed toward the ligands.

Copper(II) complexes (d^9 ion) generally are tetragonal ($\mathbf{D}_{4h}$ with four short M—X distances in a plane and two long M—X distances perpendicular to the plane), as expected from the Jahn-Teller effect. With the high-energy $d_{x^2-y^2}$ orbital occupied singly, the additional stabilization over the regular octahedron is $+\beta - 2\beta = -\beta$, as shown in Figure 7.10. The t_{2g} levels are filled, so no stabilization results from their splitting. There is no additional stabilization for a high-spin d^8 ion (Ni^{2+}), because the two e_g^* levels are occupied singly. Of course, no ligand field stabilization occurs for d^0, d^5 (high-spin), or d^{10} ions for regular or

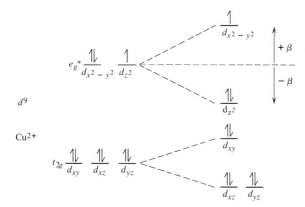

Figure 7.10 Splitting of the energy levels in an octahedral (tetragonal) field elongated along the z direction.

distorted octahedra. We expect distorted octahedral complexes for low-spin d^7 ions and for high-spin d^4 ions (one electron in d_{z^2} in each case). Slight distortion is expected for complexes involving unequal occupancy of the t_{2g} orbitals, but the effect is small because of the smaller splitting of these orbitals.

7.5.3   Planar Complexes

If the ligands along the z axis are removed completely, forming a planar complex (still $\mathbf{D}_{4h}$), the orbital energy splitting is increased greatly. The most favorable electronic configuration for planar complexes is that of a d^8 ion—for example, Ni^{2+}, Pt^{2+}, and Au^{3+}—in which the splitting is great enough to bring about pairing of all electrons. The very high energy $d_{x^2-y^2}$ orbital is left vacant (see Figure 7.11). Planar complexes are expected also for d^7 and d^9 ions. In a planar complex of a d^7 ion such as Co^{2+}, the highest energy orbital is vacant and the next-to-highest energy orbital is occupied singly. A planar copper(II) (d^9) complex has a single electron in the highest energy orbital.

The radius ratio ($r_+/r_- \geq 0.414$) required for a square planar arrangement is identical to that for an octahedral complex. Since the additional energy released from the interaction with the two extra ligands favors the octrahedral complex, the higher coordination number should be expected. The square planar arrangement is encountered only in cases where the additional splitting of the levels is most advantageous or where the planar configuration is imposed by

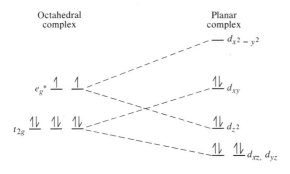

Figure 7.11 Energy levels for square planar nickel(II) complexes.

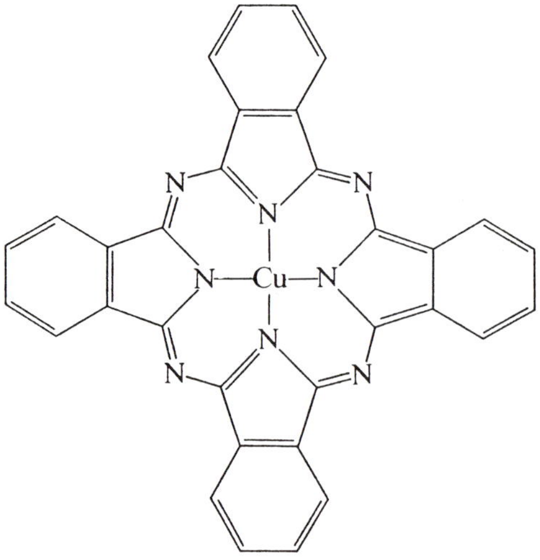

Figure 7.12 The planar copper(II) phthalocyanine complex.

the geometry of the ligand. The copper phthalocyanine complex (see Figure 7.12) can be sublimed *in vacuo* at about 580°C without decomposition and is not decomposed by strong mineral acids. The planar fused-ring system is similar to that of the porphyrins (see Section 16.1.5). The divalent metal ions that produce planar complexes with this ligand (e.g., Cu, Be, Mn, Fe, Co, Ni, and Pt) include several that normally form tetrahedral complexes.

7.5.4 Tetrahedral Complexes

A tetrahedral arrangement not only is the structure of least ligand-ligand repulsion for a four-coordinated metal ion, but also is the structure expected for a complex with a radius ratio between 0.225 and that required for an octahedral complex (0.414). The ligand field splitting is less important in tetrahedral complexes, since $10Dq$ is only four-ninths that of a regular octahedral arrangement for the same ligands. This is because there are only four ligands and they do not approach along the direction of any of the orbitals. The coordinate axes, chosen for convenience, are the two-fold symmetry axes that bisect the six edges of the tetrahedron. The high-energy orbitals are the d_{xy}, d_{xz}, and d_{yz} orbitals directed most nearly toward the ligands (see Figure 7.13). Because of the smaller ligand-field splitting, low-spin complexes generally are not encountered and tetrahedral complexes are not expected in cases where the ligand field stabilization is great for octahedral or planar complexes. Tetrahedral complexes are more likely for nontransition metals and transition metals with no ligand field stabilization (d^0, high-spin d^5, and d^{10}). Tetrahedral complexes of d^2 and d^7 ions demonstrate high ligand-field stabilization. Maximum LFSE occurs in the low-spin d^4 case, and $[Cr[N(Si(CH_3)_3)_2]_3NO]$ is a low-spin complex with a distorted tetrahedral (C_{3v}) structure.[4] It is formulated as a Cr(II) (d^4) complex containing NO^+. Large ligands or those with high negative charge favor the formation of tetrahedral complexes over planar or octahedral complexes, in order to minimize ligand–ligand repulsion. The oxo anions and chloro complexes such as those shown in Table

---

[4] D. C. Bradley, *Chem. Britain* 1975, *11*, 393.

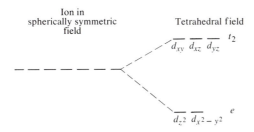

Figure 7.13 Ligand field splitting for tetrahedral complexes.

7.9 are common examples. Working out the effects of Jahn-Teller distortions of tetrahedral complexes is left as an exercise (Problem 7.5).

Table 7.9 Examples of regular tetrahedral complexes

d^0	d^2	d^5 (high-spin)	d^7 (high-spin)	d^{10}
$AlCl_4^-$	FeO_4^{2-}	$FeCl_4^-$	$CoCl_4^{2-}$	$ZnCl_4^{2-}$
$TiCl_4$				$GaCl_4^-$
MnO_4^-				
CrO_4^{2-}				

Ligand field stabilization greatly favors planar complexes for Cu(II), Ni(II) and Co(II), but $CuCl_4^{2-}$, $NiCl_4^{2-}$, and a few other complexes of these metals are tetrahedral.[5] The importance of covalent bonding, and particularly of π bonding is difficult to assess in many cases. Planar nickel complexes usually are those with an appreciable amount of π bonding (e.g., $[Ni(CN)_4]^{2-}$), and it seems likely that the necessary splitting for spin pairing requires covalent bonding. Although the pure crystal field treatment does not take covalent character into account, increased covalent character or π bonding can be handled in terms of the effect of giving a stronger ligand field.

In normal spinels, $MgAl_2O_4$ or $M^{II}M_2^{III}O_4$, the M^{III} ions are found in octahedral sites and M^{II} ions in tetrahedral sites (p. 213). When M^{II} shows strong LFSE for octahedral coordination, these ions might displace M^{III} from octahedral sites (unless they also show strong preference for octahedral coordination) to form an inverse spinel.

7.6 SPLITTING PATTERNS IN FIELDS OF VARIOUS SYMMETRIES

Krishnamurthy and Schaap's instructive approach to determining the relative energies of the *d* orbitals in fields of various symmetries is based on the additivity of effects of combined

---

[5] $CuCl_4^{2-}$ exists as flattened tetrahedra in Cs_2CuCl_4, but at high pressure the structure changes to square planar. Some hydrogen-bonded solids contain both square planar and tetrahedrally distorted square-planar $CuCl_4^{2-}$. See J. R. Ferraro and G. J. Long, "Solid-state Pressure Effects on Stereochemically Nonrigid Structures," *Acc. Chem. Res.* 1975, *8*, 171.

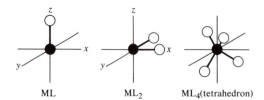

ML   ML$_2$ ML$_4$(tetrahedron)

Figure 7.14 The three primary ligand groups. (From R. Krishnamurthy and W. B. Schaap, *J. Chem. Educ.* 1969, *46*, 799.)

groupings of ligands. Adding the energies of the *d* orbitals for a square planar arrangement of four ligands (*xy* plane) and for a linear arrangement (along the *z* axis) of two ligands thus gives the relative *d* orbital energies for an octahedral field:

	d_{z^2}	$d_{x^2-y^2}$	d_{xy}	d_{xz}	d_{yz}
Square planar (*xy*)**D**$_{4h}$	$-4.28Dq$	$12.28Dq$	$2.28Dq$	$-5.14Dq$	$-5.14Dq$
Linear (*z*) **D**$_{\infty h}$	10.28	-6.28	-6.28	1.14	1.14
Octahedral **O**$_h$	$6.00Dq$	$6.00Dq$	$-4.00Dq$	$-4.00Dq$	$-4.00Dq$

Cases usually encountered require three primary ligand groups, ML, ML$_2$, and ML$_4$—as shown in Figure 7.14. Table 7.10 gives the relative energies of the *d* orbitals for these cases.

The field of a linear ML$_2$ complex is twice that of ML, so the energies of each orbital are doubled. The square planar field is twice that of the ML$_2$ primary ligand group. The field for an octahedral field is, then, $V_{O_h} = 2V_{ML} + 2V_{ML_2}$. For a square pyramid it is $V_{C_{4v}} = V_{ML} + 2V_{ML_2}$ or $V_{C_{4v}} = V_{O_h} - V_{ML}$.

The *total* potential energy for six unit charges in the *xy* plane is the same whether there are -3 units of charge along *x* and -3 units along *y*, $-1\frac{1}{2}$ units each at $\pm x$ and $\pm y$, or unit charges arranged in a planar hexagon. The latter case is treated in the reference. For three unit charges in a planar-trigonal arrangement ML$_3$, the effects would be half as great as for six units of charge in a plane. Multiplying the energies of the *d* orbitals for the ML$_2$ case by 3/2 gives us the energies for three units of charge, either $-1\frac{1}{2}$ at $+x$ and $-1\frac{1}{2}$ at $+y$, or $-3/4$ each at $\pm x$ and $\pm y$:

	d_{z^2}	$d_{x^2-y^2}$	d_{xy}	d_{xz}	d_{yz}
ML$_3$ or 3/2 × ML$_2$ (90°)	-3.21	9.21	1.71	-3.85	-3.85

The total potential energy is the same for a trigonal planar ML$_3$ arrangement, but the **D**$_{3h}$

Table 7.10 Relative *d* orbital energies for three primary geometric configurations

	Relative Energies in Units of Dq				
Configuration	d_{z^2}	$d_{x^2-y^2}$	d_{xy}	d_{xz}	d_{yz}
M–L (along *z* axis)	5.14	-3.14	-3.14	0.57	0.57
ML$_2$ (two ligands at right angles, along *x* and *y* axes)	-2.14	6.14	1.14	-2.57	-2.57
ML$_4$ (regular tetrahedron)[a]	-2.67	-2.67	1.78	1.78	1.78

[a]The original reference also gives values for distorted tetrahedra.

character table (Appendix 3.3) indicates two doubly-degenerate pairs of *d* orbitals: d_{xz} and d_{yz}, and $d_{x^2-y^2}$ and d_{xy}. The first pair is degenerate, and we can make the other pair equivalent by averaging to give the values shown for planar ML$_3$, **D**$_{3h}$.

	d_{z^2}	$d_{x^2-y^2}$	d_{xy}	d_{xz}	d_{yz}
Planar ML$_3$, **D**$_{3h}$	-3.21	5.46	5.46	-3.85	-3.85
Linear ML$_2$	10.28	-6.28	-6.28	1.14	1.14
Trigonal bipyramidal ML$_5$, **D**$_{3h}$	7.07	-0.82	-0.82	-2.71	-2.71

Adding the energies of the *d* orbitals for a linear ML$_2$ arrangement to those for planar ML$_3$ gives the energies of the *d* orbitals for a trigonal bipyramidal complex, ML$_5$, with five equidistant ligands.

Interestingly, the *d* orbitals are five-fold degenerate in an icosahedral, **I**$_h$, field—there is no splitting. Removing two ligands along an axis (choose *z*, or the linear ML$_2$ case) of the icosahedron leaves us with a pentagonal antiprism, such as the staggered arrangement of ferrocene, Fe(C$_5$H$_5$)$_2$. The corresponding treatment of the relative energies of the *d* orbitals gives

	d_{z^2}	$d_{x^2-y^2}$	d_{xy}	d_{xz}	d_{yz}
ML$_{12}$, **I**$_h$	0	0	0	0	0
ML$_2$ (linear)	10.28	-6.28	-6.28	1.14	1.14
ML$_{10}$, **D**$_{5d}$ (Pentagonal antiprism)	-10.28	$+6.28$	$+6.28$	-1.14	-1.14

We can use this approach to determine the relative energies of *d* orbitals for various geometries and for the further splittings of complexes such as *cis*- and *trans*-[MX$_4$Y$_2$] (see Problem 7.10). The original paper discusses limitations of this approach, which applies specifically to the d^1 case (no interelectron repulsion) or, using the hole formalism, to the d^9 case.

7.7 LIGAND FIELD SPECTRA

This section uses the notation and terminology of group theory but does not require the detailed manipulations of Section 3.5.

7.7.1 Energy States from Spectral Terms

We have noted already that the d^1 ion Ti(H$_2$O)$_6^{3+}$ shows a single absorption band in the visible region of the spectrum corresponding to the transition $t_{2g}^1 \rightarrow e_g^{*1}$. The energy corresponding to the band maximum is the splitting parameter $10Dq$. To examine cases with more than a single *d* electron, we must deal with spectral terms (p. 29).

The *d* orbitals are five-fold degenerate in the free ion, and the only spectral term for the d^1 case is ^2D. If we impose an octahedral field, the *d* orbitals split into a set of two, e_g, and a set of three, t_{2g}. The *D* term ($L = 2$) behaves like *d* orbitals ($l = 2$) and also splits in an **O**$_h$ field to give E_g and T_{2g} states. An *s* orbital ($l = 0$) is totally symmetrical and nondegenerate. In an **O**$_h$ field it is designated as a_{1g}, and an *S* ($L = 0$) term belongs to the representation (state) A_{1g}. The *p* orbitals ($l = 1$) remain three-fold degenerate in an **O**$_h$ field, so they are t_{1u}.

Table 7.11 Splitting of spectral terms for d^n configurations in an octahedral field

Term	Degeneracy	States in an Octahedral Field
S	1	A_{1g}
P	3	T_{1g}
D	5	$E_g + T_{2g}$
F	7	$A_{2g} + T_{1g} + T_{2g}$
G	9	$A_{1g} + E_g + T_{1g} + T_{2g}$
H	11	$E_g + T_{1g} + T_{1g} + T_{2g}$
I	13	$A_{1g} + A_{2g} + E_g + T_{1g} + T_{2g} + T_{2g}$

A P ($L = 1$) term corresponds to T_{1g} in an $\mathbf{O}_h$ field. In a centrosymmetric point group—$\mathbf{O}_h$, for example—spectral terms arising from d^n configurations give g states. Terms arising from p^n or f^n configurations give states that are u if n is odd and g if n is even ($u \times u = g$). If the point group is noncentrosymmetric—$\mathbf{T}_d$, for example—the g or u subscript is omitted. Table 7.11 lists the states arising from various spectral terms in an $\mathbf{O}_h$ field—note that the degeneracy (five-fold for D, seven-fold for F, etc.) is conserved.

7.7.2 Splitting Diagrams

There is direct correspondence between the configurations t_{2g}^1 and e_g^{*1} and the states $^2T_{2g}$ and ^2E_g for the d^1 case, but not for the general case. The separation between the ground state $^2T_{2g}$ and the excited state ^2E_g is $10Dq$, increasing with field strength (Dq) of the ligands. For a tetrahedral field the splitting is reversed, with ^2E the ground state (Figure 7.15).

We have seen that d^9 gives only one spectral term, ^2D—the electron can be missing from any one of the 10 possibilities. The transition in an octahedral field is $t_{2g}^6 e_g^{*3} \rightarrow t_{2g}^5 e_g^{*4}$—the "hole" has moved to the t_{2g} orbital. The ground state is E_g (corresponding to the partially filled level), and the excited state is T_{2g}. The splitting diagram is reversed, as on the left side of Figure 7.15. The tetrahedral case of d^9 is the same as the octahedral case of d^1 (a second reversal).

The high-spin d^6 case gives only one quintet term, ^5D, and this behaves as the d^1 case. We can assume that the half-filled configuration is unaltered and only one electron can move. Similarly, the high-spin d^4 case corresponds to d^9—there is one "hole" in the half-filled d^5 configuration. Thus a single diagram (called an Orgel diagram) in Figure 7.15 shows the change in energy with Dq for four different cases, all corresponding to one electron or one hole. It is qualitative since even for the same ligand $Dq_{\text{oct}} > Dq_{\text{tet}}$ and the energies depend on the configuration—for example, d^1 or d^4—and the charge on the ion.

The many other spectral terms for d^4 and d^6 differ in spin multiplicity. The *spin selection rule* (see Section 7.7.5) states that only transitions between states of the same multiplicity are allowed—or, that electrons do not pair or unpair during a transition. Hence we need only consider the spin-allowed transitions. Here they all involve states from the D term only. Spin-forbidden transitions sometimes are observed, if they are isolated from other transitions, but their intensities generally are very low ($\epsilon < 5$).

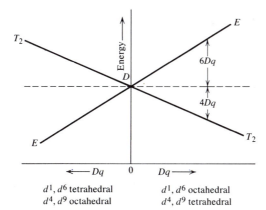

d¹, d⁶ tetrahedral
d⁴, d⁹ octahedral

d¹, d⁶ octahedral
d⁴, d⁹ tetrahedral

Figure 7.15 Orgel D term splitting diagram.

For the d^2 configuration there are two terms of maximum multiplicity, ^3F and ^3P. The ^3P is higher in energy than the ground state ^3F by 15B.[6] We can plot the energies of the free-ion spectral terms, as obtained from atomic spectroscopy, on the vertical axis at the left; on the right, we can plot the energies of the discrete configurations for the symmetry of the complex (see Section 7.6). The order of $\mathbf{O}_h$ is $t_{2g}^2 < t_{2g}^1 e_g^{*1} < e_g^{*2}$. There is no splitting of ^3P in an octahedral field (the label becomes $^3T_{1g}$), but ^3F splits to give $^3A_{2g} + {}^3T_{1g} + {}^3T_{2g}$. The axis at the left corresponds to zero applied field and the axis at the right corresponds to an infinitely strong octahedral field. We then connect the $\mathbf{O}_h$ representation derived from the spectral terms with those derived from the strong-field configurations. The states on the left are connected by lines with the corresponding states on the right, observing the noncrossing rule: tie lines for states of the same designation (including multiplicity) cannot cross.[7] This determines that the states from ^3F increase in energy in the order $^3T_{1g} < {}^3T_{2g} < {}^3A_{2g}$. The resulting correlation diagram is a qualitative energy-level splitting diagram. Figure 7.16a shows the splitting diagram for d^2 resulting from this correlation diagram, omitting the strong field configurations to the right. The diagram provides an initial approximation of the changes in relative energies of the triplet states with increasing field strength (Dq).

The ground state for a d^2 ion in an $\mathbf{O}_h$ field is $^3T_{1g}$. In a very weak field, transitions are expected to $^3T_{2g}$ (ν_1, 8Dq), $^3A_{2g}$ (ν_2, 18Dq), and $^3T_{1g}$ (P) (ν_3, 6Dq + 15B). For a stronger field, $^3T_{1g}(F) \rightarrow {}^3T_{1g}(P)$ will be at lower energy than $^3T_{1g}(F) \rightarrow {}^3A_{2g}$.

As noted for the d^1 case, splitting in a tetrahedral field is the reverse of that in an $\mathbf{O}_h$ field, resulting in the left side of the Orgel diagram for d^2 in Figure 7.16b. As the noncrossing rule also applies here, the lines for the $^3T_1$ states bend away from one another to avoid crossing $[^3A_2 \rightarrow {}^3T_1(F)$ is less than 18Dq, and $^3A_2 \rightarrow {}^3T_1(P)$ is greater than 6Dq + 15B]. This so-called configuration interaction results from mixing of the states. The interaction is greatest when the states approach one another in energy, but bending of the lines for the $^3T_{1g}$ states also occurs with increasing field strength for the $\mathbf{O}_h$ case. The hole formalism relates the d^8, d^3, and high-spin d^7 configurations, as shown in the Orgel diagram.

[6]The energy separations from interelectron repulsion are expressed in terms of the Racah parameters B and C. The energy separation of terms of maximum multiplicity can be expressed as a function of B only. These parameters are evaluated empirically from spectra.

[7]R. W. Jotham, "Why Do Energy Levels Repel One Another?" *J. Chem. Educ.* 1975, *52*, 377.

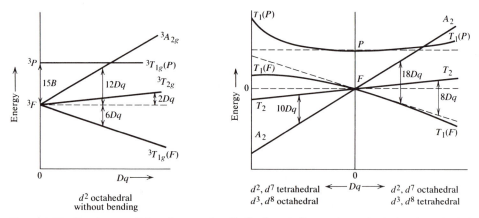

Figure 7.16   Orgel term splitting diagrams for d^2, d^3, d^7, and d^8 cases in octahedral and tetrahedral fields. (Adapted with permission from L. E. Orgel, *J. Chem. Phys.* 1955, *23*, 1004. Copyright 1955, American Chemical Society.)

Only two diagrams describe the splitting of terms of maximum spin multiplicity for all configurations except d^5, because they all have either ^nD or ^nF ground-state terms. The d^5 configuration is unique: it is its own hole counterpart, and there are no spin-allowed transitions. A single absorption band is expected for octahedral complexes of d^1, d^9, and high-spin d^4 and d^6 ions. Because of Jahn-Teller splitting of the e_g^* level, the band might be unsymmetrical whether these orbitals are occupied unequally in the ground state (d^4 and d^9) or in the excited state (d^1 and d^6). Octahedral complexes of ions with the other high-spin configurations (except d^5) should show three ligand-field absorption bands. The energies of two of the transitions are nearly the same in fields of intermediate strength, and they might appear as a single band. Octahedral high-spin d^5 ions—$Mn(H_2O)_6^{2+}$ for example—are faintly colored, since all d–d transitions are spin-forbidden. Since there are several ways to arrange four electrons of one spin-set and one of the other spin-set, however, there are four free-ion quartet spectral terms, and many quartet states arise in an octahedral field. Consequently, there are many peaks of low intensity (see p. 283).

7.7.3   The Spectrum of CoF_6^{4-}

Consider as an example the absorption spectrum of crystals of $KCoF_3$ in which Co^{2+} (d^7) is surrounded octahedrally by $6F^-$. There are absorption bands at 7.15, 15.2, and 19.2 kK. For a weak-field ligand, from Figure 7.16 we expect these to be $^4T_{1g} \rightarrow {}^4T_{2g}$, $^4T_{1g} \rightarrow {}^4A_{2g}$, and $^4T_{1g} \rightarrow {}^4T_{1g}(P)$, respectively. We could calculate Dq from v_1, which is $8Dq$, but this does not allow for configuration interaction (bending) between the T_{1g} states. The $v_2 - v_1$ difference is $10Dq$, and these excited states are not involved in configuration interaction, giving $10Dq = 15.2 - 7.2 = 8.0$, or $Dq = 0.8$ kK. From the expressions

$$v_1 = 8Dq + c \tag{7.8}$$

$$\nu_2 = 18Dq + c \tag{7.9}$$

(where c is the bending from configuration interaction), we get $c = 0.8$ kK in each case.

The Racah parameter B can be evaluated from ν_3, but the value of B in complexes (B′) is always less than the value for the free ion (B′ $\cong$ 0.7B $\rightarrow$ 0.9B), because of the decreased electron repulsion caused by greater delocalization of the d electrons in the complex. Jørgensen calls this the *nephelauxetic* ("cloud expanding") effect. The greater the covalent interaction, the more B′ will be lowered. We obtain the value of B′ empirically from spectra. Here

$$\nu_3 = 15B' + 6Dq + 2c \tag{7.10}$$

gives B′ = 0.85 kK, compared with the free-ion value for Co^{2+} of 0.97 kK. This value of B′ is reasonable for the expected small amount of covalent bonding in CoF_6^{4-}. If we do not know the value of B′, we can use ~0.8B as an approximation (a mid-range value).

7.7.4 Tanabe-Sugano Diagrams

The simple Orgel diagrams are useful for spin-allowed transitions when the number of peaks observed is not less than the number of empirical parameters (Dq, B′, and bending). Other cases require the use of Tanabe-Sugano diagrams, which differ from Orgel diagrams in having the ground state as the constant reference (it becomes the horizontal axis). Configuration interaction is included, as are states with spin multiplicity lower than that of the ground state. Figure 7.17 shows the T-S diagram for the d^2 configuration. In order to make the diagram general for various d^2 ions and ligands (both of which would change B′), the axes are in units of E/B and Dq/B. These ratios contain the actual values of B for the complex—or really, B′. For the $KCoF_3$ (d^7) case, Dq/B′ = 0.94 and the positions of the three peaks agree reasonably with the T-S diagram. Note that the spin-forbidden transition $^4T_{1g} \rightarrow {}^2E_g$ is about halfway between ν_1 and ν_2 and might appear as a very weak peak in this region. Two spin-forbidden transitions—$^4T_{1g} \rightarrow {}^2T_{1g}$, $^2T_{2g}$—are at only slightly greater energy than ν_2 and are likely to be hidden under the ν_2 band.

In the Tanabe-Sugano diagram for d^7 (Figure 7.17b), discontinuities occur at Dq/B $\cong$ 2.2, which is the ligand field strength required to overcome the pairing energy and go over to the low-spin case. The ground state becomes ^2E_g and the spin-allowed transitions are $^2E_g \rightarrow$ $^2T_{1g}$, $^2T_{2g}$ (nearly the same energy), $^2E_g \rightarrow {}^2A_{1g}$ at much higher energy, and others at energies too high to be observed ordinarily.

Compare the Tanabe-Sugano diagram for the d^5 configuration shown in Figure 7.17c with the assignments for the high-spin Mn^{2+} spectrum (Figure 7.18). Note these spin-forbidden bands have very low intensities and that all involve quartet excited states. Any transition for the d^5 case must change the number of unpaired electrons, but a two-electron transition (to give a doublet) is most improbable. Note the sharp peaks for transitions to the $^4A_{1g}$, ^4E_g (^4G) and $^4E_g(^4D)$ states. The sharpness stems from their energies' independence of ligand field strength (horizontal lines in the T-S diagram), which comes about because they involve pairing electrons within the t_{2g} ($^4A_{1g}$, ^4E_g) or $e_g[^4E_g(^4D)]$ orbitals. The ligand strength (Dq) is very sensitive to changes in the M–L distance. M–L vibrations in effect sweep a range of Dq values and cause broad bands, except for transitions that are independent of Dq. These field-indepen-

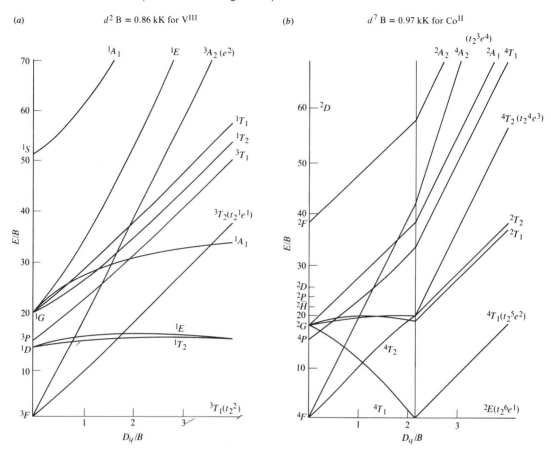

Figure 7.17   Semiquantitative energy-level diagrams for octahedral symmetry. (After Y. Tanabe and S. Sugano, *J. Phys. Soc. Japan* 1954, 9, 753.)

dent bands are useful in making assignments not only because they stand out, but also because their energies do not change with a change in ligands.

7.7.5 Selection Rules

Absorption of electromagnetic radiation is accompanied by a change in the electric dipole moment of the molecule for *electric dipole transitions* and a change in the magnetic dipole moment of the molecule for *magnetic dipole transitions*. The absorption bands for transition metal compounds, with which we are concerned here, are usually electric dipole transitions. The transitions observed for rare earth compounds are commonly for magnetic dipole transitions. Certain requirements of the ground and excited states, called *selection rules*, must be met for a transition to occur. If a selection rule is obeyed, the transition is said to be allowed; if the selection rule is violated, the transition is said to be forbidden. Since the selection rules deal with specific descriptions, often idealized, of the ground and excited states, we often observe forbidden transitions. Nevertheless, forbidden transitions have low probability and the observed intensities generally are low.

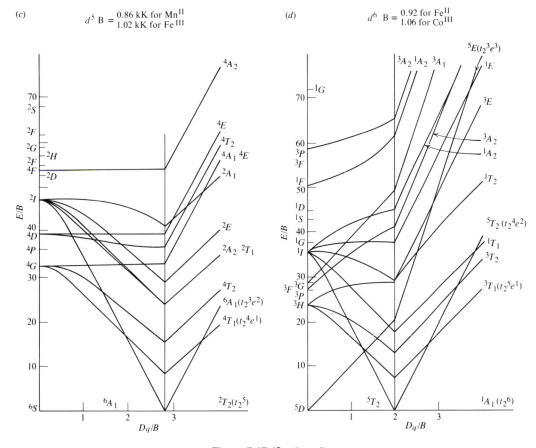

(c) d^5 $B = \begin{matrix} 0.86 \text{ kK for Mn}^{II} \\ 1.02 \text{ kK for Fe}^{III} \end{matrix}$

(d) d^6 $B = \begin{matrix} 0.92 \text{ for Fe}^{II} \\ 1.06 \text{ for Co}^{III} \end{matrix}$

Figure 7.17 *(Continued)*

Spin Selection Rule

We have noted that only transitions between states of the same multiplicity are allowed. Allowed transitions involve promoting electrons without changing spin. Spin-forbidden transitions, if observed at all, usually have intensities lower by a factor of 0.01–0.001 than those

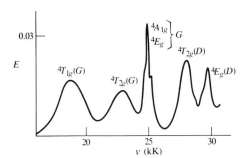

Figure 7.18 Absorption spectrum of $Mn(H_2O)_6^{2+}$. (From C. K. Jørgensen, *Acta Chem. Scand.* 1954, *8*, 1502.)

of spin-allowed transitions. We have used the Russell-Saunders coupling scheme (L–S coupling—see p. 29), since this generally works best for light elements. The j–j coupling scheme applies to heavier elements (see p. 31). For many elements, the coupling falls between these schemes. This involves some mixing of different spin states and the relaxation of the spin selection rule, which accounts for the appearance of weak spin-forbidden bands.

Symmetry Selection Rule

The d^6 case is important. The lower pairing energy means that the low-spin case is achieved at lower field strength ($Dq/B = 2.0$) than with d^5 (pairing occurs above $Dq/B \cong 2.8$). For Co(III) only CoF_6^{3-} is a well-characterized high-spin complex. The ground state for the low-spin case is $^1A_{1g}$ (the t_{2g} orbitals are filled). The spin-allowed transitions (T–S diagram, Figure 7.17) within the usual spectral range are $^1A_{1g} \rightarrow {}^1T_{1g}$ and $^1T_{2g}$, so two absorption bands are observed for $[Co(NH_3)_6]^{3+}$—very similar to the spectrum of $[Co(en)_3]^{3+}$ (see Figure 8.19). Tanabe-Sugano diagrams for d^3, d^4, and d^8 configurations are given in Figure 7.19.

Ligand field absorption bands have low intensities ($\epsilon = 5$ to 100) compared with electronically allowed transitions ($\epsilon \sim 10,000$). The d–d transitions violate the *Laporte* or *symmetry* selection rule, which states that $g \rightarrow g$ ($d \rightarrow d$) or $u \rightarrow u$ ($p \rightarrow p$) transitions are forbidden. This selection rule is relaxed for metal complexes because of the *vibronic mechanism* (the intensity-giving mechanism). In simple terms, an odd vibration (one that destroys the center of symmetry) distorts the octahedron so that the g and u designations are not applicable and there can be mixing of d and p orbitals. The vibrational and electronic transitions are coupled. The transitions involve excited vibrational states of the electronic ground and/or excited states, and this contributes to the broadening of bands. Intensities for noncentrosymmetric complexes—for example, tetrahedral $CoCl_4^{2-}$—generally are higher ($\epsilon \sim 100$) than those for octahedral complexes. The $[Co(en)_3]^{3+}$ ion is noncentrosymmetric ($\mathbf{D_3}$), but the absorption bands are not much more intense than those for $[Co(NH_3)_6]^{3+}$. We must conclude that the $\mathbf{O}_h$ selection rules apply in both cases. What seems to be important in both cases is the octahedral CoN_6 chromophore. Although the chelate rings that impose $\mathbf{D_3}$ symmetry have little effect here, their influence is seen in the circular dichroism spectrum (p. 316).

The intensity of an electronic transition is proportional to the dipole strength, D. The square root of D is an integral involving the wave functions for the ground and excited states and an operator called the dipole moment vector ($d\tau$ is a volume element). Since the integral of an odd function

$$\sqrt{D} = \int \Psi_{\text{ground}} (\text{operator}) \Psi_{\text{excited}} d\tau$$

vanishes (the positive and negative portions cancel when integrated over all space), the intensity will be nonzero only for an even function. For electric dipole transitions, the electric dipole transition moment (the operator) transforms as a vector along x, y, or z. Consequently, the irreducible representations are those of x, y, and z, which are necessarily u (odd) for a centrosymmetric group. The products $g \times g = g$ and $u \times u = g$, but $g \times u = u$ dictate that for an odd operator, (operator)$_u$, the function will be g only if the product of the wave functions also is u. This is the basis for the *symmetry selection rule*.

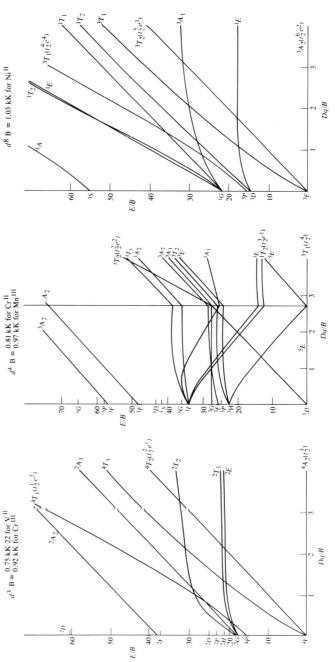

Figure 7.19 Semiquantitative energy-level diagrams for octahedral symmetry. (After Y. Tanabe and S. Sugano, *J. Phys. Soc. Japan* 1954, 9, 753.)

$$g \times u \times u = g \text{ and } u \times u \times g = g$$

but

$$g \times u \times g = u \text{ and } u \times u \times u = u$$

Only transitions that are $u \rightarrow g$ or $g \rightarrow u$ give a symmetric function, so $d_g \rightarrow d_g$ and $p_u \rightarrow p_u$ are forbidden. If there is some mixing of d and p, the g and u do not apply (strictly) and a "forbidden" transition might be observed. It is often said that here the selection rule is relaxed, but actually, the case does not conform to the ideal model upon which the selection rule is based. This situation is achieved for noncentrosymmetric point groups (to varying degrees, depending upon the *effective* symmetry) or for $d \rightarrow d$ transitions by coupling the electronic transition with an odd vibration.

Actually, the selection rules are even more restrictive. The function involved in the integral must not only be even, it must be *totally symmetric*—it must belong to the totally symmetric representation (A_{1g} for the $\mathbf{O}_h$ case). The product of two representations is A_{1g} (or contains A_{1g} for a reducible representation) only if the two representations are identical. Therefore, the product of the representations for the ground and excited states (this product is called the symmetry of the transition) must belong to the *same representation* as one of the vectors **x**, **y**, and **z.** This provides a basis for more detailed assignments of transitions.

For magnetic dipole transitions (encountered for rare earths), the operator is an even function, so the product of the ground and excited states must be even. The requirements for optically active transitions (p. 316) are very stringent, since nonzero intensity for circular dichroism or optical rotatory dispersion depends on the product of the integrals for electric and magnetic dipole transitions. This leads to the limitation of optical activity to (rotational) groups lacking any S_n axis.

For a good discussion of selection rules at an appropriate level, see M. Orchin and H. H. Jaffé, *Symmetry, Orbitals, and Spectra,* Wiley-Interscience, New York, 1971.

Lowering of Symmetry

Lowering the symmetry of a complex by substituting another ligand, as in the case of *cis-* and *trans*-[Co(en)$_2$F$_2$]$^+$, partially removes the degeneracy of the T states. The states that arise can be obtained from group theory (see Table 7.12, the correlation table for $\mathbf{O}_h$). The actual symmetry of the *cis* isomer is $\mathbf{C}_2$ ($\mathbf{C}_{2v}$ for the corresponding NH$_3$ complex) and that of the *trans* isomer is $\mathbf{D}_{2h}$ ($\mathbf{D}_{4h}$ for the corresponding NH$_3$ complex). As with [Co(en)$_3$]$^{3+}$ we can ignore the chelate rings for the absorption spectra and use the higher symmetries of the corresponding ammonia complexes. Note the somewhat higher intensity (Figure 7.20) for the noncentrosymmetric *cis* isomer—this is usually observed, and was noted above for tetrahedral complexes.

For absorption spectra the effective symmetry—the symmetry consistent with the splitting observed—is determined by the ligand atoms rather than the overall symmetry. Ligands that are trans to one another interact with the same d orbital, and the effective field along this axial direction is the average of the field strengths of the two ligands. The simplified diagram of the *trans* isomer has $\mathbf{D}_{4h}$ (tetragonal) symmetry with a weak field (2 F) along z (Figure 7.21). The

Table 7.12 Correlation table for some groups derived from O_h

O_h	T_d	D_{4h}	D_3	C_{4v}	C_{2h}	C_{2v}
A_{1g}	A_1	A_{1g}	A_1	A_1	A_g	A_1
A_{2g}	A_2	B_{1g}	A_2	B_1	B_g	A_2
E_g	E	$A_{1g} + B_{1g}$	E	$A_1 + B_1$	$A_g + B_g$	$A_1 + A_2$
T_{1g}	T_1	$A_{2g} + E_g$	$A_2 + E$	$A_2 + E$	$A_g + 2B_g$	$A_2 + B_1 + B_2$
T_{2g}	T_2	$B_{2g} + E_g$	$A_1 + E$	$B_2 + E$	$2A_g + B_g$	$A_1 + B_1 + B_2$
A_{1u}	A_2	A_{1u}	A_1	A_2	A_u	A_2
A_{2u}	A_1	B_{1u}	A_2	B_2	B_u	A_1
E_u	E	$A_{1u} + B_{1u}$	E	$A_2 + B_2$	$A_u + B_u$	$A_1 + A_2$
T_{1u}	T_2	$A_{2u} + E_u$	$A_2 + E$	$A_1 + E$	$A_u + 2B_u$	$A_1 + B_1 + B_2$
T_{2u}	T_1	$B_{2u} + E_u$	$A_1 + E$	$B_1 + E$	$2A_u + B_u$	$A_2 + B_1 + B_2$

cis isomer also can be treated as a tetragonal case, with a strong field (2 N) along z and the same weaker field along x and y $\left(\dfrac{N + F}{2}\right)$. The splitting between the levels from T_{1g} depends on the differences between the axial and in-plane field strengths and is twice as great for the *trans* isomer as for the *cis* isomer (see Problem 7.11 for verification), and in the opposite direction. In one case, the unique d_{z^2} orbital interacts with the weaker field (it will be lower in energy than $d_{x^2 - y^2}$); in the other case, it interacts with the stronger field.

Trans isomers often show distinct splitting of the first absorption band (T_{1g}) if the ligands differ appreciably in field strength. The *cis* isomers often show no distinct splitting, only, perhaps, a shoulder on the opposite side of the main band compared with the minor component of the *trans* isomer. In Figure 7.20 the *cis* isomer shows slight broadening on the low-energy side. The peak at ~22.5 kK for the *trans* isomer is the "minor" component. (Its intensity should be compared with the minima on either side; the intensity is raised by contributions of "tails," particularly from the higher energy band.) The second band (T_{2g}) usually shows no splitting.

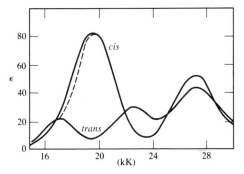

Figure 7.20 The absorption spectra of *cis*- and *trans*-[Co(en)$_2$F$_2$]NO$_3$. The dotted line outlines the main Gaussian band. (After F. Basolo, C. J. Ballhausen, and J. Bjerrum, *Acta Chem. Scand.* 1955, 9, 810.)

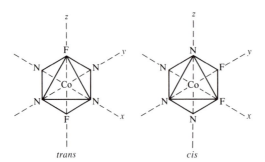

trans   *cis*   **Figure 7.21** *cis*- and *trans*-[CoN$_4$F$_2$]$^+$

7.8 CHARGE TRANSFER BANDS

At energy higher than the ligand field absorption bands, we commonly see one or more very intense bands that go off scale unless log ϵ is plotted. These normally are charge transfer bands, corresponding to electron transfer processes that might be ligand → metal or metal → ligand. Metal → ligand transitions occur for metal-ion complexes that have filled, or nearly filled, t_{2g} orbitals with ligands that have low-lying empty orbitals. These empty orbitals are ligand π^* orbitals in complexes such as those of pyridine, bipyridine, 1,10-phenanthroline, CN$^-$, CO, and NO.

The ligand → metal charge-transfer spectra have been studied more thoroughly. The intense bands are for electronically allowed transitions commonly of the (ligand) $p \rightarrow d$ (metal) type. The ionization energy for the ligand, or its ease of oxidation, determines the energy of the transition. No net oxidation-reduction occurs, because of the short lifetime of the excited state. However, many complexes are decomposed photochemically by this process, so they are not stored in strong light. Figure 7.22 shows the spectra of [Co(NH$_3$)$_5$X]$^{2+}$ ions, where X is a halide ion. The spectra are plotted on a log ϵ scale so that we can see the weak ligand field and strong charge-transfer bands. In each case the first two (lower-energy) bands are ligand field (d–d) bands. Note that these bands shift slightly toward lower energy as the field strength of X$^-$ decreases (F$^-$ highest and I$^-$ lowest). Little splitting of the first band is observed for [Co(NH$_3$)$_5$F]$^{2+}$, because the field strength of F$^-$ does not differ greatly from that of NH$_3$. A shoulder appears on this band for the Cl$^-$ and Br$^-$ complexes, because of the significantly lower field strength along the unique (N–Co–X) direction. As noted for [Co(en)$_2$F$_2$]$^+$, little splitting of the second band occurs.

The charge transfer bands (X$^- \rightarrow$ M) are shifted much more than the ligand field bands. The shift toward lower energy corresponds to the change in ease of oxidation, I$^- >$ Br$^- >$ Cl$^- >$ F$^-$. The charge transfer band merges with the second ligand-field band (a shoulder here) for [Co(NH$_3$)$_5$Br]$^{2+}$. The peak for this second band is more distinct for the I$^-$ complex, but its intensity is enhanced greatly by overlap with the more intense band. Even the lowest energy band is affected for [Co(NH$_3$)$_5$I]$^{2+}$, so the expected pronounced splitting is obscured.

The intense colors of the MnO$_4^-$ (purple), MnO$_4^{2-}$ (green), and CrO$_4^{2-}$ (yellow) ions arise from charge transfer (O → M). The source of color in metal sulfides, selenides, and tellurides also results from charge-transfer transitions (see p. 198).

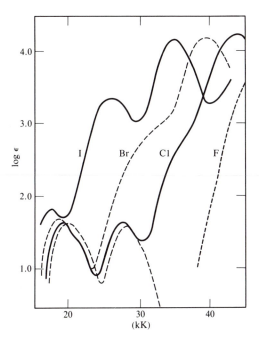

Figure 7.22 The spectra of the $[Co(NH_3)_5X]^{2+}$ ions, where X is a halide ion. (After M. Linhard and M. Weigel, *Z. Anorg. Chem.*, 1951, *266*, 49.)

7.9 PICTORIAL MOLECULAR-ORBITAL DESCRIPTION OF BONDING

We follow the same approach for a molecular orbital description of a metal complex, ML_n, that we used for other molecules, AX_n (Sections 4.2 and 4.3). Thus bonding is very much the same in $AlCl_4^-$ as in $SiCl_4$, and in AlF_6^{3-} as in SF_6. Even though metal complexes are not special cases, they often exhibit features that justify separate discussion. In transition metal complexes, the participation of the *d* orbitals is particularly important.

Consider sigma bonding in an octahedral complex, ML_6^{n+}, where M is a transition metal ion, such as Ni^{2+}, and L is a Lewis base, such as :NH_3. Each NH_3 molecule has a filled orbital directed toward the metal ion. We must identify all metal orbitals with reasonable energies for bonding and proper symmetry. Group theory helps us obtain all proper combinations of metal and ligand orbitals. In simple cases, such as this one, we can use a pictorial approach.

We recognize that the energy of the 4*s* electrons is only slightly higher than that of the 3*d* electrons, since the transition metals give stable positive oxidation numbers corresponding to removal of 3*d* electrons after the 4*s* electrons are removed. Also, at atomic numbers 19(K) and 20(Ca), the 4*s* energy level is *lower* than the 3*d* level. The energies of the 4*p* orbitals are only slightly higher than those of the 4*s* and 3*d* orbitals. Thus the orbitals we must consider are 4*s*, 3*d*, and 4*p*, for a total of nine. Now we must choose the orbitals of proper symmetry.

The 4*s* orbital is spherically symmetrical and can interact equally with all six ligands. One bonding MO corresponds to the combination of the 4*s* orbital with the σ orbital of the same sign on each of the six ligands (Figure 7.23). For convenience these six ligands' orbitals will

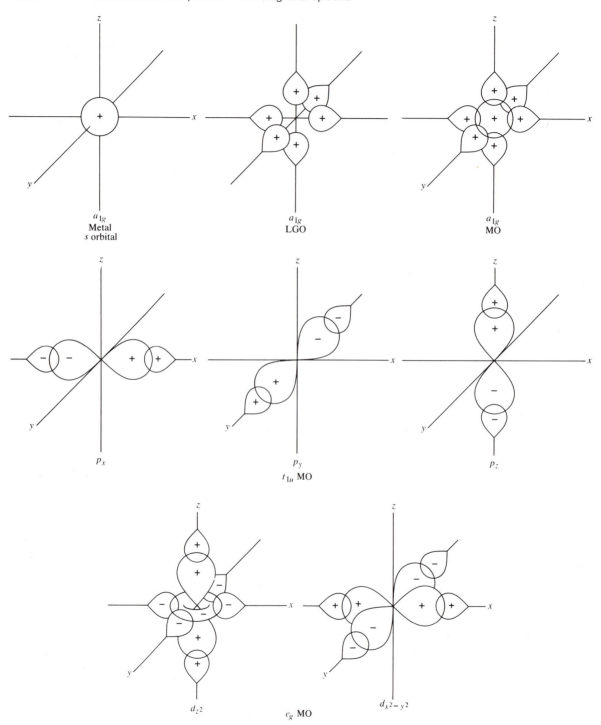

Figure 7.23   Combination of atomic orbitals and LGO to form sigma bonds in an octahedral complex.

be treated together as a ligand group orbital *(LGO)*. The combination of the $4s$ orbital with the *LGO* of opposite sign is antibonding. The $4s$ orbital and this *LGO* are totally symmetric: that is, they are symmetric with respect to every symmetry operation of the point group. For the $\mathbf{O}_h$ group, this is the A_{1g} representation. The bonding MO is designated a_{1g}, and the antibonding combination a_{1g}^*.

Each $4p$ orbital on the metal ion is directed along one of the coordinate axes and can combine with *LGO* of the two ligands lying along the axis (Figure 7.23). The p orbitals together belong to the triply degenerate representation T_{1u}. There are three bonding (t_{1u}) and three antibonding (t_{1u}^*) combinations.

Two of the d orbitals ($d_{x^2-y^2}$ and d_{z^2}) are directed along the coordinate axes and hence are suitable for σ bonding. The bonding *LGO* are those that overlap the lobes of the appropriate metal orbitals with the signs matching (Figure 7.23). These two d orbitals belong to the E_g representation, as do the corresponding *LGO*. The other three d orbitals (t_{2g}) directed at 45° to the coordinate axes are not suitable for σ bonding; they are nonbonding.

We have a total of six bonding MO, six antibonding MO, and three nonbonding MO. This total (15) must agree with the total number of orbitals combined (nine from M and six from the ligands).

7.10 GROUP THEORETICAL TREATMENT[8]

7.10.1 Sigma Bonding in Octahedral Complexes

The systematic approach is to apply group theory (see Sections 3.5, 4.3, and 4.4). We can look at transformation properties of the metal orbitals to identify their representations, but we know that the s orbital is totally symmetric (A_{1g}), the p orbitals transform as the vectors along x, y, and z (T_{1u}), and the representations for the d orbitals (E_g and T_{2g}) can be identified from the $\mathbf{O}_h$ character table (Appendix 3.3). To obtain the representations to which the *LGO* belong, we can examine the effects of all classes of symmetry operations on the individual ligand orbitals. Of importance here is not whether the ligand orbitals are atomic or hybrid orbitals, but only that there is a component along the M—L bond for σ bonding.

The ligands in Figure 7.24 are numbered in accordance with nomenclature rules. We will tabulate the effect of *one* symmetry operation (the *same* one for all ligands) of *each class* on

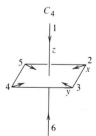

Figure 7.24 Orientation of the ligand σ orbitals in an octahedral complex.

---

[8]This section makes use of the methods presented in Chapter 3 and applied in Chapter 4. See also S. F. A. Kettle, "Ligand Group Orbitals of Octahedral Complexes," *J. Chem. Educ.* 1966, *43*, 21.

Table 7.13 Character table for the **O** point group

O	E	$6C_4$	$3C_2\,(=C_4^2)$	$8C_3$	$6C_2$		
A_1	1	1	1	1	1		
A_2	1	-1	1	1	-1		
E	2	0	2	-1	0		(z^2, x^2-y^2)
T_1	3	1	-1	0	-1	(x,y,z)	
T_2	3	-1	-1	0	1		(xy, xz, yz)

each of the ligand orbitals. Each symmetry operation will either leave an orbital unchanged (enter character 1) or transform it into another orbital (enter character 0). If we were examining the metal orbitals, with the nucleus at the origin, it would also be possible to reverse the sign of the orbital to give a character -1: for example, to reflect a p orbital through a plane of symmetry perpendicular to it. We can simplify the matter by using the **O** character table (Table 7.13, the pure rotational subgroup) instead of the $\mathbf{O}_h$ table. The results, as tabulated in Table 7.14, do not identify the representations as g or u, but that can be done separately by examining the effects of i and/or planes of symmetry.

The sums of the characters correspond to a reducible representation. Inspection (or the method discussed in Section 3.5.7) gives us the irreducible representations contained. We can obtain results more directly by recognizing that there will be an A_1 representation and keeping the metal orbitals in mind. The result is

$$\Gamma_\sigma = A_1 + E + T_1$$

The combination with all ligand orbitals having the same sign must be A_{1g} for $\mathbf{O}_h$. Since the character for the reducible representation (Γ_σ) is zero for the center of symmetry (each orbital goes into another), the representations in $\mathbf{O}_h$ symmetry, by inspection, must be

$$\Gamma_\sigma = A_{1g} + E_g + T_{1u} \tag{7.11}$$

Superimposing these on their metal counterparts in Figure 7.23, we see that since there is no LGO belonging to T_{2g}, the metal t_{2g} orbitals must be nonbonding.

Table 7.14 Effects of the symmetry operations of the **O** point group on the ligand orbitals of an octahedral complex.

O	E	C_4^z	$C_2^z(C_4^2)$	C_3	C_2	i (for $\mathbf{O}_h$)
L_1	1	1	1	0	0	0
L_2	1	0	0	0	0	0
L_3	1	0	0	0	0	0
L_4	1	0	0	0	0	0
L_5	1	0	0	0	0	0
L_6	1	1	1	0	0	0
Γ_σ	6	2	2	0	0	0

The systematic way to obtain the detailed description of the *LGO* and their *LCAO* wave functions is to apply the *projection operator* approach (p. 156). Here we tabulate all of the symmetry operations and write down the orbital obtained as a result of performing each operation on this orbital. Choosing L_1, for example: it is still L_1 after the identity operation, but it becomes L_2 after one C_3 operation, L_3 after another C_3 operation, etc. (Table 7.15). Next we multiply each orbital in the table by the character for the corresponding symmetry operation for the A_1 representation (all $+1$) and sum. In this case we get $a_1 = 4L_1 + 4L_2 + 4L_3 + 4L_4 + 4L_5 + 4L_6$; or, simplifying, $a_1 = L_1 + L_2 + L_3 + L_4 + L_5 + L_6$. This is normalized to give the *LCAO* wave function by dividing by the square root of the sum of the squares of the coefficients.

$$\Psi_{a_1} = \frac{1}{\sqrt{6}} (L_1 + L_2 + L_3 + L_4 + L_5 + L_6) \tag{7.12}$$

Multiplying the orbitals in Table 7.15 by the characters for the representation E and summing gives

$$e(1) = 4L_1 + 4L_6 - 2L_2 - 2L_3 - 2L_4 - 2L_5$$

or

$$2L_1 + 2L_6 - L_2 - L_3 - L_4 - L_5 \tag{7.13}$$

and

$$\psi_{e(1)} = \frac{1}{\sqrt{12}} (2L_1 + 2L_6 - L_2 - L_3 - L_4 - L_5) \tag{7.14}$$

This gives us one of the pair of e *LGO*—one that matches perfectly with the metal d_{z^2} orbital. We can get the other one by performing an operation that interchanges ligands along z with those in the xy plane. Using a C_2 axis such that $L_1 \rightarrow L_2$ and $L_6 \rightarrow L_4$, we get

$$2L_2 + 2L_4 - L_1 - L_3 - L_5 - L_6 \tag{7.15}$$

This is equivalent to the *LGO* already obtained, except that it corresponds to a different numbering scheme. Adding it to the one obtained just gives another equivalent combination. Keeping in mind that the other metal e orbital is $d_{x^2-y^2}$, we recognize that we want an *LGO* entirely contained in the xy plane, with L_1 and L_6 not participating. Multiplying Equation 7.15 by two and adding to the first *LGO* gives us the second e orbital.

$$3L_2 + 3L_4 - 3L_3 - 3L_5 \text{ and } \Psi_{e(2)} = \frac{1}{2} (L_2 - L_3 + L_4 - L_5) \tag{7.16}$$

Table 7.15 Results of applying all symmetry operations to L_1

E	$C_4 (1)$	$C_4 (2)$	$C_4 (3)$	$C_4 (4)$	$C_4 (5)$	$C_4 (6)$	$C_2 (1)$	$C_2 (2)$	$C_2 (3)$
L_1	L_1	L_5	L_2	L_3	L_4	L_1	L_1	L_6	L_6
$C_3 (1)$	$C_3 (2)$	$C_3 (3)$	$C_3 (4)$	$C_3 (5)$	$C_3 (6)$	$C_3 (7)$	$C_3 (8)$		
L_2	L_3	L_4	L_5	L_5	L_2	L_3	L_4		
$C_2' (1)$	$C_2' (2)$	$C_2' (3)$	$C_2' (4)$	$C_2' (5)$	$C_2' (6)$				
L_6	L_6	L_2	L_3	L_5	L_4				

We can obtain the t_1 *LGO* in similar fashion, recognizing that for each we need a pair of orbitals of opposite sign along one axis.

$$t_1 (1) = \frac{1}{\sqrt{2}} (L_1 - L_6) \tag{7.17}$$

$$t_1 (2) = \frac{1}{\sqrt{2}} (L_3 - L_5) \tag{7.18}$$

$$t_1 (3) = \frac{1}{\sqrt{2}} (L_2 - L_4) \tag{7.19}$$

Symmetry and group theory do not help in arriving at the energies of the molecular orbitals. We can get some qualitative ordering of the energy levels from the relative energies of the metal and ligand orbitals and the extent of orbital overlap. Such a qualitative energy-level diagram is shown in Figure 7.25.

7.10.2 Pi Bonding in Octahedral Complexes

The π bonding can be treated independently of σ bonding, because no group operation interchanges σ and π orbitals. With the representations for the metal orbitals known, we find the reducible representation for the ligand orbitals oriented properly for π bonding as before.

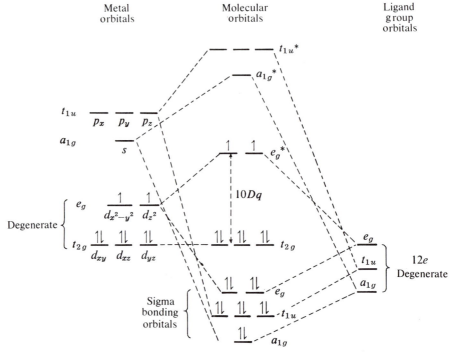

Figure 7.25 Qualitative diagram for the molecular orbitals of an octahedral d^8 complex such as $[Ni(NH_3)_6]^{2+}$ (without π bonding).

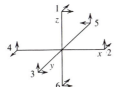

Figure 7.26 Orientation of the ligand π orbitals in an octahedral complex.

Figure 7.26 depicts these orbitals as vectors oriented so that they are interchanged by some symmetry operation. Vectors that transform in this way belong to the same set. If we were dealing with a trigonal bipyramid, there would be two independent sets—those on ligands in the equatorial plane and those in axial positions.

Using the **O** point group, we see that each vector is unchanged by the identity operation, giving a total character of 12 for this symmetry operation. For the C_2 axis along z, C_2^z, the four vectors for ligands 1 and 6 have their directions reversed—giving a character of -4. Each of the other vectors is changed into another vector by any operation (other than E), so the characters are zero.

O	E	C_3	C_2	C_4	$C_2^z(C_4^2)$	$i(\mathbf{O}_h)$
Γ_π	12	0	0	0	4	0

The reducible representation obtained can be identified by inspection as $2T_1$ and $2T_2$: that is, the sums of the characters for $2T_1$ and $2T_2$ give the characters for Γ_π. Since the character for i is zero for all 12 vectors, in the $\mathbf{O}_h$ point group

$$\Gamma_\pi = T_{1g} + T_{2g} + T_{1u} + T_{2u} \qquad (7.20)$$

Since there are no metal t_{1g} or t_{2u} orbitals, these *LGO* must be nonbonding. We can sketch the bonding *LGO* using the metal orbitals as guides (see Figure 7.27). The t_{2g} orbitals will be important for π bonding unless both ligand and metal orbitals are empty or both filled. The t_{1u} metal orbitals are used effectively for σ bonding, where they provide better overlap; they generally are not involved much in π bonding.

7.11 EFFECTS OF π BONDING

There are two types of π bonding for metal complexes. If the ligands have filled t_{2g} orbitals whose energies are close to those of the empty or only partially filled metal t_{2g} orbitals, then ligand → metal donation can occur. This situation is favored for metal ions with high oxidation numbers and few d electrons. If the metal has a few d electrons, these will go into the antibonding t_{2g}^* orbitals, since the shared ligand electrons will be in the bonding t_{2g} orbitals. If evaluated spectroscopically, $10Dq$ will be *smaller* than without π bonding (Figure 7.28). This probably is the case for complexes of F^-, Cl^-, or OH^-. The lower $10Dq$ value for OH^- (weaker-field ligand) compared with H_2O comes about because OH^- is a better π donor than H_2O. If there are as many as six t_{2g} metal electrons, the π bonding will not increase the stability of the complex.

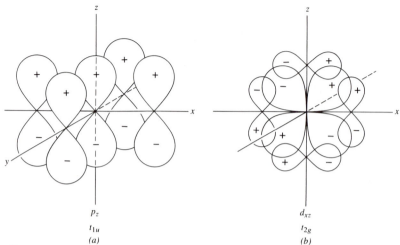

Figure 7.27 The two symmetry types of π bonds in octahedral complexes. (a) One of the t_{1u} orbitals; there are identical orbitals along the x and y directions. (b) One of the t_{2g} π orbitals; there are identical orbitals in the xy and yz planes.

The other type of π bonding involves metal → ligand donation, which is important for metals with filled or nearly filled t_{2g} orbitals and ligands with empty low-energy orbitals, such as the d orbitals of P or S or π* orbitals of CO, CN^-, NO^+, 1,10-phenanthroline, etc. (Section 7.3.2). Of course, the ligands with empty π* orbitals also have filled π orbitals and could function as π donors. The acceptor role, however, is more important, especially when there are several metal t_{2g} electrons. The metal (t_{2g}) → π* (ligand) bonding strengthens the metal–ligand bond but weakens the C—N bond in CN^-, for example, because of population of the antibonding (with respect to the C—N bond) orbital. The energy of the filled t_{2g} levels decreases as a result of the bonding—increasing the value of $10Dq$ (Figure 7.28). Most of the ligands near the top of the spectrochemical series (p. 270) are π-acceptors, whereas π donors are near the bottom of the series.

7.12 COMPARISON OF THE DIFFERENT APPROACHES TO BONDING IN COORDINATION COMPOUNDS

Research in the field of coordination compounds made rapid progress utilizing the valence bond approach for about 25 years, starting in the early 1930s. As more results were obtained, the qualitative nature of this approach became a serious limitation. Even some of the qualitative predictions of the relative stabilities of complexes based on the availability of low-energy orbitals are incorrect: for example, Cu(II) and Zn(II), with no vacant inner d orbitals, form more stable complexes than some of the metals with vacant inner d orbitals.

The crystal field theory offered the advantages of permitting the interpretation of spectra of complexes and more detailed interpretation and explanation of the magnetic behavior, stability, sterochemistry, and reaction rates of complexes. The pure crystal field approach, which

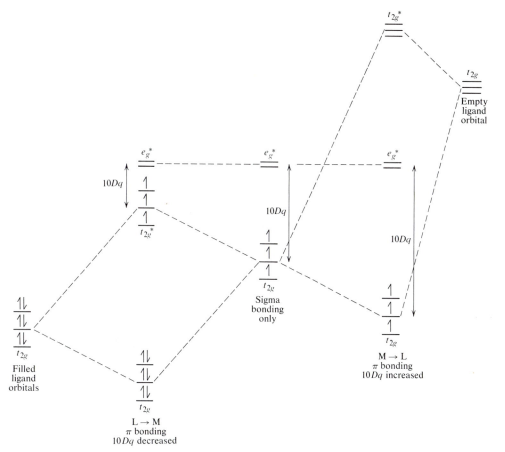

Figure 7.28 Comparison of the effects of π bonding using (a) filled low-energy π ligand orbitals for L → M donation and (b) empty ligand orbitals of π symmetry for M → L donation.

does not consider covalent bonding, might seem as limited as the extreme valence bond approach, which considers only covalent bonding. The effect of σ bonding, however, can be treated empirically as though it were the result of a very strong ligand field. Without π bonding, the results of the MO treatment are so similar to the ligand field representation that the latter is adequate for most applications. The important spectral transitions and the differences that result in ligand field stabilization involve the t_{2g} (nonbonding) and e_g^* (antibonding) levels.

For the present, the ligand field theory seems to offer the most practical approach to bonding in coordination compounds. It is essentially the crystal field theory modified to take covalent character into account when necessary. In applying the ligand field theory, we can take advantage of simple pictorial models similar to those that proved so useful in the valence bond approach. The ligand field theory lends itself to the simple qualitative applications and predictions, but it also permits quantitative applications.

The molecular orbital theory encompasses the crystal field and valence bond approaches as special cases. Applications of MO theory to metal complexes have become increasingly important with the availability of large computers.

GENERAL REFERENCES

C. J. Ballhausen, *Introduction to Ligand Field Theory*, McGraw-Hill, New York, 1962.

F. Basolo and R. G. Pearson, *Mechanisms of Inorganic Reactions*, 2nd ed., Wiley, New York, 1967. Covers more than just mechanisms.

J. K. Burdett, *Molecular Shapes*, Wiley-Interscience, New York, 1980. A good treatment of CFT, MO, and angular-overlap approaches.

J. P. Fackler, Jr., *Symmetry in Coordination Chemistry*, Academic Press, New York, 1971. Treats bonding in metal complexes.

B. N. Figgis, *Introduction to Ligand Fields*, Wiley-Interscience, New York, 1966.

S. F. A. Kettle, *Coordination Compounds*, Appleton-Century-Crofts, New York, 1969.

A. B. P. Lever, *Inorganic Electronic Spectroscopy*, Elsevier, Amsterdam, 1968. The best source of spectral data.

A. E. Martell, Ed., *Coordination Chemistry*, Vol. I, Van Nostrand-Reinhold, New York, 1971; Vol. II, ACS Monograph 174, SIS/American Chemical Society, Washington, D.C., 1978.

K. Nakamoto and P. J. McCarthy, Eds., *Spectroscopy and Structure of Metal Chelate Compounds*, Wiley, New York, 1968.

PROBLEMS

7.1   Which of the following complexes obey the rule of 18 (EAN rule)?
   a. $Cu(NH_3)_4^{2+}$, $Cu(en)_3^{2+}$, $Cu(CN)_4^{3-}$.
   b. $Ni(NH_3)_6^{2+}$, $Ni(CN)_4^{2-}$, $Ni(CO)_4$.
   c. $Co(NH_3)_6^{3+}$, $CoCl_4^{2-}$.
   d. $Fe(CN)_6^{3-}$, $Fe(CN)_6^{4-}$, $Fe(CO)_5$.
   e. $Cr(NH_3)_6^{3+}$, $Cr(CO)_6$.

7.2   Determine the number of unpaired electrons and the LFSE for each of the following.
   a. $Fe(CN)_6^{4-}$.
   b. $Fe(H_2O)_6^{3+}$.
   c. $Co(NH_3)_6^{3+}$.
   d. $Cr(NH_3)_6^{3+}$.
   e. $Ru(NH_3)_6^{3+}$.
   f. $PtCl_6^{2-}$.
   g. $CoCl_4^{2-}$ (tetrahedral).

7.3   Discuss briefly the factors working for and against the maximum spin state of d electrons in transition metal complexes.

7.4   Explain why square planar complexes of transition metals are limited (other than those of planar ligands such as porphyrins) to those of **a.** d^7, d^8, and d^9 ions and **b.** very strong field ligands which can serve as π acceptors.

7.5   For which d^n configurations would no Jahn-Teller splitting be expected for the tetrahedral case (ignore possible low-spin cases).

7.6   Give the orbital occupancy (identify the orbitals) for the Jahn-Teller splitting expected for tetrahedral complexes with high-spin d^3 and d^4 configurations. Indicate the nature of the distortions expected.

7.7   **a.** Why are low-spin complexes usually not encountered for tetrahedral coordination? **b.** Octahedral splitting is expressed as 10 Dq. What should be the splitting for ML_8 with cubic coordination? Assume the same ligands at the same distance as for the octahedral and tetrahedral cases.

7.8   Negative ions might be expected to create stronger ligand fields than neutral molecules. Explain why OH^- is a weaker-field ligand than H_2O. Why is CO such a strong-field ligand?

7.9   Calculate the relative energies of the d orbitals for an ML_6 complex with trigonal prismatic coordination ($\mathbf{D}_{3h}$), assuming that the ligands are at the same angle relative to the xy plane as for a regular tetrahedron. (*Hint:* Start with the tetrahedral case, but allow for three ligands up and down, instead of two. The degeneracy of the *d* orbitals is the same as for other $\mathbf{D}_{3h}$ complexes, such as trigonal, ML_3, and trigonal bipyramidal, ML_5, cases.)

7.10 Calculate the relative energies of the *d* orbitals for the following complexes, assuming $Dq(X) = 1.40\ Dq(Y)$ and that where X and Y are along the axis (trans to one another), the field strength is the same as for two equivalent ligands with the average field strength of X and Y. Use *z* as the unique axis.
 a. MX_5Y.
 c. cis-$[MX_4Y_2]$ (both Y's in *x,y* plane).
 b. trans-$[MX_4Y_2]$.
 d. facial-$[MX_3Y_3]$ (see Figure 8.7).

7.11 Compare the splitting of the e_g ($\mathbf{O}_h$) orbitals (difference in energies of d_{z^2} and $d_{x^2-y^2}$) for the cases in Problem 7.10 with the spectral results of p. 287.

7.12 Calculate the relative energies of the *d* orbitals (see Section 7.6) for ML_5 in the TBP($\mathbf{D}_{3h}$) and SP ($\mathbf{C}_{4v}$) geometries. For high-spin $d^6 - d^9$ metal ions, for which cases is there a significant preference for one of the two geometries?

7.13 Identify the ground state with the spin multiplicity for the following cases in *a.* octahedral complexes and *b.* tetrahedral complexes.

$$Cu^{2+},\ V^{3+},\ Cr^{3+},\ Mn^{2+},\ Fe^{2+},\ \text{and}\ Ni^{2+}$$

7.14 Explain why the ligand field *(d-d)* bands are shifted only slightly for the $[Co(NH_3)_5X]^{2+}$ ions, but the charge transfer bands are shifted greatly.

7.15 Common glass used for windows and bottles appears colorless, but when viewed through the edge it appears faintly green. Fe^{3+} causes the color. Why is it so faintly colored? Would you expect one broad, very weak absorption peak or many weak peaks?

7.16 For $Cr(NH_3)_6^{3+}$ there are two absorption bands observed at 21,500 cm^{-1} and 28,500 cm^{-1} and a very weak peak at 15,300 cm^{-1}. Assign the bands and account for any missing spin-allowed bands. Calculate Dq for NH_3 using the Orgel diagram. Account for any discrepancy between the observed position of any of the spin-allowed bands and that expected from the Orgel diagram.

7.17 The following data are available for $Ni(H_2O)_6^{2+}$ and $Ni(NH_3)_6^{2+}$.

$Ni(H_2O)_6^{2+}$	$Ni(NH_3)_6^{2+}$	
8,600 cm^{-1}	10,700 cm^{-1}	
13,500	17,500	
25,300	28,300	
15,400	15,400	Very weak peaks for both
18,400	18,400	complexes

Assign the bands. Calculate 10 Dq and the expected positions of the spin-allowed bands. Account for any discrepancy between the experimental and calculated energies of the bands. Account for the relative position of corresponding bands for the two complexes.

7.18 Interpret the following comparisons of intensities of absorption bands for transition metal complexes.
 a. Two isomers of a Co(III) complex believed to be *cis*- and *trans*-isomers give the following spectral features:

 Both give two absorption bands in the visible region. Complex A has two symmetrical bands with $\epsilon = 60$–80. The lower energy band for B is broad with a possible shoulder and has lower intensity. Assign the isomers. Explain.

 b. An octahedral complex of Co(III), with an amine and Cl$^-$ coordinated, gives two bands with $\epsilon = 60$–80, one very weak peak with $\epsilon = 2$ and a high energy band with $\epsilon = 20,000$. What is the presumed nature of these transitions? Explain.

c. Two complexes of Ni(II) are believed to be octahedral and tetrahedral. Each has three absorption bands, but complex A has $\epsilon \cong 10$ and B has $\epsilon \cong 150$. Which probably is the tetrahedral complex? Explain. Measurement of what physical property would exclude the possibility of either complex being square planar?

7.19 For a square planar complex with the ligands lying along the x and y axes, indicate all of the metal orbitals that may participate in σ bonds. Sketch one ligand group orbital that could enter into a σ_g molecular orbital and one that could enter into a σ_u molecular orbital. Repeat the above, this time for π bonds.

7.20 Give a pictorial approach to obtain the bonding molecular orbitals for σ and π bonding in square planar complexes.

7.21 Use the group theoretical approach to obtain the representations and LGO for σ bonding in square planar complexes.

7.22 The molecular orbitals formed by the six p π orbitals in benzene may be depicted as shown below (top half only shown, the sign of ψ reverses on going through the plane of the sheet of paper).

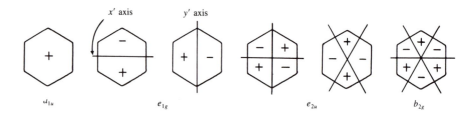

Consider a molecule of bis(benzene)chromium as having a chromium atom at the origin of a set of Cartesian axes with a benzene ring on each side, centered on, and perpendicular to the z axis. Indicate below each of the above figures the metal orbital(s) that would combine with the group orbitals of the ligands and the type of molecular orbital (σ_g, δ_u, etc.) that would be formed. Draw an energy-level diagram showing ligand orbitals, metal orbitals, and molecular orbitals with appropriate labeling.

VIII

Stereochemistry of Coordination Compounds

The previous chapter dealt with the spectra and the approaches to bonding of coordination compounds. Most of the examples considered had coordination numbers four or six. We shall now examine the structures and stereochemistry for a wider range of coordination numbers. After discussing types of isomerism that might be encountered for any coordination number, we will treat the individual coordination numbers, beginning with the most important cases—four and six. Organometallic compounds, such as ferrocene, and polynuclear carbonyls have very special structural features and are treated in later chapters.

8.1 ISOMERISM OF COORDINATION COMPOUNDS

8.1.1 Simple Types of Isomerism

Reagent chromium chloride, labeled $CrCl_3 \cdot 6H_2O$, is usually a green crystalline compound. When dissolved in water it produces a green solution, which upon standing for a week or more turns violet. Violet crystals, which have the composition $CrCl_3 \cdot 6H_2O$, can be obtained from the solution. Actually, there are three isomers with this composition. They differ as follows.

$[Cr(H_2O)_6]Cl_3$	Violet
$[Cr(H_2O)_5Cl]Cl_2 \cdot H_2O$	Blue-Green
$[Cr(H_2O)_4Cl_2]Cl \cdot 2H_2O$	Green

In the violet isomer the six water molecules are coordinated to the chromium and the three chloride ions are separate ions in the solid and in solution. In the second isomer, a chloride

Table 8.1   Some types of isomerism among coordination compounds

Examples	*Type of Isomerism*
$[Co(NH_3)_6][Cr(CN)_6]$ and $[Cr(NH_3)_6][Co(CN)_6]$	Coordination
$[Pt(NH_3)_4Cl_2]Br_2$ and $[Pt(NH_3)_4Br_2]Cl_2$	Ionization
$[Co(NH_3)_5(NO_2)]Cl_2$ and $[Co(NH_3)_5(ONO)]Cl_2$	Linkage
$[Pd(bipy)(SCN)_2]$ and $[Pd(bipy)(NCS)_2]$	

ion displaces from the coordination sphere one water molecule, which is held as a molecule of water of crystallization. In the third isomer there are two coordinated chloride ions and two molecules of water of crystallization. Commercial chromium chloride is usually a mixture of the two green isomers. This type of isomerism is known as *hydrate isomerism.*

Several other kinds of isomerism are illustrated in Table 8.1. *Coordination isomers* involve the exchange of the metal ions between the cation and anion. *Ionization isomers* differ only in the exchange of a coordinated anion for one present to maintain charge balance. Conceivably, many ligands might form *linkage isomers* similar to the nitro complex (coordination through N) and nitrito complex (coordination through O) shown. Ligands that form linkage isomers are called *ambidentate ligands.*[1] Most ligands normally coordinate through the same atom. Cyanide ion usually coordinates through C, except when it serves as a bridging group. In the clathrate compound $Ni(NH_3)_2(CN)_2C_6H_6$ (see p. 191) the six-coordinate Ni^{2+} ions are bonded to two NH_3 and the N atoms of four bridging CN^- ions; the four-coordinate Ni^{2+} ions are bonded to the C of the CN^- ions. When bonded to the less electronegative carbon, Ni should have the lower coordination number, because more negative charge is transferred to the Ni in the Ni—C bond (the electroneutrality principle, p. 263). The thiocyanate ion coordinates through the nitrogen atom in most cases—for example, $[Co(NCS)_4]^{2-}$, but it can coordinate through S, as in $[Hg(SCN)_4]^{2-}$. These are the isomers expected, since Co(II), a relatively hard Lewis acid, is bonded through N (hard Lewis base) and Hg(II), a soft Lewis acid, is bonded through S (soft Lewis base) (see p. 541). Two isomers have been obtained for $[Pd(bipy)(SCN)_2]$. Although the S-bonded isomer is stable in the solid at room temperature, it rearranges to give the N-bonded isomer at elevated temperature or in solution. The structures were assigned from infrared data.

8.1.2 Stereoisomerism

Another significant isomer type is stereoisomerism—isomers that differ only in the spatial arrangement of bonded groups. Stereoisomers bearing a mirror-image relationship to each other are termed *enantiomers.* Enantiomers have identical bond lengths and bond angles, except for the sense of the angles. These mirror-image isomers are nonsuperimposable only if the molecule has no S_n axis (see p. 89). Geometric isomers differ in the spatial arrangement of the connected atoms, other than the handedness—a pair of enantiomers is classified as the *same* geometrical isomer. The words *cis* and *trans* (Latin "on the same side" and "across," respectively) were

[1] A. H. Norbury and A. I. P. Sinha, *Quart. Rev.* 1970, *5,* 69; J. L. Burmeister, *Coord. Chem. Rev.* 1968, *3,* 225.

first used to describe isomers in van't Hoff's classic 1874 paper, in which he not only developed the molecular model for optical activity, but also *predicted* the existence of geometrical isomers of olefins. The *cis, trans* nomenclature applies also to square planar and octahedral complexes.

8.2 COORDINATION NUMBER FOUR

Coordination compounds with C.N. 4 are encountered with tetrahedral and square planar configurations.

8.2.1 Tetrahedral Complexes

Tetrahedral complexes display only optical isomerism. Because of lability, few optical isomers in which four different groups are coordinated to the metal have been prepared. Most of these are organometallic compounds of Fe, Mn, and Ti of the type shown in Figure 8.1. In this case, when the acetyl group is reduced the configuration at the Fe center is retained. When the Fe center is involved in the reaction, inversion sometimes occurs.[2] Most of the tetrahedral complexes that have been resolved contain unsymmetrical chelating agents, giving isomerism similar to that of the organic spiran compounds. Thus bis(benzoylpyruvato)beryllium(II) can be separated into enantiomers (see Figure 8.2). Other tetrahedral complexes of Be, B, Zn, and Cu(II) have been resolved. In many cases we cannot rule out the possibility that the Cu(II) and Zn(II) complexes are really six-coordinate (water or anion coordinated).

8.2.2 Square Planar Complexes

Werner (1893) assumed the planar structure for Pt(II) complexes, because this structure led to the prediction of the number of isomers encountered. The planar structures of $K_2[PtCl_4]$ and $K_2[PdCl_4]$ were confirmed by x-ray studies in 1922. Pauling (1932) predicted that Ni(II) also

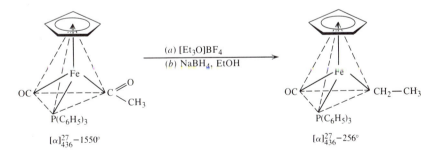

Figure 8.1 Optically active compounds of iron. The angles of rotation of plane-polarized light at 27°C and 436 nm are given. (From H. Brunner, "Optical Activity," in *The Organic Chemistry of Iron*, E. A. Koerner Von Gustorf, F.-W. Grevels, and I. Fishler, Eds., Academic Press, New York, 1978, p. 299.)

[2]H. Brunner, "Chiral Metal Atoms in Optically Active Organo-transition-metal Compounds," *Adv. Organomet. Chem.* 1980, *18*, 151.

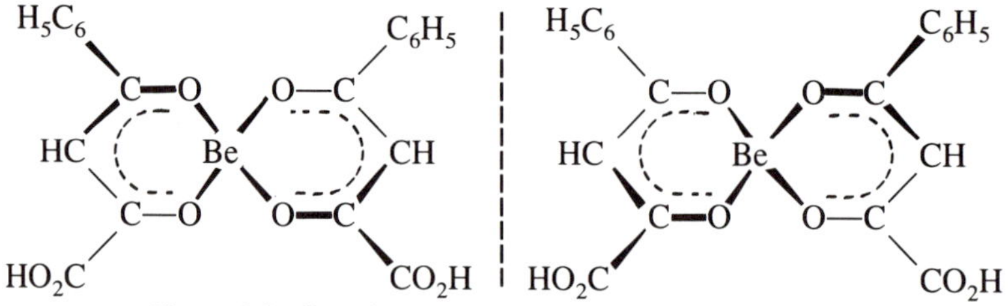

Figure 8.2 Enantiomers of bis(benzoylpyruvato)beryllium.

would form planar complexes. Many planar complexes are now known for Rh(I), Ir(I), Ni(II), Pd(II), Pt(II), Cu(II), Ag(II), Au(III), and Co(II).

The compound [Pt(NH$_3$)$_2$Cl$_2$] exists as *cis*- and *trans*-isomers. The isomers are prepared as shown in Figure 8.3, taking advantage of the *trans effect* (see p. 371). The Cl$^-$ has a greater *trans* directing influence than NH$_3$ does, so a group (circled) *trans* to Cl$^-$ is replaced in the second step. The compounds *cis*-[Pt(NH$_3$)$_2$Cl$_2$] and *cis*-[Pt(en)Cl$_2$] are effective antitumor drugs; *trans*-[Pt(NH$_3$)$_2$Cl$_2$] is not very effective.

Three isomers have been reported for [Pt(NH$_3$)(NH$_2$OH)(NO$_2$)(C$_5$H$_5$N)]NO$_2$, corresponding to the ammonia *trans* to each of the other three groups. The isomers, represented as [Pt(*abcd*)] are as follows.

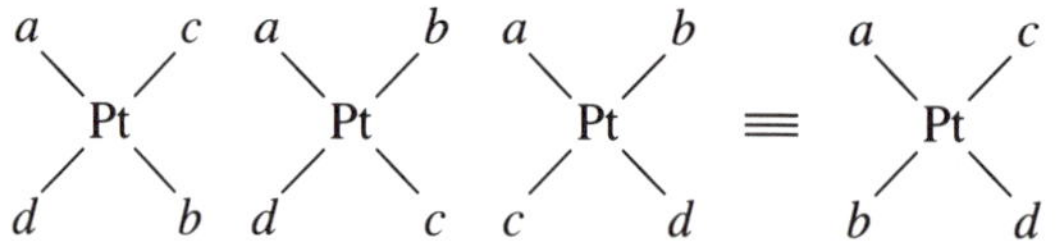

Even four different groups coordinated to the metal ion do not lead to optical isomerism, because each isomer has a plane of symmetry (the molecular plane) and hence mirror images are superimposable. Nevertheless, the ingenuity of Mills and Quibell (1935) provided an example of an optically active planar complex. The compound [Pt(NH$_2$CHC$_6$H$_5$·CHC$_6$H$_5$NH$_2$)(NH$_2$C(CH$_3$)$_2$CH$_2$NH$_2$)]Cl$_2$ should be optically active if planar, because the plane

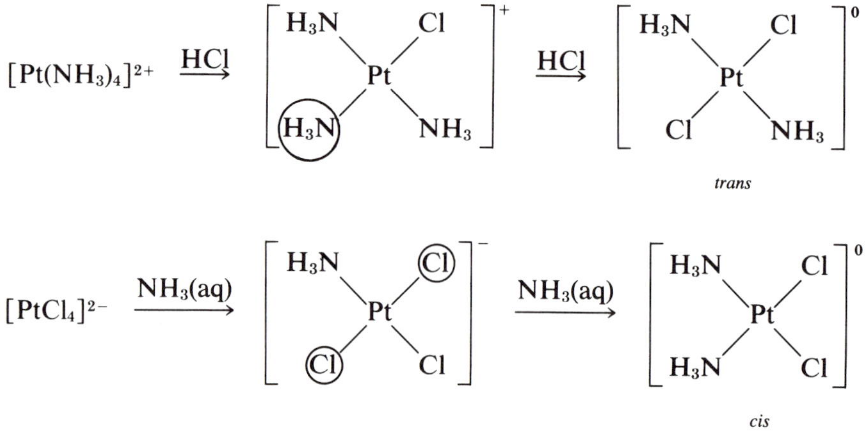

Figure 8.3 Reactions for the formation of *cis*- and *trans*-[Pt(NH$_3$)$_2$Cl$_2$].

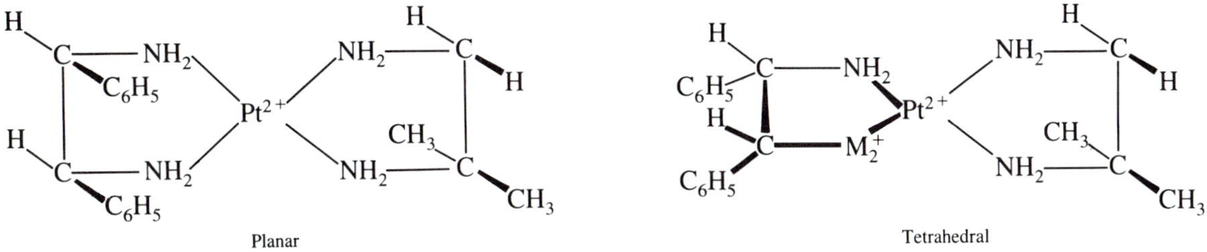

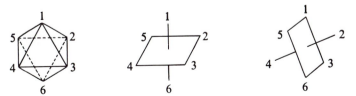

Figure 8.4 Possible configurations of a Pt(II) complex.

of symmetry is eliminated; if it were tetrahedral, on the other hand, it would possess a plane of symmetry and hence could not be optically active (Figure 8.4). In the representation of the tetrahedral configuration, the ring to the right would be in the plane of the paper, and the one to the left, perpendicular to it. The plane of the paper represents a plane of symmetry.

8.3   COORDINATION NUMBER SIX

Probably the most common and certainly the most thoroughly studied complexes are those with C.N. 6. Several possible configurations have been considered, but the weight of physical and chemical evidence strongly favors the octahedron, although a number of trigonal prismatic complexes is now known (p. 317). The octahedron is sometimes distorted by the structure of the ligands or because of the Jahn-Teller effect (p. 272).

8.3.1   Enumeration of Isomers

That the octahedron can be represented in many ways is illustrated in Figure 8.5, which emphasizes both the three-fold axis of the octahedron and the equivalence of the substitution sites. Here we shall use the first representation unless the symmetry being discussed is discerned more easily in one of the other representations.

In drawing isomers of a complex: (1) name the unique structural features; (2) draw the structure in a fashion that emphasizes the symmetry present; (3) use enumeration equations to check the results.

It is useful to draw a *cis* pair of ligands to emphasize the C_{2v} parent geometry and a *trans* pair to emphasize the D_{4h} parent geometry of the remaining positions (see Fig. 8.6). For the octahedral complex $M(abcde_2)$, the *trans*-e_2 arrangement gives rise to the three isomers having *trans ab, ac,* and *ad*. All of these are optically inactive, because of the mirror plane through $M(abcd)$.

Figure 8.5   Representations of the octahedron.

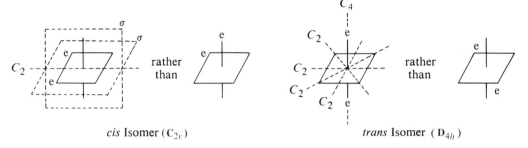

cis Isomer ($\mathbf{C}_{2v}$) • *trans* Isomer ($\mathbf{D}_{4h}$)

Figure 8.6 Convenient representations of *cis* and *trans* isomers of an octahedral complex.

For *n* different ligands the total number of stereoisomers possible is $n!/\sigma$ where σ is the symmetry number,[3] which is identical to the order of the rotational group except for linear molecules (where it is 2 for $\mathbf{D}_{\infty h}$ and 1 for $\mathbf{C}_{\infty v}$). Thus for the *trans*-e_2 arrangement just considered, the *trans*-$[M(e_2)]$ arrangement pictured in Figure 8.6 belongs to the $\mathbf{D}_{4h}$ point group and the rotational group is $\mathbf{D}_4$, a group of order 8. The number of stereoisomers is $4!/8 = 3$. For the *cis*-$[M(e_2)]$ arrangement as pictured in Figure 8.6, the point group is $\mathbf{C}_{2v}$ and the rotational group is $\mathbf{C}_2$. For $[M(abcde_2)]$ with the *cis*-e_2 arrangement, the number of stereoisomers would be $4!/2 = 12$ (where 2 is the order of the $\mathbf{C}_2$ group). Since neither of the mirror planes of the $\mathbf{C}_{2v}$ *cis*-e_2 structure remains after adding the other ligands, we have 6 *d,l* pairs. Arbitrarily picking the remaining *trans* position, we can immediately make the *trans* pairings

that is, *ab, ac, ad, bc, bd,* and *cd*—and draw these and their mirror images.

The rotational group is the group or subgroup with only rotations—no center of symmetry, planes of symmetry, or S_n axes. Thus the rotational groups lack any letter subscripts—$\mathbf{O}$, $\mathbf{D}_4$, $\mathbf{D}_3$, $\mathbf{C}_4$, $\mathbf{C}_2$, etc.

Example Draw all the isomers of M($abcd\overset{\frown}{ee}$), where $\overset{\frown}{ee}$ is a bidentate ligand.

Solution The presence of a bidentate ligand limits the number of isomers possible, because normally a chelated ligand can span only *cis* (adjacent) positions. In this case we introduce the bidentate ligand first and use it for our basic framework. For the configuration M($\overset{\frown}{ee}$), Figure 8.6 shows the $\mathbf{C}_2$ axis and σ planes. The point group is $\mathbf{C}_{2v}$ and the rotational group is $\mathbf{C}_2$, giving $4!/2 = 12$ stereoisomers (6 *d,l* pairs) for $[M(abcd\overset{\frown}{ee})]$. Placing the *a* ligand in an "out-of-plane" position in Figure 8.8 leads to three *trans* pairs *ab, ac,* and *ad*. Starting with *b* and then *c* in the "out-of-plane" position, we get *bc* and *bd* and then *cd* as the unique *trans* pairs. Each of the six geometrical isomers can exist as a pair of enantiomers for 12 stereoisomers—one pair of enantiomers is shown.

---

[3]For a group theoretical approach to determining the number of isomers, see D. H. McDaniel, *Inorg. Chem.* 1972, *11*, 2678.

1,2,3 Isomer
Facial C_{3v}.

1,2,6 Isomer
Meridional C_{2v}. **Figure 8.7** Isomers of $M(a_3b_3)$.

A set of three similar ligands may be arranged on an octahedron in an all-*cis* fashion, giving the facial or *fac* isomer, or with one pair *trans*, giving the meridional or *mer* isomer (Figure 8.7). The *fac* and *mer* isomers of an Ma_3b_3 octahedral complex are optically inactive.

> The more general nomenclature scheme uses the numbering system shown in Figure 8.5. The *cis*- and *trans*-$[M(a_2b_4)]$ isomers are 1,2- and 1,6-isomers, respectively. For $M(a_3b_3)$, the facial isomer is 1,2,3 and the meridional isomer is 1,2,6.

The known examples of optically active octahedral complexes have at least one chelate ring. Three of the five possible geometrical isomers of $[Pt(NH_3)_2(py)_2Cl_2]^{2+}$ (see Figure 8.9) have been prepared. The geometrical isomers with all like groups *cis* to one another should be optically active.

8.3.2 Optical Activity

Enantiomers Based upon the Chirality of Chelate Rings

The compounds $[Co(NH_3)_6]Cl_3$ (O_h) and $[Co(en)_3]Cl_3$ (D_3) are very similar, but only the latter can be resolved into enantiomers (Figure 8.10). We can prove the nonsuperimposability either by examining models, or, with experience, by examining planar representations as shown. The

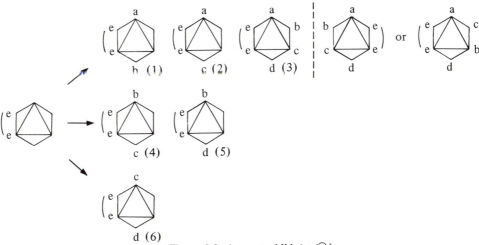

Figure 8.8 Isomers of [Mabcdee].

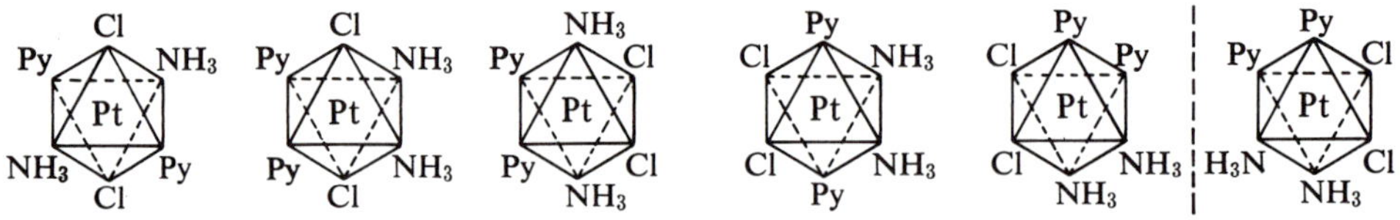

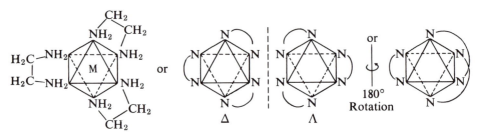

Figure 8.9 Isomers of $[Pt(NH_3)_2(py)_2Cl_2]^{2+}$

Figure 8.10 Enantiomers of $[M(en)_3]^{n+}$ (D_3).

rotation of the Λ-isomer by 180° demonstrates that this orients one chelate ring as for the Δ-isomer, whereas the other two rings are oriented differently. The other method of proving nonsuperimposability is to examine the symmetry properties. Enantiomers can exist only if there is no S_n axis (including $\sigma \equiv S_1$ and $i \equiv S_2$). The O_h point group has several of these symmetry elements, but there are none for D_3 (or any rotational group). In this case, the *chirality* (Greek word "handedness") arises from the spiral configuration of the chelate rings, providing the basis for the Δ (right) and Λ (left) notation (see p. 310). The complexes *cis*-$[Co(NH_3)_4Cl_2]Cl$ (C_{2v}, violet) and *trans*-$[Co(NH_3)_4Cl_2]Cl$ (D_{4h}, bright green) have the same colors as the corresponding complexes *cis*-$[Co(en)_2Cl_2]Cl$ (C_2) and *trans*-$[Co(en)_2Cl_2]Cl$ (D_{2h}). Of these, only *cis*-$[Co(en)_2Cl_2]Cl$ belongs to a rotational group and is optically active. The isomers of $M(a_3b_3)$ (see Figure 8.7) are not optically active, but the complex $M(\widehat{ab})_3$, with

three unsymmetrical bidentate ligands—for example, $\left[Co\left(\begin{smallmatrix} NH_2 \\ \\ O-C \end{smallmatrix} \begin{smallmatrix} \\ CH_2 \\ \end{smallmatrix} \right)_3 \right]$, tris(gly-

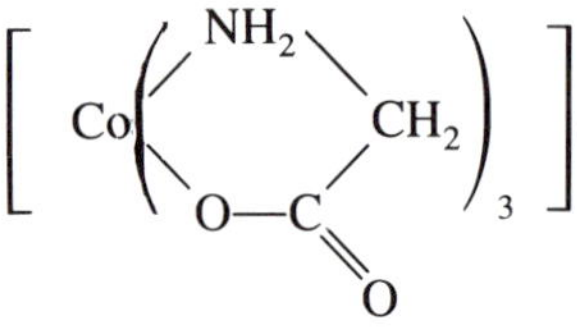

cinato)cobalt(III)—exists as facial (C_3) and meridional (C_1) isomers, each of which exists as enantiomeric pairs.

Table 8.2 lists the metals for which octahedral complexes have been resolved. Some of

Table 8.2 Metals for which octahedral complexes have been resolved

								Al	Si	
	Ti	Cr	Fe	Co	Ni	Cu	Zn	Ga	Ge	As
Y			Ru	Rh			Cd			
Gd,Nd			Os	Ir	Pt					

the gaps in the table doubtless stem from the small amount of work done with metals, other than Group VIII, beyond the first transition series.

Resolution Procedures

The physical properties of the enantiomers of a complex such as [Co(en)$_3$]Cl$_3$ are identical except for the interaction with polarized light. However, if the chloride ions are replaced by optically active ions, the resulting salts might differ in properties such as solubility. [Co(en)$_3$]$^{3+}$ is commonly resolved by replacing two chloride ions by the *d*-tartrate ion, followed by fractional crystallization. The diastereomers[4] (+)-[Co(en)$_3$]Cl*d*-tart and (−)-[Co(en)$_3$]Cl*d*-tart, are not mirror-images, because the tartrate ion has the same configuration in each salt. Upon crystallization, the (+)[Co(en)$_3$]Cl*d*-tart separates in the form of large crystals. The (−) complex is much more soluble, and the solution becomes a thick gelatinous mass before fine needles begin to crystallize. We can convert the diastereoisomers to the active chloride complexes by treating them with concentrated hydrochloric acid. The active complex is stable in solution for many weeks without racemization.

The antimony tartrate ion[5] and the *d*-α-bromocamphor-π-sulfonate ion have been useful resolving agents for cationic complexes. Anionic complexes—for example, [M(C$_2$O$_4$)$_3$]$^{3-}$— usually are resolved using an optically active cation such as those formed by the bases strychnine or brucine. Resolved complex ions such as those of [Co(edta)]$^-$, [Co(en)$_3$]$^{3+}$, or *cis*-[Co(en)$_2$(NO$_2$)$_2$]$^+$ can be used as resolving agents.

Partial resolution of active complexes also has been achieved by differential adsorption on an optically active solid such as quartz, starch, or sugar. Failure to resolve a complex by one or several means does not justify the conclusion that the complex is not dissymmetric. The method tried might not be applicable for that complex, or the complex might racemize so rapidly that the optical isomers cannot be isolated.

Since racemic mixtures contain equal amounts of the enantiomers, complete resolution would give 50% of the material as the (+) isomer. Dwyer was able to isolate nearly 100% of [Fe(phen)$_3$]$^{2+}$ as the (−) isomer by fractional crystallization of the antimony tartrate salt. The complex racemizes at an appreciable rate in solution, but very slowly in the solid. As the (−) isomer crystallizes, the (+) isomer undergoes racemization in solution. Werner (1912) observed a similar situation in the resolution of [Cr(C$_2$O$_4$)$_3$]$^{3-}$, which racemizes rapidly in solution.

The presence of an optically active substance in solution can affect greatly the equilibrium between the enantiomers of a complex. Pfeiffer noted that the optical rotation of a solution of zinc *d*-camphor-π-sulfonate changed from +0.98° to +0.09° upon addition of enough 1,10-phenanthroline to form the [Zn(phen)$_3$]$^{2+}$ ion. The complex ion isolated from the solution is inactive. Presumably, the active anion causes a shift in the equilibrium between (+)- and (−)-[Zn(phen)$_3$]$^{2+}$ in solution, perhaps through ion-pair formation. However, Dwyer has shown also that the rates of racemization of (+)-[Ni(phen)$_3$]$^{2+}$ and (−)-[Ni(phen)$_3$]$^{2+}$ differ

---

[4]Whereas enantiomers are mirror-image isomers, diastereoisomers (also called diastereomers) are not. The optical isomers (+)-[Co(L-alanine)$_3$] and (−)-[Co(L-alanine)$_3$] are diastereomers because both contain L-alanine, not one L- and one D-alanine.

[5]The antimony tartrate ion often is represented as SbO(*d*-tartrate)$^-$, but two tartrate ions bridge two antimonate ions, (+)-bis[μ-tartrato(4−)]diantimonate(2−).

when in the presence of optically active anions *or cations*. In the latter case, the unusual ion-pair interactions cannot be responsible.

Designation of Absolute Configurations

Various symbols (D or L, R or S, P or M, and Δ or Λ) have been used to designate absolute configurations of octahedral complexes. Piper suggested Δ for a right-handed spiral (and Λ for a left spiral) of chelate rings about the C_3 axis of a tris(bidentate) complex such as $[Co(en)_3]^{3+}$. Such a complex ($\mathbf{D}_3$ symmetry) also has three C_2 axes (perpendicular to C_3). For the $\Delta(C_3)$ enantiomer, two chelate rings define the opposite (Λ) chirality about each of the C_2 axes. There are many more optically active complexes with C_2 axes than with C_3 axes, and for a while the symbols Δ and Λ were used to define chirality with reference to either C_2 or C_3 axes. The $(+)$-$[Co(en)_3]^{3+}$ ion (positive optical rotation at the sodium $\mathbf{D}$ line, unless otherwise designated) is $\Lambda(C_3)$, but $\Delta(C_2)$.

   The 1970 IUPAC rules[6] avoid the use of a reference axis. Two skewed lines not orthogonal to each other define a helical system—either of them can be considered the axis of a cylinder, with the other tangent to a helix on the surface of the cylinder. Helices are designated as Δ (right) and Λ (left). For $(+)$-$[Co(en)_3]^{3+}$, using any pair of octahedral edges spanned by chelate rings (Figure 8.11) as the skewed lines, the designation is Λ—the same as relative to the C_3 axis. For $(+)$-*cis*-$[Co(en)_2Cl_2]^+$, the only pair of octahedral edges spanned by chelate rings defines a left (Λ) helix—the same as relative to the "pseudo" C_3 axis. In orienting the octahedron so that one octahedral edge is spanned on the back side and horizontal (A----A in Figure 8.12), the other edge spanned (in front) is tipped down to the right for Δ and to the left

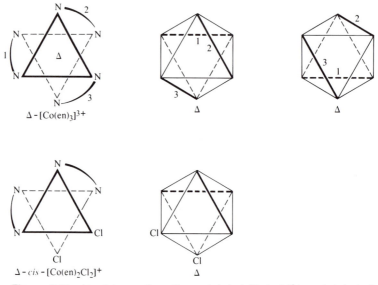

Figure 8.11 Absolute configurations of Δ-$(-)$-$[Co(en)_3]^{3+}$ and Δ-$(-)$-*cis*-$[Co(en)_2Cl_2]^+$

[6]See *Inorg. Chem.* 1970, *9*, 1; or the reference cited for Appendix B.

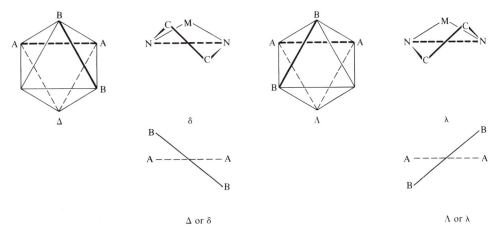

Figure 8.12 Projection of skewed lines with A---A below the plane of the paper and B—B above the paper. The corresponding elements of the Δ and Λ octahedra are shown as well as the δ and λ conformations of a chelate ring formed by ethylenediamine.

for Λ. The same system applies to the designation of enantiomeric chelate-ring conformations. An ethylenediamine chelate ring (*gauche* configuration) is viewed with the octahedral edge spanned horizontal (A----A). The C—C bond (closer to the viewer) is skewed down to the right for a δ conformation and down to the left for a λ conformation. The direction of viewing is not important, but the more distant line should be horizontal. If you view the ring from the side of the metal and tip the ring so that the more distant line (the C—C bond) is horizontal, the nearer line (the octahedral edge) is tipped down to the right for the δ isomer.

The sign of optical rotation (at the sodium **D** line, unless otherwise specified) of an optically active ligand is designated. The absolute configuration of the ligand (*R*, for *rectus;* and *S*, for *sinstra*) can be given if known: for example, *S*-(+)-pn. The symbols *D* and *L* are still in limited use, particularly for naturally occurring amino acids, all of which have the *L* (or *S*) configuration.

Stereospecific and Stereoselective Effects

The complex ion [Co(edta)]⁻ can be resolved into enantiomers, but if we use optically active (−)-propylenediaminetetraacetic acid, only $(+)_{546}$-[Co(−)-pdta]⁻ results. The reaction appears to be completely stereospecific and the complex ion does not racemize under conditions for the racemization of [Co(edta)]⁻. At first it was believed that racemic propylenediamine formed only two enantiomeric tris complexes, $(+)$-[Co(+)-pn₃]³⁺ and $(-)$-[Co(−)-pn₃]³⁺. Subsequent work by Dwyer and coworkers showed that others of the eight combinations are formed, even though these two isomers are favored. Such a reaction is stereoselective.

In a classic paper, Corey and Bailar[7] provided the first explanation of stereoselective

[7]E. J. Corey and J. C. Bailar, Jr. *J. Am. Chem. Soc.* 1959, *81*, 2620. The ring conformations were designated *k* and *k′* as shown in Figure 8.13, but an error in another figure reversed these designations and this error was carried forth in some papers, leading to confusion. The error apparently was caused by the reversal of a photographic print, since the original print supplied by Professor Bailar and used in the first edition of this book was correct. The symbols λ and δ are used now.

effects in metal complexes. The *gauche* configuration of the chelate ring in an ethylenediamine complex can be compared to the chair form of cyclohexane. The substituents on the puckered ring can be classified as approximately axial or equatorial. In a square planar complex of the type $[M(en)_2]^{n+}$, the two puckered rings can be arranged as shown in Figure 8.13. In the $\lambda\delta$ form the two en rings are mirror images. The clockwise sequence of substituents, *a* or *e*, from above in the chelate ring for λ is M*eaea* and for δ it is M*aeae*. The $\lambda\lambda$ form, with the hydrogens staggered around both rings, should be more stable than the $\lambda\delta$ form, in which the hydrogen atoms on adjacent nitrogen atoms are directly opposed.

In octahedral complexes the interactions are not so apparent, but the situation is similar. There are four different forms possible for an octahedral complex of the type $[M(en)_3]^{n+}$: $\lambda\lambda\lambda$, $\lambda\lambda\delta$, $\lambda\delta\delta$, and $\delta\delta\delta$. The four forms differ in energy, with the energies of $\lambda\lambda\delta$ and $\lambda\delta\delta$ intermediate between those of $\lambda\lambda\lambda$ and $\delta\delta\delta$. The hydrogen–hydrogen and carbon–hydrogen interactions were evaluated for the $\Delta(\lambda\lambda\lambda)$ form [or the enantiomeric $\Lambda(\delta\delta\delta)$ form] and the $\Delta(\delta\delta\delta)$ form [or the enantiomeric $\Lambda(\lambda\lambda\lambda)$ form]. In Figure 8.14 these are designated *lel* (the C—C bonds approximately paral*lel* to the C_3 axis) and *ob* (the C—C bonds *ob*lique to the C_3 axis). The *lel* isomer was found to be more stable than the *ob* isomer by about 7.5 kJ/mole (or 2.5 kJ/ligand). The hydrates of $[Cr(en)_3][Ni(CN)_5]$ contain none of the expected $\Lambda(\delta\delta\delta)$

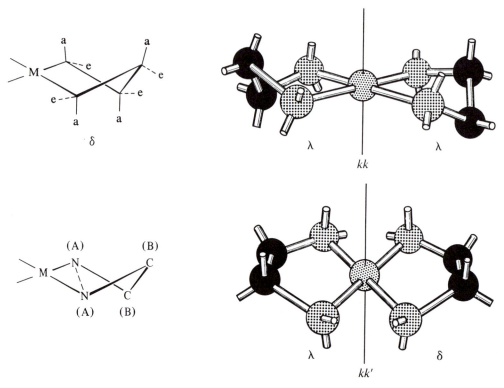

Figure 8.13   Conformation of the chelate ring in a planar ethylenediamine complex. The skewed lines (*AA* and *BB*) considered in Figure 8.12 are shown for comparison. (Adapted from E. J. Corey and J. C. Bailar, Jr., *J. Am. Chem. Soc.* 1959, *81*, 2620. Copyright 1959, American Chemical Society.)

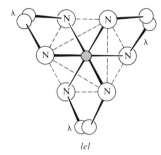
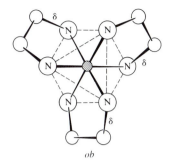

lel *ob*

Figure 8.14 The *lel* and *ob* conformations of Δ-[M(en)$_3$]$^{n+}$ viewed along the C$_3$ axis. (Adapted from Y. Saito in K. Nakamoto and P. J. McCarthy, *Spectroscopy and Structure of Metal Chelate Compounds*, Wiley, New York, 1968, p. 29.)

isomers, but instead the three conformational isomers Λ(δδλ), Λ(δλλ), and Λ(λλλ).[8] Since the energy barrier for inversion is small the conformations are not fixed in solution. The relative stabilities of the conformational isomers considered above are based on enthalpies. Free energies actually favor slightly the "mixed" conformations, because of the statistical entropy effect.

For an optically active ligand such as (+)-propylenediamine, one conformation (δ) with the methyl group equatorial (Figure 8.15) is more stable than the other by about 8 kJ/mole of pn. Thus (+)-pn has a strong preference for the δ conformation, so the preferred isomers are Λ(δδδ) (the more stable *lel* configuration with all methyls equatorial) and Δ(δδδ) (the less stable *ob* configuration, but all methyls are equatorial). Because of the opposing effects, *lel* is more stable than *ob* by only about 6.7 kJ/mole (*ca.* 2 kJ/pn). Both isomers have been isolated [as well as mixed (+)-pn, (−)-pn isomers], and the structures and absolute configurations were determined by x-ray methods for Δ-(−)-[Co(−)-pn$_3$]$^{3+}$ (*lel*) and Δ-(−)-[Co(+)-pn$_3$]$^{3+}$ (*ob*). Both are facial isomers (methyl groups on a face rather than around an edge).

Stereoselectivity can arise because of differences in thermodynamic stability (as noted), or because of kinetic effects. Metal ions might play important roles in the remarkable stereospecificity encountered in biological systems.

Circular Dichroism and Optical Rotatory Dispersion

Optical Rotatory Dispersion. Optical rotation, the rotation of the plane of polarization of plane-polarized light, is the usual test for optical activity. Fresnel (1823) explained this phenomenon as the result of different interactions of the optically active substance with the two

Figure 8.15 Enantiomeric conformations of a *gauche* chelate ring of S-(+)-propylenediamine placing the methyl group in axial (λ) and equatorial (δ) positions. Viewed along the C—C bond.

[8]K. N. Raymond, P. W. R. Corfield, and J. A. Ibers, *Inorg. Chem.* 1968, *7*, 843.

circularly-polarized components of plane-polarized light. The right circularly-polarized component's electric vector spirals to the right (E_r) along the direction of propagation (with the viewer looking toward the light source) and thus sweeps out a right spiral on the surface of a cylinder of radius equal to E_r, whereas the left circularly-polarized component spirals to the left. These two components of equal magnitude add vectorially to produce plane-polarized light (Figure 8.16). They travel with the same velocity in an optically inactive medium, but in an optically active medium one component is slowed relative to the other: that is, the indices of refraction differ, $n_\ell \neq n_r$. Since one component is slowed down more than the other, the resultant vector will be rotated through an angle α after passing through the optically active medium (Figure 8.17). A plot of angle of rotation versus wavelength (or frequency) is an *optical rotatory dispersion curve* (ORD) (Figure 8.18).

Circular Dichroism.   An optically active substance absorbs the two circularly polarized components to different extents. The differential absorption is known as *circular dichroism* (CD).

$$\Delta\epsilon = \epsilon_\ell - \epsilon_r$$

Optical rotation and circular dichroism together are known as the *Cotton effect*. We need not measure both ORD and CD, since the curves are described by the same set of three parameters: (1) the position of the electronic transition in wavelength or frequency, ν_{max}, for CD and the inversion point for ORD; (2) intensity, $\Delta\epsilon_{max}$, for CD and the trough-to-peak height for ORD; and (3) width or half-width (half-width at $\frac{1}{2}\Delta\epsilon_{max}$) for CD and trough-to-peak spread for ORD (Figure 8.18).

CD curves have the advantage of the simple shape (Gaussian), with $\Delta\epsilon$ vanishing not far from $\Delta\epsilon_{max}$. The different signs of CD peaks can be helpful in separating overlapping peaks. Extending far from the inversion center, an ORD curve's tail (a "normal" dispersion curve) is responsible for the observed optical rotation of a colorless substance such as sugar or tartaric acid in the visible region. This rotation is really a composite of the tails of "anomalous"

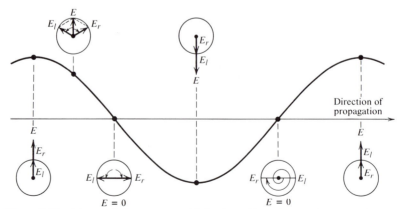

Figure 8.16   A beam of plane polarized light viewed from the side (sine wave) and along the direction of propagation at specific times (circles) where the resultant electric vector E and the circularly polarized components E_l and E_r are shown.

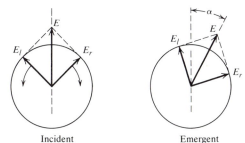

Incident	Emergent

Figure 8.17 Plane polarized light before entering and after emerging from an optically active substance.

dispersions for all optically active transitions of the molecule throughout the visible and ultraviolet regions. In ORD studies you always have to contend with an appreciable background from other transitions.

Stereochemical Information from CD. $[Co(en)_3]^{3+}$ demonstrates the value of CD studies. The absorption spectrum (Figure 8.19) is very similar to that of $[Co(NH_3)_6]^{3+}$, showing two symmetrical bands in the visible region. One would conclude that the effective field symmetry is $\mathbf{O}_h$ and that these are the $A_{1g} \rightarrow T_{1g}$ and $A_{1g} \rightarrow T_{2g}$ bands. In the region of the first (lower-energy) absorption band, the ORD curve shows a hint of a second Cotton effect near 23,000 cm^{-1}, but two CD peaks of opposite sign are unmistakable within this band. The symmetry of $[Co(en)_3]^{3+}$ is $\mathbf{D}_3$ and the triply degenerate states (for $\mathbf{O}_h$ symmetry) should split to give $T_{1g}(\mathbf{O}_h) \rightarrow A_2 + E(\mathbf{D}_3)$ and $T_{2g}(\mathbf{O}_h) \rightarrow A_1 + E\ (\mathbf{D}_3)$. We can observe both A_2 and E for the first band, but only a weak E peak for the second band, since A_1 is magnetic-dipole forbidden

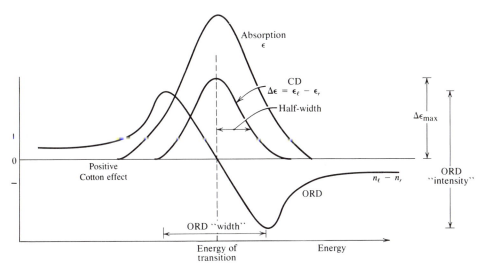

Figure 8.18 Theoretical curves for absorption, circular dichroism, and optical rotatory dispersion for a single electronic transition of one optical isomer. For the other optical isomer the Cotton effect will be negative for this transition, the CD curve will be negative, and the ORD curve will be the mirror image of the one shown, with the trough at lower energy than the peak.

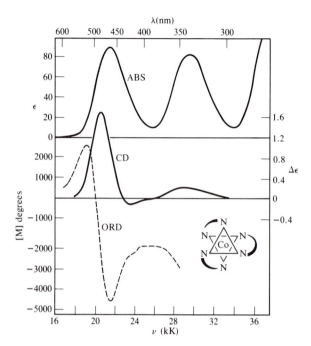

Figure 8.19 Absorption, ORD and CD curves of Λ-(+)-[Co(en)$_3$]Cl$_3$.

(p. 286).[9] The absorption spectrum does not reveal the presence of two components, because the peaks are broad and merge into one for small splittings. Mason used CD studies of single crystals of [Co(en)$_3$]$^{3+}$ to identify the major CD peak as *E*. This is an important case, since it was the first complex for which the absolute configuration was determined by x-ray crystallography. First resolved by Werner, it is used widely as a reference complex for assigning absolute configurations of closely related complexes on the basis of the signs of CD peaks for transitions that can be related to those of [Co(en)$_3$]$^{3+}$.

The absorption spectrum of [Co(edta)]$^-$ shows two bands in the visible region with no apparent splitting. This was interpreted to mean that the "effective" symmetry about Co(III) was cubic—that the field was averaged out to become effectively $\mathbf{O}_h$ instead of $\mathbf{C}_2$. The CD spectrum (Figure 8.20) shows two peaks in the region of the first band and three weak peaks in the second-band region. We probably see only two of the three peaks expected in the first band for $\mathbf{C}_2$ symmetry because of small splittings: that is, only two peaks are resolved. The expected three peaks appear for a ligand field model compound, which also has $\mathbf{C}_2$ symmetry, [Co(en)(mal)$_2$]$^-$ (mal = malonate ion). We know the absolute configurations for both complexes. CD studies have proven useful in distinguishing among geometrical isomers formed by a number of complexes, even in cases where the isomers differ only in the chelate ring conformations, such as for triethylenetetraamine. Absolute configurations can be assigned from CD spectra for closely related complexes if one of the series has a known configuration.

[9]The absorption bands for transition metal complexes involve electric dipole transitions. For ORD and CD the intensities depend on the product of integrals for the electric dipole transition *and* the magnetic dipole transition. If either of these is zero, the transition is forbidden. One of these will be zero if the molecule possesses a plane or center of symmetry or *any* S_n axis (p. 89). In the case of [Co(en)$_3$]$^{3+}$ the *E* and A_2 transitions are electric- and magnetic-dipole allowed, but the A_1 transition is magnetic-dipole forbidden (see Problem 8.14).

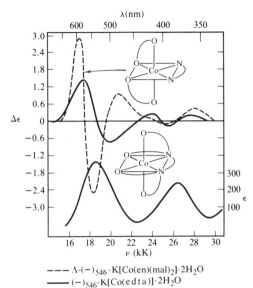

Figure 8.20 Absorption and CD curves of $(-)'_{546}K[Co(edta)]\cdot 2H_2O$ and CD curve of $(-)_{546}$-$K[Co(en)(mal)_2]\cdot 2H_2O$.

Legend to figure:
- - - - Λ-$(-)_{546}$-$K[Co(en)(mal)_2]\cdot 2H_2O$
——— $(-)_{546}$-$K[Co(edta)]\cdot 2H_2O$

Uncertainty arises whenever there is any doubt about the assignments of the electronic transitions or the basis for determining the CD sign.

8.3.3 Trigonal Prismatic Coordination[10]

General acceptance of the existence of square planar complexes (C.N. 4) was delayed by the remarkable success of the tetrahedron in organic chemistry. Ligand–ligand repulsion is minimized for tetrahedral coordination. Similarly, octahedral coordination minimizes ligand–ligand repulsion for C.N. 6, and exceptions came as surprises.[11] Trigonal prismatic coordination was first observed for the layer lattices of MoS_2 and WS_2 (p. 215), but these could be rationalized as peculiarities of the solid state. In 1965 the complex Re $\left(\begin{array}{c} S \quad C_6H_5 \\ | \quad C \\ \parallel \\ C \\ S \quad C_6H_5 \end{array} \right)_3$ was reported to have trigonal prismatic coordination of Re. Although similar complexes of Mo and V have some crystallographic differences, in spite of size differences the M–S distances are remarkably constant, varying from 232.5 pm for Re to 233.8 pm for V—even though V has the smallest radius! The tetragonal faces are almost square (as is true also for MoS_2) with very short S–S distances ($\sim$306 pm). The S–S distance along the edges spanned by the chelate ring are shorter than for the other edges but within 1 pm of the S–S distance between ligands. These short S–S

[10]R. A. D. Wentworth, *Coord. Chem. Rev.* 1972, *9*, 171.

[11]Various geometries for C.N. 6 are considered theoretically by R. Hoffmann, J. M. Howell, and A. R. Rossi in "Bicapped Tetrahedral, Trigonal Prismatic, and Octahedral Alternatives in Main and Transition Group Six-coordination," *J. Am. Chem. Soc.* 1976, *98*, 2484.

distances suggest that S–S bonding might be important in establishing the trigonal prismatic stereochemistry. The square faces are reminiscent of square planar S_4^{2+} rings observed by Gillespie.

Other 1,2-dithiolene and 1,2-ethanedithiolate complexes of these metals, as well as Fe and W, have been obtained, and more examples should be forthcoming. Researchers found that two-electron reduction of the neutral complex to produce $M[S_2C_2(CN)_2]_3^{2-}$ causes twisting about the C_3 axis, to give stereochemistry intermediate between a trigonal prism and an octahedron. Presumably, the electrons occupy orbitals that are antibonding with respect to S–S, increasing repulsion. An octahedron would result from a twist angle of 60°. One intramolecular mechanism for racemization and isomerization of tris(bidentate) complexes is Bailar's trigonal twist (p. 366), which converts a complex of left chirality to one of right chirality, going through a trigonal prismatic configuration.

Of the few reported trigonal pyramidal complexes of ligands without sulfur, most involve three relatively rigid bidentate ligands that can coordinate along the tetragonal edges of a prism, with the three ligands joined to cap one or both ends of the trigonal prism. Such a ligand is *cis, cis*-1,3,5-tris(pyridine-2-carboxaldimino)cyclohexane (the Schiff base from *cis, cis*-1,3,5-triaminocyclohexane (tach) with pyridine-2-carboxaldehyde). Figure 8.21 shows the Zn(II) complex ion $[Zn(py)_3tach]^{2+}$. The nitrogens form a slightly tapered trigonal prism only slightly twisted (4°). The Ni^{2+} complex with high octahedral LFSE (d^8) shows a twist angle of about 32°, about halfway between an octahedron and a trigonal prism. Less rigid encapsulating ligands give varying twist angles, depending on the preference of the metal ion for octahedral coordination.

Wentworth studied the loss in LFSE in going from octahedral to trigonal prismatic coordination. The most favorable cases for trigonal prismatic coordination are d^0, d^5, and d^{10} (0 LFSE); d^1; and high-spin d^7. The coordination of cobalt(II) (high-spin d^7) is trigonal prismatic in the complex ion $[Co(NH_2CH_2CH_2O)_3]_2Co^{2+}$. The $Co(NH_2CH_2CH_2O)_3$ molecules are

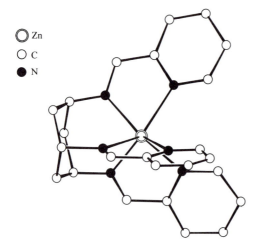

Zn
C
N

Figure 8.21 The structure of the Zn[(py)₃tach]²⁺ ion in the anhydrous perchlorate salt. (From W. O. Gillum, J. C. Huffman, W. E. Streib, and R. A. D. Wentworth, *Chem. Commun. 1969*, 843.)

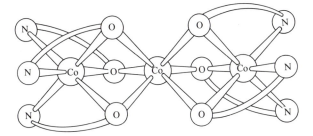

Figure 8.22 Structure of [Co(Co-(OCH$_2$CH$_2$NH$_2$)$_3$)$_2$]$^{2+}$ in the anhydrous acetate salt. (From J. A. Bertrand, J. A. Kelley, and E. G. Vassian, *J. Am. Chem. Soc.* 1969, *91*, 2394. Copyright © 1969, American Chemical Society.)

octahedral and of the same chirality. Bonded to six oxygens, the Co^{2+} ion bridges the two octahedra (Figure 8.22). For octahedral coordination of Co^{2+} there are unfavorable nonbonding interactions between hydrogens of the Co(NH$_2$CH$_2$CH$_2$O)$_3$ molecules of the same chirality, favoring trigonal prismatic coordination. The authors expect octahedral coordination of Co^{2+} to be more favorable for the meso isomer where the Co(III) complexes have opposite chirality.

8.4 COORDINATION NUMBERS OTHER THAN FOUR AND SIX

Not all of the "other" coordination numbers are rare, or even uncommon. They are grouped separately because of the lengthy considerations of C.N. 4 and 6.

8.4.1 Coordination Number Two

We encounter C.N. 2 for Cu(I), Ag(I), Au(I), and Hg(II). Cu(I) forms [Cu(NH$_3$)$_2$]$^+$ and CuCl$_2^-$, but most of the complexes of Cu(I) are tetrahedral. In solid K[Cu(CN)$_2$] the C.N. of Cu(I) is 3 and solid halide complexes usually involve tetrahedral coordination. [Ag(CN)$_2$]$^-$, [Ag(NH$_3$)$_2$]$^+$, Hg(CN)$_2$, and HgCl$_2$ involve linear coordination in the solid state. The Ag$^+$ and Hg^{2+} ions add more ligands to form tetrahedral complexes, but the remaining two ligands are bound weakly. In solution the MX$_2$ species might be solvated strongly to give an effective C.N. of 4. Only for Au(I) is C.N. 2 characteristic, with little tendency to add additional ligands.

Although ethylenediamine complexes generally are much more stable than those of ammonia, this is not true for [Ag(en)]$^+$ compared with [Ag(NH$_3$)$_2$]$^+$. We attribute the unexpectedly low stability of [Ag(en)]$^+$ to the fact that the en molecule cannot span the Ag$^+$ to permit the desired linear arrangement. In chelate compounds the five-membered rings generally are most stable. In the case of silver, larger rings permit less strained bonding and compounds containing six-, seven-, and eight-membered rings are all more stable than [Ag(en)]$^+$.

A linear arrangement usually is associated with *sp* hybridization, but more favorable bonding can be achieved by participation of the d_{z^2} orbital (see p. 165). Infrared studies of the linear HgCl$_2$, HgBr$_2$, and HgI$_2$ in the vapor state indicate that the bending force-constants are extremely low, which suggests that the bonding orbitals have only weak directional properties.

8.4.2 Coordination Number Three[12]

Apparently, no metal ions show a preference for C.N. 3, which is encountered primarily for metal ions of low oxidation number with bulky ligands or when imposed in the solid. The C.N. of Cu(I) is 3 in $K[Cu(CN)_2]$ (through bridging CN^-) but 4 in $K_2[Cu(CN)_3]$. Planar trigonal coordination occurs for HgI_3^-, $Pt[P(C_6H_5)_3]_3$, $Cu[S-P(CH_3)_3]_3$, and $Cu[SC(NH_2)_2]_3^+$ in the solid. The bulky ligand $N[Si(CH_3)_3]_2^-$ gives three-coordinate ML_3 molecules with Fe^{3+}, Cr^{3+}, and even some rare earth ions.

8.4.3 Coordination Number Five[13]

Stoichiometry often is a poor indication of the actual C.N. Thus PCl_5 exists in the solid as $[PCl_4]^+[PCl_6]^-$, and more commonly C.N. 6 is achieved through association—$MoCl_5$, $TaCl_5$, SbF_5, and $Tl_2[AlF_5]$ are examples.

Square Pyramidal Complexes

Antimony(III) forms the square pyramidal (SP) complexes SbF_5^{2-} (Figure 8.23) and $SbCl_5^{2-}$ because of the steric requirements of the lone pair of electrons. In SbF_5^{2-} the Sb is slightly below the square base of the pyramid, because of the influence of the unshared electron pair. For transition metal complexes the metal ion commonly is above the square base of the pyramid [by 34 pm for $Ni(CN)_5^{3-}$]. The apical position is not equivalent to the basal positions for the SP (C_{4v}). The vanadyl group, VO^{2+}, retains its identity in many compounds and the acetylacetonate complex, $VO(acac)_2$, is square pyramidal. The bidentate ligand $o\text{-}C_6H_4[As(CH_3)_2]_2$ (diars) forms SP complexes of divalent Ni, Pd, and Pt with an anion in the apical position, $[M^{II}(diars)_2X]^+$.

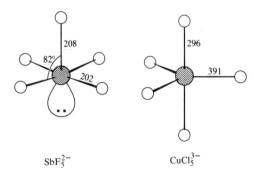

SbF_5^{2-} $CuCl_5^{3-}$

Figure 8.23 Structures of SbF_5^{2-} and $CuCl_5^{3-}$. Bond lengths are in pm.

[12]P. G. Eller, D. C. Bradley, M. B. Hursthouse, and D. W. Meek, "Three Coordination in Metal Complexes," *Coord. Chem. Rev.* 1977, *24*, 1; D. C. Bradley "Steric Control of Metal Coordination," *Chem. Britain* 1975, *11*, 393.

[13]E. L. Muetterties and R. A. Schunn, *Quart. Rev.* 1966, *20*, 245; C. Furlani, *Coord. Chem. Rev.* 1968, *3*, 141; B. F. Hoskins and F. D. Williams, *Coord. Chem. Rev.* 1973, *9*, 365; R. Morassi, I. Bertini, and L. Sacconi, *Coord. Chem. Rev.* 1974, *11*, 343; A. R. Rossi and R. Hoffmann, *Inorg. Chem.* 1975, *14*, 365; M. C. Favas and D. L. Kepert, *Prog. Inorg. Chem.* 1980, *27*, 325.

Five-coordinate transition metal complexes without constraints of chelating ligands occur with the trigonal bipyramidal (TBP) structure—for example, $CuCl_5^{3-}$ (Figure 8.23), $Co(CNCH_3)_5^+$ and $Fe(CO)_5$—and with the square pyramidal structure—for example, $Ni(CN)_5^{3-}$ (Figure 8.24). These two structures differ less than the usual representations suggest. In general, chelated complexes or those with two or more different ligands coordinated, display versions of these structures so distorted that it often is difficult to say whether the structure is closer to TBP or SP. The stabilities of the two regular structures are very similar; in fact, $[Cr(en)_3][Ni(CN)_5] \cdot 1.5H_2O$ contains *both* SP and TBP anions. The corresponding compound containing the $[Cr(1,3\text{-propanediamine})_3]^{3+}$ cation contains only the SP $[Ni(CN)_5]^{3-}$.

Trigonal Bipyramidal Complexes

Five-coordinate compounds of nonmetals—PF_5, for example—generally have the trigonal bipyramidal structure. The trigonal bipyramid (TBP, D_{3h}) has nonequivalent axial and equatorial positions. Whereas in PX_5 molecules the axial bonds are longer, in $CuCl_5^{3-}$ (Figure 8.23) the equatorial bonds are slightly longer.

That only slight displacement of ligands is required to go from SP to TBP, or *vice versa,* is illustrated in Figure 8.24, in which the regular SP is shown for $Ni(CN)_5^{3-}$ alongside the distorted TBP that occurs in the same structure. Two diagonally related ligands in the base of the SP structure are displaced upward to become the axial ligands for the TBP, and the other two basal ligands are displaced downward to equatorial positions for the TBP. In this distorted TBP, one C—Ni—C bond angle in the equatorial plane is much larger (142°) than the other two (107.3° and 111.5°). Further displacement to close the large angle would give a regular TBP. The actual C_2 configuration of the TBP $Ni(CN)_5^{3-}$ apparently is not imposed by the crystal lattice, since such distortion of the TBP geometry also is encountered for $Ni(CN)_2[P(C_6H_5)(OC_2H_5)_2]_3$ (where the CN^- ions are in the axial positions of the distorted TBP; or, alternatively, there is one long Ni—P bond that might be considered apical for a distorted SP with two CN^- and two phosphines in the nearly square base).

Several tripod ligands, such as $N[C_2H_4N(CH_3)_2]_3$ and $P[C_3H_6As(CH_3)_2]_3$, form TBP complexes with Ni(II), Co(II), and some other metals. The quadridentate umbrellalike ligands

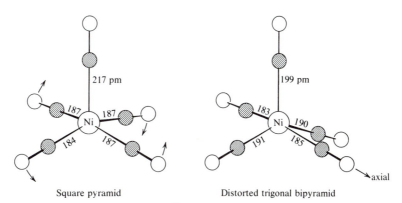

Square pyramid   Distorted trigonal bipyramid

Figure 8.24 Structures of $Ni(CN)_5^{3-}$ in $[Cr(NH_2C_2H_4NH_2)_3][Ni(CN)_5] \cdot 1.5H_2O$. Only slight shifts in the directions of the arrows are required to convert from SP → TBP.

coordinate symmetrically, with another ligand—for example, CN^-, Cl^-, or Br^-—in the remaining axial position.

Ligand–ligand repulsion alone favors the TBP arrangement, but bonding interactions, and especially π bonding, involving d orbitals can favor the SP structure. Both arrangements give a pair of d orbitals of lowest energy and one d orbital of very high energy (see Problem 7.12). Low-spin five-coordinate complexes with d^5, d^6, d^7, and d^8 configurations leave this highest-energy orbital empty.

No examples of isomerism for five-coordinate complexes have been reported, but these are unlikely because of the easy TBP $\rightleftarrows$ SP interconversion which can scramble all positions. Although diffraction studies and infrared spectroscopy show the axial and equatorial groups in $Fe(CO)_5$ and PF_5 to be nonequivalent, on the NMR time scale all five groups are equivalent. These are examples of fluxional molecules[14] (see Section 8.5).

8.4.4 Coordination Number Seven[15]

The molecule IF_7 has a pentagonal bipyramidal structure, as do $UO_2F_5^{3-}$, UF_7^{3-}, ZrF_7^{3-}, HfF_7^{3-}, and $V(CN)_7^{4-}$ (Figure 8.25). The structure of NbF_7^{2-} and TaF_7^{2-} ions derives from a trigonal prism, with a seventh F^- added to one of the tetragonal faces. Subsequent distortion to diminish anion–anion repulsion results in a configuration closer to that shown in Figure 8.25, with a tetragonal face opposite a trigonal face. The La_2O_3 structure (A-type M_2O_3) involves layers of LaO_6 octahedra with an oxygen of a neighboring octahedron directly over one triangular face, to give C.N. 7. The compound $(NH_4)_3SiF_7$ contains $[SiF_6]^{2-}$ and F^- anions rather than a seven-coordinate species. Studies have shown that quinquedentate macrocyclic and related quinquedentate ligands (Figure 8.26) form seven-coordinate complexes with a number of metal ions. The five ligand atoms are nearly in a plane, allowing plenty of room for coordination of anions (e.g., NCS^-, Cl^-, or Br^-) or water molecules above and below the

Figure 8.25 Structures of UF_7^{3-} and NbF_7^{2-}.

[14]See also J. R. Shapley and J. A. Osborn, "Rapid Intramolecular Rearrangements in Pentacoordinate Transition Metal Compounds," *Acc. Chem. Res.* 1973, *6*, 305.

[15]D. L. Kepert, "Aspects of the Stereochemistry of Seven Coordination," *Prog. Inorg. Chem.* 1979, *25*, 41; R. K. Boggess and W. D. Wiegele, "Seven-coordinate Complexes of First-row Transition Metals," *J. Chem. Educ.* 1978, *55*, 156.

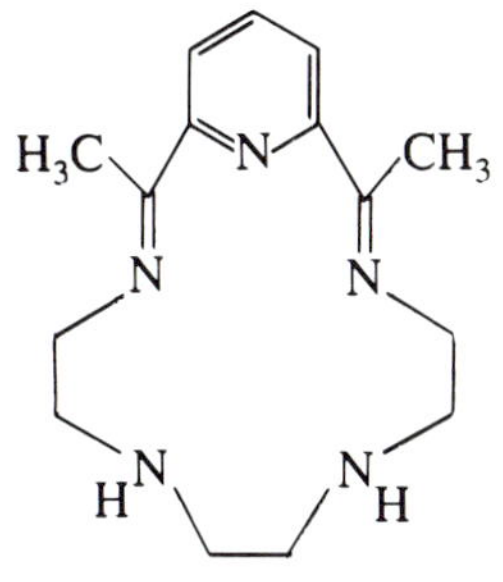

Figure 8.26 A quinquedentate macrocyclic ligand (L) capable of forming seven-coordinate complexes of the type [Fe(L)Cl$_2$]ClO$_4$.

plane, to give C.N. 7. The septidentate cryptate ligand crypt, 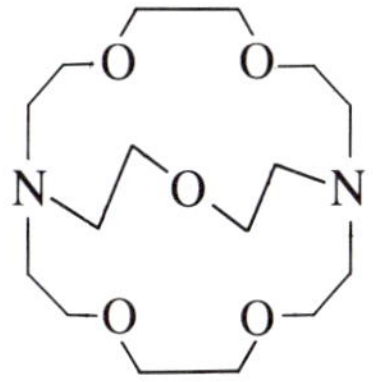 , forms a

distorted trigonal bipyramidal complex with Co(II), [Co(crypt)][Co(SCN)$_4$]. One oxygen of each of the O⌒O arms is in an axial position.

8.4.5 Coordination Number Eight[16]

After C.N. 6 and 4 the most common C.N. is 8, which is encountered in solids for large cations in the CsCl and CaF$_2$ structures, ZrSiO$_4$, and garnets. Eight-coordination is cubic in solids with close-packed anions. A cubic coordination polyhedron has been found for discrete complex ions in (Et$_4$N)$_4$[U(NCS)$_8$] and Na$_3$[PaF$_8$]. More commonly found is the dodecahedron with triangular faces (**D**$_{2d}$)[17] (Figure 8.27). Whereas the dodecahedron occurs for Mo(CN)$_8^{4-}$, Mo(CN)$_8^{3-}$, and Zr(C$_2$O$_4$)$_4^{4-}$, the square antiprism (**D**$_{4d}$) occurs for TaF$_8^{3-}$, Cs$_4$[U(NCS)$_8$], and Zr(acac)$_4$ (**D**$_4$)—in the last-named case, the acetylacetonate chelate rings span edges on the square faces. It is interesting that U(NCS)$_8^{4-}$ gives both the antiprismatic and cubic structures: the Et$_4$N$^+$ cations fit into the cube-faces to stabilize the cube, while the Cs$^+$ ions are above the triangular faces of the antiprism.

Both the dodecahedron and antiprism give one *d* orbital of lowest energy, and many examples have d^0, d^1, or d^2 configurations. The Mo(CN)$_8^{4-}$ ion is diamagnetic (d^2). Having similar energies, the two stereochemistries provide no firm basis for choice between them. As the "bite" (closeness of the donor atoms determined by the size of the chelate ring) is increased, however, the square antiprism (**D**$_{4d}$) should become more favored than the dodecahedron

[16]S. J. Lippard, *Prog. Inorg. Chem.* 1968, 8, 109; M. G. B Drew, *Coord. Chem. Rev.* 1977, 24, 179; J. K. Burdett, R. Hoffmann, and R. C. Fay, *Inorg. Chem.* 1978, *17*, 2553.

[17]There is also a dodecahedron with pentagonal faces (**I**$_h$), sometimes used for calendars. The **D**$_{2d}$ dodecahedron with triangular faces has two equivalent sets of four vertices, each set forming a trapezoid. The planes of the intersecting trapezoids are perpendicular. See Problem 3.8.

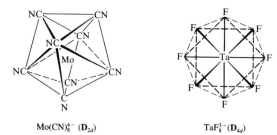

$Mo(CN)_8^{4-}$ $(\mathbf{D}_{2d})$

TaF_8^{3-} $(\mathbf{D}_{4d})$

Figure 8.27 Complexes with C.N. 8 (Mo—C bonds omitted).

(although ligands with large "bites" are likely to form bridged complexes). Physical measurements on $Mo(CN)_8^{4-}$ in solution are inconclusive with respect to effective $\mathbf{D}_{2d}$ or $\mathbf{D}_{4d}$ symmetry.

C.N. 8 is favored for large cations and small ligands—F^- and CN^- for example. Chelate ligands with a small "bite" show C.N. 8 with smaller cations, and

$Co(NO_3)_4^{2-}$, for example, both have approximately dodecahedral coordination, with the average positions of each ligand approximately tetrahedral. The spectral characteristics of $Co(NO_3)_4^{2-}$ resemble those of tetrahedral complexes. A similar dodecahedral structure (Figure 8.28) was found for tetrakis(*N*-ethylsalicylaldiminato)zirconium(IV) (the bidentate ligand is the Schiff base from ethylamine and salicylaldehyde). The actual symmetry of the complex is $\mathbf{S}_4$. This is an interesting example of a compound with no center or plane of symmetry, but it is optically inactive because of an S_4 axis. No substantiated cases of isomerism are known for C.N. 8.

C.N. 8 is achieved for compounds of dioxo cations—for example, uranyl ion, UO_2^{2+}—through hexagonal bipyramidal coordination. The $[UO_2(NO_3)_3]^-$ ion has three bidentate NO_3^- slightly twisted from the hexagonal equatorial plane. The axial positions of the uranyl oxygens and the average positions of the nitrate ions describe a trigonal bipyramid.

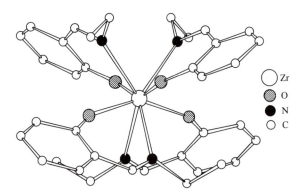

○ Zr
◉ O
● N
○ C

Figure 8.28 Structure of the eight coordinate complex $Zr(OC_6H_4CH=NC_2H_5)_4$ (S_4 symmetry). (From D. C. Bradley, M. B. Hursthouse, and I. F. Rendall, *Chem. Commun., 1970*, 368.)

8.4.6 Higher Coordination Numbers[18]

The $Nd(H_2O)_9^{3+}$ ion in $Nd(BrO_3)_3 \cdot 9H_2O$ has the Nd^{3+} ion at the center of a trigonal prism with an additional water molecule over each tetragonal face; the polyhedron is described as a tricapped trigonal prism (see Figure 14.5, p. 633). Other hydrated rare earth ions with this structure have been reported. The structures of ReH_9^{2-} and TcH_9^{2-} are similar. C.N. 9 is achieved through sharing of anions in $PbCl_2$, UCl_3, $La(OH)_3$, and other compounds having these structures. In the solid, ThF_8^{4-} achieves C.N. 9 in a capped (shared F^-) square antiprism.

Cerium(IV) shows C.N. 10 in $[Ce(NO_3)_4\{OP(C_6H_5)_3\}_2]$ where the four nitrate ions are bidentate. If we view the phosphine oxygens as occupying axial positions of an octahedron, the nitrate ions would be in the four equatorial positions, but with the coordinated oxygens tipped out of the equatorial plane. The C.N. is also 10 for $Ce(NO_3)_5^{2-}$. Here we can visualize the structure by viewing the nitrate ions as arranged about a trigonal bipyramid. The coordinated oxygens are twisted about the Ce—N lines, to minimize repulsion. The structure of $[Ce(CO_3)_5]^{6-}$ is a distorted bicapped square antiprism ($\mathbf{D}_{4d}$ for the regular polyhedron).

C.N. 12 is achieved in the compound $Ce_2Mg_3(NO_3)_{12} \cdot 24H_2O$ [containing discrete $[Ce^{III}(NO_3)_6]^{3-}$ ions] and in $(NH_4)_2Ce(NO_3)_6$, both involving bidentate nitrate ligands. The approximately octahedral positions of the nitrate ions result in a figure with a center of symmetry. Opposite pairs of nitrate ligands are twisted slightly from the xy, xz, and yz planes. The coordinated oxygens are at the 12 corners of an irregular icosahedron (Figure 8.29). The actual symmetry is nearly $\mathbf{T}_h$ for the Ce(IV) complex. Icosahedral complexes are interesting, because there is no splitting of the energies of the d orbitals in an icosahedral field. Chelated nitrate complexes have high coordination numbers because of the small ''bite'' of the bidentate nitrate ligand, and their mean positions—the positions of the N of the nitrate ion—usually correspond to a lower polyhedron (C.N./2) (See Table 8.3).

Table 8.3 Simplified shapes of some complexes based on the mean positions of bidentate ligands

Coordination Number	Figure Described by Positions of Ligand Atoms	Figure Described by Mean Position of Ligand	Examples
4	Tetrahedron ($\mathbf{D}_{2d}$)	Linear	$Be(acac)_2$, $Cu(NO_3)_2^a$
6	Octahedron[b] ($\mathbf{D}_3$)	Trigonal plane	$Co(NO_3)_3$
8	Dodecahedron ($\mathbf{D}_{2d}$)	Tetrahedron	$Ti(NO_3)_4$, $Fe(NO_3)_4$
8	Square antiprism ($\mathbf{D}_{4d}$)	Square plane	$Zr(acac)_4$
10	''Bicapped'' trigonal antiprism ($\mathbf{C}_1$)[c]	Trigonal bipyramid	$Ce(NO_3)_5^{2-}$
12	Icosahedron ($\mathbf{T}_h$)	Octahedron	$Ce(NO_3)_6^{3-}$, $Ce(NO_3)_6^{2-}$

[a]The structure of $Cu(NO_3)_2$ in the vapor is distorted tetrahedral; the O—Cu—O angle is 70°.
[b]The octahedron is distorted toward the trigonal prism because of the small ''bite'' of the ligand. The simplified shape would still be trigonal planar even if the structure were the trigonal prism.
[c]Each cap is a bidentate nitrate ligand.

[18]M. G. B. Drew, ''Structures of High Coordination Complexes,'' *Coord. Chem. Rev.* 1977, *24*, 179; R. E. Robertson ''Coordination Polyhedra with Nine and Ten Atoms,'' *Inorg. Chem.* 1977, *16*, 2735; D. L. Kepert, ''Aspects of C.N. 9, 10, and 12,'' *Prog. Inorg. Chem.* 1981, *28*, 309.

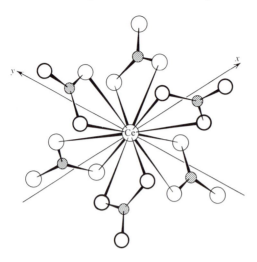

Figure 8.29   Twelve-coordination of $Ce(NO_3)_6^{2-}$ in $(NH_4)_2Ce(NO_3)_6$. (From T. A. Beineke and J. Delgaudio, *Inorg. Chem.* 1968, 7, 715. Copyright © 1968, American Chemical Society.)

In $U(BH_4)_4$ uranium has C.N. 14. Four of the six BH_4^- surrounding each U are attached by two H atoms and use their other two H to bridge neighboring U atoms. Two additional BH_4^- ions are bonded to U by three H. The idealized polyhedron formed by the hydrogens is a bicapped hexagonal antiprism.

8.5   STEREOCHEMICALLY NONRIGID AND FLUXIONAL MOLECULES[19]

For most molecules, one stereochemical arrangement is much more stable than others. The molecule will display the stable structure unless it is in an excited vibrational or electronic state or as a transition state for a reaction or stereochemical transformation. As noted above, $Ni(CN)_5^{3-}$ is found as a square pyramid or trigonal bipyramid, even in the same crystal (p. 321). Such complex ions with two structures of comparable stability are said to be *stereochemically nonrigid*. If the two (or more) configurations are chemically equivalent, the molecule is *fluxional*. For the trigonal bipyramid, the axial positions are not equivalent geometrically to the equatorial positions; this generally results in different bond lengths. However, minor deformations passing through a square pyramidal intermediate can interchange the axial and equatorial positions. Structural techniques with a slower time scale (p. 698)—NMR of PF_5, for example—"see" the five groups as equivalent (see also p. 322). The PF_5 molecule is fluxional. Observers have shown that for four-coordinate complexes of nickel(II), such as $[Ni(R_3P)_2X_2]$, the planar $\rightleftarrows$ tetrahedral transformation is accessible thermally. Since the structures are not equivalent chemically, these complexes are stereochemically nonrigid but not fluxional.

The structure of $Mo(CN)_8^{4-}$ in the solid is the dodecahedron; some studies have favored a square antiprism in solution. This case is inconclusive, with stereochemically nonrigid behavior suggested. Stereochemically nonrigid or fluxional behavior is rather common among

[19]F. A. Cotton, *Acc. Chem. Res.* 1968, *1*, 257; E. L. Muetterties, *Inorg. Chem.* 1965, *4*, 769; E. L. Muetterties, *Acc. Chem. Res.* 1970, *3*, 266; J. R. Shapley and J. R. Osborn, *Acc. Chem. Res.* 1973, *6*, 305.

organometallic molecules and cluster compounds (see p. 696 and p. 699) and accounts for the conflicting structures obtained from structural techniques with different time scales.

GENERAL REFERENCES

F. Basolo and R. G. Pearson, *Mechanisms of Inorganic Reactions,* 2nd ed., Wiley, New York, 1967. Covers more than mechanisms.

C. J. Hawkins, *Absolute Configuration of Metal Complexes,* Wiley-Interscience, New York, 1971. Covers more than implied by the title.

G. B. Kaufman, Ed., *Classics in Coordination Chemistry,* Part I, *The Selected Papers of Alfred Werner,* 1968; Part II, *Selected Papers* (1798–1899), 1976; Part III, *Twentieth Century Papers* (1904–1935), 1978, Dover, New York.

A. E. Martell, Ed., *Coordination Chemistry,* Vol. I, Van Nostrand-Reinhold, New York, 1971; Vol. II, ACS Monograph 174, SIS/American Chemical Society, Washington, D.C., 1978.

K. Nakamoto and P. J. McCarthy, Eds., *Spectroscopy and Structure of Metal Chelate Compounds,* Wiley, New York, 1968.

Y. Saito, "Absolute Stereochemistry of Chelate Compounds," *Topics in Stereochemistry* 1978, *10,* 95.

PROBLEMS

8.1 Name the following compounds according to the modified IUPAC rules (see Appendix B):

K_2FeO_4

$[Cr(NH_3)_6]Cl_3$

$[Cr(NH_3)_4Cl_2]Cl$

$K[PtCl_3(C_2H_4)]$

$K_3[Al(C_2O_4)_3]$

$K_2[Co(N_3)_4]$

$K[Co(edta)]$

$[Cr(NH_3)_2(H_2O)_3(OH)](NO_3)_2$

$Fe(C_5H_5)_2$

$(CO)_5Mn-Mn(CO)_5$

$$[(NH_3)_4 Co \underset{NH_2}{\overset{OH}{<>}} Co(en)_2]Cl_4$$

8.2 Sketch all of the possible geometrical isomers for the following complexes and indicate which of these would exhibit optical activity.

$[Co(en)(NH_3)_2BrCl]^+$.

$[Co(NH_2CH_2CO_2)_2(NH_3)Cl]^+$.

$[Pt(NH_3)BrCl(NO_2)]^-$.

$[Co(trien)Cl_2]^+$ (consider the different ways of linking trien to Co).

$$[(gly)_2 Co \underset{OH}{\overset{OH}{<>}} Co(gly)_2].$$

8.3 Draw *all* possible isomers for Ma_2bcd assuming the complex forms a square pyramid.

8.4 How might one distinguish between the following isomers?

a. $[Co(NH_3)_5Br]SO_4$ and $[Co(NH_3)_5SO_4]Br$.

b. $[Co(NH_3)_3(NO_2)_3]$ and $[Co(NH_3)_6][Co(NO_2)_6]$.

c. *cis-* and *trans-*$[CoCl_2(en)_2]Cl$.

d. *cis-* and *trans-*$NH_4[Co(NO_2)_4(NH_3)_2]$.

e. *cis-* and *trans-*$[Pt(gly)_2]$.

8.5 Give examples of the following types of isomerism:
  a. Hydrate isomerism.   *d.* Ionization isomerism.
  b. Coordination isomerism.   *e.* Geometrical isomerism.
  c. Linkage isomerism.

8.6 Draw all the isomers possible for octahedral M(*abcdef*). (Hint: Calculate the number of stereo-isomers for [M(*abcdef*)]. Write the unique pairings trans to *a* and calculate the number of possible isomers for each of these. Then consider unique *trans* pairings for the remaining groups.)

8.7 Draw all possible isomers for
  a. M(*abcdef*) ($\widehat{ef}$ is bidentate).
  b. M($\widehat{aa}bcd_2$) (*Hint:* leave $\widehat{bc}$ in a fixed position in all drawings except for enantiomers; $\widehat{aa}$ is a symmetrical bidentate ligand and $\widehat{bc}$ is an unsymmetrical bidentate ligand.)

8.8 Sketch the isomers possible for a trigonal prismatic complex M($\widehat{aa}$)$_3$, where $\widehat{aa}$ is a planar bidentate ligand. Could any of these be optically active? Assign the point groups.

8.9 Sketch octahedral and trigonal prismatic [M(NO$_3$)$_3$] complexes to show that the N atoms are in a trigonal planar arrangement in each case. (*Hint:* Sketch the complexes looking down the three-fold axes.) Assign the point groups.

8.10 Determine the number of Δ and Λ pairs of chelate rings for the isomer of [Co(edta)]⁻ shown. The pairs of rings attached at a common point and those spanning parallel edges are omitted.

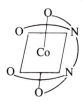

8.11 Which chelate ring conformation is favored for coordinated S-(+)-propylenediamine, and why?

8.12 For *trans*-[Co(trien)Cl$_2$]⁺ (trien = NH$_2$C$_2$H$_4$NHC$_2$H$_4$NHC$_2$H$_4$NH$_2$) there are three isomers possible, depending on the chelate ring conformations. Two are optically active and one is meso with the central diamine ring in an eclipsed "envelope" conformation. Sketch the two optically active *trans* isomers (all rings *gauche*) and assign the absolute configurations (δ or λ) to the chelate rings. (See D. A. Buckingham, P. A. Marzilli, and A. M. Sargeson, *Inorg. Chem.* 1967, *6*, 1032.)

8.13 What is (are) the number(s) of *d* electrons for metals usually encountered with the following stereochemistries?
  a. Linear.   *c.* Trigonal prismatic.
  b. Square planar.   *d.* Dodecahedral.

8.14 Selection rules for electric dipole transitions (those usually observed for transition metals) require that the symmetry of the transition (the direct product of the representations for the ground and excited states) must be the same as the representation for one of the electric dipole moment operators, and these correspond to the representations for *x*, *y*, and *z*. Selection rules for magnetic dipole transitions (observed for lanthanide metals) require that the symmetry of the transition be the same as the representation of one of the magnetic dipole moment operators, and these correspond to rotations about *x*, *y*, and *z* (R$_x$, R$_y$, and R$_z$ in the character tables). For CD and ORD, the transitions must obey the selection rules for electric *and* magnetic dipole transitions. For [Cr(NH$_3$)$_6$]$^{3+}$ (**O**$_h$) the *d-d* transitions are $^4A_{2g} \rightarrow {}^4T_{2g}$ and $^4A_{2g} \rightarrow {}^4T_{1g}$. For [Cr(en)$_3$]$^{3+}$ (**D**$_3$) the transitions are $^4A_2 \rightarrow {}^4E$ and $^4A_2 \rightarrow {}^4A_1$ (both derived from $^4A_{2g} \rightarrow {}^4T_{2g}$ by lowering symmetry), and $^4A_2 \rightarrow {}^4E$ and $^4A_2 \rightarrow {}^4A_2$ (both derived from $^4A_{2g} \rightarrow {}^4T_{1g}$ by lowering symmetry). Which of these are electric-dipole-allowed? Which are *both* electric- and magnetic-dipole-allowed?

IX
Reaction Mechanisms for Coordination Compounds

The chemistry of the transition metals involves processes in which ligands are coordinated to metals, ligands are exchanged, the coordination sphere is rearranged, oxidation and reduction occur, and the reactivity of the ligands is changed by coordination. A thorough appreciation of the chemistry of the transition metals requires a knowledge of their reaction mechanisms. The relatively simple complexes discussed here serve as models displaying the main features of reaction mechanisms of coordination compounds.

In this chapter we look first at the model for discussing reaction kinetics and mechanisms. Then we discuss substitution reactions of octahedral Werner complexes, developing the concepts needed for relating the model to experiment. After dealing briefly with racemization reactions of octahedral complexes, we treat ligand substitution reactions of square planar complexes. Next, the mechanisms of redox reactions are explicated. Finally, a brief look is taken at photochemical reactions. Reaction mechanisms for complexes containing π-acid ligands are discussed in Chapter 10.

9.1 THE KINETIC MODEL

The model that chemists currently use to rationalize and predict what goes on during chemical reactions was developed in the 1930's from transition state theory. It views the course of reactions as involving collision of sufficiently energetic reactants to form the activated complex (or *transition state*). Along the path from reactants to products, kinetic energy is converted to potential energy by bond stretching, partial bond formation, molecular distortions, etc. The exact processes that occur depend on the particular reaction. The transition state is the species of maximum potential energy resulting from these motions. All the energetically unfavorable processes have occurred by the time a transition state is reached. Thereafter, the collection of atoms rearranges toward a more stable species: old bonds are fully broken, new ones fully made, and geometric rearrangement toward a more stable configuration (relaxation) occurs. In multistep reactions a new transition state is reached for each distinguishable step.

Plots can be made (at least in principle) for the energy of a reacting system as a function of the relative atomic positions. To take a concrete example, consider the calculated[1] energy for the reaction $H_2 + F \rightarrow H + HF$ when all the reactants are confined to a straight line. The variables describing relative atomic positions are then r_{H-H} and r_{H-F}, the distances between atoms. Such a plot of energy versus two variables is three-dimensional. Figure 9.1*a* shows a two-dimensional contour-plot representation. The lines represent the shape of the intersection of a plane having energy value written on the line with the three-dimensional energy surface. (The energy zero is the energy of $H_2 + F$.) Such a contour plot is like a terrain map showing contours of constant altitude. At very large r_{H-F} (bottom right), a cut through the plot (*A-B*) is just the H–H energy versus distance curve. At very large r_{H-H}, a cut through the plot (*C-D*) gives the H–F energy versus distance curve. Figure 9.1*b* shows the computer-generated three-dimensional plots for the two different sections of the potential surface. In both cases the view is looking toward the origin. At very large distances of the atoms from each other (top right on the contour plot), the energy essentially is independent of the distance and the energy surface is a flat plateau. Any linear collection of HHF atoms would take the lowest energy path along the valleys of the contour plot. This path, called the *reaction coordinate,* involves changes in both H–H and H–F distances. It is indicated by a dashed line in Figure 9.1*a*.

The intersection of the two parts of the potential surface in Figure 9.1*b* involves a saddle point (marked * in Figure 9.1*a*), which is an energy maximum as the reaction travels along the valleys of the plot (reaction coordinate) but a minimum along a path such as *E-F*. The surface around the point is shaped like a saddle. Figure 9.1*c* is a plot of the energy of the system as it moves along the reaction coordinate. Such a plot is called a *reaction profile.*

Whether the $H_2 + F \rightleftharpoons HF + H$ reaction proceeds forward or backward, it follows the low-energy reaction coordinate. The atomic positions simply reverse themselves as the reaction is reversed, like a film shown backwards. This is a statement of the principle of microscopic reversibility.

The principal of microscopic reversibility is extremely important in mechanistic studies, because it requires that *at equilibrium* both forward and reverse reactions proceed at equal rates backward and forward and along the reaction coordinate. Hence, the study of exchange reactions where leaving and entering groups X are the same is quite important, because the mechanism of entry for the new ligand must be just the reverse of that for loss of the leaving ligand. *Insofar as some other entering ligand Y resembles the leaving one X,* the principle of microscopic reversibility places limitations on the possible mechanism for replacement of X by Y: that is, we would expect the mechanism to resemble that for replacement of X by X (exchange). This is true for each separate step of a multistep reaction.

Instead of a saddle point, some potential surfaces (although not the one we have discussed) have a potential well or depression corresponding to a slightly stable configuration of reacting

[1]C. F. Bender, S. V. O'Neill, P. K. Pearson, and H. F. Schaeffer, III, *Science* 1972, *176*, 1412.

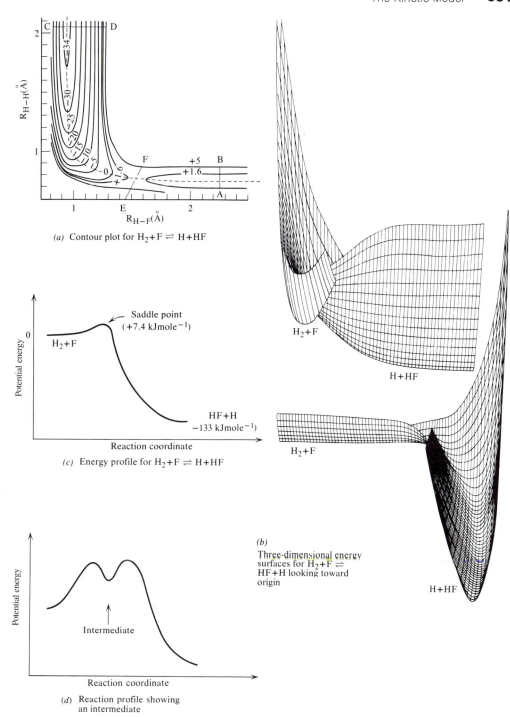

(a) Contour plot for $H_2 + F \rightleftharpoons H + HF$

(c) Energy profile for $H_2 + F \rightleftharpoons H + HF$

(b)
Three-dimensional energy
surfaces for $H_2 + F \rightleftharpoons$
$HF + H$ looking toward
origin

(d) Reaction profile showing
an intermediate

Figure 9.1 Plots for the reaction $H_2 + F \rightleftharpoons H + HF$. [(a) and (b) from C. F. Bender, S. V. O'Neill, P. K. Pearson, and H. E. Schaeffer, III, *Science* 1972, *176*, 1412.]

atoms, an *intermediate*. Slight stability means that the intermediate can persist for some small length of time. However, it eventually will follow the reaction coordinate down into a valley. Figure 9.1*d* shows a reaction profile for a system with one intermediate. (Of course, several successive intermediates are possible in multistep reactions.)

The width of the low energy valley on the energy versus atomic position plot is related to the entropy requirements for the reaction. A steeply sloping and narrow energy valley is associated with very negative entropies of activation involving constriction of the atomic motions of reactants. Conversely, a wide valley is associated with relative freedom in the motion and orientation of the reactants (more positive entropy of activation).

More complex reactions are hard to depict in plots of energy as a function of all the relative atomic positions. A nonlinear molecule of N atoms has $(3N - 6)$ interatomic distances and angles that could be varied to change the energy [$(3N - 5)$ if the molecule is linear]. Thus, a plot of energy versus each coordinate would require a $(3N - 5)$-dimensional hyperspace for plotting [or $(3N - 4)$ for a linear system]. Also, calculating such a multidimensional energy surface would be tremendously difficult and time-consuming—especially when our interest lies along the path of minimum energy from reactants to products. Hence we use a two-dimensional reaction profile. All the complexities of many variables are lumped into the definition of the reaction coordinate, a parameter that collects all the bond distances and angles that must change as the reactants rearrange to products. The reaction profile of energy versus reaction coordinate is used as a qualitative picture, and experiments are designed to show what reacting species are present in the transition state and what factors are important in determining the energy required to reach this state. Also of interest is whether intermediates exist along the reaction path.

The rate of reaction is determined by the energy required to surmount the activation barrier. The higher the barrier, the fewer the number of reactant molecules that will have this much energy at a given temperature, and thus, the slower the reaction. Fast reactions are those with low activation energies.

It would be a good idea to review the essential features of transition state theory as they bear on this chapter.[2] A major theme of this chapter is how features of the model provided by transition state theory can be verified experimentally. After all, only those concepts that can be investigated by a properly designed experiment have any scientific meaning. These considerations are quite important in elucidating the nature of the transition state, which is (by definition) an energy maximum and hence too short-lived for direct observation. Even intermediates (which are at relative minima on the energy surface) seldom last long enough for direct detection. Hence, designing and interpreting experiments bearing on these points is a nontrivial and subtle endeavor.

9.2 INTRODUCTION TO LIGAND SUBSTITUTION REACTIONS

Chemical reactions in which the composition of the first coordination sphere around a metal changes are called ligand substitution reactions. The rates of such reactions vary widely,

---

[2]See, for example W. J. Moore, *Physical Chemistry*, 4th ed., Prentice-Hall, Englewood Cliffs, N. J., 1972, pp. 376–387; K. Laidler, *Chemical Kinetics*, 2nd ed., McGraw-Hill, New York, 1965.

ranging from completion within the time for mixing of reactants (such as the formation of blue-purple $[Cu(NH_3)_4]^{2+}$ discussed in Section 7.1) to days (such as the substitution by H_2O on $[Co(NH_3)_6]^{3+}$). Taube[3] called complexes having $t_{\frac{1}{2}} < 30$ sec for substitution *labile* and those with longer $t_{\frac{1}{2}}$ *inert*. The study of reactions of labile complexes requires special experimental techniques such as stopped-flow,[4] P-jump[5] or T-jump[6] (in which a system at equilibrium is perturbed by a sudden change in pressure or temperature and its relaxation to a new equilibrium monitored).

Slower reactions can be monitored by conventional techniques (including NMR, UV-Vis spectroscopy and polarimetry) using convenient time periods. Hence, considerably more information is available on reactions of inert species.

Lability and *inertness* are kinetic terms referring to how quickly a reacting system reaches equilibrium. They are unrelated to the thermodynamic stability of the complexes. Table 9.1 gives exchange rates and formation constants, a measure of thermodynamic stability (See Sec. 12.3.3), for several cyano complexes. There is no necessary connection between the magnitude of the equilibrium constant for complex formation and the rate of exchange with labeled cyanide ion in aqueous solution. Thus $[Hg(CN)_4]^{2-}$ is both thermodynamically stable and kinetically labile. In solution, it exchanges ligands many times each second, but the entering ligand is always cyanide instead of the more abundant water. Inertness is sometimes referred to as kinetic stability—as differentiated from thermodynamic stability, which is given by the formation constant.

We shall study the ligand substitution reactions of octahedral and of square-planar complexes in turn. Many of the concepts involved in studying octahedral complexes also apply to square-planar ones.

Table 9.1 Formation constants and exchange rates of some cyano complexes[a]

Complex	Formation Constant[b] (K_f)	Exchange Rate
$[Ni(CN)_4]^{2-}$	10^{30}	Very fast
$[Hg(CN)_4]^{2-}$	10^{42}	Very fast
$[Fe(CN)_6]^{3-}$	10^{44}	Very slow
$[Fe(CN)_6]^{4-}$	10^{37}	Very slow
$[Pt(CN)_4]^{2-}$	$\sim 10^{40}$	$t_{\frac{1}{2}} = 1$ min

[a]From F. Basolo and R. Pearson, *Mechanisms of Inorganic Reactions*, 2nd ed., Wiley, New York, 1967.

[b]$K_f = \dfrac{[M(CN)_n]^{y-}}{[M^{(n-y)+}][CN^-]^n}$ for the reaction $M^{(n-y)+} + n\,CN^- \rightleftharpoons [M(CN)_n]^y$

[3]H. Taube, *Chem. Rev.* 1952, *50*, 69.

[4]See, for example, G. Dulz and N. Sutin, *Inorg. Chem.* 1963, *2*, 917.

[5]C. F. Bernasconi, *Relaxation Kinetics*, Academic Press, New York, 1976, p. 227; *ibid*, p. 181.

[6]G. H. Czerlinski and M. Eigen, *Z. Elektrochem.* 1959, *63*, 652; G. G. Hammes, *Acc. Chem. Res.* 1968, *1*, 321.

9.3 OCTAHEDRAL COMPLEXES–LIGAND SUBSTITUTION REACTIONS

9.3.1 Kinetics of Water Exchange

The simplest substitution reaction is exchange of coordinated water around a metal ion in aqueous solution with bulk solvent (in the absence of other ligands). Figure 9.2 displays values of log k for the exchange reaction.[7] The relative values of k show how such features as ionic charge and size affect exchange rates. Langford and Gray[8] have divided metal ions into four classes, based on exchange rates.

Class I   Very fast (diffusion-controlled) exchange of water occurs; $k \geq 10^8$ sec^{-1}. The ions in this class are those of the alkali metals and alkaline earths (except for Be^{2+} and Mg^{2+}), Group IIB (except for Zn^{2+}), Cr^{2+}, and Cu^{2+}.

Class II   Exchange-rate constants are between 10^4 and 10^8 sec^{-1}. The divalent first-row transition metal ions (except for V^{2+}, Cr^{2+}, and Cu^{2+}), as well as Mg^{2+} and the trivalent lanthanide ions, are members of this class.

Class III   Exchange-rate constants are between 1 and 10^4 sec^{-1}. This class includes Be^{2+}, V^{2+}, Al^{3+}, Ga^{3+}, and several trivalent first-row transition metal ions.

Class IV   Ions in this class are inert in Taube's sense; their rate constants for exchange fall between 10^{-6} and 10^{-3} sec^{-1}. Members of the set are Cr^{3+}, Co^{3+}, Rh^{3+}, Ir^{3+}, and Pt^{2+}.

Obviously, one factor determining exchange rates for ions with the noble gas and pseudo–noble gas electron configurations is the ionic potential q/r (see Section 7.4.1). Those ions with high ionic potential exchange relatively slowly, suggesting that a principal contribution to the ac-

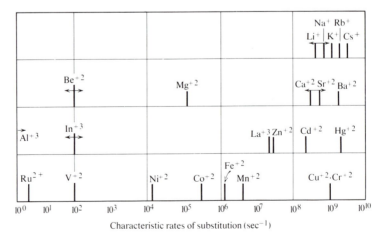

Figure 9.2 Exchange rates for metal aqua complexes. (From M. Eigen, *Pure Appl. Chem.* 1963, *6*, 105.)

[7]In this chapter, k is used for rate constants and K for equilibrium constants.

[8]H. B. Gray and C. H. Langford, *Chem. Eng. News.* 1968, *April 1*, p. 68.

tivation energy is the breaking of the bond to the leaving molecule. With the transition metals the behavior is more complex. One factor involved is still the ionic potential, since trivalent ions tend to exchange more slowly than divalent ones. However, the d-electron configuration also plays an important role. Rapid exchange by Cr^{2+} and Cu^{2+} (which places them in Group I) is related to the Jahn-Teller distortion in the ground state (see Section 7.5.2), which makes axial bonds longer than equatorial bonds. The geometry of the ground state for these aquated ions thus is close to that of the transition state for water exchange, and axial water ligands can exchange more rapidly than for undistorted octahedral complexes.

9.3.2 Lability and *d*-Electron Configuration

One way of viewing the problem of electron configuration and its relation to lability is to note that inert complexes are the ones with very high LFSE: d^3, low-spin d^4, low-spin d^5, and low-spin d^6. The d^8 configuration is borderline. The octahedral d^8 complexes encountered are the weak-field complexes of Ni(II). Low-spin complexes of d^8 ions—for example Rh(I), Ir(I), Ni(II), Pd(II), Pt(II), and Au(III)—are square planar. The weak-field (high-spin) complexes of nickel(II) usually react much more rapidly than those of d^3 or d^6 ions, but slower than those of Cu(II), Co(II), and Zn(II) complexes.

Two extreme possibilities for substitution mechanisms in octahedral complexes come to mind. A dissociative path would involve prior departure of the leaving ligand followed by attack of the entering ligand on the resulting five-coordinate species. An associative path involving prior attack of the entering ligand and giving a seven-coordinate species also is possible. This would be followed by departure of the leaving ligand.

Ligand field calculations (see Section 7.6) indicate a large loss in LFSE energy for d^3, low-spin d^4, low-spin d^5, low-spin d^6, and d^8 complexes in going from six- to either five- or seven-coordinate species. For both dissociative and associative pathways, these configurations experience a ligand field contribution to the activation energy (LFAE). Hence these configurations are kinetically inert. Other configurations lose little or no LFSE and are labile.[9]

Note that for a dissociative mechanism the activation energy must be equal to or greater than the bond dissociation energy. The LFSE is only one factor in determining this energy and, when the change in LFSE is small, other factors might be more important. This should apply particularly to complexes involving π bonding.

9.3.3 Mechanisms for Ligand Substitution Reactions

We now will consider in more detail dissociative and associative pathways for ligand substitution. Henceforth, for consistency in notation we shall call the leaving group X and the entering group Y. In particular cases, X and/or Y may be identical with the other ligands in the complex. Any of these ligands may be water or other solvent. Thus, the general reaction (omitting charge) may be written

$$L_5MX + Y \rightarrow L_5MY + X \tag{9.1}$$

[9]F. Basolo and R. G. Pearson, *Mechanisms of Inorganic Reactions*, 2nd ed., John Wiley, New York, 1967, pp. 141–158.

Two issues are important in ligand substitution reactions: first, whether the main contribution to the activation energy (and hence the factor controlling the rate) is the breaking of the bond to the leaving group (dissociative activation) or the making of the bond to the entering group (associative activation); and second, what the sequence of elementary steps leading from reactants to products is. Langford and Gray have labeled these as the *intimate* and *stoichiometric mechanisms,* respectively.

Dissociative (*d*) Mechanisms

To understand this distinction better, consider first the case of a ligand substitution in which the activation energy is determined primarily by the energy required to break the bond to the leaving group, a so-called dissociative or *d* reaction. At the start of the reaction the complex ion with its coordination sphere of L and X is surrounded by an outer sphere of loosely held solvent molecules, as well as of Y and other species present in solution. In one possible *d* mechanism, two elementary steps are detectable: the complex accumulates enough energy to break completely the M—X bond, leaving a five-coordinate intermediate; then, this intermediate reacts with Y (which could be solvent) in the second coordination sphere.

$$L_5MX \underset{k_{-1}}{\overset{k_1}{\rightleftharpoons}} L_5M + X$$

$$L_5M + Y \xrightarrow{k_2} L_5MY$$

(9.2)

The rate law for this mechanism is

$$\frac{d\,[L_5MY]}{dt} = k_2\,[L_5M]\,[Y]$$

(9.3)

Applying the steady-state approximation to L_5M,

$$\frac{d\,[L_5M]}{dt} = 0 = k_1[L_5MX] - k_{-1}\,[L_5M]\,[X] - k_2\,[L_5M]\,[Y]$$

(9.4)

$$[L_5M] = \frac{k_1\,[L_5MX]}{k_{-1}\,[X] + k_2\,[Y]}$$

(9.5)

And in terms of experimentally measurable concentrations the rate becomes

$$\frac{d\,[L_5MY]}{dt} = \frac{k_1 k_2\,[L_5MX]\,[Y]}{k_{-1}\,[X] + k_2\,[Y]}$$

(9.6)

If we are to show that any reaction does in fact proceed via this mechanism, L_5M must have a sufficiently long lifetime to be experimentally detectable. (Later, we will see some ways to do this.)

Consider now the case in which L_5M does *not* have an appreciable lifetime. L_5MX accumulates sufficient thermal energy into the M–X vibration to begin to break the M—X bond. Before the bond can be broken fully, M begins to form a bond with whatever species Y happens to be in a suitable geometric position to enter the first coordination sphere. This is just the limiting case as the lifetime of an intermediate becomes shorter and shorter and finally does not survive long enough to be at all detectable—corresponding to the relative energy minimum in Figure 9.1*d* becoming shallower and shallower until it no longer exists. The likelihood of any Y being present in the outer sphere will be proportional to [Y]. Also involved is the equilibrium constant for outer sphere complex formation. The magnitude of K depends on the ionic charge, being larger when Y and the complex are oppositely charged. This possible mechanism can be written as follows where the presence of Y in the outer sphere (the formation of an outer sphere complex) is indicated explicitly in the first step.

$$L_5MX + Y \overset{K}{\rightleftharpoons} (L_5MX,Y)$$

$$(L_5MX,Y) \overset{k}{\rightarrow} (L_5MY,X) \tag{9.7}$$

$$(L_5MY,X) \overset{fast}{\rightarrow} L_5MY + X$$

The experimental parameter most accessible is the initial concentration of the reactant complex $[L_5MX]_0$, which (in a solution containing Y) exists partly in the form of the outer sphere complex.

$$[L_5MX]_0 = [L_5MX] + [(L_5MX,Y)]$$

$$= [L_5MX] + K[L_5MX][Y] \tag{9.8}$$

Hence

$$[L_5MX] = \frac{[L_5MX]_0}{1 + K[Y]} \tag{9.9}$$

and the rate is

$$\frac{d[L_5MX]}{dt} = \frac{kK[L_5MX]_0[Y]}{1 + K[Y]} \tag{9.10}$$

For each of two mechanistic possibilities for ligand substitution reactions we have just discussed, the energy necessary for bond breaking mainly determines the activation energy and hence the rate. The intimate mechanism in both cases is d (dissociative). The two cases differ in that an intermediate of reduced coordination number may or may not be detectable experimentally. These two possibilities lead to differing sequences of elementary steps (different stoichiometric mechanisms). When the five-coordinate intermediate is detectable, the stoichiometric mechanism is labeled D (dissociative); the other case is labeled I_d (dissociative interchange). Table 9.2 shows the relationship between intimate and stoichiometric mechanism

Table 9.2 Relation between intimate and stoichiometric reaction mechanisms

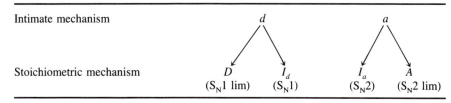

Intimate mechanism		d		a	
Stoichiometric mechanism		D (S_N1 lim)	I_d (S_N1)	I_a (S_N2)	A (S_N2 lim)

(associative mechanisms are discussed below) and presents notation relating the stoichiometric mechanisms to the S_N1 and S_N2 labels of organic chemistry.

An important point about the *d* mechanisms just discussed is that the rate laws given in Equations 9.6 and 9.10 both give second-order kinetics under certain circumstances (see below). Thus the preeminence of bond breaking in the intimate mechanism does not necessarily lead to first-order kinetics for inorganic complexes, and so the molecularity of the reactions does not necessarily indicate the mechanism.

Associative (*a*) Mechanisms

Bond making to an entering group also might be the main factor determining the size of the activation energy. Such reactions are given the *a* designation, for an associative intimate mechanism. Again, two types of stoichiometric mechanism are possible. In the first, an intermediate of higher coordination number survives long enough to be detected; this is the so-called *A* (or S_N2 lim) stoichiometric mechanism.

$$L_5MX + Y \underset{k_{-1}}{\overset{k_1}{\rightleftharpoons}} L_5MXY$$

$$L_5MXY \overset{k_2}{\to} L_5MY + X \tag{9.11}$$

An *A* mechanism would lead to a rate law of the form rate $= \dfrac{k_1\,[L_5MX]\,[Y]}{k_{-1} + k_2}$. At present, no firmly established examples of *A* mechanisms have been observed for octahedral substitution reactions. The other possibility is the I_a mechanism, the associative counterpart of I_d. The outer-sphere complex between the reactants is formed first. The bond between the metal and the entering group begins to form before the bond to the leaving group begins to break. In the transition state, both the entering and leaving group are bonded to the metal, but the extent of bond making exceeds that of bond breaking. The two transition states could be diagrammed as follows for I_a and I_d, where ". . ." denotes the less important bond energy contribution.

$$\begin{array}{cc} Y{-}M \cdots X & Y \cdots M{-}X \\ I_a & I_d \end{array} \tag{9.12}$$

The steps in the stoichiometric mechanism and the form of the rate law resemble those in Equations 9.7 and 9.10. The difference is that in the I_a mechanism, the magnitude of *k* is controlled by the energy of bond formation. Another way of viewing the I_a mechanism is as the limiting case of an *A* mechanism in which bond breaking begins too soon for detection of a seven-coordinate intermediate.

The relationship among these four models for reaction mechanism is shown in Table 9.2. As with any models, a crucial question is what experimental evidence can be used to distinguish among them. (Analysis of experimental results also leads to refinement of models themselves in favorable cases. Such considerations led to the notion that the S_N1 and S_N2 models of reactivity did not suffice for inorganic complexes.) Plainly, the form of the rate law will not serve to distinguish—*D*, I_d and I_a mechanistic possibilities lead to a rate law of the form

$$\text{rate} = \frac{a\,[L_5MX]\,[Y]}{1 + b\,[Y]} \tag{9.13}$$

Under certain circumstances (for $b[Y] \gg 1$), the reaction could be first-order for any of the mechanisms. If $b[Y] \ll 1$, the reaction would be second-order. We would see mixed-order kinetics when $b[Y] \approx 1$. As long as the final step is not reversible, second-order kinetics would result for A reactions.

9.3.4 Experimental Tests of Mechanism

No reaction mechanism ever can be really "proven." Evidence in favor of one mechanism can be advanced and experiments designed to refute or confirm the proposed reaction path. The accepted mechanism turns out to be the one with the least evidence against it.

Intimate Mechanism

For octahedral ligand substitution reactions, the most straightforward distinction to make is that between d and a activation (the intimate mechanism). Most of the data so far obtained involve reactions of inert ions such as Co(III) (especially), Cr(III), Rh(III), Ir(III), Pt(IV) and the borderline Ni(II). Two frequently studied reactions (along with solvent exchange) are aquation (sometimes called acid hydrolysis),

$$L_5MX^{n+} + H_2O \rightarrow L_5M(OH_2)^{n+1} + X^- \qquad (9 14)$$

and anation,

$$L_5M(OH_2)^{n+1} + Y^- \rightarrow L_5MY^{n+} + H_2O \qquad (9.15)$$

Sensitivity to the Nature of the Entering or Leaving Group. Comparing the rates of reaction of a series of complexes with different leaving groups and the same entering group affords us evidence about intimate mechanisms. Table 9.3 reports rates of aquation for the series of complexes $[Co(NH_3)_5X]^{2+}$. The rates depend heavily on the nature of X, the leaving

Table 9.3 Rate constants for acid aquation of some octahedral complexes[a] of Co(III)[b] at 25°C

Complex	$k(sec^{-1})$
$[Co(NH_3)_5OP(OCH_3)_3]^{3+}$	2.5×10^{-4}
$[Co(NH_3)_5(NO_3)]^{2+}$	2.7×10^{-5}
$[Co(NH_3)_5I]^{2+}$	8.3×10^{-6}
$[Co(NH_3)_5(H_2O)]^{3+}$	5.8×10^{-6}
$[Co(NH_3)_5Cl]^{2+}$	1.7×10^{-6}
$[Co(NH_3)_5SO_4]^{+}$	1.2×10^{-6}
$[Co(NH_3)_5F]^{2+}$	8.6×10^{-8}
$[Co(NH_3)_5N_3]^{2+}$	2.1×10^{-9}
$[Co(NH_3)_5NCS]^{2+}$	5.0×10^{-10}

[a]Leaving ligand is written last.
[b]F. Basolo and R. Pearson, *op. cit.*

Table 9.4 Limiting rate constants for anation by Y^{n-} and water exchange (k_e) at 45°C[a] of $[Co(NH_3)_5(H_2O)]^{3+}$

Y^{n-}	$k(sec^{-1})$	k/k_e
H_2O	100×10^{-6}	—
N_3^-	100×10^{-6}	1.0
SO_4^{2-}	24×10^{-6}	0.24
Cl^-	21×10^{-6}	0.21
NCS^-	16×10^{-6}	0.16
H_2O	5.8×10^{-6} [b]	—
$H_2PO_4^-$	7.7×10^{-7} [b]	0.13

[a]From R. G. Wilkins, *The Study of Kinetics and Mechanism of Reactions of Transition Metal Complexes*, Allyn and Bacon, Boston, 1974.
[b]At 25°C.

group, and vary over some six orders of magnitude. In contrast, the rates of anation of $[Co(NH_3)_5(H_2O)]^{3+}$ (shown in Table 9.4) display relatively little sensitivity to Y^-. Clearly, these reactions are much more sensitive to the kind of bond being broken than to the kind being formed. This constitutes evidence of *d* activation. Further support comes from the aqueous chemistry of $[Co(NH_3)_5X]^{2+}$ complexes where no direct replacement of X^- by Y^- is observed. *Instead, aquation occurs, followed by anation with* Y^-. Such behavior indicates that the energetically significant process is Co—X bond breaking and that bond making has so little energetic significance that whatever species is present in greatest concentration (in this case, the solvent water) serves as the initial entering group independent of its identity.

The substitution rate constants for $[Ru(edta)(H_2O)]^-$ reported in Table 9.5 stand in striking contrast, indicating a thousandfold rate variation with the nature of the entering group and hence an *a* intimate mechanism. The rate constants for aquation of $[Ru(edta)L]^{n-}$ complexes span a lesser range (by a factor of 10), indicating the smaller sensitivity to the identity of the leaving group. (You may have noticed that the $[Ru(edta)L]^{n-}$ complexes react much faster

Table 9.5 Rate constants for substitution by L (k_L) of $[Ru(edta)(H_2O)]^-$ and for aquation k_{aq} of $[Ru(edta)L]^{n-}$ at 25°C[a,b]

L	$k_L(M^{-1}sec^{-1})$	$k_{aq}(sec^{-1})$
Pyrazine	2.0×10^4	2.0
Isonicotinamide	8.3×10^3	0.7
Pyridine	6.3×10^3	0.061
Imidazole	1.86×10^3	—
NCS^-	2.7×10^2	0.5
CH_3CN	3.0×10	3.2

[a]T. Matsubara and C. Creutz, *Inorg. Chem.* 1979, *18*, 1956.
[b]edta is quinquedentate.

than would be expected from low-spin d^5 species. This interesting point is discussed in the reference.)

Steric Effects of Inert Ligands.   Crowding around the metal ion would be expected to retard the rates of reactions that occur by an *a* mechanism and to speed up those occurring via a *d* mechanism. The data in Table 9.6 show that aquation reactions of *trans*-[Co(N—N)$_2$Cl$_2$]$^+$

Table 9.6   Effect of nonleaving ligands on acid hydrolysis rates of some Co(III) complexes

a. *trans*-[Co(N—N)$_2$Cl$_2$]$^+$ + H$_2$O → [Co(N—N)$_2$(H$_2$O)Cl]$^{2+}$ + Cl$^-$ [a,b]

N—N	$k(sec^{-1})$
NH$_2$CH$_2$CH$_2$NH$_2$	3.2×10^{-5}
NH$_2$CH$_2$CH(CH$_3$)NH$_2$	6.2×10^{-5}
d,ℓ-NH$_2$CH(CH$_3$)CH(CH$_3$)NH$_2$	1.5×10^{-4}
meso-NH$_2$CH(CH$_3$)CH(CH$_3$)NH$_2$	4.2×10^{-4}
NH$_2$CH$_2$C(CH$_3$)$_2$NH$_2$	2.2×10^{-4}
NH$_2$C(CH$_3$)$_2$C(CH$_3$)$_2$NH$_2$	Instantaneous
NH$_2$CH$_2$CH$_2$NH(CH$_3$)	1.7×10^{-5}

b. *trans*-[Co(N$_4$)LCl]$^{n+}$ + H$_2$O → *trans*-[Co(N$_4$)L(H$_2$O)]$^{(n+1)+}$ + Cl$^-$ [a,c,d]

Complex	$k(sec^{-1})$
trans-[Co(cyclam)Cl$_2$]$^+$	1.1×10^{-6}
trans-[Co(cyclam)(NCS)Cl]$^+$	1.1×10^{-9}
trans-[Co(cyclam)(CN)Cl]$^+$	4.8×10^{-7}
trans-[Co(tet-*b*)Cl$_2$]$^+$	9.3×10^{-4}
trans-[Co(tet-*b*)(NCS)Cl]$^+$	7.0×10^{-7}
trans-[Co(tet-*b*)(CN)Cl]$^+$	3.4×10^{-4}
trans-[Co(dmgH)$_2$Cl$_2$]$^-$	2.7×10^{-4}
trans-[Co(trans[14]diene)(N$_3$)Cl]$^+$	8.8×10^{-3}

c. [Co(en)$_2$LCl]$^{n+}$ + H$_2$O → [Co(en)$_2$L(H$_2$O)]$^{(n+1)+}$ + Cl$^-$ [a,e]

Complex	$k(sec^{-1})$
cis-[Co(en)$_2$(OH)Cl]$^+$	0.012
trans-[Co(en)$_2$(OH)Cl]$^+$	1.60×10^{-3}
cis-[Co(en)$_2$Cl$_2$]$^+$	2.4×10^{-4}
trans-[Co(en)$_2$Cl$_2$]$^+$	3.5×10^{-5}
cis-[Co(en)$_2$(NCS)Cl]$^+$	1.1×10^{-5}
trans-[Co(en)$_2$(NCS)Cl]$^+$	5×10^{-8}
cis-[Co(en)$_2$(NH$_3$)Cl]$^{2+}$	5×10^{-7}
trans-[Co(en)$_2$(NH$_3$)Cl]$^{2+}$	3.4×10^{-7}
cis-[Co(en)$_2$(H$_2$O)Cl]$^{2+}$	1.6×10^{-6}
trans-[Co(en)$_2$(H$_2$O)Cl]$^{2+}$	2.5×10^{-6}
cis-[Co(en)$_2$(CN)Cl]$^+$	6.2×10^{-7}
trans-[Co(en)$_2$(CN)Cl]$^+$	8.2×10^{-5}
cis-[Co(en)$_2$(NO$_2$)Cl]$^+$	1.1×10^{-4}
trans-[Co(en)$_2$(NO$_2$)Cl]$^+$	9.8×10^{-4}

[a] At 25°C.
[b] R. G. Pearson, C. R. Boston and F. Basolo, *J. Am. Chem. Soc.* 1953, *75*, 3089.
[c] N$_4$-quadridentate amine ligand; see Figure 9.3 for structures.
[d] From the work of Poon, Tobe, and others.
[e] From the work of Chan, Tobe, and others.

(where N—N is a bidentate amine ligand) exhibit steric acceleration. This behavior is indicative of d activation, as is the rate increase on going from cyclam complexes to those of the hexamethyl analogue tet-b. (The ligand structures are shown in Figure 9.3.)

Electronic Effects of Inert Ligands. Along with steric effects, inert (nonleaving) ligands can also exert electronic effects on reaction rates. Comparing the rates of acid hydrolysis of a series of compounds containing the progressively more unsaturated ligands cyclam, dimethylglyoximate (dmgH) and trans[14]diene (Table 9.6b and Figure 9.3) reveals rate increases of 10^2–10^3 for similar leaving groups. The steric requirements of these ligands do not differ much, and the rate increase is considered to arise from enhanced possibilities for electron delocalization into unsaturated ligands, making the metal center softer (see Section 12.3) and stabilizing the transition state. Another characteristic of dmgH complexes is the importance of the identity of the *trans* ligand in determining the reaction rate. The rates of dissociative Br$^-$ anation of *trans*-$[(MeO)_3PCo(dmgH)_2X]$ vary over a range of 10^5, depending on X.[10] We do not see this effect in Co(III) complexes with harder ligands.

The data on acid aquation of $[Co(en)_2LX]^{n+}$ complexes (Table 9.6c) is explained nicely by a dissociative mechanism. Consider the *cis*-$[Co(en)_2LX]^{n+}$ series. When the leaving group X departs, the d^2sp^3 hybrid orbital that bound it to the Co is empty. If the *cis* ligand is a good π-donor, it can supply electrons to the electron-deficient Co, thereby stabilizing the transition state and lowering the activation energy (see Figure 9.4). As a consequence, *cis* complexes where L is a good π-donor react more rapidly than *cis* complexes of π-acceptor ligands (such as CN$^-$) or σ-donor-only ligands (such as NH$_3$). When L is *trans* to the leaving group X, no π-donation into the vacant Co orbital can occur without rearrangement to a trigonal bipyramid. The energy of rearrangement causes the activation energy for aquation of *trans*-$[Co(en)_2LX]^{n+}$ to be larger and the rates lower than those of the corresponding *cis* complexes. The reactivity pattern of *cis*- and *trans*-$[Co(en)_2LX]^{n+}$ complexes would be difficult to explain assuming an a intimate mechanism. (The argument here does not require VB hybrid orbitals; it also could be understood in terms of MO's.)

Comparison of Rates of Anation and Water Exchange. Water exchange is an important reaction. Because the concentration of water (or any solvent) cannot be varied, its presence in the transition state (or lack thereof) cannot be determined kinetically. The form of the rate law for Reaction 9.1 (X = Y = H$_2$O) is always

$$\text{rate} = k_{\text{obs}} [\text{L}_5\text{MX}] \qquad (\text{since } [\text{H}_2\text{O}] = [\text{Y}] = 55.5\ M) \qquad (9.16)$$

We might consider a complex having X = H$_2^{18}$O placed in H$_2$O = Y so that solvent exchange could be detected by entry of H$_2$O of natural isotopic abundance into the coordination sphere. For a D mechanism, $k_2[\text{H}_2\text{O}] \gg k_{-1} [\text{X}]$ in Equation 9.6, and the rate becomes $k_1[\text{L}_5\text{MX}]$. Thus k_{obs}, the measured rate constant, can be interpreted as the rate constant for breaking of the M—O bond. In the I_d mechanism, $K[\text{H}_2\text{O}] \gg 1$ and k_{obs} is identified with k, the rate constant for interchange of a water molecule between the outer and inner coordination spheres. For a mechanisms, similar limiting forms of the rate law are operative for aquation. For the A pathway, the observed first-order rate constant represents the rate of attack by Y = H$_2$O to

---

[10]P. J. Toscano and L. G. Marzilli, *Inorg. Chem.* 1979, *18*, 421.

NH$_2$CH$_2$CH$_2$NHCH$_2$CH$_2$NH$_2$

dien (1,4,7-tetraazaheptane)

NH$_2$CH$_2$CH$_2$NHCH$_2$CH$_2$NHCH$_2$CH$_2$NH$_2$

trien (1,4,7,10-tetraazadecane)

edda (ethylenediaminediacetate)

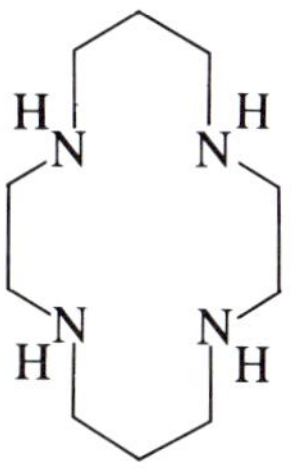

cyclam (1,4,8,11-tetraazacyclotetradecane)

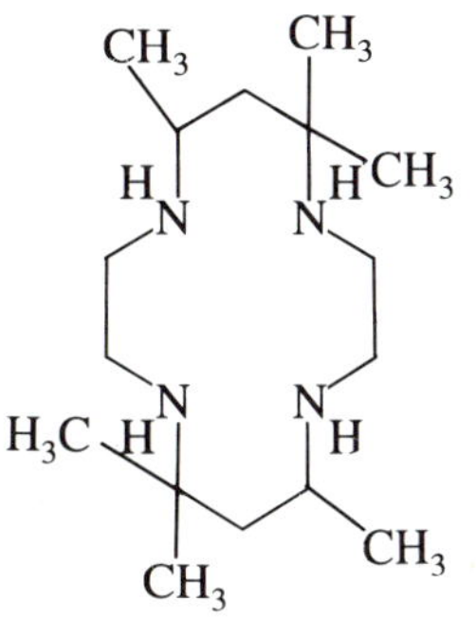

tet-*b*(*d,l*-1,4,8,11-tetraaza-5,5,7,12,12,14-hexamethylcyclotetradecane)

tet-*a*(*meso*-1,4,8,11-tetraaza-5,5,7,12,12,14-hexamethylcyclotetradecane)

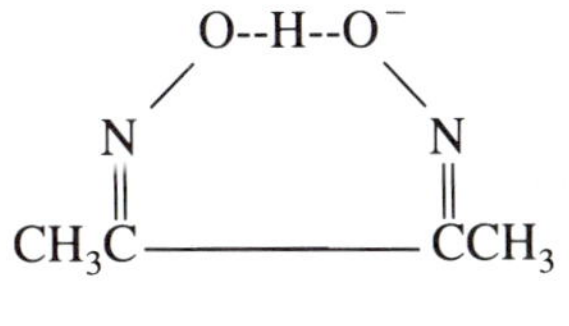

dmgH (dimethylglyoximate)

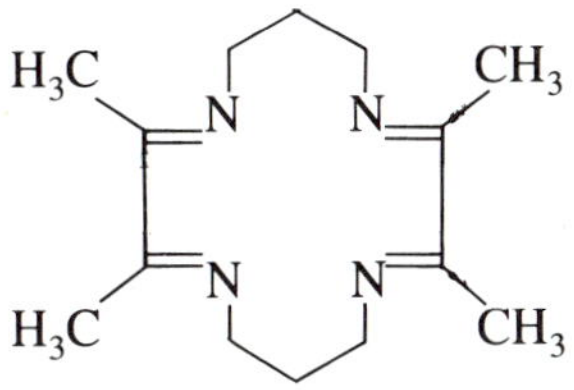

trans[14]diene (2,3,9,10-tetramethyl-1,4,8,11-tetraazacyclotetradeca-1,3,8,10-tetraene)

Figure 9.3 Structures of some chelating ligands.

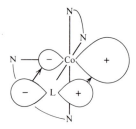

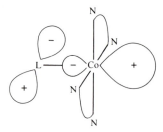

(a) π-Donation by a *cis* ligand
into the *p* component of an
empty d^2sp^3 hybrid.

(b) π-Orbital of *trans* L is
orthogonal to d^2sp^3 and
no donation occurs.

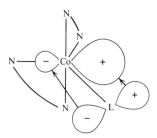

(c) π-Donation from *trans* L can occur
upon rearrangement to trigonal
bipyramid

Figure 9.4

form the seven-coordinate intermediate L_5MXY. For the I_a mechanism, k_{obs} is the rate of associative interchange of Y = H_2O between the outer and inner coordination sphere.

For ligands Y other than water, a similar limiting rate is often reached if [Y] is made large enough. The reaction then becomes first-order in L_5MX and k_{obs} is in units of sec^{-1}. When [Y] is very small (impossible when Y = solvent) the rate laws have the limiting forms summarized in Table 9.7. The results are obtained from Equations 9.6, 9.10, and 9.13. In the most general case, second-order kinetics are observed at low [Y]; at high [Y], we see the limiting case of first-order behavior. Figure 9.5 shows a plot of k_{obs} versus [Y] for two reactions. In one reaction, the limiting first-order behavior is not seen at high [Y], because either k_2 or K is too small.

Comparing the values of the limiting first-order rate constants for anation and for solvent exchange allows us to choose between a and d intimate mechanisms. In the case of d activation, the main contribution to the activation energy is the breaking of the M—X bond. Bond formation to the entering group tends to take place rather unselectively with any Y that is in correct position in the second coordination sphere. Because of tremendous concentration advantage, water is the most likely entering group in aqueous solution, and so we can expect that the rate of anation should never be greater than the rate of water exchange for d activation. The data presented in Table 9.4 show that the anation reactions of $[Co(NH_3)_5(H_2O)]^{3+}$ are subject to d activation. In cases of a activation, we can expect to see selectivity displayed toward the entering ligand and, it is entirely possible that some ligands could be better entering groups than water. Table 9.8's data indicate an a mechanism. The fact that limiting first-order rates are reached indicates the mechanism to be I_a rather than A. Although a limiting first-order rate constant for anation larger than that for exchange is a sufficient criterion for establishing

Table 9.7 Limiting forms of rate laws and significance of rate constants.

Stoichiometric Mechanism	Conditions	Rate Law	k_{obs}	Comments
D	$k_2[Y]$ very large	$k_1[L_5MX]$	k_1	k_{obs} represents rate of M—X dissociation.
D	$k_2[Y]$ very small	$\dfrac{k_1 k_2[L_5MX][Y]}{k_{-1}[X]}$	$\dfrac{k_1 k_2}{k_{-1}[X]}$	If X = solvent, [X] is constant; otherwise, can observe mass law retardation.
I_d	$K[Y]$ very large	$k[L_5MX]$	k	k_{obs} represents interchange rate.
I_d	$K[Y]$ very small	$kK[L_5MX][Y]$	kK	k_{obs} is composite.
I_a	$K[Y]$ very large	$k[L_5MX]$	k	k_{obs} represents rate of ligand interchange.
I_a	$K[Y]$ very small	$kK[L_5MX][Y]$	kK	k_{obs} is composite.
A A	$\left.\begin{array}{l} [Y] \text{ very large} \\ [Y] \text{ very small} \end{array}\right\}$	$\dfrac{k_1[L_5MX][Y]}{k_{-1} + k_2}$	$\dfrac{k_1}{k_{-1} + k_2}$	Kinetics always second-order if second step is irreversible ($k_{-2} = 0$).

an I_a mechanism, it is not a necessary one. Even though the anation reactions of $[Cr(H_2O)_6]^{3+}$ are thought to proceed via the I_a mechanism, no soft base is as effective as H_2O for attacking the hard acid Cr^{3+}.[11,12]

Effect of Charge on Reaction Rate. All other things being equal, increased positive charge on a reacting complex should make the breaking of bonds between nucleophiles and metals more difficult. Hence we expect the reaction rate to decrease with increasing positive charge for complexes displaying d activation. The fact that the water exchange reactions of a number

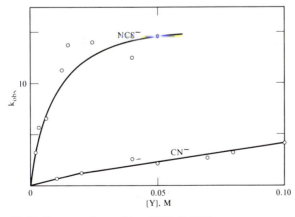

Figure 9.5 Plots of k_{obs} versus concentration of entering ligand for the anation reactions

$[Co(III)hematoporphyrin(IX)(H_2O)_2] + Y^-$

showing almost linear behavior for $Y^- = CN^-$ and transition from second- to first-order behavior for $Y^- = NCS^-$. (Reprinted with permission from E. B. Fleischer, S. Jacobs, and L. Mestichelli. *J. Am. Chem. Soc.* 1968, 90, 2527. Copyright 1968, American Chemical Society.)

[11]J. H. Espenson, *Inorg. Chem.* 1969, 8, 1554.

[12]A metal ion with its normal complement of water ligands is sometimes written as M^{n+}

Table 9.8 Limiting rate constant[a] for anation by Y^{n-} and water exchange (k_e).

Complex	Y^{n-}	$k(sec^{-1})$	k/k_e
$[Rh(NH_3)_5(OH_2)]^{3+\,b,d}$	Br^-	7.9×10^{-3}	4.9
	Cl^-	4.2×10^{-3}	2.6
	SO_4^{2-}	1.7×10^{-3}	1.0
	H_2O	1.6×10^{-3}	—
$[Ir(NH_3)_5(OH_2)]^{3+\,c,e}$	Cl^-	9.2×10^{-4}	~4
	H_2O	2.2×10^{-4}	—

[a]See Table 9.7 for explanation.
[b]F. Monacelli, *Inorg. Chim. Acta*, 1968, 2, 263.
[c]E. Borghi, F. Monacelli, and T. Prosperi, *Inorg. Nucl. Chem. Lett.* 1970, 6, 667.
[d]At 65°C.
[e]At 85°C.

of main-group metals mentioned in Section 9.3.1 follow this trend, coupled with the similarity of substitution to aquation rates, provides evidence for dissociative (*d*) activation. With transition metal complexes, LFSE considerations (Section 9.3.2) are superimposed on those of charge. Moreover, there is no way to vary the overall charge on the complex without introducing differently charged ligands, which leads to changes in σ and π bonding. Even so, an increase in rate as charge decreases often has been accepted as evidence of *d* activation. Table 9.9 shows rate constants for anation of $[Cr(H_2O)_6]^{3+}$ and $[Cr(OH)(H_2O)_5]^{2+}$. The increase in rate for the hydroxo complexes implies a *d* mechanism. It may be that electron density can be supplied to the metal by OH^-, thus compensating for loss of the electron pair during the bond-breaking step.

Table 9.9 Rate constants for anation by Y^{n-} at 25°C for $[Cr(OH_2)_6]^{3+}$ and $[Cr(OH)(OH_2)_5]^{2+\,a}$

Y^{n-}	$k(M^{-1}sec^{-1})$ for $[Cr(OH_2)_6]^{3+}$	$k(M^{-1}sec^{-1})$ for $[Cr(OH)(OH_2)_5]^{2+}$
SO_4^{2-}	1.1×10^{-5}	3.9×10^{-4}
H_2O	2.5×10^{-6b}	$<1.0 \times 10^{-4b}$
NCS^-	1.8×10^{-6}	4.9×10^{-5}
Cl^-	2.9×10^{-8}	2.8×10^{-5}
SCN^-	4×10^{-9}	5×10^{-6}
I^-	8×10^{-10}	2.6×10^{-6}

[a]From R. G. Wilkins, *The Study of Kinetics and Mechanism of Reactions of Transition Metal Complexes*, Allyn and Bacon, Boston, 1974, p. 194.
[b]sec^{-1}

The expectations for *a* mechanisms as the positive charge on the complex increases are not clear. Although bond making should be enhanced (thus increasing rates for *A* reactions), bond breaking should be retarded, leading to opposing effects in I_a pathways.

Activation Parameters.[13]   A remarkable feature of several substitution reactions is the existence of a correlation between rate and ΔG^0: more thermodynamically favored reactions proceed faster. Figure 9.6 shows a plot of log k as a function of log $K (=-\Delta G^0/2.3\ RT)$ for aquation reactions of $[Co(NH_3)_5X]^{n+}$. The physical reason for the relationship lies in Hammond's postulate,[14] which states that two consecutively occurring states that have similar energy along a reaction coordinate will involve only small structural reorganization. Hence, the plot of Figure 9.6, which has a slope of 1.0, is considered evidence of a *d* mechanism in which the transition state (or intermediate) lies late along the reaction coordinate and so *resembles the products* in geometry (having free solvated X). Hence ΔG^* will be related to ΔG^0, the overall free-energy change of the reaction.

For *a* reactions the transition state resembles the reactants in having X still bound to the metal. Hence, it occurs earlier along the reaction coordinate. For $[Cr(H_2O)_5X]^{2+}$, a linear plot of log k versus log K has a slope of 0.56, indicating[13] a lesser resemblence of the transition state geometry to product geometry and hence an *a* mechanism. Swaddle shows in detail why linear plots of log k versus log K can be associated with a *d* mechanism when the slope is 1.0, and an *a* mechanism when it is around 0.5.

Recently, Stranks and his coworkers studied pressure dependence of reaction rates.[15] Their measurements permit us to calculate the volume of activation $\Delta V^{\ddagger}$. For a *d* mechanism, $\Delta V^{\ddagger}$ should (to a first approximation) approach the total volume change in going from ML_5X to L_5M and free X, and should be positive. For *a* mechanisms, we would expect $\Delta V^{\ddagger}$ might be negative. Table 9.10 shows values of $\Delta V^{\ddagger}$ for some aquation reactions. $\Delta V^{\ddagger}$ values for the water exchange reactions of $[M(NH_3)_5(H_2O)]^{3+}$ (M = Co, Cr) show that when a neutral ligand is released into solution, our expectations are realized. Much other evidence indicates that reactions of Co(III) complexes are *d*.[16] The negative values of $\Delta V^{\ddagger}$ for some Co(III) species

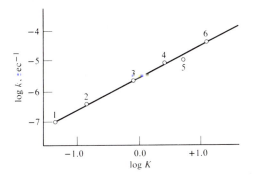

Figure 9.6 Plot of log rate constant versus log of equilibrium constant for the acid hydrolysis reaction of $[Co(NH_3)_5X]^{+2}$ ions. Measurements made at 25.0°. Points are designated: $1, X^- = F^-$; $2, X^- = H_2PO_4^-$; $3, X^- = Cl^-$; $4, X^- = Br^-$; $5, X^- = I^-$; and $6, X^- = NO_3^-$. (Reprinted with permission from C. H. Langford, *Inorg. Chem.*, 1965, *4*, 265. Copyright 1965, American Chemical Society.)

[13]T. W. Swaddle, *Coord. Chem. Rev.* 1974, *14*, 217.

[14]G. S. Hammond, *J. Am. Chem. Soc.* 1955, *77*, 334.

[15]G. A. Lawrance and D. R. Stranks, *Acc. Chem. Res.* 1979, *12*, 403.

[16]C. H. Langford and V. S. Sastri, *M. T. P. International Review of Science, Inorganic Chemistry, Series One*, Vol. 9, M. L. Tobe, Ed., University Park Press, Baltimore, 1972, p. 210.

Table 9.10 Volumes of activation for aquation and water exchange reactions of some octahedral complexes[a]

Complex	$\Delta V^{\ddagger}$ (cm^{3}mole^{-1})
$[Co(NH_3)_5Cl]^{2+}$	-10.6 ± 0.4
$[Co(NH_3)_5SO_4]^+$	-18.5 ± 0.7
$[Co(NH_3)_5(OH_2)]^{3+}$	$+1.2 \pm 0.1$
$[Cr(NH_3)_5Cl]^{2+}$	-10.8 ± 0.3
$[Cr(NH_3)_5(OH_2)]^{3+}$	-5.8 ± 0.2
$[Co(NH_3)_5(urea)]^{3+}$	$+1.3 \pm 0.5$
$[Ni(phen)_3]^{2+}$	-1.2 ± 0.2
$[Rh(NH_3)_5(H_2O)]^{3+}$ [b]	-4.1

[a]From G. A. Lawrance and D. A. Stranks, *Acc. Chem. Res.* 1979, *12*, 403.
[b]T. W. Swaddle and D. A. Stranks, *J. Am. Chem. Soc.* 1972, *94*, 8357.

could be attributed to solvent electrostriction by anions released into solution. Another factor not yet assessed is the contribution by the nonlabile ligands whose bond strengths and bond distances from M will surely change on going to the transition state.[17,18]

The relation of $\Delta H^{\ddagger}$ and $\Delta S^{\ddagger}$ to mechanism in octahedral substitution reactions has been discussed (see footnote 13).

Summary of Results on Intimate Mechanism.   In this section we have developed models for mechanisms of ligand substitution reactions within the transition state theory. Equally importantly, we have discussed what kinds of experimental data can be used to distinguish among possible pathways. The rate law for any reaction gives only the chemical composition of the transition state. The energetic importance of various structural features and of reactant identity must be assessed by examining series of similar complexes, a point of view exemplified in the preceding material on octahedral complexes.

Real situations often are more complicated than our discussion has indicated. For example, the aquation reactions of $[Cr(H_2O)_5X]^{2+}$ complexes display some or all of the terms in the rate expression

$$\text{rate} = k_{H^2}[Cr(H_2O)_5X^{2+}] [H^+]^2 + k_H[Cr(H_2O)_5X^{2+}] [H^+]$$

$$+ k[Cr(H_2O)_5X^{2+}] + \frac{k_{OH}[Cr(H_2O)_5X^{2+}]}{[H^+]} \quad (9.17)$$

depending on the acidity of the solution and the nature of X$^-$. As usual, a multiterm rate law indicates the simultaneous operation of several reaction mechanisms. Terms in k_{H^2} and k_H are observed only when X is protonated readily and indicate activated complexes of composition

[17]C. H. Langford, *Inorg. Chem.* 1979, *18*, 3288.
[18]K. E. Newman and A. E. Merbach, *Inorg. Chem.* 1980, *19*, 2481.

$[Cr(H_2O)_5XH_2]^{4+}$ or $[Cr(H_2O)_5X^{2+}, 2H^+]$ and $[Cr(H_2O)_5XH]^{3+}$ or $[Cr(H_2O)_5X^{2+}, H^+]$, respectively.[19] Whereas k_H terms arise for $X^- = Cl^-$, CN^-, etc.,[20] a k_{H2} term also is seen when $X^- = NO_2^-$.[21] These pathways are known as acid-assisted aquation. The k_{OH} term implies an activated complex of composition $[Cr(OH)(H_2O)_4X]^+$ or $[Cr(H_2O)_5X^{2+}, OH^-]$. The former is preferred, since the enhanced rate for the k_{OH} path relative to the k path points to a *d* mechanism for a species of reduced charge (Section 9.3.4).

The choice of *a* or *d* activation as a function of metal is not understood completely. All complexes of Co(III) investigated so far have been found to react via a *d* mechanism. This seems to be the most usual mechanism for octahedral complexes. Aside from Co(III), the most studied metal ion has been Cr(III). It now seems clear that Cr(III) complexes can react with either *d* or *a* activation. The aquation reactions of $[M(NH_3)_5X]^{2+}$, as well as the anation reactions of $[M(NH_3)_5(H_2O)]^{3+}$, proceed via *d* mechanisms for M = Cr, Co. However, water exchange of $[Cr(H_2O_6]^{3+}$ and acid hydrolysis of $[Cr(H_2O)_5X]^{2+}$ exhibit *a* activation. It may be that Co(III) aqua complexes parallel those of Cr(III) in reactivity, but the highly oxidizing nature of $[Co(H_2O)_6]^{3+}$ (which oxidizes water) has prevented resolution of this question. A considerable number of Ru(III) complexes react via *a* pathways and a claim to the discovery of an *A* mechanism recently has been advanced[22] on the basis of the very negative $\Delta V^{\ddagger}$ for both the forward and reverse of Equation 9.18.

$$Ru(NH_3)_5Cl^{2+} + H_2O \rightleftharpoons Ru(NH_3)_5(H_2O)^{3+} + Cl^- \tag{9.18}$$

Table 9.11 summarizes presently available information on intimate mechanism as a function of metal ion. As more experiments are done, we can anticipate that evidence for *a* activation will be found for more of the species now in the *d* column (especially those with high positive charge).

Table 9.12 provides representative data on acid hydrolysis reactions of some inert metals

Table 9.11 Intimate mechanisms for substitution reactions of octahedral complexes

a	*a and d*	*d*
Ti(III)?, V(III)?	Ta(V), Cr(III)	Mo(V), V(II)?
Mo(III), Ga(III)	Ru(III), Rh(III)	Mn(III), Fe(III) high-spin,
In(III)	Ir(III), Al(III)	Fe(III) low-spin, Fe(II) high-spin, Fe(II) low-spin, Ru(II), Co(III)
		Co(II), Ni(II), Cu(II), Mg(II)

[19]See E.L. King, *Chemistry*, Painter Hopkins, Sausalito, Cal., 1979, Chapter 15, for a discussion of the relation between the rate law and composition of the activated complex.

[20]R. G. Wilkins, *The Study of Kinetics and Mechanism of Reactions of Transition Metal Complexes*, Allyn and Bacon, Boston, 1974, p. 203.

[21]T. C. Matts and P. Moore, *J. Chem. Soc. A* 1969, 1997.

[22]M. T. Fairhurst and T. W. Swaddle, *Inorg. Chem.* 1979, *18*, 3241.

Table 9.12   Acid hydrolysis rate constants for some octahedral complexes of inert metals[a,b]

Complex	$T(°C)$	$k(sec^{-1})$
$[Cr(NH_3)_5(ONO_2)]^{2+}$	25	7.0×10^{-4}
$[Cr(NH_3)_5F]^{2+}$	25	1×10^{-7}
$[Cr(NH_3)_5Cl]^{2+}$	25	7.3×10^{-6}
$[Cr(NH_3)_5I]^{2+}$	25	1.0×10^{-3}
$[Cr(H_2O)_5(ONO_2)]^{2+}$	25	7.35×10^{-5}
$[Cr(H_2O)_5Cl]^{2+}$	25	3.5×10^{-7}
$[Cr(H_2O)_5Br]^{2+}$	25	5.14×10^{-6}
$[Cr(H_2O)_5I]^{2+}$	25	1.10×10^{-4}
$[Ru(NH_3)_5(O_2CCH_3)]^{+}$	25	5.0
cis-$[Ru(NH_3)_4(H_2O)Cl]^{+}$	25	4
trans-$[Ru(NH_3)_4(H_2O)Cl]^{+}$	25	<0.1
$[Ru(NH_3)_5Cl]^{2+}$	80.1	3.28×10^{-4}
$[Ru(NH_3)_5Br]^{2+}$	80.1	3.99×10^{-4}
$[Ru(NH_3)_5I]^{2+}$	80.1	1.64×10^{-4}
$[Rh(NH_3)_5Cl]^{2+}$	50	1.13×10^{-6}
$[Rh(NH_3)_5Br]^{2+}$	50	1.00×10^{-6}
$[Rh(NH_3)_5I]^{2+}$	50	2.4×10^{-7}
trans-$[Ir(en)_2Cl_2]^{+}$	105	5.9×10^{-6}
trans-$[Ir(en)_2BrCl]^{+}$	105	9.0×10^{-6}
trans-$[Ir(en)_2ICl]^{+}$	105	1.59×10^{-4}

[a]Leaving group written last in formula.
[b]From the work of Monacelli, Taube, King, Swaddle, Linck, Kane-Maguire, Poë, Basolo, and others.

besides Co(III). Some points worth noting include the rather similar reactivity of Cr(III) as compared with Co(III) complexes (Tables 9.3, 9.4, 9.6) and the low reactivity of Rh(III) and Ir(III) complexes, presumably because of the large LFAE. In addition, large cations such as Ru(III), Rh(III), and Ir(III) likely will have bond making of greater energetic importance than in Co(III), even when the intimate mechanism is still *d*. Note, for example, the unusual reactivity order of the $[Ru(NH_3)_5X]^{2+}$ complexes. One feature of the behavior of Cr(III) ammine complexes not apparent from the data quoted here is the greater tendency for Cr—N bond rupture, which complicates kinetic studies.

Stoichiometric Mechanisms

The previous section outlined methods for distinguishing between dissociative *(d)* and associative *(a)* intimate substitution mechanisms. A further goal of mechanistic studies is to discover the sequence of elementary steps that leads from reactants to products: that is, to discover the stoichiometric mechanism for a reaction. We now address the problem of how to differentiate between D (Equation 9.2) and I_d (Equation 9.7) and between A (Equation 9.11) and I_a (Equation 9.7) mechanisms. The distinction in each case rests on the ability to detect the existence of a five-coordinate (for D) or seven-coordinate (for A) intermediate. Detection of such intermediates is a fairly rigorous requirement that can be fulfilled only if the intermediate's lifetime is sufficiently long for us to observe it somehow. If not, the reaction must be regarded as concerted (I_d or I_a). Thus, by virtue of being labeled a or d, a reaction is automatically at least I_a or I_d. Further experimental evidence must be amassed if the A or D label is to be applied.

A versus I_a. Although a seven-coordinate intermediate ML_5XY could survive long enough for spectroscopic detection if X were a very poor leaving group,[23] so far no such intermediates have been detected. The only claim for an A mechanism (Section 9.3.4) relies on $\Delta V^{\ddagger}$ measurements that do not depend on activation energies. So until further evidence is accumulated, the presently known a reactions all must be regarded provisionally as I_a.

D versus I_d. An intermediate of reduced coordination number need not be detected spectroscopically, but can be recognized by so-called kinetic tests involving discrimination in reaction with various entering groups or product stereochemistry.

Second-Order Limiting Rate Constant. Just as the relative values of rate constants under first-order limiting conditions enable us to recognize a and d mechanisms, so the values of rate constants under second-order limiting conditions (low $[Y]$) permit a choice between D and I_d stoichiometric mechanisms. We can think of ligand interchange between outer and inner spheres in the I_d path as occurring so rapidly that the outer coordination sphere lacks sufficient time to reorganize (relax) before the reaction is done. Consequently, whatever ligand happens to be in the proper geometric orientation in the outer sphere enters ("accidental bimolecularity"). In the D mechanism there is sufficient time for rearrangement in the outer sphere before a new group enters the inner sphere, so the coordinatively unsaturated intermediate can react with preferred ligands over others.

Under second-order limiting conditions, k_{obs} can be identified with kK, the product of the dissociation rate constant and the equilibrium constant for outer-sphere complexation in the I_d mechanisms. K often can be measured from spectroscopic experiments on nonreacting systems or calculated from theoretical expressions.[24] If we compare a series of reactions in which a variety of entering groups displaces the same leaving group, we would expect any differences in k_{obs} to stem from differences in K if the mechanism is I_d. Table 9.13 presents data for ligand replacement reactions of $[Ni(H_2O)_6]^{2+}$. Values of k obtained by dividing k_{obs} by a calculated K are essentially the same, even though the charge on the entering group varies from -2 to $+1$, as expected for an I_d mechanism.[25]

[23]For an A reaction the second step in Equation 9.11 could be reversible: $k_{-2} \neq 0$. A departure from strict second-order kinetics would be seen. This kinetic detection could occur even if the intermediate were too short-lived for spectroscopic observation.

[24]D. B. Rorabacher, *Inorg. Chem.* 1966, *5*, 1891.

[25]Although it may be possible to demonstrate the presence of an outer-sphere complex, such a demonstration does not necessarily mean that it is an intermediate along the reaction path. See J. Halpern, *J. Chem. Educ.* 1968, *45*, 372.

Table 9.13   Observed rate constants (k_{obs}), calculated outer sphere complexation equilibrium constants (K), and resulting values of $k\left(= \dfrac{k_{obs}}{K}\right)$ for $[Ni(H_2O)_6]^{2+}$ reactions with ligands Y^{n-} at 25°C[a]

Y^{n-}	$k_{obs}(M^{-1}sec^{-1})$	$K(M^{-1})^b$	$k = \dfrac{k_{obs}}{K}\ (sec^{-1})$
CH_3COO^-	1×10^5	3	3×10^4
SCN^-	6×10^3	1	6×10^3
F^-	8×10^3	1	8×10^3
HF	3×10^3	0.15	2×10^4
H_2O			3×10^3
NH_3	5×10^3	0.15	30×10^3
$NH_2(CH_2)_2N(CH_3)_3^+$	4×10^2	0.02	20×10^3

[a]From R. G. Wilkins, *Acc. Chem. Res.* 1970 3, 408.
[b]Calculated from the expression

$$K = \frac{4\,\pi N a^3}{3000}\,e^{-U(a)/kT}$$

where

$$U(a) = \frac{Z_1 Z_2 e^2}{aD} - \frac{Z_1 Z_2 e^2 \kappa}{D(1 + \kappa a)}$$

(representing the sum of the attractive and repulsive potentials between two ions of charge $Z_1 e$ and $Z_2 e$ at a distance a in a medium of dielectric constant D)

$$\kappa^2 = \frac{8\pi^2 N e \mu}{1000\ DkT}$$

and a is taken as 500 pm.

For any reaction proceeding via a D mechanism, the second-order limiting value of k_{obs} can be identified as $\dfrac{k_1 k_2}{k_{-1}[X]}$ (Table 9.7). When X = solvent, [X] will be a constant. Moreover, in comparing a series of reactions featuring a solvent leaving group and different entering groups Y, k_1/k_{-1} will be constant and differences in k_2 will control the changes in k_{obs} on varying Y. The variation in k_{obs} with Y will reflect the ability of the five-coordinate intermediate to discriminate among Y. Indeed, a truly five-coordinate intermediate, no matter what its origin, will always display the same selectivity toward Y. It is a species with no memory of the leaving group. Hence, we attribute any differences in k_{obs} to the identity of the entering group.

We might expect such behavior to be apparent if outer sphere complexation between the complex and entering reagent (required for the I_d path) could be avoided—a situation likely to obtain in reactions of two anionic species such as

$$Co(CN)_5(H_2O)^{2-} + Y^- \rightarrow Co(CN)_5Y^{3-} + H_2O \tag{9.19}$$

Kinetic measurements of this reaction fit well the applicable form of Equation (9.6)

Table 9.14 Relative reactivities (k_2/k_{-1}) of entering groups Y^- toward the five-coordinate intermediate $[Co(CN)_5]^{2-}$ at 40°C[a]

Y^-	Relative Reactivity
H_2O	1
N_3^-	0.53
SCN^-	0.34
I^-	0.19
Br^-	0.1
OH^-	3×10^3

[a]Reprinted with permission from A. Haim, R. J. Grassi, and W. K. Wilmarth, *Adv. Chem. Ser.* 1965 *49*, 31. Copyright 1965, American Chemical Society.

corrected for the possibility of reversibility in the second step of the mechanism of Equation (9.2). Table 9.14 reports ratios k_2/k_{-1}, which represent the reactivities of various reagents with the five-coordinate intermediate compared with the reactivity of water. Note that the values quoted do represent evidence for a *d* intimate mechanism since (1) the range they span is only one power of 10 if OH^- is excluded and (2) the rates are all smaller than that for H_2O exchange. (The rate enhancement by OH^- is discussed below.)

Product Analysis. Sometimes the relative reactivity of an intermediate toward different reagents can be obtained by product analysis. If the same ratio of products is obtained by running a reaction under various sets of conditions, we can conclude that the same intermediate must have been generated in all cases.

Table 9.15 reports data for the spontaneous and induced (or assisted) reactions of

Table 9.15 Values of $[Cr(H_2O)_5(CH_3OH)^{3+}]/[\text{Total Cr(III)}]$ in the Cr(III) products produced from $[Cr(H_2O)_5X]^{2+}$ [a]

	Z (= mole fraction CH_3OH)			
Reaction	0.28	0.46	0.64	0.87
$[Cr(H_2O)_5I]^{2+}$ + solvent	0.21	0.29	0.49	—
$[Cr(H_2O)_5I]^{2+}$ + Tl^{3+}	0.20	0.32	0.46	0.74
$[Cr(H_2O)_5I]^{2+}$ + Hg^{2+}	0.20	0.33	0.47	0.72
$[Cr(H_2O)_5Cl]^{2+}$ + Hg^{2+}	0.19	0.34	0.47	0.72
$[Cr(H_2O)_5I]^{2+}$ + Ag^+	0.21	0.32	0.47	0.74
$[Cr(H_2O)_5N_3]^{2+}$ + HNO_2	0.20	0.32	0.48	0.72

[a]Reprinted with permission from S. P. Ferraris and E. L. King, *J. Am. Chem. Soc.* 1970, *92*, 1215. Copyright 1970, American Chemical Society.

$[Cr(H_2O)_5X]^{2+}$ with water containing CH_3OH. The fraction of the products that appears as $[Cr(H_2O)_5(CH_3OH)]^{3+}$ depends only on the mole fraction of methanol—not on the identity of X^- or on the way of assisting removal of X^-. This is evidence that the same intermediate (which must be $[Cr(H_2O)_5]^{3+}$) is always produced. The last entry in Table 9.15 is an example of "induced" or "assisted" aquation brought about by HNO_2 oxidation of an azido ligand to a species that is an excellent leaving group.

So far, rather few d reactions can be characterized as D. Lack of definitive evidence for the unsaturated intermediate forces us to label them I_d.[26]

9.3.5 Base Hydrolysis

The replacement of a ligand by OH^- is called base hydrolysis. (Equation 9.1, $Y = OH^-$). In Co(III) complexes containing amine ligands, the rate of base hydrolysis is very much faster than acid hydrolysis or aquation. Table 9.16 lists second-order rate constants for base hydrolysis of several complexes. These results should be compared with the aquation rate constants in Tables 9.4, 9.6, and 9.12. Considering first the Co(III) complexes, two points are noteworthy: the much larger rate constants for base hydrolysis and the fact that reactivity orders are not the same as for acid aquation. Base hydrolysis reactions ordinarily are second-order and never seem to reach the mixed first- and second-order behavior (rate saturation) of the rate laws for the D and I_d mechanisms (Section 9.3.3). Base hydrolysis of $[Co(NH_3)_5Cl]^{2+}$ obeys the rate expression

$$\text{rate} = k[Co(NH_3)_5Cl^{2+}][OH^-] \tag{9.20}$$

up to $[OH^-] = 1.0\,M$. These facts seem to point to a special mechanism for base hydrolysis. In fact, it was argued for some time that base hydrolysis provided an example of an associative reaction involving rate-determining attack by OH^-. However, it would be extremely surprising if OH^- were the *only* nucleophile in aqueous solution capable of attack. Currently, this fact together with other evidence is considered to point to the mechanism first proposed by Garrick in 1937:

$$[Co(NH_3)_5Cl]^{2+} + OH^- \overset{\text{fast}}{\rightleftharpoons} [Co(NH_3)_4NH_2Cl]^+ + H_2O \tag{9.21}$$

$$[Co(NH_3)_4(NH_2)Cl]^+ \rightarrow [Co(NH_3)_4(NH_2)]^{2+} + Cl^- \tag{9.22}$$

$$[Co(NH_3)_4(NH_2)]^{2+} + H_2O \overset{\text{fast}}{\rightarrow} [Co(NH_3)_5OH]^{2+} \tag{9.23}$$

The first step involves the removal of a proton by OH^- in a rapid acid–base equilibrium, giving a complex ion of lower charge that then loses a Cl^- more rapidly than the starting complex. The last step is relatively fast, so the second step is rate-determining. This is referred to as a *D-CB* mechanism,[27] indicating that it involves a D reaction of the conjugate base *(CB)* of the starting complex (see Figure 9.7).

---

[26]One seemingly simple piece of evidence of a D mechanism would be the observation of the mass law retardation by X predicted in Equation 9.6. However, it has been shown that outer-sphere complexation by X released in the reaction leads to a rate law of the same form. See J. A. Ewen and D. J. Darensbourg, *J. Am. Chem. Soc.* 1976, *98*, 4317.

[27]This terminology has not yet been adopted widely in the literature. The mechanism is most often referred to using older terminology as S_N1CB.

Table 9.16 Rate constants for base hydrolysis of some octahedral complexes[a,b]

Complex	T(°C)	$k(M^{-1}sec^{-1})$
$[Co(NH_3)_5NCS]^{2+}$	25	8.0×10^{-4}
$[Co(NH_3)_5NO_2]^{2+}$	25	4.2×10^{-6}
$[Co(NH_3)_5F]^{2+}$	25	1.2×10^{-2}
$[Co(NH_3)_5Cl]^{2+}$	25	0.85
cis-$[Co(en)_2(CN)Cl]^+$	0	8.9×10^{-3}
trans-$[Co(en)_2(CN)Cl]^+$	0	0.13
cis-$[Co(en)_2(NH_3)Cl]^{2+}$	0	0.50
trans-$[Co(en)_2(NH_3)Cl]^{2+}$	0	1.25
cis-$[Co(en)_2(NCS)Cl]^+$	0	1.40
trans-$[Co(en)_2(NCS)Cl]^+$	0	0.35
cis-$[Co(en)_2(NO_2)Cl]^+$	0	3×10^{-2}
trans-$[Co(en)_2(NO_2)Cl]^+$	0	8.0×10^{-2}
cis-$[Co(en)_2(OH)Cl]^+$	0	0.37
trans-$[Co(en)_2(OH)Cl]^+$	0	1.7×10^{-2}
cis-$[Co(en)_2Cl_2]^+$	0	15.1
trans-$[Co(en)_2Cl_2]^+$	0	85
$[Cr(NH_3)_5Cl]^{2+}$	25	1.8×10^{-3}
$[Cr(NH_3)_5Br]^{2+}$	25	7.1×10^{-2}
$[Cr(NH_3)_5I]^{2+}$	25	3.7
$[Ru(NH_3)_5Cl]^+$	25	4.95
$[Ru(NH_3)_5Br]^+$	25	11.3
$[Ru(NH_3)_5I]^+$	25	5.11
$[Rh(NH_3)_5Cl]^{2+}$	25	4.06×10^{-4}
$[Rh(NH_3)_5Br]^{2+}$	25	3.37×10^{-4}
$[Rh(NH_3)_5I]^{2+}$	25	7.26×10^{-5}
$[Ir(NH_3)_5Cl]^{2+}$	25	4.83×10^{-8}
$[Ir(NH_3)_5Br]^{2+}$	25	2.15×10^{-8}
$[Ir(NH_3)_5I]^{2+}$	25	8.26×10^{-9}

[a]Leaving ligand written last in formula.
[b]From the work of Lalor, Chan, Bailar, Taube, Basolo, Pearson, Wallace, Kane-Maguire, and others.

The rate law for the base hydrolysis is

$$\text{Rate} = k[\text{complex}][\text{base}] \tag{9.24}$$

We can easily show this to be consistent with the *D-CB* mechanism if the first step in Equation 9.21 is a rapidly established equilibrium. The rate law would be first-order in conjugate base $[Co(NH_3)_4(NH_2)Cl^+]$, but the concentration of the conjugate base can be related to the concentrations of the initial complex and OH^-. Consider the hydrolysis reaction

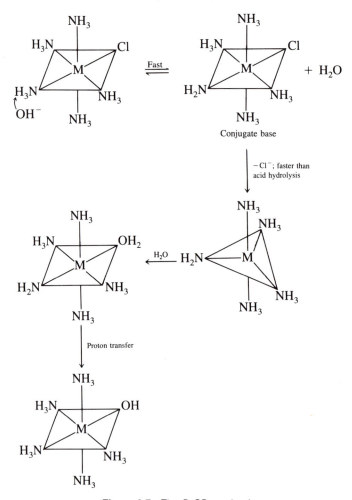

Figure 9.7 The *D-CB* mechanism.

Complex + OH$^-$ ⇌ Conjugate base + H$_2$O, K_h

$$K_h = \frac{K_a}{K_w} = \frac{\text{[Conjugate base]}}{\text{[Complex][OH}^-\text{]}}$$

$$\text{[Conjugate base]} = \frac{K_a}{K_w}\text{[Complex] [OH}^-\text{]} \qquad (9.25)$$

Since

$$\text{Rate} = k'\,\text{[Conjugate base]}$$

$$\text{Rate} = k'\,\frac{K_a}{K_w}\text{[Complex] [OH}^-\text{]} \qquad (9.26)$$

Equation 9.26 is the observed rate law (Equation 9.24) where

$$k = k' \frac{K_a}{K_w} \tag{9.27}$$

The *D-CB* mechanism requires a moderately acidic proton in the starting complex. A complex without an acidic proton should react with OH^- much more slowly, and the rate of the reaction would be expected to be independent of OH^-. And this, in fact, is what is observed for the base hydrolysis of $[Co(CN)_5Br]^{3-}$ and *trans*-$[Co(py)_4Cl_2]^+$. However, $[Pt(CN)_4Br_2]^{2-}$ [28] as well as *cis*-$[Co(bipy)_2Cl_2]^+$, *cis*-$[Co(phen)_2Cl_2]^+$, *cis*-$[Cr(bipy)_2Cl_2]^+$, and *cis*-$[Cr(phen)_2Cl_2]^+$ [29] all undergo base hydrolysis in spite of a lack of acidic protons, for reasons that are not well understood.

The important features of the Garrick mechanism (Equations 9.21, 9.22, and 9.23), including acidity of ammine ligand protons in complexes [30] (see also Problem 9.1), have been well documented in the literature. [31] The conjugate base is believed to be about 10^6 more reactive to substitution than the parent. A true five-coordinate intermediate is produced having trigonal bipyramidal geometry (unless the ligands are too rigid to rearrange). The trigonal bipyramid is stabilized by the π-donor NH_2^-. (See Figure 9.4 where $L = NH_2^-$.) On rearrangement to a trigonal bipyramid, ligands both *cis* and *trans* to the leaving group occupy equivalent positions. Hence the reactivities of *trans*-$[Co(en)_2XCl]^+$ reflect π-donor abilities of X. The low reactivity of $[Co(en)_2(NO_2)Cl]^+$ and $[Co(en)_2(CN)Cl]^+$ to base hydrolysis is related to the fact that NO_2^- and CN^- are σ donors and π acceptors. Although (as pointed out in Section 9.3.4) a good π donor *cis* to the leaving ligand can donate electrons without the necessity for rearrangement to a trigonal bipyramid, it seems likely that the five-coordinate intermediates of base hydrolysis survive long enough to rearrange. Note that both *cis*- and *trans*- $[Co(en)_2LX]^{n+}$ display similar reactivities in base hydrolysis—in contrast to acid hydrolysis, where the *cis* complexes are more reactive. Further evidence supporting this view lies in the stereochemical course of base hydrolysis discussed in Section 9.3.6.

If a five-coordinate intermediate exists, it must be able to discriminate among various entering groups (Section 9.3.4). In aqueous solution in the presence of Y^- the intermediate could be captured either by Y^- or by H_2O.

$$[Co(NH_3)_4(NH_2)]^{2+} \xrightarrow[\quad]{+H_2O} [Co(NH_3)_5(OH)]^{2+} \tag{9.28}$$

$$\xrightarrow[\quad]{+Y^-} [Co(NH_3)_4(NH_2)Y]^+ \xrightarrow{H_2O} [Co(NH_3)_5Y]^{2+} + OH^- \tag{9.29}$$

The competition ratio $R = \dfrac{[Co(NH_3)_5Y^{2+}]}{[Y^-][Co(NH_3)_5OH^{2+}]}$ obtained from separation and isolation of products should depend on the nature of Y^- and should be the same no matter how

[28] C. E. Skinner and M. M. Jones, *J. Am. Chem. Soc.* 1969, *91*, 1984.

[29] J. Josephson and C. E. Schäffer, *Chem. Comm.* 1970, 61.

[30] See, for example, D. A. Buckingham, P. A. Marzilli, and A. M. Sargeson, *Inorg. Chem.* 1969, *8*, 1595; E. Grunwald and D. W. Fong, *J. Am. Chem. Soc.* 1972, *94*, 7371; D. M. Goodall and M. J. Hardy, *Chem. Comm.* 1975, 919.

[31] An excellent overall review is provided by M. L. Tobe, *Acc. Chem. Res.* 1970, *3*, 377.

Table 9.17 Competition ratios $\dfrac{[Co(NH_3)_5Y^{2+}]}{[Y^-][Co(NH_3)_5OH^{2+}]}$ from the base hydrolysis of $[Co(NH_3)_5X]^{2+}$ in the presence of Y^- [a]

Y^-	I^-	Br^-	Cl^-	NO_3^-
			X^-	
NO_2^-	4.5	5.0	4.2	5.1
NCS^-	6.3	6.1	5.5, 5.4	7.1, 6.9, 7.2
N_3^-	9.9, 9.8	8.7	8.5	10.6, 10.2

[a]D. A. Buckingham, I. I. Olsen, and A. M. Sargeson, *J. Am. Chem. Soc.* 1966, *88*, 5443.

$[Co(NH_3)_4(NH_2)]^{2+}$ is generated. Data in Table 9.17 give measured competition ratios for base hydrolysis of several complexes $[Co(NH_3)_5X]^{2+}$ in the presence of Y^-. The competition ratios show very little dependence on the nature of the leaving group X^-, but do depend on the entering group Y^-, being different for different X^-. *The course of the base hydrolysis in the presence of competing ions differs from reactions in acid solution where $[Co(NH_3)_5X]^{2+}$ first undergo aquation, then anation.*

In the Garrick *D-CB* mechanism, the hydroxide ion simply acts to deprotonate the starting complex to the more labile conjugate base. Hydroxide (being the lyate ion of the solvent) is the strongest base that can exist in aqueous solution and thus is uniquely effective. The other possibility, that OH^- is the best entering group and nucleophile (which would support interpretation of the results via an *A* mechanism) has been ruled out in a classic labeling experiment.[32]

Other evidence against OH^- as an entering group includes competition experiments with HO_2^- [33] and reactions in nonaqueous solvents containing added OH^-.[34]

In summary, base hydrolysis in complexes with acidic protons occurs via a *D-CB* mechanism. The five-coordinate intermediate involved often has a trigonal bipyramidal geometry (see Section 9.3.6). Water, rather than OH^-, is the entering group.

9.3.6 Stereochemistry of Octahedral Substitution Reactions

General

Any satisfactory theory of octahedral substitution reactions must explain their stereochemical course. Werner believed that substitution stereochemistry was determined by the geometry of the second coordination sphere. According to Werner, if the incoming ligand were located opposite to the departing one, it entered across the complex and a *trans* product resulted. On the other hand, if the incoming group were located in the outer sphere next to the departing ligand, a *cis* product was formed. As more data were accumulated, it became apparent that

[32]M. Green and H. Taube, *Inorg. Chem.* 1963, *2*, 948.

[33]R. G. Pearson and D. N. Edgington, *J. Am. Chem. Soc.* 1962, *84*, 4607.

[34]R. G. Pearson, H. H. Schmidtke, and F. Basolo, *J. Am. Chem. Soc.* 1960, *82*, 4434.

additional factors must be involved. In particular, the existence of variable stereochemistry when solvent (which occupies all positions in the second coordination sphere) is the entering group, provides evidence that other factors, including the geometry in the transition state, must be of significance.

Figure 9.8 depicts possible geometries of the five-coordinate intermediate arising in D mechanisms. These may also be regarded as models for activated complexes in I_d mechanisms. The figures represent complexes of the type $[M(N—N)_2LX]^{n+}$, where N—N is a chelating ligand and L a stereochemical marker. The fact that (N—N) is bidentate permits the existence of optical isomers of cis-$[M(N—N)_2LX]^{n+}$, of which only one is shown. Analyzing product stereochemistry provides some insight into the activated complex geometry.

We can see from Figure 9.8 that reactions occurring through a square pyramid lead to retention of both geometric and optical configuration, whereas a trigonal bipyramid permits rearrangement. A square pyramid and a trigonal bipyramid have very similar energies and readily can be interconverted by one of the vibrational motions, if the species is sufficiently long-lived. The motion involves the opening up of the angle between two of the equatorial ligands E and the closing of the angle between two axial ligands A to give a square pyramidal species in which the two A ligands and two of the E ligands are equivalent.

If the angle between the A pair closes and the angle between the other E pair opens, a trigonal bipyramid again will form. This time, two of the formerly equatorial ligands are now axial, and vice-versa. This motion, providing a mechanism for exchanging all equatorial and axial ligands, is called *pseudorotation*, because if all ligands are equivalent, the net effect appears to be rotation of the molecule to give new spatial orientation.[35] (See Sections 8.5 and 14.9.4.)

In Figure 9.8 direct substitution of Y for X in *trans*-$[M(N—N)_2LX]$ gives a *trans* product. Each of the trigonal bipyramids **A** and **B** from the square pyramid gives one *trans* and two *cis* products. On a statistical basis *trans*-$[M(N—N)_2LX]$ could produce a 5:4 ratio of *trans:cis* product. If pseudorotation can occur, then the statistical ratio of products is not changed. Starting with *cis*-$[M(N—N)_2LX]$, the statistical product isomer mixture is 1:5 *trans:cis*. This is the expectation whether or not pseudorotation interconverts **C** and **D**, since both can be produced from the square pyramid. These considerations lead to the prediction (borne out by the data in Tables 9.18, 9.19, and 9.20) that the amount of *cis* product from the *cis* isomer is always greater than the amount of *cis* product from the *trans* isomer of $[M(N—N)_2LX]$.

---

[35]Another possible motion that interconverts axial and equatorial ligands is the turnstile rotation, in which two ligands (say, A_1 and E_2) are held constant while the other three undergo a three-fold rotation around an axis that is not a symmetry element.

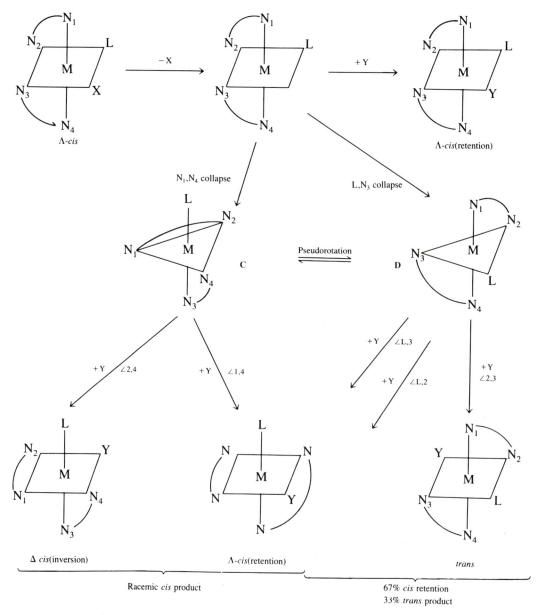

Figure 9.8 Stereochemical course of substitution in *cis*- and *trans*-[M(N-N)$_2$LX]$^{n+}$ complexes (After *Ligand Substitution Processes*, 1966 by C. H. Langford and H. B. Gray with permission of the publisher, Benjamin/Cummings, Inc, Reading, Mass.)

(Continued)

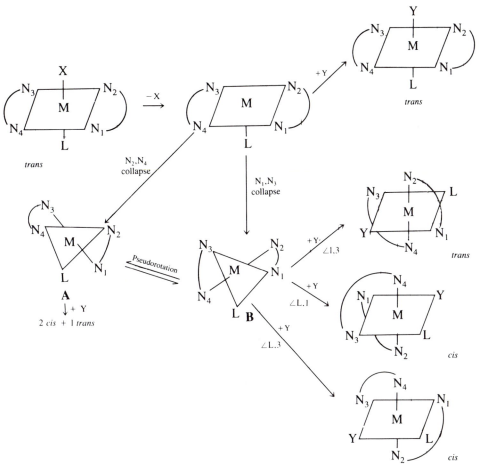

Figure 9.8 *(Continued)*

The product stereochemistry is, of course, affected not only by statistical probability, but also by steric and electronic factors that stabilize a square pyramid or one or both of the trigonal bipyramids and by whether these are intermediates or activated complexes

Stereochemistry of Acid Hydrolysis and Anation

Tables 9.18 and 9.19 record stereochemical consequences of acid aquation and anation of some representative complexes. The results for *trans*-$[Co(en)_2LX]^{n+}$ are consistent with a trigonal bipyramidal species along the reaction coordinate. This presumably occurs because of stabilization by a π-donor L (See Section 9.3.4). Retention of configuration is the rule for *cis*-$[Co(en)_2LX]^{n+}$ and for complexes containing stereochemically rigid quadridentate chelating ligands. Presumably, a square pyramidal species is involved in these reactions. Except for complexes of the quadridentate chelates, a correlation exists between rearrangement and positive $\Delta S^{\ddagger}$.

Table 9.18   Stereochemical results of acid hydrolysis of some Co(III) and Cr(III) complexes[a,b]

Complex	$\Delta S^{\ddagger}$ (J/K mole)	Percent Retention
trans-[Co(NH$_3$)$_4$(ND$_3$)Br]$^{2+}$	-17	100
trans-[Co(NH$_3$)$_4$Cl$_2$]$^+$	$+36$	45 ± 10
cis-[Co(en)$_2$(CN)Cl]$^+$	-20	100
trans-[Co(en)$_2$(CN)Cl]$^+$	-8	100
trans-[Co(en)$_2$(NH$_3$)Cl]$^{2+}$	-46	100
cis-[Co(en)$_2$(NCS)Cl]$^+$	-59	100
trans-[Co(en)$_2$(NCS)Cl]$^+$	$+38$	40 ± 10
cis-[Co(en)$_2$(OH)Cl]$^+$	-42	100
trans-[Co(en)$_2$(OH)Cl]$^+$	$+84$	25
cis-[Co(en)$_2$Cl$_2$]$^+$	-20	100
trans-[Co(en)$_2$Cl$_2$]$^+$	$+59$	65
trans-[Co(cyclam)(OH)Cl]$^+$	-29	100
cis-[Co(cyclam)Cl$_2$]$^+$	-25	100
trans-[Co(cyclam)Cl$_2$]$^+$	-13	100
cis-[Cr(en)$_2$(OH)Cl]$^+$	$-$	96
trans-[Cr(en)$_2$(OH)Cl]$^+$	$-$	13–16
cis-[Cr(en)$_2$(H$_2$O)Cl]$^{2+}$	$-$	100
cis-[Cr(en)$_2$Cl$_2$]$^+$	-25	≥ 99
trans-[Cr(en)$_2$Cl$_2$]$^+$	$+12$	≥ 86

[a]Leaving ligand written last in formula.
[b]From the work of Linck, Tobe, Chan, Poon, Garner, Bailar, Buckingham, and others.

Table 9.19   Stereochemical results of substitution reactions of Ru(II), Ru(III), Rh(III)[a]

Complex	Reagent	Product	Percent Retention
cis-[Ru(NH$_3$)$_4$Cl$_2$]	py	*cis*-[Ru(NH$_3$)$_4$py$_2$]$^{2+}$	>98
trans-[Ru(NH$_3$)$_4$Cl$_2$]	py	*trans*-[Ru(NH$_3$)$_4$py$_2$]$^{2+}$	>98
cis-[Ru(NH$_3$)$_4$Cl$_2$]$^+$	H$_2$O	*cis*-[Ru(NH$_3$)$_4$(H$_2$O)Cl]$^{2+}$	100
trans-[Ru(NH$_3$)$_4$Cl$_2$]$^+$	H$_2$O	*trans*-[Ru(NH$_3$)$_4$(H$_2$O)Cl]$^{2+}$	100
trans-[Rh(en)$_2$(NO$_2$)$_2$]$^+$	HCl	*trans*-[Rh(en)$_2$(NO$_2$)Cl]$^+$	100
cis-[Rh(en)$_2$(NO$_2$)$_2$]$^+$	HCl	*cis*-[Rh(en)$_2$(NO$_2$)Cl]$^+$	100
cis-[Rh(en)$_2$Cl$_2$]$^+$	NH$_3$	*cis*-[Rh(en)$_2$(NH$_3$)Cl]$^{2+}$	100
trans-[Rh(en)$_2$Cl$_2$]$^+$	NH$_3$	*trans*-[Rh(en)$_2$(NH$_3$)Cl]$^{2+}$	100

[a]From the work of Ford, Kane-Maguire, Broomhead, Basolo, Pearson, Klabunde, and others.

Table 9.20   Stereochemical results of base hydrolysis of some M(III) complexes[a,b]

Complex	$\Delta S^{\ddagger}$ (J/K mole)	Percent trans	% R cis[c]	% I cis[c]
trans-[Co(NH$_3$)$_4$(^{15}NH$_3$)Cl]$^{2+}$	—	60	40	
cis-[Co(en)$_2$(CN)Cl]$^+$	+75	10 ± 3	90 ± 3	
trans-[Co(en)$_2$(CN)Cl]$^+$	+92	100	0	
cis-[Co(en)$_2$(NH$_3$)Cl]$^{2+}$	—	22 ± 2	63 ± 2	15 ± 2
trans-[Co(en)$_2$(NH$_3$)Cl]$^{2+}$	—	36 ± 2	64 ± 2	
cis-[Co(en)$_2$(NCS)Cl]$^+$	—	20 ± 2	56 ± 4	24 ± 4
trans-[Co(en)$_2$(NCS)Cl]$^+$	+113	24 ± 2	76 ± 2	
cis-[Co(en)$_2$(OH)Cl]$^+$	+88	<3	61 ± 3	36 ± 3
trans-[Co(en)$_2$(OH)Cl]$^+$	+71	6 ± 2	94 ± 2	
cis-[Co(en)$_2$Cl$_2$]$^+$	+155	63 ± 2	21 ± 2	16 ± 2
trans-[Co(en)$_2$Cl$_2$]$^+$	+146	>95	<5	
cis-[Co(cyclam)Cl$_2$]$^+$	—	0	100	
trans-[Co(cyclam)Cl$_2$]$^+$	—	100	0	
cis-[Ru(en)$_2$Cl$_2$]$^+$	—	0	100	0
trans-[Rh(en)$_2$(OH)Cl]$^+$	−2 ± 4	100	0	
cis-[Rh(en)$_2$Cl$_2$]$^+$	—	0	100	0
trans-[Rh(en)$_2$Cl$_2$]$^+$	+88 ± 17	0		100
cis-[Rh(cyclam)Cl$_2$]$^+$	+63	0		100
trans-[Rh(cyclam)Cl$_2$]$^+$	+42	100		0

[a]Leaving ligand written last in formula.
[b]From the work of Tobe, Basolo, Pearson, Green, Buckingham, Fitzgerald, and others.
[c]R = same configuration as starting material; I = inverted configuration.

Stereochemistry of Base Hydrolysis

The extensive rearrangement observed in base hydrolysis reactions of both *cis*- and *trans*-[Co(en)$_2$LX]$^{n+}$ complexes (see Table 9.20) is consistent with the production of a five-coordinate intermediate that survives sufficiently long to convert to a trigonal bipyramid and perhaps to undergo pseudorotation. In accord with this picture are the large positive values of $\Delta S^{\ddagger}$ and considerable optical inversion in the base hydrolysis of *cis* complexes. Identity of nonleaving ligands around Co is important, complexes of the quadridentate cyclam ligand reacting with complete retention. For Rh(III), Ir(III), and Ru(III), as well as for Cr(III), configuration generally is retained. With the Rh, Ir, and Ru complexes, the identity of the *trans* ligand is quite important in determining the rate of base hydrolysis, suggesting a dissociative reaction proceeding via a square pyramid.

9.3.7   Substitution without Breaking the Metal–Ligand Bond

In acid solution, carbonato complexes such as $[Co(NH_3)_5CO_3]^+$ are converted to aqua complexes with the release of CO_2. When the reaction is carried out in the presence of oxygen-18 labeled water, none of the ^{18}O is found in the resulting aqua complex. Hence the Co—O bond must be retained during the reaction. The reaction presumably involves protonation of the oxygen attached to the Co, followed by removal of CO_2. The complex $[Co(NH_3)_4CO_3]^+$, in which the CO_3^{2-} is bidentate, is converted to *cis*-$[Co(NH_3)_4(H_2O)_2]^{3+}$ in acid solution. If the reaction is carried out in the presence of $H_2^{18}O$, the product is found to have derived half its oxygen from the solvent. The first step in the reaction involves[36] breaking the chelate ring, with an H_2O or H_3O^+ substituting for the O of the CO_3^{2-} ion. The second step involves removal of CO_2 without rupture of the Co—O bond, as in the case of $[Co(NH_3)_5CO_3]^+$ (see Figure 9.9). Another example is the aquation of nitrito complexes $[Cr(NH_3)_5(ONO)]^{2+}$ and $[Co(NH_3)_5(ONO)]^{2+}$ catalyzed by Cl^- and Br^- ions. The mechanism is thought to proceed through species such as

$$M-O-N\begin{matrix} H & X \\ | & \\ & \\ & O \end{matrix}$$

and to involve elimination of NOX (which is rapidly hydrolyzed with N—O bond cleavage).

9.4   RACEMIZATION REACTIONS

Octahedral complexes can undergo substitution reactions involving geometrical isomerization and/or racemization.[37] We will not discuss these reactions here, but instead treat racemization of metal tris-chelate complexes such as $[Co(en)_3]^{3+}$, $[Fe(phen)_3]^{3+}$, $[Cr(C_2O_4)_3]^{3-}$, etc.[38,39] Because chelating ligands are more difficult to remove than unidentate ones, we might expect that intramolecular pathways for racemization would be more important for tris-chelate complexes. And although this generally is true, the complexes $[Ni(phen)_3]^{2+}$ and $[Ni(bipy)_3]^{2+}$ provide striking exceptions. The rates of racemization and ligand exchange[40] (as well as the activation parameters) are the same within experimental error for both complexes. Hence the mechanism must be

$$\Lambda\text{-}[Ni(phen)_3]^{2+} \rightleftharpoons [Ni(phen)_2]^{2+} + phen \rightleftharpoons \Delta\text{-}[Ni(phen)_3]^{2+} \qquad (9.30)$$

[36]T. P. Dasgupta and G. M. Harris, *J. Am. Chem. Soc.* 1969, *91*, 3207.

[37]See F. Basolo and R. G. Pearson, *Mechanisms of Inorganic Reactions*, 2nd ed., Wiley, New York, 1967.

[38]J. J. Fortman and R. E. Sievers, *Coord. Chem. Rev.* 1971, *6*, 331; N. Serpone and D. G. Bickley, *Prog. Inorg. Chem.* 1972, *17*, 391.

[39]L. H. Pignolet, *Top. Curr. Chem.* 1975, *56*, 91.

[40]For comparison with k exchange, the rate constant for racemization must be divided by a factor of two, since the process 2 *d*-$[Ni(phen)_3]^{2+} \rightarrow$ *d*-$[Ni(phen)_3]^{2+}$ + *l*-$[Ni(phen)_3]^{2+}$ cancels the optical rotation provided by 2 *d* complex ions.

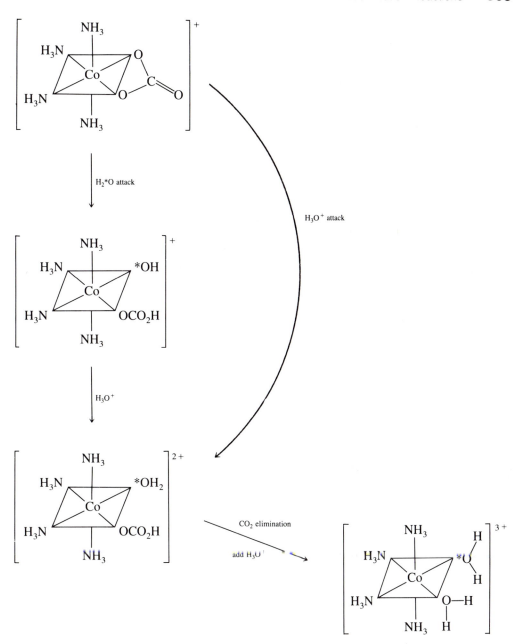

Figure 9.9 Aquation of $[Co(NH_3)_4CO_3]^+$

The bis-chelate intermediate (which may contain coordinated solvent molecules) either is symmetric or loses optical activity faster than it recombines with phen.

A number of tris-chelate complexes of both inert and labile metals racemize faster than they exchange ligands. These racemizations must be intramolecular. Most classical studies have been on inert complexes ($k_{rac} < 10^{-2} sec^{-1}$) requiring isomer separation and polarimetric measurements. NMR techniques now are being employed[39] to investigate racemizations of labile complexes ($k_{rac} \sim 10^{-2}$ to $10^6 sec^{-1}$) in which rearrangement often occurs too fast to permit isolation of optical isomers. Figure 9.10 shows the intramolecular pathways that may lead to racemization for [M(chel)$_3$] complexes. Pathways a and b involve the dissociation of one end of a chelate to produce a trigonal bipyramid and a square pyramid, respectively. Attack by the starred (dissociated) end at the position indicated by arrows leads to optical inverstion. Attack at the equivalent position not designated by the arrow simply leads back to the starting isomer. Pathway c is known as the Bailar (or trigonal) twist. The figure depicts an alternative view of the chelate complex looking down the C_3 axis, which goes through the center of an octahedral face. Clockwise twisting of the triangle of chelate atoms in the upper plane (dotted

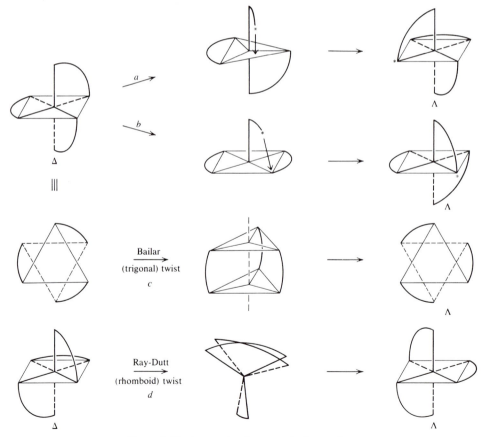

Figure 9.10 Intramolecular racemization pathways.

lines indicate a plane below that of the page) leads to an eclipsed trigonal prism after 60°. If the rotation is continued yet another 60°, the Λ configuration results. The bottom left view shows the chelate rotated such that we are looking down an "imaginary" C_3 axis of the complex. A 60° rotation around this axis (Ray-Dutt twist) leads to the intermediate shown, and a further 60° produces a complex of Λ configuration. We must now consider how to decide whether one (or both) of the dissociative mechanisms (*a* or *b*) or one (or both) of the twist mechanisms (*c* or *d*) is operative for a particular intramolecular racemization.

A dissociative pathway has been postulated[41] in the racemization of optically active

The rate constant for linkage isomerism to give

is reported to be 3.0×10^{-5} sec^{-1} at 90° in chlorobenzene. Under the same conditions the rate constant for racemization is 4.7×10^{-4} sec^{-1}. It was argued that both processes involve the dissociation of one end of the chelate to give a five-coordinate species. The linkage isomerism is slower because it requires rotation around a C—C bond, which does not happen as readily as the reattack to give optical inversion. In another investigation, the ^{18}O exchange for the oxygens not bonded to the metal in $[Co(C_2O_4)_3]^{3-}$ was found to occur much faster than racemization, whereas six more O's exchanged much more slowly and at a rate similar to that for racemization.[42] Presumably these last two processes go through a common intermediate that must involve dissociation of one O for exchange.

Direct evidence for twist mechanisms is difficult to amass. Some criteria have included a rate far in excess of ligand exchange and a low $\Delta S^{\ddagger}$. To distinguish between the Bailar (*c*) and the Ray-Dutt (*d*) twists, we must investigate the behavior of complexes of unsymmetric chelates such as $C_5H_5C(O)CHC(O)CH_3^-$. In these situations, both optical and geometric isomerism are possible, giving a greater number of parameters for tracing the stereochemical course.[41,43]

[41]A. Y. Girgis and R. C. Fay, *J. Am. Chem. Soc.* 1970, *92*, 7061.

[42]L. Damrauer and R. M. Milburn, *J. Am. Chem. Soc.* 1968, *90*, 3884.

[43]J. G. Gordon and R. H. Holm, *J. Am. Chem. Soc.* 1970, *92*, 5319; D. G. Bickley and N. Serpone, *Inorg. Chem.* 1979, *18*, 2200.

9.5 SQUARE PLANAR COMPLEXES— LIGAND SUBSTITUTION REACTIONS

9.5.1 General Features

Metal ions with the d^8 configuration [Au(III), Ni(II), Pd(II), Pt(II), Rh(I), Ir(I)] usually form four-coordinate square planar complexes, especially with strong-field ligands. We might anticipate (as turns out to be true) that the lack of steric crowding and availability of an empty p orbital perpendicular to the molecular plane would lead to enhanced importance for associative (a) mechanisms in ligand substitution reactions. Two general characteristics have emerged after study of a large number of reactions: the mechanisms are associative and the rate law has two terms. For the reaction

$$ML_2AX + Y \rightarrow ML_2AY + X \tag{9.31}$$

the rate $= (k_1 + k_2[Y])[ML_2AX]$. Moreover, the substitutions are stereospecific: a *trans* reactant gives a *trans* product ML_2AY, and a *cis* reactant gives a *cis* product.

The associative (a) character of the intimate mechanism is shown by the fact that the nature of the entering group is important in determining the reaction rate. We can visualize the course of the reactions as shown in Figure 9.11, proceeding through a trigonal bipyramidal intermediate. Obviously, the mechanism depicted accounts nicely for the stereospecificity of the substitution. In some cases, for example in the reactions of $[AuCl_4]^-$ with NCS^-,[44] five-coordinate intermediates actually have been detected. Also, many stable five-coordinate d^8 species are known to exist. For example, $[Ni(CN)_4]^{2-}$ reacts with excess CN^- to give $[Ni(CN)_5]^{3-}$, which can be isolated by crystallization of its $[Cr(en)_3]^{3+}$ salt.

Quite likely, at some time during the substitution process both entering and leaving groups will be coordinated to the central metal to give a five-coordinate species. This species may be an intermediate (A mechanism) or a transition state (I_a mechanism), depending on the particular reaction. It is convenient to treat substitution reactions of square planar complexes *as if* they proceeded via the A mechanism depicted in Figure 9.11 and involved the trigonal bipyramidal intermediate **C.** Much of the rest of this section consists of rationalizing kinetic data in terms of a mechanism involving a trigonal bipyramid. The vast majority of data have been accumulated on complexes of Pt(II), since these react slowly enough for spectroscopic monitoring. The reactivity order is Ni(II) > Pd(II) > Pt(II). Other complications in studying reactions include the tendency of Ni(II) to form octahedral complexes with weak-field ligands and the

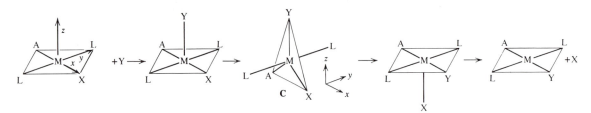

Figure 9.11 Steric course of square planar substitution.

[44]A. J. Hall and D. P. N. Satchell, *J. Chem. Soc., Dalton Trans.* 1977, 1403.

ease with which Pd(II) is reduced to Pd(0). Relatively few kinetic studies have been done on complexes of Rh(I), Ir(I), and Au(III).

9.5.2 Significance of the Rate Law

The two-term rate law indicates two parallel reaction paths. These reactions are studied conveniently under pseudo-first-order conditions. A large excess of the entering ligand Y is introduced so that [Y] will remain effectively constant throughout the reaction. The rate is measured at several values of [Y], and the rate constant k_{obs} is plotted as a function of [Y]. Figure 9.12 shows a plot of k_{obs} versus the concentration of several different entering nucleophiles replacing Cl^- in *trans*-[Pt(py)$_2$Cl$_2$]. The straight-line plot shows that $k_{obs} = k_1 + k_2[Y]$. Values of k_2 differ for different nucleophiles and can be found from the slope of the straight line. The differing values of k_2 reflect the nucleophilicity of various entering ligands. A particularly noteworthy feature of the plot is that all the lines have the same intercept—that is, the value of k_1 is the same for all the nucleophiles. This implies an attack by the same species in all these different reactions. The only other species present in all the cases and capable of nucleophilic attack is the solvent methanol. Indeed, we generally find that the value of k_1 is the same for all nucleophiles and depends only on the solvent. Hence the two terms in the rate law represent nucleophilic attack by the solvent and by the entering group Y as parallel paths to the product.[45]

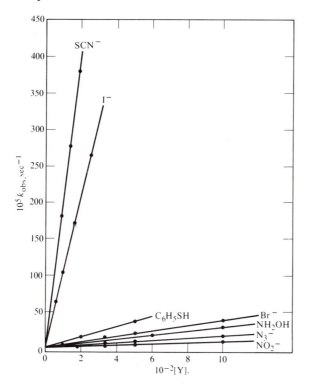

Figure 9.12 Rates of reaction of *trans*-[Pt(py)$_2$Cl$_2$] in methanol at 30° as a function of the concentrations of entering nucleophile. (Reprinted with permission from U. Belluco, L. Cattalini, F. Basolo, R. Pearson and A. Turco, *J. Am. Chem. Soc.* 1965, *87*, 241. Copyright 1965 American Chemical Society.)

[45]In some reactions, one of the rate constants is large enough to swamp the other and only a one-term rate law is found.

$$(9.32)$$

In order for the mechanism of 9.32 to hold, displacement of solvent S by Y must be much faster than displacement of X by S, and this is generally the case.

9.5.3 Effect on Rates of Entering and Leaving Ligands

If the intimate mechanism of square planar substitution reactions is indeed associative, we might expect the nature of both entering and leaving ligands to affect rates substantially. The data given in Table 9.21 show that this is so. Values of k_2 for various ligands replacing Cl^-

Table 9.21 Rate constants for ligand displacement in some square planar Pt(II) complexes

a. *trans*-[Ptpy$_2$Cl$_2$] + Y → *trans*-[Ptpy$_2$ClY] + Cl^{-a}

Y	T(°C)	$k_2(M^{-1}sec^{-1})$	n_{Pt}
CH$_3$OH	25	2.7×10^{-7}	0.00
OCH$_3^-$	25	very slow	<2.4
Cl$^-$	30	4.5×10^{-4}	3.04
NH$_3$	30	4.7×10^{-4}	3.07
N$_3^-$	30	1.55×10^{-3}	3.58
I$^-$	30	1.07×10^{-1}	5.46
CN$^-$	25	4.00	7.14
PPh$_3$	25	249	8.93

b. [Pt(dien)X]$^+$ + py → [Pt(dien)py]$^{2+}$ + X^b

X	$k_{obs}(M^{-1}sec^{-1})^c$
CN$^-$	1.7×10^{-8}
SCN$^-$	3.0×10^{-7}
I$^-$	1.0×10^{-5}
Cl$^-$	3.5×10^{-5}
H$_2$O	1.9×10^{-3d}
NO$_3^-$	very fast

^{a}In CH$_3$OH. From U. Belluco, L. Cattalini, F. Basolo, R. G. Pearson, and A. Turco, *J. Am. Chem. Soc.* 1965, *87*, 241; R. G. Pearson, H. Sobel, and J. Songstad, *J. Am. Chem. Soc.* 1968, *90*, 319.
^{b}At 25° in water. From F. Basolo, H. B. Gray, and R. G. Pearson, *J. Am. Chem. Soc.* 1960, *82*, 4200; H. B. Gray and R. J. Olcott, *Inorg. Chem.* 1962, *1*, 481.
^{c}[py] = 5.9×10^{-3}.
^{d}[py] = 5×10^{-3}.

in *trans*-[Ptpy$_2$Cl$_2$] provide a measure of the nucleophilicity of the attacking ligands Y. As you can see, the identity of Y affects the rate over a range of $\sim 10^9$ in this series of reactions.

A nucleophilic reactivity constant n_{Pt} can be defined as

$$n_{Pt} = \log \frac{k_2(Y)}{k(CH_3OH)} = \log k_2(Y) - \log k(CH_3OH) \qquad (9.33)$$

This constant for a particular Y is a measure of the nucleophilicity of Y relative to that of the solvent methanol. The factors making for good nucleophilicity toward Pt complexes differ from those making for activity in nucleophilic displacements on CH$_3$I and for basicity. The factors involved include both basicity toward Pt(II) and oxidation potential of Y.[46] Values of n_{Pt} are given in Table 9.21*a*. The same nucleophilicity order holds for other Pt complexes in other solvents.

The results of classic studies by Gray on the effect of leaving groups on reaction rates involve py as an entering group with [Pt(dien)X]$^+$ complexes. The effect of the nature of X is large, spanning six orders of magnitude here (see Table 9.21*b*).

The diagrams of Figure 9.13 show that in a particular case, either the entering or the leaving group could make the overriding contribution to activation energy. Hence for separation of the two effects in the [Ptpy$_2$Cl$_2$] studies, Cl$^-$ was chosen because it is a relatively good leaving ligand (that is, the energy profile in Figure 9.13*b* applies) and the reaction rate reflects the height of the first activation barrier.

The nucleophilicities of ligands lie in the same order toward most Pt(II) complexes, but the spread in reactivity is not always the same: that is, the reaction rates depend not only on the nucleophilicity of Y, but also on the electrophilicity of the Pt complex. In other words, different Pt(II) substrates display different abilities to discriminate among nucleophiles. For reactions of nucleophiles Y with other Pt(II) complexes in other solvents (S) besides methanol, the relative reactivity of Y and S is related to n_{Pt}.

$$\frac{\log k_2(Y)}{\log k_2(S)} = sn_{Pt} \qquad (9.34)$$

where *s* is the so-called nucleophilic discriminating factor and depends on the Pt complex used.[47] Similar treatments recently have been extended to Pd(II) complexes. For Au(III), however, no such correlation is possible. The relative values of n_{Au} are very dependent on the complex undergoing substitution.

9.5.4 The *Trans* Effect

An extremely significant aspect of square planar substitution reactions is the *trans effect,* the effect on the rate of replacement of some ligand by the identity of the *trans* ligand.[48] This

---

[46]For many organic and inorganic reactions, nucleophilicities can be related explicitly to basicity and oxidation potential by the Edwards equation $\log \frac{k_2(Y)}{k_1} = \alpha E + \beta H$, where E is the oxidation potential for Y and H is related to its proton basicity. (J. O. Edwards, *J. Am. Chem. Soc.* 1954, *76*, 1540; J. O. Edwards, *Inorganic Reaction Mechanisms,* W. A. Benjamin, New York, 1964). However, this seems not to be the case for substitution reactions of Pt(II) (R. G. Pearson, H. Sobel, and J. Songstad, *J. Am. Chem. Soc.* 1968, *90*, 319).

[47]U. Belluco, L. Cattalini, F. Basolo, R. G. Pearson, and A. Turco, *J. Am. Chem. Soc.* 1965, *87*, 241.

[48]The effects of the identity of the *cis* ligands are smaller by a factor of $\sim 10^2$.

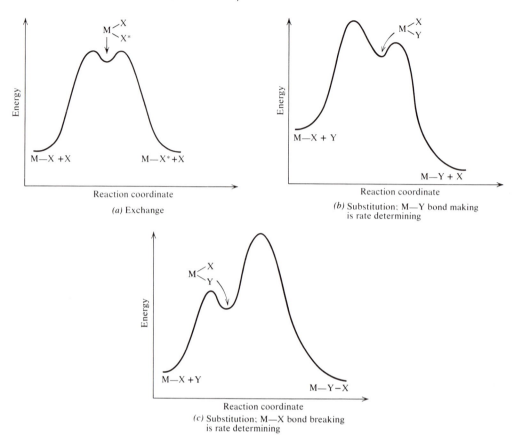

Figure 9.13 Reaction profiles for square planar substitution. (Reprinted from *Ligand Substitution Processes,* 1966, by C. H. Langford and H. D. Gray with permission of the publisher, Benjamin/Cummings, Inc. Reading, Mass.)

Table 9.22 Effect of *trans* ligand L on the rate of the reactions *trans*-[Pt(PEt$_3$)$_2$LCl] + py → *trans*-[Pt(PEt$_3$)$_2$L(py)]$^+$ + Cl$^{-\ a,b}$

L	$k_1(sec^{-1})$	$k_2(M^{-1}sec^{-1})$	$T(°C)$
H$^-$	1.8×10^{-2}	4.2	0
CH$_3^-$	2×10^{-4}	7×10^{-2}	25
C$_6$H$_5^-$	2×10^{-5}	2×10^{-2}	25
Cl$^-$	1.0×10^{-6}	4.0×10^{-4}	25

[a]In ethanol.
[b]From F. Basolo, J. Chatt, H. B. Gray, R. G. Pearson, and B. L. Shaw, *J. Chem. Soc.* 1961, 2207.

effect was recognized rather early on, in the preparation of the two geometric isomers of $Pt(NH_3)_2Cl_2$. (See Section 8.2.2.)

Data in Table 9.22 give some idea of the magnitude of the *trans* effect. Allowing for the difference in temperature, you can see that varying the *trans* ligand can change the speed of replacement over a range of 10^6! The ordering of *trans* labilization is not precisely the same for every substrate. However, a sort of "average" order is

$$CN^-, CO, C_2H_4 > PR_3, H^- > CH_3^-, SC(NH_2)_2 >> C_6H_5^-, NO_2^-,$$

$$I^-, SCN^- > Br^- > Cl^- > py > NH_3 > H_2O \tag{9.35}$$

$$>> OH^-$$

The *trans* effect has great utility in the synthesis of square planar complexes. In the synthesis of the three isomers of $[Pt(NH_3)(py)BrI]$, the *trans* effect and the bond-strength order ($Pt—NH_3 > Pt—py > Pt—Br > Pt—Cl$) govern the results.

$$\begin{bmatrix} & Cl & \\ & | & \\ Cl—&Pt&—Cl \\ & | & \\ & Cl & \end{bmatrix}^{2-} \xrightarrow[-Cl^-]{+NH_3} \begin{bmatrix} & NH_3 & \\ & | & \\ Cl—&Pt&—Cl \\ & | & \\ & Cl & \end{bmatrix}^- \xrightarrow[-Cl^-]{+Br^-} \begin{bmatrix} & NH_3 & \\ & | & \\ Cl—&Pt&—Br \\ & | & \\ & Cl & \end{bmatrix}^- \xrightarrow[-Cl^-]{+py} \begin{bmatrix} & NH_3 & \\ & | & \\ py—&Pt&—Br \\ & | & \\ & Cl & \end{bmatrix} \tag{9.36}$$

When Br^- is added, one of the ligands *trans* to Cl^- is replaced faster. The product turns out to be $[Pt(NH_3)BrCl_2]$ rather than $[PtBrCl_3]^-$, because of the greater $Pt—NH_3$ bond strength.

$$\begin{bmatrix} & Cl & \\ & | & \\ Cl—&Pt&—Cl \\ & | & \\ & Cl & \end{bmatrix}^{2-} \xrightarrow[-Cl^-]{+py} \begin{bmatrix} & py & \\ & | & \\ Cl—&Pt&—Cl \\ & | & \\ & Cl & \end{bmatrix}^- \xrightarrow[-Cl^-]{+Br^-} \begin{bmatrix} & py & \\ & | & \\ Cl—&Pt&—Br \\ & | & \\ & Cl & \end{bmatrix}^- \xrightarrow[-Cl^-]{+NH_3} \begin{bmatrix} & py & \\ & | & \\ H_3N—&Pt&—Br \\ & | & \\ & Cl & \end{bmatrix} \tag{9.37}$$

The reversal in the order of py and NH_3 addition reverses their positions in the product in (9.37).

$$\begin{bmatrix} & Cl & \\ & | & \\ Cl—&Pt&--Cl \\ & | & \\ & Cl & \end{bmatrix}^{2-} \xrightarrow[-Cl^-]{+2py} \begin{bmatrix} & py & \\ & | & \\ Cl—&Pt&—py \\ & | & \\ & Cl & \end{bmatrix}^- \xrightarrow[-Cl^-]{NH_3} \begin{bmatrix} & py & \\ & | & \\ Cl—&Pt&—py \\ & | & \\ & NH_3 & \end{bmatrix}^- \xrightarrow[-py]{+Br^-} \begin{bmatrix} & py & \\ & | & \\ Cl—&Pt&—Br \\ & | & \\ & NH_3 & \end{bmatrix}^+ \tag{9.38}$$

The second step in Equation 9.38 may seem at first to contradict the notion of the *trans* effect, but this is not so. The *trans* effect does *not* predict that *any* ligand *trans* to Cl^- should be replaced faster than *any* ligand *trans* to py. Rather, it leads us to expect that a Cl^- *trans* to a Cl^- should be replaced faster than a Cl^- *trans* to a py if there were a choice (here there is not). The lesser bond strength of $Pt—Cl$ can serve as a rationale for Cl^- replacement. Events in the final step are not rationalized easily. This reminds us that the *trans* effect is an empirical correlation that cannot necessarily be used to predict what reactions will occur. Given the stoichiometry of a reaction (that is, what the entering and leaving groups are), the *trans* effect permits us to assign product geometry—a not-inconsiderable accomplishment. In the final step of Equation 9.38, if we know that Br^- is the entering group and py the leaving group, the *trans* effect leads us to expect that the py *trans* to Cl^- has been replaced.

9.5.5 Theories of the *Trans* Effect

The *trans* effect is essentially an empirical statement of fact. However, numerous efforts have been made to explain the origin of this effect in electronic structural terms. The presence of certain ligands in *trans* position labilizes a particular ligand to substitution. In other words, the activation energy for the substitution is lowered relative to the activation energy when some other ligand is in the *trans* position (see the previous section for examples). The activation energy for any reaction can be lowered in either of two ways: by raising the energy of the ground state of the reactants or by lowering the energy of (stabilizing) the transition state. Good *trans* directing ligands must act in one or both of these ways.

Ground-State Effects

A good *trans* directing ligand could operate by weakening the bond between the metal and the *trans* ligand. This ground-state weakening is, of course, a thermodynamic effect that has been given the name *trans influence*,[49] to distinguish it from the kinetic *trans* effect. There have been several theories as to how a ligand might weaken the bond *trans* to itself. One of the earliest was the polarization theory of Grinberg, in which a good *trans* directing ligand was visualized as being very polarizable and forming an ionic bond to the metal. A dipole in the opposite sense is induced in the metal (see Figure 9.14). This dipole then serves to repel the *trans* ligand, thus weakening the bond. This could account for the presence of H^- and CH_3^- high in the *trans* effect series.

A second view emphasizes the importance of good σ-donor ligands as effective at weakening the *trans* bond. A good σ-donor will claim a larger share of the metal σ-orbitals, leaving a lesser share to bond the *trans* ligand.

Studies indicate that the nature of a ligand A can have a substantial effect on the strength of the bond to the *trans* group. For example, the Pt—Cl distance in *trans*-$[Pt(PEt_3)_2Cl_2]$ has been determined by x-ray crystallography to be 229.4 pm.[50] In *cis*-$[Pt(PMe_3)_2Cl_2]$ where Cl^- is trans to PMe_3, the Pt—Cl distances are lengthened to 236.4 and 238.8 pm.[51] Thus (presuming a relation between bond length and bond strength) a phosphine ligand in the *trans* position weakens a Pt—Cl bond as compared with a chloride in *trans* position. Not every ligand that is an effective *trans* director displays a large *trans* influence. In particular, π-acid ligands have low *trans* influence. This is not unexpected, since reduction in activation energy for substitution can occur in ways other than through the *trans* influence. Several other techniques have been

Figure 9.14 Polarization of a metal by a *trans* ligand L. (From F. Basolo and R. Pearson, *Mechanisms of Inorganic Reactions*, 2nd ed., Wiley, New York, 1967.)

[49]A recent review of the theory and experimental evidence for the *trans* influence is T. G. Appleton, H. C. Clark, and L. E. Manzer, *Coord. Chem. Rev.* 1973, *10*, 335.

[50]G. G. Messmer and E. L. Amma, *Inorg. Chem.* 1966, *5*, 1775.

[51]G. G. Messmer, E. L. Amma, and J. A. Ibers, *Inorg. Chem.* 1967, *6*, 725.

employed to demonstrate the dependence of bond strengths on the identity of the *trans* ligand,[52] including IR, ^{1}H, ^{31}P and ^{19}F NMR (examination of both chemical shifts and coupling constants).[49,52]

Transition-State Effects

Substitution in square planar complexes proceeds via a trigonal bipyramid, as shown in Figure 9.11. If the mechanism is I_a, this trigonal bipyramid is the activated complex in the transition state, and any features that stabilize it will lower the activation energy barrier and speed the reaction. If the mechanism is *A,* this trigonal bipyramid will represent an intermediate. However, Hammond's postulate tells us that the geometry of the transition state (which is not far away along the reaction coordinate) will be quite similar. So in this case, too, we expect features that stabilize a trigonal bipyramid to lower the activation energy barrier. Ligands capable of a high degree of σ donation or π-acceptance can confer such stability and are good *trans* directors. Langford and Gray discuss the theoretical reasons for this.[53]

The role of good *trans* directing ligands in stabilizing the transition state is evident from the data given in Table 9.21. Ligands high in the *trans* effect series are both good entering groups and poor leaving groups.

9.5.6 Steric Effects of Inert Ligands[47,54,55]

We can make no clear-cut prediction about the effect of bulky ligands on the rates of substitution by associative mechanisms. For, whereas the attack by an entering ligand should be retarded, the departure of the leaving group should be assisted. Nevertheless, the steric retardation observed in several cases provides good evidence in favor of an associative mechanism, since there is no way to rationalize such retardation on the assumption of a dissociative mechanism.

Table 9.23 shows data for substitution of Cl^- in a series of progressively more hindered complexes *trans*-[Pt(PEt$_3$)$_2$RCl], where R = phenyl, *o*-tolyl, and mesityl. With less nucleophilic entering groups such as NO_2^-, Br^-, and I^- the rate law displays only a first-order term, rate = k_1 [Pt(PEt$_3$)(mesityl)Cl]. Moreover, k_1 is the same for all the entering groups. This is evidence that the severe steric hindrance has blocked associative pathways completely and that a dissociative mechanism is now operative. The even lower nucleophilicity of methanol rules out the possibility that rate-determining solvent attack occurs when the nucleophilic strength of entering groups is low. MO studies of a three-coordinate intermediate have been reported.[56]

The existence of both *a* and *d* mechanisms in the reactions of square planar complexes should not be surprising or disconcerting. Many kinds of reaction pathways always are theoretically available. Those actually observed are those with the lowest activation energy (fastest rate), which generally are similar for similar complexes. If these ordinary pathways are blocked for some reason, then ordinarily unseen mechanisms take over.

[52]There also is evidence for a *trans* effect in octahedral complexes. However, its relation to kinetics of substitution reactions is not yet clear. However, see R. R. Whittle, Ph.D. Dissertation, University of Cincinnati, 1981.

[53]C. H. Langford and H. B. Gray, *Ligand Substitution Processes,* Benjamin/Cummings, Reading, Mass., 1966, pp 24–29.

[54]G. Faraone, V. Ricevuto, R. Romeo, and M. Trozzi, *Inorg. Chem.* 1969, 8, 2207; *ibid.* 1970, 9, 1525.

[55]R. Romeo, D. Minniti, and M. Trozzi, *Inorg. Chem.* 1976, 15, 1134.

[56]S. Komiya, T. A. Albright, R. Hoffmann, and J. K. Kochi, *J. Am. Chem. Soc.* 1976, 98, 7255.

Table 9.23 Rate constants for substitution of some Pt(II) complexes at 30°C[a]

		$k_2(M^{-1}sec^{-1})$	
		trans-[Pt(PEt$_3$)$_2$RCl] + Y → *trans*-[Pt(PEt$_3$)$_2$RY] + Cl$^-$	
R	*Solvent*	*Y = CN$^-$*	*Y = SC(NH$_2$)$_2$*
C$_6$H$_5$	methanol	3.61	6.30
C$_6$H$_5$	dimethylsulfoxide	very fast	0.532
o-C$_6$H$_4$CH$_3$	methanol	0.234	0.652
o-C$_6$H$_4$CH$_3$	dimethylsulfoxide	3.54	0.106
2,6-C$_6$H$_3$(CH$_3$)$_2$	methanol	8.49×10^{-3}	4.94×10^{-2}
2,6-C$_6$H$_3$(CH$_3$)$_2$	dimethylsulfoxide	3.17×10^{-2}	9.52×10^{-3}

[a]From G. Faraone, V. Ricevuto, R. Romeo, and M. Trozzi, *Inorg. Chem.* 1970, 9, 1525.

9.6 LIGAND SUBSTITUTION REACTIONS IN OTHER GEOMETRIES

Far less information is available on reaction mechanisms for tetrahedral, trigonal bipyramidal, and square pyramidal complexes than on those for octahedral and square planar ones. Practically no studies have been done on species of C.N. >6. In general, both *a* and *d* mechanisms have been observed for tetrahedral and for five-coordinate species. Evidence seems to be accumulating that associative mechanisms can occur when the metal has a vacant orbital capable of bonding to the incoming ligand. In contrast, dissociative mechanisms dominate when all the metal orbitals are filled with electrons or used in bonding to the coordinated ligands.

For example, in the tetrahedral complex Ni(CO)$_2$(PPh$_3$)$_2$, all *d* orbitals of Ni are filled with electrons and *sp^3* hybrids are used to bond to the ligands. The phosphine replacement reactions[57] are first-order in complex and zero-order in free ligand, implying a *d* mechanism.

$$\text{Ni(CO)}_2\text{(PPh}_3)_2 + \text{P(OPh)}_3 \rightarrow \text{Ni(CO)}_2\text{(PPh}_3)\text{(P(OPh)}_3) + \text{PPh}_3 \quad (9.39)$$

On the other hand, tetrahedral complexes with unoccupied *d* orbitals display second-order kinetics reflecting an associative pathway.[58]

$$\text{MBr}_2\text{(PAr}_3)_2 + \text{PAr}_3' \rightarrow \text{MBr}_2\text{(PAr}_3)\text{(PAr}_3') + \text{PAr}_3 \quad (9.40)$$

$$M = \text{Fe, Ni, Co}$$

$$\text{CoCl}_3\text{(solv)}^- + \text{Cl}^- \rightarrow \text{CoCl}_4^{2-} + \text{solv} \quad (9.41)$$

$$\text{solv} = \text{acetone, DMF, CH}_3\text{NO}_2, \text{etc.}$$

[57]L. S. Meriwether and M. L. Fiene, *J. Am. Chem. Soc.* 1959, *81*, 4200.

[58]L. H. Pignolet, *et al.*, *J. Am. Chem. Soc.* 1968, *90*, 922; U. Mayer, *Pure Appl. Chem.* 1975, *41*, 291.

The five-coordinate trigonal bipyramidal complexes

$$\left[M\!\!\left(O\!-\!N\!\!\begin{array}{c}CH_3\\ \\ \\CH_3\end{array}\!\!\right)_5 \right]^{2+} \quad (M = Co, Ni),^{59}\ [NiX_2(PMe_3)_3]\quad (X = Cl, Br, I, CN)^{60}$$

and $[NiP_3X_2]$ (P = phosphine, phosphite, X = Cl, Br, CN)[61] all exchange neutral ligands by a dissociative process in organic solvents. On the other hand, we can view the square pyamidal species[62]

$$M = Fe, Co$$

as intermediate between square planar and octahedral complexes having one vacant metal orbital. These complexes react by parallel a and d pathways to substitute the phosphine ligand with other nucleophiles. $[Co(tmc)(OH_2)]^{2+}$ [63] has $\Delta S^{\ddagger} = -34$ J/K mole for water exchange and presumably exhibits only an associative pathway. Charge can also make a difference to mechanism; the ligand substitution reactions of square pyramidal $[M(S_2C_2(CF_3)_2)_2L]^{z}$ [64] (M = Fe, Co; L = phosphine, phosphite) in which L is replaced by another phosphine or phosphite are dissociative when $z = -1$ and associative when $z = 0$.

Too little work has been done on complexes having C.N. > 6 for any general trends to emerge.

9.7 CATALYSIS OF SUBSTITUTION BY REDOX PROCESSES

As pointed out in Section 9.3.2, the substitution reactions of octahedral low-spin d^6 complexes should proceed quite slowly. Experimental data for Co(III), Rh(III), and Ir(III) (see Table 9.12) indicated that this was the case. Pt(IV) also forms octahedral complexes that undergo slow substitution. In the presence of catalytic amounts of square planar Pt(II) complexes, however, substitution rates are accelerated by factors of 10^4–10^5. The mechanism of these catalytic reactions is thought to be[65]

[59]P. M. Enriquez, S. S. Zumdahl, and L. Forshey, *Chem. Comm.* 1970, 1527; P. M. Enriquez and S. S. Zumdahl, *Inorg. Chem.* 1973, *12*, 2475.

[60]P. Meier, *et al.*, *J. Am. Chem. Soc.* 1976, *98*, 6402.

[61]C. G. Grimes and R. G. Pearson, *Inorg. Chem.* 1974, *13*, 970.

[62]D. A. Sweigart and D. G. DeWit, *Inorg. Chem.* 1970, *9*, 1582; D. Sweigart, *Inorg. Chim. Acta* 1974, *8*, 317.

[63]P. Meier, A. Merbach, S. Burki, and T. A. Kaden, *Chem. Comm.* 1977, 36. (tmc = 1,4,8,11-tetramethyl-1,4,8,11-tetraazacyclotetradecane)

[64]D. A. Sweigart, *Inorg. Chim. Acta* 1976, *18*, 179.

[65]W. R. Mason, *Coord. Chem. Rev.* 1972, *7*, 241.

$$Pt^{II}L_4 + Y \overset{fast}{\rightleftharpoons} Pt^{II}L_4Y$$

$$Pt^{II}L_4Y + X\overset{*}{P}t^{IV}L_4Z \rightleftharpoons X\overset{*}{P}t^{IV}L_4Z\text{-}Pt^{II}L_4Y$$

$$X\overset{*}{P}t^{IV}L_4Z\text{-}Pt^{II}L_4Y \rightleftharpoons X\overset{*}{P}t^{II}L_4\text{-}ZPt^{IV}L_4Y \qquad (9.42)$$

$$X\overset{*}{P}t^{II}L_4\text{-}ZPt^{IV}L_4Y \rightarrow X\overset{*}{P}t^{II}L_4 + ZPt^{IV}L_4Y$$

$$X\overset{*}{P}t^{II}L_4 \rightleftharpoons X + \overset{*}{P}t^{II}L_4$$

In the third step of the mechanism, the Pt^{II} is oxidized to Pt^{IV} while the $\overset{*}{P}t^{IV}$ is reduced to Pt^{II}. The two metals exchange roles via a redox process. The relatively fast substitution on Pt^{II} enables the reaction to occur rapidly, and the product $ZPt^{IV}L_4Y$ contains a metal that originally was Pt^{II}. We also find redox catalysis in Pd^{II}/Pd^{IV}, Cr^{II}/Cr^{III} and Ru^{II}/Ru^{III} substitutions where one oxidation state is more labile than the other and substitution on the labile center can be followed by rapid oxidation or reduction to an inert product.

9.8   REDOX REACTIONS

Redox reactions involving transition metals having more than one stable oxidation state are of wide importance in chemistry. Many classical analytical methods are based on rapid redox reactions, including Fe(II) determinations by titration with $HCrO_4^-$, as well as oxidations of substances such as $C_2O_4^{2-}$ by MnO_4^-. The role of transition metal ions in life processes (see Chapter 16) depends on their ability to participate selectively in electron transfer reactions in complexes. Redox reactions involving two transition-metal complexes generally occur fairly rapidly. Thus, the values of E^0 for reactions are a rather good guide to the chemistry that actually will occur commonly on a convenient time scale in the laboratory. This is true because redox processes involving metals often occur by electron transfer. In contrast, redox reactions of nonmetals occur more often via group transfer, which often requires high activation energy. Atom transfers are also known for metal systems.

For example, the oxidation–reduction of I_2 in base changes the oxidation number from 0 to $+I$ and $-I$

$$I_2^0 + OH^- \rightarrow I^- + I^I(OH) \qquad (9.43)$$

The mechanism involves a displacement whereby the OH^- group is transferred to I^+. The aqueous chemistry of nonmetallic species often involves group transfer, and this may make even thermodynamically feasible reactions kinetically slow.

Below we consider models for redox reactions of two transition-metal-containing complexes. Oxidations of nonmetallic species by metallic ions have been reviewed.[66]

[66]A. McAuley, *Coord. Chem. Rev.* 1970, *5*, 245; J. K. Beattie and G. P. Haight, *Prog. Inorg. Chem.* 1972, *17*, 93; D. Benson, "Mechanisms of Oxidation by Metal Ions," in *Reaction Mechanisms in Organic Chemistry*, Vol. 10, Elsevier, New York, 1976.

9.8.1 Inner- and Outer-sphere Reactions

Two important models for the course of redox reactions of metal complexes currently are considered operative. In the inner-sphere mechanism, the transition state for electron transfer involves interpenetrating coordination spheres of the two metals. A bridging ligand is coordinated to both the oxidant and the reductant and forms part of the first coordination sphere of each. In the outer-sphere mechanism, the two separate inner coordination spheres remain intact.

The first definitive evidence for an inner-sphere process was provided by Taube and Myers, who investigated the Cr^{2+} reduction of $[Co(NH_3)_5Cl]^{2+}$.[67] The final products of the reaction in acid solution are Co^{2+}, Cr^{3+}, $CrCl^{2+}$, Cl^-, and NH_4^+. It is possible to imagine both inner- and outer-sphere mechanisms that would lead to these products.

Inner Sphere

$$Co(NH_3)_5Cl^{2+} + Cr^{2+} \rightleftharpoons Co(NH_3)_5ClCr^{4+}$$

$$Co(NH_3)_5ClCr^{4+} \rightarrow CrCl^{2+} + Co(NH_3)_5^{2+} \qquad (9.44)$$

$$Co(NH_3)_5^{2+} + 5H^+ \rightleftharpoons Co^{2+} + 5NH_4^+$$

Outer Sphere

$$Co^{III}(NH_3)_5Cl^{2+} + Cr^{2+} \rightarrow Co^{II}(NH_3)_5Cl^+ + Cr^{3+}$$

$$Co(NH_3)_5Cl^+ + 5H^+ \rightarrow Co^{2+} + 5NH_4^+ + Cl^- \qquad (9.45)$$

$$Cr^{3+} + Cl^- \rightleftharpoons CrCl^{2+}$$

However, the inner-sphere pathway must be the one operative, because Cr^{3+} is inert to substitution on the time scale for the reaction. At 25°C, the second-order rate constant for Cl^- anation of Cr^{3+} is $2.9 \times 10^{-8} M^{-1}sec^{-1}$, whereas that for reduction is $6 \times 10^5 M^{-1}sec^{-1}$. Hence, the $CrCl^{2+}$ could not have arisen by substitution with free Cl^-. The electron transfer step converts labile Cr^{2+} into substitution-inert Cr(III), which simply retains the bridging chloride in its coordination sphere when the bridged species breaks up. Notice also that Cr^{2+} is sufficiently labile ($k_{exch} \sim 10^8 sec^{-1}$ at 25°C) that substitution of the bridging Cl^- into the coordination sphere is not rate-limiting.

A sufficient condition for establishing that a reaction is inner-sphere is that one of the products be substitution-inert and retain the bridging ligand that was coordinated originally to the other reactant. This means that the oxidant and reductant must be chosen so that one is inert while the other is labile and the products are labile and inert in the opposite sense. Not every inner-sphere reaction fulfills this requirement (that is, the condition is not a necessary one). Later, we shall devote attention to other ways of deciding whether a reaction is inner sphere.

The above discussion reveals that a sufficient condition for the outer-sphere mechanism to be operative is that the redox rate be much faster than substitution on either metal center. An example is the reduction of $[Ru(NH_3)_5Br]^{2+}$ by V^{2+}. The second-order rate constant for

---

[67]In discussing redox reactions, M^{n+} ordinarily is taken to represent the metal ion with its first coordination sphere occupied by aqua ligands. Thus $Cr^{2+} \equiv Cr(H_2O)_6^{2+}$. ML^{n+} represents the complex in which one aqua ligand is replaced by L.

the reduction is $5.1 \times 10^3 M^{-1} sec^{-1}$, whereas aquation reactions on the Ru(III) and V(II) centers have first-order rate constants of 20 and 40 sec^{-1}, respectively.

Before discussing other ways of assigning mechanism, we will review some general features of outer- and inner-sphere reactions.

Outer-sphere Reactions

A well known outer-sphere reaction is the so-called self-exchange. Take, for example,

$$Fe(H_2O)_6^{3+} + \overset{*}{Fe}(H_2O)_6^{2+} \rightleftharpoons \overset{*}{Fe}(H_2O)_6^{3+} + Fe(H_2O)_6^{2+} \qquad (9.46)$$

The equilibrium constant for this reaction is 1 and $\Delta G^0 = 0$. The activation energy for the process is ~32 kJ/mole and $k = 3.0$ $M^{-1}sec^{-1}$ at 25°. Two major contributions to the activation energy involve reorganization of the first and second coordination spheres. Before an electron can be transferred, the Fe—OH$_2$ bond lengths must distort. Those of the Fe(III) complex lengthen to a distance halfway between the Fe(II)–OH$_2$ and Fe(III)–OH$_2$ distance. Those of the Fe(II) complex contract to the same distance. Energy must be expended to accomplish this. The solvation shell (outer coordination sphere) will then rearrange to accommodate the new dimensions of the reactants; this also costs energy. Only then can electron transfer take place and the products relax back to their equilibrium geometries. This model for the outer-sphere process is in accord with the Franck-Condon principle, which states that electron transfer is much faster than nuclear motion. As far as the electrons are concerned, the nuclear positions are frozen during the time period required for electron transfer.

Notice that this reorganization energy (sometimes called Franck-Condon energy) *must* be expended. Otherwise the initial products of self-exchange would be $[\overset{*}{Fe}(H_2O)_6]^{3+}$ with bond distances characteristic of $[\overset{*}{Fe}(H_2O)_6]^{2+}$ and $[Fe(H_2O)_6]^{2+}$ with bond distances characteristic of $[Fe(H_2O)_6]^{3+}$. The energy released when these species relax back to the stable geometry would be created from nothing, in violation of the first law of thermodynamics.

The reaction profile[68] for self-exchange is shown in 'Figure 9.15a. The energy is the total

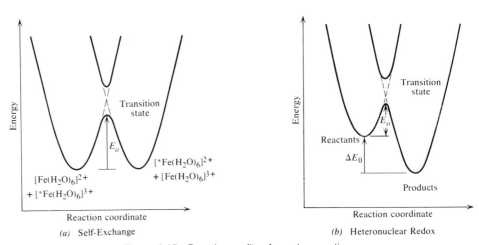

Figure 9.15 Reaction profiles for redox reactions.

[68]N. A. Lewis, *J. Chem. Educ.* 1980, *57*, 478.

potential energy for a pair of complex ions. The reaction coordinate represents all changes in bond lengths and angles in the first and second coordination spheres for the *pair* of complex ions. The parabolas are identical, since products and reactants are identical. At the value of the reaction coordinate appropriate for the transition state, the wave functions mix, creating two separate energy states instead of the parabolas crossing. Dotted lines show the course of the parabolas if no mixing occurred. The path of the self-exchange reaction is the motion of a point from the reactant-well up to the energy maximum (the Franck-Condon barrier) and on down into the product-well.

Most redox reactions of interest are heteronuclear—they involve two different metals. Nevertheless, each metal complex must undergo separately the kind of rearrangements described above. Thus, we might expect that the activation energy (and hence the value of the rate constant) for the heteronuclear reaction will be related to the self-exchange rates of each of the reactants. Figure 9.15b shows a reaction profile for a heteronuclear redox reaction that is thermodynamically favorable. As we can see from the figure, the height of the activation barrier depends on the relative placement of the reactant and product potential energy curves— which is equivalent to saying that the value of the rate constant also depends on the free-energy change for the reaction. This, in turn, is related to the emf (see Chapter 11).

$$\Delta G^0 = -n\mathcal{F}E^0 = -RT\ln K_{eq} \tag{9.47}$$

Marcus Theory

The considerations discussed above have been put on a firmer theoretical basis by Marcus, who derived[69] an equation for predicting rate constants for heteronuclear outer-sphere redox reactions from the self-exchange rate constants for each partner and the overall equilibrium constant. The Marcus equation is

$$k_{12} = \sqrt{k_{11}k_{22}K_{12}f_{12}} \tag{9.48}$$

where

$$\log f_{12} = \frac{(\log K_{12})^2}{4 \log\left(\dfrac{k_{11}k_{22}}{Z^2}\right)}$$

and Z is the number of collisions per second between particles in solution. The Marcus equation shows that the rates of redox reactions depend on intrinsic factors (through k_{11} and k_{22}, the self-exchange constants) and a thermodynamic factor (through K_{12}). The rate is related to the driving force, and more thermodynamically favorable reactions are faster. This equation is an example of a linear free-energy relationship (LFER).

To predict the rate constant for the reduction of, for example, $[Co(bipy)_3]^{3+}$ by $[Co(terpy)_2]^{2+}$, we require self-exchange constants.

$$[Co(bipy)_3]^{2+} + [\overset{*}{C}o(bipy)_3]^{3+} \rightleftharpoons [Co(bipy)_3]^{3+} + [\overset{*}{C}o(bipy)_3]^{2+}$$

$$k_{11} = 9.0 M^{-1}sec^{-1} \text{ at } 0°C \tag{9.49}$$

[69]R. A. Marcus, *Ann. Rev. Phys. Chem.* 1964, *15*, 155; T. W. Newton, *J. Chem. Educ.* 1968, *45*, 571.

$$[Co(terpy)_2]^{2+} + [\overset{*}{C}o(terpy)_2]^{3+} \rightleftharpoons [Co(terpy)_2]^{3+} + [\overset{*}{C}o(terpy)_2]^{2+}$$

$$k_{22} = 48 M^{-1} sec^{-1} \text{ at } 0°C \tag{9.50}$$

Reduction potentials for $[Co(terpy)_2]^{3+}$ and $[Co(bipy)_3]^{3+}$ are $+0.31$ and $+0.34$ volt, respectively. Hence,

$$\log K_{12} = \frac{n\mathscr{F}E^0}{2.303\,RT} = \frac{9.65 \times 10^4 \text{ coul} \times 0.03 \text{ volt}}{2.303(1.99 \text{ cal K}^{-1})(273 \text{ K})} \times \frac{1 \text{ cal}}{4.184 \text{ J}} \times \frac{1 \text{ J}}{\text{volt coul}} = 0.553$$

$$K_{12} = 3.57$$

$$\log f_{12} = \frac{(0.553)^2}{4 \log\left(\dfrac{9.0 \times 48.0}{10^{22}}\right)} = -3.95 \times 10^{-3} \tag{9.51}$$

$$f_{12} = 0.99$$

$$k_{12} = \sqrt{(9.0 \text{ M}^{-1}\text{sec}^{-1})(48.0 \text{ M}^{-1}\text{sec}^{-1})(3.57)(0.99)} = 39 \text{ M}^{-1}\text{sec}^{-1}$$

This calculated value compares favorably with the measured value of 64 $M^{-1}sec^{-1}$.[70] The agreement between calculated and measured rate constants generally is good, but in many cases the predicted outer-sphere rate constants do not agree well with experiment. Many of these disagreements occur for Co(III) complexes. The data provided in Table 9.24 are fairly typical. In the case of an unknown mechanism, agreement between the observed rate constant and that calculated from the Marcus equation is regarded as good evidence for an outer-sphere pathway.[71]

Suppose we compare the outer-sphere reactions of reagent 1 with two different reagents, 2 and 3. Then,

$$\frac{k_{12}}{k_{13}} = \frac{\sqrt{k_{11}k_{22}K_{12}f_{12}}}{\sqrt{k_{11}k_{33}K_{13}f_{13}}} \tag{9.52}$$

If the redox potentials of reagents 2 and 3 are not too different, and if the f values are similar, then

$$\frac{k_{12}}{k_{13}} \approx \sqrt{\frac{k_{22}}{k_{33}}} \tag{9.53}$$

and by the same reasoning, the ratio of outer-sphere rate constants for reaction with reagent 4 should also be

$$\frac{k_{42}}{k_{43}} \approx \sqrt{\frac{k_{22}}{k_{33}}} \tag{9.54}$$

[70]R. Farina and R. G. Wilkins, *Inorg. Chem.* 1968, *7*, 514.

[71]For a critical review of the applicability of the Marcus equation, see M. Chou, C. Creutz, and N. Sutin, *J. Am. Chem. Soc.* 1977, *99*, 5615.

Table 9.24 Comparison between calculated and observed rate constants for outer-sphere reactions[a]

Reaction	k_{12} obsd. $(M^{-1} sec^{-1})$	k_{12} calcd. $(M^{-1} sec^{-1})$	k_{11} $(M^{-1} sec^{-1})$	k_{22} $(M^{-1} sec^{-1})$	$E_1 V(K_{12})$	f
$[Fe(CN)_4]^{4-}-[IrCl_6]^{2-}$	3.8×10^5	1×10^6	7.4×10^2	2.3×10^5	$0.24(1.2 \times 10^4)$	5.1×10^{-1}
$[Fe(CN)_6]^{4-}-[Fe(phen)_3]^{3+}$	$>10^8$	$>1 \times 10^8$	7.4×10^2	$>3 \times 10^7$	$0.39(3.5 \times 10^6)$	1.2×10^{-1}
$[Fe(CN)_6]^{4-}-MnO_4^-$	1.7×10^5	6×10^4	7.4×10^2	3×10^3	$0.20(2.5 \times 10^3)$	6.6×10^{-1}
$[Fe(CN)_6]^{4-}-Ce(IV)$	1.9×10^6	6×10^6	7.4×10^2	4.4	$0.76(5.8 \times 10^{12})$	4.7×10^{-3}
$[Fe(phen)_3]^{2+}-Ce(IV)$	1.4×10^5	$>7 \times 10^6$	$>3 \times 10^7$	4.4	$0.36(1.1 \times 10^6)$	2.1×10^{-1}
$[Fe(phen)_3]^{2+}-MnO_4^-$	6.1×10^3	>3	$>3 \times 10^7$	3×10^3	$-0.50(3.4 \times 10^{-9})$	2.4×10^{-2}
$Fe^{2+}-[Fe(phen)_3]^{3+}$	3.7×10^4	$>5 \times 10^6$	4.0	$>3 \times 10^7$	$0.35(7.6 \times 10^5)$	2.4×10^{-1}
$Fe^{2+}-Ce(IV)$	1.3×10^6	5×10^5	4.0	4.4	$0.71(8.3 \times 10^{11})$	2.0×10^{-2}
$Fe^{2+}-[IrCl_6]^{2-}$	3.0×10^6	2×10^4	4.0	2.3×10^5	$0.16(5.0 \times 10^2)$	8.9×10^{-1}
$Cr^{2+}-Fe^{3+}$	2.3×10^3	$<1 \times 10^6$	$\leq 2 \times 10^{-5}$	4.0	$1.18(6.6 \times 10^{19})$	1.7×10^{-4}

[a]From D. A. Pennington in *Coordination Chemistry*, Vol. 2, A. E. Martell, Ed., American Chemical Society, Washington, 1978. k_{11} and k_{22} are the rate constants for self-exchange for reductant and oxidant, respectively, for the given redox pair.

That is, the relative rates of reduction (or oxidation) by a pair of reagents is expected to be the same for reactions with a series of oxidants (or reductants), as long as the mechanism does not change.

Table 9.25 shows data for outer-sphere reduction of some Co(III) and Ru(III) complexes by Cr^{2+}, Eu^{2+}, and V^{2+}. For these reactions, we can obtain average values of $k_{Cr^{2+}}/k_{V^{2+}}$ of 0.021 and $k_{Eu^{2+}}/k_{V^{2+}}$ of 0.35. These ratios can be used to determine the structure of transition states for pairs of reactions when one member of the pair is known to be outer-sphere. For example, some very fast reductions ($k \sim 40 \ M^{-1}sec^{-1}$) by V^{2+} are known to be outer-sphere. The reduction of Fe^{3+} by V^{2+} has $k_{V^{2+}} = 1.8 \times 10^4$. If the reductions by Cr^{2+} and Eu^{2+} were also outer-sphere, their predicted rate constants would be

Table 9.25 Rate constants $(M^{-1}sec^{-1})$ for outer-sphere reductions of some Co(III) and Ru(III) complexes at 25°C[a]

Oxidant	Reductant		
	Cr^{2+}	Eu^{2+}	V^{2+}
$[Co(NH_3)_6]^{3+}$	8.9×10^{-5}	2×10^{-2}	3.7×10^{-3}
$[Co(en)_3]^{3+}$	2×10^{-5}	5×10^{-3}	2×10^{-4}
$[Co(NH_3)_5py]^{3+}$	4.3×10^{-3}	8.3×10^{-2}	2.4×10^{-1}
$[Co(NH_3)_5(3\text{-}CH_3py)]^{3+}$	4.3×10^{-3}	3.8×10^{-2}	—
$[Co(NH_3)_5(DMF)]^{3+}$	7.2×10^{-3}	6.7×10^{-2}	—
$[Co(tet\text{-}a)(NH_3)_2]^{3+}$	4.8×10^{-3}	—	1.7×10^{-1}
$[Ru(NH_3)_6]^{3+}$	2×10^2	—	8×10^1
$[Ru(NH_3)_5py]^{3+}$	3.4×10^3	—	1.2×10^5

[a]From R. G. Linck in *M.T.P. International Review of Science, Inorganic Chemistry*, Series Two, Vol. 9, M. L. Tobe, Ed., University Park Press, Baltimore, 1974.

$$k_{Cr^{2+}} \sim 0.021 k_{V^{2+}} = 3.8 \times 10^2 \ M^{-1} sec^{-1}$$

$$k_{Eu^{2+}} \sim 0.35 k_{V^{2+}} = 6.3 \times 10^3 \ M^{-1} sec^{-1}$$

(9.55)

These correspond quite well with some of the measured values shown in Table 9.26. Thus, reductions by Cr^{2+} and Eu^{2+} can be assigned an outer-sphere mechanism. Table 9.26 shows some other assignments of mechanism based on this rate-ratio method. If the predicted value of k is too small [as for the Eu^{2+} reduction of Co(III) complexes[72]], the mechanism probably is inner-sphere.

Inner-sphere Reactions

A generalized mechanism for inner-sphere electron-transfer reactions involves three separate steps.

$$M^{II}L_6 + XM'^{III}L_5' \rightleftharpoons L_5M^{II}-X-M'^{III}L_5' + L$$

$$L_5M^{II}-X-M'^{III}L_5' \rightleftharpoons L_5M^{III}-X-M'^{II}L_5'$$

(9.56)

$$L_5M^{III}-X-M'^{II}L_5' \rightarrow products$$

The first step involves substitution of a ligand into the coordination sphere of the labile reactant (usually the reductant[73]) by the bridging group X, to form the precursor complex. In the second step, the precursor complex undergoes the same kind of reorganization described for outer-sphere pathways, followed by electron transfer to give the successor complex (sometimes rather inelegantly called the "postcursor" complex). Thus events in this step are related to those in

Table 9.26 Rate-ratio predictions for reductions by Cr^{2+} and Eu^{2+} based on known rates for V^{2+} [a]

Oxidant	Reductant	Predicted $k(M^{-1}sec^{-1})$	Observed $k(M^{-1}sec^{-1})$
Fe^{3+}	Cr^{2+}	3.8×10^2	$\leq 5.7 \times 10^2$
Fe^{3+}	Eu^{2+}	6.3×10^3	$\leq 3.4 \times 10^3$
$Fe(NCS)^{2+}$	Eu^{2+}	2.3×10^5	3.2×10^5
$FeCl^{2+}$	Eu^{2+}	1.6×10^5	2.0×10^6
V^{3+}	Eu^{2+}	3.5×10^{-3}	9.0×10^{-3}
$[Co(NH_3)_5F]^{2+}$	Eu^{2+}	9.1×10^{-1}	2.6×10^4
$[Co(NH_3)_5Cl]^{2+}$	Eu^{2+}	2.7	3.9×10^2
$[Co(NH_3)_5Br]^{2+}$	Eu^{2+}	8.8	2.5×10^2
$[Co(NH_3)_5I]^{2+}$	Eu^{2+}	4.2×10^1	1.2×10^2
$[Co(NH_3)_5NCS]^{2+}$	Eu^{2+}	2.2×10^{-3}	$\sim 7 \times 10^{-1}$

[a]From R. G. Linck in *M.T.P. International Review of Science, Series Two*, Vol. 9, M. L. Tobe, Ed., University Park Press, Baltimore, 1974.

[72]In fact, inner-sphere mechanisms seem to be prevalent for Eu^{2+}.

[73]The bridging ligand may be carried by the reductant, as in the $[Fe(CN)_6]^{4-}$ reduction of $HCrO_4^-$.

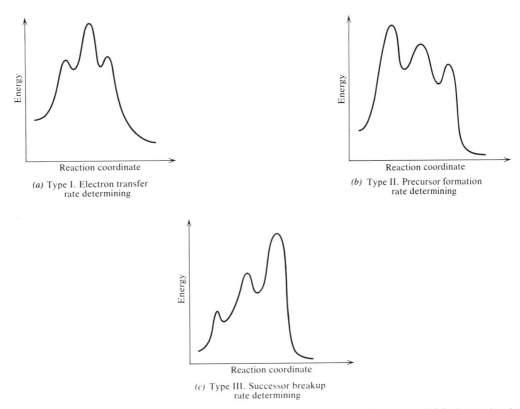

(a) Type I. Electron transfer
rate determining

(b) Type II. Precursor formation
rate determining

(c) Type III. Successor breakup
rate determining

Figure 9.16 Reaction profiles for inner-sphere redox reactions. (From R. G. Linck in *M.T.P. International Review of Science, Chemistry,* vol. 9 series one, M. L. Tobe, Ed. University Park Press, Baltimore, 1972.)

outer-sphere reactions. In the last step the successor complex breaks up to give the products. If M^{III} is inert to substitution and M'^{II} is labile, transfer of the bridging group occurs, affording $M^{III}XL_5$ as one of the products—which provides evidence of the inner-sphere mechanism. This is not always the case, however. For example, in Reaction 9.57 no ligand transfer occurs, since both Fe complexes are inert.

$$Co(edta)^{2-} + Fe(CN)_6^{3-} \rightarrow Co(edta)^- + Fe(CN)_6^{4-} \tag{9.57}$$

In principle, any one of the three steps could be rate-determining. Figure 9.16 shows potential-energy diagrams corresponding to each possible case.

Type I. Apparently, most redox reactions of inner-sphere mechanism correspond to the case in which the electron transfer step (with attendant bond-length and solvation changes) that transforms the precursor into the successor complex is slowest. The inner-sphere reactions mentioned so far all are examples of this type. Table 9.27 presents rate constants for some of these reactions. Notice that the rates vary considerably with the metal ions and bridging ligands involved. Inner-sphere redox reactions are placed in this class if no evidence exists for assigning them to either of the other types.

Table 9.27 Rate constants for some case-I inner-sphere redox reactions[a]

Reaction	$T(°C)$	$k(M^{-1}sec^{-1})$
$CrF^{2+} + Cr^{2+}$	0	2.4×10^{-3}
$CrCl^{2+} + Cr^{2+}$	0	9
$CrNCS^{2+} + Cr^{2+}$	25	1.4×10^{-4}
$CrNNN^{2+} + Cr^{2+}$	25	6.1
$CrSCN^{2+} + Cr^{2+}$	25	40
$FeCl^{2+} + Fe^{2+}$	0	5.4
$FeNCS^{2+} + Fe^{2+}$	0	4.2
$FeNNN^{2+} + Fe^{2+}$	0	1.9×10^3
$[Co(NH_3)_5F]^{2+} + Cr^{2+}$	25	2.5×10^5
$[Co(NH_3)_5Cl]^{2+} + Cr^{2+}$	25	6×10^5
$[Co(NH_3)_5NCS]^{2+} + Fe^{2+}$	25	$<3 \times 10^{-6}$
$[Co(NH_3)_5NNN]^{2+} + Fe^{2+}$	25	8.8×10^{-3}
$[Co(NH_3)_5SCN]^{2+} + Fe^{2+}$	25	1.2×10^{-1}

[a]From the work of Sutin, Moore, Espenson, Haim, and Halpern.

Because the electron transfer step involves a reorganization barrier similar to the outer-sphere case, a Marcus-type relationship applies to inner-sphere reactions.[74] The study of binuclear complexes such as $[(bipy)_2ClRu^{II}—N\bigcirc N—Ru^{III}(bipy)_2Cl]^{3+}$ as model systems for inner-sphere precursor complexes has enabled evaluation of Franck-Condon barriers, solvation energies, etc.[75] Reaction profiles such as those shown in Figure 9.15 are applicable to the electron transfer step in inner-sphere reactions.

Type II. When formation of the precursor complex is rate-determining (Figure 9.16b), the reaction rate is controlled by the rate at which the bridging ligand is substituted into the coordination sphere of the labile reactant. Table 9.28 gives kinetic parameters for some reductions by V^{2+}, a number of which are substitution-controlled. The rates of all these reductions are quite similar, and the kinetic parameters closely match those for substitution on V^{2+}. For water exchange on V^{2+}, $k = 100$ sec^{-1}, $\Delta H^{\ddagger} = 68.6$ kJ/mole, $\Delta S^{\ddagger} = -23$ J/mole K. For NCS$^-$ substitution, $k = 28$ $M^{-1}sec^{-1}$, $\Delta H^{\ddagger} = 56.5$ kJ/mole and $\Delta S^{\ddagger} = -29$ J/mole K. That the rates of reduction should be so similar for a wide variety of bridging ligands indicates that the process is controlled by the dissociative activation of the substitution (see Section 9.3.4).

A corollary of these considerations is that any V^{2+} reduction with a rate constant substantially in excess of the substitution rate constant must be outer-sphere (for example, the Fe^{3+} reduction and others on which the predictions in Table 9.26 are based).

[74]R. G. Wilkins, *The Study of Kinetics and Mechanism of Reactions of Transition Metal Complexes,* Allyn and Bacon, Boston, 1974, p. 282.

[75]T. J. Meyer, *Acc. Chem. Res.* 1978, *11*, 94.

Table 9.28 Rate parameters for some reductions by V^{2+} at 25°C[a]

Oxidant	$k\ (M^{-1}sec^{-1})$	$\Delta H^{\ddagger}\ (kJ/mole)$	$\Delta S^{\ddagger}\ (J/mole\ K)$
$CrSCN^{2+}$	8.0	50.2	-46
$[Co(NH_3)_5C_2O_4H]^{2+}$	12.5	51.0	-54
$[Co(NH_3)_5C_2O_4]^+$	45.3	—	—
$[Co(NH_3)_5N_3]^{2+}$	13	49.0	-58
cis-$[Co(en)_2(N_3)_2]^+$	32.7	—	—
cis-$[Co(en)_2(NH_3)(N_3)]^{2+}$	10.3	52.7	-50
cis-$[Co(en)_2(H_2O)(N_3)]^{2+}$	16.6	50.6	-50
$trans$-$[Co(en)_2(N_3)_2]^+$	26.6	51.0	-46
$trans$-$[Co(en)_2(H_2O)(N_3)]^{2+}$	18.1	46.0	-67
$[Co(NH_3)_5SCN]^{2+}$	30	—	—
Cu^{2+}	26.6	46.7	-57.7

[a]From the work of Taube, Espenson, Sutin, Linck, and others.

Substitution-controlled reductions by Cr^{2+}, Fe^{2+}, Cu^+, and Eu^{2+} also are known.

Type III. You might expect the breakup of the successor complex to be rate-determining when the electron configuration of both metals after the electron transfer leads to substitution inertness. Usually, the existence of a binuclear complex in equilibrium with reactants is indicated by the form of the rate law—as, for example, in the Cr^{2+} reduction of $[Ru(NH_3)_5Cl]^{2+}$. A complex having both Ru and Cr is kinetically detectable. Because the successor complex would contain inert Ru^{II} and Cr^{III}, we assume that this must be the species detected.[76] The mechanism involved is

$$Ru^{III}(NH_3)_5Cl^{2+} + Cr^{2+} \rightleftharpoons [Ru(NH_3)_5ClCr]^{4+}, K_{eq} \tag{9.58}$$

$$[Ru(NH_3)_5ClCr]^{4+} \xrightarrow{k_1} Ru^{II} + CrCl^{2+}$$

This is consistent with the observed rate law

$$\frac{-d[Ru^{III}]}{dt} = \frac{k_1 K_{eq}}{1 + K_{eq}[Cr^{2+}]}\ [Ru^{III}]_0[Cr^{2+}] \tag{9.59}$$

which you should verify as an exercise.[77]

Sometimes the successor complexes are sufficiently stable that they can be prepared independently and their breakup followed. An example is the successor complex from the re-

[76]It is true that the mere detection of a bridged species does not mean that it is involved in the redox reaction. Other evidence, including comparison with related systems, is needed to establish this point. See, for example, L. Rosenheim, D. Speiser, and A. Haim, *Inorg. Chem.* 1974, *13*, 1571; D. H. Huchital and J. Lepore, *Inorg. Chem.* 1978, *17*, 1134.

[77]*Hint:* The rate = $k_{obs}[Cr^{2+}][Ru^{III}]_0$ where $[Ru^{III}]_0$ represents the concentration of $[Ru(NH_3)_5Cl]^{2+}$ used in preparing the reacting solution. However, this is now partitioned between two solution species, $[Ru(NH_3)_5Cl]^{2+}$ and the Ru—Cr dimer.

duction of VO^{2+} by V^{2+}, $[V^{III}(OH)_2V^{III}]^{4+}$ [78] which can be prepared also by hydrolysis of V(III) solutions. Comparatively few redox reactions have been shown to belong to Type III.

The Bridging Ligand.[79]   Even in reactions in which precursor complex formation is not rate-determining, the nature of the bridging ligand can be quite important. Even in Type I or Type III reactions, the activation barrier for precursor formation may represent a substantial fraction of the overall activation energy. Also, the nature of the bridging ligand may influence the mechanism for the electron transfer step. This section points out some aspects of the role of the bridging ligand and their mechanistic consequences.

If a ligand is to function in a bridging capacity, an unshared pair of sufficiently basic electrons is required. Thus complexes such as $[Co(NH_3)_6]^{3+}$ and $[Co(NH_3)_5py]^{3+}$ react only via outer-sphere paths. Current evidence also makes it seem likely that unshared pairs on water are not sufficiently basic to allow aqua complexes an inner-sphere path.

Evidence for an inner-sphere mechanism in selected cases follows from changes in rate as a function of bridging ability of ligands. Table 9.29 contains rate data for several reductions as the bridging ligand is changed from H_2O to OH^- and from N_3^- to NCS^-. The $[Ru(NH_3)_6]^{2+}$ reductions must be outer-sphere (o.s.) since the reductant is both inert to substitution on the time scale of the redox reaction and without electron pairs for bridging. The rates of reaction with aqua and hydroxo complexes are quite similar. The large rate enhancement on going from H_2O to OH^- complexes when Cr^{2+} is the reductant supplies strong evidence that OH^- is a good bridging ligand and that reduction of the hydroxo complexes is inner sphere (i.s.).

Remote Attack.   Table 9.29's information on reduction of thiocyanato and azido complexes raises several interesting points. First, notice that all the products are consistent with transfer of the bridging group in an inner-sphere mechanism. Another interesting feature is the fact that lone pairs for bridging are not always available on the atom adjacent to the oxidant.

Table 9.29   Rate constants for some redox reactions at 25°C[a]

Oxidant	Reductant	$k(M^{-1}sec^{-1})$	Mechanism
$[Co(NH_3)_5(H_2O)]^{3+}$	Cr^{2+}	≤ 0.1	probably o.s.
$[Co(NH_3)_5(OH)]^{2+}$	Cr^{2+}	1.5×10^6	i.s.
$[Co(NH_3)_5(H_2O)]^{3+}$	$[Ru(NH_3)_6]^{2+}$	3.0	o.s.
$[Co(NH_3)_5(OH)]^{2+}$	$[Ru(NH_3)_6]^{2+}$	0.04	o.s

Reaction	Product	$k(M^{-1}sec^{-1})$
$[Co(NH_3)_5NCS]^{2+} + Cr^{2+}$	$CrSCN^{2+}$	19
$[Co(NH_3)_5N_3]^{2+} + Cr^{2+}$	CrN_3^{2+}	3×10^5
$[Co(NH_3)_5SCN]^{2+} + Cr^{2+}$	71% $CrNCS^{2+}$ + 29% $CrSCN^{2+}$	1.9×10^5

[a]From the work of Haim, Linck, Taube, Endicott, and others.

[78]T. W. Newton and F. B. Baker, *Inorg. Chem.* 1964, *3*, 569; L. Pajdowski and B. Jezowska-Trzebiatowska, *J. Inorg. Nucl. Chem.* 1966, *28*, 443.

[79]A. Haim, *Acc. Chem. Res.* 1975, *8*, 264.

Thus, for example, the only site available for bridging in $[Co(NH_3)_5NCS]^{2+}$ is on the S atom remote from the metal center. The product, $CrSCN^{2+}$, indicates that remote attack occurs on S. Of course, the symmetry of the azido ligand makes such a distinction experimentally impossible without isotopic labeling. However, by drawing the Lewis structure you can convince yourself that only the remote N can act as a bridge. In $[Co(NH_3)_5SCN]^{2+}$, both adjacent and remote attack occur. On standing, the $CrSCN^{2+}$ produced rearranges to the more stable $CrNCS^{2+}$ isomer. Because Cr^{2+} is a hard acid, it prefers to bond with the hard N end of the thiocyanato ligand. This preference is reflected in a lesser stability of the $Cr—SCN—Co$ precursor and accounts for the rate differences observed between the S- and N-bonded isomers.[80] The latter rates are about the same as those for azido complexes. (This is the same kind of effect seen for OH^- and H_2O.) Finally, these data show that electron transfer can ocur through a multi-atomic bridge.

One such electron transfer involves remote attack

by Cr^{2+} on O in $\left[(NH_3)_5Co—N\bigcirc—C\overset{\displaystyle \ddot{O}}{\underset{NH_2}{\diagdown}} \right]^{3+}$. The electron pair on N apparently is

not sufficiently basic to provide a site for attack. This seems to be a general feature of $—NH_2$ groups. Whereas $[(NH_3)_5Co \leftarrow NH_2—\underset{\underset{\displaystyle :O:}{\|}}{C}—H]^{3+}$ undergoes reduction with Cr^{2+} via remote

attack on the carbonyl O, its linkage isomer, $\left[(NH_3)_5Co\ddot{O}=C\overset{\displaystyle H}{\underset{\ddot{N}H_2}{\diagdown}} \right]^{3+}$, follows an outer-

sphere path. (Presumably, steric factors block adjacent attack.)

Organic Bridging Ligands.[81] Organic bridging ligands having conjugated π-systems display further interesting effects on redox chemistry. A study of the Cr^{2+} reduction of

$[(NH_3)_5Co—C(O)—\bigcirc—\overset{\overset{\displaystyle O}{\|}}{C}H]^{2+}$ shows that remote attack occurs on the carbonyl O and

gives Co^{2+} as the product. However, reduction of $[(NH_3)_5Co—C(O)—\bigcirc—NO_2]^{3+}$ gives

no Co^{2+}. Instead, the nitro group on the ligand is reduced. The Cr^{2+} reduction of

$\left[(NH_3)_5Co—OC(O)—\langle\bigcirc\rangle\overset{\displaystyle N}{\underset{N}{\diagup}} \right]^{2+}$ has been found to involve two detectable intermediates.

The first decays to give the successor complex, which then affords Co^{2+} as the product.[82] Apparently, in some cases (where the organic ligand is reducible to a fairly stable radical)

---

[80]S seems to be an especially good mediator of electron transfer. Thus, attack at S sites often occurs kinetically, even when the product is not the most thermodynamically stable.

[81]H. Taube and E. S. Gould, *Acc. Chem. Res.* 1969, *2*, 321.

[82]E. S. Gould, *J. Am. Chem. Soc.*, 1972, *94*, 4360.

electron transfer can first reduce the bridging ligand, which then may or may not in turn reduce the other metal. This mode of electron transfer, represented in Mechanism 9.60 is called the *chemical* or *radical-ion mechanism.*

$$M^{III}—L + Cr^{2+} \rightleftharpoons M^{III}—L—Cr^{2+}$$

$$M^{III}—L—Cr^{2+} \rightleftharpoons M^{III}—\dot{L}^-—Cr^{III}$$

$$M^{III}—\dot{L}^-—Cr^{III} \rightarrow M^{II}—L—Cr^{III}$$

$$M^{II}—L—Cr^{III} \rightarrow M^{II} + Cr^{III}L$$

(9.60)

We also can recognize the occurrence of the radical-ion mechanism by the insensitivity of reaction rates to the identity of the metal in the oxidant. For example, Co(III) complexes typically are reduced in inner-sphere reactions about 10^5-10^7 times as fast as Cr(III) complexes with the same ligands. However, for the complexes $[ML_5N\bigcirc—C(O)NH_2]^{3+}$, the reaction is only about 10 times as fast for $ML_5 = Co(NH_3)_5$ as for $ML = Cr(H_2O)_5$.

Just because the organic ligand could be reduced in a radical-ion mechanism does not mean it always is. In our model of electronic structures, electrons must always exist in orbitals. The electrons transferred to unsaturated organic ligands are presumed to occupy π^*-orbitals. They are transferred into e_g orbitals of σ symmetry when Cr(III) and Co(III) are reduced. The reorganization energy required for distortion of the complex to achieve overlap of the ligand π and oxidant σ orbitals slows down the reduction of the metal center. The lifetime of the radical-ion intermediate is prolonged. However, when the metal orbital is t_{2g} of π-symmetry, the electron transfer can occur immediately via a *resonance mechanism.*

$[Ru^{III}(NH_3)_5—N\bigcirc—C(O)NH_2]^{3+}$ has the $(t_{2g})^5$ configuration and is reduced by Cr^{2+} 30,000 times as fast as the Co(III) complex. The resonance transfer mechanism is the one most usually observed in redox chemistry and the only possible mechanism for saturated bridging ligands. This is consistent with the bridging inefficiency of multiatom saturated ligands. Such complexes react by outer-sphere pathways.

A rather surprising role for organic bridging ligands recently has been found[83]: in some cases, the bridging group simply holds the oxidant and reductant in proximity while the actual electron transfer occurs via an outer-sphere process. At least this seems the most plausible explanation for relative rates of electron transfer in a series of long-lived precursor complexes $[(NC)_5Fe—L—Co(NH_3)_5]$ where $L = N\bigcirc—X—\bigcirc N$ and $X = CH_2$, CH_2CH_2, $(CH_2)_3$, and C=O. Since the complex with $X = (C=O)$ has a conjugated π system, we would expect it to undergo resonance electron transfer, which would occur faster than in the other complexes where X is saturated. Instead, it reacts an order of magnitude slower. This implies that the flexible, saturated X groups simply hold the two reactants together and permit them to achieve a suitable conformation for outer-sphere electron transfer!

Multiple Bridging.   Product analysis reveals several instances in which more than one ligand

[83]J.-J. Jwo, P. L. Gaus, and A. Haim, *J. Am. Chem. Soc.* 1979, *101*, 6189.

acts as bridge in an inner-sphere transition state. For example, the reaction of $^{51}Cr^{2+}$ with *cis*-$[Cr(H_2O)_4(N_3)_2]^+$ gives exchange of $^{51}Cr^{2+}$ 30 times faster than the catalyzed aquation occurs giving $[^{51}Cr(H_2O)_5N_3]^{2+}$.[84]

Other Issues in Inner-sphere Reactions

The nonbridging ligands play a role in rates of redox reactions because they affect the relative energies of donor and acceptor orbitals. Also, the important question of ionic strength and medium effects greatly affects the determined form of the rate law.[85]

Table 9.30, which summarizes methods for distinguishing between inner- and outer-sphere models for the transition state, is presented here because this information is interspersed throughout the general discussion of the characteristics of inner- and outer-sphere reactions.

Table 9.30 Methods for distinguishing between inner- and outer-sphere redox reactions

Method	Comments
Form of rate law	Usually, no distinction possible; most have rate = k[oxidant][reductant]. Can sometimes detect stable complex. (See Section 9.8.1.)
Analysis of products	Detection of product with transfer of bridging ligand is evidence of i.s. mechanism. Product must be substitution-inert on experiment time scale.
Rate prediction from Marcus equation	Agreement indicates o.s. May be applicable in i.s. cases.
Rate ratio	Agreement with prediction from ratio for known o.s. mechanism implies o.s. mechanism (see Section 9.8.1). May also apply to i.s. mechanism.
Rate enhancement by bridging ligand	Applicable to limited class of ligands; shows i.s. behavior. N_3^- vs. NCS^-; OH^- vs. H_2O; not applicable to halides.
Insensitivity of rate to possible bridging ligand	Comparison to reactions of known mechanism must be made carefully. Could represent o.s. or substitution-controlled or radical-ion type i.s.
Value of rate constant and activation parameters	If these are similar to values for substitution, i.s. substitution-controlled mechanism applies. If rate is $\gg$ substitution rate for either reactant, mechanism is o.s.
Detection of binuclear species	If this is a kinetically detected intermediate, i.s. mechanism applies. If spectroscopically detected, most likely i.s.
Sensitivity to metal ions	Probably results from thermodynamic factors in o.s. or i.s. case. Main significance is in cases where separate evidence for i.s. or o.s. behavior exists.
Ligand identity	If no ligand has a lone pair for bridging, must be o.s. Even H_2O and NH_2 probably do not allow i.s. path.
$\Delta V^{\ddagger}$	Positive for the few inner-sphere reactions investigated.

[84]R. Snellgrove and E. L. King, *J. Am. Chem. Soc.* 1962, *84*, 4609.

[85]See, for example, D. E. Pennington, in *Coordination Chemistry,* Vol. 2, A. E. Martell, Ed., American Chemical Society, Washington, 1978. See also the references at the end of the chapter.

9.9 PHOTOCHEMICAL REACTIONS

In *photochemistry,* an area of rapidly growing interest in inorganic chemistry, energy is imparted to molecules by irradiating them with visible and ultraviolet light. The quantities of energy available from light quanta range from ~170 to ~590 kJ/mole—up to 3.5 times that available thermally at ordinary temperatures. Only certain wavelengths of light are absorbed by transition metal complexes—namely, those corresponding to differences between electronic energy states (see Sections 7.7 and 7.8).

The excited complexes can undergo several kinds of energy transfer. The most straightforward is luminescence, in which light is emitted as the excited state returns to the ground state (a radiative process). Alternatively, the energy may be converted to vibrational energy and dissipated thermally to the environment as the complex returns to the ground state (internal conversion). Of more interest to chemists is the fact that relatively long-lived excited states of complexes may display very different reaction chemistry from the thermally activated ground state. Figure 9.17 diagrams the possible fates of a photochemically activated complex and compares them with thermally activated processes. The photochemically excited state may luminesce with a rate constant k_r, undergo internal conversion with a rate constant k_q, or react (perhaps with other species present in solution) with rate constants k_A, k_B, . . . , etc. The figure also indicates schematically how products that are thermodynamically inaccessible *(A)* or kinetically inaccessible *(C)* in thermal reactions might result from photochemical excitation.

A further complicating possibility is so-called intersystem crossing, in which a transition occurs to a different excited state, which can then undergo any of the fates depicted in the figure.

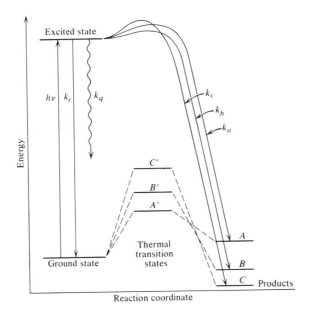

Figure 9.17 Comparison of thermal and photochemical activation. (After Balzani and Carassiti, reference 90).

Four types of photochemically excited states can be achieved for transition metal complexes. Ligand field (LF) excited states, which result from *d–d* transitions and generally place electron density in $e_g\sigma^*$ orbitals, are expected to lead to weakened L—M bonds and to favor ligand labilization. Charge-transfer excited states may be L → M or M → L. In the latter, more common case, the reactivity of the excited state should parallel that of $\dot{L}^-$ or involve electron transfer to the oxidized metal center. Transitions localized on the ligands result in intraligand (IL) excited states. Insofar as anything is known about the reactivity to be expected, ligand rearrangement and ligand reaction with other substrates may be involved. Finally, a charge transfer to solvent (CTTS) state involving an oxidized metal ion and a solvated e_s could lead to electron transfer to the oxidized metal and reduction of solvent or other substrate by e_s. As one example, the photoaquation of $[Co(NH_3)_6]^{3+}$ to give $[Co(NH_3)_5(H_2O)]^{3+}$ occurs readily upon irradiating the lower-wavelength ligand field band of the hexaammine. In contrast, an acid solution of $[Co(NH_3)_6]^{3+}$ can be boiled for days without a reaction occurring. The photochemically excited state is $\sim 10^{13}$ as reactive as the ground state.

Irradiation of $[Ru(bipy)_3]^{2+}$ and related complexes produces a CT excited species, $[\overset{*}{Ru}(bipy)_3]^{2+}$, with an oxidized Ru and an electron localized on a bipy ligand. This excited molecule is both a better oxidizing agent and a better reducing agent than $[Ru(bipy)_3]^{2+}$, as the reduction potentials in Table 9.31 show. $[\overset{*}{Ru}(bipy)_3]^{2+}$ is luminescent; quenching of this luminescence by other species in solution Q is evidence of a reaction that may be oxidation, reduction, or energy-transfer.

$$[\overset{*}{Ru}(bipy)_3]^{2+} + Q \rightarrow [Ru(bipy)_3]^{3+} + Q^- \qquad (9.61)$$

$$[\overset{*}{Ru}(bipy)_3]^{2+} + Q \rightarrow [Ru(bipy)_3]^{1+} + Q^+ \qquad (9.62)$$

$$[\overset{*}{Ru}(bipy)_3]^{2+} + Q \rightarrow [Ru(bipy)_3]^{2+} + {}^*Q \qquad (9.63)$$

The oxidation and reduction reactions represent ways of converting light energy into chemical energy. Energy storage could also be achieved if a suitable quencher could convert the excited species into complexes such as the excellent oxidizing agent $[Ru(bipy)_3]^{3+}$ and the excellent reducing agent $[Ru(bipy)_3]^+$. (The first of these has been shown to oxidize OH^- to O_2—representing a possible step in the decomposition of water to H_2 and O_2 by solar energy.) One such set of quenching reactions that is thermodynamically feasible is

$$[\overset{*}{Ru}(bipy)_3]^{2+} + [Ru(NH_3)_6]^{2+} \rightarrow [Ru(bipy)_3]^{3+} + [Ru(NH_3)_6]^+ \qquad (9.64)$$

Table 9.31 Reduction potentials for some Ru complexes at 25°C[a]

Reaction	E^0 (volts)
$[Ru(bipy)_3]^{3+} + e \rightarrow [Ru(bipy)_3]^{2+}$	1.26
$[\overset{*}{Ru}(bipy)_3]^{2+} + e \rightarrow [Ru(bipy)_3]^{1+}$	0.84
$[Ru(bipy)_3]^{3+} + e \rightarrow [\overset{*}{Ru}(bipy)_3]^{2+}$	−0.84
$[Ru(bipy)_3]^{2+} + e \rightarrow [Ru(bipy)_3]^{1+}$	−1.28

[a]From the work of Creutz and Sutin.

$$[\overset{*}{Ru}(bipy)_3]^{2+} + [Ru(NH_3)_6]^{2+} \rightarrow [Ru(bipy)_3]^{1+} + [Ru(NH_3)_6]^{3+} \qquad (9.65)$$

Unfortunately, both thermodynamics and kinetics favor the back-reaction of the products to regenerate $[Ru(NH_3)_6]^{2+}$, dissipating the stored energy to the solution. A possible solution to this problem is to use an organic quencher such that Q^- or Q^+ reacts further to produce stable products, leaving energy stored either as $[Ru(bipy)_3]^{3+}$ or $[Ru(bipy)_3]^+$. Gray and coworkers have prepared isocyanide complexes such as $[Rh_2(\mu_2\text{-}NCCH_2CH_2CH_2CN)_4]^{2+}$, which on ir-radiation reduces H^+ to H_2. Both kinds of complexes do have potential application to solar energy storage and splitting of water. However, our main focus here is on the fact that pho-tolytic activation can lead to chemistry quite different from that caused by thermal activation.

An interesting connection exists between photochemistry and the redox reactions just discussed in this chapter. Taube, Creutz, Meyer, and others have prepared a series of mixed-

valence compounds, of which $[(bipy)_2ClRu^{II}\text{—}N\bigcirc N\text{—}Ru^{III}(bipy)_2Cl]^{3+}$ is an example.

These complexes contain the same metal in two different oxidation states and serve as good models for the precursor complex in inner-sphere electron transfer reactions when the bridging ligand contains some unsaturation.[86] We can study the rates of electron transfer apart from considerations relating to the precursor stability.

Mixed-valence complexes usually are intensely colored, because of an absorption in the low-energy region. The absorption is not present in the spectrum of $[Ru(bipy)_2Clpy]^{n+}$ ($n = 2$ or 3), and results from electron transfer from one metal to the other—the intervalence transfer (IT). Figure 9.18a makes clear the origin of this band. This figure is like Figure 9.15a in that the reactants and products are identical. However, E_{op} (the IT energy) differs from E_{th} (the thermal energy for electron transfer) because of the Franck-Condon principle. Most of the complex ions in solution are at the equilibrium position for the ground state when the photon energy is absorbed. Because nuclear motions are so slow compared with electronic transitions, the energy absorbed in going to the upper state is E_{op}, corresponding to a vertical transition. For symmetrical complexes, we can show that $E_{op} = 4E_{th}$.[87]

Example Show that $E_{op} = 4E_{th}$ for a symmetrical mixed-valence complex.

Solution Suppose a = the distance between energy minimum and thermal transition state. Both curves are parabolas. Hence, at the thermal transition state, $E_{th} = k(a)^2$, where k is the proportionality constant. At the value of reaction coordinate corresponding to the reactants, the distance from the product curve minimum is $2a$. Hence, the energy on the upper parabola is $E_{op} = k(2a)^2$. Hence, $E_{op} = 4E_{th}$.

At least theoretically, measurements of E_{op} should lead to estimates of self-exchange in sym-metrical complexes. Figures 9.18b and 9.18c show alternative energy possibilities for unsym-metrical mixed-valence complexes that might contain different metals or different ligands (for

example, $[(NH_3)_4ClRu^{III}\text{—}N\bigcirc N\text{—}Ru^{II}(bipy)_2Cl^{3+}]$).

Unsaturated bridging ligands such as pyrazine give mixed-valence complexes that are

---

[86]See T. J. Meyer, *Acc. Chem. Res.* 1978, *11*, 94.

[87]N. S. Hush, *Prog. Inorg. Chem.* 1967, *8*, 391; *Chem. Phys.* 1975, *10*, 361; J. J. Hopfield, *Proc. Nat'l Acad. Sci., U.S.A.* 1974, *71*, 3640.

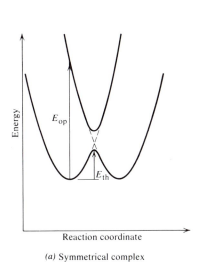

(a) Symmetrical complex

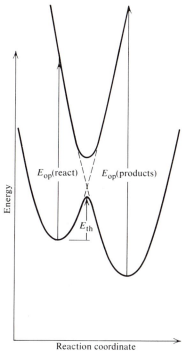

(b) Unsymmetrical complex; products more stable than reactants

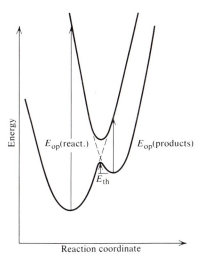

(c) Unsymmetrical complex; products less stable than reactants

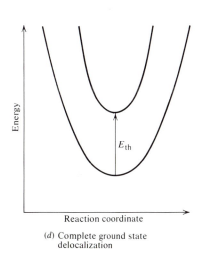

(d) Complete ground state delocalization

Figure 9.18   Reaction profiles for intervalence transfer.

good models for inner-sphere precursor complexes, because of orbital overlap between π^* ligand orbitals and each separate metal orbital. If the bridging ligand is saturated and sufficiently long, electronic coupling between redox sites is essentially from outer-sphere orbital overlap. Such a complex is $[(bipy)_2ClRu^{II}Ph_2PCH_2CH_2PPh_2Ru^{III}Cl(bipy)_2]^{3+}$ which also exhibits an IT band, but one of much lower intensity, because of unfavorable orbital overlap.[88] Thus complexes with low-intensity IT bands are models for outer-sphere electron-transfer reactions.

Redox chemistry can often be brought about by irradiating the IT band.

$$[(NH_3)_5Co^{III}N\equiv C-Ru^{II}(CN)_5]^- \xrightarrow{h\nu} 5NH_3 + Co^{2+} + [Ru^{III}(CN)_6]^{3-} \qquad (9.66)$$

A third group of complexes also may be prepared from metals in two different oxidation states. Where the bridging ligand has delocalized orbitals and the metal orbitals can overlap with particular effectiveness (because of a short physical separation, low oxidation state, etc.), the two metal centers may "communicate" electronically in the ground state. In these cases, delocalized molecular orbitals exist and the electrons are distributed equivalently over both metals in the ground state. Examples include $[(NH_3)_5Ru-N\bigcirc N-Ru(NH_3)_5]^{5+}$ and $[(NH_3)_5Ru-O-Ru(NH_3)_5]^{5+}$,[89] where each Ru is assigned 5.5 d electrons. This corresponds to the more-or-less complete coalescence of the curves in Figures 9.18a, 9.18b, and 9.18c to give a curve like the one in Figure 9.18d in the symmetrical case. Notice that a band corresponding to a transition between two delocalized MO's now is expected. The delocalized case can be recognized from electrochemical, visible-UV spectral and magnetic measurements.

Inorganic photochemistry is a comparatively new research area. Important questions demand investigation: What is the exact electronic state that undergoes photoreaction? (This is the photochemists' equivalent of stoichiometric mechanism). What are the reaction products (the equivalent of intimate mechanism)? What fraction of the light quanta absorbed leads to a particular photoreaction? (This is the quantum yield for the reaction and corresponds to the thermal rate constant.) Besides photoreaction, what other processes do excited molecules undergo? We cannot do justice to this new and exciting field here. For further information, consult the references.[90]

GENERAL REFERENCES

F. Basolo and R. Pearson, *Mechanisms of Inorganic Reactions,* 2nd ed., Wiley, New York, 1967. The classic text in the field.

M. L. Tobe, *Inorganic Reaction Mechanisms,* Nelson, London, 1972.

R. G. Wilkins, *The Study of Kinetics and Mechanism of Reactions of Transition Metal Complexes,* Allyn

[88] B. P. Sullivan and T. J. Meyer, *Inorg. Chem.* 1980, *19,* 752; C. A. Stein and H. Taube, *J. Am. Chem. Soc.* 1978, *100,* 1635.

[89] J. A. Baumann and T. J. Meyer, *Inorg. Chem* 1980, *19,* 345.

[90] V. Balzani and V. Carassiti, *Photochemistry of Coordination Compounds,* Academic Press, London, 1970; A. W. Adamson and P. D. Fleischauer, *Concepts of Inorganic Photochemistry,* Wiley, New York, 1975; M. S. Wrighton, Ed., *Inorganic and Organometallic Photochemistry, Adv. in Chem. Ser. No. 168,* American Chemical Society, Washington, 1978; G. L. Geoffroy and M. S. Wrighton, *Organometallic Photochemistry,* Academic Press, New York, 1979.

and Bacon, Boston, 1974. Excellent discussions of rate laws and their interpretation and experimental establishment.

Substitution Reactions

L. Cattalini, *Prog. Inorg. Chem.* 1970, *13*, 263. Intimate mechanism of square planar substitution reactions.

C. H. Langford and H. B. Gray, *Ligand Substitution Processes*, Benjamin, Cummings, Reading, Mass., 1966. A very readable account detailing the development of the D, A, I_d, I_a nomenclature.

C. H. Langford and V. S. Sastri, in *M.T.P. International Review of Science, Inorganic Chemistry*, Series one, Vol. 9, M. L. Tobe, Ed., University Park Press, Baltimore, 1972. An authoritative general account of octahedral substitution.

D. W. Margerum, G. R. Cayley, D. C. Weatherburn, and G. K. Pagenkopf in *Coordination Chemistry*, Vol. 2, A. E. Martell, Ed., *ACS Monograph 174*, American Chemical Society, Washington, 1978. A discussion of ligand-exchange and substitution reactions, focusing on labile ions and multidentate ligands.

M. L. Tobe, *Acc. Chem. Res.* 1970, *3*, 377. Mechanism of base hydrolysis.

Redox Reactions

W. L. Reynolds and R. W. Lumry, *Mechanisms of Electron Transfer*, Ronald Press, New York, 1966.

H. Taube, *Electron Transfer Reactions of Complex Ions in Solution*, Academic Press, New York, 1970.

R. G. Linck in *M.T.P. International Review of Science, Inorganic Chemistry*, Series one, Vol. 9, and Series two, Vol. 9, M. L. Tobe, Ed., University Park Press, Baltimore, 1972 and 1974. Insightful accounts of mechanistic developments in redox reactions.

D. E. Pennington in *Coordination Chemistry*, Vol. 2, American Chemical Society, ACS Monograph 174, Washington, 1978. A recent and complete account of mechanistic investigations in redox chemistry.

PROBLEMS

9.1 In 1.0 M OH^- solution, at least 95% of $[Co(NH_3)_5Cl]^{2+}$ that undergoes base hydrolysis must be present in the pentaammine form. If 5% or more were present as the conjugate base $[Co(NH_3)_4(NH_2)Cl]^+$, it could be detected by departure from second-order kinetics which is *not* observed to happen.

    a. Using this fact together with expressions in Equation 9.25, show that for $[Co(NH_3)_5Cl]^{2+}$, $K_a \leq 5 \times 10^{-16}$.

    b. Show the form of the rate that would result if the conjugate base were present in appreciable amount. (*Hint:* The concentration of $[Co(NH_3)_5Cl]^{2+}$ used to make up the solution will be partitioned between the complex and its conjugate base.)

9.2 Explain how the data on water exchange given in this chapter (see p. 334) suggest a dissociative mechanism.

9.3 Account for the difference in rate constants for the following two reactions.

$$[Fe(H_2O)_6]^{2+} + Cl^- \rightarrow [Fe(H_2O)_5Cl]^+ + H_2O, \; k(M^{-1}sec^{-1}) = 10^6$$
$$[Ru(H_2O)_6]^{2+} + Cl^- \rightarrow [Ru(H_2O)_5Cl]^+ + H_2O, \; k(M^{-1}sec^{-1}) = 10^{-2.0}$$

9.4 Distinguish between the intimate and stoichiometric mechanism of a reaction.

9.5   What is the significance of the following facts for the mechanism of substitution at Co(III) in aqueous solution?
  a. The rates of aquation are always given by the expression rate $= k_{aq}[Co(NH_3)_5X]^{2+}$.
  b. No direct replacement of X^- by Y^- is ever observed. Instead, water enters first and is subsequently replaced by Y^-.

9.6   The following data have been obtained at 50°C for aquation of $[Cr(NH_3)_5X]^{2+}$ (k_{aq}) and anation by Y^- of $[Cr(NH_3)_5(H_2O)]^{3+}$ (k_{an}).

Y^-	$k_{aq}(sec^{-1})$	$k_{an}(M^{-1}sec^{-1})$
NCS^-	0.11×10^{-4}	4.16×10^{-4}
$CCl_3CO_2^-$	0.37×10^{-4}	1.81×10^{-4}
$CF_3CO_2^-$	0.50×10^{-4}	1.37×10^{-4}
Cl^-	1.75×10^{-4}	0.69×10^{-4}
Br^-	12.5×10^{-4}	—
I^-	102×10^{-4}	—
$HC_2O_4^-$	—	6.45×10^{-4}
H_2O	13.7×10^{-4} (exchange)	—

What can you say about the intimate mechanism of these reactions? (See T. Ramasani and A. G. Sykes, *Chem. Comm.* 1976, 378.)

9.7   The following data have been reported for aquation at 298.1 K of complexes *trans*-$[Co(N_4)LCl]^{n+}$, where N_4 is a quadridentate chelate.

	$k(sec^{-1})$	
L	$(N_4) = $ cyclam	$(N_4) = $ tet-b
Cl^-	1.1×10^{-6}	9.3×10^{-4}
NCS^-	1.1×10^{-9}	7.0×10^{-7}
NO_2^-	4.3×10^{-5}	4.1×10^{-2}
N_3^-	3.6×10^{-6}	2.1×10^{-2}
CN^-	4.8×10^{-7}	3.4×10^{-4}
NH_3	7.3×10^{-8}	—

What do these data suggest about the intimate mechanism for aquation? (See W.-K. Chau, W.-K. Lee, and C. K. Poon, *J. Chem. Soc., Dalton Trans.* 1974, 2419.)

9.8   The reactions $[Cr(NCS)_6]^{3-} + $ solv $\rightarrow [Cr(NCS)_5(solv)]^{2-} + NCS^-$ have been investigated and found to have the following rate constants near 70° C.

Solvent	$k(sec^{-1})$
dimethylacetamide	9.5×10^{-5}
dimethylformamide	12.4×10^{-5}
dimethylsulfoxide	6.2×10^{-5}

What do these values suggest about the intimate mechanism of these reactions? (See S. T. D. Lo and D. W. Watts, *Aust. J. Chem.* 1975, *28*, 1907.)

9.9   Figure 9.19 shows plots of k_{obs} vs. $[X^-]$ for the anation reactions

$$[Co(en)_2(NO_2)(DMSO)]^{2+} + X^- \rightarrow [Co(en)_2(NO_2)X]^+ + DMSO$$

All three reactions are presumed to have the same mechanism.
  a. What is the significance of the shapes of the curves?

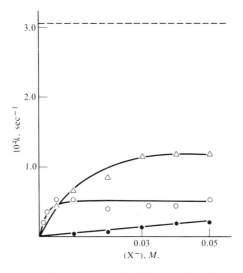

Figure 9.19 Rate of anation of *cis*-[Co(en)$_2$-NO$_2$dmso]$^{2+}$ as a function of the concentration of the entering anion X$^-$: △,X$^-$ = NO$_2^-$; ○,X$^-$ = Cl$^-$; ●,X$^-$ = SCN$^-$. The broken line shows the rate of the dmso-exchange reaction. (Reprinted with permission from W. R. Muir and C. H. Langford, *Inorg. Chem.* 1968, *7*, 1032. Copyright 1968, American Chemical Society.)

b. What is the significance of the fact that the first-order limiting rate constants are smaller than that for DMSO exchange?

c. If the mechanism were *D*, to what would the limiting rate constants correspond?

d. If the mechanism were *I$_d$*, to what would the limiting rate constants correspond?

e. The limiting rate constants are 0.5×10^{-4}sec^{-1} and 1.2×10^{-4}sec^{-1} for Cl$^-$ and NO$_2^-$, respectively. For NCS$^-$, the limiting rate constant can be estimated as 1×10^{-4}sec^{-1}. Do these values constitute evidence for a *D* or an *I$_d$* mechanism?

(See W. R. Muir and C. H. Langford, *Inorg. Chem.* 1968, *7*, 1032.)

9.10 The following data were obtained for the base hydrolysis of $(+)_{589}$-[Co(en)$_2$(NH$_3$)X]$^{n+}$ in the presence of added N$_3^-$ (1.0 *M*) and are reproducible to ± 0.50%.

	Products					
	[Co(en)$_2$(NH$_3$)OH]$^{2+}$, %			[Co(en)$_2$(NH$_3$)N$_3$]$^{2+}$, %		
Complex	*trans*	*cisR*	*cisI*	*trans*	*cisR*	*cisI*
[Co(en)$_2$(NH$_3$)Cl]$^{2+}$	17	48	11	7	13	4
[Co(en)$_2$(NH$_3$)Br]$^{2+}$	17	48	12	7	12	4
[Co(en)$_2$(NH$_3$)(NO$_3$)]$^{2+}$	17	46	11	8	13	5
[Co(en)$_2$(NH$_3$)(DMSO)]$^{3+\,a}$	16	45	9	10	17	4
[Co(en)$_2$(NH$_3$)(TMP)]$^{3+\,b}$	16	45	8	10	18	3

^{a}DMSO = dimethylsulfoxide.
^{b}TMP = trimethylphosphate.

What do these data reveal about the stoichiometric mechanism of these reactions? (See D. A. Buckingham, I. I. Olsen, and A. M. Sargeson, *J. Am. Chem. Soc.* 1968, *90*, 6654; D. A. Buckingham, C. R. Clark, and T. W. Lewis, *Inorg. Chem.* 1979, *18*, 1985.)

9.11 The complexes *trans*-[Rh(en)$_2$LX]$^+$ react with various Y$^-$, giving [Rh(en)$_2$LY]$^+$. The rate law for the appearance of the product is rate = (k$_1$ + k$_2$[Y$^-$])[Rh(en)$_2$LX]$^+$. Do these data constitute evidence for an *a* mechanism? Why or why not? (See A. J. Poë and C. P. J. Vuik, *Inorg. Chem.* 1980, *19*, 1771.)

9.12   The anation of *trans*-[Rh(en)$_2$(H$_2$O)$_2$]$^{3+}$ with chloride recently has been studied. Two plots of k_{obs} versus concentration of species in solution are reproduced in Figure 9.20.
   a. Account for the linearity of the plots in Figure 9.20*a*.
   b. What does the nonzero intercept for each of the two lines in Figure 9.20*a* indicate about the reaction mechanism?
   c. Account for the shape of the two curves in Figure 9.20*b*. Do these curves alone allow you to distinguish what the stoichiometric mechanism is?

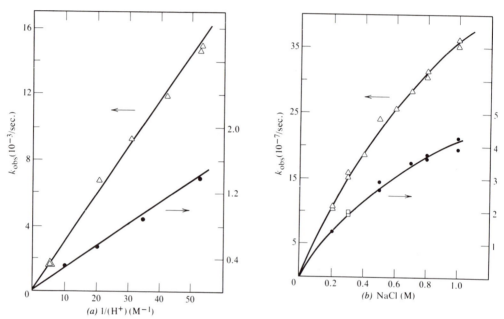

(a) 1/(H$^+$) (M^{-1})

(b) NaCl (M)

Figure 9.20   (a) Dependence of k_{obs} on reciprocal concentration of H$^+$, with I = 1.00 *M*, (complex) = 0.001 *M:* △,65°,(NaCl) = 0.801 *M;* ○,50°,(NaCl) = 0.641 *M.* Arrows indicate the corresponding ordinate. (b) Dependence of k_{obs} on concentration of chloride with T = 40°, I = 1.00 *M*, and (complex) = 0.001 *M:* △,(H$^+$) = 0.0190 *M;* ○,(H$^+$) = 0.199 *M.* Arrows indicate the corresponding ordinate. (Reprinted with permission from M. J. Pavelich, *Inorg. Chem.* 1975, *14*, 982. Copyright 1975, American Chemical Society.)

9.13   Rate constants for the reaction *trans*-[Ptpy$_2$Cl$_2$] + XC$_6$H$_4$SC$_6$H$_4$Y → *trans*-[Ptpy$_2$Cl(XC$_6$H$_4$SC$_6$H$_4$Y)] + Cl$^-$ were measured in methanol at 30°C. From these data calculate, n_{Pt} for the following ligands.

X	Y	$k_2(M^{-1}sec^{-1})$
NH$_2$	NH$_2$	7.25 × 10^{-3}
OH	OH	2.72 × 10^{-3}
CH$_3$	CH$_3$	1.84 × 10^{-3}
NH$_2$	NO$_2$	0.78 × 10^{-3}
NO$_2$	NO$_2$	0.096 × 10^{-3}

(See J. R. Gaylor and C. V. Senoff, *Can. J. Chem.* 1971, *49*, 2390.)

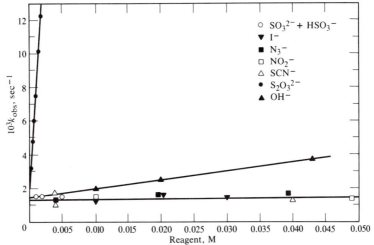

Figure 9.21 Plot of k_{obs} vs. concentration of entering nucleophile for anation of [Pd(Et$_4$dien)Br]$^+$ in water at 25°C. (Reprinted with permission from J. B. Goddard and F. Basolo, *Inorg. Chem.* 1968 *7*, 936. Copyright 1968, American Chemical Society.)

9.14 A plot of k_{obs} versus [X$^-$] is shown in Figure 9.21 for the reactions [Pd(Et$_4$dien)Br]$^+$ + X$^-$ → [Pd(Et$_4$dien)X]$^+$ + Br$^-$. Account for the shape of the plot. In particular, what mechanism can you propose to account for the zero slope when X$^-$ = N$_3^-$, I$^-$, NO$_2^-$, SCN$^-$?

9.15 Base hydrolysis, a well-known reaction of octahedral complexes, is generally about 10^6 times faster than acid hydrolysis. Given the following facts, what can you say about the existence and importance, if any, of a special mechanism for base hydrolysis of square planar complexes?

a. Au(III) complexes, which are also d^8, typically react about 10^4 times as fast as Pt(II) complexes. The following data are representative.

$$[M(dien)Cl]^{n+} + Br^- \xrightarrow{H_2O} [M(dien)Br]^{n+} + Cl^-$$

M	n	$k_1(sec^{-1})$	$k_2(M^{-1}sec^{-1})$
Pt	1	8.0×10^{-5}	3.3×10^{-3}
Au	2	0.5	154

b. The reactivity of [Au(dien)Cl]$^{2+}$ and [Pt(dien)Cl]$^+$ toward anions increases in the order OH$^-$ << Br$^-$ < SCN$^-$ < I$^-$.

c. The replacement of chloride in [Au(Et$_4$dien)Cl]$^+$ by anions Y$^-$ is independent of [Y$^-$]. The rate constant (25°C) increases with pH from $k = 1.9 \times 10^{-6}$ to a maximum of $1.3 \times 10^{-4}sec^{-1}$.

9.16 The following activation parameters have been measured for the reactions

$$[PtL_2ClX] + py \xrightarrow[\text{methanol}]{30°C} [PtL_2Clpy]^+ + X^-$$

Complex	$k_2(M^{-1}sec^{-1})$	$\Delta H_2^{\ddagger}$ (kJ/mole)	$\Delta S_2^{\ddagger}(J/K\ mole)$	$\Delta V^{\ddagger}(cm^3/mole)$
trans-[Ptpy$_2$Cl(NO$_2$)]	7.35×10^{-3}	49.3	−94	−8.8
cis-[Ptpy$_2$Cl(NO$_2$)]	0.150×10^{-3}	55.2	−110	−19.8
trans-[Pt(PEt$_3$)$_2$Cl$_2$]	0.53×10^{-3}	53.9	−100	−13.6

What mechanistic information can be extracted from these values? (See M. Kotowski, D. A. Palmer, and H. Kelm, *Inorg. Chem.* 1979, *18*, 2555.)

9.17 Show that for a redox reaction of the type $A^+ + B \overset{K}{\rightleftharpoons} \text{Intermediate} \overset{k}{\rightarrow} A + B^+$, where B is present in excess, the rate law will be of the form

$$\text{rate} = \frac{a[A^+]_0[B]}{1 + b[B]}$$

whether the intermediate is a precursor or successor complex. (*Hint:* A_0 will be partitioned between free A and the intermediate.)

9.18 Calculate predicted rate constants for the following outer-sphere redox reactions from the information provided. Measured values of k_{12} are given for comparison.

Reaction	$k_{11}(M^{-1}sec^{-1})$	$k_{22}(M^{-1}sec^{-1})$	ΔE° *(volts)*	$k_{12}^{meas}(M^{-1}sec^{-1})$
$Cr^{2+} + Fe^{3+}$	$\leq 2 \times 10^{-5}$	4.0	1.18	2.3×10^3
$[W(CN)_8]^{4-} + Ce(IV)$	$> 4 \times 10^4$	4.4	0.90	$> 10^8$
$[Fe(CN)_6]^{4-} + MnO_4^-$	7.4×10^2	3×10^3	0.20	1.7×10^5
$[Fe(phen)_3]^{2+} + Ce(IV)$	$> 3 \times 10^7$	4.4	0.36	1.4×10^5

9.19 Assign an outer- or inner-sphere mechanism for each of the following:

a. The main product of the reaction between $[Cr(NCS)F]^+$ and Cr^{2+} is CrF^{2+}. (See F. N. Welch and D. E. Pennington, *Inorg. Chem. 1976, 15*, 1515.)

b. When $[VO(edta)]^{2-}$ reacts with $[V(edta)]^{2-}$, a transient red color is observed. (See F. J. Kristine, D.R. Gard, and R. E. Shepherd, *Chem. Comm.* 1976, 944.)

c. The rates of reduction of $[Co(NH_3)_5py]^{3+}$ by $[Fe(CN)_6]^{4-}$ are insensitive to substitution on py. (See A. J. Miralles, R. E. Armstrong, and A. Haim, *J. Am. Chem. Soc.* 1977, *99*, 1416.)

d. The rate of reduction of $[Co(NH_3)_5NCS]^{2+}$ by Ti^{3+} is 36,000 times smaller than the rate of $[Co(NH_3)_5N_3]^{2+}$ reduction. (See J. P. Birk, *Inorg. Chem.* 1975, *14*, 1724.)

e. Activation parameters for some reductions by V^{2+} are

Complex	$\Delta H^\ddagger$ *(kJ/mole)*	$\Delta S^\ddagger$ *(J/K mole)*
$[Co(NH_3)_5F]^{2+}$	46.4	-77.4
$[Co(NH_3)_5Cl]^{2+}$	31.4	-120
$[Co(NH_3)_5Br]^{2+}$	30.1	-115
$[Co(NH_3)_5I]^{2+}$	30.5	-103
$[Co(NH_3)_5N_3]^{2+}$	48.9	-58.5
$[Co(NH_3)_5SO_4]^+$	48.5	-54.8

(See M. R. Hyde, R. S. Taylor and A. G. Sykes, *J. Chem. Soc., Dalton Trans.* 1973, 2730.)

f. A series of Co(III) carboxylato complexes is known to be reduced by Cr^{2+} in an inner sphere mechanism. The rate constants for Eu^{2+} give the log–log plot versus $k_{Cr^{2+}}$ shown in Figure 9.22. (See F.-R. Fan and E. S. Gould, *Inorg. Chem.* 1974, *13*, 2639.)

g. Reduction of *(a)* by Cr^{2+} is much slower than reduction of *(b)*.

$$\left[(en)_2Co \overset{O}{\underset{O}{\diagdown}} \begin{matrix} CH_3 \\ \diagup \\ H \\ \diagdown \\ CH_3 \end{matrix} \right]^{2+} \qquad \left[(en)_2Co \overset{O}{\underset{O}{\diagdown}} \begin{matrix} CH_3 \\ \diagup \\ CHO \\ \diagdown \\ CH_3 \end{matrix} \right]^{2+}$$

(a)   (b)

(See R. J. Balahura and N. A. Lewis, *Can. J. Chem.* 1975, *53*, 1154.)

h. The reduction rates of $[(NH_3)_5Co-O-C(O)R]^{2+}$ (R = Me, Et) by Eu^{2+}, V^{2+}, and Cr^{2+} decrease as the pH decreases. (See J. C. Thomas, J. W. Reed and E. S. Gould, *Inorg. Chem.* 1975, *14*, 1696.)

i. On mixing $\quad [(NH_3)_5\,CoO-\overset{O}{\overset{\|}{C}}-\langle N \rangle]^{2+}$ and Cr^{2+}, a transient ESR signal could be observed.

The *g*-value indicated that the odd electron resided mainly in the aromatic ring. (See H. Spiecker and K. Wieghardt, *Inorg. Chem* 1977, *16*, 1290.)

9.20 The complex $[(bipy)_2ClRu-N\langle\bigcirc\rangle N-Ru(bipy)_2Cl]^{5+}$ displays an intervalence transfer absorption band at $7.69 \times 10^3 \text{ cm}^{-1}$. Estimate the energy barrier to thermal electron transfer. If E_{th} can be very approximately equated to $\Delta G^{\ddagger}$ for thermal electron transfer, estimate the rate constant. The thermal rate constant has been reported to be $<3 \times 10^{10} \text{ sec}^{-1}$.

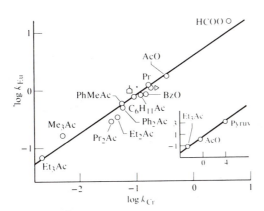

Figure 9.22 Log-log plot comparing specific rates of reductions of carboxylatopentaammine-cobalt(III) complexes $R(NH_3)_5Co^{2+}$ by Eu^{2+} and Cr^{2+} at 25° (Reprinted with permission from F.-R. Fan and E. S. Gould, *Inorg. Chem.* 1974, *13*, 2639. Copyright 1974, American Chemical Society.)

---

k_{obs} is defined by kineticists in the most convenient way and this is not always consistent. For example, the plots of k_{obs} vs [Y] in Figure 9.20*b* and in Figure 9.5 indicate that k_{obs} is defined as a function of [Y]. On the other hand, the discussion of k_{obs} for I_d reactions on page 352 compares values at constant [Y], in effect defining k_{obs} as independent of [Y].

X

Organometallic Chemistry

This chapter deals with the chemistry of compounds containing metal–carbon bonds. The simplest organometallic compounds are CO compounds that exist only for transition metals of low oxidation number. Their stability depends on the existence of π* orbitals that are empty in free CO and are of sufficiently low energy to accept electrons from a metal— so-called π-acid behavior. Hence we first discuss the synthesis, structure, and stoichiometry (which is governed by the EAN rule—see Section 7.3.2) of carbonyl complexes. Reaction mechanisms for carbonyls then are treated.

Next, the bonding of organic groups to metals via their π-systems is introduced through a treatment of olefin complexes, which are typical of π-donor complexes. We then briefly consider the application of the EAN rule to complexes of more complicated π-donor ligands, without dealing explicitly with their chemistry. This allows us to explain and predict the stoichiometry of a wide variety of organometallic complexes.

Experimental evidence bearing on the measurement of π-acidity in ligands is then presented through a consideration of the difficulties of defining accurately the oxidation number in organometallic complexes and a discussion of the IR spectra of carbonyl complexes. π-donor complexes are examined systematically in order of increasing number of C atoms bonded to the metal, from two (alkyne) to eight (cyclooctatetraene). The chemistry of complexes with M—C σ-bonds then is reviewed.

Finally, some important general reactions of organometallic compounds are presented, and the chemistry given is applied to select catalytic processes of commercial importance.

10.1 INTRODUCTION

Organometallic chemistry is the chemistry of compounds containing metal–carbon bonds. The metals involved are those of the main group and the transition series. Very recently, the organometallic chemistry of the lanthanide and actinide metals also has been studied.[1]

Although organometallic compounds have been known since 1827, and Frankland developed the chemistry of alkyl zinc compounds in the mid and late nineteenth century,[2] the field has undergone a renaissance in the last thirty years. Much interest has centered recently on organometallic chemistry, because of its importance to homogeneous catalysis. The transformations in organic molecules that can be effected on laboratory and industrial scales often involve catalysis by metals. Catalysts act by bonding organic species to themselves (at least transiently), providing a low-energy reaction pathway. Synthetic, spectroscopic, and kinetic techniques have been brought to bear on processes that occur with reactants in the same phase (homogeneously). Investigation of the fundamental chemistry of organometallic complexes has uncovered a tremendous number of intriguing reactions and modes of ligand attachment. In many instances the results have been applied to improving the product yield and selectivity of catalyzed reactions. A variety of novel and synthetically useful stoichiometric reactions also has been developed.

10.2 CARBONYL COMPLEXES

10.2.1 CO—the Most Important π-Acid Ligand

Our first topic is the bonding of CO to a metal as an example of how π-acid ligands bond. As the valence bond structure of carbon monoxide indicates, an electron pair is available for donation to the metal.

$$:C \equiv O:$$

The molecular orbital diagram of Figure 4.9 shows that this is the nonbonding σ-electron pair localized on C. Figure 10.1*a* shows the bonding resulting from overlap of the filled C σ orbital and an empty metal σ orbital. CO has a pair of empty, mutually perpendicular π* orbitals that also are indicated in Figure 4.9 and whose shapes appear in Figure 4.10. The bonding between these empty orbitals and a filled metal orbital is depicted in Figure 10.1*b*. The CO π* orbitals help to drain excess negative charge from the metal onto the ligands. Because these ligands act as acceptors of electrons from filled metal orbitals into their empty orbitals of π symmetry, they behave as Lewis acids and are known as π acids. The metal-to-ligand electron donation involved is referred to as back bonding.

The most important energetic component of the bonding is ligand-to-metal σ donation. Since, however, the back-bonding component assumes greater relative importance when the metal has many electrons to dissipate, low oxidation states are stabilized by π-acid ligands.

Both types of bonding reinforce each other: the greater the donation of electrons from the filled ligand σ orbital to the metal, the greater the partial positive charge on the ligand and the more stable the π* orbitals become, making them better acceptors. This kind of mutual rein-

---

[1]See T. J. Marks, *Prog. Inorg. Chem.* 1978, *24*, 51; M. Tsutsui, N. Ely, and R. Dubois, *Acc. Chem. Res.* 1976, *9*, 217; E. C. Baker, G. W. Halstead, and K. N. Raymond, *Struct. Bonding* 1976, *25*, 23; T. J. Marks, *Acc. Chem. Res.* 1976, *9*, 223; T. J. Marks and R. D. Fischer, Eds., *Organometallics of the f-Elements*, Reidel, Boston, 1979.

[2]An excellent presentation of the history of organometallic chemistry is given in J. S. Thayer, *Adv. Organomet. Chem.* 1976, *13*, 1.

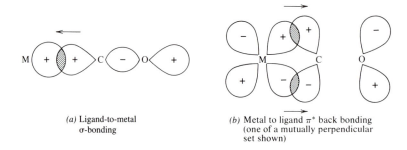

(a) Ligand-to-metal
σ-bonding

(b) Metal to ligand π^* back bonding
(one of a mutually perpendicular
set shown)

Figure 10.1   Orbital overlap in M—CO bonding.

forcement is called *synergism*. Evidence for the importance of synergic bonding lies in the fact that CO forms a very large number of complexes with transition metals in low oxidation states, even though it is an extremely poor Lewis base toward other species.

Bonding of other π-acid ligands displays features similar to CO bonding. Organic ligands that also act as two-electron donors and have empty orbitals of π symmetry include isocyanides CNR and carbenes :C(X)(Y). Several other ligands have similar bonding features but do not contain C, including NO^+ (isoelectronic with CO), phosphines R_3P, arsines R_3As, stibines R_3Sb, bipyridine, and 1,10-phenanthroline. Both CNR and NO^+ have a set of mutually perpendicular π^* orbitals, like CO. Bipy and phen possess a single, delocalized π^* orbital perpendicular to the molecular plane. The Group V ligands have a single, empty d orbital, and carbenes a single, empty C $2p_z$ of proper symmetry for back-donation. Accordingly, complexes containing all these ligands are mentioned in this chapter, even though, strictly speaking, not all of their complexes are organometallic. (The description of back bonding given in this section is the MO equivalent of the VB formulation in Section 7.3.2.)

10.2.2 Binary Carbonyl Complexes

The simplest class of π-acid complexes is the binary carbonyls. Table 10.1 lists some neutral carbonyls, which often are used as reactants in the preparation of other compounds. Most are available commercially.

Synthesis of Carbonyls

There are two ways to synthesize simpler transition-metal carbonyls: direct reaction of a metal with CO, and reductive carbonylation in which a metal salt is allowed to react with CO in the presence of a reducing agent (which also may be CO).

Only Fe and Ni react directly with CO under nonvigorous conditions to give $Fe(CO)_5$ and $Ni(CO)_4$, both of which are very toxic. The extreme ease of thermal decomposition for $Ni(CO)_4$ is the basis of the Mond process for purification of Ni. The impure metal reacts with CO to give gaseous $Ni(CO)_4$, which is separated easily and converted back to Ni. The CO can be recycled.

A synthetic method for making carbonyls from metal salts involves reduction of an acetate

Table 10.1 Binary metal carbonyls

V(CO)₆	*Cr(CO)₆*	*Mn₂(CO)₁₀*	*Fe(CO)₅*	*Co₂(CO)₈*	*Ni(CO)₄*
Blue-green, $\mathbf{O}_h$, dec. 60–70°C, air-sensitive, pyrophoric	Colorless, $\mathbf{O}_h$, m.p. 154–5°C, air-stable	Golden yellow, $\mathbf{D}_{4d}$, m.p. 153–154°C, heat- and air-sensitive	Light yellow, $\mathbf{D}_{3h}$, m.p. −20°C, b.p. 103°C	Dark orange, dec. 51–52°C, $\mathbf{D}_{3d}$ (solution), $\mathbf{C}_{2v}$ (solid), heat- and air-sensitive	Colorless, $\mathbf{T}_d$, m.p. −25°C, toxic, dec. to Ni + 4CO w/heat
	Mo(CO)₆	*Tc₂(CO)₁₀*	*Ru(CO)₅*	*Rh₂(CO)₈*	
	Colorless, $\mathbf{O}_h$, dec. 150°C, air-stable	Colorless, $\mathbf{D}_{4d}$	Colorless, $\mathbf{D}_{3h}$ (solution), m.p. −25°C, difficult to purify	Stable only at low T and high CO pressure	
	W(CO)₆	*Re₂(CO)₁₀*	*Os(CO)₅*	*Ir₂(CO)₈*	
	Colorless, $\mathbf{O}_h$, dec. (?) 150°C	Colorless, $\mathbf{D}_{4d}$, m.p. 177°C, air-stable	Colorless, $\mathbf{D}_{3h}$ (solution), difficult to purify	Stable only at low T and high CO pressure	
			Fe₂(CO)₉	*Co₄(CO)₁₂*	
			Yellow orange, $\mathbf{D}_{3h}$, dec. 100°C, more reactive than Fe(CO)₅	Black, $\mathbf{C}_{3v}$, air-sensitive	
			Ru₂(CO)₉	*Rh₄(CO)₁₂*	
				Red, $\mathbf{C}_{3v}$, dec. 150°C, air-stable	
			Os₂(CO)₉	*Ir₄(CO)₁₂*	
				Yellow, $\mathbf{T}_d$, dec. 230°C, air-stable	
			Fe₃(CO)₁₂		
			Green-black, $\mathbf{C}_{2v}$ (solid), sol'n structure different, dec. 140°C, air-sensitive		
			Ru₃(CO)₁₂		
			Orange, $\mathbf{D}_{3h}$, dec. 154 6°C, air stable		
			Os₃(CO)₁₂		
			Yellow, $\mathbf{D}_{3h}$, m.p. 224°C, air-stable		

hydrate by H_2 in the presence of CO and sufficient acetic anhydride to convert the water of hydration to acetic acid.

$$2Co(H_2O)_4(CH_3CO_2)_2 + 8(CH_3CO)_2O + 8CO + 2H_2 \qquad (10.1)$$

$$\rightarrow Co_2(CO)_8 + 20CH_3CO_2H$$

With an anhydrous salt, Al is often the reducing agent.

$$CrCl_3 + Al + 6CO \xrightarrow[C_6H_6]{AlCl_3} Cr(CO)_6 + AlCl_3 \tag{10.2}$$

CO may also function as a reducing agent.

$$Re_2O_7 + 17CO \rightarrow Re_2(CO)_{10} + 7CO_2 \tag{10.3}$$

Higher carbonyls can be synthesized by thermolysis of the lower ones. The reaction often happens at ambient temperature, rendering the lower carbonyls unstable—as in the case of $Os(CO)_5$.

$$3Os(CO)_5 \rightarrow Os_3(CO)_{12} + 3CO \tag{10.4}$$

Such reactions occur because metal—CO bond cleavage produces unsaturated fragments that combine into higher nuclearity species. Photochemical bond cleavage also is employed, as in the synthesis of diiron enneacarbonyl.

$$2Fe(CO)_5 \xrightarrow{h\nu} Fe_2(CO)_9 + CO \tag{10.5}$$

Molecular and Electronic Structures of Carbonyls—The EAN Rule

Complexes of the transition metals with π-acid ligands, as well as their organometallic complexes, generally obey the EAN rule (see Section 7.3.2). Thus their stoichiometries and molecular structures usually can be predicted and understood as arising from a tendency to surround each metal with a full complement of 18 electrons. This is just another way of saying that the nd, $(n + 1)s$ and $(n + 1)p$ orbitals all are valence orbitals in the transition series, and that all their bonding capacity is used when the 18-electron configuration is reached.

Figure 10.2 shows the changes in energy for the $3d$, $4s$, and $4p$ orbitals of the first-transition-series atoms. As you can see, all the orbitals become more stable as the effective atomic number increases across the series. Around the central part (V—Co), all the orbitals are of fairly similar energy, and so all are available for bonding and the EAN rule is expected to apply. As the end of the series is approached, the $3d$ and $4s$ orbitals drop in energy faster than the $4p$, until at Zn the $3d$'s are part of the atomic core and too stable to participate in bonding. Hence, we expect that Zn (and the representative metals beyond) should not obey the EAN rule. For Cu the prediction is similar. For Ni(0) it makes no difference whether or not we consider the d's to be part of the core, since they are filled completely whereas the $4s$ and $4p$ orbitals are still available for bonding, leading to a prediction of 18 electrons around Ni(0). As some orbitals disappear into the core toward the end of the transition series, 16- and even 14-electron species become more stable. Early in the series [before V(0)] the d orbitals are not all energetically available for bonding, and hence the EAN rule is not expected to hold. For the second and third transition series, similar trends are anticipated. In short, we expect the EAN rule to be valid toward the middle of the series for metals with low or negative oxidation numbers. Increasing the oxidation number changes the slopes of the binding energy versus atomic number curves and increases their separation. So at sufficiently positive metal oxidation number ($\geq$ II), the EAN rule no longer obtains. Within its region of applicability, however, the EAN rule is a rather reliable guide to molecular structure and stoichiometry.

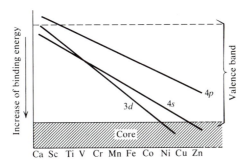

Figure 10.2 The change in energy of the $3d$, $4s$ and $4p$ orbitals of the first transition series. (After C. S. G. Phillips and R. J. P. Williams in *Inorganic Chemistry*, Vol. II, Oxford, Oxford, 1966.)

The EAN rule can be applied to the stoichiometry and structures of the binary metal carbonyls. In counting electrons, follow these simple (but arbitrary) rules:

1. Arbitrarily assign the metal a positive oxidation number equal to the sum of the negative charges on the ligands and the total charge on the complex. Count the number of electrons it contributes.

2. Each linearly bonded CO contributes two electrons. (Assignments for other ligands are made later.)

3. Each metal–metal bond contributes one electron.

Example The electrons in $Mo(CO)_6$ are counted as follows: CO has no charge and the complex is uncharged; hence the oxidation state for Mo is $0 + 0 = 0$.

$$Mo(0) + 6 \times CO$$
$$6e + 6 \times 2e = 18e$$

This result is consistent with the known structure of $Mo(CO)_6$, in which Mo is surrounded octahedrally by the six CO ligands. The rules given above only enable us to count electrons; they provide no information about the actual density distribution.

Example Give the electron count for $Re_2(CO)_{10}$, which has the $\mathbf{D}_{4d}$ structure with staggered M—C—O axes as shown in Figure 10.3. For each Re, we have

$$Re(0) + 5 \times CO + 1 \text{ Re—Re bond}$$
$$7e \quad + 5 \times 2e \quad + 1e = 18e$$

The odd electron Re atoms can only achieve the 18-electron configuration if two $Re(CO)_5$ fragments dimerize to form a Re—Re bond. If no Re—Re bond existed, the molecule would be paramagnetic, which is not found.

The x-ray structure of $Fe_3(CO)_{12}$ (see Figure 10.3) reveals a feature that we have not encountered before—CO's that bridge two metal atoms. In molecular formulas, these doubly-bridging carbonyls are written as μ-CO or μ_2-CO. Hence, a formula for triirondodecacarbonyl that conveys structural information is $Fe_3(\mu_2\text{-CO})_2(CO)_{10}$.[3]

The doubly bridging CO unit is considered to resemble (at least formally) the ketonic

$$\text{C}=\text{O}.$$ C is considered to be sp^2-hybridized and to contribute one hybrid orbital and one

electron to each of the two metals.

[3]C. H. Wei and L. F. Dahl, *J. Am. Chem. Soc.* 1969, *91*, 1351.

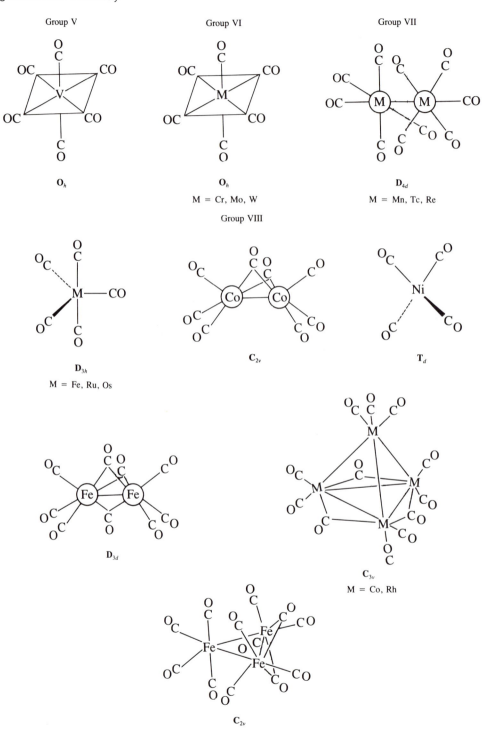

Figure 10.3 Solid-state structures of some neutral binary carbonyls.

Group VIII *(Continued)*

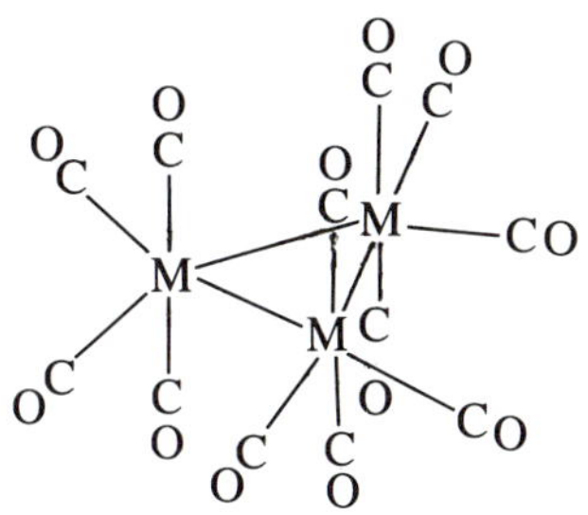

$\mathbf{D}_{3h}$

M = Ru, Os

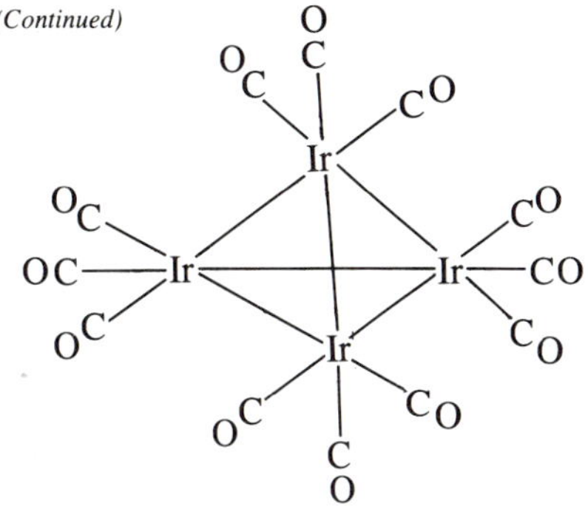

$\mathbf{T}_d$

Figure 10.3 *(Continued)*

Example Electrons in $Fe_3(CO)_{12}$ are counted as follows.

Two bridged Fe

$$Fe(0) + 3 \times CO + 2 \times \mu_2\text{-}CO + 2\ Fe\text{—}Fe \text{ bonds}$$
$$8e \quad + 3 \times 2e + 2 \times 1e \quad + 2 \times 1e = 18e$$

Unique Fe

$$Fe(0) + 4 \times CO + 2\ Fe\text{—}Fe \text{ bonds}$$
$$8e \quad + 4 \times 2e + 2 \times 1e = 18e$$

Notice that we cannot *predict* via the EAN rule the presence of the μ_2-CO's. Once this information is known from structural studies, the structure can be rationalized on the basis of the rule. In fact, a molecular structure having no μ-CO's and containing three $Fe(CO)_4$ units would also satisfy the EAN rule. Both $Ru_3(CO)_{12}$ and $Os_3(CO)_{12}$ adopt such a structure containing only nonbridging (called *terminal*) CO's. Presumably, this occurs because the distances between the larger atoms are too great for effective bridge-bond formation. (See Problem 10.1.) A similar situation exists in the $M_4(CO)_{12}$ (M = Co, Rh, and Ir), where Co and Rh give a $\mathbf{C}_{3v}$ structure whereas Ir forms a $\mathbf{T}_d$ one (Figure 10.3).

The EAN rule is not applicable universally. The known exceptions include $V(CO)_6$ (a 17-*e* species) and $Rh_6(CO)_{16}$ (discussed in Chapter 15). Sometimes, different structures consistent with the EAN rule occur in different environments. For example, $Co_2(CO)_8$ has a structure with two μ_2-CO's in the solid state. However, no fewer than three isomers are known to exist in solution: the bridged form, a form containing all terminal CO's, and a third form

of unknown structure.[4] The structures M——M and M—M are equivalent in electron count,

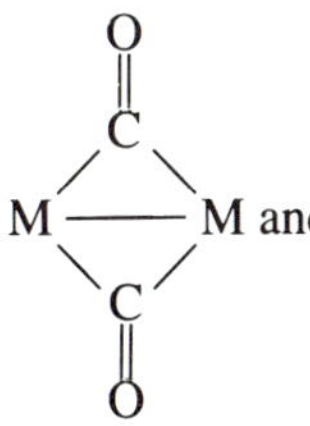

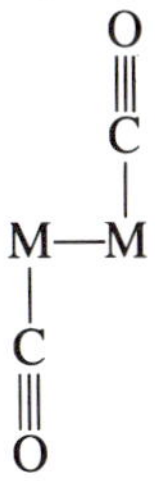

and bridge-terminal tautomerism occurs frequently.

[4]G. Bor and K. Noack, *J. Organomet. Chem.* 1974, *64*, 367; S. Onaka and D. F. Shriver, *Inorg. Chem.* 1976, *15*, 915.

10.2.3 Substituted Carbonyls and Carbonylate Anions

The EAN rule allows us to predict that stable species can result when CO is replaced by another two-electron donor ligand or when electrons are added or removed to form charged species. Thus, for example, replacing CO by Br^- in $Mo(CO)_6$ gives $[Mo(CO)_5Br]^-$, which also obeys the rule of 18 for Mo and the octet rule for Br. Likewise, replacing Mo by iso-electronic Mn^+ leads to $[Mn(CO)_6]^+$, which obeys the EAN rule. The anion $[Fe(CO)_4]^{2-}$ also is a known 18-electron species. Table 10.2 gives electron counts for selected ligands.

Substituted Carbonyls

Neutral Lewis bases or anionic two-electron donors can replace carbonyl ligands. The three most common preparative routes are direct replacement (Equations 10.6, 10.7, 10.8), oxidation by halogens (Equations 10.11 and 10.12), and reactions between metal halides and CO (Equation 10.14).

$$Cr(CO)_6 + CNPh \xrightarrow{-CO} Cr(CO)_5(CNPh) \qquad (10.6)$$

$$Ir_4(CO)_{12} + 2PPh_3 \xrightarrow{-CO} Ir_4(CO)_{10}(PPh_3)_2 \qquad (10.7)$$

$$Mo(CO)_6 + Br^- \xrightarrow{-CO} Mo(CO)_5Br^- \qquad (10.8)$$

Photochemical activation sometimes is employed to aid in breaking the M—CO bond, but many reactions are carried out thermally. Substitution usually stops after two or three carbonyls have been replaced for all but the best π-acid ligands, because the remaining CO ligands become saturated in their ability to withdraw electron density from the metal. With good π-acids, however, all the CO's can be replaced, giving compounds such as $Ni(PF_3)_4$, $Cr(CNPh)_6$, etc.

Since NO is a three-electron donor, $2NO = 3CO$ in electron count. Hence direct displacement with NO gives products of different stoichiometry.

$$Fe(CO)_5 + 2NO \xrightarrow{-CO} Fe(CO)_2(NO)_2 \qquad (10.9)$$

$$Co(CO)_4^- + NO_2^- \xrightarrow{H^+} Co(CO)_3(NO) \qquad (10.10)$$

Carbonyl halides result from the oxidizing action of halogens on carbonyls.

$$Mn_2(CO)_{10} + Cl_2 \rightarrow 2Mn(CO)_5Cl \qquad (10.11)$$

Table 10.2   Electron counts for some ligands

Ligand	Number of Valence Electrons Contributed
$H^-, F^-, Cl^-, Br^-, I^-, CN^-, NCS^-, CO, CNR, NO^+$ $PR_3, P(OR)_3, AsR_3, NR_3, SbR_3, SR_2,$ $:C(X)(Y), R^-, C^-(O)R, Ar^-$	2
NO	3

$$Fe(CO)_5 + Br_2 \xrightarrow{-CO} Fe(CO)_4Br_2 \tag{10.12}$$

A formally related reaction is

$$Co_2(CO)_8 + H_2 \rightarrow 2HCo(CO)_4 \tag{10.13}$$

It seems strange to think of an "oxidation" by H_2! However, it is an arbitrary convention to consider all M—H compounds as containing hydridic H. In fact, the chemical behavior of $HCo(CO)_4$ is consistent with the presence of protonated $[Co(CO)_4]^-$ containing $Co(-I)$. (See Section 10.4.1).

Typical of the third class of reactions is

$$2RhCl_3 + 4CO \xrightarrow[94-95°]{} [Rh(CO)_2Cl]_2 + 2Cl_2 \tag{10.14}$$

This carbonyl has the halogen-bridged structure

Bridging in carbonyl halides is usually through a noncarbonyl ligand. Some monomeric compounds form bridged dimers on heating to induce loss of CO.

$$2Mn(CO)_5Br \underset{+CO,\ pressure}{\overset{\Delta,\ -CO}{\rightleftharpoons}} Mn_2(\mu\text{-Br})_2(CO)_8 \tag{10.15}$$

The μ-Cl species is the direct product of the reaction of $Re_2(CO)_{10}$ with Cl_2, analogous to Equation 10.11. Bridged species often can be cleaved by Lewis bases—for example,

$$[Rh(CO)_2Cl]_2 + 4PPh_3 \xrightarrow{-CO} 2\ trans\text{-}[Rh(PPh_3)_2(CO)Cl] \tag{10.16}$$

Although metal carbonyls undergo many different kinds of reactions, substitutions are among the most important. In later sections we shall encounter examples in which carbonyls are substituted by organic groups.

Carbonylate Anions

We know of a very large number of carbonyl anions but very few cations. This is not very surprising in view of the considerable π-acidity of CO, which stabilizes low oxidation states.

Three common methods are utilized for preparing carbonyl anions, which often are made *in situ* and used for reaction without isolation.

1. Reaction of metal carbonyl derivatives with base. Examples include

$$13Mn_2(CO)_{10} + 40OH^- \rightarrow 24[Mn(CO)_5]^- + 2Mn^{2+} + 10CO_3^{2-} + 20H_2O \tag{10.17}$$

$$Fe_2(CO)_9 + 4OH^- \rightarrow [Fe_2(CO)_8]^{2-} + 2H_2O + CO_3^{2-} \tag{10.18}$$

$$Co_2(CO)_8 + 5RNC \rightarrow [Co^I(CNR)_5]^+ + [Co^{-I}(CO)_4]^- + 4CO \tag{10.19}$$

Reactions 10.17 and 10.19 amount to base-induced disproportionations. Although convenient, this method often involves conversion of part of the metal to a cationic product.

2. Reaction of metal carbonyls with reducing agents. Alkali metal amalgams, hydride reagents, and Na/K alloy in basic solvents, including liquid ammonia, have been used. Examples include

$$Fe(CO)_5 + 2Na \xrightarrow[NH_{3(l)}]{} Na_2[Fe(CO)_4] \tag{10.20}$$

$$Cr(CO)_6 \xrightarrow[\text{boiling THF}]{\text{Na/Hg}} Na_2[Cr(CO)_5] \tag{10.21}$$

$$Co_2(CO)_8 + 2Li[HB(C_2H_5)_3] \xrightarrow[\text{THF}]{25°} 2Li[Co(CO)_4] + 2B(C_2H_5)_3 + H_2 \tag{10.22}$$

$$Mn_2(CO)_{10} + 2KH \xrightarrow[\text{THF}]{25°} 2K[Mn(CO)_5] + H_2 \tag{10.23}$$

$$Cr(CO)_6 \xrightarrow[NH_{3(l)}]{NaBH_4} Na_2[Cr_2(CO)_{10}] \tag{10.24}$$

3. Reactions of metal carbonyls with anions. With carbonylate anions, heteronuclear carbonyls result; other reactions are substitutions. An example is

$$[Mn(CO)_5]^- + Re(CO)_5Br \rightarrow MnRe(CO)_{10} + Br^- \tag{10.25}$$

Also, Reaction 10.8 and

$$Re(CO)_5Cl + 2KCN \xrightarrow[\text{methanol}]{100°} K[Re(CO)_4(CN)_2] \tag{10.26}$$

Two especially important reactions of carbonylate anions are those with alkyl or acyl halides to give organic derivatives (Equations 10.27 and 10.28) and protonation to afford metal hydrides (Equations 10.29 and 10.30).

$$[Re(CO)_5]^- + CH_3I \rightarrow CH_3Re(CO)_5 + I^- \tag{10.27}$$

$$[Mn(CO)_5]^- + CH_3\overset{O}{\overset{\|}{C}}Cl \rightarrow CH_3\overset{O}{\overset{\|}{C}}Mn(CO)_5 + Cl^- \tag{10.28}$$

The above reactions are analogous to S_N2 displacements on organic halides.

$$[Mn(CO)_5]^- + H^+ \rightarrow HMn(CO)_5 \tag{10.29}$$

$$[Fe(CO)_4]^{2-} + H^+ \rightarrow [HFe(CO)_4]^- \xrightarrow{H^+} H_2Fe(CO)_4 \tag{10.30}$$

The hydride ligands in the above neutral products occupy positions in the coordination sphere, giving octahedral species.[5] Some known hydrides involve bridging H: for example, protonation of $[Cr_2(CO)_{10}]^{2-}$ gives $[Cr(CO)_5—H—Cr(CO)_5]^-$, in which the Cr—H—Cr bond (which is slightly nonlinear) may be regarded as a three-center/two-electron bond or as a protonated metal–metal bond.

[5]For a discussion of structures of transition-metal hydride complexes, see R. Bau, R. G. Teller, S. W. Kirtley, and T. F. Koetzle, *Acc. Chem. Res.* 1979, *12*, 176.

10.2.4 Reaction Mechanisms for Substitution in Carbonyl Complexes

Mechanistic studies on organometallic compounds are considerably rarer than those on coordination compounds (see Chapter 9). However, some studies have been made, including work on substitution reactions of the octahedral d^6 carbonyls with Lewis bases.[6] For the reaction

$$LM(CO)_5 + L' \rightarrow LL'M(CO)_4 + CO \tag{10.31}$$

(where L may also be CO), a two-term rate law is observed.

$$\text{rate} = k_1[LM(CO)_5] + k_2[LM(CO)_5][L'] \tag{10.32}$$

As usual, such a rate law reflects two parallel paths for substitution. For some reactions the second-order path is undetectably slow ($k_2 \ll k_1$). It is generally agreed that the first-order path is D (Section 9.3) in character and involves dissociation of CO as the rate-determining step. For substitution reactions of Group VIA hexacarbonyls, the relative values of k_1 are Mo:Cr:W, 100:1:0.01—paralleling M—C bond strengths. The identity of L also is important in determining the rate and site for CO dissociation. A study of ^{13}CO exchange with $Mn(CO)_5Br$[7] showed that equatorial carbonyls exchanged much faster than axial ones. This is understandable on the basis of the ground-state properties of the molecule. Since the Br^- ligand has filled π orbitals, both the d_{xz} and d_{yz} orbitals on Mn that point toward Br^- and the *trans* carbonyl will be able to act as π-electron donors only toward this CO ligand, thus strengthening the Mn—C bond. The d_{xz} orbital is shared between a pair of equatorial CO's that are *trans* to one another (likewise for d_{yz}). Hence, the bond-strengthening to the equatorial CO's located *cis* to Br^- is smaller, and these carbonyls might be expected to be more labile.

Ground-state effects cannot account, however, for the observation that replacement of CO by any L which is a worse π acid increases the first-order reaction rate. The argument given above might lead you to expect that *cis*-carbonyl ligands in $Mn(CO)_5Br$ should be less labile to substitution than those in $[Mn(CO)_6]^+$, which has no π-donor ligands. Instead, the bromide is many orders of magnitude more reactive, much beyond the difference accounted for by the smaller charge. As another example, $Cr(CO)_5(PPh_3)$ is about 100 times as reactive to substitution as $Cr(CO)_6$, whereas $[Cr(CO)_5Br]^-$ is 10^7 as reactive. Obviously, the identity of L is quite important in influencing rates of CO dissociation. A good deal of evidence indicates that L exerts its influence by stabilizing a square pyramid resulting from dissociation of a *cis* CO.

$$\tag{10.33}$$

Ligands L most likely to stabilize the 16-e transition state when in the equatorial position are

[6]G. R. Dobson, *Acc. Chem. Res.* 1976, *9*, 300.

[7]J. D. Atwood and T. L. Brown, *J. Am. Chem. Soc.* 1975, *97*, 3380.

those that are good π donors.[8] Hence for L $=$ I$^-$, py, Br$^-$, Cl$^-$, NO$_3^-$, etc., rate enhancement of *cis* ligand dissociation is anticipated, because of transition state stabilization. The final products are usually (but not always) the *cis* complexes indicated in Equation 10.33, since the intermediate may last long enough to rearrange before reacting with L$'$. This is why CO exchange studies are so important in establishing the pattern of *cis* CO activation: the principle of microscopic reversibility requires entry of ^{13}CO into the position vacated by the leaving carbonyl, whereas no such requirement exists for L$' \neq$ CO. Even when *cis*-M(CO)$_4$LL$'$ is the kinetic product, rearrangement often occurs to the thermodynamically favored *trans* product.

The second-order term in Equation 10.32 may represent the operation of an *a* mechanism. Two possibilities arise for sites of attack: the metal itself or one of the carbonyl ligands. The ligands that are good π-acceptors probably react through attack on the metal, presumably via an I_a process. The electrophilic properties of the incoming ligand could serve to drain off excess charge and thus to stabilize the transition state. Data are rather few, however—those in Table 10.3 imply a less spectacular difference in rates as a function of incoming ligand than might be expected for an I_a mechanism. Hence it is quite likely that the second-order pathway may be an I_d mechanism with considerable bond-making in the transition state.

For substitution by basic reactants, it recently has been shown that attack on a carbonyl ligand is responsible for a second-order term in the rate law.[9] The CO ligand is polarized—in the sense shown below—when coordinated to cationic or neutral metals (see Section 10.2.1)

$$M^- \leftarrow C^+ \equiv O$$

Sufficiently nucleophilic reactants, then, can attack at the positive C. Such reactions are known for carbonyls with several kinds of nucleophiles.

Table 10.3   Rate data for reactions of carbonyl complexes with Lewis bases[a]

Complex	*Lewis Base*	*T(°C)*	*Solvent*	$k_1 \times 10^4 (sec^{-1})$	$k_2 \times 10^4 (M^{-1}sec^{-1})$
Cr(CO)$_6$	PPh$_3$	130.7	decalin	1.38	0.450
W(CO)$_6$	PPh$_3$	165.7	decalin	1.15	0.888
Mo(CO)$_6$	PPh$_3$	112.0	decalin	2.13	1.77
Mo(CO)$_6$	(*n*-Bu)$_3$P	112.0	decalin	2.13	20.5
Mo(CO)$_6$	C$_6$H$_5$CH$_2$NH$_2$	112.0	decalin	2.44	4.4
Mo(CO)$_6$	Br$^-$	55.0	C$_6$H$_5$Cl	—	32.4
Mo(CO)$_5$py	P(OCH$_2$)$_3$CCH$_3$	47.9	ClCH$_2$CH$_2$Cl	9.3	~4.8
[Mo(CO)$_5$Br]$^-$	PPh$_3$	19.6	diglyme	1.34	4.1
[Mo(CO)$_5$I]$^-$	PPh$_3$	29.8	diglyme	1.13	1.8
[Mo(CO)$_5$I]$^-$	P(*p*-C$_6$H$_4$F)$_3$	29.8	diglyme	1.13	2.8
[Mo(CO)$_5$I]$^-$	P(*p*-C$_6$H$_4$Cl)$_3$	29.8	diglyme	1.13	3.6

[a]From the work of Dobson, Angelici, and others.

[8]J. D. Atwood and T. L. Brown, *J. Am. Chem. Soc.* 1976, *98*, 3160; D. L. Lichtenberger and T. L. Brown, *loc. cit.* 1978, *100*, 366.

[9]T. L. Brown and P. A. Bellus, *Inorg. Chem.* 1978, *17*, 3727; P. A. Bellus and T. L. Brown, *J. Am. Chem. Soc.* 1980, *102*, 6020.

$$[Mn(CO)_6]^+ + LiCH_3 \xrightarrow{THF} Mn(CO)_5\overset{\overset{\displaystyle O}{\|}}{C}CH_3 \tag{10.34}$$

$$[Mn(PPh_3)(CO)_5]^+ + 2CH_3NH_2 \xrightarrow{THF} \text{\textit{cis}-}Mn(PPh_3)(CO)_4\overset{\overset{\displaystyle O}{\|}}{C}NHCH_3 + NH_3CH_3^+ \tag{10.35}$$

The products are substituted more readily than the parent compounds, because of the activation of dissociation of *cis* CO's by the reduced π-acidity of the ligands discussed above.

By employing a base that could not undergo deprotonation leading to a stable carbamoyl product, Bellus and Brown were able to demonstrate that the second-order term in the rate law for reaction of $[Mn(CO)_5(CH_3CN)]^+$ with py represents ligand attack on CO. The cation is known to be stable in CH_3NO_2 for a week or more and not to undergo exchange with ^{13}CO in $C_2H_5NO_2$ over at least 72 hr at 30°C. On addition of excess py, however, the following reaction occurs rapidly.

$$[Mn(CO)_5(CH_3CN)]^+ + CH_3NO_2 + 3\ py \rightarrow \textit{fac-}[Mn(CO)_3(py)_3]^+$$

$$+\ CH_3CN + CO + CO_2 + CH_2NOH \tag{10.36}$$

The mechanism proposed is shown in Figure 10.4. The rate-determining step is dissociation of the first CO ligand from the neutral intermediate. The dissociations of the other two carbonyls are rapid, because of *cis* activation by the noncarbonyl ligands. This pathway represents the equivalent of base hydrolysis for carbonyl complexes. By way of contrast, the reaction of $[Mn(CO)_5(CH_3CN)]^+$ with nucleophilic (but weakly basic) PPh_3 is 10^2 slower than with py (in CH_3CN).

A free-radical mechanism is likely in the base-induced disproportionations of metal carbonyls to produce carbonylate anions (Equations 10.17 and 10.19). The mechanism requires the d^7 radicals $\cdot Co(CO)_4$ to be labile to substitution (Y is a Lewis base):

$$(CO)_4Co\!-\!Co(CO)_4 + \overset{\cdot}{Y}Co(CO)_3 \rightarrow YCo(CO)_3^+ + \overset{\cdot}{C}o_2(CO)_8^-$$

$$\overset{\cdot}{C}o_2(CO)_8^- \rightarrow Co(CO)_4^- + \overset{\cdot}{C}o(CO)_4$$

$$\overset{\cdot}{C}o(CO)_4 + Y \xrightarrow{\text{fast}} \overset{\cdot}{C}o(CO)_3Y + CO \tag{10.37}$$

$$Y\overset{\cdot}{C}o(CO)_3 + Y \xrightarrow{\text{fast}} Y_2\overset{\cdot}{C}o(CO)_3, \text{ etc.}$$

An interesting feature of this mechanism is the transfer of one electron from $\overset{\cdot}{Y}Co(CO)_3$ to give the radical anion $\overset{\cdot}{C}o_2(CO)_8^-$, which gives the anion $Co(CO)_4^-$ and the propagating radical $\overset{\cdot}{C}o(CO)_4$.

Studies have established several patterns of reactivity for organometallic compounds: (1) The production by ligand dissociation of intermediates with reduced coordination number; such coordinately unsaturated species are encountered more often in organometallic chemistry than with more conventional coordination compounds. (2) Activation of *cis* ligands toward dissociation when CO's are replaced by non-π-acid ligands. The group that dissociates is the one with the weakest bond to the metal. (3) Nucleophilic attack on carbonyl (and other) ligands

$$py + CH_3NO_2 \rightleftharpoons {}^-CH_2NO_2 + pyH^+$$

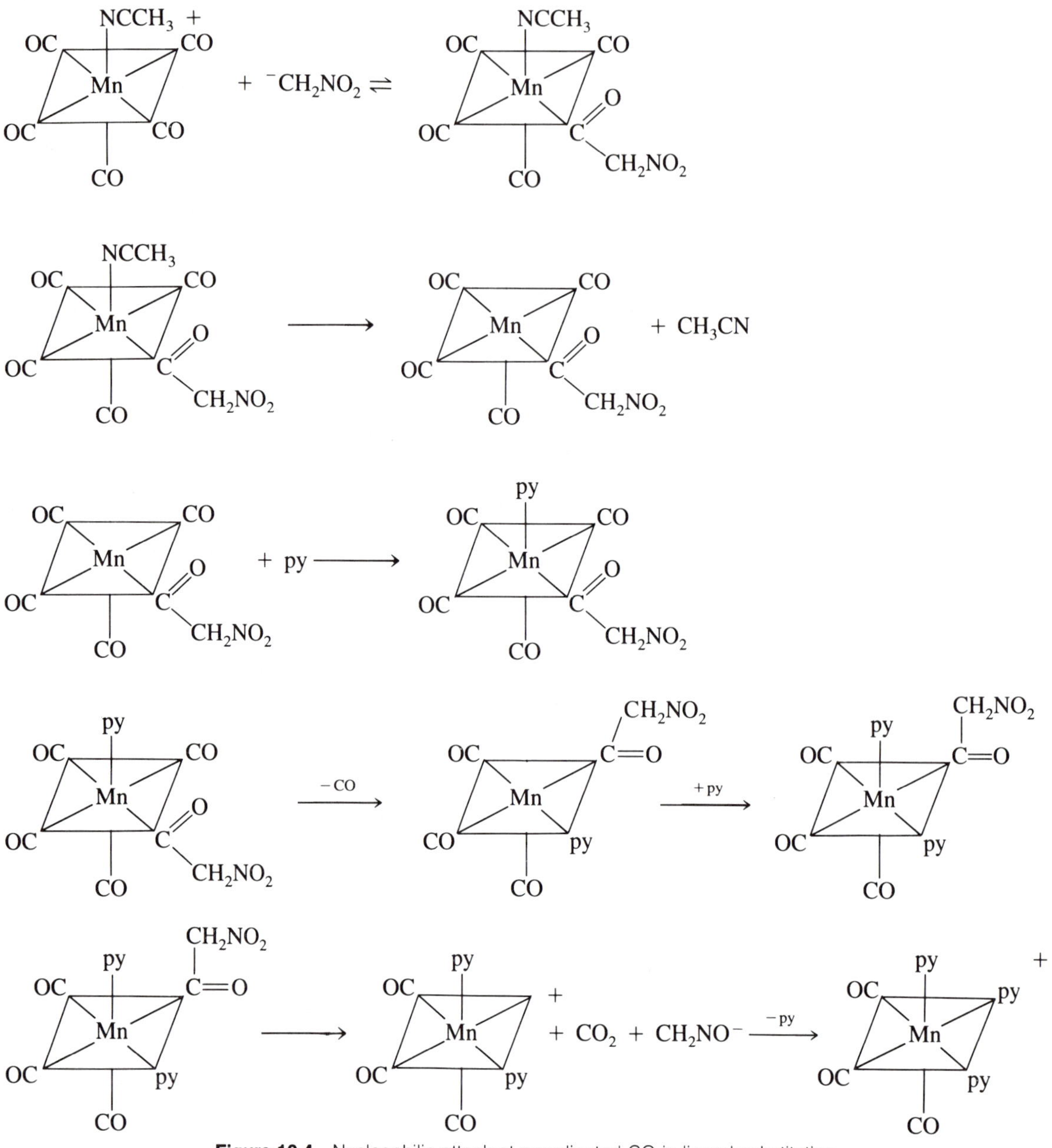

Figure 10.4 Nucleophilic attack at coordinated CO in ligand substitution.

polarized by coordination to give stable compounds or reactive intermediates. (4) Free-radical mechanisms, sometimes involving one-electron transfer processes.[10]

10.3 BONDING OF ORGANIC LIGANDS TO METALS

10.3.1 Olefin Complexes—The Paradigmatic π-Donor Complexes

The first organometallic compound was prepared in 1827 by Zeise, who heated a $PtCl_2$ + $PtCl_4$ mixture in ethanol, evaporated the solvent, and treated the residue with aqueous KCl. The product was Zeise's salt $K[PtCl_3(C_2H_4)]$, whose nature was not understood until x-ray studies in the early 1950's showed that the ethylene molecule was coordinated approximately perpendicular to the molecular plane of the anion, with the hydrogens bent away from the metal. The arrangement is symmetrical with respect to the $PtCl_3$ plane.

A molecular orbital approach to the bonding in this complex, first developed by Dewar, Chatt, and Duncanson, serves as the paradigm for bonding in π complexes. As you can see by drawing its valence bond structure, ethylene (C_2H_4), unlike CO, has no lone pairs for ligand-to-metal charge donation. However, when the C_2H_4 molecule is oriented as shown above, the filled π and empty π* orbitals are situated properly for overlap with metal orbitals (see Figure 10.5). The ethylene–metal bond has a component of σ symmetry in which electrons are donated

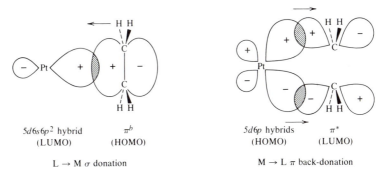

Figure 10.5 The Chatt-Dewar-Duncanson picture of bonding in a Pt olefin complex.

[10]For an extensive treatment of free-radical processes, see J. K. Kochi, *Organometallic Mechanisms and Catalysis*, Academic Press, New York, 1978. See also P. R. Jones, *Adv. Organomet. Chem.* 1977, *15*, 273.

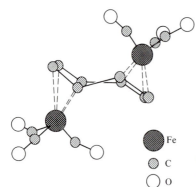

Figure 10.6 Structure of $(\eta^4,\eta'^4\text{-cot})Fe_2(CO)_6$. (Reproduced with permission from B. Dickens and W. N. Lipscomb, *J. Am. Chem. Soc.* 1961, *83*, 489. Copyright 1961, American Chemical Society.)

from the filled ethylene π orbital into an empty metal $5d6s6p^2$ hybrid orbital. In the component of π symmetry, electrons from the platinum are shared with the olefin through the overlap of a filled hybridized $5d6p$ orbital (a hybridized orbital is used to provide better overlap than occurs with d_{xz} or d_{yz} orbitals) with the antibonding π^* orbital of the olefin. The π^* orbital is antibonding with respect to the bonding in the olefin molecule. Its use should weaken the C—C bond, but its overlap with a Pt orbital would strengthen the Pt—olefin bond.

The two components of the olefin–metal bond reinforce one another synergically. Olefins, like CO, behave as two-electron donors. The ''difference'' between bonding in organometallic π-complexes and that in carbonyl complexes is just that the ligand orbital involved in the σ component of the metal–ligand bond is one of π symmetry in the ligand itself.

Polyolefins with isolated double bonds also form complexes. One such ligand is 1,5-cyclooctadiene (1,5-cod), which forms compounds in which it functions as a 2×2-electron donor. Another is 1,3,5,7-cyclooctatetraene (cot), which may function as a 2×2-electron donor to one or two different metals (see Figure 10.6) or a 3×2-electron donor to a single metal (also see Table 10.4). Metal–olefin bonds often are represented by arrows, since the ligands are formally π-donors (see Problem 10.2).

Coordinated olefins (like coordinated CO) are activated toward attack by strong nucleophiles, because electron density is withdrawn by donation to the metal—especially if the complex is positively charged and/or the metal has a high positive oxidation state.

$$\begin{array}{c} \text{Cl} \quad\quad \text{PPh}_3 \\ \diagdown \quad \diagup \\ \text{Pt} \diagup^{\text{CH}_2} \\ \diagup \quad \diagdown\!\!\!| \\ \text{Cl} \quad\quad \text{CH}_2 \end{array} + \text{Et}_2\ddot{\text{N}}\text{H} \rightarrow \begin{array}{c} \text{Cl} \quad\quad \text{PPh}_3 \\ \diagdown \quad \diagup \\ \text{Pt}^- \\ \diagup \quad\quad \diagdown \\ \text{Cl} \quad\quad \text{CH}_2\text{CH}_2\overset{+}{\text{N}}\text{HEt}_2 \end{array} \quad\quad (10.38)$$

In Reaction 10.38, the nitrogen lone pair attacks an olefin carbon, converting the olefin to a substituted alkyl ligand. This mode of reaction contrasts with the behavior of free olefins, which on account of the electron-rich double bond, ordinarily are attacked by electrophiles.

Other examples include[11]

---

[11] The Cp = C_5H_5 ligand is discussed in the next section. The point here is the behavior of the coordinated olefin.

$$\left[CpFe(CO)_2 - \overset{CH_2}{\underset{CH_2}{\|}} \right]^+ + CN^- \longrightarrow CpFe(CO)_2CH_2CH_2CN \qquad (10.39)$$

$$\left[CpW(CO)_3 - \overset{CH_2}{\underset{CH_2}{\|}} \right]^+ + PPh_3 \longrightarrow [CpW(CO)_3CH_2CH_2PPh_3]^+ \qquad (10.40)$$

10.3.2 The EAN Rule for π-Donor Complexes

Besides olefins, other organic species can act as π donors. The chemistry of some of these compounds along with special features of the bonding, is discussed later in the chapter. All complexes with π-donor ligands exhibit features similar to the bonding in olefin complexes, including electron donation from filled ligand π orbitals to the metal and donation from the metal into empty π^* orbitals on the ligand. The most stable of such compounds obey the EAN rule. As the ligands grow more complicated, so does the number and symmetry of the π orbitals. For now, we will focus simply on electron-counting rules for π-donor ligands, as given in Table 10.4. Note that some ligands can bond to a metal with varying numbers of π electrons: for example, cyclooctatetraene can donate electrons from two or three of its double bonds. This is reflected by the η^n nomenclature, where n denotes the number of carbons bonded to the metal. η is from the Greek prefix *hapto* (derived from *haptein*, "to fasten"). Thus η^3-C_3H_5 is read "trihaptoallyl" and denotes that all three carbons are bonded to the metal by the π orbitals that extend over all three C atoms; η^5-C_5H_5 (often abbreviated Cp) is read "pentahaptocyclopentadienyl," and so on. Index numbers of the C atoms bound to the metal must be specified whenever confusion could arise—as, for example, in the η^4 cyclooctatetraene complex shown in Table 10.4, in which alternate (instead of adjacent) double bonds are attached.

Arrows sometimes are drawn from the π system to represent formal donation of a pair of electrons. In drawing structural formulas, the totality of bonds usually is represented by a single line—under the assumption that the number of bonded C's is that needed to satisfy the EAN rule.

Olefin ligands and arenes simply are named as such. A more complex and arbitrary situation arises with ligands having an odd number of C's and delocalized π systems, which are named as if they were odd-electron radicals. For example, C_5H_5 (cyclopentadienyl) has the contributing structures

Since it has both olefin *(-ene)* and radical *(-yl)* functionality, it is named as an *-enyl*. The same is true of other ligands bonded through an odd number of carbons. For electron-counting purposes, however, we treat π-donor ligands as existing in a closed-shell electron configuration. For cyclic π systems, we assume that the anions have the $(4n + 2)$ π electrons required

Table 10.4 Electron counting for π-donor ligands

Electrons Contributed	Ligand	Structure	Example
2	η^2-C_2H_4	$H_2C\!=\!CH_2$	$[PtCl_3(C_2H_4)]^-$ trichloro(ethylene)platinate($1-$)
4	η^3-$C_3H_5^-$ η^3-allyl		 bis(η^3-allyl)di-μ-bromodipalladium
4	η^4-C_4H_8 η^4-butadiene		 η^4-butadienetricarbonyliron
4	η^4-C_5H_6 η^4-cyclopentadiene		 dicarbonylbis(η^4-cyclopentadiene)- molybdenum
4	η^4-C_8H_8 η^4-cyclooctatetraene (cot)		 (1,2,5,6-η^4-cyclooctatetraene)(η^5- cyclopentadienyl) cobalt
6	η^4-$C_4H_4^{2-}$ η^4-cyclobutadiene		 tricarbonyl(η^4-cyclobutadiene)iron
6	η^5-$C_5H_5^-$ η^5-cyclopentadienyl (Cp)		 tricarbonylchloro(η^5- cyclopentadienyl)molybdenum

(Continued)

Table 10.4 Electron counting for π-donor ligands *(Continued)*

Electrons Contributed	Ligand	Structure	Example
6	η^5-$C_5H_6^-$ η^5-pentadienyl		tricarbonyl(η^5-pentadienyl)iron(1+)
6	η^6-C_6H_6 η^6-benzene		bis(benzene)chromium
6	η^7-$C_7H_7^+$ η^7-tropylium		tricarbonyl(η^7-tropylium)- molybdenum(1+)
6	η^6-C_7H_8 η^6-cyclo- heptatriene		tricarbonyl(η^6- cycloheptatriene)- molybdenum
6	η^6-C_8H_8 η^6-cyclooctatetraene (cot)		tricarbonyl(η^6-cyclooctatetraene)- chromium

for aromaticity. Although this convention is completely arbitrary, it has some connection with the chemistry of π-donor complexes. For example, $Na^+C_5H_5^-$ can be prepared and used to synthesize cyclopentadienyl compounds. Noncyclic π donors are also treated as closed-shell species for electron counting since their compounds are not fundamentally different from those of cyclic ligands. Finally, it is worth noting that some ligands are related to those in Table

10.4 by closing a chain compound with CH_2 groups. For example, η^3-$C_5H_7^-$ is just an allyl ligand closed by two CH_2 groups.

The following examples demonstrate the application of the EAN rule to π-donor complexes using the rules given on p. 409 and the electron counts provided in Tables 10.2 and 10.4.

Example For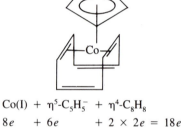

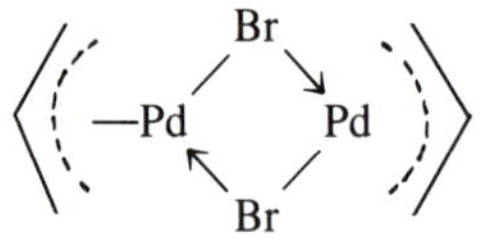

For each metal the oxidation number is II; each Br forms one dative bond that does not contribute to the charge.

$$Pd(II) + Br^- + \mu\text{-}Br + allyl^-$$
$$8e + 2e + 2e + 4e = 16e$$

This species thus does not obey the EAN rule—but that is not too surprising, since Pd is near the end of the second transition series.

Example

$$Co(I) + \eta^5\text{-}C_5H_5^- + \eta^4\text{-}C_8H_8$$
$$8e + 6e + 2 \times 2e = 18e$$

Example

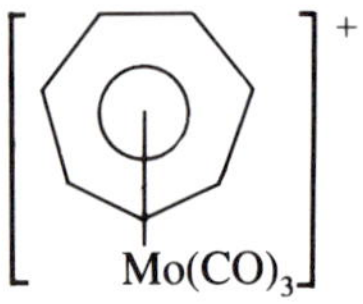

The oxidation number of Mo is $-1 + 1 = 0$.

$$Mo(0) + 3 \times CO + tropylium$$
$$6e + 3 \times 2e + 6e = 18e$$

Now that you are somewhat familar with applying the EAN rule to several kinds of ligands, you can understand and predict the formulas and some structural features of organometallic compounds. Hence we can discuss compounds containing several different kinds of ligands, even though we may wish to focus on the chemical and structural features associated with only one.

10.3.3 σ Bonds between C and Metals

We have already discussed in some detail the bonding between the σ-donor CO and metals. Other organic species also behave as two-electron sigma donors: isocyanides, :CNR, and carbenes, $:C\overset{\diagup X}{\diagdown Y}$ —both of which also have π orbitals capable of accepting back-donation.

Table 10.2 includes electron counts for some organic ligands that are not π acids. Alkyl, aryl, and acyl ligands are capable only of σ donation to metals. Notice that we again consider all ligands to contain sufficient electrons for a closed-shell configuration. So even though alkyl and aryl ligands are named as radicals, for electron-counting purposes we consider them to be negatively charged carbanions.

10.4 EXPERIMENTAL EVIDENCE FOR BACK DONATION

10.4.1 The Problem of Oxidation Numbers

Computing the oxidation number for a metal by using the procedures given above is straightforward. However, the concept of oxidation number has little physical meaning for organometallic complexes, since many of the ligands both donate and receive electrons (see Section 10.3.1). Thus the charge on each ligand is assumed arbitrarily in the calculation as that necessary to give a closed-shell electron configuration where possible, and it may bear little resemblance to the actual electron distribution in a complex. Indeed, it may not be the same for all complexes, depending on the ability of the metal to back-donate electrons. Consider the case of olefin complexes. The formal effect of back donation into the π^* orbitals of the olefin is to destroy the π component of the C=C bond. If the back donation were as great as one electron pair, the ligand would be represented better as the carbanion $H_2\overset{(-)}{C}-\overset{(-)}{C}H_2$ capable (formally) of bonding to the metal as two σ-bonded alkyl groups. The two extreme situations can be represented as

$$M^{n+}\leftarrow\overset{CH_2}{\underset{CH_2}{\|}} \qquad \text{versus} \qquad M^{(n+2)}\overset{\diagup CH_2}{\underset{\diagdown CH_2}{\big|}}$$

(a) Metal-olefin   *(b)* Metallacyclopropane

where the oxidation number of the metal changes by two units from the metal–olefin formulation to the metallacyclopropane formulation. The real situation likely will fall somewhere between the two and correspond to the electroneutrality principle (see Section 2.2.2).

We can estimate the contribution of structure b by noting that a carbanion would have sp^3 hybridization around each C ($109.47°$ bond angles), whereas an olefin would have sp^2 hybridization ($120°$ bond angles). Also, the C—C bond distance would be longer in the metallacyclopropane. Hence x-ray structural and NMR evidence can be brought to bear. On coordination, the best π-acid olefins should exhibit distortions toward a metallacyclopropane structure. A good example is the structure of the $Ir(PPh_3)_2(CO)(Br)(tcne)$, where tcne is tetracyanoethylene shown in Figure 10.7. The substituents on the olefinic carbons are bent away from the metal, and the coordinated molecule is no longer planar. The C—C bond length is ~17 pm longer

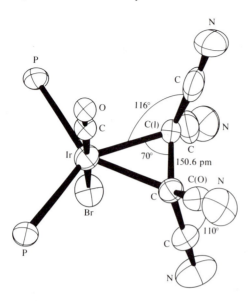

Figure 10.7 The molecular structure of IrBr(CO)(PPh$_3$)$_2$(tcne). (From L. Manojlovic-Muir, K. W. Muir, and J. A. Ibers, *Disc. Faraday Soc.* 1969, *47*, 84.)

than that in free tcne. Even for the C$_2$H$_4$ ligand in Zeise's anion, the hydrogens are somewhat bent away from Pt, with a 3.5 pm lengthening of the C—C bond. The extent of bending has been observed in a series of olefin complexes to increase with the length of the C—C bond,[12] and is thus related to the π-acid capacity of the olefin. ^{19}F NMR studies[13] on compounds such as Pt(PPh$_3$)$_2$(C$_2$F$_4$) show that chemical shifts typically are more characteristic of F bonded to sp^3 than to sp^2 hybridized C. These experimental results confirm the importance of back donation and indicate the difficulties it introduces in employing the concept of oxidation number.

An example from the chemistry of carbonyl hydrides also serves to point up the difficulty of defining a meaningful oxidation number in organometallic compounds. The series of compounds Mn(CO)$_5$Cl, Mn(CO)$_5$Br, Mn(CO)$_5$I can be viewed as containing halide ligands and Mn (being a metal, which, by definition, tends to lose electrons) with an oxidation number of I. Hence we can consider Mn(CO)$_5$H as a hydride complex by analogy. Consider the series of hydride complexes Mn(CO)$_5$H, H$_2$Fe(CO)$_4$, and HCo(CO)$_4$. Table 10.5 indicates that the hydrogens show NMR signals in the high-field region typical of hydrides. However, these compounds become progressively stronger as acids until HCo(CO)$_4$ behaves as a strong acid in water, dissociating into H$^+$ and [Co(CO)$_4$]$^-$!

ESR studies[14] on V(CO)$_6$, the only simple paramagnetic carbonyl, show that the unpaired electron is not confined to V, but is delocalized throughout the molecule.

Plainly, the calculation of oxidation numbers for these compounds is only a device for counting electrons according to arbitrary, if plausible and consistent, rules. Our convention has been to compute ligand charges and metal oxidation numbers by considering that all ligands

[12]J. K. Stalick and J. A. Ibers, *J. Am. Chem. Soc.* 1970, *92*, 5333.

[13]F. G. A. Stone, *Pure Appl. Chem.* 1972, *30*, 551.

[14]K. A. Rubinson, *J. Am. Chem. Soc.* 1976, *98*, 5188.

Table 10.5 Some properties of carbonyl hydrides[a]

	$HV(CO)_6$	$HMn(CO)_5$	$H_2Fe(CO)_4$	$HCo(CO)_4$
Color	—	colorless	colorless solid, yellow liquid	yellow
M.p.(°C)	—	−24.6	−70	−26.2
$\tau(M{-}H)$	—	17.50[b]	20.8[b]	20.7 ± 2[b]
K_a	strong	8×10^{-8} (20°)	4×10^{-5} (17.5°) = K_1 4×10^{-14} (17.5°) = K_2	1
		$HRe(CO)_5$		
Color		colorless		
M.p.(°C)		12.5		
$\tau(M{-}H)$		15.66[c]		
K_a		very weak		

[a]From A. P. Ginsberg, *Transition Metal Chemistry* 1965, *1*, 112.
[b]Neat.
[c]In THF.

are present in closed-shell configurations. (It is also possible to use a convention that considers ligands as uncharged radicals[15] and the metals as possessing the charge on the complex.)

10.4.2 Infrared Spectra of Carbonyl Complexes[16]

CO Stretching Frequencies and Electron Density

The extent of back donation in carbonyl complexes is monitored conveniently by IR spectroscopy. CO stretching vibrations usually are quite intense and occur in a region (2100 cm^{-1} to 1500 cm^{-1}) fairly well isolated from other types of vibrations likely to be present.

One use for IR spectra is in structural diagnosis. Terminal C≡O ligands in neutral molecules have stretches in the 2100–1850 cm^{-1} region, whereas μ_2-CO's absorb between 1860 and 1700 cm^{-1} and μ_3-CO's below 1750 cm^{-1}. The quantities of energy involved in these stretches reflect a progressively decreasing C—O bond order. Thus, for example, two different structures for $Mn_2(CO)_{10}$ are consistent with the EAN rule:

$$(OC)_5Mn{-}Mn(CO)_5 \quad \text{and} \quad (OC)_4Mn \underset{\overset{C}{\underset{O}{}}}{\overset{\overset{O}{\overset{C}{}}}{\quad\quad}} Mn(CO)_4$$

(a) (b)

[15]See, for example, J. E. Ellis, *J. Chem. Educ.* 1976, *53*, 2.

[16]See P. S. Braterman, *Structure and Bonding* 1976, *26*, 1; a more complete and rigorous treatment is given in P. S. Braterman, *Metal Carbonyl Spectra*, Academic Press, New York, 1975.

The IR spectrum in the carbonyl stretching region shows only bands in the region for terminal CO, ruling out *b*. On the other hand, the compound $(\eta^5\text{-}C_5H_5)_2Fe_2(CO)_4$ was shown to exist in several geometries in solution, depending on solvent.[17]

In compounds containing terminal carbonyl ligands, the electron-donor properties of other parts of the molecule are reflected in the frequencies of the CO stretching vibrations. The greater the electron density supplied by the metal and other ligands, the greater can be the back donation into π^* orbitals of CO, and hence the lower will be the CO bond order and the stretching frequency. As the data in Table 10.6 demonstrate, decreasing the positive charge on the metal lowers v_{CO}—the carbonyl stretching frequency—in a series of hexacarbonyls and indicates an increase in the extent of back bonding.

The electron-donating abilities of other ligands can be ascertained from data on the two sets of Ir complexes. In general, the more electronegative the other ligands, the less electron density available for back donation to CO and the higher the v_{CO}. In the first group of $[\text{Ir}(PPh_3)_2(CO)Cl(X)(Y)]$ compounds, the most electronegative (poorest donor) X and Y ligands lead to the highest values of v_{CO}. In the second group of $\text{Ir}(PPh_3)_2(CO)L$ compounds (whose structures resemble that shown in Figure 10.7), the ligands L have their own empty π^* orbitals, which can compete with those of CO for back-donated electrons. The most effective competitors (best π-acids) lead to the highest values for v_{CO}.

Another point worth noting is that v_{CO} is lower for carbonyl ligands than for free CO, indicating that back-bonding populating the π^* orbitals actually does occur.

Table 10.6 v_{CO} for some carbonyl compounds

Compound	$v_{CO}(cm^{-1})$
$[V(CO)_6]^-$	1858^a
$Cr(CO)_6$	1984^a
$[Mn(CO)_6]^+$	2094^a
$Ir(PPh_3)_2(CO)Cl_2H$	2046^b
$Ir(PPh_3)_2(CO)Cl(I)_2$	2067^b
$Ir(PPh_3)_2(CO)Cl(Br)_2$	2072^b
$Ir(PPh_3)_2(CO)Cl_2I$	2074^b
$Ir(PPh_3)_2(CO)Cl_3$	2075^b
$Ir(PPh_3)_2(CO)Cl(O_2)$	2015^b
$Ir(PPh_3)_2(CO)Cl[(CN)HC{=}CH(CN)]$	2029^b
$Ir(PPh_3)_2(CO)Cl(C_2F_4)$	2049^b
$Ir(PPh_3)_2(CO)Cl(tcne)$	2054^b
CO	2143^b

[a]From P. S. Braterman, *Metal Carbonyl Spectra*, Academic Press, New York, 1975.
[b]From the work of Vaska and Shriver.

[17]A. R. Manning, *J. Chem. Soc. A 1968*, 1319; P. A. McArdle and A. R. Manning, *loc. cit. 1969*, 1498; J. G. Bullitt, F. A. Cotton and T. J. Marks, *J. Am. Chem. Soc.* 1970, 92, 2155.

Normal Modes

Compounds with more than one carbonyl ligand often have more than one CO stretch active in IR measurements. Each stretching frequency must correspond to the symmetry of the complex: that is, each frequency corresponds to a linear combination of stretching and compression of CO bonds that transforms like some irreducible representation of the molecular point group. Such a linear combination is called a *normal vibrational mode* (and also is important, of course, for M—C, etc. stretches). The C—O stretching modes belong to the same symmetry species as the M—C σ bonds. $M(CO)_6$ would have stretching modes A_{1g}, E_g, T_{1u}, etc. Only normal modes that have the symmetry properties of (transform like) Cartesian coordinates *x, y, z* are active in the IR.

Table 10.7 contains a listing of expected IR-active bands in the carbonyl region for several common geometries. Note that the varying numbers of IR-allowed bands permits us to distinguish among isomeric forms of the same compound. In general, the more symmetric species have fewer IR-active bands. However, distortions from the idealized point group may lower the symmetry sufficiently to permit additional bands to appear. A case in point is that of the two isomers of $Mo(CO)_4$-$[P(OPh)_3]_2$ whose IR spectra appear in Figure 10.8. The more symmetric *trans* compound has the simpler spectrum. However, the nonallowed A_{1g} and B_{1u} modes (for $\mathbf{D}_{4h}$ symmetry) exhibit weak bands, because of distortions from $\mathbf{D}_{4h}$ symmetry brought about by bulky $P(OPh)_3$ ligands.[18]

10.5 SURVEY OF THE CHEMISTRY OF π-DONOR COMPLEXES

This section considers π-donor ligands having two–eight C atoms bonded to a metal. The bonding, the methods of synthesis, and some typical reactions are given for each kind of ligand. Because of the truly enormous number of π complexes known, only some of the main features can be emphasized.

10.5.1 Alkyne Complexes

Alkynes, RC≡CR′, form π complexes with a variety of transition metals. Alkynes have two mutually perpendicular sets of π and π* orbitals, each of which can engage in the same kind of bonding with metal orbitals as ethylene. In mononuclear complexes,[19] alkynes often function essentially like olefins. An example is the Pt(II) compound *trans*-[PtCl$_2$(*p*-toluidine)(tBuC≡CtBu)], whose structure is shown in Figure 10.9a. The acetylenic ligand is perpendicular to the molecular plane (like ethylene in Zeise's salt), and the *t*-Bu groups are bent back by ~20°. The C—C bond is lengthened from ~120 pm in the parent acetylene to 124 pm in the complex, and $\nu_{C≡C}$ lowered from that of free tBuC≡CtBu. Although caused by π back bonding, these effects are sufficiently small that we can best represent the bonding as a π-bonded alkyne *(a)*, rather than as a metallocyclic alkene *(b)*.

[18]Values of CO force constants can be extracted from normal stretching frequencies; see F. A. Cotton and C. S. Kraihanzel, *J. Am. Chem. Soc.* 1962, *84*, 4432. These can be used to evaluate quantitatively the donor properties of other ligands present; see W. A. G. Graham, *Inorg. Chem.* 1968, *7*, 315.

[19]S. Otsuka and A. Nakamura, *Adv. Organomet. Chem.* 1976, *14*, 245.

Table 10.7 Infrared active modes for some carbonyl complexes

Complex			Point Group	Symmetry of IR-active CO Normal Modes	Number v_{CO} Expected
1. M(CO)$_6$			O_h	T_{1u}	1
2. M(CO)$_5$L			C_{4v}	$A_1 + E \ (+A_1)$	2 or 3
3. M(CO)$_4$L$_2$	*trans:*		D_{4h}	E_u	1
	cis:		C_{2v}	$A_1 + B_1 + B_2 \ (+A_1)$	3 or 4
4. M(CO)$_3$L$_3$	*mer:*		C_{2v}	$2A_1 + B_2$	3
	fac:		C_{3v}	$A_1 + E$	2
5. M(CO)$_5$			D_{3h}	$A_2'' + E'$	2
6. M(CO)$_4$L	*ax:*		C_{3v}	$2A_1 + E$	3
	eq:		C_{2v}	$2A_1 + B_1 + B_2$	4
7. M(CO)$_3$L$_2$			D_{3h}	E'	1
			C_s	$2A' + A''$	3
8. M(CO)$_4$			T_d	T_2	1

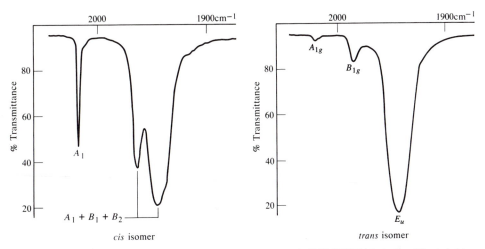

Figure 10.8   Infrared spectra of the geometric isomers of $Mo(CO)_4[P(OPh)_3]_2$ in the CO stretching region. (From M. Y. Darensbourg and D. J. Darensbourg, *J. Chem. Educ.* 1970, *47*, 33.)

However, in the Pt(0) complex $Pt(PPh_3)_2(PhC\equiv CPh)$, shown in Figure 10.9b, the C—C bond is lengthened to 128 pm and the phenyl groups are bent back by ~40°. The C≡C stretching frequency is lowered from ~2200 cm^{-1} in free $PhC\equiv CPh$ to 1750 cm^{-1}. The increased back bonding of Pt(0) over Pt(II) indicates that the structure is best represented as (b).

Alkynes form a number of binuclear complexes,[20] in which each of the two perpendicular π systems may be considered formally as bonding to one of the metals. The structure of $(tBuC\equiv CtBu)Co_2(CO)_6$, depicted in Figure 10.9c, is fairly typical. The C—C bond length and bond angles indicate rehybridization to give some σ character to the M—C bonds (see Figure 10.9d and 10.9e).

Alkyne complexes usually are prepared by adding or displacing other ligands from complexes, as in the following reactions.

$$IrCl(CO)(PPh_3)_2 + CF_3C\equiv CCF_3 \rightarrow IrCl(CO)(PPh_3)_2(CF_3C\equiv CCF_3) \quad (10.41)$$

$$Co_2(CO)_8 + PhC\equiv CPh \xrightarrow{-CO} Co_2(CO)_6(PhC\equiv CPh) \quad (10.42)$$

$$(olefin)Pt(PPh_3)_2 + PhC\equiv CPh \rightarrow (PhC\equiv CPh)Pt(PPh_3)_2 + olefin \quad (10.43)$$

Acetylenes with at least one substituted C are most likely to form isolable complexes. Especially with metals having several *d* electrons, acetylenes tend to dimerize to form metal-

[20]E. L. Muetterties *et al., J. Am. Chem. Soc.* 1978, *100*, 2090.

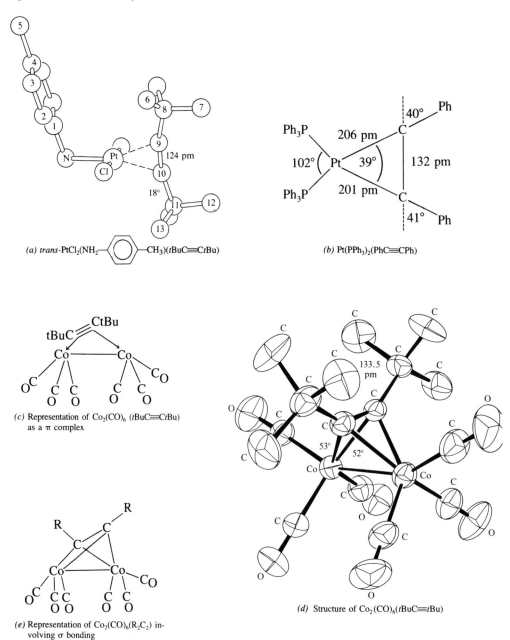

(a) *trans*-PtCl₂(NH₂—⟨benzene⟩—CH₃)(tBuC≡CtBu)

(b) Pt(PPh₃)₂(PhC≡CPh)

(c) Representation of Co₂(CO)₆ (tBuC≡CtBu) as a π complex

(d) Structure of Co₂(CO)₆(tBuC≡tBu)

(e) Representation of Co₂(CO)₆(R₂C₂) involving σ bonding

Figure 10.9 Structures of some alkyne complexes. (*a* is from G. R. Davies, W. Hewertson, R. H. B. Mais, P. G. Owston, and C. G. Patel, *J. Chem. Soc. A* **1970**, 1873. *b* is from J. O. Glanville, J. M. Stewart, and S. O. Grim, *J. Organomet. Chem.* **1967**, *7*, P9. *d* is reproduced with permission from F. A. Cotton, J. D. Jamerson, and B. R. Stults, *J. Am. Chem. Soc.* **1976**, *98*, 1774. Copyright 1976, American Chemical Society.)

lacyclopentadienes, to trimerize to form substituted benzenes[21] or to incorporate CO to give cyclic dienones. Hence relatively few alkyne complexes have a single π-bonded ligand.

$$Fe(CO)_5 + xs\ RC{\equiv}CR \rightarrow \qquad\qquad\qquad\qquad (10.44)$$

$$(R = H,\ Ph,\ CF_3)$$

$$Fe(CO)_5 + CH_3C{\equiv}CCH_3 \xrightarrow{h\nu} \qquad\qquad\qquad (10.45)$$

$$+ xs\ RC{\equiv}CR \rightarrow \qquad\qquad\qquad\qquad (10.46)$$

$$R = CO_2CH_3$$

$$Ph_3Cr(THF)_3 + xs\ CH_3C{\equiv}CCH_3 \rightarrow \qquad\qquad (10.47)$$

The coupling reactions obviously are preparative routes to alkene and arene complexes, many of which also can be made directly from the free ligands.

[21] K. P. C. Vollhardt, *Acc. Chem. Res.* 1977, *10*, 1.

10.5.2 Allyl Complexes

The allyl group $\overset{(-)}{C}H_2$—CH=CH$_2$ can bond to metals in a σ (or η^1) fashion (see Section 10.3.2), or it can behave as a four-electron π-donor ligand with η^3 coordination. The VB representation indicates a delocalized π-system:

Figure 10.10 shows the molecular orbitals for the allyl anion. Ψ_1 and Ψ_2 are filled with a total of four electrons, whereas Ψ_3 is empty and available for back donation of metal electrons. Although allyl complexes exist for many transition metals, the most important ones are those of Ni, Pd, and Pt, which have an extensive organic chemistry and are useful catalysts.[22]

A typical structure of π-allyl complexes is that of the dimer (η^3-C$_3$H$_5$PdCl)$_2$, pictured in Figure 10.11. The plane of the allyl group is tilted at a 111.5° angle with respect to the Pd$_2$Cl$_2$ plane (instead of being perpendicular). Moreover, the Pd$_2$Cl$_2$ plane is closer to the end carbons than to the central one. As would be expected from the MO description, both C—C distances are the same (136 pm).

Three preparative methods commonly are employed to synthesize allyl complexes: reaction of allyl Grignard reagents with metal halides, reaction of allyl halides with carbonylate anions, and hydrogen addition to or abstraction from olefin complexes.

Bis(allyl) complexes are prepared conveniently via the Grignard route.

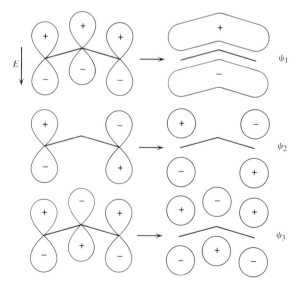

Figure 10.10 MO's for C$_3$H$_5^-$.

[22]R. Baker, *Chem. Rev.* 1973, *73*, 487.

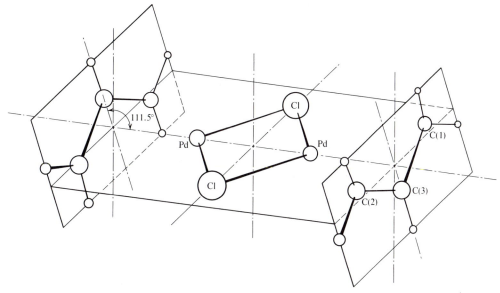

Figure 10.11 Structure of $[(\eta^3\text{-}C_3H_5)PdCl]_2$. (From A. F. Smith, *Acta Cryst.* 1965, *18*, 331.)

$$2C_3H_5MgBr + NiCl_2 \xrightarrow[-10°]{\text{diethyl ether}} \quad Ni \quad + 2MgBrCl \qquad (10.48)$$

Reaction of allyl halides with carbonylate anions usually gives the η^1-allyl, which can be converted to an η^3-allyl by loss of CO.

$$[Mn(CO)_5]^- + CH_2\text{=}CHCH_2Cl \rightarrow (CO)_5MnCH_2CH\text{=}CH_2 \xrightarrow[-CO]{\Delta} \left(\text{—Mn(CO)}_4\right) \quad (10.49)$$

Several hydrogen transfer reactions involving olefin complexes generate allyl ligands.

$$2CH_2\text{=}CHCH_3 + 2PdCl_2 \rightarrow \left[\begin{array}{c} \underset{H}{\overset{\displaystyle H \quad CH_3}{\underset{\displaystyle}{C}}} \\ \end{array} \right]$$

$$\rightarrow \left(\text{—Pd} \underset{Cl}{\overset{Cl}{<}} Pd\text{—} \right) + 2HCl \qquad (10.50)$$

The second step involves elimination of an allylic proton from propene, leaving (formally) a $C_3H_5^-$ ligand. With conjugated diene complexes, a formal hydride addition saturates one of the olefinic carbons, giving an allyl.

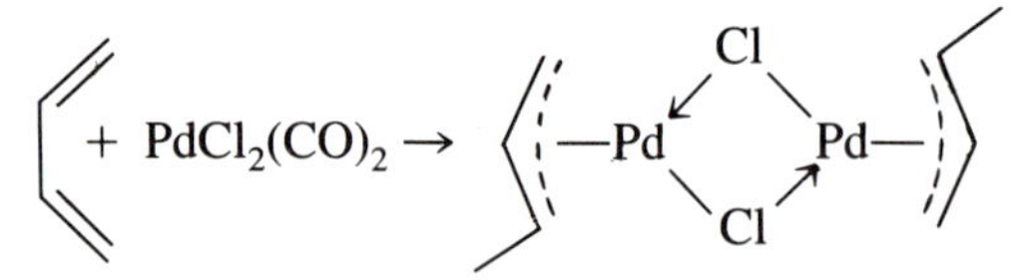

$$(10.51)$$

Presumably, the reaction proceeds by way of an intermediate diene-Pd complex; the hydrogen comes from the solvent. In some reactions the hydride may come from the addition of a metal hydride to a diene.

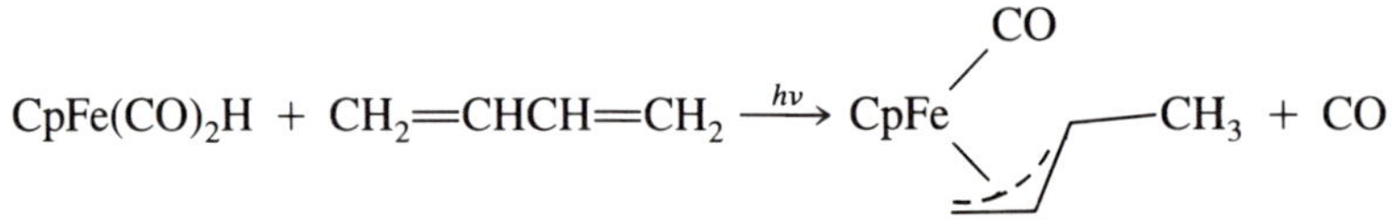

$$CpFe(CO)_2H + CH_2{=}CHCH{=}CH_2 \xrightarrow{h\nu} CpFe \qquad\qquad {-}CH_3 + CO \qquad (10.52)$$

Allyls substituted on an end C are capable of isomerism; the substituent may be on the same side as the center C (*syn* isomer), or on the opposite side (*anti* isomer). This same feature leads to nonequivalence of the terminal protons. In solution, interconversion of terminal groups sometimes can occur in the presence of basic ligands; the mechanism is thought to be that shown in Figure 10.12. Excess ligand displaces an olefinic bond, converting the η^3-allyl ligand to an η^1-allyl. Free rotation around the M—C single bond interconverts *syn* and *anti* isomers. The process can be followed in solution by NMR. (See Section 15.10.3.)

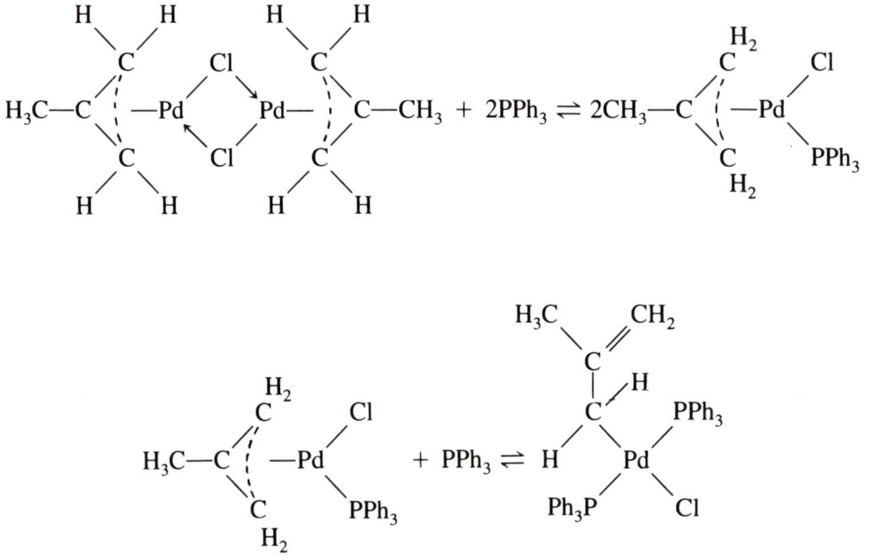

Figure 10.12 Mechanism for interconversion of *syn* and *anti* substituents in η^3-allyl complexes.

10.5.3 Butadiene Complexes

Butadiene and substituted butadienes, which can be treated as four-electron donor ligands, form an especially large number of complexes with the $Fe(CO)_3$ group. Since these ligands are conjugated (have alternating double bonds), they may be described in terms of the following contributing VB structures.

Hence, two contributing structures are possible for metal complexes.

The C—C bond lengths in $C_4H_6Fe(CO)_3$ are identical within experimental error, indicating a significant contribution from structure *f*. However, the ^{13}C—H coupling constants for terminal H suggest sp^2, rather than sp^3, hybridization at the C. Structural data on a variety of butadiene complexes do indicate some bending out of plane for the terminal groups. Hence the butadiene ligand is different from nonconjugated polyene ligands, which have isolated double bonds.

The MO bonding description[23] resembles that for allyl complexes (Section 10.5.2). Equalization of C—C bond lengths is viewed as resulting from ligand-to-metal π donation from butadiene orbitals that are bonding between C_1—C_2 and C_3—C_4, thereby reducing the bond order. Metal-to-ligand back donation occurs into an empty π^* orbital that is bonding between C_2 and C_3. Even though the best description of the bonding would be structure *(d)*, emphasizing delocalization for the C_4H_6 ligand, it is usual to draw structure *(a)*, to emphasize the four-electron donor ability of the ligand. Some known complexes do display the alternating C—C bond lengths implicit in this formulation.

Like olefin and acetylene complexes, butadiene complexes generally are prepared by direct reaction between the ligand and a metal complex. Often, CO is displaced.

$$Co_2(CO)_8 + 2C_4H_6 \xrightarrow{-CO}$$ (10.53)

[23]See G. E. Coates, M. L. H. Green, and R. Wade, *Organometallic Compounds,* Chapman and Hall, London, 1968, Vol. II, p. 70.

$$Fe(CO)_5 + C_4H_6 \xrightarrow[-CO]{} \text{[structure]} \quad (10.54)$$

Coordinated butadiene does not undergo the hydrogenation and Diels-Alder reactions of the free molecule.

10.5.4 Cyclic π Complexes

Several cyclic organic species form complexes involving transition metals. Complete electron delocalization around the ring ordinarily occurs. We consider that the cyclobutadiene, cyclopentadienyl, and arene ligands all behave as six-electron donors, possessing in their π-systems the $(4n + 2)$ electrons necessary for aromaticity. Of course, all these ligands are bonded by donation of electrons from their filled π orbitals to the metal and by back donation by the metal into empty ligand π* orbitals.[24]

Cyclobutadiene Complexes[25]

Unlike the straight chain C_4H_6, cyclobutadiene is not known in the free state, since it would have only four electrons rather than the six required for aromaticity. An early consideration (1956) of orbital symmetries led Orgel and Longuet-Higgens to predict that cyclobutadiene could be stablized by coordination to a metal. Shortly thereafter, a tetramethylcyclobutadiene complex was prepared by Criegee. Petit and coworkers prepared an unsubstituted analogue by a similar reaction.

$$\text{[structure]} + Fe_2(CO)_9 \xrightarrow{-CO} \text{[structure]} \quad (10.55)$$

Cyclobutadiene complexes often result from coupling reactions of substituted acetylenes. As shown in Equation 10.56, several coupling products often result.

$$CpCo(PPh_3)_2 + xs\ PhC\equiv CPh \xrightarrow{-CO} \text{[structure]}$$

$$+ \text{[structure]} + CpCo \text{[structure]} \quad (10.56)$$

---

[24]For orbital pictures, see G. E. Coates, M. L. H. Green, and K. Wade, *Organometallic Compounds*, Vol. II, Chapman and Hall, London, 1968.

[25]A. Efraty, *Chem. Rev.* 1977, *77*, 691.

Justification for considering the cyclobutadiene ring as aromatic arises from the variety of electrophilic substitution reactions that occur on the ring, as depicted in Figure 10.13.

Oxidation of a cyclobutadiene complex with Ce^{4+} releases the free ligand, which can be trapped by a Diels-Alder reaction.

$$\text{(cyclobutadiene)Fe(CO)}_3 + \text{MeO}_2\text{CC}\equiv\text{CH} \xrightarrow{Ce^{4+}} \text{product–CO}_2\text{Me} \tag{10.57}$$

Cyclopentadienyl Complexes

Current interest in organometallic chemistry began in the early 1950's, when two research groups, working independently, prepared the astonishing Fe hydrocarbon derivative $(\eta^5\text{-C}_5\text{H}_5)_2\text{Fe}$, by reaction of $FeCl_2$ with C_5H_5MgBr and by reaction of reduced Fe with cyclopentane under nitrogen at $300°$ in the presence of K_2O. The product, now known as ferrocene, is an orange solid melting at 172.5 to 174°C with sublimation. Ferrocene is soluble in alcohol, ether, and benzene and insoluble in water, 10% NaOH solution, and concentrated HCl. It is oxidized easily to a blue cation, $[\text{Fe}(C_5H_5)_2]^+$. Ferrocene is diamagnetic and the cation obtained by oxidation contains one unpaired electron. Cyclopentadienyl compounds have now been prepared for many of the transition metals (Table 10.8). The crystal structure of ferrocene (Figure 10.14) showed that the cyclopentadienyl rings are planar, with all

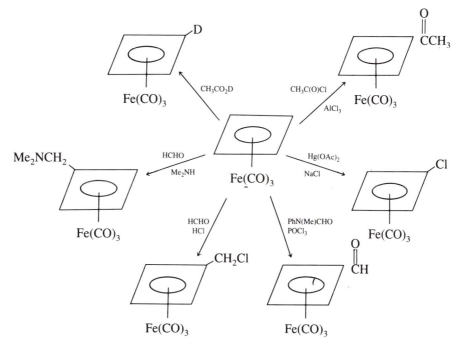

Figure 10.13 Some electrophilic substitution reactions of cyclobutadiene iron tricarbonyl. (After Tsutsui *et al.*, *Introduction to Metal Pi Complex Chemistry*, Plenum, New York, 1970.)

Table 10.8 Some bis(cyclopentadienyl) complexes of transition metals

Compound	Appearance	m.p.(°C)	No. Unpaired Electrons	Structure and Comments
"Cp$_2$Ti"	Dark green crystals	Dec. 200	0	Dimer with μ-H
	Gray-black powder		Paramagnetic	Dimer containing M—M bond and (η^1, η^5-C$_5$H$_5$)
(Me$_5$C$_5$)$_2$Ti	Yellow-orange	Dec. 60	2	Monomer, tilted rings
"Cp$_2$Zr"	Purple-black crystals	Dec. 300	0	Structure similar to Ti μ-H dimer
Cp$_2$V	Purple crystals	167–68	3	Air-sensitive; reacts with CO, halogens, alkyl halides
"Cp$_2$Nb"	Yellow solid		0	Dimer with μ-H and μ(η^1, η^5-C$_5$H$_5$)
"Cp$_2$Ta"			0	Isomorphous with "Cp$_2$Nb"
Cp$_2$V$^+$	Blue-green (as Reineckate salt)		2	
Cp$_2$V^{2+}	Green (as picrate salt)		1	
Cp$_2$Cr	Scarlet crystals	172–73	2	Air-sensitive
Cp$_2$Mo?	Yellow			Not well-characterized
Cp$_2$W				Exists only as very reactive intermediate
Cp$_2$Cr$^+$	Brown-black (as CpCr(CO)$_3^-$ salt)		3	
Cp$_2$Mn	Brown solid		5	Stable form ≤ 159°, chain structure
	Light orange crystals		5	Stable form > 159°; isomorphous with ferrocene (but ionic)
Cp$_2$Fe	Orange crystals	173	0	Staggered configuration in solid, eclipsed in gas phase; thermally stable >500°; air-stable

(Continued)

Table 10.8 Some bis(cyclopentadienyl) complexes of transition metals *(Continued)*

Compound	*Appearance*	*m.p.(°C)*	*No. Unpaired Electrons*	*Structure and Comments*
Cp_2Ru	Light yellow crystals	199–201	0	Eclipsed configuration in solid; most thermally stable metallocene (>600°)
Cp_2Os	Colorless crystals	229–30	0	Eclipsed configuration in solid
Cp_2Fe^+	Blue (as PF_6^- salt)		1	Prepared by oxidation of Cp_2Fe
	Dichroic in aqueous solution (blue-green to blood red)			($Cp_2Fe^+ + e \rightarrow Cp_2Fe$ $E^0 = -0.3$ V vs S.C.E)
Cp_2Ru^+	Light yellow (as ClO_4^- salt)		1	Prepared by oxidation of Cp_2Ru with I_2 or $HClO_4$ in ethanol
$Cp_2Os(OH)^+$	Orange-red (as PF_6^- salt)		0	Product of $FeCl_3$ oxidation of Cp_2Os
Cp_2Co	Purple-black crystals	173–74	1	19e species; easily oxidized by air to Cp_2Co^+
Cp_2Rh	Brown-black solid		Paramagnetic	19e species; monomeric $\leq -196°$
	Yellow-orange crystals	Dec. 140	0	Dimer linked through rings

Cp_2Ir	Colorless crystals		Paramagnetic	19e species; monomeric $\leq -196°$
	Yellow crystals	Dec. 230	0	Same structure as Rh dimer
Cp_2Co^+	Yellow (with colorless anion)		0	Prepared by oxidation of Cp_2Co or directly from metal halide $+ C_5H_5^-$
Cp_2Rh^+	Yellow (as PF_6^- salt)		0	Prepared as above
Cp_2Ir^+	Yellow (as PF_6^- salt)		0	Prepared as above
Cp_2Ni	Green crystals	Dec. 173–74	2	20e species; toxic
Cp_2Pd?	Red			Not well-characterized
Cp_2Ni^+	Yellow-orange crystals (with colorless anion)		1	Unstable

C—C distances equal, and are arranged in the staggered sandwich configuration ($\mathbf{D}_{5d}$), rather than with eclipsed carbons ($\mathbf{D}_{5h}$). Both ruthenocene and osmocene have the eclipsed configuration in the crystal—as does ferrocene itself in the gas phase. This indicates that the energy barrier to ring free rotation is quite small.

The sandwich structure of the metallocenes gave the first clue that organic ligands could be bonded to metals via their π systems. The preparation of ferrocene opened up the study of the whole field of compounds with metal–carbon bonds.

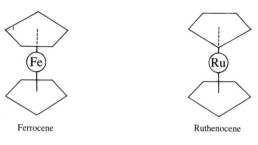

Ferrocene   Ruthenocene

Staggered ($\mathbf{D}_{5d}$)   Eclipsed ($\mathbf{D}_{5h}$)   **Figure 10.14**   Metallocene structures.

Although the structure of ferrocene was known to involve the ligand π system, formulation of the electronic structure presented a considerable challenge. Because of the number of VB structures that can be written for the cyclopentadienyl radical (see Section 10.3.2) or the cyclopentadienide anion, a VB description of ferrocene requires 560 contributing structures![26] In this situation the MO description is more appealing.

The MO's of C_5H_5 are depicted in Figure 4.25. In describing the electronic structure of ferrocene, we construct on the two separate rings linear combinations of the MO's that conform to the molecular symmetry ($\mathbf{D}_{5d}$). Figure 10.15 depicts these linear combinations along with the metal orbitals with which they can overlap. Figure 10.16 presents an energy diagram for ferrocene and other metallocene complexes. The most energetically significant bonding interaction appears to be the overlap of the $e_{1g}(d_{xz}, d_{yz})$ metal orbitals with the ligand e_{1g} pair. Although metal e_{1u} (p_x, p_y) orbitals overlap favorably with ligand e_{1u}, the energy difference is too large for effective MO formation. Hence the $1e_{1u}$ orbitals are localized largely on the ligands. Since the metal e_{2g}'s $(d_{x^2 - y^2}, d_{xy})$ and a_{1g} (d_{z^2}) do not overlap very effectively with ring orbitals, they are nonbonding and localized on the metal. The 18 electrons of ferrocene exactly fill the energy levels through $2a_{1g}$, above which a substantial energy gap exists. This gap, together with effective occupation by Cp rings of all the coordination positions around Fe, accounts for the remarkable kinetic and thermal stability of ferrocene.

The energy-level diagram in Figure 10.16 is applicable to the metallocene sandwich compounds in Table 10.8. It can account for their magnetic properties, if the reasonable assumption is made that electron-pairing energy sometimes is greater than the $1e_{2g} - 2a_{1g}$ energy separation (see Problem 10.4).

Manganocene is unique in exhibiting the maximum-possible number of unpaired electrons (five). Its bonding (in the high-temperature form) has been described as being more ionic than that in other metallocenes, with a large contribution from the structure $(Cp^-)_2Mn^{2+}$. This is borne out to some extent in the chemistry of manganocene, which is hydrolyzed easily and reacts with CO_2 to give carboxylic acid salts, in contrast to other metallocenes. However, the compound is soluble in organic solvents and gives solutions of low conductivity.

As noted in Table 10.8, many bis(cyclopentadienyl) compounds do not have the metallocene sandwich structure. Some are dimeric and/or do not contain planar, parallel Cp rings. This occurs especially when such rearrangement can lead to an 18-electron structure (for example, "Cp$_2$Nb" and "Cp$_2$Rh"). By referring to the energy-level diagram, we see that

---

[26]This number can be reduced by considering only *d*-orbitals of proper symmetry for metal–ring bonding.

Cp_2Co, Cp_2Rh, Cp_2Ir, and Cp_2Ni contain electrons in the $2e_{1g}$ orbital that is antibonding. Hence it is not surprising that these compounds undergo reactions leading to 18-electron species. (See Reactions 10.58–10.61.)

$$2Cp_2Rh \xrightarrow{\;>\,-196°C\;} [CpRh(\eta^4\text{-}C_5H_4)]_2 \tag{10.58}$$

$$Cp_2Rh \;+\; NaBH_4 \longrightarrow \tag{10.59}$$

$$Cp_2Rh \xrightarrow{\;Br_2\;} [Cp_2Rh]^+Br_3^- \tag{10.60}$$

Nickelocene, a 20-electron species, undergoes addition reactions such as 10.61, converting one ring to an allyl and giving and 18-electron complex.

$$Cp_2Ni \;+\; C_2F_4 \longrightarrow CpNi \tag{10.61}$$

Some bis(cyclopentadienyl) compounds of main-group metals also are known. In Cp_2Be and Cp_2Mg, no d orbitals are available and so the bonding is likely to be ionic. Both compounds feature a sandwich structure, but the Be compound is especially interesting in that one of the rings is considerably closer to Be than the other.

Main group cyclopentadienides such as NaCp and TlCp exhibit "open face" sandwich structures with planar Cp rings.

Cyclopentadienyl Complexes Containing Other Ligands.   A very large number of organometallic complexes, including Cp_2TiCl_2, Cp_2ReH, and Cp_2NbOCl, contains two cyclopentadienyl rings along with other ligands. An important feature of their structures is the bending back of the planar Cp rings to make available for bonding to other ligands the $d_{x^2-y^2}$ and d_{xy} orbitals. Of course, many complexes contain only one Cp ring acting as a planar six-electron donor occupying three coordination positions. Examples include $CpTi(CO)_2$, $[CpCr(CO)_3]_2$, $[CpFe(CO)_2]_2$, and $CpMo(O)_2Cl$.

In general, cyclopentadienyl rings are less reactive than other ligands and can be thought of as inert ligands that take up three coordination positions. This point is illustrated by the summary of the chemistry of $Cp_2Fe_2(CO)_4$ given in Figure 10.17. The chemistry of the Cp ring is discussed later in this section.

Synthesis of Cyclopentadienyl Compounds.   The usual way to add a Cp ring is to allow alkali metal cyclopentadienides to react with metal complexes in THF or another ether. The cyclopentadienides are prepared *in situ* from cyclopentadiene.

$$2Na \;+\; 2C_5H_6 \longrightarrow 2C_5H_5^- \;+\; 2Na^+ \;+\; H_2 \tag{10.62}$$

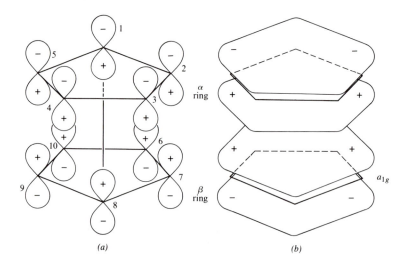

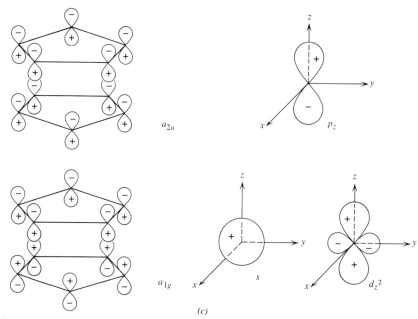

Figure 10.15 *(a)* Orientation of orbitals for consideration of bonding in ferrocene. *(b)* a_{1g} orbitals for the two rings. *(c)* Symmetry combinations of Cp orbitals (in order of increasing energy) and their metal counterparts in the D_{5d} symmetry of ferrocene.

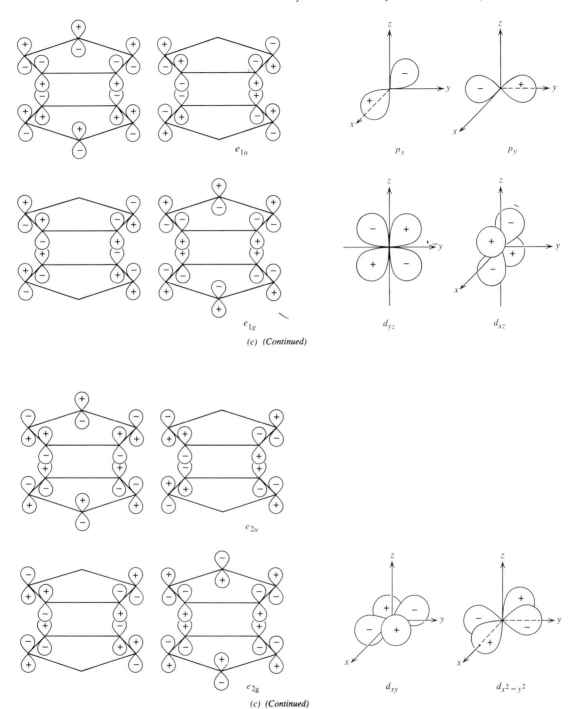

e_{1u} p_x p_y

e_{1g} d_{yz} d_{xz}

(c) *(Continued)*

e_{2u}

e_{2g} d_{xy} $d_{x^2 - y^2}$

(c) *(Continued)*

Figure 10.15 *(Continued).*

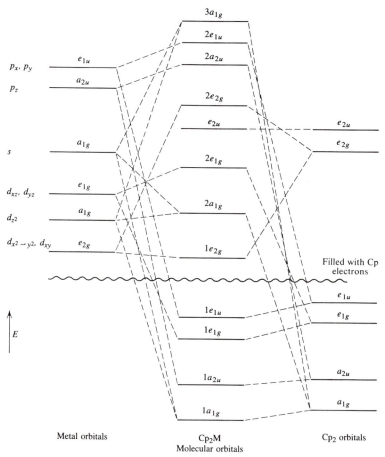

Figure 10.16 Molecular orbital energy diagram for metallocene complexes. (Reproduced with permission from Y. S. Sohn, D. N. Hendrickson, and H. B. Gray, *J. Am. Chem. Soc.* 1971, *93*, 3603. Copyright © 1971, American Chemical Society.)

$$VCl_3 + 3NaC_5H_5 \rightarrow Cp_2V \tag{10.63}$$

In the above reaction, sodium cyclopentadienide reduces V(III) to V(II).

$$W(CO)_6 + NaC_5H_5 \rightarrow Na^+[CpW(CO)_3]^- + 3CO \tag{10.64}$$

Sometimes $Tl(C_5H_5)$ is used, because it can be prepared separately and stored.

$$TiCl_4 + 2Tl(C_5H_5) \rightarrow Cp_2TiCl_2 + 2TlCl \tag{10.65}$$

A second method is direct reaction between cyclopentadiene and metal complexes, often in the presence of a base acting as a proton acceptor.

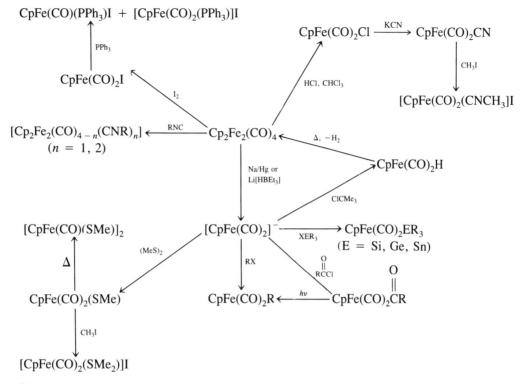

Figure 10.17 The chemistry of $Cp_2Fe(CO)_4$. (After M. L. H. Green in G. E. Coates, M. L. H. Green, and K. Wade, *Organometallic Compounds*, Vol. II, Chapman and Hall, London, 1968.)

$$2C_5H_6 + 2Fe(CO)_5 \rightarrow [2CpFe(CO)_2H] \rightarrow Cp_2Fe_2(CO)_4 + H_2 \qquad (10.66)$$

$$NiCl_2 + 2C_5H_6 \xrightarrow{Et_2NH} Cp_2Ni + 2[Et_2NH_2]^+Cl^- \qquad (10.67)$$

The Chemistry of Ferrocene Derivatives.[27] All neutral metallocenes are soluble in organic solvents and sublimable. In general, they behave like covalent compounds—except for manganocene, which contains five unpaired electrons and exhibits behavior more akin to the ionic main-group compounds, such as Cp_2Mg.

We must discuss briefly the enormous chemisty of ferrocene and its derivatives, since it is sufficiently stable to survive a variety of reaction conditions and its reactions can be regarded as fairly typical of coordinated Cp rings.[28] The Cp ligands display the chemistry of aromatic rings (lending credence to our formulation of the ligand as $C_5H_5^-$). Indeed, ferrocene is more

[27]See M. Rosenblum, *Chemistry of the Iron Group Metallocenes*, Interscience, New York, 1965.

[28]Note also the reactions of "electron-rich" metallocenes in this section.

reactive than benzene toward electrophilic reagents. It undergoes Friedel-Crafts acylation, giving $(\eta^5\text{-}C_5H_4C(O)CH_3)(\eta^5\text{-}C_5H_5)Fe$ with equimolar acetyl chloride. With excess $CH_3C(O)Cl$, two disubstituted isomers are produced.

$$Cp_2Fe + xs\ CH_3C(O)Cl \xrightarrow{AlCl_3} \quad \quad \quad \quad \quad \quad \quad \quad (10.68)$$

60   .   1

The entry of the acetyl group, which is electron-withdrawing, deactivates the ring to further substitution. Hence the major product is 1,1'-diacetylferrocene. (The prime denotes substitution on the second ring–heteroannular substitution).

Alkylation also occurs under Friedel-Crafts conditions. Dialkyl and polyalkyl products result. Because alkyl groups are electron-donating, dialkylation occurs homoannularly. Several aromatic substitution reactions of ferrocene are presented in Figure 10.18.

Because ferrocene is oxidized easily to the ferricinium ion $[Cp_2Fe]^+$, direct halogenation and nitration cannot be carried out. Instead, ferrocenyl lithium is used to prepare these and other derivatives, as shown in Figure 10.19.

Arene Complexes[29]

Neutral arenes form complexes with transition metals. Some contain only arene ligands, whereas others also contain carbonyl, phosphine, or other ligands. The first arene complexes were prepared in 1919 by Hein, who treated $CrCl_3$ with C_6H_5MgBr in ether. However, these were not recognized as π complexes until 1954. Upon hydrolysis, Hein's reaction mixture afforded $(\eta^6\text{-}C_6H_6)_2Cr$, $(\eta^6\text{-}C_6H_5\text{—}C_6H_5)(\eta^6\text{-}C_6H_6)Cr$, and $(\eta^6\text{-}C_6H_5\text{—}C_6H_5)_2Cr$.

The x-ray structure of bis(benzene)chromium (Figure 3.11) shows eclipsed rings and $\mathbf{D}_{6h}$ symmetry. The arene ligands behave as six-electron donors and each can be considered to occupy three coordination positions around the metal. An approximate MO diagram for bis(benzene)chromium is presented in Figure 10.20. The most important bonding interaction is that between the d_{xz}, d_{yz} orbitals and the ligand combination of e_{1g} symmetry. The six d electrons of Cr fill the levels through a_{1g} (d_{z^2})—which accounts for the diamagnetism of $(\eta^6\text{-}C_6H_6)_2Cr$. This energy-level scheme probably also applies to other bis(arene) complexes.

[29]W. E. Silverthorn, *Adv. Organomet. Chem.* 1975, *13*, 47.

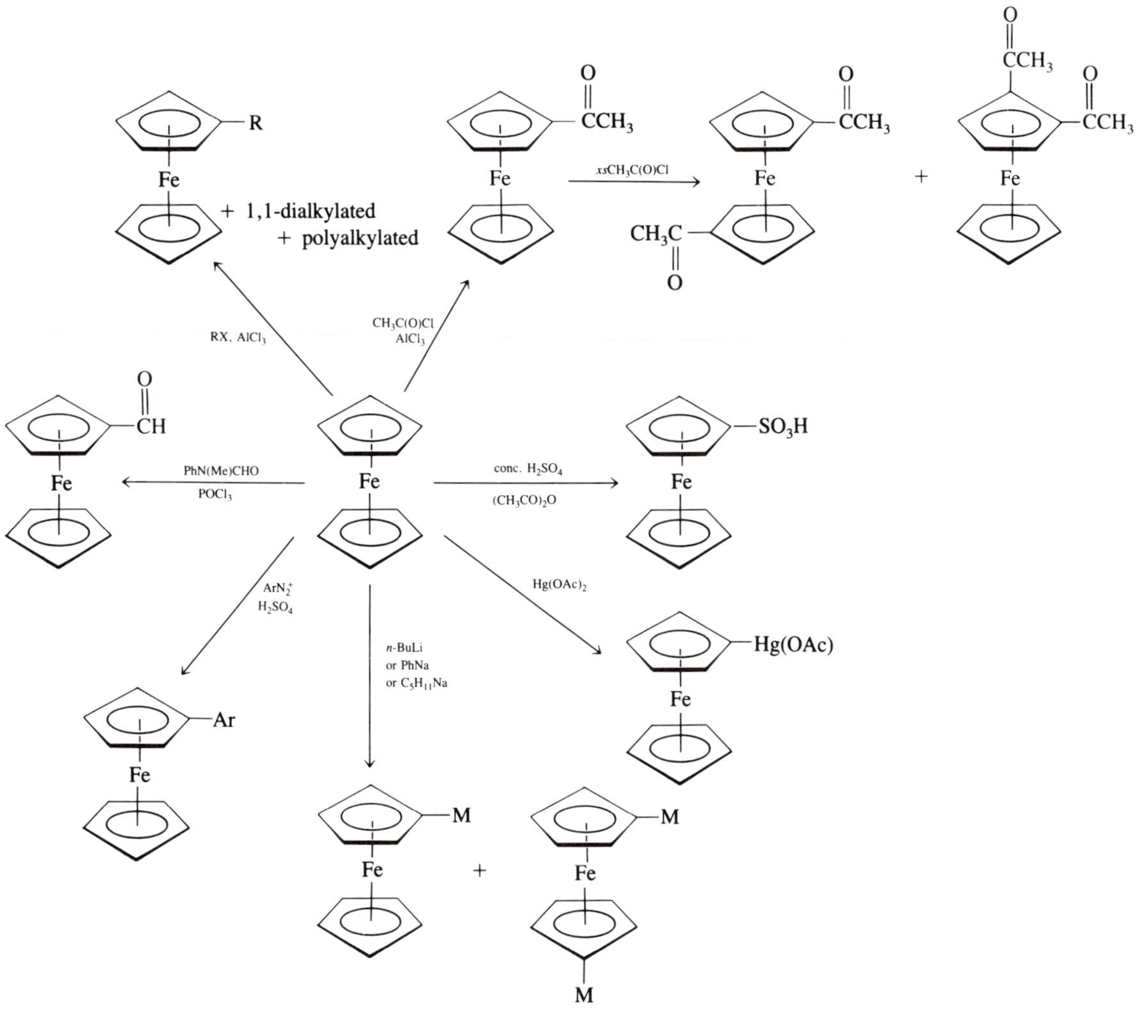

Figure 10.18 Some aromatic substitution reactions of ferrocene.

Not all bis(arene) complexes that seem to violate the EAN rule actually do so. For example, $(Me_6C_6)_2Ru$ would be a 20-e species having two unpaired electrons if both arene rings were coordinated in planar geometry. Instead, the complex is diamagnetic and features a structure[30] in which one hexamethylbenzene ligand behaves as an η^4-four electron donor and is nonplanar, thus enabling the metal to attain an 18-electron configuration. Several electron-rich metals form complexes featuring bending away of the uncoordinated portion of the arene ring or the destruction of ring aromaticity, as shown by alternating C—C bond lengths.

Most neutral bis(arene) complexes are soluble in organic solvents, can be sublimed, and are oxidized by air to mono cations. The strength of the metal–ring bond is less than that in ferrocene and is even weaker in the 19- and 20-electron compounds.

[30]G. Huttner and S. Lange, *Acta Cryst., Sect. B* 1972, *28*, 2049.

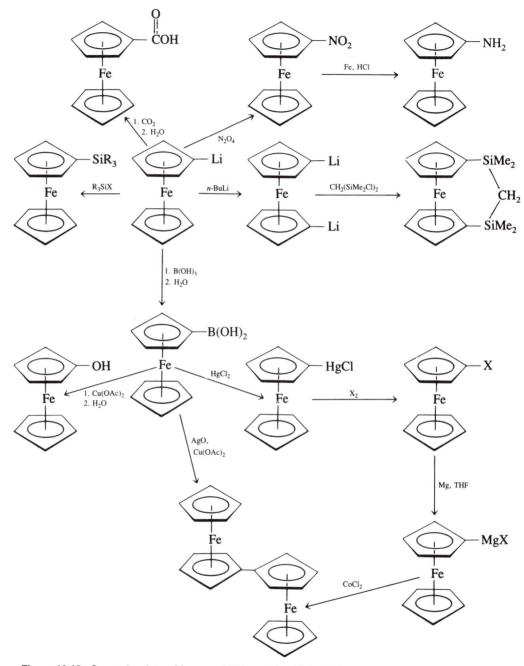

Figure 10.19 Some chemistry of ferrocenyl lithium. (After M. L. H. Green in *Organometallic Compounds*, Vol. II, *The Transition Elements*, Chapman and Hall, London, 1968.)

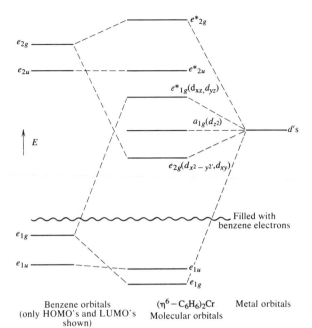

Figure 10.20 A qualitative MO diagram for $(\eta^6\text{-}C_6H_6)_2M$ complexes.

Benzene orbitals
(only HOMO's and LUMO's
shown)

$(\eta^6\text{-}C_6H_6)_2Cr$
Molecular orbitals

Metal orbitals

Coordination of an arene ligand usually is achieved by the Fischer-Hafner method, cyclic condensation of alkynes, or the replacement of other ligands. The Fischer-Hafner method, one of the earlier systematic ways of synthesizing arene complexes, is applicable to most of the transition metals [except for Ti(0)] and generally leads to bis(arene) products. A metal halide is allowed to react with the arene in the presence of $AlCl_3$ or $AlBr_3$ and Al metal, which acts as a reducing agent.

$$3CrCl_3 + 2Al + AlCl_3 + 6 \text{ arene} \rightarrow 3[(\text{arene})_2Cr]^+[AlCl_4]^- \qquad (10.69)$$

Cationic compounds can be reduced to neutral ones by aqueous dithionite $(S_2O_4^{2-})$. If no reduction of the metal is required, Al is omitted. The reaction conditions are those of the Friedel-Crafts reaction. The method is not applicable to aryl halides, which are dehalogenated. The cyclic condensations of three alkyne units to form arenes has been mentioned previously (Section 10.5.1).

Carbonyl ligands can be replaced by other donors, including arenes.

$$M(CO)_6 + \text{arene} \rightarrow (\text{arene})M(CO)_3 + 3CO \qquad (10.70)$$

$$M = Cr, Mo, W$$

The yields in Reaction 10.70 are low. Better starting materials have more easily replaced ligands.

$$(CH_3CN)_3Cr(CO)_3 + C_6H_6 \rightarrow (\eta^6\text{-}C_6H_6)Cr(CO)_3 + 3CH_3CN \qquad (10.71)$$

Halides are removed along with CO's by $AlCl_3$.

$$Mn(CO)_5Cl + arene + AlCl_3 \rightarrow [(arene)Mn(CO)_3]^+[AlCl_4]^- \qquad (10.72)$$

Carbonyl displacement is a good way to make mixed complexes.

$$(\eta^4\text{-}Ph_4C_4)Co(CO)_2Br + arene + AlBr_3 \rightarrow [(\eta^6\text{-}arene)(\eta^4\text{-}Ph_4C_4)Co]^+[AlBr_4]^- \qquad (10.73)$$

 In discussing the reactivity of coordinated arenes, we will concentrate on complexes in which the ligand is coordinated through its entire π system and is planar. A major organic reaction of free arenes is electrophilic attack—often leading to substitution. Coordination to a metal, as you might expect, draws away charge from the arene ring and deactivates the organic ligand to electrophilic attack. Although most bis(arene) complexes do not survive Friedel-Crafts reaction conditions, $(\eta^6\text{-}C_6H_6)Cr(CO)_3$ can be acetylated, although not very readily. Other evidence of the "electron-poor" state of coordinated arenes includes the fact that benzoic acid is a weaker acid than $(\eta^6\text{-}C_6H_5CO_2H)Cr(CO)_3$ and that $(\eta^6\text{-}C_6H_5NH_2)Cr(CO)_3$ is a weaker base than aniline.

 In contrast to free arenes, coordinated arenes are activated toward attack by nucleophiles, as the following reactions demonstrate.

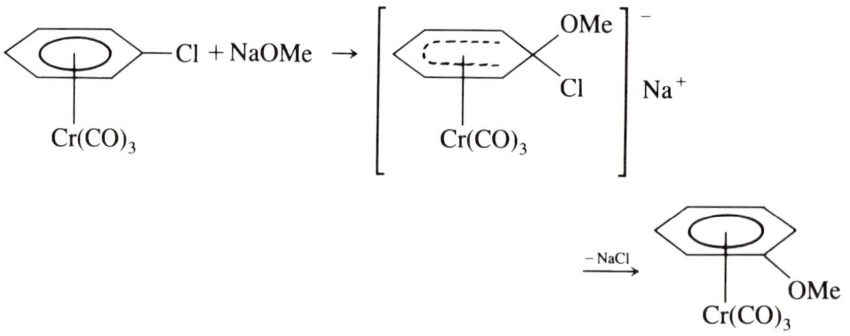

$$(10.74)$$

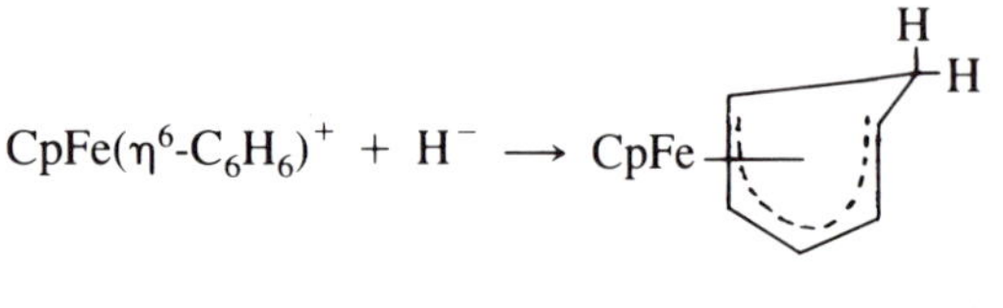

$$CpFe(\eta^6\text{-}C_6H_6)^+ + H^- \longrightarrow CpFe{-}\!\!\!\!\!\!\!\!\quad \qquad (10.75)$$

The other common reaction is replacement of the arene ligand with either other arenes or donor ligands. For example, arene ligands may be exchanged in the presence of $AlCl_3$:

$$[(\eta^6\text{-}1,3,5\text{-}Me_3C_6H_3)_2Cr]^+ + 2C_6H_6 \xrightarrow{AlCl_3} [(\eta^6\text{-}C_6H_6)_2Cr]^+ + 2(1,3,5\text{-}Me_3C_6H_3) \qquad (10.76)$$

 As a rule, bis(arene) complexes are somewhat less reactive than mono(arene) complexes.

Cycloheptatrienyl and Cyclooctatetraene Complexes

The seven-carbon ligand C_7H_8 (cycloheptatriene) behaves as a 3 × 2–electron donor toward metals. The coordination geometry shown in Figure 10.21, in which the saturated C is bent away from the metal, is typical. Such complexes with three isolated double bonds can undergo

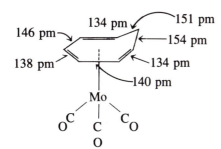

Figure 10.21 Molecular structure of $(C_7H_8)Mo(CO)_3$. (From J. D. Dunitz and P. Pauling, *Helv. Chim. Acta* 1960, *43*, 2188.)

hydride abstraction to bring the remaining C into conjugation, thus producing the planar delocalized tropylium or cycloheptatrienyl ligand $C_7H_7^+$.

$$(C_7H_8)Mo(CO)_3 + Ph_3C^+ \longrightarrow \left[\begin{array}{c} \\ Mo(CO)_3 \end{array} \right]^+ + Ph_3CH \qquad (10.77)$$

Attack by nucleophiles on tropylium complexes occurs on the ring and converts them back to cycloheptatriene species.

Most of the known cycloheptatrienyl complexes of the later transition metals are mixed-sandwich complexes, rather than those containing other ligands such as CO, Cl, or NO. This may be because complexes obeying the 18-electron rule would have formulas such as $(\eta^7\text{-}C_7H_7)Co(CO)$, and have one side open for attack, thus reducing kinetic stability.

Cyclooctatetraene (cot), like cyclobutadiene, is *not* a $(4n + 2)$ aromatic system in the uncoordinated state. In contrast to cyclobutadiene, cyclooctatetraene most often behaves toward transition metals as a ligand containing isolated double bonds. Examples in Table 10.4 show its functioning as a 2×2- or 3×2-electron donor featuring nonplanar geometry with nonbonded carbons bent away from the metal. With $Fe(CO)_3$ groups it acts as a 1×4- or 2×4-electron donor, behaving as one or two butadiene ligands (see Figure 10.6). It does, however, act as planar aromatic (10π electrons) $[C_8H_8]^{2-}$ in the complex $Ti(C_8H_8)Cp$, whose structure is shown in Figure 10.22. Ti is coordinated to one cot^{2-} planar ring, as well as to a Cp.

It is probable that the Ti–ring bonding has considerable ionic character and that, in general, d orbitals are not sufficiently diffuse and lack proper symmetry for effective overlap with the MO's of planar aromatic cot^{2-}. Hence the transition metal complexes of cyclooctatetraene seldom display planar $[C_8H_8]^{2-}$ rings. However, f orbitals in actinides are suitable for such overlap and the compound uranocene $(\eta^8\text{-}C_8H_8)_2U$ (Figure 3.11) has been prepared from UCl_4 and $[C_8H_8]^{2-}$ in THF. The green product is pyrophoric in air and not very soluble in organic solvents. In contrast to transition metal complexes containing aromatic rings, uranocene does not undergo electrophilic substitutions or metalation. Thorocenes also are known.[31,32]

[31] A. Streitweiser *et al.*, *J. Am. Chem. Soc.* 1973, *95*, 8644.

[32] C. Levanda and A. Streitweiser, *Inorg. Chem.* 1981, *20*, 656.

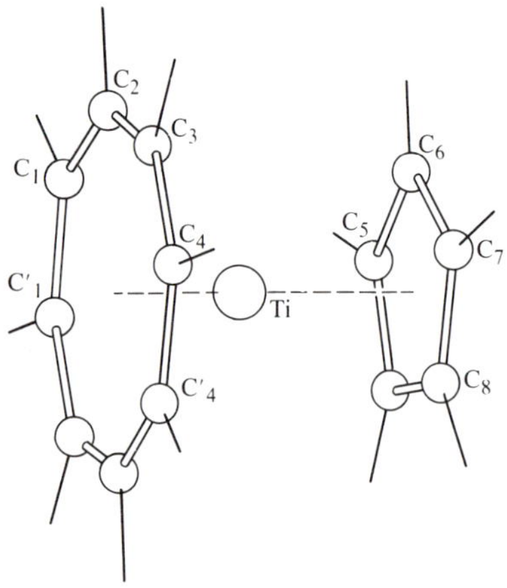

Figure 10.22 Molecular structure of $(C_8H_8)(C_5H_5)Ti$. (From P. A. Kroon and R. B. Helmholdt, *J. Organometal. Chem.* 1970, *25*, 451.)

10.6 COMPOUNDS WITH METAL–CARBON σ BONDS

10.6.1 Metal Alkyls[33]

In contrast to many organic π-donor ligands, alkyl groups bond to main-group metals as well as transition metals. In fact, the alkyls of the main-group metals are more numerous and more industrially and synthetically useful than those of the transition metals.

The method of choice for the preparation of the various σ-bonded organometallic compounds depends on the activity of the metal and of the organic moiety that provides the alkyl or aryl group (usually a halide). When the reactivity of these is very great, as in the case of Na and RX, coupling of the organic radical becomes the major path (termed the Wurtz reaction when the coupling produced is an alkane and the Fittig reaction when the product is aromatic). When the reactivity is too low, direct synthesis is either too slow or not possible.

Metal reactivity parallels the electropositive character, with the periodic trends as indicated.

$$
\begin{array}{lll}
\text{Li} \longleftarrow & & \text{Si} \\
\quad\downarrow \quad \text{Reactivity} & \uparrow \text{Cu} \quad \text{Zn} \quad \text{Ge} \\
& \text{Ag} \quad \text{Cd} \quad \text{Sn} \\
\text{Cs} & \text{Au} \quad \text{Hg} \quad \text{Pb}\downarrow
\end{array}
$$

For the alkyl or aryl halides, the order of reactivity is I > Br > Cl; and allyl or benzyl > alkyl > aryl. The temperature at which direct reaction of the metal with methyl chloride normally is carried out increases with a decrease in metal activity, and may be lowered by the presence of a Lewis base solvent or catalyst. Although a large number of organometallic compounds may be prepared by direct synthesis from the metal and an organic halide,[34] laboratory

[33]R. R. Schrock and G. W. Parshall, *Chem. Rev.* 1976, *76*, 243; P. J. Davidson, M. F. Lappert, and R. Pearce. *Chem. Rev.* 1976, *76*, 219.

[34]E. G. Rochow, *J. Chem. Eauc.* 1960, *43*, 58.

syntheses usually start with either a Grignard reagent, RMgX, or an organolithium reagent, RLi.

More active metals react with organometals of less active metals, displacing them from their compounds. On the other hand, the organometallic compounds of the less reactive metals may be formed by metathesis reactions between the halides or alkoxides of the less active metal and the alkyl or aryl derivatives of the more active metal. These reactions are driven by the large lattice energies of the inorganic products that often precipitate from the organic solvent. Typical reactions include

$$2 \text{ Na} + \text{R}_2\text{Hg} \xrightarrow{\text{hexane}} 2 \text{ NaR} + \text{Hg} \tag{10.78}$$

$$2 \text{ RMgX} + \text{HgX}_2 \xrightarrow{\text{ether}} \text{R}_2\text{Hg} + 2 \text{ MgX}_2 \tag{10.79}$$

$$3 \text{ PhLi} + \text{AlCl}_3 \cdot \text{OEt}_2 \xrightarrow{\text{ether}} \text{Ph}_3\text{Al} \cdot \text{OEt}_2 + 3 \text{ LiCl} \tag{10.80}$$

$$4 \text{ MeLi} + \text{TiCl}_4 \xrightarrow[-78°]{\text{THF}} \text{Me}_4\text{Ti} + 4 \text{ LiCl} \tag{10.81}$$

With Al alkyls, transfer is not always complete, leading to loss of efficiency.

$$3 \text{ Et}_3\text{Al} \cdot \text{OEt}_2 + 2 \text{ BF}_3 \cdot \text{OEt}_2 \xrightarrow{\text{ether}} 2 \text{ Et}_3\text{B} + 3 \text{ EtAlF}_2 + 5 \text{ Et}_2\text{O} \tag{10.82}$$

In the presence of alkyl aluminum, the following reaction occurs *only* for terminal olefins (compare with hydroboration, Section 15.8.3).

$$\text{Al} + 3/2 \text{ H}_2 + \text{RCH}{=}\text{CH}_2 \rightarrow (\text{RCH}_2\text{CH}_2)_3\text{Al} \tag{10.83}$$

The preparation of trialkyl aluminums from olefins and H_2 is very significant industrially, since the products are liquids and can serve as solvents as well as alkylating agents for use in Ziegler-Natta catalysts (see Section 10.7.1).

Transition metal alkyls can be made from olefins and hydrides.

$$\text{Pt(H)(PEt}_3)_2\text{Cl} + \text{C}_2\text{H}_4 \rightarrow \text{Pt(C}_2\text{H}_5)(\text{PEt}_3)_2\text{Cl} \tag{10.84}$$

A more widely used and convenient procedure is reaction of carbonylate anions with alkylating agents: reaction 10.28 is an example, as is

$$[\text{CpFe(CO)}_2]^- + \text{C}_6\text{H}_5\text{CH}_2\text{Cl} \rightarrow \text{CpFe(CO)}_2\text{CH}_2\text{C}_6\text{H}_5$$

Transition metals can also undergo *oxidative coupling* reactions such as

$$\text{Fe(CO)}_3(\text{C}_2\text{F}_4)_2 \rightarrow (\text{CO})_3\text{Fe}\underset{\text{CF}_2-\text{CF}_2}{\overset{\text{CF}_2-\text{CF}_2}{\big\langle}} \tag{10.85}$$

Since the alkyl metals of the most electropositive main-group elements have too few electrons and too many valence orbitals to obey the octet rule as monomers, these electron-deficient species form aggregates in which the "sigma" electrons of the alkyl group are delocalized in multicentered bonds. The tendency to form such bonds is proportional to the electropositive character of the metal; it decreases as the steric requirements of the alkyl group increase. In the solid state, LiMe exists as a tetramer having a tetrahedron of Li atoms, with each face triply bridged by a methyl group. The tetramer persists in diethyl ether, but hexamers form in hydrocarbon solvents.

The other alkali-metal alkyls perhaps are best considered as ionic compounds. They are rather intractable white solids that decompose on heating and are insoluble in virtually all solvents.

Both $BeMe_2$ and $MgMe_2$ exist in the solid as long-chain polymers in which the metals are tetrahedrally coordinated by bridging methyls. In contrast, R_2Cd and R_2Hg exist in solution as linear monomers.

Of the organometal halides, Grignard reagents are best known. The product that crystallizes from an ether solution of EtMgBr displays distorted tetrahedral geometry, with two moles of diethyl ether completing the octet around Mg. The species present in solution correspond to the following equilibria.

$$
R-Mg\overset{X}{\underset{X}{\diamond}}Mg-R \rightleftharpoons 2\,RMgX \rightleftharpoons R_2Mg + MgX_2 \rightleftharpoons \overset{R}{\underset{R}{}}Mg\overset{X}{\underset{X}{\diamond}}Mg \quad (10.86)
$$

The position of the various equilibria depends on R, X, and solvent. The basicity of various solvents and ligands with respect to alkyl magnesium is NR_2^-, $OR^- > py > THF$, $Cl^- > N$-methylpyrrole, 2-MeTHF, 1,4-dioxane $> Et_2O >$ tetrahydrothiophene $> n$-$C_6H_{11}F$. Br^- and I^- lie between Et_2O and THF.[35]

$(AlMe_3)_2$ is a dimer with two μ-methyl groups even in the vapor phase. In solution, straight-chain Al alkyls are also dimers, but i-Pr_3Al and t-Bu_3Al are monomers for steric reasons. Steric bulk also accounts for the monomeric structures of $M[CH(SiMe_3)_2]_2$ where M = Ge, Sn and Pb. Structures of electron-precise organometals (those obeying the octet or EAN rule) are those expected from related halides, etc.

Homoleptic[36] alkyls of the transition metals often are electron-deficient also. Aggregation is discouraged by bulky alkyl groups or if d electrons are unavailable for M—M bond formation. Hence $TiMe_4$ and WMe_6 are monomers. However, there are some known dimers containing multiple metal–metal bonds, such as $Mo_2(CH_2CMe_3)_6$ and $[Cr_2Me_8]^{4-}$.

The generally high states of aggregation exhibited by organo compounds of the most electropositive main-group metals lead to high melting and boiling points. Reactivity also parallels metal electropositivity, because of the resulting carbanionic character of organo groups and the energetic availability of vacant metal orbitals. Group IA, IIA, and IIIB organometals are all quite reactive with O_2, and most inflame spontaneously in air. Except for BR_3 compounds, hydrolysis to give R—H and M—OH occurs, often with explosive vigor. Reactivity toward CO_2 is confined to the most electropositive metal compounds, but these often inflame spontaneously in a carbon dioxide atmosphere. Organo compounds of the less electropositive metals are less reactive in general, although several still inflame spontaneously in air, including $CdMe_2$ and $ZnMe_2$. In contrast, $HgMe_2$ is stable to air and water. The organometals of Group IVB are the least reactive, since their more accessible valence orbitals already are employed

---

[35]H. Woltermann, Ph.D. thesis, University of Cincinnati, 1972.

[36]Homoleptic alkyls contain only metals and alkyl groups.

in bonding. However, they do react with more active reagents such as halogen and HX. Aryl compounds are less reactive than alkyl. For a given metal, the moisture- and oxygen-sensitivity of an organometallic halide parallels the sensitivity of the corresponding homoleptic alkyl.

The ability to form organo bridges provides a facile mechanism for alkyl exchange, making the more active metal compounds good alkylating agents. Availability of metal orbitals leads to Lewis acid behavior by Groups IA, IIA, and IIIB compounds. Many complexes are formed, such as $Me_3B \leftarrow NMe_3$, $Et_3Al \leftarrow OEt_2$, $Li[ZnMe_4] \cdot OEt_2$, and $Ph_2Mg \cdot OEt_2$. This can lead to interesting solvent effects on reactivity. For example, N,N,N',N'-tetramethylethylenediamine (tmed) complexes very strongly with Li^+, occupying two of the coordination sites. The added electron density considerably enhances the carbanionic character and hence the reactivity of the alkyl group in *n*-BuLi. In contrast, ethers form strong complexes with trialkylaluminums, blocking the one remaining coordination site and reducing the reactivity.

A very significant reaction of alkyl complexes containing β-H is so-called β-elimination.

$$M—CH_2CH_2R \rightleftharpoons M—\overset{\displaystyle H}{\underset{\displaystyle \underset{CH_2}{\overset{CH_2}{\|}}}{|}} \rightleftharpoons M—H + CH_2{=}CHR \qquad (10.87)$$

Olefin is eliminated (it may remain coordinated if M is a transition metal) and the β-H remains bonded to the metal. This reaction affords a very facile path for decomposition for M = Li, Na, Mg, Be, B, Al and for transition metals. The occurrence of the β-elimination led to such kinetic instability for transition metal alkyls that the inability to prepare them for a long time was ascribed to the weakness of the transition metal–C bond.[37] Recently, kinetically stable alkyls have been isolated through the use of alkyl groups such as CH_3, $CH_2C_6H_5$, and CH_2CMe_3, which have no β-H, thus supressing the elimination.

Thermochemical data indicate[38] that transition metal–carbon bonds are considerably stronger (100–200 kJ/mole) than had been realized earlier, though still somewhat weaker than M—F, M—OR, or M—Cl bonds (300–400 kJ/mole). Since the instability of transition metal alkyls is of kinetic rather than thermodynamic origin, these species can be stabilized by blocking reaction pathways. Because a vacant metal coordination site is required to form a hydride olefin complex from an alkyl, ligands that are strongly bonded and occupy all coordination positions act to stabilize alkyls. Examples include Cp, CO, and PR_3, as well as NH_3 in the substitution-inert $[Rh(NH_3)_5(C_2H_5)]^{2+}$. The bulky alkyls CH_2CMe_3 and CH_2SiMe_3 form stable homoleptic alkyls such as $Cr(CH_2SiMe_3)_4$ because they effectively block all coordination positions.

10.6.2 Metal Acyl Complexes

Metal acyls (which are unknown for all main-group metals except Si) are related to alkyl complexes. They are prepared from carbonylate anions and acyl halides

$$[CpFe(CO)_2]^- + C_6H_5\overset{\displaystyle O}{\overset{\|}{C}}Cl \rightarrow CpFe(CO)_2\overset{\displaystyle O}{\overset{\|}{C}}C_6H_5 \qquad (10.88)$$

[37]G. Wilkinson, *Science* 1974, *185*, 109.

[38]J. A. Connor, *Top. Curr. Chem.* 1977, *71*, 71.

and by insertion reaction (see Section 10.7.1). Their chemistry is discussed further in later sections of the chapter.

10.6.3 Compounds with Multiple Metal–Carbon Bonds

Carbenes and Alkylidenes

Species containing divalent carbon called *carbenes*, :C(XR)(R') where X is a heteroatom such as O or N, can be stabilized by coordination to transition metals. The first such complex was prepared in 1964 by Fischer via a route involving nucleophilic attack on coordinated CO, followed by alkylation of the acyl.

$$
W(CO)_6 + PhLi \rightarrow \left[(CO)_5W \overset{\overset{\displaystyle O}{\parallel}}{\underset{}{C}} Ph \right]^{-} Li^{+} \xrightarrow{Me_4N^+Cl^-}
$$

$$
\left[(CO)_5W \overset{\overset{\displaystyle O}{\parallel}}{\underset{}{C}} Ph \right]^{-} Me_4N^{+} \xrightarrow{H^+} (CO)_5W{=}C\overset{OH}{\underset{Ph}{\diagdown}}
$$

$$
\xrightarrow{CH_2N_2} (CO)_5W{=}C\overset{OCH_3}{\underset{Ph}{\diagdown}} \tag{10.89}
$$

It is more usual now to carry out the alkylation with trialkyloxonium salts or methyl fluorosulfonate.

Three canonical structures can be written for the metal—carbene bond

$$
M{=}C\overset{XR}{\underset{R'}{\diagup}} \leftrightarrow \overset{-}{M}{\leftarrow}\overset{+}{C}\overset{XR}{\underset{R'}{\diagup}} \leftrightarrow \overset{-}{M}{-}C\overset{\overset{+}{X}R}{\underset{R'}{\diagup}} \equiv M{\overset{.}{=}}C\overset{'XR}{\underset{R'}{\diagup}}
$$

$$
\quad\; (a) \qquad\qquad (b) \qquad\qquad (c)
$$

The third of these structures is thought to make the largest contribution.

These Fischer-type carbene complexes have metals in low oxidation states and are characterized by electrophilicity of the carbene C. The main feature of their chemistry is attack by nucleophiles.

$$(CO)_5W{=}C\underset{Ph}{\overset{OMe}{\big<}} \xrightarrow[-78°]{PhLi} (CO)_5W^{-}{-}\underset{Ph}{\overset{OMe}{\underset{|}{\overset{|}{C}}}}{-}Ph \xrightarrow[-78°]{HCl} (CO)_5W{=}CPh_2 \qquad (10.90)$$

Metallocarbene intermediates are thought to be attacked by olefins in the catalysis of olefin metathesis.[39]

Formally similar complexes of Nb and Ta in high oxidation states have recently been prepared by Schrock.[40] The route would be expected to result in the production of $Ta(CH_2CMe_3)_5$, since $TaMe_5$ is prepared in this way.

$$M(CH_2CMe_3)_3Cl_2 + 2Me_3CCH_2Li \rightarrow (Me_3CCH_2)_3M{=}C\underset{CMe_3}{\overset{H}{\big<}} + CMe_4 + 2LiCl$$

$$M = Nb, Ta \qquad (10.91)$$

The absence of β-hydrogens and the presence of such bulky ligands is thought to lead to intramolecular abstraction of an α-hydrogen and elimination of Me_4C from an $M(CH_2CMe_3)_4Cl$ or $M(CH_2CMe_3)_5$ intermediate.

The crystal structure of a related complex is shown in Figure 10.23a. In $Cp_2Ta(CH_2)CH_3$ a comparison can be made between the $Ta{-}CH_2$ and $Ta{-}CH_3$ distances. The ~20 pm shortening of the M—C distance for the methylene group compared to the methyl indicates the importance of the metal–carbon double-bond formulation. Moreover, NMR studies indicate a rather substantial (60–80 kJ/mole) barrier to rotation about the Ta=C bond. Since these features offer a striking contrast to the molecular and electronic structures of the Fischer-type carbene complexes, we distinguish them by calling them alkylidenes. Several of these complexes do not obey the EAN rule.

The presently known chemistry is also quite different, involving behavior of the alkylidene C as a *nucleophile*.

$$Cp_2TaMe(CH_2) + AlMe_3 \rightarrow Cp_2TaMe(CH_2AlMe_3) \qquad (10.92)$$

Alkylidene ligands evidently behave as dianions CRR'^{2-}. These Group VA compounds display several analogies with phosphorus ylids such as $Ph_3P{=}CH_2$.

Carbynes and Alkylidynes

Complexes containing a metal–carbon triple bond M≡CR are related to carbene and alkylidene complexes, respectively. Treatment of carbene complexes with electrophilic $BX_3(X = Cl, Br, I)$ removes one of the substituents on the carbene carbon, to give a metal carbyne.[41]

[39]T. J. Katz, *Adv. Organomet. Chem.* 1977, *16*, 283; N. Calderon, J. P. Lawrence, and E. A. Ofstead, *loc. cit.* 1979, *17*, 449.

[40]R. R. Schrock, *Acc. Chem. Res.* 1979, *12*, 98.

[41]E. O. Fischer, *Adv. Organomet. Chem.* 1976, *14*, 1.

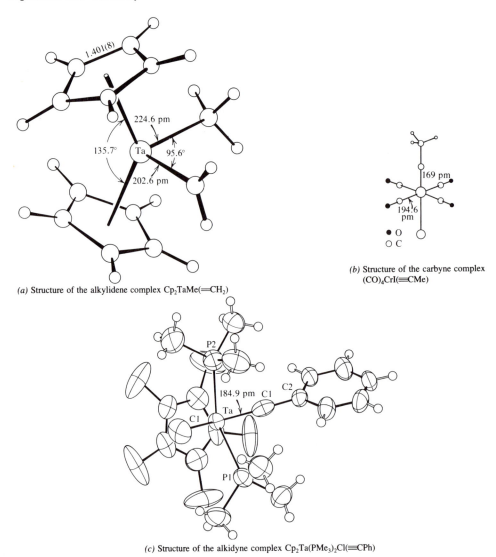

(a) Structure of the alkylidene complex Cp₂TaMe(═CH₂)

(b) Structure of the carbyne complex (CO)₄CrI(≡CMe)

(c) Structure of the alkidyne complex Cp₂Ta(PMe₃)₂Cl(≡CPh)

Figure 10.23 Structures of some complexes with M—C multiple bonds. (a is reprinted with permission from L. H. Guggenberger and R. R. Schrock, *J. Am. Chem. Soc.* 1975, *97*, 6578. Copyright 1975, American Chemical Society. *b* is from G. Huttner, H. Lorenz, and W. Gartzke, *Angew. Chem., Int'l Ed. Engl.* 1974, *13*, 609. *c* is reprinted with permission from S. J. McLain, C. D. Wood, L. W. Messerle, R. R. Schrock, F. J. Hollander, W. J. Youngs, and M. R. Churchill, *J. Am. Chem. Soc.* 1978, *100*, 5962. Copyright 1978, American Chemical Society.)

$$(CO)_5M{=}C\underset{R}{\overset{OCH_3}{<}} \;+\; BX_3 \xrightarrow[\substack{-\,BX_2OMe \\ -\,CO}]{} (CO)_4XM{\equiv}CR \qquad (10.93)$$

$$M = Cr,\ Mo,\ W \quad R = Me,\ Et,\ Ph \quad X = Cl,\ Br,\ I$$

The metal—C bond length is quite short, as we expect for a triple bond such as shown in Figure 10.23*b*.

The α-hydrogen of an alkylidene ligand can be removed by adding nucleophilic reagents.

$$CpTaCl(CH_2CMe_3)(=CHCMe_3) + 2\ PMe_3 \rightarrow$$

$$CpTa(PMe_3)_2Cl(\equiv CCMe_3) + Me_4C \quad (10.94)$$

Figure 10.23*c* shows the structure of the alkylidyne complex $CpTa(PMe_3)_2Cl(\equiv CPh)$. The linear $\equiv C$—Ph ligand forms a very short Ta$\equiv$C bond of 184.9 pm—as compared with a Ta$=$C bond distance of 203.0 pm in $Cp_2TaCl(=CHCMe_3)$.

10.7 SOME STOICHIOMETRIC REACTIONS OF ORGANOMETALLICS

10.7.1 Insertion Reactions

A large class of reactions involves insertion of small molecules X—Y into metal ligand bonds. As shown in Equations 10.95 and 10.96, both 1,1- and 1,2-insertions are possible.

$$M\text{—}L + X\text{—}Y \rightarrow M\text{—}\underset{|}{\overset{Y}{X}}\text{—}L \quad (10.95)$$

$$M\text{—}L + X\text{—}Y \rightarrow M\text{—}X\text{—}Y\text{—}L \quad (10.96)$$

The term *insertion* describes only the result of the reaction—it has no mechanistic significance. Many insertion reactions are reversible, and the reverse reaction is called *extrusion* or *elimination*. Table 10.9 shows a representative sampling of insertion reactions involving transition metals.

A good deal of what is known about CO insertion (or *carbonylation*)[42] comes from studies on $RMn(CO)_5$. In an early study, $CH_3Mn(CO)_5$ was shown to react with ^{14}CO to give $CH_3C(O)Mn(CO)_4(^{14}CO)$: that is, the inserted CO is one previously coordinated to the metal, not the labeled one added to the complex. This suggests that other Lewis bases L also can bring about insertion. The generally accepted mechanism is

$$RMn(CO)_5 + S(= solvent) \rightleftharpoons RC(O)Mn(CO)_4S$$
$$RC(O)Mn(CO)_4S + L \underset{\Delta}{\rightleftharpoons} cis\text{-}RC(O)Mn(CO)_4L \rightleftharpoons trans\text{-}RC(O)Mn(CO)_4L \quad (10.97)$$

An acyl intermediate is formed with solvent occupying one coordination position (if the solvent is basic enough). The Lewis base L enters to give the *cis* complex, which then establishes equilibrium with the *trans* isomer. Complexes of Fe, Co, Rh and Ir behave similarly.

The labeling experiments depicted in Figure 10.24 show that either the alkyl or one of the coordinated CO's may move during carbonylation, depending on the identity of the metal and other ligands. Many acyl complexes may be decarbonylated either thermally or photochemically.

---

[42]A. Wojcicki, *Adv. Organomet. Chem* 1973, *11*, 87; F. Calderazzo, *Angew. Chem., Int'l. Ed. Engl.* 1977, *16*, 299; E. J. Kuhlmann and J. J. Alexander, *Coord. Chem. Rev.* 1980, *33*, 195.

Table 10.9   Some representative insertion or group-transfer reactions[a] [From F. A. Cotton and G. Wilkinson, *Advanced Inorganic Chemistry*, 4th ed., Wiley, New York, 1980]

"Inserted" Molecule	*Bond*	*Product*
CO	M—CR$_3$	MCOCR$_3$
	M—OH	MCOOH
	M—NR$_2$	MCONR$_2$
CO$_2$	M—H	MCOOH
	M—C	MC(O)OR or MO$_2$CR
	M—NR$_2$	MOC(O)NR$_2$
	M—OH	MOCO$_2$H
CS$_2$	M—M	MSC(S)M
	M—H	MS$_2$CH and MSC(S)H
C$_2$H$_4$	M—H	MC$_2$H$_5$
C$_2$F$_4$	M—H	MCF$_2$CF$_2$H
RC≡CR′	M—H	MC(R)=CH(R′)
CH$_2$=C=CH$_2$	M—R	M(η^3-allyl)
SnCl$_2$	M—M	MSn(Cl)$_2$M
RNC	M—H	MCH=NR^a
	M—R′	MCR′=NR
	M—η^3-C$_3$H$_5$	MC(=NR)(CH$_2$CH=CH$_2$)
RNCS	M—H	M(RNCHS)^b
RN=C=NR	M—H	M(RNCHNR)^b
SO$_2$	M—C	MS(R)O$_2$ or MOS(O)R
	M—M^c	MOS(O)M
	M—(η^3-C$_3$H$_3$)	MSO$_2$CH$_2$CH=CH$_2$
SeO$_2$	M—C	MSeO$_2$R
SO$_3$	M—C	M—OSO$_2$R
O$_2$	M—CR$_3$	M—OOH
	M—CR$_3$	M—OOCR$_3$

[a]The *N*-aryl or alkyl formimidoyl ligand can also act as a bridge between three Os atoms; see R. D. Adams and N. M. Golembeski, *J. Am. Chem. Soc.*, 1978, *100*, 4622.
[b]S. D. Robinson et al., *Inorg. Chem.*, 1977, *16*, 2722, 2728.
[c]Pd, L. S. Benner et al., *J. Organomet. Chem.*, 1978, *153*, C31.

In contrast to CO, SO$_2$ inserts directly into M—C bonds, often producing an *S*-sulfinate.[43]

$$\text{CpFe(CO)}_2\text{R} + \text{SO}_2(l) \xrightarrow{-10°C} \text{CpFe(CO)}_2\underset{\underset{\text{O}}{\|}}{\overset{\overset{\text{O}}{\|}}{\text{S}}}\text{R}$$ (10.98)

[43]Some other products can be formed with other metals, as Table 10.9 indicates.

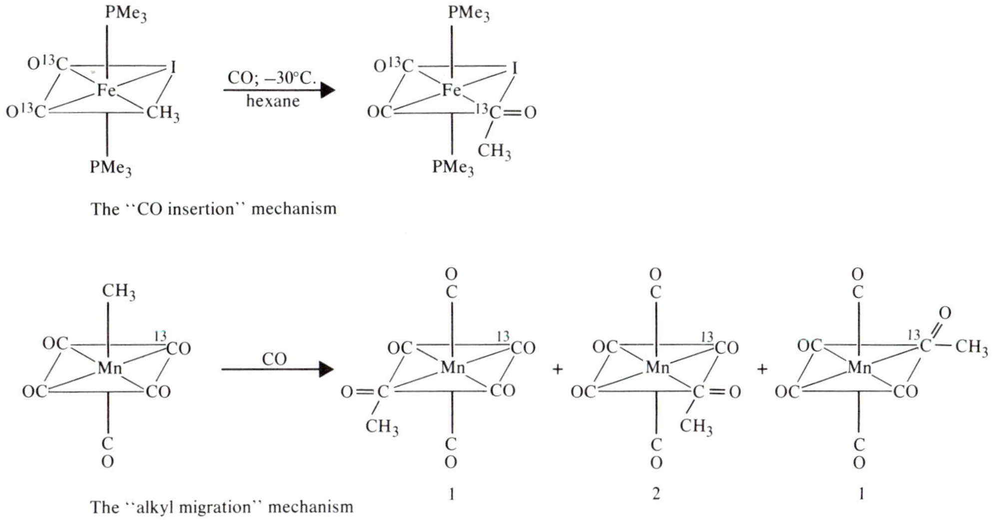

The "CO insertion" mechanism

The "alkyl migration" mechanism

Figure 10.24 Stereochemical course of carbonylation.

Kinetic studies have shown that SO_2 behaves as a Lewis acid, attacking the alkyl ligand rather than the metal.[44] The mechanism is depicted in Figure 10.25.

Numerous examples of insertions into a metal—H bond are known for olefins.[45] The most investigated reactions involve Pt—H bonds—for example,

$$\textit{trans-}[Pt(PEt_3)_2H(acetone)]^+ + CH_2\!\!=\!\!CH_2 \rightleftharpoons \textit{trans-}[Pt(PEt_3)_2(C_2H_5)(acetone)]^+ \quad (10.99)$$

Olefin insertions probably involve π coordination of olefin to metal *cis* to the M—H bond, followed by hydride migration.

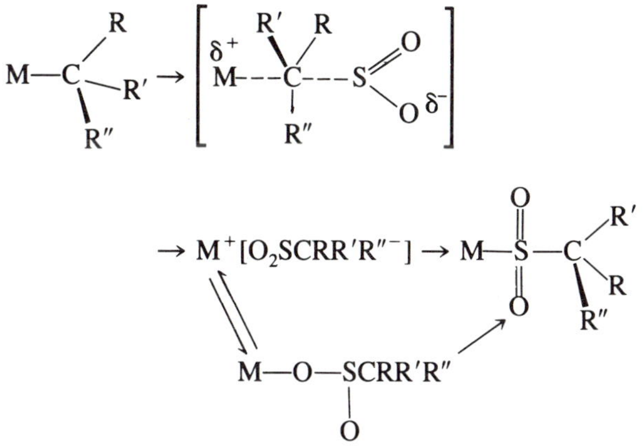

Figure 10.25 Mechanism of SO_2 insertion.

[44]A. Wojcicki, *Adv. Organomet. Chem.* 1974, *12*, 31.

[45]G. H. Olivé and S. Olivé, *Top. Curr. Chem.* 1976, *67*, 107.

$$M \leftarrow \|\begin{array}{c} H \\ | \\ C\beta \\ \| \\ C\alpha \end{array}\begin{array}{c} H \\ \diagup \\ \diagdown \\ H \end{array} \rightleftharpoons \left[\begin{array}{c} H---C\beta \\ | \quad | \\ M---C\alpha \end{array}\begin{array}{c} H \\ \diagdown \\ H \\ \diagdown \\ H \end{array}\right]^{\ddagger} \rightleftharpoons M-C\alpha\begin{array}{c} H \\ | \\ \beta \\ H \end{array}\begin{array}{c} H \\ \diagup \\ \diagdown \\ H \end{array}$$

(10.100)

The forward reaction converts the hydrido-olefin complex into an alkyl complex by migration of hydride to the β-carbon of the olefin, and in the process vacates a coordination position. Note that the oxidation state of the metal does not change. The metal and the hydride add in *cis* fashion to the olefin.[46]

The reverse of olefin insertion into an M—H bond is, of course, the β-elimination reaction of metal alkyl complexes, which can occur only if a vacant coordination position is available and β-hydrogen is present. If these conditions are fulfilled, a facile route exists for the decomposition of metal alkyl complexes. (See Section 10.6.1)

A recent study[47] of the thermal decomposition of $CpFe(CO)(PPh_3)(n\text{-}C_4H_9)$ according to Equation 10.101 revealed that the reaction

$$CpFe(CO)(PPh_3)(n\text{-}C_4H_9) \rightarrow CpFe(CO)(PPh_3)H + \text{butene} \qquad (10.101)$$

was severely retarded by added PPh_3. This suggests that dissociation of PPh_3 to create a vacant coordination position on Fe is a necessary step in the decomposition via β-elimination.

Few examples of olefin insertion into M—C bonds are known. Ethylene polymerization catalyzed by Ti(III) salts and Al alkyls (Ziegler-Natta catalysts)[48] may involve such an insertion. R_3Al could alkylate the surface of a $TiCl_3$ crystal. Coordination of ethylene to a vacant Ti site may be followed by insertion, coordination of ethylene, etc.

$$R_3Al + TiCl_3 \rightarrow R\text{—}Ti \xrightarrow[\;\;C_2H_4\;\;]{CH_2=CH_2} R\text{—}Ti \rightarrow RCH_2CH_2Ti \xrightarrow{C_2H_4}$$

$$RCH_2CH_2\text{—}Ti \xrightarrow[\;\;]{CH_2=CH_2} R(CH_2CH_2)_2Ti \xrightarrow{C_2H_4} \text{etc.} \qquad (10.102)$$

However, the usual preference for olefin insertion is into M—H, rather than M—C, bonds.

$$Cp_2Nb\begin{array}{c} H \\ \diagup \\ \diagdown \\ CH_2 \\ \| \\ CH_2 \end{array} + C_2H_4 \xrightarrow{60°} Cp_2\,Nb\begin{array}{c} CH_2 \\ \diagdown \| \\ CH_2 \\ \diagdown \\ C_2H_5 \end{array} \xrightarrow{C_2H_4} N.R. \qquad (10.103)$$

---

[46]This insertion is related to attack on coordinated olefins by nucleophiles (Section 10.7.4), except that here the hydride nucleophile is precoordinated to the metal.

[47]D. L. Reger and E. C. Culbertson, *J. Am. Chem. Soc.* 1976, *98*, 2789.

[48]See H. Sinn and W. Kaminsky, *Adv. Organomet. Chem.* 1980, *18*, 99.

In spite of the reactivity of M—H bonds toward olefin insertion, no examples of CO insertion to give an isolable MC(O)H formyl complex have yet been found. This likely results from the greater M—H than M—C bond strength.

10.7.2 Oxidative Addition[49]

A ubiquitous reaction in organometallic chemistry is *oxidative addition,* in which a low-valent transition metal complex reacts with a molecule XY to yield a product in which both the oxidation number and coordination number of the metal are increased. An example is the reaction of Vaska's compound $Ir(CO)Cl(PPh_3)_2$ with Cl_2,[50] shown in Equation 10.104.

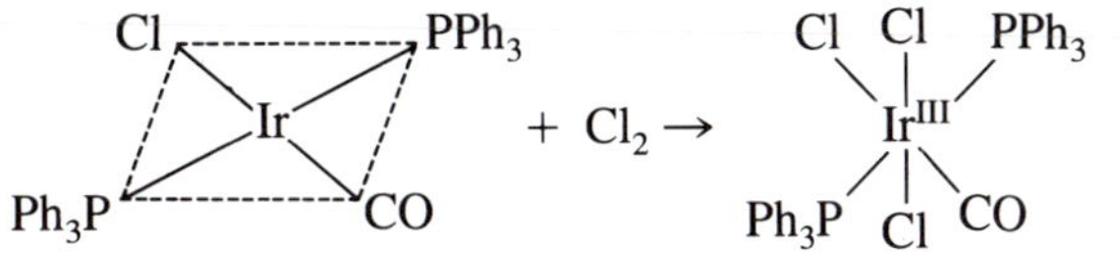

$$\text{(structure: } Cl \cdots PPh_3 / Ir / Ph_3P \cdots CO) + Cl_2 \rightarrow \text{(structure: } Cl\ Cl\ PPh_3 / Ir^{III} / Ph_3P\ Cl\ CO) \tag{10.104}$$

The d^8, four-coordinate, 16-electron starting complex is converted into a d^6, six-coordinate, 18-electron product by addition of Cl_2. The oxidation number of Ir increases from one to three. Consistent with this change, $\nu_{C\equiv O}$ increases from 1967 cm^{-1} to 2075 cm^{-1}. The Cl—Cl bond is broken and two Ir—Cl bonds are formed (in this case, in the *trans* position). Various molecules (see Table 10.10) can react in this fashion, including XY = Cl_2, Br_2, I_2, HX, RX, RC(O)X, H_2, R_3SiH, and R_3GeH.[51] As Table 10.6 shows, oxidative addition products of carbonyl complexes exhibit increased values of $\nu_{C\equiv O}$. The reverse of oxidative addition is termed *reductive elimination.*

Other Group VIII complexes having d^8 metals also undergo oxidative addition; the tendency to become oxidized to d^6 increases on going down a triad or to the left, as shown in Figure 10.26. Lower-valent metals in Group VIII tend to form five-coordinate complexes that obey the 18-electron rule (coordinatively saturated). Hence their oxidative addition reactions also involve ligand loss, as shown in Equation 10.105.

$$Fe(CO)_5 + Br_2 \rightarrow cis\text{-}[Fe(CO)_4Br_2] + CO \tag{10.105}$$

The mechanisms for these oxidative additions differ (see below) from that for coordinatively

Increasing Tendency to Five-Coordination ←

Fe(0)	Co(I)	Ni(II)
Ru(0)	Rh(I)	Pd(II)
Os(0)	Ir(I)	Pt(II)

Increasing ease of oxidative addition ↓

Figure 10.26 Reactivity of Group VIII d^8 metal atoms and ions.

[49]J. P. Collmann and W. R. Roper, *Adv. Organomet. Chem.* 1968, 7, 53.

[50]See L. Vaska. *Acc. Chem. Res.* 1968, 1, 335.

[51]Oxidative addition reactions could be thought of as insertions into the X—Y bond. However, the term *insertion* conventionally is reserved for reactions involving no change in metal oxidation number.

Table 10.10   Substances that can be added oxidatively[a] [From F. A. Cotton and G. Wilkinson, *Advanced Inorganic Chemistry*, 4th ed., Wiley, New York, 1980.]

Atoms Separate	Atoms Remain Attached
$X—X$	O_2
H_2,	$\overset{\cdot}{S}O_2$
Cl_2, Br_2, I_2, $(SCN)_2$, $RSSR^b$	
	$CF_2{=}CF_2$, $(CN)_2C{=}C(CN)_2$
$C—C$	
	$RC{\equiv}CR'$
$Ph_3C—CPh_3$, $(CN)_2$ C_6H_5CN,	$RNCS$
$MeC(CN)_3$	
	$RNCO$
$H—X$	
	$RN{=}C{=}NR'$
HCl, HBr, HI, $HClO_4$, C_6F_5OH, C_6H_5SH, H_2S, H_2O^c,	
CH_3OH	
$C_6F_5NH_2$ 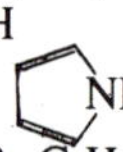 ⟨NH, $HC{\equiv}CR$, C_5H_6, CH_3CN, HCN,	
HCO_2R, C_6H_6, C_6F_5H, $HSiR_3$, $HSiCl_3$	
$H{-}B_{10}C_2HPMe_2$, $H{-}B_5H_8$	
	$RCON_3$
$C—X$	
	$R_2C{=}C{=}O$
CH_3I, C_6H_5I; CH_2Cl_2, CCl_4	
CH_3COCl, $C_6H_5CH_2COCl$	
C_6H_5COCl, CF_3COCl	
	CS_2
$M—X$	
	$(CF_3)_2CO$, $(CF_3)_2CS$, CF_3CN
Ph_3PAuCl, $HgCl_2$, $MeHgCl$, R_3SnCl, R_3SiCl, $RGeCl_3$	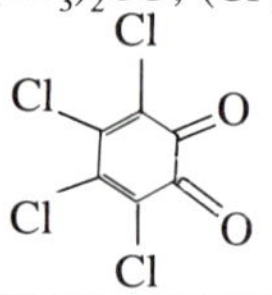 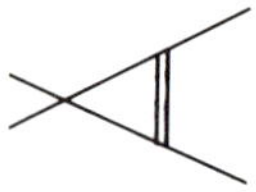
H_8B_5Br, Ph_2BX	
Ionic	
PhN_2^+ BF_4^-, Ph_3C^+ BF_4^-	

[a]Oxidative addition can occur with neutral, anionic, and cationic complexes. Additions to more than one metal center are also possible. See R. Poilblanc, *Nouv. J. Chim.*, 1978, 2, 145.
[b]A. W. Gal et al., *Inorg. Chim. Acta*, 1979, 32, 235.
[c]T. Yoshida et al., *J. Am. Chem. Soc.*, 1979, 101, 2027; B. N. Chaudret et al., *J.C.S. Dalton*, 1977, 1546.

unsaturated 16-electron complexes. As you might expect, increasing the basicity of the ligands in a complex increases its reactivity toward oxidative addition. Steric effects also play a role, with bulky ligands retarding attack by the oxidant.

Pearson[52] has pointed out that reactions proceeding with reasonably low activation energies involve the flow of electrons between orbitals on each reactant with the same symmetry properties. For the oxidative addition reactions under discussion, electrons must flow from a filled metal orbital into an antibonding X—Y orbital; this then allows the X—Y bond to be broken and new bonds to the metal to be formed. As shown in Figure 10.27, the X—Y antibonding orbital must overlap in phase with a filled metal orbital. The figure shows three possibilities for such overlap, leading to three different mechanisms for oxidative addition, of X—Y to a square planar ML_4 molecule.

[52]R. G. Pearson, *Symmetry Rules for Chemical Reactions*, Wiley, New York, 1976.

Both *cis* and *trans* addition have been observed. However, these stereochemical results at the metal are not mechanistically diagnostic, since there is no proof that the geometries are controlled kinetically.

Because many d^8 complexes are coordinatively saturated species, their reaction mechanisms must be different from those of four-coordinate species. Figure 10.28 depicts possible pathways for oxidative addition. Pathway *d* involves nucleophilic attack by the metal on XY to afford an $[ML_5X]^+$ species that may or may not undergo subsequent attack by Y^- to displace L. A probable case of mechanism *e* is Equation (10.106), which occurs only under conditions vigorous enough to promote ligand dissociation—in contrast to H_2 addition to square planar complexes, which takes place at room temperature and atmospheric pressure.

$$H_2 + Os(CO)_5 \xrightarrow[80\ atm]{100°C} H_2Os(CO)_4 + CO \qquad (10.106)$$

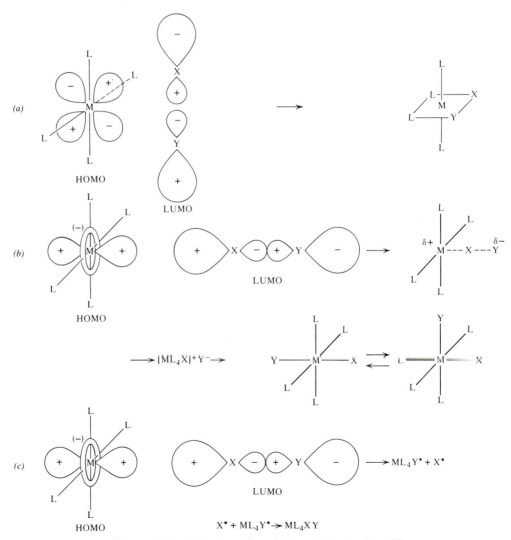

Figure 10.27 Mechanisms for oxidative addition to planar ML_4.

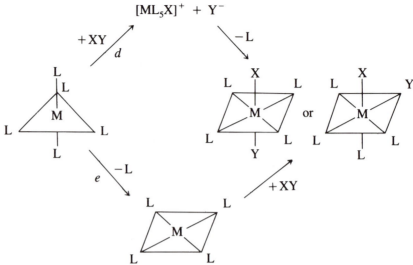

Figure 10.28   Mechanisms for oxidative addition to ML_5. (Reproduced with permission from R. G. Pearson, *Acc. Chem. Res.* 1971, *4*, 152. Copyright 1971, American Chemical Society.)

Other Electron Configurations

Oxidative-addition and reductive-elimination reactions involving other electron configurations also are known.

$d^{10} \rightarrow d^8$

$$Pt(PPh_3)_3 + RBr \rightarrow \textit{trans}\text{-}RPt(PPh_3)_2Br + PPh_3 \qquad (10.107)$$

Free-radical mechanisms are known to occur more often in oxidative additions of RBr and RI to Pt(0) and Ir(I) than to Pd(0) and Rh(I). The reactivity order for d^{10} complexes in general parallels ease of oxidation by two units.

$$Pt(0) \sim Pd(0) \sim Ni(0) > Au(I) > Cu(I) \gg Ag(I)$$

$d^8 \rightarrow d^7$

Some dinuclear complexes of Rh undergo oxidative addition at two centers.

$$Rh_2[\mu_2\text{-}CN(CH_2)_3NC]_4^{2+} + 4Br_2 \rightarrow [BrRh\mu_2\text{-}(CN(CH_2)_3NC)_4RhBr](Br_3)_2 \quad (10.108)$$

$d^7 \rightarrow d^6$

$[Co(CN)_5]^{2-}$ undergoes free-radical oxidative addition in one-electron steps.

$$\overset{\cdot}{Co}(CN)_5^{2-} + RX \rightarrow Co(CN)_5X^{2-} + \overset{\cdot}{R}$$
$$\underline{\overset{\cdot}{R} + \overset{\cdot}{Co}(CN)_5^{2-} \rightarrow RCo(CN)_5^{2-}}$$
$$2[Co(CN)_5]^{2-} + RX \rightarrow [RCo(CN)_5]^{2-} + [Co(CN)_5X]^{2-} \qquad (10.109)$$

$$2[Co(CN)_5]^{2-} + H_2 \rightarrow 2[HCo(CN)_5]^{2-} \qquad (10.110)$$

$d^6 \rightarrow d^4$

$$W(CO)_4(bipy) + SnCl_4 \rightarrow W(CO)_3(bipy)(SnCl_3)Cl \qquad (10.111)$$

We can view reactions of the main-group elements as oxidative additions—for example, the addition of F_2 to BrF_3 and PF_3, yielding BrF_5 and PF_5.

Reductive Elimination

Many species can undergo reductive elimination. Sometimes the molecule eliminated is different from the one added oxidatively.

An interesting oxidative addition reaction is the so-called ortho-metallation, in which the *ortho* C—H bond of a phenyl group on a coordinated aromatic phosphine or phosphite adds to the metal to yield an aryl-hydrido complex. In the presence of D_2, the reductive elimination of a C—D bond can lead to complete ortho deuteration, because the addition is reversible.

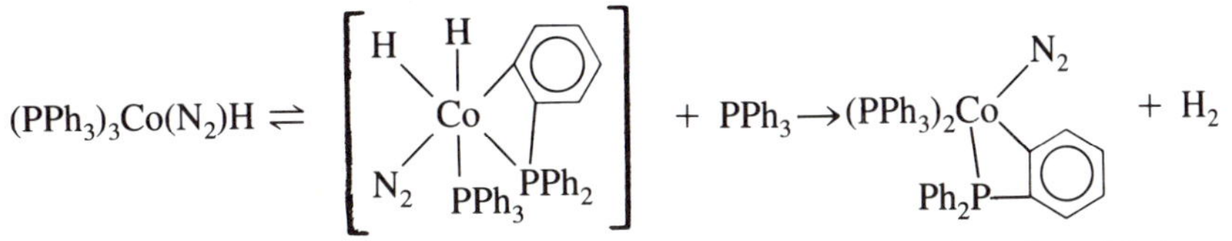

$$(PPh_3)_3Co(N_2)H \rightleftharpoons \left[\begin{array}{c} \text{complex structure} \end{array} \right] + PPh_3 \rightarrow (PPh_3)_2Co \quad + H_2 \qquad (10.112)$$

The high C—H bond energy ($\sim$470 kJ/mole) makes this activation via oxidative addition very significant.

Addition of Oxygen—An Intermediate Case

Some molecules X=Y undergo oxidative addition without cleavage of the X—Y bond (see Table 10.10). A good example is the addition of O_2 to $Ir(PPh_3)_2(CO)X$. The adduct could be described formally as either of the two valence-bond structures shown below.

(a)   *(b)*

Structure *(a)* involves a six-coordinate Ir(III) complex of O_2^{2-}, whereas *(b)* involves a five-coordinate Ir(I) complex of neutral O_2. As usual, structural data can help make the distinction. For the X = Cl complex, the O—O distance is 130 pm—comparable to that (128 pm) in the superoxide ion O_2^-. This indicates transfer of electron density from the metal into the π^* orbitals of O_2. When X = I, the O—O distance is 151 pm, very near that (149 pm) in organic peroxides R—O—O—R. Obviously, as the metal complex becomes a better base, an entire pair of electrons is donated to the oxygen ligand, giving an Ir(III) complex. These results show that all degrees of electron donation are possible. Table 10.6 presents several adducts of $Ir(PPh_3)_2(CO)Cl$ with neutral molecules in which electron density is transferred from the metals, as shown in $\nu_{C\equiv O}$ values. Because all X—Y bonds are not broken in the adduct, we would *formally* label these complexes of Ir(I).[53] The actual extent of electron transfer can be deter-

---

[53]Another way to think of these reactions is as the addition of neutral Lewis *acid* ligands.

mined only by structural and spectroscopic data. This is, of course, the same consideration alluded to in the discussion of the π-acid properties of olefins (see Section 10.4.1).[54]

10.8 HOMOGENEOUS CATALYSIS BY SOLUBLE TRANSITION-METAL COMPLEXES

Much of the impetus for the development of organometallic chemistry over the past 50 or so years has arisen from the ability of transition metals to catalyze various kinds of organic transformations. The function of a catalyst is, of course, to provide an alternative, low-activation-energy path by which a reaction can occur. The catalytic species is regenerated and can be used repeatedly, rather than being consumed stoichiometrically. The best catalysts are both effective (producing high product yields rapidly) and selective (producing mainly the desired product). The catalytic process is thought to involve several of the stoichiometric reactions that have been discussed in this chapter: coordination of ligands to metals, insertion of olefins into M—C and M—H bonds, insertion of CO into M—C bonds, and oxidative addition—as well as the reverse of each of these reactions (dissociation, β-elimination, decarbonylation, and reductive elimination, respectively). Under reaction conditions as practiced industrially (usually involving high temperatures and pressures), organometallic species often are too labile kinetically to be detected spectroscopically. (After all, if a catalyst is to be useful, any reaction intermediates ought to react rather rapidly to form later intermediates and products). Hence many of the steps in catalytic sequences are postulated to occur by analogy with reactions that are known to take place with related but more stable systems and/or under milder reaction conditions.

Catalysis may occur homogeneously (catalyst, reactants, and products all in the same phase) or heterogeneously (usually with the catalyst in a solid phase and reactants and products in a solution phase).[55] This section focuses on homogeneously catalyzed reactions (which also serve as models for possible mechanisms in heterogeneous catalysis), because organometallic species thought to be present resemble those treated in this chapter. Evidence for their presence can be amassed by solution spectroscopic and kinetic studies on model systems and, sometimes, even under actual industrial conditions.[56] Such studies have revealed considerable information about the mechanisms of homogeneous catalytic processes and so have contributed to the design of better catalysts.

10.8.1 Feedstocks for Chemical Industry

Substances available in large quantity directly from natural sources or involving only minimal treatment of naturally obtained mixtures are the starting materials, or *feedstocks,* for industrial synthesis. Over the past 25 years the heavy organic chemicals industry has come to rely almost entirely on feedstocks derived from petroleum. Crude petroleum is distilled initially into a

---

[54]For a review of the chemistry of transition-metal dioxygen complexes, see J. S. Valentine, *Chem. Rev.* 1973, *73,* 235.

[55]Heterogeneous catalysis is used extensively in chemical industry. See Section 15.10.1.

[56]See, for example, D. E. Morris and H. B. Tinker, *Chem. Tech.* 1972, *2,* 554.

fraction volatile below 400°C and a nonvolatile residual oil. The volatile fraction is then redistilled into several cuts: *(a)* C_1—C_4 hydrocarbons, *(b)* light gasoline, *(c)* naptha (or heavy gasoline), *(d)* kerosene, and *(e)* light gas oil in order of increasing boiling temperature. The residual oil is distilled under reduced pressure to yield heavy gas oils and a second high-boiling residue, which can be utilized for fuel oil. Heavy oils are "cracked"—that is, broken up into lower-molecular-weight hydrocarbons suitable for fuel use. A dual-function heterogeneous catalyst permits concurrent "reforming," the conversion of straight-chain to branched-chain hydrocarbons for smoother burning.

Thermal (noncatalytic) cracking of the naptha fraction yields olefins and acetylene—the exact product mixture depending on the temperature. This method is used in Europe for olefin production. In the United States, natural gas (which consists of CH_4 and other volatile hydrocarbons) is used to produce olefins from cracking the heavier fractions. These olefins (ethylene, propylene, C_4 and C_5 unsaturated hydrocarbons) along with methane and synthesis gas (CO + H_2) constitute the principal building blocks for the heavy organic chemicals industry. Also used are higher alkanes and arenes, most of which are oxidized catalytically by O_2.

Synthesis gas is prepared from the combustion of CH_4 or other hydrocarbons to give a mixture of CO and H_2O. These are equilibrated catalytically with H_2 and CO_2 in the water gas shift reaction.

$$CO + H_2O \rightleftharpoons CO_2 + H_2 \qquad (10.113)$$

Condensation of water and CO_2 leaves H_2 and CO. Alternatively, coke (C) prepared by heating coal in the absence of air can be allowed to react with hot steam in the water-gas reaction:

$$C + H_2O \rightarrow CO + H_2 \qquad (10.114)$$

The requisite heat is supplied by admitting air and burning coke to CO_2 ($\Delta H^0 = -394$ kJ/mole); after the CO_2 is vented, steam is admitted to the hot coke bed and the endothermic water gas reaction ($\Delta H^0 = +131$ kJ/mole) cools the reactants. Admission of air converts coke to CO_2 and heat, and begins another cycle.

The task of the organic chemicals industry is to convert the available feedstocks—olefins, alkanes, arenes, and synthesis gas—to compounds possessing additional functional groups that confer desired properties or reactivity. Among the industrially important compounds produced are alcohols, aldehydes, acids, and polymers. This chapter briefly discusses only a few of the many industrial reactions that are catalyzed homogeneously.[57]

As petroleum grows more scarce and more expensive, it becomes increasingly desirable to make optimal use of petroleum and to replace petroleum feedstocks. Synthesis gas can be obtained from cheaper coal, as well as from combustion of petroleum hydrocarbons. Direct combination of CO and H_2 could produce CH_3OH and other alcohols. At least in theory, hydrocarbons could be produced by hydrogenation of CO! Synthesis-gas chemistry promises to increase in importance. Carbon monoxide hydrogenation can be catalyzed heterogeneously by solid oxides of transition metals (the Fischer-Tropsch synthesis), but the process is not yet very economically feasible. Heterogeneous catalysis and its relation to inorganic chemistry are discussed more thoroughly in Section 15.9.2.[58]

---

[57]For a more complete treatment and many excellent references, see G. W. Parshall, *Homogeneous Catalysis*, Wiley-Interscience, New York, 1980.

[58]Significant chemistry related to homogeneous stoichiometric reduction of CO has been discovered by Bercaw and coworkers. See P. T. Wolczanski and J. E. Bercaw, *Acc. Chem. Res.* 1980, *13*, 121.

10.8.2 Hydroformylation

Olefins can add the elements of H_2 and CO across the double bond to give aldehydes in the presence of a Co (or Rh) catalyst, the net reaction being

$$RCH{=}CH_2 + H_2 + CO \xrightarrow{\;Co\;} RCH_2CH_2\overset{\displaystyle O}{\overset{\|}{C}}H + \underset{\underset{HC=O}{|}}{RCHCH_3} \qquad (10.115)$$

The relative amounts of straight- and branch-chain aldehydes produced depend on the identity of R and other constituents of the reaction mixture. Straight-chain aldehydes, the more desirable products, usually are hydrogenated, affording straight-chain alcohols, or self-condensed, affording more complex aldehydes.

The process is run at high pressure (~200 atm) and temperature (~90 to 140°) in the presence of a solvent. Finely divided Co metal or any Co salt can be used as the catalyst precursor, which is converted to $Co_2(CO)_8$ by reduction and CO coordination. The active catalyst is $HCo(CO)_4$, formed via the reaction

$$Co_2(CO)_8 + H_2 \rightarrow 2HCo(CO)_4 \qquad (10.13)$$

High pressures of CO are required to prevent decomposition of the active catalyst. The first step in the catalytic process involves CO dissociation followed by olefin coordination.

$$HCo(CO)_4 \rightleftharpoons HCo(CO)_3 + CO \qquad (10.116)$$

$$HCo(CO)_3 + RCH{=}CH_2 \rightleftharpoons H(CO)_3\,Co\!-\!\!\overset{R\;\;\;H}{\underset{H\;\;\;H}{\diagup\!\!\diagdown}} \qquad (10.117)$$

Subsequently, the coordinated olefin is inserted into the Co—H bond and the resulting alkyls are stabilized by CO coordination, giving 18-electron species.

$$\underset{H\;\;\;H}{\overset{R\;\;\;H\;\;H}{\diagup\!\!\diagdown}}\!\!\overset{C}{\underset{C}{\|}}\!\!-\!Co(CO)_3 \rightleftharpoons RCH_2CH_2Co(CO)_3 + CH_3\underset{R}{CH}Co(CO)_3 \qquad (10.118)$$

$$RCH_2CH_2Co(CO)_3 + CO \rightleftharpoons RCH_2CH_2Co(CO)_4 \qquad (10.119)$$

$$CH_3\underset{R}{CH}Co(CO)_3 + CO \rightleftharpoons CH_3\underset{R}{CH}Co(CO)_4 \qquad (10.120)$$

Note that the hydride can migrate to either of the two olefin β-carbons, yielding two possible alkyl products. Now CO can be inserted to give acyl complexes

$$RCH_2CH_2Co(CO)_4 \rightleftharpoons RCH_2CH_2\overset{\displaystyle O}{\overset{\|}{C}}Co(CO)_3 \qquad (10.121)$$

$$CH_3CHCo(CO)_4 \rightleftharpoons CH_3CH\overset{\overset{\displaystyle O}{\|}}{C}Co(CO)_3 \qquad (10.122)$$
$$\qquad R \qquad\qquad\qquad R$$

These 16-electron Co(I) species now can undergo oxidative addition with H_2, giving 18-electron Co(III) dihydrides.

$$RCH_2CH_2\overset{\overset{\displaystyle O}{\|}}{C}Co(CO)_3 + H_2 \rightleftharpoons RCH_2CH_2\overset{\overset{\displaystyle O}{\|}}{C}\overset{\displaystyle H}{\underset{\displaystyle H}{Co(CO)_3}} \qquad (10.123)$$

$$CH_3CH\overset{\overset{\displaystyle O}{\|}}{C}Co(CO)_3 + H_2 \rightleftharpoons CH_3CH\overset{\overset{\displaystyle O}{\|}}{C}\overset{\displaystyle H}{\underset{\displaystyle H}{Co(CO)_3}} \qquad (10.124)$$
$$\quad R \qquad\qquad\qquad\qquad R$$

If this is followed by reductive elimination of the aldehyde product, $HCo(CO)_3$ is left and the cycle can be reentered with a new olefin molecule at Equation 10.117.

$$RCH_2CH_2\overset{\overset{\displaystyle O}{\|}}{C}\overset{\displaystyle H}{\underset{\displaystyle H}{Co(CO)_3}} \rightarrow RCH_2CH_2\overset{\overset{\displaystyle O}{\|}}{C}H + HCo(CO)_3 \qquad (10.125)$$

$$CH_3CH\overset{\overset{\displaystyle O}{\|}}{C}\overset{\displaystyle H}{\underset{\displaystyle H}{Co(CO)_3}} \rightarrow CH_3CH\overset{\overset{\displaystyle O}{\|}}{C}H + HCo(CO)_3 \qquad (10.126)$$
$$\quad R \qquad\qquad\qquad\qquad R$$

An alternative cleavage reaction (written here only for the straight-chain isomer) also occurs for the acyl.

$$RCH_2CH_2\overset{\overset{\displaystyle O}{\|}}{C}Co(CO)_3 + HCo(CO)_4 + CO \rightarrow RCH_2CH_2\overset{\overset{\displaystyle O}{\|}}{C}H + Co_2(CO)_8 \qquad (10.127)$$

The information in Equations 10.116–10.127 is summarized conveniently in the diagram shown in Figure 10.29, which also emphasizes the cyclic nature of the process.

The yield of straight-chain product can be increased by adding $(n\text{-}Bu)_3P$ to the reaction mixture, presumably because steric interactions destabilize the $CH_3CH(R)Co(CO)_3 [P(n\text{-}Bu_3)]$ intermediate relative to the straight-chain alkyl complex.

10.8.3 Hydrogenation of Olefins

At present, the homogeneous catalytic hydrogenation of olefins is not of much industrial importance. However, a relatively small-scale operation is conducted by the Monsanto Company in order to produce L-dopa (for treatment of Parkinson's disease) from a substituted

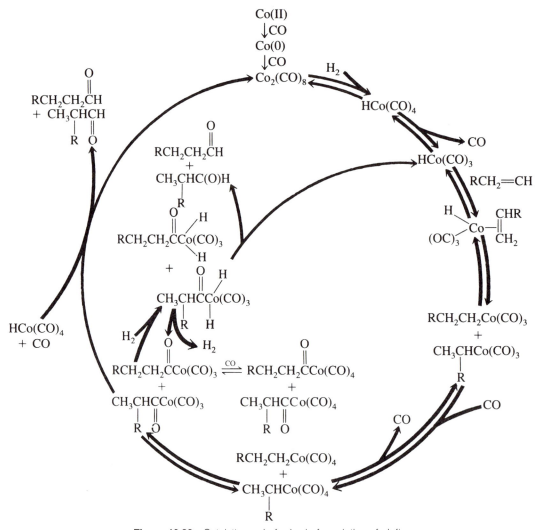

Figure 10.29 Catalytic cycle for hydroformylation of olefins.

cinnamic acid using an Rh catalyst. We mention it here because the selectivity for synthesis of just one optical isomer from an optically inactive precursor provides a good example of the kinds of advantages offered by homogeneous catalysis. The catalytic mechanism probably is related to the one that is operative for $Rh(PPh_3)_3Cl$ (Wilkinson's catalyst), depicted in Figure 10.30. The process can be run at low H_2 pressures.

A similar catalyst can be generated by starting with $[(diene)Rh(PPh_3)_2]^+$. In this case, the diene is hydrogenated, leaving $[Rh(PPh_3)_2(solvent)_2]^+$, the active catalyst. In the L-dopa process, the catalyst precursor is $[(diene)Rh(P\!\!-\!\!P)]^+$, where (P—P) is the optically active chelating phosphine, shown below.

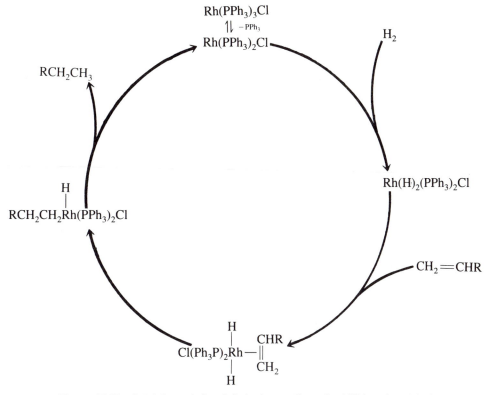

Figure 10.30 Catalytic cycle for olefin hydrogenation using Wilkinson's catalyst.

This optically active phosphine forms a "chiral hole" in the coordination sphere of the Rh, so that the prochiral olefin

presents one face preferentially to the complex when it coordinates. This leads to stereospecificity in the migratory insertion into the Rh—H bond, giving one optical isomer of the alkyl ligand, and subsequently, of the alkane product.

10.8.4 The Wacker Process

Another way of producing aldehydes (especially acetaldehyde) is by way of the Wacker process, in which olefins are oxidized by O_2 rather than carbonylated. Hence the aldehyde product has the same number of carbons as the starting olefin, in contrast to the hydroformylation reaction. The overall process is given by Equation 10.128, and the catalytic cycle is diagrammed in Figure 10.31.

$$CH_2\!\!=\!\!CH_2 + \tfrac{1}{2}O_2 \xrightarrow[120°, \ 4 \ \text{atm}]{PdCl_2, \ CuCl_2} CH_3\overset{\overset{\displaystyle O}{\|}}{C}H \qquad\qquad (10.128)$$

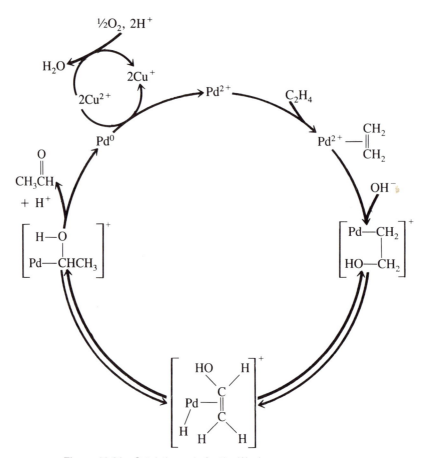

Figure 10.31 Catalytic cycle for the Wacker process.

At the beginning of the cycle, square planar $PtCl_4^{2-}$ undergoes a ligand displacement by ethylene, giving $[PdCl_3(C_2H_4)]^-$, an analogue of Zeise's salt. Other ligands around Pd^{2+} also are replaced (since the reaction is inhibited by Cl^-), but the exact composition of the coordination sphere is uncertain and so the catalytically active species is indicated as $Pd(C_2H_4)^{2+}$. Cu^{2+} in the solution reoxidizes the Pd(0) to Pd^{2+}, which reenters the cycle. O_2 from air converts Cu^+ back to Cu^{2+}, making the process catalytic.

A similar cycle is used to manufacture acetone from propylene (see Problem 10.7). If OH^- is replaced by acetate, vinyl acetate can be made.

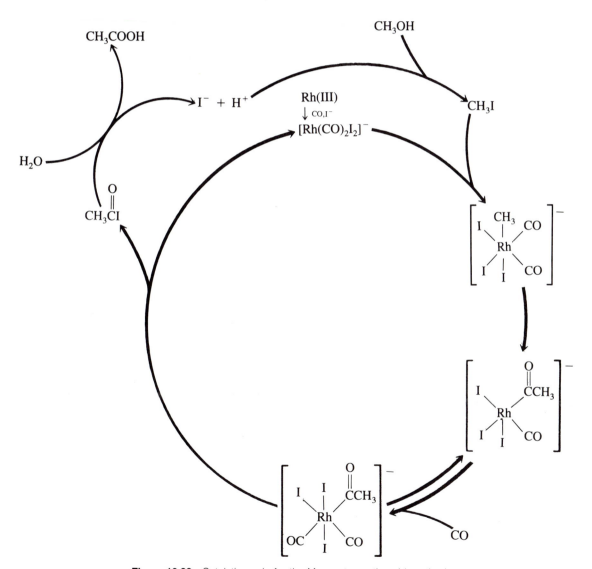

Figure 10.32 Catalytic cycle for the Monsanto acetic acid synthesis.

10.8.5 The Monsanto Acetic Acid Synthesis[59]

Chemists at Monsanto recently developed a catalytic process for the production of acetic acid by the direct carbonylation of methanol in the presence of Rh and I^-.

$$CH_3OH + CO \xrightarrow[180°, \ 30-40 \ atm]{Rh, \ I^-} CH_3COOH \tag{10.129}$$

The acetic acid produced is employed in the manufacture of vinyl acetate; cellulose acetate; other acetate esters, used as solvents; and in pesticides and other chemicals. An advantage of the synthesis is the relatively low pressure required. Figure 10.32 depicts the catalytic cycle involved. Rh can be added as the readily available $RhCl_3 \cdot 3H_2O$ at a concentration of $\sim 10^{-3}$ M. During an incubation period, the Rh(III) compound is reduced by CO to Rh(I) and converted to $[Rh(CO)_2I_2]^-$, the catalytically active species.

10.8.6 The 16- and 18-Electron Rule in Homogeneous Catalysis

All the mechanisms given in this section involve two-electron changes in metal oxidation state to give either 16- or 18-electron species, which the EAN rule predicts as most stable. There is considerable evidence for the importance[60] of species with these electron counts in homogeneous catalytic processes. Under vigorous catalytic reaction conditions, it may be that one-electron processes which give free radical species are involved sometimes. (See Section 10.7.2). A recent book interprets many organometallic reactions in terms of one-electron reactions (see footnote 10).

GENERAL REFERENCES

B. J. Aylett, *Organometallic Compounds*, 4th ed., Chapman and Hall, New York: Vol. 1, Part II, *The Main Group Elements, Groups IV and V*, 1979; Vol. 2, *The Transition Elements*, projected for 1982. The authoritative treatise on bonding, structure, synthesis, and reactions of organometallic compounds.

G. E. Coates, M. L. H. Green, and R. Wade, *Organometallic Compounds*, 3rd ed., Chapman and Hall, London: Vol. 1, *The Main Group Elements*, 1967; Vol. 2, *The Transition Elements*, 1968.

R. B. King, *Transition Metal Organometallic Chemistry*, Academic Press, New York, 1969. A good, readable introduction.

R. F. Heck, *Organotransition Metal Chemistry, A Mechanistic Approach*, Academic Press, New York, 1974. Emphasizes mechanisms of reactions related to catalysis.

D. S. Matteson, *Organometallic Reaction Mechanism*, Academic Press, New York, 1974. Deals with nontransition elements.

For specific topics, consult the following series.
Advances in Organometallic Chemistry
Progress in Inorganic Chemistry
Specialist Periodical Reports of The Chemical Society

---

[59]See D. S. Forster, *Adv. Organomet. Chem.* 1979, *17*, 255.
[60]C. A. Tolman, *Chem. Soc. Rev.*, 1972, *1*, 337.

PROBLEMS

10.1 Show that the EAN rule is obeyed by $Os_3(CO)_{12}$, whose molecular geometry appears in Figure 10.3.

10.2 Give an electron count for the complex in Figure 10.7. Does it obey the EAN rule?

10.3 Show that each of the metal-containing products in Reactions 10.53, 10.54 obeys the EAN rule.

10.4 Using Figure 10.16, write down electron configurations that account for the magnetic properties (given in Table 10.8) of Cp_2V, Cp_2V^+, Cp_2Cr, Cp_2Mn, Cp_2Fe, Cp_2Fe^+, Cp_2Co, Cp_2Co^+, Cp_2Ni, and Cp_2Ni^+.

10.5 ***a.*** The butene product from Reaction 10.101 consists of 1-butene as well as *cis*- and *trans*-2-butene. Keeping in mind the reversibility of the β-hydride elimination, write a mechanism to account for this.

b. When Reaction 10.101 is run with $CpFe(CO)_2(CD_2CH_2Et)$, complete scrambling of D occurs. Write a mechanism to account for this observation.

10.6 ***a.*** A reaction thought to be important in catalysis is the so-called 1,3-hydride shift. An alkyl C—H bond in a propene π-bonded ligand adds oxidatively to the metal, giving an allyl hydride complex. Show how the reversibility of this process can afford a mechanism for shifting H from C_1 to C_3.

b. Account for the following observation by a mechanism involving the 1,3-hydride shift: in the complex

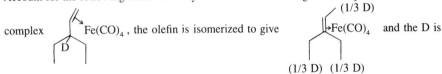

the olefin is isomerized to give $Fe(CO)_4$ and the D is

scrambled to all terminal methyl groups. (C. P. Casey and C. R. Cyr, *J. Am. Chem. Soc.* 1973, *95*, 2248).

10.7 Write a catalytic cycle for the production of acetone from propylene via the Wacker process.

10.8 Give the valence electron count for the following species. Which ones obey the EAN rule?

$Co_2(\mu_2\text{-}CO)_2(CO)_6$ $Mn(CO)_5\,(CH_2C_6H_5)$ $HRh(CO)_4$
$W(CO)_6$ $Fe(CO)_4Br_2$ $Ru_3(CO)_{12}$
$Cr(CNMe)_6$ $[Mn(CO)_5]^-$ $[Co(CN)_5]^{2-}$
$[Fe(CN)_6]^{4-}$ $Ni(PPh_3)_4$ $H_2Fe(CO)_4$
$[Cr(CO)_5]^{2-}$

10.9 Name each of the species in Problem 10.8.

10.10 Give the valence electron count for the following species. Which ones obey the EAN rule?

$Cp_2Ru_2(CO)_4$ $[(\eta^4\text{-cod})Rh(\mu\text{-Cl})]_2$ $Rh(C_2H_4)(PPh_3)_2Cl$
$[\eta^3\text{-}CH_2C(CH_3)CH_2]_2Ni$ $Ru(Me_2PCH_2CH_2PMe_2)_2$ $(\eta^6\text{-}C_6H_6)_2Mo$
$(\eta^4\text{-cot})Fe(CO)_3$ $CpMo(PPh_3)(CO)_2CH_3$ $HIr(CO)_3(PPh_3)$
$CH_3C(O)Re(CO)_4(CNMe)$ $[CpTa(CO)_4]$ $Rh_2(CO)_4(\mu\text{-Cl})_2$

10.11 Name each of the species in Problem 10.10.

10.12 The following kinetic data have been reported for the reactions $Mo(CO)_5(\text{amine}) + PPh_3 \xrightarrow{35°C}$ $Mo(CO)_5(PPh_3) + \text{amine}$.

Amine	$k_1(sec^{-1})$	$k_2(M^{-1}sec^{-1})$
Quinuclidine	0.9×10^{-5}	3.0×10^{-5}
Piperidine	1.8×10^{-5}	1.24×10^{-3}
Cyclohexylamine	9.9×10^{-5}	3.9×10^{-3}

Considering these data together with those in Table 10.3, what can you say about the *a* or *d* character of the k_2 path for substitution at Mo? Are the data given here consistent with a *d* path for the k_1 term? (W. D. Covey and T. L. Brown, *Inorg. Chem.* 1973, *12*, 2820.)

10.13 Rationalize the trends in the following sets of IR-active CO stretching frequencies (in cm^{-1}).

    a. $(\eta^6\text{-}C_6H_6)Cr(CO)_3$ 1980, 1908

         $CpMn(CO)_3$ 2027, 1942

    b. $CpV(CO)_4$ 2030, 1930

         $CpMn(CO)_3$ 2027, 1942

         $[CpFe(CO)_3]^+$ 2120, 2070

    c. $W(CO)_5(Pn\text{-}Bu_3)$ 2068, 1936, 1943

         $W(CO)_5(PPh_3)$ 2075, 1944, 1944

         $W(CO)_5P(OBu)_3$ 2079, 1947, 1957

         $W(CO)_6$ *ca* 2000

    d. $Ni(CO)_4$ 2046

         $Co(CO)_4^-$ 1883

         $Fe(CO)_4^{2-}$ 1788

10.14 *a.* Draw at least two possible structures of $Os_3(CO)_9(PPh_3)_3$.

    b. The IR spectrum of this compound in CH_2Cl_2 has CO stretches at 1962 and 1917 cm^{-1}. How does this knowledge help to narrow the possible structures?

10.15 The IR spectrum of $Rh_2I_2(CO)_2(PPh_3)_2$ has CO stretches at 2061 and 2005 cm^{-1}. Suggest a structure consistent with this.

10.16 Explain in your own words how the VB and MO descriptions of the electronic structure of $C_3H_5^-$ are equivalent.

10.17 Coordination of ligands to transition metals often makes them subject to nucleophilic attack. Make a list of reactions given in this chapter involving nucleophilic attack on coordinated ligands.

10.18 Give the formula of the most stable compound of the type M(olefin) (CO)$_x$ expected for each of the following metals, with each olefin listed.

Metals	Olefins
Cr, Mn, Fe	$C_3H_5^-$, Cp^-, C_6H_6, $C_7H_7^+$

10.19 Identify the following reactions by type and predict the products.

$Re_2(CO)_{10} + Na/Hg$ $Rh(PPh_3)_3Br + Cl_2$

$W(CO)_6 + (n\text{-}Bu_4N)I$ $Pt(PPh_3)_3 + CF_3I$

$CH_3Mn(CO)_5 + C_6H_5NH_2$ $RuH_2(PPh_3)_4 + C_2H_4$

$CpCo(CO)_2 + PPh_3$ $HIrCl_2(PPh_3)_2 \xrightarrow{}$

$[Rh(CO)_2Cl_2]^-, + CH_3I$ $Ge(OMe)_2 + CH_3I$

$CH_3Mn(CO)_5 + SO_2$ $RuH(CO)_2(PPh_3)_2C(O)R \xrightarrow{}$

$EtMgI + SnCl_4$ $Ru(PPh_3)_2(CO)_3 + H_2$

$Me_2Hg + Cp_2TiCl_2$ $PtMe_2Cl(PMe_2Ph)C(O)CH_3 \xrightarrow{\Delta}$

$MeLi(excess) + CuCl_2$ $AlMe_3 + CrCl_3$

$[CpFe(CO)_2]^- + C_6H_5Cl$ $LiMe + WCl_6$

$(\eta^1\text{-}C_3H_5)Co(CO)_4 \xrightarrow{}$ $CpMo(CO)_3Me + CN^-$

$Cp_2Fe + n\text{-}BuLi$ $CpFe(CO)_2CH_2CH{=}CH_2 \xrightarrow{h\nu}$

$Mo(CO)_5py + PPh_3$

10.20 $CpMoH_2$ reacts with diphenylacetylene to give $CpMo(PhC_2Ph)$ and *cis*-$PhCH{=}CHPh$. Propose a mechanism for this reaction. (A. Nakamura and S. Otsuka, *J. Am. Chem. Soc.* 1972, *94*, 1886.)

10.21 The rate of reaction of O_2 with *trans*-$[IrX(CO)(PPh_3)_2]$ in benzene decreases in the order $X = NO_2 > I > ONO_2 > Br > Cl > N_3 > F$. Explain this observation. (L. Vaska, L. S. Chen, and C. V. Senoff, *Science* 1971, *174*, 587.)

10.22 Propose a mechanism for the stoichiometric decarbonylation of $C_6H_5CH_2C(O)Cl$ by $Rh(PPh_3)_3Cl$, giving benzyl chloride. Keep in mind the 16- and 18-electron rule.

10.23 Explain the trends depicted in Figure 10.26.

10.24 Write a catalytic cycle for the production of ethyl acetate via the Monsanto acetic acid synthesis.

10.25 $HNi[P(OEt)_3]_4^+$ is known to be a catalyst for olefin isomerization. Write a catalytic cycle for isomerization of 1-butene catalyzed by this species. Keep in mind the 16- and 18-electron rule. How would you formulate in the most reasonable way the electronic structure of the Ni cation?

10.26 Give electron counts for all the species postulated to be involved in the catalytic cycle for the hydroformylation reaction (Figure 10.29).

10.27 Kinetic studies indicate that the hydroformylation reaction rate is enhanced by an increase in H_2 pressure and inhibited by an increase in CO pressure. How is the mechanism proposed in this chapter consistent with these observations?

10.28 Propose a mechanism for the following reaction [J. Chatt and B. L. Shaw, *Chem. Ind.* (London) 1960, 931]

$$IrCl_3(PEt_3)_3 + C_2H_5OH + KOH \rightarrow HIrCl_2(PEt_3)_3 + CH_3CHO + KCl + H_2O$$

10.29 What products would you expect from the hydroformylation of $C_3H_7C(CH_3)DCH{=}CH_2$? Show how each is obtained.

10.30 Explain how the facts that $(\eta^6\text{-}C_6H_5CO_2H)Cr(CO)_3$ is a stronger acid than benzoic acid and that $(\eta^6\text{-}C_6H_5NH_2)Cr(CO)_3$ is a weaker base than aniline show that the $Cr(CO)_3$ group withdraws electrons from the aromatic rings.

10.31 What is the point group of the Li_4Me_4 unit described in Section 10.6.1?

XI

Prediction and Correlation of Oxidation–Reduction Reactions

> This chapter deals primarily with the use of half-cell electromotive force data in predicting and correlating chemical reactions involving oxidation and reduction. Since emf data comprise only one facet of thermodynamic data in general, we will begin with a brief review of thermodynamic functions.
>
> Tables of thermodynamic data provide one of the most compact ways of storing chemical information. A single entry may summarize months of investigation. Thermodynamic data may be used not only to reconstruct much of the information from which the data came, but also to make predictions about unknown reactions. Readily available are extensive tables listing standard free energies and heats of formation, and entropies; free energy functions, and half-cell electromotive force data.[1]
>
> Of particular interest to the chemist is the free-energy change accompanying a reaction under a given set of conditions, for this indicates the direction of the reaction in approaching equilibrium.

11.1 CONVENTIONS COVERING STANDARD STATES

At equilibrium the free-energy change accompanying a reaction at a fixed temperature and pressure is zero. Thus, from the equilibrium partial pressures at 1000 K of 0.564 atm SO_2, 0.102 atm O_2, and 0.333 atm SO_3, we may write

[1]See the general references listed at the end of this chapter.

$$2SO_2(0.564 \text{ atm}) + O_2(0.102 \text{ atm}) \rightarrow 2SO_3(0.333 \text{ atm}) \quad \Delta G_{1000 \text{ K}} = 0$$

Assuming ideal gas behavior for the above system, we then may write

$$K_p = \frac{(P_{SO_3})^2}{(P_{SO_2})^2(P_{O_2})} = 3.42 \text{ atm}^{-1} \quad \text{or} \quad 4.5 \times 10^{-3} \text{ torr}^{-1}$$

We can discover the ΔG value for the above reaction under conditions other than the equilibrium value by means of the following equation, known as the van't Hoff reaction isotherm.

$$\Delta G_T = RT \ln \frac{Q}{K} \tag{11.1}$$

where Q is an activity quotient similar in form to the equilibrium constant but with arbitrary activity values. Thus at 1000 K and with $P_{SO_2} = 5$ atm, $P_{O_2} = 2$ atm, and $P_{SO_3} = 10$ atm,

$$Q = \frac{(10)^2}{(5)^2(2)} = 2 \text{ atm}^{-1}$$

$$K = 3.42 \text{ atm}^{-1}$$

and

$$\Delta G_{1000 \text{ K}} = 2.303 \times 8.314 \times 1000(\log 2 - \log 3.42) = -4461 \text{ J}$$

Under the stated conditions, equilibrium is approached by the formation of SO_3.

The units used in expressing Q and K must be the same, of course. When the arbitrary activity of unity is selected for both the reactants and the products, the free-energy change for the reaction is called the "standard" free-energy change, designated by superscript zero: that is, ΔG^0.

From the relationship $\Delta G^0 = -RT \ln K$, we calculate a "standard" free energy change for the SO_2 oxidation as follows.

$$2SO_2 (1 \text{ atm}) + O_2 (1 \text{ atm}) \rightarrow 2SO_3 (1 \text{ atm}) \quad \Delta G^0_{1000 \text{ K}} = -9372 \text{ J}$$

or[2]

$$2SO_2 (1 \text{ torr}) + O_2 (1 \text{ torr}) \rightarrow 2SO_3 (1 \text{ torr}) \quad \Delta G^0_{1000 \text{ K}} = +44980 \text{ J}$$

This example reveals that the equilibrium constant and the standard free-energy change depend upon the standard states selected for the reactants and the products.

For thermodynamic equilibrium constants, the quantities used in the equilibrium product should be activities rather than concentrations. We define the standard state of any substance as a state of unit activity. According to convention, the standard state for *gases* is unit fugacity: that is, the gas behaving ideally at 1 atm pressure. For most gases the standard state may thus be taken as 1 atm. For *pure liquids* and *solids,* which occur as separate phases in reactions, the standard state is defined conventionally as the pure solid or pure liquid at 1 atm pressure. Conventions covering *solutions* are not as uniform as those for gases and pure liquids or solids.

---

[2]The standard state of one torr assumed here is not the conventional one, and this ΔG would normally not be designated as ΔG^0 (see below).

We usually define the standard state for the solvent as that of the pure liquid. For the solute the standard state may be taken as an *activity, a* (thermodynamic concentration), of one molal. For approximate calculations *concentrations, m,* in units of molality may be used instead of activities. These two quantities may be related to each other through the use of activity coefficients (γ).

$$a = \gamma m$$

At high dilution, γ approaches 1. Note that using activities in terms of molarities or mole fractions involves different standard states from those selected by the use of molalities.

Example A standard free-energy change of 45.6 kJ is calculated from the K_{sp} for $PbSO_4$. Write the reaction for which this ΔG^0 applies.

Solution $PbSO_4(s) \rightarrow Pb^{2+}$ ($\gamma^{\pm}m = 1$) + SO_4^{2-} ($\gamma^{\pm}m = 1$)

Example At 20°C, K_N is 4 for the esterification reaction of acetic acid with ethanol. Mole fractions are used as concentrations in the expression for K_N. Calculate ΔG^0 and write the hypothetical reaction to which it applies. What data would be needed to calculate K_p?

Solution $\Delta G^0 = -RT \ln K_N = -3430\,J$
for the reaction

$$CH_3CO_2H(l) + C_2H_5OH(l) \rightarrow CH_3CO_2C_2H_5(l) + H_2O(l)$$

This must also be the free-energy change for the above reaction when each of the substances is in the gas phase with a partial pressure equal to the vapor pressure of the pure liquid. We thus can find K_p by using the ΔG^0 as ΔG in Equation 11.1 and evaluating Q from the vapor pressures.

11.2 METHODS OF DETERMINING CHANGES IN FREE ENERGY

Changes in free energy, enthalpy, and entropy are related through the equation

$$\Delta G = \Delta H - T\Delta S \tag{11.2}$$

Knowing any two of these quantities allows us to calculate the other, and knowing all three enables us to check the assumptions involved in the calculation and/or in the experimental methods. The most common methods used to obtain one or more of these pieces of information are

1. Direct calorimetric measurement of heats of reaction.
2. Heat capacity measurements.
3. Direct determination of equilibrium constants.
4. Galvanic cell potentials.
5. Statistical mechanics and appropriate spectral data.
6. Approximation methods.

11.3 CALORIMETRIC DATA

The enthalpy, entropy, or free-energy change accompanying a reaction is independent of the reaction path. Accordingly, if we know the ΔH (or ΔG or ΔS) values for all but one step in a cycle, we readily can obtain the value for the last step. Thus knowledge of the enthalpy changes for the combustion of C to CO_2 and CO to CO_2 allows us to calculate the heat of combustion of C to CO: that is

$$
\begin{array}{lll}
(1) & O_2(g) + C(s) \rightarrow CO_2(g) & \Delta H^0 = -393.5 \text{ kJ} \\
(2) & \tfrac{1}{2} O_2(g) + CO(g) \rightarrow CO_2(g) & \Delta H^0 = -283.0 \text{ kJ} \\
\hline
& \tfrac{1}{2} O_2(g) + C(s) \rightarrow CO(g) \ \Delta H_3^0 = \Delta H_1^0 - \Delta H_2^0 = -110.5 \text{ kJ}
\end{array}
$$

It is convenient to *define* the standard heat of formation at a particular reference temperature as the heat of reaction of the elements to form a compound, both reactants and product being in their standard states. Under this definition elements in the stable form at a particular reference temperature have ΔH_f^0 values of zero. Heats of formation can be found in extensive tables that allow the calculation of ΔH^0 at 25°C by the simple equation

$$
\Delta H^0 = \Sigma \Delta H_f^0 \text{ (products)} - \Sigma \Delta H_f^0 \text{ (reactants)}
$$

Coupled with heat capacity data and heats of transition for phase changes, ΔH^0 may be calculated at other temperatures.

The standard enthalpy of a solute includes not only the heat of formation of the isolated solute, but also its heat of solution. By adopting the *convention* that the H^+ ion in aqueous solution at 25°C be assigned a ΔH_f^0 of zero, we can assign ΔH_f^0 values to other ions in aqueous solution.

Example Calculate the heat of reaction of

$$
AgNO_3 \text{ (1 molar)} + NaCl(s) \rightarrow AgCl(s) + NaNO_3 \text{ (1 molar)}
$$

Solution $\Delta H = \Delta H_{AgCl}^0 + \Delta H_{Na^+(aq)}^0 - \Delta H_{Ag^+(aq)}^0 - \Delta H_{NaCl}^0$
Using values from Latimer,[3]

$$
\begin{aligned}
\Delta H &= -127.0 - 239.7 - (+105.9) - (-411.0) \\
&= -61.6 \text{ kJ}
\end{aligned}
$$

Standard free energies of formation of compounds and ions follow the same conventions as ΔH_f^0. If we know the standard free energies of formation for both reactants and products, we can calculate the standard free-energy change of the reaction.

$$
\Delta G^0 = \Sigma \Delta G_f^0 \text{ (products)} - \Sigma \Delta G_f^0 \text{ (reactants)}
$$

Example Calculate ΔG^0 for

$$
AgNO_3 \text{ (1 molar)} + NaCl(s) \rightarrow AgCl(s) + NaNO_3 \text{ (1 molar)}
$$

[3]W. M. Latimer, *The Oxidation States of the Elements and Their Potentials in Aqueous Solution*, 2nd ed., Prentice-Hall, New York, 1952.

Solution   Using values from Latimer,

$$\Delta G^0 = -109.7 + (-261.9) - (77.1) - (-384.0)$$
$$\Delta G^0 = -64.7 \text{ kJ}$$

The convention covering entropy differs from that for enthalpy and free energy in that most pure crystalline substances are assigned an entropy of zero at 0 K. Not assigned zero values at absolute zero are those substances with a degree of randomness in their crystalline lattice, and hence a residual entropy. This convention is based on the third law of thermodynamics. Entropies for the substance at temperatures above 0 K may be obtained from heat capacity data coupled with heats of transition and the entropy at 0 K.

$$S_T = \int_0^T C_p \, d \ln T + \sum \left(\frac{\Delta H_{tr}}{T_{tr}} \right) + R \ln \sigma \tag{11.3}$$

The integral may be evaluated graphically; the $\ln \sigma$ term gives the 0 entropy (σ is the number of distinguishable orientations of the molecule in the crystal at 0 K). An exception to the above convention is the hydrogen ion, which is assigned a molal entropy of zero at 25°C at infinite dilution in aqueous solution. This allows us to assign entropy values to other ions.

   Values of ΔH_f^0, ΔG_f^0, and S^0 at 25°C have been tabulated for many substances. These tabulations permit predictions to be made about reactions occurring at 25°C and 1 atm. ΔH_f^0 and ΔG_f^0 values at higher temperatures are increasingly becoming available.[4]

Example   Calculate the equilibrium constant at 2500 K for

$$\text{SiO}_2(\text{l}) + \text{Si}(\text{l}) \rightarrow 2\text{SiO}(\text{g})$$

Solution   From Elliott and Gleiser, the ΔG_f^0 values at 2500 K are -267.4 and 424.7 kJ/mole, respectively,

$$\Delta G^0 = 2(-267.4) - (-424.7) = -110.1 \text{ kJ}$$

$$K_p = \exp_{10} \left(\frac{\Delta G^0}{2.3RT} \right) = \exp_{10} \left(\frac{110,100}{2.3 \times 8.314 \times 2500} \right) = 10^{2.31} = 200 \text{ atm}^2$$

SiO is thermodynamically unstable at room temperature, but a quenched sample is kinetically stable. SiO condensed from the gas on optical parts provides a protective coating.

   We can obtain approximate values for ΔG at temperatures other than 25°C by assuming ΔH and ΔS to be temperature-independent: then $\Delta G_T = \Delta H_{298 \text{ K}} - T\Delta S_{298 \text{ K}}$. This is equivalent to assuming that ΔC_p for the reaction is zero. More accurate values for ΔG at temperatures other than 298 K may be made from free energy function data.[5]
   Displacement reactions of the type

$$\text{AB}_x + y\text{C} \rightleftharpoons \text{AC}_y + x\text{B}$$

---

[4]See particularly Elliott and Gleiser under general references. This volume contains such data for carbides, nitrides, oxides, phosphides, silicides and sulfides as well as C_p, S, free energy functions and enthalpy functions for a number of elements.

[5]For a discussion of free energy functions, see J. L. Margrave, *J. Chem. Educ.*, 1955, 32, 520.

Table 11.1 Some heats of formation in kJ per equivalent

LiF	Li$_2$O	Li$_3$N
609	297	63
NaF	Na$_2$O	Na$_3$N
569	207	—
CaO	CaS	CaSe
318	238	184

generally occur if ΔH_f of AC$_y$ per equivalent is much greater than that of AB$_x$ per equivalent (that is, the reaction is exothermic). When the heats of formation are almost the same, entropy effects may cause endothermic reactions, as in the case of

$$2Ag(s) + Hg_2Cl_2(s) \rightleftharpoons 2AgCl(s) + 2Hg(l)$$

Van Arkel[6] has generalized on data such as that in Table 11.1 to construct the displacement chart for the nonmetals in ionic compounds shown in Figure 11.1.

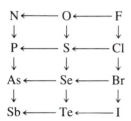

Figure 11.1 Displacement chart for the nonmetals.

Except for halogen displacement, these reactions generally are slow at room temperature. At higher temperatures more complex reactions might occur. Free-energy functions are used to make better predictions since (1) entropy changes are not ignored and (2) data are available in this form for many materials at high temperatures.

11.4 SCHEMATIC REPRESENTATION OF GALVANIC CELLS

Electrodes—An electrode consists of a metallic conductor in contact with (or a part of) a phase boundary across which a difference in electrical potential occurs. The phase boundary is represented by a single vertical line, $|$, separating the components of the phases. The schematic representation of the electrode signifies a half-reaction in which positive ions move from left to right and negative ions move from right to left. The representations below thus imply the half-reactions that follow.

[6]A. E. Van Arkel, *Molecules and Crystals in Inorganic Chemistry*, translated by J. C. Swallow, 2nd ed., Wiley-Interscience, New York, 1946, Chapter 5.

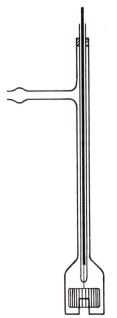

Figure 11.2   The hydrogen electrode. (Reprinted with permission from J. H. Hildebrand, *J. Am. Chem. Soc.*, 1913, *35*, 847. Copyright 1913, American Chemical Society.)

Electrode	*Half-Reaction*
Ag $\mid$ Ag$^+$(aq)	Ag(s) $\rightarrow$ Ag$^+$(aq) $+$ e
Ag$^+$(aq) $\mid$ Ag(s)	Ag$^+$(aq) $+$ e $\rightarrow$ Ag(s)
Pt(s) $\mid$ Fe^{2+}(aq), Fe^{3+}(aq)	Fe^{2+}(aq) $\rightarrow$ Fe^{3+}(aq) $+$ e
Ag(s), AgCl(s) $\mid$ Cl$^-$(aq)	Ag(s) $+$ Cl$^-$(aq) $\rightarrow$ AgCl(s) $+$ e
Pt(s), H$_2$(g) $\mid$ H$^+$(aq)	$\frac{1}{2}$H$_2$(g) $\rightarrow$ H$^+$(aq) $+$ e

The last electrode listed above is called the hydrogen electrode, which under standard conditions has a defined electrical potential of zero. It consists of a platinum electrode coated with platinum black in contact with a solution saturated with hydrogen. An early design by Hildebrand, still widely used, is shown in Figure 11.2.

11.5   CONVENTIONS REGARDING CELLS

To construct a galvanic cell we must be able to carry out an oxidation half-reaction and a reduction half-reaction in separate places (that is, at the electrodes). Electrolytic conduction (current carried by ions) must be possible internally and metallic conduction (current carried by electrons) externally. For some systems the electrolyte may be the same throughout the cell and the cell represented by the appropriate combination of electrodes, as in the following example.

$$\textit{Cell} \quad \text{Pt, H}_2\text{(g)} \mid \text{HCl(aq)} \mid \text{AgCl(s), Ag}$$
$$\textit{Reaction} \quad \text{AgCl(s)} + \tfrac{1}{2}\text{H}_2 \rightarrow \text{Ag(s)} + \text{HCl(aq)}$$

To construct a cell in which the reaction $Zn + 2HCl \rightarrow H_2 + ZnCl_2$ may be carried out reversibly, we must prevent the hydrogen ion from coming into direct contact with the Zn. One way of doing this is to insert a salt bridge, designated by two parallel vertical lines, $\parallel$, that prevents the direct mixing of the electrolytes of the two electrodes. (The salt bridge may introduce liquid junction potentials, which will not be discussed here.) The reaction of hydrogen ion and Zn could thus be carried out reversibly in the following cell.

$$Zn(s) \mid ZnCl_2(aq) \parallel HCl(aq) \mid H_2(g), Pt(s)$$

11.5.1 Electrode Potentials

When two electrodes are coupled to make a cell, the voltage developed by the cell will be simply the difference in the electrical potential of the two electrodes. If we assign the hydrogen electrode under standard conditions a potential of zero, then we can assign the potential of any electrode relative to hydrogen. By this convention the potential of the Zn electrode in contact with Zn ions under standard conditions is -0.76 V. Note that the sign of the *potential* of the Zn electrode is independent of the cell in which the Zn electrode occurs whether oxidation or reduction takes place at the Zn electrode.

11.5.2 Relationship Between Cell Voltage and the Free-Energy Change for the Cell Reaction

To obtain the free-energy change for a reaction occurring in a galvanic cell, we simply multiply the potential difference of the cell electrodes by the number of coulombs passing through the circuit, or

$$\Delta G = -n\mathscr{F}E \tag{11.4}$$

where

ΔG is the free-energy change for a redox reaction

n is the number of equivalents oxidized or reduced (that is, the number of electrons exchanged per formula unit)

$\mathscr{F}$ is the coulombs/Faraday $= 96,500$

E is the voltage produced by the galvanic cell under reversible conditions

By convention, the voltage for a spontaneous reaction is positive whereas the free-energy change for a spontaneous reaction is negative. If the reactants and products are all present at unit activity, then E is E^0 and ΔG is ΔG^0.

11.5.3 Half-Cell emf Values

Under the conventions stated earlier, the electromotive force for the reversible operation of the cell

$$Zn(s) \mid Zn^{2+}(aq) \parallel H^+(aq) \mid H_2(g), Pt(s)$$

is related to the free-energy change for the reaction

$$Zn(s) + 2H^+(aq) \rightarrow Zn^{2+}(aq) + H_2(g) \tag{11.5}$$

through the relationship

$$\Delta G^0 = -n\mathscr{F}E^0$$

Since the conventions assign ΔG_f^0 values of zero for hydrogen and hydrogen ion, the ΔG_f^0 value obtained may be assigned entirely to the free energy change for the oxidation half-reaction

$$Zn \rightarrow Zn^{2+} + 2e \qquad E^0 = +0.76$$

Since Reaction 11.5 occurs spontaneously under standard conditions, ΔG^0 must be negative and, accordingly, E^0 positive. For the reverse of Reaction 11.5,

$$H_2(g) + Zn^{2+}(aq) \rightarrow Zn(s) + 2H^+(aq)$$

the signs of ΔG^0 and E^0 are reversed and accordingly E^0 for the reduction half-reaction

$$Zn^{2+}(aq) + 2e \rightarrow Zn(s) \qquad E^0 = -0.76$$

is reversed. Note that the standard emf for a reduction half-reaction is identical in sign and magnitude to the value of the standard potential for the electrode at which the half-reaction is carried out. As pointed out above, the electrode potential is independent of whether an oxidation or reduction reaction is taking place at the electrode. Accordingly, terms such as oxidation potentials or reduction potentials are inappropriate in talking about electromotive force values for half-cell oxidation or reduction reactions. Nevertheless, the International Union of Pure and Applied Chemistry has agreed to the use of the word "potential" as synonymous with the emf value for the *reduction* half-cell reaction. Because of lack of adherence to a single convention in many sources of emf data, you should remember the sign of a half-cell emf reaction such as given in Reaction 11.5 and check the sign given this half-reaction in any tabulation you use.

11.6 CALCULATIONS USING emf DIAGRAMS

Given below is an emf diagram for the standard half-cell reduction reactions in acid solution at 25°C.

$$E_A^0 \quad MnO_4^- \xrightarrow{0.564} MnO_4^{2-} \xrightarrow{2.26} MnO_2 \xrightarrow{0.95} Mn^{3+} \xrightarrow{1.51} Mn^{2+} \xrightarrow{-1.19} Mn$$

The arrows indicate the direction of the half-reactions. Except for writing the half-reactions as reductions, the above diagram is identical to those given in Latimer's[7] extensive treatment. Our diagram abbreviates the half-reactions by showing only the substances that change in oxidation number. Thus the E^0 of 2.26 is for the half-reaction[8]

---

[7]W. M. Latimer, *The Oxidation States of the Elements and Their Potentials in Aqueous Solution*, 2nd ed., Prentice-Hall, New York, 1952.

[8]We can write the half-reactions readily by the following sequence: (1) Balance the equation with respect to the atoms undergoing change in oxidation number. (2) Obtain a balance of oxygen atoms by adding H_2O to the appropriate side of the equation. (3) Balance the equation with respect to hydrogen atoms by adding H^+ to the appropriate side of the equation. (4) Balance the equation with respect to charge by adding electrons to the appropriate side of the equation.

$$2e + 4H^+ + MnO_4^{2-} \rightarrow MnO_2 + 2H_2O \qquad E^0 = 2.26$$

The superscript zero implies that all substances shown in the equation are in their standard states; E_A^0 specifically indicates that the hydrogen ion activity is unity, and E_B^0 indicates that the hydroxide ion activity is unity. For many practical purposes, we can predict a reaction directly, without evaluating activity coefficients, using *formal emf values*. The formal emf, E^F, is the emf of a half-reaction with the *concentrations* of oxidant and reductant equal to unity, and with arbitrarily chosen concentrations of other electrolytes, including acids. Unfortunately, however, formal emf values vary with the ionic strength of the solution, and a given list applies to one ionic strength only.

If formal emf values are available for a particular medium, we can make reaction predictions based on them conveniently. In many practical situations the necessary activity coefficients needed to permit the use of the standard emf data are not easily available. Formal emf values in $1M$ perchloric acid solution often are used as close approximations of E_A^0 values.

11.6.1 Combination of Half-Cell emf Values to Obtain Other Half-Cell emf Values (the Use of Volt Equivalents)

From the emf diagram given for manganese, we can find the half-cell emf values for any half-reaction that occurs between a pair of the species given in the diagram. Since the E^0 refers to the voltage per electron, evaluating the E^0 for the new half-reaction corresponds to calculating a weighted average. Volt equivalents—that is, the product of the number of electrons involved in the half-reaction and the E^0 of the half-reaction—are used. That this gives the same E^0 as the use of the more familiar handling of ΔG^0 values can be seen from the following example.

Reaction	E^0	$\Delta G^0 = -nFE^0$	nE^0 (volt equivalent)
$Mn^{2+} + 2e \rightarrow Mn$	-1.19	$2F1.19$	-2.38
$Mn^{3+} + e \rightarrow Mn^{2+}$	1.51	$-F1.51$	1.51
$Mn^{3+} + 3e \rightarrow Mn$	$\Delta G^0 = 0.87F$		-0.87

$$E^0 = \frac{-(0.87F)}{3F} = -0.29$$

In summary, we can obtain the E^0 for a couple by multiplying each of the E^0 values of intervening couples by the number of electrons involved, adding the products, and dividing the sum by the total number of electrons involved in the new half-reaction. Thus the E_A^0 value for $MnO_4^- \rightarrow Mn^{2+}$ is given by

$$\frac{(1)(1.51) + (1)(0.95) + (2)(2.26) + (1)(0.564)}{5} = 1.51 \text{ V}$$

11.6.2 Effect of Concentration on Half-Cell emf Values

Appropriately combining Equations 11.1 and 11.4 leads to the well-known Nernst equation, which gives the effect of variation of activity (thermodynamic concentration) on the emf of a cell or half-cell: that is,

$$E = E^0 - \frac{RT}{n\mathscr{F}} \ln Q$$

where Q is an activity product identical in form to the equilibrium constant, but in which the activities need not represent equilibrium values. If there are no changes in the species involved, the Nernst equation readily permits calculation of the effect of pH on the emf values.

Example   Given E_A^0 for the half-reaction $MnO_4^{2-} \rightarrow MnO_2(s)$, express the E value as a function of pH.

Solution   From the Nernst equation and the balanced half-reaction, we may write

$$E = E_A^0 - \frac{0.059}{2} \log \frac{1}{[H^+]^4 [MnO_4^{2-}]}$$

where the constant 0.059 incorporates a temperature of 298 K and the conversion factor from log to $\ln$; 1 is used for the activity of the solid phase. If the manganate activity is kept at 1, the pH dependence is

$$E = E_A^0 - \frac{(0.059)(4)}{2} pH = 2.26 - 0.118pH$$

E_B^0 is used to indicate that the potentials are standard potentials for base solution—all reactants and products, including hydroxide ion, are at unit activity. In the above example, we find the E_B^0 value by using a pH value of 14 (since $pH + pOH = pK_w = 14$), giving $E_B^0 = 0.60$ volts.

Example   Given E_A^0 for the couple $MnO_4^- \rightarrow MnO_4^{2-}$, what is E_B^0?

Solution   Since hydrogen is not involved in the half-reaction E_B^0 and E_A^0 will be identical.

When a change of pH results in the formation of an insoluble hydroxide, we may calculate the emf value for the half-cell at the new pH, if the value of K_{sp} is known, by application of the Nernst equation, as the following free-energy cycle demonstrates.

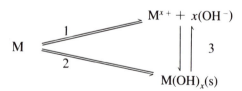

If M^{x+} exists in a solution in contact with $M(OH)_x$ (that is to say, in equilibrium), ΔG for step 3 must be zero and ΔG for steps 1 and 2 must be identical.

Example   Given $E_A^0 = -1.19$ for $Mn^{2+} + 2e \rightarrow Mn$ and K_{sp} for $Mn(OH)_2 = 2 \times 10^{-13}$, what is E_B^0 for the $Mn(OH)_2(s) \rightarrow Mn$ couple?

Solution K_{sp} is the equilibrium constant for the reaction

$$Mn(OH)_2(s) \rightleftharpoons Mn^{2+} + 2(OH^-)$$

$$K_{sp} = [Mn^{2+}][OH^-]^2$$

In 1 molar base $(OH^-) = 1$ and thus $(Mn^{2+}) = K_{sp} = 2 \times 10^{-13}$. From the Nernst equation,

$$E = E^0 - \frac{0.059}{2} \log \frac{1}{[Mn^{2+}]}$$

$$= -1.19 - \frac{0.059}{2} \log \frac{1}{2 \times 10^{-13}}$$

$$= -1.56$$

The calculation's result is that expected qualitatively on the basis of LeChatelier's principle. As the metal ion concentration decreases, it becomes easier to form the ion from the metal. Similar considerations apply to the effect of complexing of metal ions by ligands such as cyanide, chloride, ammonia, etc. The relative effectiveness of such ligands in decreasing the free-metal ion concentration in solution is revealed in the following emf values.

$$Zn(H_2O)_w^{2+} + 2e \rightarrow Zn + wH_2O \qquad E^0 = -0.76$$

$$Zn(NH_3)_4^{2+}(aq) + 2e \rightarrow Zn + 4NH_3 \qquad E^0 = -1.03$$

$$Zn(CN)_4^{2-}(aq) + 2e \rightarrow Zn + 4CN^- \qquad E^0 = -1.26$$

We examine in more detail the use of emf data to obtain both stoichiometries and stability constants of complex ions on page 506.

11.6.3 The Use of Half-Cell emf Data in the Prediction of Chemical Reactions

Since $\Delta G = -n\mathscr{F}E$, a positive value of E means that the reaction will be "spontaneous" in a thermodynamic sense—that is to say, that the reactants and products are not in equilibrium and equilibrium will be approached by the formation of more product. The major advantage using emf data rather than free-energy data is that with emf data the stoichiometries need not be taken into account, because the half-cell emf data are given in terms of volts per electron transferred. Thus the sum of the emf values for any oxidation half-reaction and any reduction half-reaction supplies an answer as to whether such a reaction may occur "spontaneously."

When a substance is added to water, it may be unstable for one of the following reasons: it may undergo disproportionation, or auto-oxidation-reduction, in which species of higher and lower oxidation states are produced; it may react with water to evolve hydrogen; or it may react with water to evolve oxygen.

Example Is Mn^{2+} stable in solution in water?

Solution Referring to the emf diagram for manganese, we can obtain a sum for the half-reactions as follows.

$$
\begin{array}{ll}
 & E^0 \\
Mn^{2+} + 2e \rightarrow Mn & -1.19 \\
\underline{2(Mn^{2+} \rightarrow Mn^{3+} + e)} & \underline{-1.51} \\
3Mn^{2+} \rightarrow Mn + 2Mn^{3+} & -2.70
\end{array}
$$

Since E^0 is negative, the reaction does not occur and Mn^{2+} does not disproportionate.

Note that the emf for the oxidation half-reaction

$$Mn^{2+} \rightarrow Mn^{3+} + e$$

is simply the negative of the emf for the reduction half-reaction given on the emf diagram. In this case, we need not consider disproportionation to higher oxidation states than Mn^{3+}, since inspection of the diagram shows that the half-cell emf for the oxidation of Mn^{2+} to any higher state will be negative. Using calculations similar to those given above, we can verify that Mn^{3+} or MnO_4^{2-} would be unstable with respect to disproportionation in acid solution. In order to determine whether Mn^{3+} and MnO_4^{2-} would be stable in water, a new emf diagram should be calculated for pH = 7. For

$$MnO_2(s) + 4H^+ + e \rightarrow 2H_2O + Mn^{3+}$$

$$E = E_A^0 - \frac{0.059}{1} \log \frac{1}{(10^{-7})^4} = -0.73$$

Similar calculations should be made for the $MnO_4^{2-} \rightarrow MnO_2$ couple. These new emf values are the emf values in neutral water, E_N. The E_N diagram for Mn is:

$$MnO_4^- \xrightarrow{0.564} MnO_4^{2-} \xrightarrow{1.43} MnO_2 \xrightarrow{-0.73} Mn^{3+} \xrightarrow{1.51} Mn^{2+} \xrightarrow{-1.19} Mn$$

$$(2.86 + 0.56)/3$$

Inspection of this diagram indicates that MnO_4^{2-} will disproportionate in water ($E_N = 1.43 - 0.564$) and Mn^{3+} will disproportionate ($E_N = 1.51 + 0.73$). All other species are stable with respect to disproportionation in water. Disproportionation cannot occur for terminal species (Mn, or MnO_4^-), since such species cannot undergo both oxidation and reduction.

Hydrogen Evolution

E_A^0 for the half-reaction

$$2H^+ + 2e \rightarrow H_2$$

by convention is given a value of zero. Using the Nernst equation, we obtain $E_N = -0.414$ and $E_B^0 = -0.828$. For the reaction of manganese with acid, an E_A^0 of $+1.19$ is obtained. For gas evolution reactions to occur as reasonable rates, the emf for the reaction generally must be of the order of $+0.4$ V or more, because of overvoltage effects, which are specific for both the metal and the gas evolved. Since the E_A^0 for the reaction of manganese with acid exceeds the gas evolution overvoltage, reaction should take place. Similarly, E_N for hydrogen evolution by manganese is $+0.77$ and hydrogen evolution should occur.

Oxygen Evolution

We write the following half-cell emf value for oxygen evolution.

$$2H_2O = O_2 + 4H^+ + 4e \qquad E = -E_A^0 = -1.229$$

From this, we may obtain $E_N = -0.815$ and $E_B = -0.401$.

Example Will permanganate ion evolve oxygen in acid solution?

Solution For the reaction

$$2MnO_4^- + H_2O \rightarrow 2MnO_4^{2-} + \tfrac{1}{2}O_2 + 2H^+$$

$$E = +0.564 - 1.229 = -0.665$$

Hence this reaction does not occur. However, for the reaction

$$2MnO_4^- + 2H^+ \rightarrow 2MnO_2 + H_2O + \frac{3}{2}O_2$$

$$E_A^0 = 1.695 - 1.229 = 0.466 \text{ V}$$

Since this value is barely in excess of the overvoltage needed for oxygen evolution, we find that oxygen evolution of permanganate solutions does take place very slowly in acid. The E_N value shows that oxygen evolution will not take place in neutral solution unless a suitable catalyst, such as platinum, is present to reduce the overvoltage.

11.6.4 Overvoltage

In order to carry out an electrolytic reaction resulting in gas evolution, a potential difference, which exceeds that calculated for a reversible galvanic cell, must be applied to the electrodes. This excess potential, called overvoltage, increases as the current density at the electrodes increases. The overvoltage depends on the gas evolved and on the material used for the electrode. Thus at a current density of 0.01 A/sq cm, shiny platinum has a hydrogen overvoltage of 0.07 V, whereas the oxygen overvoltage is 0.40 V. Under the same conditions, nickel has a hydrogen overvoltage of 0.56 V, whereas the oxygen overvoltage is 0.35 V. Metal deposition does not show an overvoltage effect. Hence, by working at high current density and low temperatures (where overvoltages are highest), we can electroplate from aqueous solution metals, such as nickel, that lie above hydrogen in the activity series.

The overvoltage effect is not limited to reactions carried out in an electrolytic cell, but occurs in many reactions producing a gas in solution—particularly those producing hydrogen and oxygen, as noted above.

11.7 POURBAIX (OR PREDOMINANCE AREA) DIAGRAMS

Figure 11.3 shows the potentials of some manganese couples as a function of pH. The activity of all solution species containing manganese, and involved in the couples whose potentials are plotted, is taken as unity. The potentials of couples unaffected by pH, such as Mn/Mn^{2+}, appear as horizontal lines. To the right of the intersection of the Mn/Mn^{2+} line with the $Mn/Mn(OH)_2$ line, the Mn/Mn^{2+} line is dashed, since a concentration of one-molal Mn^{2+} ion cannot be maintained at high pH values because of precipitation of $Mn(OH)_2$. To the left of this intersection, the $Mn/Mn(OH)_2$ line is dashed, since the solubility of $Mn(OH)_2$ is greater than one molal at low pH values. For similar reasons the Mn^{2+}/Mn_3O_4 line is dashed to the right of the intersection with the $Mn(OH)_2/Mn_3O_4$ line, and the latter is dashed to the left of

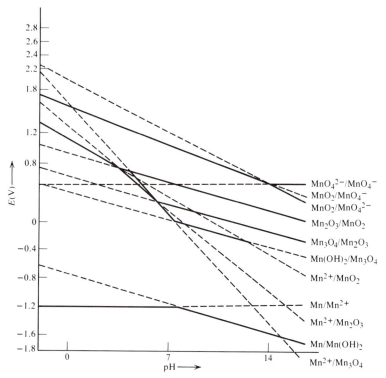

Figure 11.3   Variation with pH of the potentials of some Mn couples.

the intersection. Species undergoing disproportionation are either omitted entirely (Mn^{3+}, for example) or, in the case of couples involving such species, are shown as dotted lines in the region of instability of one of the species involved (MnO_4^{2-}, for example).

Figure 11.4 presents the same data as Figure 11.3, except that only the solid lines are shown for manganese couples and several vertical lines have been added to represent acid–base equilibria not involving electron transfer. In addition, the H_2O/H_2 and H_2O/O_2 potentials at 1 atm pressure are shown by dotted lines. This type of figure is known as a Pourbaix diagram,[9] or potential–pH diagram, or predominance area diagram. The latter name reflects the fact that the species listed in any area is the predominant species of that element within the E—pH range given by its boundaries. The oxidation state always increases with increasing E on crossing a boundary. For solution species, the boundaries will be concentration-dependent, in accordance with the Nernst equation. Geologists and corrosion engineers frequently use diagrams with solution concentrations of 10^{-6} molal. Lowering the concentration of a solution species has the effect of increasing the predominance area of the solution species at the expense of solid species. Some predominance area diagrams use dotted lines to separate solution species, such as MnO_4^{2-} and MnO_4^-, and solid lines to indicate the existence of phase boundaries. Note that information not included in the construction of the diagram (species for which data may not be available, etc.) cannot be obtained from the diagram. For temperatures other than 25°C, you should construct a new Pourbaix diagram using the temperature coefficients of the poten-

[9]See work cited under "M. Pourbaix," in General References.

tials of the appropriate couples.[10] Finally, remember that the Pourbaix diagrams give the results expected *at equilibrium*. Frequently, equilibrium is achieved too slowly to make the conclusions valid—particularly in cases involving gas evolution, as pointed out earlier. The area between the dotted lines in Figure 11.4 delineates the region of stability of substances in contact with water, but as noted already, overvoltage effects permit the existence of species within a range of 0.4 volts above or below the indicated area for periods of time sufficient to carry out experiments in the laboratory.

What is the meaning of a point within the predominance area of a solution species? As indicated earlier, the predominant solution species is taken to have a fixed activity—usually, unity—so other species are set by the E and pH of the point selected, and can be determined if the appropriate half-cell emf data and acid-base equilibria data are available. At a pH of 4.0 and an E of -0.2, we would find (from the Nernst equation and the standard potential for Mn^{2+}/Mn^{3+}) that the Mn^{3+} concentration is 10^{-29} molal; and from the literature value of $10^{-3.9}$ for the instability constant of $Mn(OH)^+$ and the given pH, we calculate an $Mn(OH)^+$ concentration of 10^{-6} m. If the E is raised to $+0.4$, the Mn^{3+} concentration rises to 10^{-19} m.

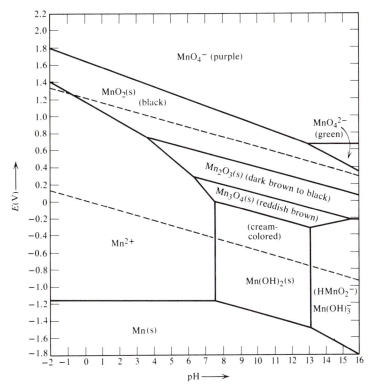

Figure 11.4 Pourbaix diagram for Mn species. (After M. Pourbaix, *Atlas of Electrochemical Equilibria in Aqueous Solution*, Trans. by J. A. Franklin, Pergamon, London, 1966.)

[10]See also H. E. Townsen, Jr., *Corrosion Science* 1970, *10*, 343–358 for extrapolations based on 25°C entropy values.

The Mn^{2+} and Mn^{3+} concentrations become equal at an E of 1.51, but for a 1 m concentration of solution species, at low pH values, $MnO_2(s)$ is the predominant species (but see below).

The fact that the Mn(II) ion is stable over an extremely wide range of E values has been attributed in part to the stability of the half-filled d orbital configuration, because of spin exchange interactions. On increasing the pH of a deaerated 1 m solution of Mn(II) ion (very pale pink), the white $Mn(OH)_2$ begins to precipitate at a pH of around 8. If air is allowed to contact the solution, the E gradient will vary from about 600 mV at the air interface to about 0 at the solution-precipitate interface, and in several hours the precipitate will develop striations that are cream colored on the bottom, through reddish-brown, dark brown, or black where the surface is in contact with the solution. After standing several days, only the black MnO_2 will be present. The reddish-brown Mn_3O_4, a mixed valence oxide containing both Mn^{2+} and Mn^{3+} ions in a slightly distorted normal spinel structure (see p. 212 for spinels), may be made directly by roasting any oxide of manganese in air at 1000°C. If we add iodine crystals to an Mn(II) solution, the solution will be ''buffered'' at an E around 600 mV. Very slowly adding base will precipitate $Mn(OH)_3$ ($K_{sp} = 10^{-36}$), which, in the absence of air goes over to a hydrated Mn_2O_3 that is brown in color. The Mn^{2+}/Mn^{3+} couple may be added to the Pourbaix diagram to give a horizontal line at 1.51 V. In the usual pH range of 0 to 14, and with concentrations (activities) of solution species set at one, the Mn^{2+}/Mn^{3+} couple has a higher potential than the Mn^{2+}/MnO_2 couple—thus Mn^{3+} is not stable under the conditions depicted in Figure 11.4. If the range of acidity is extended, however, the Mn^{2+}/Mn^{3+} line eventually will fall below the Mn^{2+}/MnO_2 line, and the Mn^{3+} species will exhibit a predominance domain. One hundred percent H_2SO_4 has an effective pH of -12 (or more precisely, an H_0 value, see p. 523). When MnO_2 is heated in concentrated sulfuric acid, $Mn_2(SO_4)_3$ forms and O_2 evolves. We also can produce Mn(III) by dissolving Mn_2O_3 in concentrated acids, or by reducing MnO_2 or MnO_4^- in concentrated acid. (Note, however, that permanganates are dehydrated in concentrated sulfuric acid and produce the very explosive Mn_2O_7, a dark greenish-brown oil). Decreasing the solution concentrations on the Pourbaix diagram will increase the Mn^{3+} domain at the expense of the MnO_2 domain. Complexing agents, such as F^- and CN^-, reduce the concentration of free Mn^{3+} and can stabilize this oxidation state even at moderate total concentration of Mn^{3+} and moderate acidities.

The Mn(I)/Mn(II) couple must have a potential less than that of Mn/Mn(II), but if the free manganese solution concentrations are low enough, the Mn(I) species has a domain between Mn and Mn(II). This occurs with the cyano complexes, and reduction of Mn(II) in the presence of cyanide gives the Mn(I) complex.

As might be surmised from our discussion of air oxidation of $Mn(OH)_2$, manganese ions in air-saturated natural waters are oxidized to Mn(IV) and precipitated as MnO_2, pyrolusite. The name of this mineral (Greek *pyro*, ''fire''; and *louein*, ''to wash'') derives from its use in glassmaking, in which it decolorizes green glass by oxidizing the Fe^{2+} impurity to Fe^{3+} while itself being reduced to Mn^{2+}. Both of the product ions are high-spin d^5 ions that are virtually transparent in the visible region, because all d-d transitions are spin-forbidden as well as LaPorte-forbidden (see p. 282). High concentration of Fe^{2+} ion in glass can reduce solar transmission by as much as 10% of the total energy.

MnO_2 can coexist with MnO_4^- except in very basic solution. Since the coexistence boundary lies above the H_2O/O_2 boundary over the entire pH range of coexistence, permanganate ion cannot be produced directly from MnO_2 by air oxidation. It can be produced, however, by an indirect air-oxidation route. At very high pH, MnO_2 has a coexistence boundary with the manganate ion (MnO_4^{2-}) that crosses the H_2O/O_2 boundary at a pH = 17.4 for 1 m solution concentrations; for 10^{-7} m MnO_4^{2-}, the boundaries cross at pH = 14. MnO_4^{2-} salts are isomorphous with sulfates, and air oxidation of MnO_2 in $Ba(OH)_2$ solution results in the formation of $BaMnO_4$. Acidifying manganates results in disproportionation to MnO_2 and MnO_4^-. Industrially, KOH is used instead of $Ba(OH)_2$, and the last step is carried out using chlorine or ozone, or electrolytically, to complete the conversion of the manganate to permanganate.

The effective pH of 10 M KOH (actually the H_- value) is 19. In this pH range the blue MnO_4^{3-} hypomanganate ion [or the tetraoxomanganate(V) ion] has a predominance domain with an E between that of MnO_4^{2-} and MnO_2. It can be prepared by reducing the permanganate ion with Na_2SO_3 in 25 to 30% NaOH.

Permanganate ion is a strong oxidizing agent that is thermodynamically unstable in water at all pH values. Because of overvoltage effects, only in acid solutions is the oxidation of water to oxygen of any consequence. The reaction is catalyzed by MnO_2, particularly in the presence of light. For this reason, solutions of $KMnO_4$ used as analytical reagents should be completely free of dust or organic matter and stored in dark glass bottles. The reduction product in acid solution is the almost colorless Mn^{2+} ion, and the endpoint of permanganate titrations in acid readily may be detected by the disappearance of the red-violet color of the MnO_4^- ion. In neutral or alkaline solutions, the reduction product of MnO_4^- is the insoluble black MnO_2.

11.8 GEOCHEMISTRY OF MANGANESE[11]

We already have seen several examples of the increase in a domain area as the solution activity of an oxidation state is decreased. From solubility products or free energies of formation, geochemists have been able to construct Pourbaix-type diagrams showing the E and pH domains of minerals for a set of conditions for other dissolved substances. Thus, setting the Mn^{2+} solution concentration at 10^{-6} m and the concentration of total dissolved carbonate species at $10^{-1.4}$ m (corresponding to a system first saturated with CO_2 under 1 atm pressure and then closed) gives us Figure 11.5. With the total dissolved sulfur species set at 10^{-1} m and the partial pressure of CO_2 at 10^{-4} atm (corresponding to an open system in marine sediment), Figure 11.6 is obtained. The small fields for pyrochroite and alabandite indicate the very limited conditions under which these minerals form and reflect their rarity. The ratio of manganese to other transition metals occurring in ocean-floor nodules has been explained[12] in terms of both the ease of mobility by ionic diffusion of the lower oxidation states of the pairs below and the E boundary between these states at pH = 8.

Element	Ni	Co	Mn	Cr	Cu	V	Fe	U
	II/III	II/III	II/IV	III/VI	I/II	IV/V	II/III	IV/VI
E (volts)	+0.8	+0.6	+0.5	+0.3	+0.1	0.0	−0.2	−0.3

[11]R. M. Garrels and C. L. Christ *Solutions, Minerals, and Equilibria*, Freeman, Cooper, San Francisco, 1965.

[12]E. Bonatti, D. E. Fisher, O. Joensuu, and H. S. Rydell, *Geochimica et Cosmochimica Acta* 1971, *35*, 189–201.

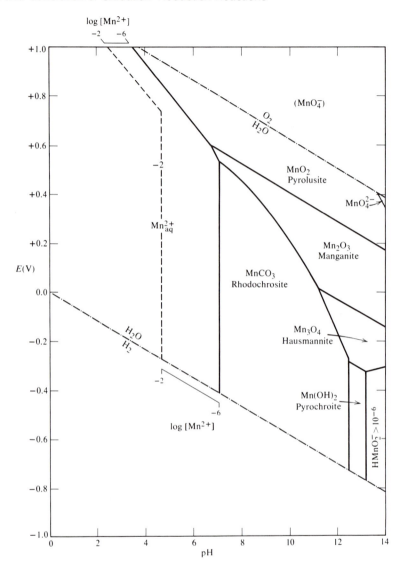

Figure 11.5 Stability relations among some manganese compounds in water at 25°C and 1 atmosphere total pressure. Total dissolved carbonate species = $10^{-1.4}$. (From R. M. Garrels and C. L. Christ, *Solutions, Minerals, and Equilibria*, Freeman, Cooper, San Francisco, 1965.)

"Elements with the higher redox potential are more easily reduced. Considering that all of these elements (except Cr, V and U) when in reduced state tend to be in solution and to precipitate when oxidized, it follows that, if the environment is reducing, the elements with higher redox potential should tend to go in solution. As a result, a

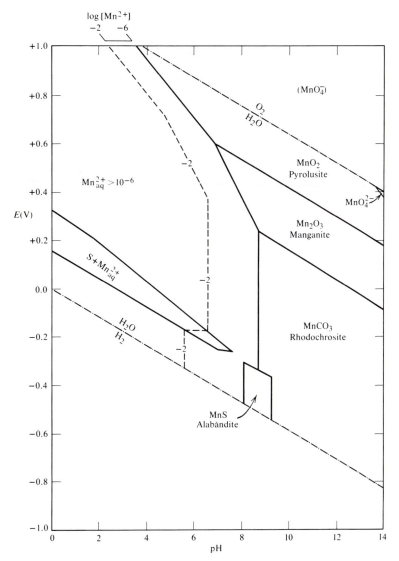

Figure 11.6 Stability relations among some manganese compounds in water at 25°C and 1 atmosphere total pressure. Total dissolved sulfur species = 10^{-1}, $P_{CO_2} = 10^{-14}$. (From R. M. Garrels and C. L. Christ, *Solutions, Minerals, and Equilibria*, Freeman, Cooper, San Francisco, 1965.)

concentration gradient should be established in the pore solution at the Eh interface whereby these elements should become concentrated in the pore solution below the interface and depleted in the solution above it. Upward migration of the elements should follow, mainly by ionic or molecular diffusion.''[12]

11.9 PERIODIC TRENDS AMONG THE TRANSITION ELEMENTS[13]

The two Pourbaix diagrams shown in Figures 11.7 and 11.8 indicate the general trends in redox behavior among the transition element species in aqueous media. Going from Mn to Tc, we find an increase in nobility of the metal—Tc does not liberate H_2 from acid, neutral, or basic solution, but it may be oxidized by dissolved oxygen in water. This nobility of the heavier transition metals extends to Re and through the transition elements to the right of Tc and Re—Ru, Rh, Pd, and Ag in the second transition series and Os, Ir, Pt, and Au in the third. We also find that TcO_4^- is stable in aqueous solution. This trend also is general in the transition metallates—the higher oxidation states are relatively more stable for the heavier elements in a given transition metal group, whereas the lower states are either unstable or exhibit relatively narrow stability domains. Thus the higher states are much milder oxidizing agents for the heavier elements, and the lower states may behave as reducing agents. (The chemistry of Tc has excited much interest recently, because of the use of ^{99m}Tc in radiophar-

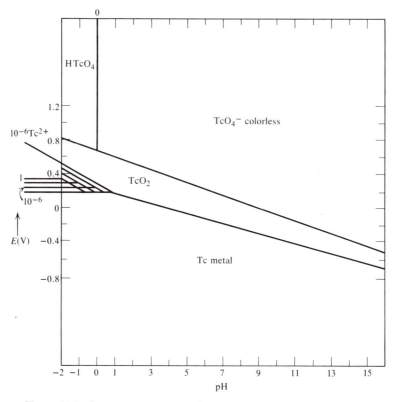

Figure 11.7   Pourbaix diagram for Tc species. (From M. Pourbaix, *op. cit.*)

[13]Campbell and Whiteker have constructed a large number of Pourbaix diagrams for the stability domain range of water, which you are urged to examine—particularly if access to Pourbaix's work is not available. J. A. Campbell and R. A. Whiteker, *J. Chem. Educ.* 1969, *46*, 90.

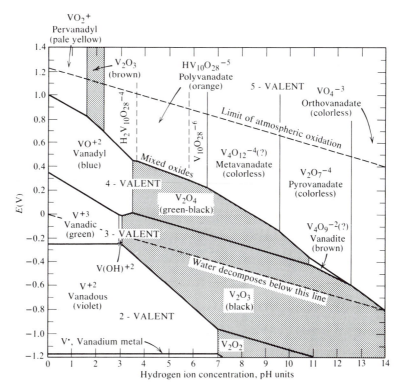

Figure 11.8 Stability of some vanadium compounds and ions in water at 25°C and 1 atmosphere total pressure. (From R. M. Garrels, *op cit.*)

maceuticals as heart- and bone-scanning agents. ^{99m}Tc is a short-lived gamma emitter that arises from the β-decay of ^{99}Mo and is separated from it by column chromatography.)

From Figure 11.8's Pourbaix diagram we see that, except in very strong acid, the V(V) species are relatively weak oxidizing agents. As the atomic number increases on crossing a transition metal series, the oxidizing ability of the highest oxide or oxoanion increases. Thus chromates are better oxidizing agents than vanadate, permanganate is still better, and FeO_4^{2-} is too strong an oxidizing agent to be in contact with water. On the other hand, V^{2+} is a reducing agent, thermodynamically unstable with respect to reducing water; because of overvoltage effects it can be prepared in aqueous solution. The pH-dependent polymerization equilibria of the highest oxidation state are perhaps most pronounced with vanadium, but are seen to a lesser extent with chromates and molybdates.

11.10 CHEMICAL REACTIONS

If two Pourbaix diagrams are superimposed, species can coexist only if their predominance areas overlap; otherwise, a reaction will take place and the resulting products will have overlapping predominance areas. The Latimer line diagrams connecting species with a value given for the potential of the couple under standard conditions gives the boundary points on a Pourbaix diagram at a pH of 0 for E_A^0 values, or a pH of 14 for E_B^0 values. The chemistry of aqueous solutions at the appropriate pH often can be understood by studying such line potential diagrams. Consider the following data for chlorine species.

$$E_A^0 \qquad ClO_4^- \xrightarrow{1.19} ClO_3^- \xrightarrow{1.15} ClO_2 \xrightarrow{1.27} HClO_2 \xrightarrow{1.65} HClO \xrightarrow{1.63} Cl_2 \xrightarrow{1.36} Cl^-$$

$$E_B^0 \qquad ClO_4^- \xrightarrow{0.36} ClO_3^- \xrightarrow{-0.5} ClO_2 \xrightarrow{1.16} ClO_2^- \xrightarrow{0.66} ClO^- \xrightarrow{0.4} Cl_2 \xrightarrow{1.36} Cl^-$$

In the laboratory, we can produce elemental chlorine from chloride ion sources using a variety of oxidizing agents in acid solution—$KMnO_4$, $K_2Cr_2O_7$, $KClO_3$, MnO_2, PbO_2, etc. Pertinent E_A^0 values are

$$MnO_4^- \xrightarrow{1.51} Mn^{2+} \qquad ClO_3^- \xrightarrow{1.47} Cl_2$$

$$MnO_2 \xrightarrow{1.23} Mn^{2+} \qquad PbO_2 \xrightarrow{1.46} Pb^{2+}$$

$$Cr_2O_7^{2-} \xrightarrow{1.33} Cr^{3+}$$

MnO_2 and $K_2Cr_2O_7$ require a concentrated acid with a hydrogen ion activity greater than 1— in actual practice, sulfuric acid usually is used since sulfuric acid itself is too weak an oxidizing agent to oxidize the Cl^- ion, it serves, if used by itself, simply to form hydrogen chloride.

$$E_A^0 \qquad SO_4^{2-} \xrightarrow{0.17} H_2SO_3 \xrightarrow{0.45} S$$

Although Cl_2 is produced by the action of concentrated nitric acid on chlorides, lower oxides of nitrogen are produced also. A mixture of concentrated nitric and hydrochloric acids (aqua regia) does serve as a strong oxidizing medium for many metals, because of the complexing action of the chloride ion with the metal ion produced.

Commercially, the electrolysis of brine solutions is used to produce chlorine. Although the evolution of oxygen should take place at lower voltages than those needed for chlorine

$$E_N \qquad H_2O \xrightarrow{-0.815} O_2$$

the overvoltage for oxygen at high current densities is much greater than that for chlorine.

Chlorine reacts slowly with water to liberate oxygen.

$$Cl_2 + H_2O \rightarrow 2HCl + O_2 \qquad E_N = 0.545$$

More immediate, however, is a disproportionation reaction that is self-repressed by the hydrogen ion formed.

$$Cl_2 + H_2O \rightarrow HOCl + H^+ + Cl^- \qquad E_N = +0.14$$

$$E_{pH4.5} = 0$$

In basic solution, hydrolysis is complete ($E_B^0 = +0.96$) to give hypochlorite and chloride ions. In one commercial procedure a brine solution is stirred during electrolysis, so that the anode and cathode products mix and hypochlorite ions are formed. More commonly, chlorine is allowed to react directly with "slaked lime" to form bleaching powder or "chloride of lime."

$$Cl_2 + Ca(OH)_2 \rightarrow CaCl(OCl) + H_2O$$

Although the couples $Cl_2 \xrightarrow{1.36} Cl^-$ and $O_2 \xrightarrow{1.23} H_2O$ indicate that in acid solution, chlorine and oxygen should have about the same ability as oxidizing agents, chlorine actually is much stronger. This is because hydrogen peroxide is an intermediate when oxygen is reduced to water and the $H_2O_2 \rightarrow O_2$ couple is the effective emf available in oxygen oxidations.

$$E_A^0 \quad O_2 \xrightarrow{0.69} H_2O_2 \xrightarrow{1.77} H_2O$$
$$\underset{1.23}{\underline{\hspace{4cm}}}$$

Accordingly, O_2 cannot be used to displace bromine from bromides. Instead, Cl_2 is used to displace bromine from bromide in sea water with the pH adjusted to 3.5. The pH adjustment prevents disproportionation of Cl_2 or Br_2 to the halide and hypohalite.

As the emf data indicate, chlorous acid is unstable with respect to disproportionation in acid solution. In base, a mixture of chlorites and chlorates is obtained from the disproportionation of ClO_2.

$$2ClO_2 + 2OH^- \rightarrow ClO_2^- + ClO_3^- + H_2O \qquad E_B^0 = 1.56$$

Reaction of ClO_2 with sodium peroxide in aqueous solution produces pure chlorites.

$$OH^- + 2ClO_2 + O_2H^- \rightarrow 2ClO_2^- + O_2 + H_2O \qquad E_B^0 = 1.24$$

In hot basic solution the disproportionation of chlorine yields chlorates and chlorides.

$$3Cl_2 + 6(OH^-) \rightarrow 5Cl^- + ClO_3^- + 3H_2O \qquad (E_B^0 = +0.89 \text{ at } 25°)$$

We also can carry out this reaction by allowing mixing of the anode and cathode compartments during electrolysis of a hot brine solution.

Perchlorates are prepared from chlorates by disproportionation of the dry alkali metal salt.

$$4KClO_3 \xrightarrow{\Delta} 3KClO_4 + KCl$$

Upon heating above 400°, the $KClO_4$ decomposes to O_2 and KCl. We can prepare perchlorates in solution by electrolytic oxidation. Perchloric acid may be obtained from perchlorates by acidifying. If concentrated $NaClO_4$ and HCl are used, NaCl is precipitated and the excess HCl removed by warming. Sulfuric acid may be used and the perchloric acid purified by vacuum distillation. Hot concentrated perchloric acid is an extremely explosive oxidizing agent. In basic media the perchlorates are mild oxidizing agents. Magnesium perchlorate commonly is used as a drying agent. Great care should be taken to assure nonacidic conditions when using magnesium perchlorate as a drying agent.

11.10.1 Numerical Calculations Using emf Data

In addition to qualitative predictions about the spontaneity of a reaction, the E^0 value for a reaction permits us to calculate the equilibrium constant for the reaction. At equilibrium, $E = 0$ and $Q = K_{eq}$, the Nernst equation then gives

$$E^0 = \frac{0.059}{n} \log K_{eq}$$

Example   Calculate the equilibrium constant for the reaction

$$Pb + Sn^{2+} \rightarrow Pb^{2+} + Sn$$

Solution   From the emf data, we calculate $E^0 = 0.126 - 0.136 = -0.01$ and hence $K_{eq} = 10^{-2(0.010)/0.059} = 0.46$.

Example   Utilizing standard emf data, calculate K_p for the reaction

$$2PbO(s) \rightarrow 2Pb(s) + O_2$$

Solution

$$
\begin{array}{lr}
4e + 2PbO + 2H_2O \rightarrow 2Pb + 4(OH^-) & E_B^0 = -0.54 \\
4(OH^-) \rightarrow O_2 + 2H_2O + 4e & E_B^0 = -0.40 \\
\hline
2PbO(s) \rightarrow 2Pb(s) + O_2 & E_B^0 = -0.94
\end{array}
$$

$$K_p = 10^{4(-0.94)/0.059} = 10^{-63.6} = 2.5 \times 10^{-64} \text{ atm at } 25°C$$

We often can obtain solubility products of slightly soluble substances using cell measurements, by reversing the calculation procedure indicated on page 493.

Stability constants can be evaluated from emf data by procedures closely resembling those used to obtain solubility products.

Example   Calculate the equilibrium constant (β_6) for $Fe^{2+}(aq) + 6CN^-(aq) \rightarrow [Fe(CN)_6]^{4-}(aq)$ from the E^0 values.

$$Fe \underset{E^0}{\overset{0.41}{\rightarrow}} Fe^{2+}$$

and

$$Fe \underset{E_c^0}{\overset{1.5}{\rightarrow}} [Fe(CN)_6]^{4-}$$

Solution   The E_c^0 value is the *standard* emf for

$$Fe(s) + 6CN^-(aq) \rightarrow [Fe(CN)_6]^{4-}$$

that is, for a solution containing CN^- and $[Fe(CN)_6]^{4-}$ at unit activity. Accordingly, the Fe^{2+} activity is given by

$$(Fe^{2+}) = \frac{1}{\beta_6} \frac{(Fe(CN)_6^{4-})}{(CN^-)^6} = \frac{1}{\beta_6}$$

Iron in equilibrium with $[Fe(CN)_6]^{4-}$ and CN^- ion at unit activity thus is also in equilibrium with Fe^{2+} at an activity of $\frac{1}{\beta_6}$; and from the Nernst equation we obtain

$$E_c^0 = E^0 - \frac{0.059}{n} \log (Fe^{2+})$$

$$1.5 = +0.41 - \frac{0.059}{2} \log \frac{1}{\beta_6}$$

$$\beta_6 = 10^{37}$$

If the stoichiometry of the complex had not been known, we could have found it by measuring E_c at several different CN^- concentrations, provided the total cyanide concentration exceeds that required by the stoichiometry of the complex. Under these conditions the Nernst equation becomes

$$E_c = E^0 + \frac{0.059}{n} \log \beta + \frac{0.059}{n} c \log (CN^-)$$

where c is the number of ligands per metal atom. Taking derivatives yields

$$\frac{d(E_c)}{d \log (CN^-)} = \frac{0.059}{n} c$$

$d(E_c)/d \log (CN^-)$ is simply the slope of E_c plotted against $\log (CN^-)$. If the complexes of other stoichiometries are present over the ligand concentration studied, a plot of E_c versus $\log$ (ligand concentration) will not be linear.

11.11 PERIODIC TRENDS IN HALF-CELL emf VALUES AND EXTRATHERMODYNAMIC RELATIONSHIPS

The heat release (Q) during the oxidation of an element to its monoatomic ion in aqueous solution can be taken as the difference between the heat of solution of the gaseous ion and the sum of its sublimation energy and ionization energy.

$$
\begin{array}{ccc}
& I_1 + I_2 + \ldots I_x & \\
M(g) & \longrightarrow & M^{x+}(aq) + xe \\
\uparrow S & & \downarrow H_+ \\
& Q & \\
M(s) & \longrightarrow & M^{x+}(aq) + xe
\end{array}
$$

$$Q = (-\Delta H) = -S - \underset{x}{\Sigma I} + H_+$$

Assuming the free-energy change to be proportional to the enthalpy change, E^0 will parallel Q/x: that is, the emf for the oxidation will increase as the reaction becomes more exothermic. The heat of solution and the ionization energy usually will vary in a parallel fashion. In most cases the variations in I and S determine the major variation in E^0. An exception to the preceding generalization is provided by lithium ion, whose very high heat of solution causes the Li/Li^+ couple to have a higher voltage than the Na/Na^+ or K/K^+ couples.

The decrease in emf for M/M^{x+} with increasing x within a period can be attributed almost entirely to the $\Sigma I/x$ factor. Crossing the first transition series in going from M to M^{x+}, the increased penetration of the $4s$ electrons is reflected in the decreasing E^0 values. Within a transition metal family, both S and I increase with atomic number, resulting in a decrease in ease of oxidation of M to M^{x+}.

Increasing emf for oxidation half-reaction

Na/Na$^+$	Mg/Mg^{2+}	Al/Al^{3+}						
K/K$^+$	Ca/Ca^{2+}	Sc/Sc^{3+}	Ti/Ti^{2+}	V/V^{2+}	Cr/Cr^{2+}	Mn/Mn^{2+}	Fe/Fe^{2+}	Co/Co^{2+}
Rb/Rb$^+$	Sr/Sr^{2+}					Tc/Tc^{2+}	Ru/Ru^{2+}	
						Re/Re^{2+}	Os/Os^{2+}	

The actinides show a regular decrease in the formal emf for the oxidation of M^{3+} to M^{4+}; the values for U, Np, Pu, and Am are, respectively, 0.631, -0.155, -0.982, and -2.6 V. Curium can be oxidized to Cm(IV) in aqueous solution only if a high concentration of fluoride ion is present to complex the product. The high stability of Cm(III) may be associated with the stability of a half-filled f shell. Oxoanions utilizing outer d orbitals (SO$_4^{2-}$, SeO$_4^{2-}$) demonstrate an increasing ability to serve as oxidizing agents with increasing atomic number, whereas those using inner d orbitals (MnO$_4^-$, TcO$_4^-$) show decreasing ability with increasing atomic number.

The coupling reaction

$$HX \rightarrow \tfrac{1}{2}X_2 + H^+ + e$$

may be related to electronegativity if the hydrogen half-reaction is added to give

$$HX \rightarrow \tfrac{1}{2}X_2 + \tfrac{1}{2}H_2$$

for which $\Delta H = -96{,}500\,(\chi_L - \chi_H)^2$ J if solvation is ignored (χ is the electronegativity). Further, if we ignore entropy changes, then for the oxidative coupling

$$-E^0 = (\chi_L - \chi_H)^2 \tag{11.6}$$

We can use this to rationalize the trends shown below.

H$_2$NNH$_2$	H$_2$O$_2$	F$_2$	
H$_2$PPH$_2$	H$_2$S$_2$	Cl$_2$	Oxidizing ability
		Br$_2$	
		I$_2$	

CH$_4$	NH$_3$	H$_2$O	HF	
SiH$_4$	PH$_3$	H$_2$S	HCl	Reducing ability
	AsH$_3$	H$_2$Se	HBr	
			HI	

Equation 11.6 has been used to evaluate group electronegativities from E^0 data (that is, χ for HO$_2$CO—, HO—, O$_2$NO—, etc).[14]

GENERAL REFERENCES

See Appendix C for a summary of half-cell emf data.

M. Pourbaix, *Atlas of Electrochemical Equilibria in Aqueous Solution*, trans. by J. A. Franklin, Pergamon, London, 1966.

[14]D. H. McDaniel, and A. Yingst, *J. Am. Chem. Soc.*, 1964, 86, 1334.

G. Milazzo and S. Caroli, *Tables of Standard Electrode Potentials,* Wiley, New York, 1978.

W. M. Latimer, *The Oxidation States of the Elements and Their Potentials in Aqueous Solution,* 2nd ed., Prentice-Hall, New York, 1952.

G. Charlot, *Selected Constants Tables No. 8: Oxydo-Reduction Potentials,* Pergamon, New York, 1958.

A. J. deBethune and N. A. S. Loud, *Standard Aqueous Electrode Potentials and Temperature Coefficients at 25°C,* Clifford A. Hempel, Skokie, Illinois, 1964.

J. F. Elliott and M. Gleiser, *Thermochemistry for Steelmaking,* Addison-Wesley, Reading, 1960.

Landolt-Bornstein. Zahlenwerte und Funktionen aus Physik, Chemie Astronomie, Geophysik und Technik. 6th Auflage. II Band. Eigenschaften der Materie in Ihren Aggregatzustanden. 4 Teil. Kalorische Zustandsgrossen. K. Schafer and E. Lax, Eds., Springer-Verlag, West Berlin, 1961.

PROBLEMS

11.1   Write balanced ionic half-reactions for

    a. $HgO \xrightarrow{\text{base}} Hg$.

    b. $VO^+ \xrightarrow{\text{acid}} HV_2O_5^-$.

    c. $HGeO_3^- \xrightarrow{\text{base}} Ge$.

    d. $S_2O_3^{2-} \xrightarrow{\text{acid}} S$.

    e. $F_2O \xrightarrow{\text{acid}} HF$.

    f. $C_{12}H_{22}O_{11} \xrightarrow{\text{acid}} CO_2$.

11.2   The following are somewhat more challenging equations to balance.

    a. $H_2O + P_2I_4 + P_4 \rightarrow PH_4I + H_3PO_2$.

    b. $ReCl_5 + H_2O \rightarrow Re_2Cl_9^{2-} + ReO_4^- + Cl^- + H^+$.

    c. $B_{10}H_{12}CNH_3 + NiCl_2 + NaOH \rightarrow Na_4[B_{10}H_{10}CNH_2]_2Ni + NaCl + H_2O$.

    d. $ICl + H_2S_2O_7 \rightarrow I_2^+ + I(HSO_4)_3 + HS_3O_{10}^- + HSO_3Cl + H_2SO_4$.

11.3   Should the value of the heat of a reaction calculated from the variation of K_p with temperature be affected by the units in which K_p is expressed? Under what circumstances would ΔH be unaffected and under what circumstances affected by the selection of units for the K_{eq}?

11.4   Calculate E^0 values for the following cells and write the reactions for which these apply.

    a. Pt, H_2 | HCl ‖ KCl | $Hg_2Cl_2(s)$, Hg. (E^0 $Hg_2Cl_2(s) \xrightarrow{0.2682} Hg + Cl^-$)

    b. Cu | Cu^{2+} ‖ I^-, CuI(s) | Cu.

    c. Pt, CuI(s) | Cu^{2+} ‖ I^- | CuI(s), Cu. (E^0 CuI(s) $\xrightarrow{-0.1852}$ Cu + I^-)

    From parts *(b)* and *(c)*, comment on the necessity of knowing the half-cells involved in making predictions of (1) the spontaneity of a reaction and (2) the free-energy change of a reaction.

11.5   Calculate the emf values for the following cells.

    a. Pt, H_2 | $H^+(a = 0.1)$ ‖ $H^+(a = 10^{-7})$ | H_2, Pt.

    b. Zn | $Zn^{2+}(a = 1)$ ‖ $Cu^{2+}(a = 10^{-4})$ | Cu.

    c. Fe, $Fe(OH)_2$ | $OH^-(a = 0.1)$ ‖ $Fe^{2+}(a = 1)$ | Fe.

11.6   Calculate the approximate half-cell emf values for the following reactions from known electro-negativities and Equation 11.6.

    a. $2H_2O \rightarrow H_2O_2 + 2H^+ + 2e$.

    b. $2HF \rightarrow F_2 + 2H^+ + 2e$.

    c. $2PH_3 \rightarrow H_2PPH_2 + 2H^+ + 2e$.

    Use Huggins' assignment of 2.2 for the electronegativity of hydrogen.

11.7   Describe the conditions of acidity most appropriate for the following processes.

    a. $Mn^{2+} \rightarrow MnO_4^-$.

    b. $CrO_4^{2-} \rightarrow Cr_2O_7^{2-}$.

    c. $Fe^{3+} \rightarrow FeO_4^{2-}$.

    d. $ClO_4^- \rightarrow ClO_3^-$.

    e. $C_2O_4^{2-} \rightarrow 2CO_2$.

    f. $H_2O_2 \rightarrow H_2O$.

    g. $H_2O_2 \rightarrow O_2$.

11.8   By extrapolation from known data for Cr, W, and Mo, Seaborg estimates the following emf values for the couples involving possible species for the currently unknown element 106:

$$E_A^0 \qquad MO_3 \xrightarrow{-0.5} M_2O_5 \xrightarrow{-0.2} MO_2 \xrightarrow{-0.7} M^{3+} \xrightarrow{0.0} M$$

Predict the results of the following.
a. M is placed in $1M$ HCl.
b. MCl_3 is placed in an acidic solution containing $FeSO_4$.
c. MO_2 and M are in contact in acid solution.
d. M^{3+} and MO_3 are in contact in acid solution.
e. M is treated with excess concentrated HNO_3.

11.9   Given the following half-cell emf diagram for osmium,

$$E_A^0 \qquad OsO_4(s) \xrightarrow{1.0} OsCl_6^{2-} \xrightarrow{0.85} OsCl_6^{3-} \xrightarrow{0.4} Os^{2+} \xrightarrow{0.85} Os$$

a. Which of the above species, if any, would be unstable in $1M$ HCl? Give balanced equations for any reactions that occur.
b. Which couple(s) would remain unchanged in their emf value on altering the pH?
c. Which couple(s) would remain unchanged in their emf value on altering the chloride ion concentration?
d. Calculate the value of E_A^0 for $OsO_4(s) \rightarrow Os$
e. Predict the results of mixing excess osmium with solid OsO_4 in contact with $1M$ HCl.

11.10   *a.* Use data from Appendix C to construct a Latimer diagram (p. 490) for Cl_2 in acid and base solution, showing only species stable with respect to disproportionation.
b. Construct a Pourbaix diagram for Cl with unit activity for predominant species.

XII
Acids and Bases

The continuing evolution of acid-base concepts reflects the continuing evolution of chemistry itself. This chapter first examines factors affecting proton affinities, since such data eliminate many complications arising from effects such as solvation, lattice energies, and multiple bonding between acid and base. We then follow the effect of solvation on acid strength and some chemical implications. The Lewis and Usanovitch definitions make virtually all chemical reactions interpretable as acid-base interactions, or with more currency the interaction of frontier orbitals.

12.1 HISTORICAL BACKGROUND

Acids as a class of compounds were well known to the alchemists, who noted their sour taste (Latin *acidus*, ''sour''), their ability to dissolve many water-insoluble substances, and their action on various vegetable dyes. When Priestly announced his discovery of ''dephlogisticated air'' in 1775, Lavoisier, whose experiments on combustion had started the overthrow of the phlogiston theory, concluded that in combination with nonmetals the newly discovered substance was the common constituent of acids. Accordingly, he named the new substance oxygen (French *oxys*, ''sharp'' or ''acid,'' plus *genesis*). The German name *Sauerstoff* is a translation of the French name. Davy proved that not all acids contain oxygen and proposed that hydrogen was the common constituent of acids. Liebig firmly established the protonic concept of acids and described an acid as a substance composed of a replaceable hydrogen and an acid radical.

Arrhenius and Ostwald's theory of electrolytic dissociation led to the present-day view of acid-base equilibria in water. Here the theory focuses on the self-dissociation or autoprotolysis of water.

$$HOH \rightleftharpoons H^+(aq) + OH^-(aq)$$

At 25°C about one in every half-billion water molecules is dissociated into a hydrogen ion and a hydroxide ion. An aqueous solution is said to be neutral when the concentrations of hydrogen ions and hydroxide ions are equal, acidic when there are more hydrogen ions, and basic when there are more hydroxide ions. We can increase the hydrogen ion concentration either by adding a substance that provides additional hydrogen ions for the system, such as hydrogen chloride, or by adding a substance that will remove hydroxide ion from the system, such as boric acid.

$$B(OH)_3 + OH^- \rightleftharpoons B(OH)_4^-$$

Such substances are termed acids. Similarly, the hydroxide ion concentration may be increased by direct dissociation to produce hydroxide ions, as in the case of sodium hydroxide, or by combination with hydrogen ions in solution, as in the case of ammonia. Combination processes of the latter type often are referred to as hydrolysis (p. 528). Neutralization reactions in water consist of combining hydrogen ions and hydroxide ions to form water.

Many solvents other than water undergo autoprotolysis, and acid-base equilibria in these solvents may be treated on a conceptual basis similar to the Arrhenius-Ostwald picture. Thus in liquid ammonia, a substance producing ammonium ions would be an acid, a substance producing amide ions would be a base, and a neutralization reaction would consist of the reaction of ammonium ion and amide ion to produce ammonia. A more general solvent theory defines an acid as a substance that produces positive solvent ions and a base as a substance that produces a negatively charged solvent ion. Neutralization consists of combining these ions to produce solvent.

A protonic picture that does not require solvent participation was presented by Brønsted and Lowry in 1923. According to their views, an acid is a proton donor and a base is a proton acceptor. In an acid-base reaction, a proton is transferred from an acid to a base to produce another acid, termed the conjugate acid of the original base, and another base, termed the conjugate base of the original acid.

$$HCl + NH_3 \rightarrow NH_4^+ + Cl^-$$

   acid base conjugate conjugate
          acid   base

A general theory of acids and bases covering all the preceding cases and extending the definition to some substances not included above was set forth by G. N. Lewis, who defined an acid as an electron pair acceptor and a base as an electron pair donor. Neutralization in the Lewis theory consists of the formation of a new covalent bond between an electron *pair* donor and an electron *pair* acceptor.

$$BF_3 + (CH_3)_3N: \rightarrow (CH_3)_3N:BF_3$$

   acid base

This chapter deals mainly with the factors that affect the relative strength of acids and bases. Protonic acids are treated first since there are fewer factors involved than with other Lewis acids and, in the case of protonic acids, these factors are more easily isolated. The pK_a—that is, $-\log K_a$—will be used often here; the higher the numerical value of the pK_a, the weaker the acid under discussion. For the acid HA, the dissociation constant, K_a, is defined by the activity quotient $K_a = [H^+aq][A^-]/[HA]$.

Usanovitch has given the most general definition of an acid: "An acid is any material which forms salts with bases through neutralization, gives up cations, combines with anions, or electrons." All of these characterizations are useful: the one selected should be the one simplest to apply to the system under study. Thus in aqueous solutions of protonic acids, chemists use the dissociation constant as a measure of acid strength. The bulk of the data currently available are such K_a values. By emphasizing *proton* transfer as the acid–base re-action, the Brønsted-Lowry picture makes acid–base reactions analogous to oxidation–reduction reactions, in which the *electron* transfer is the essential defining process. Just as a total oxidation–reduction reaction can be split into half-reactions involving the oxidized and reduced form of a given reagent, a proton transfer reaction can be broken into half-reactions involving the protonated and unprotonated form of a given base: $B + H^+ \rightarrow BH^+$. The heat released in such a gas phase reaction is referred to as the proton affinity *(PA)* of B.

The choice of the appropriate acid–base model depends on the intended use. The Ostwald dissociation picture serves for most situations involving hydrogen ion equilibria in aqueous solution. The Brønsted-Lowry picture sets limits on the intrinsic protonic acidity or basicity through the proton affinity scale and suggests the degree of attenuation of protonic acidity scales in different solvents. Dissection of factors influencing the strength of acids and bases is best carried out using proton affinity data. Comparison with solution data then permits us to assess solvation effects. The many parallelisms between protonic acids and Lewis acids allow us to extend our picture of protonic acids to a broader range of systems, such as coordination complexes. However, such an increase in generality requires that greater care be exercised with regard to the possible changes in multiple-bond character, steric effects, chelate effects, etc., that may accompany the acid–base reaction. Finally, the Usanovitch definition's extreme generality allows us to use knowledge about any one of the factors covered, such as the ability to combine with electrons, to reveal something about acidity in a "more conventional"—that is, a Lewis—sense.

12.2 PROTONIC ACIDS

12.2.1 Proton Affinities

Within the last decade, it has become possible to utilize special types of mass spectrometry to measure equilibria of the type

$$AH^+(g) + B(g) \rightarrow BH^+(g) + A(g)$$

and

$$AH(g) + B^-(g) \rightarrow BH(g) + A^-(g)$$

As indicated earlier, these are analogous to measuring a complete redox reaction, and we need a reference point for a known reaction of the type

$$A(g) + H^+(g) \rightarrow AH^+(g)$$

to establish an absolute proton affinity scale. Fortunately, sufficient data are available from spectroscopic experiments to establish a number of absolute proton affinities from the following cycle.

$$X(g) + H^+(g) \xrightarrow{\quad \Delta H_1 = -PA \quad} XH^+(g)$$

$$\Delta H_4 = -IE(X) \qquad\qquad \Delta H_2 = D_{X^+ - H}$$

$$X^+(g) + H^+(g) + e \xleftarrow{\quad \Delta H_3 = IE \quad} X^+(g) + H(g) \qquad (12.1)$$

$$PA(X) = D_{X^+ - H} + IE(H) - IE(X)$$

Combining reference points from the above cycle with data from ion cyclotron resonance spectroscopy or high-pressure mass spectroscopy yields results such as the basicities of the binary hydrides shown in Table 12.1. Using Equation 12.1, we can rationalize the trends in basicities observed in these binary hydrides. The ionization energies for the hydrides increase smoothly crossing a period, since the irregularities in spin multiplicity that cause irregularities in the atomic ionization energies have disappeared in the diamagnetic molecules. Since the ionization energy changes generally are much greater than the changes in the bond dissociation energies, the former dominates the *PA* behavior. Thus within a given period or a given group, the *PA* varies conversely with the *IE* of the binary hydride, which varies in much the same fashion as the electronegativity of the central atom. For PH_3 and NH_3, and H_2S and H_2Se, the *IE* differences are slight and the greater bond energy of the smaller central atom gives that particular molecule the higher *PA*.

For the negative ions, the energy required to remove an electron—that is, the electron affinity—generally is smaller and changes less in going down a group than the bond energy. Accordingly, within a group the trend in *PA*'s for the anions will parallel the trend in bond energies. On going from one group to another, the *EA*'s change more rapidly than the bond energies and the higher the *EA* (or χ), the lower the *PA*. Table 12.2 demonstrates these trends.

Inductive Effects

On going from NH_3 to NF_3, the ionization energy increases by 268 kJ/mole—as expected in view of the electron-withdrawing *inductive effect* of the fluoro groups. On substituting F for

Table 12.1   Order of basicities of the binary hydrides—proton affinities in kJ/mole

CH_4	<	NH_3	>	H_2O	>	HF
527		845		711		481
		$\vee$		$\wedge$		$\wedge$
		PH_3	>	H_2S	>	HCl
		774		728		565
		$\vee$				$\wedge$
		AsH_3		H_2Se		HBr
		711		—		590
						$\wedge$
						HI
						611

Table 12.2 Acidity order of the binary hydrides—proton affinities of their anions in kJ/mole[a]

Acidity Order				Proton Affinities			
CH_4 <	NH_3 <	H_2O <	HF	CH_3^-	NH_2^-	OH^-	F^-
∧	∧	∧	∧	1743	1689	1632	1554
SiH_4 <	PH_3 <	H_2S <	HCl	SiH_3^-	PH_2^-	SH^-	Cl^-
∧	∧	∧	∧	1554	1548	1474	1395
GeH_4 ≈	AsH_3 <	H_2Se <	HBr	GeH_3^-	AsH_2^-	SeH^-	Br^-
			∧	1509	1500	1420	1355
			HI				I^-
							1315

[a]From J. E. Bartmess and R. T. McIver, Jr., "The Gas-Phase Acidity Scale," in *Gas-Phase Ion Chemistry*, M. T. Bowers, Ed., Academic Press, New York, 1978.

H, the more electronegative F atom obtains a greater share of the bonding electrons and induces a partial positive charge on N relative to its state in NH_3. This not only makes ionizing one of the lone-pair electrons of the N atom more difficult, but also makes nitrogen less willing to share its electrons with a proton, and so the proton affinity drops by 259 kJ/mole on going from NH_3 to NF_3. In the case under discussion, the substituent is engaged only in sigma bonding with the reactive center. Coupling of the substituent with the reactive center through a saturated hydrocarbon chain generally assures that the substituent effect will be transmitted by sigma-bond polarization—in other words, that the substituent effect will be a pure inductive effect. The inductive effect decreases in magnitude as the substituent is moved farther from the reactive center.

Resonance Effects

Substituting F for H in PH_3 increases the ionization energy by only 121 kJ/mole. Here we might have expected the $\Delta\chi$ of P and F to lead to an even greater electron-withdrawing inductive effect than that found with NF_3, but in the case of the phosphorus compound, the low-lying empty *d*-orbitals permit the positive charge on the HPF_3^+ to be delocalized over the entire molecule.

Resonance, or conjugative, effects may be transmitted through systems of alternating (or conjugated) double bonds, or from the *ortho* and *para* positions of a phenyl group. For the effect to be transmitted in this fashion, direct resonance must be possible when the groups are directly attached.

A rather striking effect involving aromatic or antiaromatic behavior is shown by the pK$_a$'s of the following hydrocarbons, or the *PA*'s of their anions.[1]

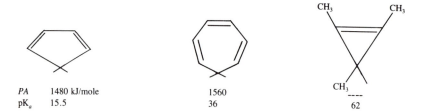

PA	1480 kJ/mole	1560	
pK$_a$	15.5	36	62

The cyclopentadiene loses a proton to give a $4n + 2$ aromatic ion, according to Hückel's rules. This ion is of course, common as a ligand in such complexes as ferrocene. The cyclo-heptatriene is only slightly destabilized, whereas the cyclopropene system is highly destabilized (both these ions are antiaromatic). The reverse order should obtain for the hydride-donating ability of the last-named two, since they become $4n + 2$ electron systems whereas the Cp$^+$ becomes antiaromatic as a $4n$ system.

Using the data given in Tables 12.1 and 12.2, or in other tables of proton affinities, to predict the direction of proton transfer upon mixing two substances, we find that the proton will go to the species with the highest proton affinity. Thus if we were somehow to mix in the gas phase the hydride ion and water, we would end up with hydroxide ion and elemental hydrogen. In mixing water and the fluoride ion in the gas phase, there is no proton transfer, since the fluoride ion has a lower proton affinity (1554 kJ/mole, compared with 1632 kJ/mole for hydroxide ion), and so the hydroxide ion hangs onto its proton: that is, water and fluoride ion remain in their initial states. Looking at the proton affinity of some other species, we find that PH$_3$ mixed with NH$_4^+$ gives no reaction, whereas (CH$_3$)$_3$P mixed with NH$_4^+$ results in the formation of (CH$_3$)$_3$PH$^+$ and NH$_3$—reactants and products all being in the gas phase.

Descending to Solvents—Leveling Effect of Solvents

Consider the following type of experiment: into a low-pressure system containing a relatively large number of monomolecular water molecules, we introduce a variety of ionic Brønsted acids and bases and then evaluate the acidity or basicity of the system. (For the moment, solvation effects are ignored.) If protonated dinitrogen, N$_2$H$^+$, is introduced, proton transfer will take place immediately and N$_2$ and H$_3$O$^+$ will be formed—the *PA* of water now determines whether future proton transfers may occur. On the other hand, if H$^-$ ion is introduced, H$_2$ and OH$^-$ will be formed, and the *PA* of OH$^-$ will determine the basicity of the system for future reactions. A third situation arises when a species such as NH$_3$ or HCl is added to the H$_2$O(g) system: no reaction takes place, but the NH$_3$ would control the basicity or the HCl would control the acidity. These and a few other examples are summarized in Table 12.3, which demonstrates that the *PA* of species engaged in proton transfer in a gaseous system with excess

[1]*PA* values are from J. E. Bartmess and R. T. McIver, Jr., "The Gas-Phase Acidity Scale," in *Gas-Phase Ion Chemistry*, M. T. Bowers, Ed., Academic Press, New York, 1978. The pK$_a$ values are from O. A. Reutov, I. P. Beletskaya, and K. P. Butin, *CH–Acids* (T. R. Crompton, Translation Ed.), Pergamon, London, 1978.

Table 12.3 Proton availability in gaseous systems containing $H_2O(g)$.

Reagent Added	Serving as	PA of Base or Conjugate Base	"PA of System"[a]	Reaction
N_2H^+	Proton donor	481 kJ/mole	711	$N_2H^+ + H_2O \rightarrow H_3O^+ + N_2$
CH_3FH^+	Proton donor	586	711	$CH_3FH^+ + H_2O \rightarrow H_3O^+ + CH_3F$
H_2I^+	Proton donor	619	711	$H_2I^+ + H_2O \rightarrow H_3O^+ + HI$
CH_3ClH^+	Proton donor	669	711	$CH_3ClH^+ + H_2O \rightarrow H_3O^+ + MeCl$
H_3S^+	Proton donor	728	728	No reaction
PH_4^+	Proton donor	791	791	No reaction
NH_4^+	Proton donor	845	845	No reaction
$(CH_3)_3PH^+$	Proton donor	954	954	No reaction
HI	Proton donor	1315	1315	No reaction
HCl	Proton donor	1395	1395	No reaction
PH_3	Proton donor	1548	1548	No reaction
H_2	Proton donor	1674	1636	No reaction
CH_4	Proton donor	1745	1636	No reaction
CH_3^-	Proton acceptor	1745	1636	$CH_3^- + H_2O \rightarrow OH^- + CH_4$
H^-	Proton acceptor	1674	1636	$H^- + H_2O \rightarrow OH^- + H_2$
PH_2^-	Proton acceptor	1548	1636	No reaction

[a]The energy required to remove one mole of H^+ ions from the system. To obtain a proton, another molecule must have a PA in excess of the one listed.

monomolecular water molecules must lie between the *PA*'s of H_2O (711 kJ/mole) and OH^- (1636 kJ/mole). Thus, in gas-phase water, nothing can be more acidic than H_3O^+ nor more basic than OH^-, and acids or bases outside this range are leveled to these values upon being added to the gaseous water system.

Now consider the effect of increasing the water activity in our system by small increments (experiments of this type have been carried out by Kebarle and coworkers using high-pressure mass spectrometry). As the water activity (pressure) increases, the charged species become increasingly hydrated. Instead of just H_3O^+, we will find increasing amounts of $H_5O_2^+$, $H_7O_3^+$, $H_9O_4^+$, etc.; and instead of just OH^-, we will find increasing amounts of $H_3O_2^-$, $H_5O_3^-$, etc. These species are hydrogen-bonded, and in addition to the energy released in protonating an H_2O molecule, energy will be released as the H_3O^+ forms these hydrogen-bonded clusters. Thus the energy released on hydrating a proton increases with the number of water molecules added to the cluster, approaching a limit of ΔH_{vap}/molecule.

$$H^+(g) + H_2O(g) \rightarrow H_3O^+(g) \qquad -\Delta H = PA_{H_2O} = 711 \text{ kJ/mole}$$
$$H^+(g) + 2H_2O(g) \rightarrow H_5O_2^+(g) \qquad -\Delta H = 914 \text{ kJ/mole } H^+$$
$$H^+(g) + 3H_2O(g) \rightarrow H_7O_3^+(g) \qquad -\Delta H = 1137$$
$$H^+(g) + 4H_2O(g) \rightarrow H_9O_4^+(g) \qquad -\Delta H = 1371$$
$$H^+(g) + H_2O(l) \rightarrow H^+(aq) \qquad -\Delta H = 1130$$

The effective "proton affinity of liquid water" is thus 1130 kJ/mole H^+. What happens to the proton affinity of OH^- as it becomes solvated? Here, the hydrogen bonds form in the $OH^- \cdot nH_2O$ cluster and are broken when the cluster disintegrates upon protonation.

$$H^+(g) + OH^-(g) \rightarrow H_2O(g) \qquad -\Delta H = PA_{OH^-} = 1632 \text{ kJ/mole } H^+$$
$$H^+(g) + HOHOH^- \rightarrow 2H_2O(g) \qquad -\Delta H = 1532 \text{ kJ/mole } H^+$$
$$H^+(g) + OH^-(aq) \rightarrow H_2O(l) \qquad -\Delta H = 1188$$

Accordingly, we may conclude that in aqueous solution, the effective "*PA*" range is 1130–1188 kJ/mole. No species more acidic than $H^+(aq)$ can exist in aqueous solution, since it would transfer its proton to water and form $H^+(aq)$. Likewise, no base stronger than $OH^-(aq)$ can exist in aqueous solution, since it would remove a proton from water and form $OH^-(aq)$.

12.2.2 Solvent Effects

If no solvation were to take place for any species other than H^+ and OH^-, all negative ions would form $OH^-(aq)$ when introduced into water, since the lowest *PA* for a negative ion is above 1250 kJ/mole, and all positive ions bearing a hydrogen atom would form $H^+(aq)$, since the highest *PA* for any neutral molecule is below 1050 kJ/mole. But this does not occur since the other ions themselves are solvated to some degree. The solvation of negative ions releases much more energy than the solvation of their neutral conjugate acids, thereby lowering the effective *PA* of the negative ions in aqueous solution. Because solvation of the positive ions is much more important than that of the neutral species produced when a proton is given up, the effective *PA* of the neutral molecules increases as they go into aqueous solution. The anions of strong acids have sufficient solvation energy that the effective proton affinity of their anions is well below 1130 kJ/mole, and thus in aqueous solution they are completely dissociated, to give $H^+(aq)$ and the solvated anion. On the other hand, solvation of anions such as H^- and CH_3^- is insufficient to lower the effective *PA* of these substances enough to permit hydrogen or methane to show acidic properties in aqueous solution. Indeed, hydride or methanide ion will remove a proton from water quantitatively to form $OH^-(aq)$ and molecular hydrogen or methane. Species such as N_2H^+, KrH^+, etc. are insufficiently stabilized by solvation to exist in aqueous solution and would transfer a proton to water immediately.

There is, of course, a group of acids and bases whose range of effective *PA* allows them to be studied in aqueous solution. Note that on going from the gas phase to an aqueous system, the effective *PA* range has been compressed by over an order of magnitude.

Another way of stating what we have just discussed is to say that acid species that are stronger than the hydronium ion in water will transfer their proton to water and the acidity will be "leveled" to the acidity of the hydronium ion. Likewise, proton acceptors (bases) that are stronger than the aquated hydroxide ion will form the hydroxide ion in water. Thus the range of acidities available in water must fall between the aquated hydronium ion–neutral water couple and the neutral water–aquated hydroxide ion couple. We have just noted the enormous compression of the acidity scale that takes place on going from the gas phase to the aqueous solution phase. This compression stems from the large degree of solvation of the protonated solvent species, the *lyonium* ion, and of the deprotonated solvent species, the *lyate* ion. The acidity range available in any solvent falls between that of the lyonium ion and that of the lyate ion. The more solvation stabilizes the lyonium and lyate ions, the greater the solution acidity scale is compressed compared with the proton affinity differences found in the gas phase. The most important specific solvation effect is hydrogen bonding, and the extent of stabilization of the lyonium or lyate ions depends primarily on the number and strength of the hydrogen bonds that may be formed. In addition to the compression of the available range of acidities that takes place on going from the gas phase to *any* solution, inversion in the order of acidities of individual pairs of acids is not uncommon on going from the gas phase to solution, or from one solution to another.

In referring to the gas phase, we have used a difference in *PA*'s—that is, an enthalpy change—as the criterion for the direction of proton transfer. We have assumed negligible entropy effects for the gas-phase proton transfer (and indeed the entropy effects have been shown to be small[2]) and thus $\Delta G^0 \approx \Delta H^0$. In solution, entropy effects are far less likely to cancel, and hence we must use ΔG^0 values. The acidity range available in a solvent may be obtained from $\Delta G^0 = -RTlnK_S$, where K_S refers to the autoprotolysis constant of the solvent.

Alcohols

Whereas the *PA* of alcohols increases with their degree of alkylation—H_2O (711 kJ/mole), CH_3OH (761), C_2H_5OH (778), *i*-PrOH (791), *t*-BuOH (799)—the *PA* of alkoxides decreases with their degree of alkylation—OH^- (1636), CH_3O^- (1586), $C_2H_5O^-$ (1573), *i*-PrO$^-$ (1565), *t*-BuO$^-$ (1561). The stabilization of the charged species (ROH_2^+, RO^-) by alkylation may be accounted for primarily by the large polarizability of the alkyl group relative to hydrogen.[3] In solution, the lyonium ion of the alcohols can form only two primary hydrogen bonds (compared with three for that of water), and both the primary H-bonds and those in the secondary solvation sphere will be weakened by steric strain caused by interference between the alkyl groups. Although the alkoxide ions, like hydroxide ions, can form three primary hydrogen bonds,

---

[2]Bartmess and McIver, *op. cit.*

[3]For a discussion of the separation of the inductive effect and the polarization effect from data such as those given here, see R. W. Taft, *et al.*, *J. Am. Chem. Soc.* 1978, *100*, 7765.

steric effects will weaken these bonds. So although the *PA* range in the gas phase between alcohols and their alkoxide ions is less than that of water and the hydroxide ion, the acidity range in solution between the lyonium and lyate ions is greater for alcohols than for water. This is reflected in the ability to titrate both ammonium salts in ethanol with alcoholic KOH (which is in equilibrium with the alkoxide ion) and carboxylate salts with alcoholic HCl. In part, the greater acidity range of alcohols arises from the smaller value of the dielectric constant of alcohols compared with water (32.6 for methanol, 24.3 for ethanol, and 78.5 for water—all at 25°C).

Liquid Ammonia

To place a solvent accurately on a chart such as Figure 12.1, we must know two facts: the absolute enthalpy (or better, the free energy) of solvation of the proton in the solvent and the autoprotolysis constant of the solvent. Although in the last decade much progress has been made toward ascertaining the absolute enthalpies of solvation of the proton and other ionic species, extrathermodynamic assumptions are still required to arrive at such values. Even so, estimates are available for relatively few solvents. The difference in the proton's free energy of solvation in water and in liquid ammonia has been estimated as 100 kJ/mole H^+, based on the electrode potentials for Rb/Rb^+ and H_2/H^+ in the two solvents and the assumption that the solvation energy of Rb^+ is the same in both solvents. (A different set of assumptions yields 67 kJ/mole H^+.)[4] The value of the autoprotolysis constant of NH_3 is 10^{-33} at $-50°C$, at which temperature the dielectric constant is 25.

The greater basicity of liquid ammonia vis-à-vis water, coupled with the greater range of basicity stemming from its lower autoprotolysis constant, permits acidic behavior to be observed for many compounds that display virtually no acidic properties in water. This is advantageous for synthesis, as the following examples show.

$$PH_3 + NaNH_2 \xrightarrow{NH_3(l)} NaPH_2 + NH_3$$
$$NaPH_2 + CH_3Cl \rightarrow CH_3PH_2 + NaCl$$
$$C_2H_2 + NaNH_2 \rightarrow NaC_2H + NH_3$$
$$NaC_2H + BuCl \rightarrow Bu\!-\!C\!\equiv\!CH + NaCl$$

In these examples, the negatively charged ion produced by deprotonation takes part in a displacement reaction with the alkyl halide. Further alkylation may be carried out by repeating the process.

Negative ions—particularly, multiply charged monatomic ions—need a basic environment in order to be stable in protic solvents. If the solvent is too acidic, protons are transferred to the negative species. In liquid ammonia we can have appreciable concentrations of ions such as S^{2-}, Se^{2-}, As^{3-}, and PH^{2-}. We might consider the ultimate in negative ions to be solvated electrons or electron pairs. When hydrogen is introduced at high pressure into an amide solution in liquid ammonia, solvated electrons and possibly even hydride ions are produced.

$$H_2 + NH_2^- \rightarrow NH_3 + e\ (NH_3)$$
$$H_2 + NH_2^- \rightarrow NH_3 + H^-$$

[4]W. L. Jolly, *J. Chem. Educ.* 1956, *33*, 512.

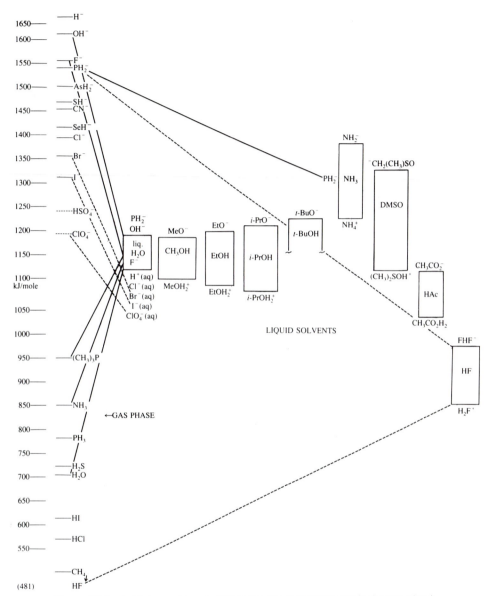

Figure 12.1   Acid–base range available in various solvents and in the gas phase.

Solvated electrons are produced more readily by the solution of alkali metals in liquid ammonia (see Section 14.2.5). The dilute solutions are deep blue in color, whereas the concentrated solutions are bronze. The volume of the solution is greater than the sum of its components—expansion occurs because of cavities in the liquid, which contain electrons. Transitions from blue to bronze solutions occur above a concentration of approximately $0.5\,M$ and are interpreted as a transition to a metallic state. Although solutions of alkali metals in ammonia are thermodynamically unstable with respect to hydrogen evolution and amide formation, in the absence of a catalyst, such as Fe_2O_3, the reaction is extremely slow. Accordingly, one of the best ways to obtain anhydrous ammonia is to distill it from its sodium solution. As might be expected, solutions of alkali metals in liquid ammonia are very good reducing agents. Among the reactions observed have been

$$K_2Ni(CN)_4 + 2K \rightarrow K_4Ni(CN)_4$$
$$Fe(CO)_5 + 2Na \rightarrow Na_2Fe(CO)_4 + CO$$
$$Pt(NH_3)_4Br_2 + 2K \rightarrow Pt(NH_3)_4 + 2KBr$$
$$9Pb + 4Na \rightarrow Na_4Pb_9$$

The plumbide ion in the last reaction is a "naked" metal cluster ion whose geometry is that of a C_{3v} tricapped trigonal prism. Adding lead iodide to the ammonia solution precipitates metallic lead.

$$Pb_9^{4-} + 2Pb^{2+} \rightarrow 11Pb$$

A number of polynuclear anions of the Group IV, V, and VI elements have been prepared using alkali metal solutions in liquid ammonia.[5]

Franklin[6] pointed out the closely analogous behavior of certain nitrogen-containing species in liquid ammonia and the corresponding oxygen species in water.

$$NH_4^+ \quad NH_3 \quad NH_2^- \quad NH^{2-} \quad N^{3-}$$
$$H_3O^+ \quad H_2O \quad OH^- \quad O^{2-}$$

Thus nitrides solvolyze in liquid ammonia to produce basic solutions of amides, just as oxides in water produce basic solutions of hydroxides. Ammonium ion reacts with active metals, liberating H_2.

One of the properties of aqueous acids is the ability to dissolve active metals with the evolution of hydrogen. Similar properties are shown by solutions containing ammonium ion in liquid ammonia. Aluminum in the form of an amalgam also can dissolve in liquid ammonia solution containing amides. Amphoteric character is found also with zinc ions, as indicated by the following sequence of reactions.

$$ZnI_2 \text{ (solvated)} + 2KNH_2 \rightarrow Zn(NH_2)_2(s) + 2KI$$
$$Zn(NH_2)_2(s) + 2KNH_2 \rightarrow K_2[Zn(NH_2)_4]$$

The tetraamidozincate ion formed corresponds to the tetrahydroxozincate ion formed in aqueous solution. Adding ammonium ion to the amidozincate ion reprecipitates zinc amide, but further ammonium ion redissolves the precipitate.

The ammono acid corresponding to acetic acid is acetamidine, $CH_3C \overset{\displaystyle NH}{\underset{\displaystyle NH_2}{\big\langle}}$, which

[5]For details and original-literature references, see H. J. Emeléus and A. G. Sharpe, *Modern Aspects of Inorganic Chemistry*, 4th ed. New York, Halsted Press, A Division of John Wiley, 1973, p. 24.

[6]E. C. Franklin, *The Nitrogen System of Compounds*, Reinhold, New York, 1935.

behaves as a weak acid in liquid ammonia. Acetonitrile, CH_3CN, is the ammono equivalent of acetic anhydride. The following equations show some of the utility of the water–ammonia analogy.

$$2H_2O + CO_2 \rightarrow H_3O^+ + HOCO_2^-$$

$$2NH_3 + CO_2 \rightarrow NH_4^+ + H_2NCO_2^-$$
$$\underset{\text{ammonium}}{} \underset{\text{carbamate}}{}$$

$$Cl_2SO_2 + 2H_2O \rightarrow H_2SO_4 + 2HCl$$

$$Cl_2SO_2 + 4NH_3 \rightarrow (NH_2)_2SO_2 + 2NH_4Cl$$
$$\underset{\text{sulfuryl}}{} \underset{\text{sulfamide}}{}$$
$$\underset{\text{chloride}}{}$$

$$Cl_2 + H_2O \rightarrow HOCl + HCl(aq)$$

$$Cl_2 + 2NH_3 \rightarrow H_2NCl + NH_4Cl$$

$$CH_3CO_2H \xrightarrow{\text{electrolysis}} CO_2 + CH_4$$

$$CH_3C(NH)NH_2 \xrightarrow[\text{cyanamide}]{\text{electrolysis}} NH_2CN + CH_4$$

Sulfuric Acid

In aqueous solutions, pH may be determined spectrophotometrically, using indicators of known pK_a, by means of the relationship

$$pH = pK_a + \log \frac{I}{IH} + \log \frac{\gamma_I}{\gamma_{IH}}$$

(the charge on the indicator or protonated indicator temporarily is ignored here). In studying concentrated acid or alkaline solutions, we neglect the activity coefficient ratio and the effective pH is termed an H_0 function if the unprotonated indicator is neutral or an H_- function if the deprotonated indicator is negatively charged.[7] A given indicator may be used to measure H functions over a range of approximately two units: that is, from I/IH 0.1 to I/IH 10. By using indicators that are all structurally related, it was felt that the activity coefficient ratios effectively would cancel and the H functions could be referenced to dilute aqueous systems.[8] The H functions provide a guide, in the absence of information about the free energy of proton solvation, in setting the acidity or basicity of a solvent, or of a solvent containing a known concentration of the anion or cation derived from autoprotolysis.

The H_0 function for pure sulfuric acid has a value of -12.1; the dielectric constant of H_2SO_4 is 100 and its autoprotolysis constant is 2.7×10^{-4} at 25°C. Sulfuric acid is thus a very acidic solvent with a very small acid–base range.

We can prepare aqueous sulfuric acid solutions with H_0 values that range from aqueous pH values up to the H_0 of pure sulfuric acid of -12.1, permitting us to determine pK_a values of very weak bases. Since phosphoric acid is 50% protonated in approximately 80% H_2SO_4, which has an H_0 value of -7.4, the pK_a of $P(OH)_4^+$ may be taken as approximately -7.4.

Sulfuric acid has a freezing-point depression constant of 6.12°C kg/mole. When nitric acid is dissolved in sulfuric acid, the freezing point depression indicates that each mole of

[7]For a complete treatment of these Hammett acidity functions, consult C. H. Rochester, *Acidity Functions*, Academic Press, New York, 1970.

[8]For a treatment not including this assumption, see J. F. Bunnett and F. P. Olsen, *Can. J. Chem.* 1967, *45*, 911.

nitric acid produces four moles of solute species. Coupled with conductance data, this is interpreted by the equation

$$HNO_3 + 2H_2SO_4 \rightarrow NO_2^+ + H_3O^+ + 2HSO_4^-$$

There is strong evidence that NO_2^+ is the active species in nitration reactions with aromatic compounds. In similar fashion, the species SO_3H^+ has been proposed as playing a role in sulfonation reactions.

The behavior of mesitoic acid in lowering the freezing point of sulfuric acid is very similar to that of nitric acid: that is,

$$2,4,6\text{-}(CH_3)_3C_6H_2CO_2H + 2H_2SO_4 \rightarrow (CH_3)_3C_6H_2CO^+ + H_3O^+ + 2HSO_4^-$$

Pouring a solution of mesitoic acid in sulfuric acid into an alcohol produces esterification in 100% yield—a reaction difficult to accomplish even in low yield by direct reaction of an alcohol with this highly hindered acid. Benzoic acid shows a lowering of the freezing point of sulfuric acid consistent with the production of two moles of solute species per mole of benzoic acid,

$$C_6H_5CO_2H + H_2SO_4 \rightarrow C_6H_5CO_2H_2^+ + HSO_4^-$$

The differing behavior of these two aromatic acids may be explained by the resonance stabilization of the protonated benzoic acid, whereas steric inhibition of resonance in the mesitoic acid makes the protonated form much higher in energy.

Formic acid decomposes in 95% to 100% sulfuric acid to give carbon monoxide, providing a convenient laboratory source of CO. The HCO^+ species is a probable intermediate in this reaction and rapidly would transfer a proton to H_2SO_4 or HSO_4^-. Oxalic acid also decomposes in sulfuric acid, yielding in this case an equimolar mixture of CO and CO_2.

Most organic compounds, except for the saturated hydrocarbons, are protonated by and dissolve in sulfuric acid. The insolubility of saturated hydrocarbons in sulfuric acid can be used to analyze mixtures containing them.

Sulfuric acid itself is an oxidizing agent, its strength increasing with its concentration. Up to about 12 *M*, the hydrogen ion is the oxidizing agent; above 12 *M*, the sulfuric acid is reduced to SO_2 during the oxidation of solutes. Just as basic media permit the existence of a variety of reduced species that are not stable in water, acidic media allow the existence of a variety of positive species such as S_8^{2+}, Se_8^{2+}, Se_4^{2+}, etc. I_3^+ is apparently produced by the reaction

$$HIO_3 + 7I_2 + 8H_2SO_4 \rightarrow 5I_3^+ + 3H_3O^+ + 8HSO_4^-$$

Dilution of the sulfuric acid media results in the destruction of these species by disproportionation reactions such as

$$16H_2O + 16Se_8^{2+} \rightarrow 120Se + 8SeO_2 + 32H^+$$

(See Chapter 13, under the individual elements, for a further discussion of these species.)

Few substances behave as acids in sulfuric acid. Perchloric acid, chlorosulfuric acid ($HClSO_3$), and fluorosulfuric acid (HSO_3F) are among the protonic acids that show weak acidic character in sulfuric acid. The Lewis acids sulfur trioxide and boron tri(hydrogen sufate) behave as moderately strong acids in sulfuric acid, increasing the lyonium ion by reducing the concentration of the lyate ion.

$$SO_3 + 2H_2SO_4 \rightarrow H_3SO_4^+ + HS_2O_7^-$$
$$B(HSO_4)_3 + 2H_2SO_4 \rightarrow H_3SO_4^+ + B(HSO_4)_4^-$$

Alternatively, we may think of these substances as forming protonic acids by reaction with sulfuric acid, with the lyonium ion subsequently being increased by the dissociation of these acids.

$$SO_3 + H_2SO_4 \rightarrow H_2S_2O_7$$
$$H_2S_2O_7 + H_2SO_4 \rightarrow H_3SO_4^+ + HS_2O_7^-$$
$$B(HSO_4)_3 + H_2SO_4 \rightarrow HB(HSO_4)_4$$
$$HB(HSO_4)_4 + H_2SO_4 \rightarrow H_3SO_4^+ + B(HSO_4)_4^-$$

Solutions of SO_3 in H_2SO_4 are called "fuming sulfuric acid" or "oleum." As the SO_3 concentration increases, higher polysulfuric acid species are formed—$HO(SO_3)_nH$—the acidity increasing slightly as n increases. The hydrogen tetra(hydrogen sulfato)borate is formed *in situ* by dissolving either boric acid or boric oxide in sulfuric acid and adding sufficient fuming sulfuric acid to react with the hydronium ion formed.

$$H_3BO_3 + 6H_2SO_4 \rightarrow B(HSO_4)_4^- + 3H_3O^+ + 2HSO_4^-$$
$$\downarrow 3H_2S_2O_7$$
$$HB(HSO_4)_4 + 8H_2SO_4$$

12.2.3 Aqueous Behavior of the Binary Oxides

Most binary oxides will dissolve in either acids or alkali and thus may be classed as basic or acidic oxides, respectively. Alternatively, if in an aqueous solution the element in question forms oxoanions, the oxide is classed as acidic; if the element forms cations (with or without oxygen content), the oxide is classed as basic. Upon dissolving in water the central atom coordinates with water and a proton transfer takes place from the aqua ligand to the oxo ligand, producing $XO_n(OH)_m$. The final coordination number of X is determined primarily by the size of X, the number of lone pairs on X, and to a lesser extent, by the pH. In only a few known ions do oxo and aqua ligands coexist: examples include the uranyl ion, UO_2^{2+} (*trans*-dioxo-tetraaquauranium(VI) ion); the vanadyl ion, VO^{2+} (oxopentaaquavanadium(IV) ion); and the dioxovanadium(V) ion, VO_2^+ (*cis*-dioxotetraaquavanadium(V) ion).

The periodic trends in acid–base behavior may be noted in the following sequence of oxides, where B indicates base, A indicates acid, Am indicates amphoteric, and S and W are strong and weak.

Na_2O	MgO	Al_2O_3	SiO_2	P_4O_{10}	SO_3	Cl_2O_7
SB	WB	Am	WA	A	SA	VSA

We can rationalize this trend readily on an electrostatic model by assuming that the charge on the central atom parallels the oxidation number. Similarly, the variation of basicity or acidity on descending a given group in the table may be rationalized on the basis of an electrostatic model and the size variation of the central atom. Table 12.4 lists the dissociation constants of some of the acids formed by these oxides. (The base constants are given in the following section, on hydrolysis.) Note that many of the acids for which dissociation constants are given

Table 12.4 pK_a values of some oxoacids

	H_2CO_3	H_2SO_4	H_2SeO_4	H_6TeO_6	H_3BO_3	
pK_1	6.73	(-3)	(-3)	7.68	9.14	
pK_2	10.33	1.92	1.92	11.29		
	HNO_3	H_3PO_4	H_3AsO_4	H_2SO_3	H_2SeO_3	H_2TeO_3
pK_1	-1.4	2.16	2.25	1.81	2.46	2.48
pK_2		7.21	6.77	6.91	7.31	7.70
pK_3		12.3	11.6			
	$HClO$	$HClO_2$	$HClO_3$	$HClO_4$	$HBrO$	HIO
pK_1	7.4	2.0	-1	(-10)	8.7	11

cannot be isolated as anhydrous compounds. The K_1 constants for "H_2CO_3" and for "H_2SO_3" apply to the equilibria

$$CO_2 \text{(dissolved)} + H_2O \rightleftharpoons H^+ + HCO_3^-$$

and

$$SO_2 \text{(dissolved)} + H_2O \rightleftharpoons H^+ + HSO_3^-$$

Part of the apparent difference in the acidity of H_2CO_3 and H_2SO_3 resides in the greater proportion of SO_2 existing in hydrated form, compared with CO_2. Pauling, among others, has formulated an empirical rule for estimating the strength of oxo acids: for an acid of the general formula $XO_n(OH)_m$, the value of pK_1 varies in approximate steps of five units and is related to n, the number of nonhydroxylic oxygen atoms, in the following approximate manner.

$$pK_1 = 7 - 5n$$

Because of the electrostatic effect, pK_2 will be approximately five units greater than pK_1. Pauling's rule may be interpreted as indicating the resonance stabilization of the anion that is afforded by each oxo ligand.

The oxo acids of phosphorus provide further interesting dissociation constants (see Table 12.5).

The pK_1 value for phosphoric acid is consistent with the predicted value of 2. The value of 2.00 for phosphorous acid suggests that the structure of this acid is

a structure recently confirmed by an analysis of the NMR spectrum of phosphorous acid. In contrast, arsenious acid has a pK value of 9.2, which is consistent with the structure

Table 12.5 Strength of the oxo acids of phosphorus[a]

Acid	Formula	pK_1	pK_2	pK_3	pK_4	pK_5	pK_6
(ortho) phosphoric[b]	H_3PO_4	2.161	7.207	12.325			
phosphorous[b]	H_3PO_3	*2.00*	*6.58*				
hypophosphorus[b]	H_3PO_2	~1					
pyrophosphoric[b]	$H_4P_2O_7$	1.52	2.36	6.60	9.25		
triphosphoric[c]	$H_5P_3O_{10}$		*1.06*	2.30	6.50	9.24	
tetraphosphoric[c]	$H_6P_4O_{13}$			*1.36*	2.23	7.38	9.11
trimetaphosphoric[b]	$H_3P_3O_9$	2.05					
tetrametaphosphoric[b]	$H_4P_4O_{12}$	2.74					
hypophosphoric[b]	$H_4P_2O_6$	<2.2	*2.81*	7.27	*10.03*		

[a]At 25° and zero ionic strength—except for italicized values, which are for other ionic strengths and/or temperatures.
[b]From J. Bjerrum, G. Schwarzenbach, and L. G. Sillén, "Stability Constants, Part II: Inorganic Ligands," *Chem. Soc. (London) Spec. Publ.,* 1958, 7.
[c]From J. I. Watters, P. E. Sturrock, and R. E. Simonaitis, *Inorg. Chem.* 1963, 2, 765.

Hypophosphorous acid is monoprotic, and its pK value is consistent with the structure

The pK_2 and pK_3 values of H_3PO_4 show the expected decrease in acidity with increasing negative charge on the anion which the proton is leaving. It is somewhat surprising that the effect of increasing charge is so slight for the successive dissociation constants of the polyphosphoric acids. The phosphorus units in a polyphosphoric acid appear to behave almost independently. Thus there is one fairly acidic hydrogen for each phosphorus atom in a polyphosphoric acid. The terminal units carry an additional weakly acidic hydrogen, with a pK comparable to the pK_2 of phosphoric acid. The difference in the last two pK values for polyphosphoric acids indicates some interaction between the ends of the chain, arising from its flexibility.

We can determine the ratio of terminal units to total units in a polyphosphoric acid from the ratio of weakly acidic hydrogen atoms to moderately acidic hydrogen atoms (that is, from the titration curves). Largely from this type of data, Van Wazer and Holst[9] concluded that no branched polyphosphates exist in aqueous solution.

[9]J. G. Van Wazer and K. A. Holst, *J. Am. Chem. Soc.* 1950, 72, 639.

Hydrolysis and Aqua Acids[10]

We can relate the dissociation constants of the metal hydroxides in water to the dissociation constants of the cationic aqua acids (or hydrolysis constants) through the equation $K_a K_b = K_w$, or $pK_a + pK_b = pK_w$. Accordingly, hydrolysis will be slight for cations forming strong bases, and *vice versa*.

For cations with a noble-gas configuration, the interaction between the metal ion and the hydroxide ion is essentially electrostatic[11]—that is, the pK_b varies in a linear fashion with z^2/r, where z is the charge on the ion and r is the ionic radius, and pK_b refers to the loss of the last hydroxide group. Thus the basicity of LiOH is the lowest of the alkali metal hydroxides, and the hydrolysis of lithium salts is greater than that of the other alkali metal salts.

$$LiOH(aq) = Li^+(aq) + OH^-(aq) \qquad pK_b = 0.18$$

$$\text{or} \quad Li(H_2O)_3OH(aq) + H_2O = Li(H_2O)_4^+(aq) + OH^-(aq)$$

Hence

$$Li^+(aq) + H_2O = LiOH(aq) + H^+(aq)$$

$$\text{or} \quad Li(H_2O)_4^+ = Li(H_2O)_3OH(aq) + H^+(aq) \qquad pK_a = 13.82$$

In each pair, the first equation in no way implies the degree of solvation occurring, and the second equation indicates speculation regarding the primary hydration of the lithium ion and the lithium hydroxide ion pair. The increase in ionic radius of Na^+ over Li^+ results in NaOH being a stronger base ($pK_b = -0.7$) than LiOH. KOH, RbOH, and CsOH are even stronger bases that appear to approach the limiting pK_b value of -1.7—in other words, K^+, Rb^+, and Cs^+, are not measurably hydrolyzed in aqueous solution.

The ionic radii of Be^{2+}, Mg^{2+}, Ca^{2+}, Sr^{2+}, and Ba^{2+} are about 30 pm less than the corresponding isoelectronic alkali metal ions, and the ionic charge is, of course, twice as great. Accordingly, all these ions are hydrolyzed more highly in solution than the lithium ion, the pK_a values varying from about 13.35 for Ba^{2+} to 11.42 for Mg^{2+} for the reaction $M^{2+}(aq) + H_2O = MOH^+(aq) + H^+(aq)$. The Be^{2+} ion is still more extensively hydrolyzed, forming polynuclear species in solution. Appreciable concentrations of the unhydrolyzed Be^{2+} ion occur only in dilute solutions under fairly acid conditions. Figure 12.2 shows the distribution of Be^{2+} among some possible species, as worked out by Sillén, *et al.*, for a total concentration of Be^{2+} of 0.1 molar. The fraction of Be^{2+} existing as a particular species at a given pH_c is given by the fraction of the vertical line appearing in the field of that species. Thus at a pH_c of 4, an 0.1 M solution of Be(II) would be approximately 20% Be^{2+}, 5% Be_2OH^{3+}, and 75% $Be_3(OH)_3^{3+}$. These formulas do not show hydrated water and do not distinguish between $Be(OH)_2$ and BeO.

Sillén suggests that the species $Be_3(OH)_3^{3+}$ may be

[10]L. G. Sillén, *Quart. Rev.* 1959, *13*, 146.

[11]C. W. Davis, *J. Chem. Soc.* 1951, 1256.

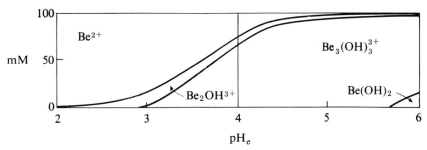

Figure 12.2 Field distribution diagram of Be²⁺ (Taken from H. Kakihana and L. G. Sillén, *Acta Chem. Scand.* 1956, *10*, 985.)

that is, a cyclic structure with hydration giving tetrahedrally coordinated Be.

Extending the electrostatic approach beyond the alkaline earth metal ions gives less satisfactory results, as Figure 12.3 indicates. Thus, although we might expect a decrease in hydrolysis of the ions in the series Mg^{2+}, Zn^{2+}, Cu^{2+}, Cd^{2+}, and Hg^{2+}, because of the increasing size of the respective ions, we actually find the reverse: Mg^{2+} is the least hydrolyzed ($pK_a = 11.42$) and Hg^{2+} the most hydrolyzed ($pK_a = 3.70$). This inversion of order results from the increasing polarizability of the metal ions in the series given, along with their increasing electronegativity (Mg^{2+} 1.3 to Hg^{2+} 2.0). Both factors lead to an increase in the covalence of the M—O bond. The hydrolysis of Cu^{2+} may also be slightly favored by ligand field stabilization energy (see Chapter 7). A somewhat perplexing set of data is provided by Sillén *et al.*, who conclude that the second dissociation constant of the aqua mercuric cationic acid is *greater* than the first.

$$Hg(H_2O)_2^{2+} \rightleftharpoons Hg(OH)(H_2O)_y^+ + H^+ \qquad pK_a = 3.70$$
$$HgOH(H_2O)^+ \rightleftharpoons Hg(OH)_2 + H^+ \qquad pK_a = 2.60$$

Accounting for these data probably is a change in the coordination number of mercury during hydrolysis. The difference between pK_1 and pK_2 of other cationic aqua acids is smaller than the five units generally found for aqua acids (that is, Zr^{4+} $pK_1 = 0.22$, $pK_2 = 0.62$, $pK_3 = 1.05$, $pK_4 = 1.17$). This indicates that the charge on the metal ion is not greatly affected by the loss of a proton from a coordinated water molecule. In contrast, changing a unit charge on the metal brings about a large change in hydrolysis [for Fe^{2+}(aq), $pK_a = 9.5$; for Fe^{3+}, $pK_a = 3.05$]. We can conclude that the loss of a proton by a water molecule is not transmitted, but a change in charge of M affects all M—O bonds and profoundly affects the acidity.

Weakly basic oxides will begin to polymerize and eventually precipitate from solution as the solution is made more alkaline—that is, as the pH increases. Weakly acidic oxides will begin to polymerize and eventually precipitate from solution as the solution becomes more acidic. Amphoteric oxides will precipitate from solution and then redissolve as the pH is increased, starting with the cationic species, or decreased, starting with the anionic species. Table 12.6 gives the equilibria involved in acidifying a vanadate solution.

Separating one element from another frequently requires judicious use of acid–base and/or redox properties. In the processing of vanadium ore, after roasting to convert the sulfoxide to the pentoxide the V_2O_5 is purified by heating with Na_2CO_3 to form $NaVO_3$, which is water-soluble at high pH but precipitates V_2O_5 on acidifying. In the production of vanadium steels, the V_2O_5 usually is reduced directly with iron oxide.

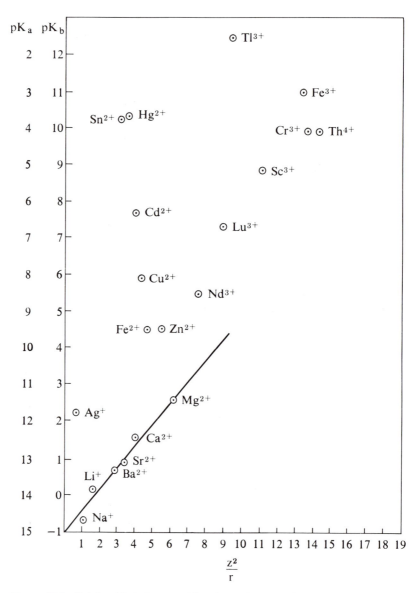

Figure 12.3 Relationship between acidity of aqua ions and the ionic charge squared to radius ratio.

Cerium can be separated from the other rare earth elements by oxidizing it to Ce(IV) while leaving the other rare earths in the III state. Increasing the pH precipitates the very weak base $Ce(OH)_4$ before any of the $RE(III)(OH)_3$ start to separate.

It is likely that most oxides exhibit amphoteric character; however, normal conditions in aqueous solution give an insufficient range of acidity to show both acidic and basic properties.

Table 12.6 Some vanadium(V) equilibria in aqueous solution at $25°$ [a]

	pK	$pH_{50\%}$ [b]
$2VO_4^{3-} + 2H^+ \rightleftharpoons V_2O_7^{4-} + H_2O$	-27.0	13.5
$2V_2O_7^{4-} + 4H^+ \rightleftharpoons V_4O_{12}^{4-} + 2H_2O$	-40.0	10.0
$5V_4O_{12}^{4-} + 8H^+ \rightleftharpoons 2V_{10}O_{28}^{6-} + 4H_2O$	-58.0	7.2
$V_{10}O_{28}^{6-} + H^+ \rightleftharpoons HV_{10}O_{28}^{5-}$	-5.8	5.8
$HV_{10}O_{28}^{5-} + H^+ \rightleftharpoons H_2V_{10}O_{28}^{4-}$	-3.6	3.6
$H_2V_{10}O_{28}^{4-} + 4H^+ \rightleftharpoons 5V_2O_5 nH_2O + (3 - n)H_2O$	-12.0	3.0
$V_2O_5 nH_2O + 2H^+ \rightleftharpoons 2VO_2^+ + (1 + n)H_2O$	$+2.4$	-1.2

[a]From, C. S. G. Phillips and R. J. P. Williams, *Inorganic Chemistry*, Oxford, Oxford, 1965.
[b]$pH_{50\%}$ is the pH at which the two species are present in equal amounts at molar concentration and pK is $-\log$(equilibrium constant of the reaction).

Thus N_2O_3 forms NO_2^-, nitrite ion, in basic solution, whereas in highly acid media NO^+, the nitrosyl cation is formed. Likewise, N_2O_5 forms NO_3^- in basic media, but NO_2^+, the nitryl cation, in concentrated sulfuric acid. Scholder and coworkers isolated from basic solution a series of salts containing $Zn(OH)_4^{2-}$, $Zn(OH)_6^{4-}$, $Cu(OH)_4^{2-}$, $Cu(OH)_6^{4-}$, $Pb(OH)_4^{2-}$, $Pb(OH)_6^{4-}$, etc.[12] In a medium in which bases of greater strengh than OH^- are available it should be possible to remove the remaining protons in these complexes. The oxide ion is much more basic than the hydroxide ion, but in water it is rapidly converted to the hydroxide ion. In fused alkali or alkaline earth metal oxides, however, we can prepare compounds such as Li_8PbO_6 or Ba_3MoO_6[13]

12.3 NONPROTONIC CONCEPTS OF ACID-BASE REACTIONS

12.3.1 Lux Concept

In the reaction sequence

$$BaO + H_2O \rightarrow Ba(OH)_2(aq)$$
$$\rightarrow BaCO_3(s) + H_2O$$
$$CO_2 + H_2O \rightarrow H_2CO_3(aq)$$

the last stage is clearly an acid-base reaction. The reaction may be carried out directly with BaO and CO_2: that is,

$$BaO + CO_2 \rightarrow BaCO_3$$

We know of many other examples of the direct reaction between acidic and basic anhydrides in the absence of water or of hydrogen ions.

[12]See Table 12.12 in F. Basolo, in *The Chemistry of the Coordination Compounds*, edited by J. C. Bailar (New York: Reinhold Publishing Corp., 1956). ''Acids, Bases, and Amphoteric Hydroxides,'' Chapter 12.
[13]R. Scholder, *Angew. Chem.*, 1958, *70*, 583.

$$3Na_2O + P_2O_5 \rightarrow 2Na_3PO_4$$
$$CaO + SiO_2 \rightarrow CaSiO_3$$

The Brønsted definition of an acid by the equation

$$\underset{\text{acid}}{AH} \rightleftharpoons \underset{\text{base}}{B^-} + \underset{\text{proton}}{H^+}$$

is not applicable. Lux has proposed for oxide systems the defining equation

$$\underset{\text{base}}{B} \rightleftharpoons \underset{\text{oxide ion}}{O^{2-}} + \underset{\text{acid}}{A}$$

According to this definition, a base is an oxide ion donor and an acid is an oxide ion acceptor. This view applies in particular to high-temperature chemistry, as in the fields of ceramics and metallurgy. Thus the ores of Ti, Ta, and Nb may be brought into solution around 800° in sodium pyrosulfate ($Na_2S_2O_7$) or potassium pyrosulfate.

$$\underset{\text{base}}{TiO_2} + \underset{\text{acid}}{Na_2S_2O_7} \rightarrow Na_2SO_4 + Ti(SO_4)_2 \text{ or } (TiO)SO_4$$

Displacement of the more volatile acid from solution is seen in the following reactions of sand or clay.

$$3SiO_2 + Ca_3(PO_4)_2 \underset{\Delta}{\rightarrow} 3CaSiO_3 + P_2O_5 \uparrow$$
$$\underset{\text{acid}}{SiO_2} + \underset{\text{salt}}{CaSO_4(\text{gypsum})} \underset{\Delta}{\rightarrow} \underset{\text{salt}}{CaSiO_3} + \underset{\text{salt}}{SO_3} \uparrow$$

By this same scheme, we may classify substances as amphoteric if they show a tendency both to take up or to give up oxide ions depending on the circumstances—that is,

$$\underset{\text{base}}{ZnO} + S_2O_7^{2-} \rightarrow Zn^{2+} + 2SO_4^{2-}$$
$$Na_2O + \underset{\text{acid}}{ZnO} \rightarrow 2Na^+ + ZnO_2^{2-}$$

Flood and Forenson have pointed out a parallelism between the thermodynamic stabilities of sulfates and carbonates toward evolution of SO_3 and CO_2, respectively, at high temperature. The order of decomposition temperatures is $Ba^{2+} > Li^+ > Ca^{2+} > Mg^{2+} > Mn^{2+} > Cd^{2+} > Pd^{2+} > Co^{2+} > Ag^+ > Fe^{2+} > Ni^{2+} > Cu^{2+} > Fe^{3+} > Be^{2+}$. This order of stability is approximately the order of base strength of the hydroxides.

The Lux oxide transfer picture of acid-base reactions can be extended to any negative ion—the halides, sulfides, or even carbanions! The following illustrates these reactions.

Negative ion donor		Negative ion acceptor	high temp	
$3NaF$	$+$	AlF_3	$\longrightarrow$	$3Na^+ + AlF_6^{3-}$
Na_2S	$+$	CS_2	$\longrightarrow$	$2Na^+ + CS_3^{2-}$
$EtNa$	$+$	Et_2Zn	$\longrightarrow$	$Na^+ + ZnEt_3^-$
base		acid		acid base

12.3.2   Lewis Acids and Bases[14]

In 1938, G. N. Lewis proposed the following operational definition of acids and bases in terms of the reactions they could undergo.

(a)   Neutralization—acids and bases react rapidly to neutralize each other.

(b)   Displacement—a strong acid will displace a weaker acid from its compounds; a strong base will displace a weaker one from its compounds.

(c)   Titration—an indicator may be used to determine the neutralization endpoint.

(d)   Catalysis—acids (and bases) may catalyze reactions.

These properties already had been widely associated with protonic acid–base reactions. Lewis then proceeded to point out that a number of nonprotonic compounds could exhibit these properties—SO_3, $SnCl_4$, and BF_3, among others. He proposed that the fundamental quality shared by these substances is a *vacant orbital* that could accept an electron pair in covalent bond formation. A base is a compound with an unshared pair of electrons. The Lewis acid picture covers (with occasional stretching) the ion transfer process as a special case and extends the concept of acids and bases to nonionic compounds, while focusing attention on the strength of the new covalent bond formed.

Strength of Lewis Bases

Comparing the strengths of Lewis acids and bases is somewhat difficult because of the many different methods various workers have used to establish orders. These methods include gas-phase dissociation data, calorimetric heats of reaction, competition experiments between several competing acids (or bases) for an insufficient amount of base (or acid), displacement methods, and studies of the volatility of addition compounds formed (the more volatile, the less stable the adduct within a series). Nevertheless, we can draw conclusions within a given series, and overlap of several series permits some generalizations, although caution should be exercised.

Position in the Periodic Table.   With regard to position in the periodic table, the strength of Lewis bases toward Lewis acids parallels their basicity toward a proton, provided that only σ bonds are formed and steric factors are not great. The following examples illustrate this.

	Molar heat of reaction with BMe_3	*The product of reaction with* HCl
Me_3N	-74 kJ	Sublimes at $250°$
Me_3P	-67 kJ	Sublimes at $125°$
Me_3As	Exists only at $-80°$	Unstable at room temperature
Me_3Sb	No compound formed	Unstable at $-80°C$

Likewise, the base strength decreases in going from Me_3N to Me_2O to MeF. Trimethylamine readily displaces dimethyl ether from boron trifluoride methyl etherate. The variation in strength

---

[14]Data largely from H. C. Brown, *J. Chem. Soc.*, *1956*, 1248.

of Lewis acids according to position in the periodic table is somewhat more complex, but for oxygen and nitrogen bases the trends parallel the tendency toward hydrolysis discussed earlier (that is, B > Be > Li; Be > Mg > Ca, etc.). Variations within the transition metals will be discussed more fully in the section dealing with stability of coordination compounds.

Effect of Substituents.   Often, we may predict the effect of substituents on the strength of Lewis acids and bases from a study of the inductive, resonance, and steric effects of the substituents. Thus an electron-withdrawing inductive effect will make an acid stronger (more willing to accept an electron pair) and a base weaker (less willing to donate an electron pair). So the base strength decreases in the series $Me_3N > NH_3 > NF_3$, whereas the acid strength increases in the series $Me_3B < BH_3$[15] $< BF_3$. This order is precisely that expected based on the inductive effect of these groups. The order of acid strength $BF_3 < BCl_3 < BBr_3$ is not the order expected from the inductive effects, but this can be explained by the greater resonance stabilization of BF_3 as compared with BCl_3.

The weakness as an acid of methylborate, $(MeO)_3B$, as compared with trimethylboron may be attributed to resonance.

Steric Effects.   In general, steric effects only slightly influence ion transfer equilibria, especially when the ions differ by only a proton. In Lewis acid-base reactions, steric effects may be quite large. Thus whereas 2-methylpyridine is a stronger base toward a proton than pyridine (as would be expected from the inductive effect), it is a much weaker base than pyridine toward Lewis acids.

Trimethylboron does not react at all with 2,6-dimethylpyridine, even though this base has a higher pK_a than the 2-methylpyridine or pyridine itself. We can attribute the effect of these *ortho*-methyl groups in reducing the stability of the Lewis acid adduct to the strain between nonbonded groups attached to different atoms that have conflicting steric requirements. Such strain is termed F-strain[16] (the "F" standing for strain at the *front* of the molecule). When triethylaluminum reacts with 2,6-dimethylpyridine, the heat released is only 13 kJ/mole less than with pyridine itself (88 kJ/mole py).[17] At first this may seem anomalous, since the Al

[15]After correction for bridge breaking.
[16]D. F. Hoeg, S. Liebman, and L. Schubert, *J. Org. Chem.* 1963, *28*, 1554.
[17]H. C. Brown, *Rec. Chem. Prog.*, 1953, *14*, 83.

atom is larger than the boron atom, but it is precisely the greater N—Al bond length, compared with the N—B bond length, that decreases the overlap of the van der Waals radii of the methyl and methylene groups.

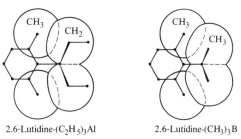

2.6-Lutidine-$(C_2H_5)_3Al$ 2.6-Lutidine-$(CH_3)_3B$

We would expect the inductive effects to be the same in triethylamine and quinuclidine, but the heat of reaction of quinuclidine is much greater than that of Et_3N with BMe_3, because of the greater F-strain in the triethylamine complex.

Triethylamine ΔH with $B(CH_3)_3 \approx -42$ kJ

Quinuclidine ΔH with $B(CH_3)_3 \approx -84$ kJ

A change in hybridization during the formation of a molecular addition compound may also bring about a change in steric requirement. Thus boron changes from sp^2 to sp^3 hybridization during acid-base reactions of trialkylboron compounds. Branching of the alkyl groups greatly reduces the acid strength of trialkylboron. Trimesitylboron is inert even to such strong bases as methoxide ion, since there would be too much crowding if the mesityl groups were forced from a trigonal planar to a tetrahedral bonding arrangement.

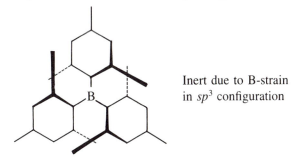

Inert due to B-strain
in sp^3 configuration

Some Inverted Orders. Although BF_3 is "normally" a stronger acid than diborane, B_2H_6, the latter forms a compound with CO—that is, H_3BCO—but the former does not. Likewise, the compound dimethylsulfide is a stronger base than dimethylether toward BH_3, although the reverse is true with BF_3. Both of these may be explained on the basis of hyperconjugation of the BH_3 group, stabilizing the addition compound: that is,

$$
\begin{array}{ccc}
\underset{\underset{\text{H}}{|}}{\overset{\overset{\text{H}}{|}}{\text{H}-\text{B}^-}}-\text{C}{\equiv}\overset{+}{\text{O}}{\scriptstyle|} & \leftrightarrow & \underset{\underset{\text{H}}{|}}{\overset{\overset{\text{H}^+}{|}}{\text{H}-\text{B}^-}}{=}\text{C}{=}\overset{..}{\text{O}}{\scriptstyle|} \\
& & \text{3 eq structures}
\end{array}
$$

$$
\begin{array}{ccc}
\underset{\underset{\text{H}\;\;\;\text{Me}}{|\;\;\;\;\;|}}{\overset{\overset{\text{H}}{|}}{\text{H}-\text{B}^-}}-\overset{+}{\bar{\text{S}}}-\text{Me} & \leftrightarrow & \underset{\underset{\text{H}\;\;\;\text{Me}}{|\;\;\;\;\;|}}{\overset{\overset{\text{H}^+}{|}}{\text{H}-\text{B}^-}}{=}\text{S}-\text{Me} \\
& & \text{3 eq structures}
\end{array}
$$

The latter involves the $3d$ orbitals of S and hence would not be important for oxygen derivatives. These inversions caused by multiple-bond character are much more common in the later transition metal compounds and will be discussed more fully later.

12.3.3 Stability of Coordination Compounds

The coordination process can be regarded as an acid-base reaction. In general, an increase in basicity of the ligand and/or an increase in the acidity of the metal enhances the stability of the complexes formed. The general properties of metals and ligands, which contribute to the stability of the complexes formed, are identified most clearly in the ligand field approach (see p. 268). Now we will consider some additional factors, which contribute to stability, and some specific effects that do not conform to the general pattern.

To speak of an organic compound such as ethanol as being stable might refer to the thermodynamic stability with respect to the elements carbon, hydrogen, and oxygen, or it might refer to the kinetic stability—the fact that ethanol can be handled under ordinary circumstances without decomposition. Ethanol is kinetically stable in air, even though the products of the reaction with O_2, (CO_2 and H_2O) are more stable thermodynamically. The stability of a metal complex, MX_m^{n+}, is usually expressed in terms of the formation constant, β_m, for the reaction

$$
M^{n+} + mX = MX_m^{n+} \qquad \beta_m = \frac{[MX_m^{n+}]}{[M^{n+}][X]^m}
$$

The β_m expresses the thermodynamic stability, since it is an equilibrium constant from which we can calculate a free-energy change for the formation of the complex. Considering the constants for each step in the formation of the complex often is worthwhile.

$$
M^{n+} + X = MX^{n+} \qquad K_1 = \frac{[MX^{n+}]}{[M^{n+}][X]}
$$

$$
MX^{n+} + X = MX_2^{n+} \qquad K_2 = \frac{[MX_2^{n+}]}{[MX^{n+}][X]}
$$

$$
MX_{m-1}^{n+} + X = MX_m^{n+} \qquad K_m = \frac{[MX_m^{n+}]}{[MX_{m-1}^{n+}][X]}
$$

$$\beta_2 = K_1 K_2$$
$$\beta_m = K_1 K_2 \cdots K_m$$

Most ligands are bases—for example, $NH_2C_2H_4NH_2$—or anions of weak acids—for example, $C_2O_4^{2-}$. For consistency, we express the basicity of the ligand in terms of pK ($-\log K$) of the acid dissociation constant for its conjugate acid. The symbols HL, HL^+, H_2L, etc. are used to indicate the species to which the pK refers.

$$H_2C_2O_4 = H^+ + HC_2O_4^- \qquad K(H_2L) = \frac{[H^+][HC_2O_4^-]}{[H_2C_2O_4]}$$

$$HC_2O_4^- = H^+ + C_2O_4^{2-} \qquad K(HL) = \frac{[H^+][C_2O_4^{2-}]}{[HC_2O_4^-]}$$

$$^+NH_3C_2H_4NH_3^+ = H^+ + NH_2C_2H_4NH_3^+ \qquad K(H_2L^{2+}) = \frac{[H^+][NH_2C_2H_4NH_3^+]}{[NH_3C_2H_4NH_3^{2+}]}$$

$$NH_2C_2H_4NH_3^+ = H^+ + NH_2C_2H_4NH_2 \qquad K(HL^+) = \frac{[H^+][NH_2C_2H_4NH_2]}{[NH_2C_2H_4NH_3^+]}$$

The formation constants are evaluated by the usual methods for determining equilibrium constants: potentiometric titrations, spectrophotometric methods, polarography, electrode potential measurements, solubility measurements, ion exchange procedures, etc.

Statistical Factor

Even if there were no change in the ΔH^0 of reaction for the successive stepwise addition of ligands, the change of symmetry of the complex would bring about a change in ΔS^0, and since $\Delta G^0 = \Delta H^0 - T\Delta S^0$, the free-energy change would differ. This would, of course, cause K_n to differ from K_{n+1}. The equation given below shows this variation of K_n and K_{n+1}, which is called the statistical factor.[18]

$$\frac{K_n}{K_{n+1}} = \frac{(N - n + 1)(n + 1)}{(N - n)n}, \text{ where } N = \text{C.N.}$$

This equation assumes all available points to be equivalent. More precisely, the relationship is found in terms of the symmetry of $ML_n(H_2O)_{m-n}$ and $ML_{n+1}(H_2O)_{m-n+1}$.[19]

Effect of Metal Ion

The stabilities of the complexes formed by various metals follow some regular trends, such as those involving size and charge effects. Although metals display pronounced differences in their tendencies to form complexes with various ligand atoms (p. 538), groups of similar metals show some helpful trends.

Generally, the stability of complexes decreases with increasing atomic number for the

[18]J. Bjerrum, *Metal Ammine Formation in Aqueous Solution*, P. Haase and Son, Copenhagen, 1941.

[19]See S. W. Benson, *J. Am. Chem. Soc.* 1958, **80**, 5151 for a more complete discussion of statistical factors.

electropositive metals—for example, Group IIA—and increases with increasing atomic number for the more noble metals, following the general trend for the ionization energies as well as the trend expected from the ligand field theory. The ionic charge density is the determining factor for electropositive metal ions (d^0), but the magnitude of the ligand field splitting must be considered for $d^1 - d^9$ ions (see Section 14.8.3), and the polarizabilities of the cations must be considered for d^{10} ions.

Effect of Ligand Atom

Classification of Metal Acceptor Properties.   As previously mentioned, for the more electropositive metals the order of stability of the halide complexes is F > Cl > Br > I, but for highly polarizing (and also polarizable) metal ions such as Hg^{2+}, we see the reverse order. The most electropositive metals show a great preference for forming complexes with ligands such as F or oxygen-containing ligands. As the electropositive character of the metals decreases, the nitrogen complexes increase in stability with respect to those of oxygen. Still more-electronegative (or more noble) metals show a preference for S and P over O and N, respectively. The noble metals show the greatest tendency to form stable olefin complexes.

Table 12.7 shows metals classified according to their acceptor properties. The metals of class *(a)* show affinities for ligands that are roughly proportional to the basicities of the ligands. The class *(b)* acceptors are the ones that form stable olefin complexes (see p. 419). The border regions are not well defined in all cases and, of course, the classification depends on the oxidation state of the metal ion. Copper(I) is a class *(b)* acceptor, but copper(II) is in the border region. Class *(a)* acceptors form the most stable complexes with the ligand atom in the second period (N, O, or F), whereas class *(b)* acceptors form their most stable complexes with ligand atoms from the third or a later period. Thus, trimethyl gallium shows the following

Table 12.7   Classification of metals according to acceptor properties[a]

[a]S. Ahrland, J. Chatt, and N. R. Davies, *Quart. Rev.* 1958, *12*, 265.
[b]Lanthanides.
[c]Actinides.

tendencies toward complex formation involving alkyls of the ligand atoms from Groups V and VI: N > P > As > Sb and O > S < Se > Te. On the other hand, for platinum(II) the order is apparently N << P > As > Sb; O <<< S >> Se < Te and F < Cl < Br < I.

Basicity and Structure of Ligand.[20] Although class *(a)* metal atoms display a preference for different ligand atoms than class *(b)* metals, with a given ligand atom the stability constants generally increase as the basicity of the ligand increases. Table 12.8 shows the correlations for a number of metal ions between the log K for metal complexation and the pK_a of the ligand according to the equation

$$\log K = a(pK_a - 7) + c$$

Table 12.8 Correlations of Brønsted basicity and Lewis basicity toward metal ions in aqueous solution[a,f]

Acid	Bases	a	c
Ag^+	$(RCH_2)_2NH^b$	0.31	1.73
Ag^+	RCH_2NH_2	0.29	2.28
Ag^+	Pyridines	0.28	2.47
Ni^{2+}	Pyridines	0.27	2.32
Cd^{2+}	Pyridines	0.32^c	1.80^c
Zn^{2+}	Pyridines	0.35^c	1.67^c
Cu^{2+}	Pyridines	0.45	3.26
CoL_5^d	Pyridines	0.21^e	3.58^e
CoL_5^d	RCH_2NH_2	0.38^e	2.22^e
CoL_5^d	RCH_2S^-	0.18^e	4.92^e
Fe^{3+}	ArO^-	0.94^c	5.99^c

^{a}At 25° C in the presence of 0.5 *M* potassium nitrate unless otherwise noted.
^{b}Including cyclic amines.
^{c}In the presence of 0.1 *M* sodium perchlorate.
^{d}Methylaquacobaloxime.
^{e}In the presence of 1.0 *M* potassium chloride.
^{f}*Source:* From J. Hine, *Structural Effects on Equilibria in Organic Chemistry*, Wiley-Interscience, 1975.

Deviations from such correlations often reveal structural differences in the complexes being formed. Thus the 2-hydroxymethylpyridine fits the correlation in Table 12.8 for Ag^+ but gives a value for Cu^{2+} and Ni^{2+} that is an order of magnitude too low for K. This indicates that the Ag ion coordinates to the N atom and not to the O. Further, the Cu^{2+} and Ni^{2+} must coordinate to both atoms, forming a chelate ring. The normal *sp* hybridization of Ag^+ prevents it from participating in the chelate formation.[21] Most metals other than silver do not tend to form

[20]See D. P. N. Satchell and R. S. Satchell, *Quart. Rev.* 1971, *25*, 171–199; *Chem. Rev.* 1969, *69*, 251–278. Also, J. R. Chipperfield, Chapter 7 in *Advances in Linear Free Energy Relationships*, Edited by N. V. Chapman, and J. Shorter, Plenum Press, New York, 1972; J. Hine, *Structural Effects on Equilibria in Organic Chemistry*, Wiley-Interscience, New York, 1975.

[21]J. Hine, *ibid.* p. 244.

linear bonds, and their complexes of chelate groups—for example, $NH_2C_2H_4NH_2$—are much more stable than those of unidentate groups—for example, NH_3. This enhanced stability is referred to as the *chelate effect*. The greater the number of chelate rings, the greater the stability of the complex. The log of the formation constant for the addition of four ammonia molecules to Cu^{2+} is 11.9; for the addition of two en molecules, 20.0; and for triethylenetetraamine (four coordinated nitrogen atoms with three chelate rings), 20.4.

The chelate rings of greatest stability are generally five-membered, because one of the bond angles in the ring is 90° for planar or octahedral complexes.

Six-membered rings, most often encountered in organic compounds, are more stable than five-membered rings in complexes of heterocyclic ligands or ligands involving conjugation in the chelate ring. Rings of more than six members are rare and only a few ligands—CO_3^{2-}, for example—give four-membered rings.

The chelate effect is predominantly an entropy effect for nontransition metal ions, but there may be an enthalpy effect for transition metal ions[22] where ligand field stabilization effects are important. After one end of an ethylenediamine molecule has attached to a metal ion, the "effective concentration" of the other —NH_2 group is increased, because its motion is restricted to a small volume in the vicinity of the metal ion. Each ethylenediamine molecule replaces two water molecules, increasing the total number of particles in the system, and hence the entropy.

Adamson[23] has suggested that the chelate effect is more apparent than real, since it practically disappears if one changes the standard states for the solute from the hypothetical one molal state to the hypothetical mole fraction unity state. Nevertheless, as concentrations and stability data usually are handled, the effect is important and useful.

Although most alkaline earth metal complexes are not very stable, ethylenediaminetetraacetate ion (edta) forms quite stable complexes with metals such as Ca^{2+}. The edta can wrap around a metal ion, displacing as many as six coordinated water molcules. These complexes of the alkaline earth metal ions form primarily because of entropy effects resulting from the increase in the number of particles in the system and the neutralization of charge on M^{2+} and $edta^{4-}$. Charge-neutralization brings about a very favorable entropy change in a polar solvent, because it removes a large part of the ordering of the solvent molecules caused by the ions.

Forced Configurations. A ligand such as porphyrin or phthalocyanine (see pp. 274, 724), which has a completely fused planar ring system, forms extraordinarily stable complexes with metal ions that tend to give planar complexes—for example, Cu^{2+}. These ligands impose planar configurations on metal ions that show no tendency to form planar complexes with unidentate ligands—for example, Be^{2+} and Zn^{2+}.

The ligands tris(2-aminoethyl)amine, $N(CH_2CH_2NH_2)_3$, tren, and triethylenetetraamine, $(NH_2C_2H_4NHCH_2—)_2$, trien, are both quadridentate, but only trien can assume a planar or nearly planar configuration. From the pK values of the ligands, we would expect tren to form more stable complexes—as indeed is the case for Zn^{2+}, which can form tetrahedral complexes (Table 12.9). The Ni^{2+} complex of tren is only slightly more stable than that of trien, and Cu^{2+}, which tends to form planar complexes or octahedral complexes with two long bonds,

[22]C. G. Spike, and R. W. Parry, *J. Am. Chem. Soc.* 1953, *75*, 2726, 3770.

[23]A. W. Adamson, *J. Am. Chem. Soc.* 1954, *76*, 1578.

Table 12.9 Some formation constants

	Tren	*Trien*
Ni^{2+} (log K_1)	14.8	14.0
Cu^{2+} (log K_1)	19.1	20.4
Zn^{2+} (log K_1)	14.65	12.1
pK (HL^+)	10.29	9.92

gives a slightly more stable complex with trien in spite of its lower basicity. Although it has sometimes been assumed that tren would give tetrahedral complexes even with Pt(II), the ligand cannot span the tetrahedral positions around a large cation. For large cations it more likely will fill four positions in an octahedral complex or form polymeric complexes by serving as a bridging ligand.

Other Steric Effects

Bulky groups substituted on the ligand atom—for example, *N*-alkylethylenediamines—or adjacent to the ligand atom—for example, 8-hydroxyquinoline with substituents in the 2 position—can result in complexes of lower stability, or even prevent complex formation. Thus, substitution of methyl or phenyl groups in 8-hydroxyquinoline has little effect on the stability of the complexes formed—except when the substitution is in the 2 position, which prevents chelation with aluminum. The effect of the substituent in the 2 position is very important for the small Al^{3+} ion but less important for even slightly larger ions, so that chelation still occurs for Cr^{3+}, Fe^{3+}, Ga^{3+}, Cu^{2+}, and Zn^{2+}. Such steric considerations can be valuable in obtaining specific or selective organic analytical reagents.

Substitution in the 6 and 6′ positions of 2,2′-bipyridine lowers the stability of the complexes formed, because the substituents crowd the metal ion. Substitution in the 5,5′ or 4,4′ positions only slightly affects the stability of the complexes formed, and in the way expected from the effect on basicity. Substitution in both the 3 and 3′ positions causes steric hindrance, because the two substituents prevent the pyridine rings from lying in the same plane for favorable π-orbital overlap among the two pyridine rings and the chelate ring.

2.2′-bipyridine (dipyridyl)

Such π bonding greatly enhances the stability of the 2,2′-bipyridine and 1,10-phenanthroline complexes such as those of Fe^{2+}

12.3.4 The Hard-Soft Acid–Base Model

G. N. Lewis's criterion for the strength of a base (or acid) is that a strong base should be able to displace a weaker base from its compounds (the criterion for acid strength is similar).

$$:B_{strong} + A_{acid}{:}B_{weak} \rightarrow A_{acid}{:}B_{strong} + :B_{weak}$$

For aqueous solutions, using the hydrogen ion as the reference acid, the order of base strength is the same as the order of pK_a of the BH^+ species: that is,

$$NH_3(aq) + HF(aq) \rightleftharpoons NH_4^+(aq) + F^-(aq)$$

$$pK_a \qquad 3.45 \qquad 9.25$$

The $\log K_{eq}$ for the displacement reaction above is simply the difference in the pK_a values of the ammonium ion and of hydrofluoric acid—5.80. Here we may consider the pK_a value to be the logarithm of the formation constant of the protonated species, either K_1 or β_1 (see p. 536).

The stability constants for a metal ion can give the order of base strength for a series of ligands (and the stability constants with a fixed ligand for a series of metal ions gives the order of acid strength of the metal ions with respect to the particular ligand and solvent system.) G. N. Lewis himself noted that the order of base strengths obtained using metal ions as the reference is not necessarily the same as that obtained using the hydrogen ion as the reference acid. With ammonia and the fluoride ion as the ligands, Ag^+ gives the same order of base strengths as H_{aq}^+, whereas the Ca^{2+} ion gives the opposite order.

$$NH_3(aq) + CaF^+(aq) = CaNH_3^{2+}(aq) + F^-(aq) \qquad (I)$$

$$\log K_1 \qquad\qquad 0.51 \qquad\qquad -0.2$$

$$\log K_{displacement} \qquad = -0.2 - (0.51) = -0.7$$

$$NH_3(aq) + AgF(aq) = Ag(NH_3)^+(aq) + F^-(aq) \qquad (II)$$

$$\log K_1 \qquad\qquad 0.36 \qquad\qquad 3.32$$

$$\log K_{displacement} \qquad = 3.32 - 0.36 = 2.96$$

Thus, in aqueous solution, ammonia is a stronger base toward silver ion than fluoride ion and a weaker base toward calcium ion than fluoride ion. In the gas phase, or a suitably nonsolvating nonaqueous medium, the F^- ion would be more basic toward both metal ions, because of the strong electrostatic interaction.

Using the stability constants given above, we obtain the following order of acid strength in aqueous solution.

$$H^+ > Ag^+ > Ca^{2+} \qquad \text{toward } NH_3(aq)$$
$$H^+ > Ca^{2+} > Ag^+ \qquad \text{toward } F^-(aq)$$

(Including $I_{(aq)}^-$ as a ligand would have given us a still different order, $Ag^+ > Ca^{2+} > H^+$). The acidity orders using different reference bases should become much more uniform, but still not identical, if all reactions were carried out in the gas phase.

Let us now consider a metathesis, or interchange, reaction involving two pairs of acid–base adducts. The equilibrium constant for the interchange reaction may be obtained in a straightforward fashion from the equilibrium constants for the displacement reactions, or from the appropriate stability constants. Thus, for the interchange between AgF and $Ca(NH_3)^{2+}$, we have

$$AgF + Ca(NH_3)^{2+} = Ag(NH_3)^+ + CaF^+$$
$$\log K_{interchange} = \log K(II) - \log K(I) = 3.62$$

where $K(I)$ and $K(II)$ refer to the displacement reactions I and II.

Pearson has proposed that *in an interchange reaction*[24] *a hard acid will prefer to combine with a hard base and a soft acid will prefer to combine with a soft base.* We will use this rule as an operational definition of hardness or softness when the assignment of hard and soft to a reference pair of acids (or bases) seems relatively unambiguous.

Although we readily can calculate $\log K_{interchange}$ from the stability of the acid–base pairs, an affinity diagram is useful when a large amount of data must be inspected. Here the affinity diagram would be a plot in which the abscissa is the affinity of a base (or acid) for various acids (as measured by $\frac{1}{n} \log \beta_n$) and the ordinate is a similar scale for a second base (or acid).

Placing unit slope lines through points on such a diagram makes the lateral (or vertical) displacement of the lines a measure of the $\log K$ (or, depending on the data used, ΔH or ΔG) for the interchange reaction. The $\log K$ for the interchange can be obtained from the difference in the intercepts on either axis. The $\log K$ will be positive when the product acid–base adduct is represented by the abscissa and the line with the greater intercept with the abscissa (the other product acid–base adduct, of course, will be represented by the ordinate and the line with the greater intercept with the ordinate). Figure 12.4 depicts this diagram for the Ag^+ and Ca^{2+} ions with NH_3 and F^-.

Pearson suggested that the *(b)* metals of the Chatt, Ahrland, Davies classification should have ions of soft acid character [see page 538 for the class *(a)* and class *(b)* metals].

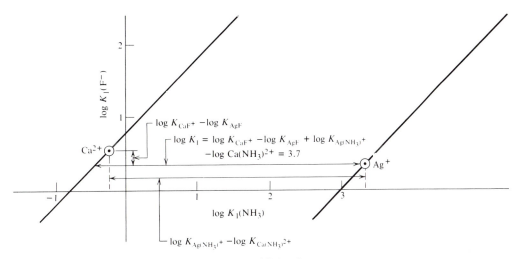

Figure 12.4 Affinity diagram.

[24]The phrase "in an interchange reaction" is an editorial insertion. The papers of Pearson listed in the general references should be consulted for examples of the original, less restricted version.

G. Schwarzenbach[25] proposed that the methyl mercury(II) ion would provide an ideal soft ion to compare with the hard hydrogen ion, since both form 1:1 acid:base complexes as their primary product in aqueous solution. Figure 12.5 shows an affinity plot of pK_a vs log $K_{CH_3Hg^+}$. From the order of the x-intercepts, and assigning the H^+ as hard and the CH_3Hg^+ as soft, we obtain the following order of softness for the bases.

$$I^- > Br^- > Cl^- > {-S_2O_3^{2-}} > S^{2-} > (C_6H_5)_3P > CN^- > {-SCN} > SO_3^{2-} >$$

$$HPO_4^{2-} > NH_3 > F^- > OH^-$$

Although the use of an interchange reaction removes much of the effect of solvation, the position of NH_3 in the above order undoubtedly stems from the high solvation energy of the NH_4^+ ion. Likewise, the ability of OH^- to fit into the water structure places OH^- as a harder base than it would otherwise fall.

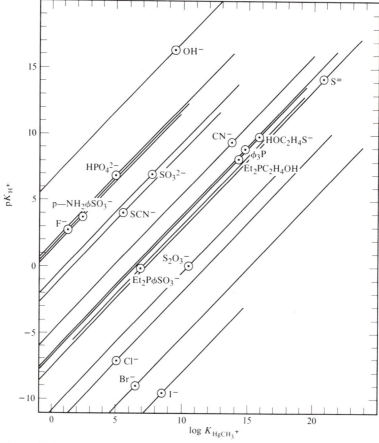

Figure 12.5 Affinity plot for various ligands. (Data of G. Schwarzenbach, *Chem. and Eng. News*, May 31, 1965, p. 92.)

---

[25]In *Chemical and Engineering News*, pp. 90–103, 1965. This article is a condensation of 12 symposium papers on the HSAB concept.

Avoiding solvation entirely, Phillips and Williams use heats of formation data for pairs of metal ions of comparable size to construct affinity plots.[26] Using Ca^{2+} compounds as hard acid complexes, and Cd^{2+} as soft acid complexes, the order of decreasing softness found is

$$Te^{2-} > Se^{2-} > S^{2-} > I^- > Br^- > O^{2-} > Cl^- > OH^- > CO_3^{2-} >$$

$$\text{hydrated ion} > NO_3^- > SO_4^{2-} > F^-.$$

In the gas phase, all metal ions, hard or soft, prefer to bind to the fluoride ion rather than the iodide ion. In an interchange reaction between metal fluorides and metal iodides the class *(b)* metals prefer iodide and the class *(a)* prefer fluoride. Assigning the F^- as a hard base and the I^- as a soft base, Williams and Hale arrived at the following order of hardness from interchange reactions.[27]

SOFT $Au^+ < Ag^+ < Hg^{2+} < Cu^+ < Cd^{2+} < Cs^+ < Rb^+ < K^+ < Na^+ < Cu^{2+} <$

$Pb^{2+}, Pb^{4+}, Ca^{2+} < Tl^+, Mn^{2+} < Zn^{2+} < In^+ < Fe^{2+} < Co^{2+}, Ni^{2+}, Sn^{4+} <$

$Cr^{2+}, Sn^{2+}, Ge^{2+}, Ba^{2+}, Sr^{2+} < In^{3+} < Ni^{3+}, Fe^{3+} < Co^{3+}, Bi^{3+} < Li^+ <$

$Mn^{3+} < Ga^{3+} < Ti^{2+} < Cr^{3+} < Sc^{2+} < Sb^{3+} < Ge^{4+} < Mg^{2+} < V^{3+} < V^{2+} <$

$Ga^+ < Y^{3+}, Ti^{3+} < Sc^{3+} < As^{3+} < Ti^{4+}, La^{3+} < Nb^{5+} < Zr^{4+} < Al^{3+} <$

Be^{2+} HARD

Finally, before we look at some interpretations and applications of the HSAB concept, note Pearson's classification of some Lewis acids and bases as given in Table 12.10.

Nucleophilicity is the term used by organic chemists to describe the *rate* at which Lewis-base attack occurs on an organic substrate (now, it is used generally for rates of basic ligand substitution reactions). *Electrophilicity* refers to rate of attack by a Lewis acid. For some time, chemists generally believed that the difference between the nucleophilicity of a group and its pK_a value represented a fundamental difference in the nature of bonds to C and bonds to H. Primarily responsible for the difference is the contribution to the activation energy of the desolvation of the nucleophile.[28] This desolvation term is greater for the more basic nucleophiles—hence the inversion in aqueous media of the order of nucleophilicity of the halide ions compared with their basicity. This effect is dramatically illustrated by the change in rate of reaction of Cl^- with CH_3Br on going from water to acetone to the gas phase: the rate increases by 3×10^5 for the first change and 3×10^9 for the second change, for an overall increase of 15 orders of magnitude in the gas phase. This is also of some significance to inorganic chemists. One order of softness has been derived from the rates of substitution reactions of various nucleophilic ligands with *trans*-$Pt(py)_2Cl_2$.[29] Again, the desolvation energy probably is the major contributor to the activation energy in these reactions!

[26]C. S. G. Phillips and R. J. P. Williams, *Inorganic Chemistry*, Vol. II, Chap. 31, Oxford, Oxford, 1966. Their approach is not actually the affinity-plot approach, but is mathematically equivalent to it.

[27]R. J. P. Williams and J. D. Hale in *Structure and Bonding*, Vol. 1, 1966, 255ff, Ed. by C. K. Jørgensen, Springer Verlag.

[28]R. T. McIver, Jr., *Scientific American* 1980, *243*, 186–196.

[29]R. G. Pearson in *C and E News, op. cit.* 1966.

Table 12.10   Pearson's classification of hard and soft behavior

---

Hard Acids

H^+ and HX in hydrogen-bonding molecules

Li^+ Be^{2+} BF_3 CO_2

Na^+ Mg^{2+} Al^{3+}, $Al(CH_3)_3$, $AlCl_3$, AlH_3 SO_3

K^+ Ca^{2+} Sc^{3+} Ti^{4+} VO^{2+} Cr^{3+} $Cr(VI)$

    Sr^{2+} Zr^{4+} MoO^{3+}

      La^{3+} Hf^{4+} WO^{4+}

      Lu^{3+}

      Ce^{3+} Th^{4+} UO_2^{2+} Pu^{4+}

Borderline acids

$B(CH_3)_3$ Fe^{2+} Co^{2+} Ni^{2+} Cu^{2+} Zn^{2+} $Ga(CH_3)_3$

      Ru^{2+} Rh^{2+} Sn^{2+} Sb^{3+}

         Ir^{3+} Pb^{2+} Bi^{3+}

Soft acids

$(BH_3)_2$ $Co(CN)_5^{3-}$ Cu^+ $Ga(CH_3)_3$, GaX_3

       Pd^{2+} Ag^+ Cd^{2+} $X = Cl, Br, I$

       $Pt^{2+,4+}$ Au^+ Hg_2^{2+}, Hg^{2+} Tl^+

Hard bases

NH_3 H_2O OH^- CO_3^{2-} F^-

RNH_2 ROH RO^- SO_4^{2-} PO_4^{3-}

       R_2O $CH_3CO_2^-$ Cl^-

N_2H_4 ClO_4^-

            NO_3^-

Borderline bases

C_5H_5N N_2 $:NO_2^-$

$C_6H_5NH_2$ N_3^- $:SO_3^{2-}$ Br^-

Soft bases

H^-

R^- CN^- C_2H_4 RNC   CO

                      R_3P, $(RO)_3P$ SCN^- $S_2O_3^{2-}$

                      R_3As RS^-, RSH, R_2S

                                         I^-

Shielding Efficiencies

An empirical shielding efficiency (S_{eff}) function for cations developed by Ahrens in the early 1950's[30] was designed to show the extent to which the valence electrons are shielded from the

[30]L. H. Ahrens, *Nature* 1954, *174*, 644.

nuclear charge by the intervening core electrons; its converse may be taken as a measure of the polarizing power of the cation. This function is

$$S_{eff} = \frac{5Q^{1.27}}{(I_n)R^{0.5}} \qquad (12.2)$$

where Q represents the unit charges on the cation, I_n is the nth ionization energy of the atom in eV and for $n = Q$, and R is the Pauling crystal radius of the cation in Ångström units. Figure 12.6 gives the S_{eff} values for several cations in the maximum oxidation states. S_{eff} varies with atomic number depending on the nature of the orbital from which the last electron is removed and the nature and occupancy of the immediately underlying orbitals. For an s^1 overlying a $1s^2$ configuration, S_{eff} is about 1.1; for an s^1 overlying an s^2p^6 configuration, S_{eff} is 1.0; for an s^1 overlying a d^{10} configuration, the S_{eff} is very low but increases smoothly with atomic number (see Cu^I through Br^{VII}). We interpret the latter as showing that as the nuclear charge increases, the d electrons are pulled in more tightly, and the penetration of the s electron thereby diminished. For a d^1 electron, S_{eff} decreases regularly with an increase in atomic number (see Sc^{III} through Mn^{VII}) and, excluding Group VIII, the screening efficiency increases with an increase in underlying electrons (on descending within a given group). Underlying f electrons are very efficient in screening, and the S_{eff} values are one or greater with any underlying f population.

Using other sets of crystallographic radii for metal ions in their common coordination states, we can use Equation 12.2 to calculate further S_{eff} values. Some trends may be seen in the following: Mn^{2+}, 0.86; Fe^{2+} or Fe^{3+}, 0.84; Co^{2+}, 0.82; Ni^{2+}, 0.80; Rh^{3+}, 0.72; Pd^{2+}, 0.70; Pt^{2+}, 0.66; Tl^+, 0.67—compared with Tl^{3+}, 0.70.

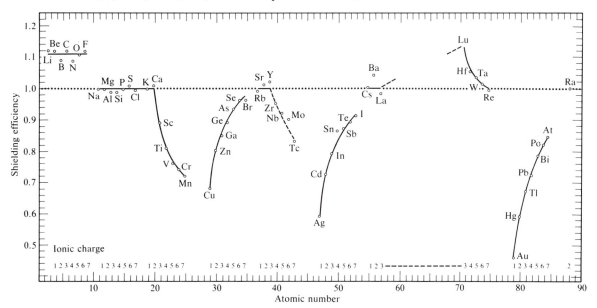

Figure 12.6 The quantity $5Q^{1.27}/I_nR^{0.5}$, defined as the shielding efficiency of a cation, is related to the atomic number and shows various regularities. The ionic charge for each cation is indicated just above the abscissa scale. (From L. H. Ahrens, *Nature* 1954, *174*, 644.)

Equating a low value of shielding efficiency to softness, we can make the following generalizations. Soft character is maximum for an *atomic* electronic configuration of $d^{10}s^1$ and decreases on going to either side—d^{10-n} or s^2p^n, the decrease being proportional to n. Increasing the ionic charge slightly, increases the softness of the transition metal ions, and slightly decreases the softness of the post-transition cations. The alkali metal ions, the alkaline earth metal ions, the lanthanide ions and the actinide ions are all hard.

Size and Charge Effects in Exchange Reactions

The preceding discussion of polarization and polarizing power may have implied that increased covalent character in soft–soft interactions is *the* driving force for exchange reactions. We readily can show that in the products in an exchange reaction between ion-pairs, the highly charged anions will pair with highly charged cations, or, if charge is kept constant, the smallest cation will pair with the smallest anion.

> Although not a necessary consequence of the above, the empirical generalization has been made that "metal complex ions which are difficult to isolate often can be isolated as salts of large ions having an equal but opposite charge"[31] (see Section 6.3.5). Table 12.11 shows a number of especially interesting compounds whose isolation supports this generalization.

Huheey and Evans have shown that the preference of small ions for other small ions does not necessarily support an electrostatic model for the acid–base pair, since covalent interactions

Table 12.11   Examples of counterion stabilization [from Burmeister and Basolo, *op. cit.*]

Compound	*Point of Interest*
$[Os(CO)_3(P(C_6H_5)_3)_2H][PF_6]$	Evidence of intermediate hydride formation in oxidative addition of HX to d^8 osmium(0) complex
$[As(C_6H_5)_4][Au(C\overset{N}{\underset{N-N}{\diagdown}}N)_4]$ $H_{11}C_6$	Possible Au—C bond
$[B(\gamma\text{-pic})(N(CH_3)_3)(Cl)H]PF_6$	First boron cation resolved into optically active isomers
$[As(C_6H_5)_4][MF_5]$	M = Si(IV), Ge(IV); abnormal coordination numbers
$[Co(tetren)Cl][ZnCl_4]$	Optically active cation containing quinquedentate amine
$[Co(pn)_3][MCl_6]$	M = Cr(III), Mn(III), Fe(III); hexachlorocomplexes stabilized in solid state
$[Cr(NH_3)_6][CuCl_5]$	Trigonal bipyramidal structure with axial Cu—Cl bond distances shorter than equatorial

Abbreviations: γ-pic = 4-methylpyridine, tetren = tetraethylenepentaamine, pn = 1,2-propanediamine.

[31] J. L. Burmeister and F. Basolo, "Synthesis of Coordination Compounds" in *Preparative Inorganic Reactions*, Vol. 5, W. L. Jolly, Ed., Interscience Pub., New York, 1968.

also will favor the matching of small with small.[32] They make a convincing argument that the matching of small with small (that is, hard with hard) is a major factor in the HSAB principle. Certainly, the shielding efficiencies of the cations give little aid in obtaining a hardness (or softness) ordering among cations of s^2p^6 configuration, whereas size variation provides a logical ordering. Increasing the charge on the cation serves to intensify the electrostatic q_1q_2/r term, as well as to decrease the ionic radius. This helps justify classifying I^{7+}, Mn^{7+}, Cr^{6+}, etc. as hard acids. But the S_{eff} values would place in the hard category only the I^{7+}, the other ions being soft or borderline. Unfortunately, there appear to be insufficient thermodynamic data to settle the question of Mn^{7+} and Cr^{6+} by the operational criteria of the exchange reactions, since most unambiguously soft bases undergo oxidation in the presence of these acids. They have been classified as hard presumably on the basis of electrostatic arguments and on the nonexistence of complexes with soft bases!

Frontier Orbitals[33]

The highest occupied molecular orbital (HOMO) and the lowest unoccupied molecular orbital (LUMO) of a molecule, termed the "frontier" orbitals, are the orbitals normally used by a Lewis base and a Lewis acid, respectively, in the formation of an adduct. If these orbitals were used invariably, the same atom in a polyatomic species would always participate in adduct formation as the base (or acid). In the case of the thiocyanate ion, coordination may occur through either the S or the N atom—hard-metal ions preferring to coordinate with the hard N end, and soft-metal ions with the soft S end.[34] The N end of SCN^- should have the greater charge density, and essentially electrostatic interactions will prefer to interact with this end of the ligand. S would lose its electrons more readily—in other words, it makes the major contribution to the HOMO—and adducts in which the LUMO of the acid is close in energy to this HOMO will prefer the bond to sulfur. Klopman calls these two types of reaction "charge controlled" and "orbital controlled," indicating which energy interaction is more important in the adduct. Hard–hard interactions are thought to be charge controlled on the basis of this model, whereas soft–soft would be orbital controlled.

Optical Electronegativities

C. K. Jørgensen used the charge transfer bands of metal complexes to assign "optical electronegativities" to both the metal ions and the ligands.[35] The energy of these charge transfer bands is essentially equal to the difference between the HOMO of the ligand and the LUMO of the metal ion as perturbed by the formation of the MO of the complex. Hard bases such as F^- and H_2O have high optical electronegativity, whereas soft bases such as I^- and $(C_2H_5O)_2PS_2^-$ have low optical electronegativity. The optical electronegativity of the metals

[32]J. E. Huheey and R. S. Evans, *J. Inorg. Nucl. Chem.* 1970, *32*, 383; *Chem. Comm.* 1969, 968.

[33]See W. B. Jensen, *The Lewis Acid–Base Concepts*, Wiley, New York, 1980, for an extensive discussion of frontier orbitals and acid–base chemistry, and for literature references to this topic.

[34]Such ligands are termed ambidentate; a clever way to prepare the nonthermodynamically favored complex is to transfer the ambidentate ion by an inner-sphere bridging mechanism (see p. 379, and also Burmeister and Basolo, *op. cit.*).

[35]C. K. Jørgensen, *Orbitals in Atoms and Molecules*, Academic Press, New York, 1962.

increases with charge and, generally, with *d*-orbital occupancy. Compounds of metals that exhibit color not caused by *d-d* transitions are generally of the soft acid–soft base type, in which the low energy of the charge transfer band places it in the visible region (see Section 7.8).

Symbiosis and Innocent Ligands

As noted above, the same metal sometimes can function as a class *(a)* (hard) or a class *(b)* (soft) acid. Cu(I), for example, is soft, whereas Cu(II) is borderline. For complex ions, assigning oxidation numbers is not always easy. Considerable ambiguity surrounds the complexes of NO in which the NO may function as NO^-, NO, or NO^+. In balancing redox equations, we normally assume that the addition of a hydrogen ion does not affect the oxidation state of atoms in a species (SO_4^{2-}, HSO_4^-, H_2SO_4, etc.). Even a hydrogen ion is not always "innocent," however. In the following reaction, which may be considered as an oxidative addition reaction, the Pt goes from Pt(0) to Pt(II).

$$Pt[P(C_2H_5)_3]_3 + H^+ = HPt[P(C_2H_5)_3]_3^+$$

Jørgensen introduced the terms "innocent" and "noninnocent" to describe ligands forming complexes in which the assignment of oxidation states is *rigorous* and *ambiguous,* respectively.[36] The probability of noninnocent behavior is related to the softness of the ligand. Since softness is related to the excitation band gap, it is not surprising that the greater the number of soft ligands surrounding a metal ion, the smaller the band gap—the more soft ligands on a metal ion, the softer the metal ion becomes as an acid; conversely, the more hard ligands around a metal ion, the harder the metal ion becomes as an acid. Jørgensen called this phenomenon *symbiosis*[37]—a term used in biology to describe organisms that live together in harmony. The effect of soft ligands softening a metal ion can be seen in the preference of $Co(NH_3)_5X^{2+}$ for $X = F^-$, but of $Co(CN)_5X^{3-}$ for $X = I^-$—the fluoride being unknown for the latter. Other examples of symbiosis include the ability of the metal carbonyls to form carbonyl hydrides and the disproportionation of HBF_3^- into BH_4^- and BF_4^- in ether solution (the HBF_3^- ion being an intermediate in the reaction of BF_3 with NaH or with the BH_4^- ion). Interestingly, although a complex forms between BH_4^- and BF_3 in the solid state, using $(C_2H_5)_4N^+$ as the cation, it decomposes into its starting materials in the absence of solvent. The concept has been used extensively in the synthesis of complexes of ambidentate ligands—those with a soft S end and a hard N end, such as SCN^-. Farona and Wojcicki proposed that when good pi-bonding ligands are present, the high oxidation states of the metal prefer the S end of SCN and the low oxidation states prefer the N end. This proposal was extended to the CNS^- complexes by Sloan and Wojcicki.[38]

---

[36]Initially, the term "innocent" was used to describe ligands in colorless complexes—color being defined as a transition below 20,000 cm⁻¹. C. K. Jørgensen, *Structure and Bonding* 1966, *1*, 234.

[37]C. K. Jørgensen, *Inorg. Chem.* 1964, *3*, 1201.

[38]See T. E. Sloan and A. Wojcicki, *Inorg. Chem.* 1968, *7*, 1268, and references therein.

The Drago-Wayland Equation

In 1965, Drago and Wayland introduced an empirical four-parameter equation to describe and predict the enthalpy change accompanying the reaction of weak neutral acids with weak neutral bases in poorly solvating solvents or in the gas phase.[39] This equation takes the form

$$-\Delta H = E_A E_B + C_A C_B$$

where E_A and C_A are parameters for the acid (arbitrarily set equal to 1.00 for I_2) and E_B and C_B are parameters for the base (initially, with E_B made proportional to the dipole moment for NH_3 and several amines and C_B made proportional to the polarizabilities of the amines). It was felt initially that the $E_A E_B$ term reflected the electrostatic part of the acid–base interaction and that the $C_A C_B$ term reflected the covalent part, and vestiges of this interpretation remain in the literature. The equation is similar in form to a number of other equations in the literature, such as the Edwards equation,[40] but differs in that the others correlate independently observable quantities such as stability constants with Brønsted acidity and oxidative coupling constants. The beauty of the D.-W. equation lies in its minimum of constraints, which in turn allows the best possible fitting of the enthalpy data by an equation of its form. Accordingly, it can do a better job (or at least as good a job) of fitting a given set of data than any equation of similar form that uses observable properties as parameters. It thus sets a limit to the degree to which any equation of its form can fit a set of data. The price paid for the reduction of constraints is an increase in the uncertainty of interpretation of the parameters.

With the constants listed in Table 12.12, we can predict the enthalpy change for 900 reactions between an acid and a base. If an interchange reaction between two acid-base adducts is considered, we can predict approximately one million such enthalpies by the equation

$$\Delta H = \Delta E_A \Delta E_B + \Delta C_A \Delta C_B$$

where the Δ terms are all consistent (either all A and B subscripted 1 subtracted from those subscripted 2, or vice versa). We can use such data to construct affinity diagrams, and it can be shown that a consistent ordering of hardness or softness cannot be achieved with the set of acids and bases for which we have E and C parameters.

Along with predicting enthalpy data, the Drago-Wayland equation can be used to select solvents that have nearly the same extent of acid–base interactions with solutes by matching the E and C parameters. Thus dimethyl sulfoxide and dimethylformamide would show virtually identical acid–base behavior toward solutes, since they have almost identical E and C values.

The deviations observed from the predictions of the D.-W. equation frequently can be used to deduce unusual steric effects, or specific solvation effects, etc., in a reaction.[41]

[39] R. S. Drago and B. B. Wayland, *J. Am. Chem. Soc.* 1965, *87*, 3571.

[40] J. O. Edwards, *J. Am. Chem. Soc.*, 1954, *76*, 1540.

[41] See, for example, R. S. Drago *et. al.*, *Inorg. Chem.* 1979, *101*, 2879.

Table 12.12 *E* and *C* parameters for various acids and bases[a, b, d]

Formula	E	C	Formula	E	C
Acids			**Bases**		
1 I_2	1.00	1.00	31 NH_3	1.15	4.75
2 ICl	5.10	0.830	32 CH_3NH_2	1.30	5.88
3 C_6H_5SH	0.99	0.198	33 $(CH_3)_2NH$	1.09	8.73
4 C_6H_5OH	4.33	0.422	34 $(CH_3)_3N$	0.808	11.54
5 p-$CH_3C_6H_4OH$	4.18	0.404	35 CH_3CN	0.886	1.34
6 p-FC_6H_4OH	4.17	0.446	36 $(CH_3)_2NCN$	1.10	1.81
7 m-FC_6H_4OH	4.42	0.506	37 $CH_3CON(CH_3)_2$	1.32	2.58
8 p-ClC_6H_4OH	4.34	0.478	38 $CH_3COOC_2H_5$	0.975	1.74
9 m-$CF_3C_6H_4OH$	4.48	0.530	39 $(CH_3)_2CO$	0.937	2.33
10 $(CH_3)_3COH$	2.04	0.300	40 $(C_2H_5)_2O$	0.936	3.25
11 CH_3CH_2OH	3.88	0.451	41 $O(CH_2)_4O$	1.09	2.38
12 $(CF_3)_2CHOH^c$	5.93	0.623	42 $(CH_2)_4O$	0.978	4.27
13 C_4H_4NH	2.54	0.295	43 $HC(S)N(CH_3)_2$	0.76	8.19
14 $CHCl_3$	3.02	0.159	44 $(CH_3)_2SO$	1.34	2.85
15 $(CH_3)_3SnCl$	5.76	0.03	45 $(CH_2)_4SO$	1.38	3.16
16 $BF_3(g)$	9.88	1.62	46 $(CH_3)_2S$	0.343	7.46
17 $B(CH_3)_3(g)$	6.14	1.70	47 $(C_2H_5)_2S$	0.339	7.40
18 $Al(CH_3)_3$	16.9	1.43	48 C_5H_5NO	1.34	4.52
19 SO_2	0.920	0.808	49 4-$CH_3C_5H_4NO$	1.36	4.99
20 $Cu(hfac)_2$	3.46	1.32	50 $(CH_3)_3P$	0.838	6.55
21 H_2O	1.64	0.571	51 C_6H_6	0.280	0.590
22 $CH_3Co(dmg)_2$	9.14	1.53	52 $C_9H_{18}NO$ (tmpno)	0.915	6.21
23 $Zn\{N[Si(CH_3)_3]_2\}_2$	5.16	1.07	53 $HC(C_2H_4)_3N$	0.700	13.2
24 $Ni(tfaccam)_2$	3.38	0.640	54 $C_6H_{10}O$ (bridged ether)	1.08	3.76
25 Ni(smdpt)	3.94	0.500	55 $(CH_3)_2Se$	0.217	8.33
26 π-allyl PdCl^c	3.41	0.980	56 $C_2H_5C(CH_2O)_3P$	0.548	6.41
27 Rh(cod)Cl^c	4.93	1.25	57 $[(CH_3)_2N]_3PO$	1.52	3.55
28 $Rh(CO)_2Cl^c$	8.72	2.02	58 C_5H_5N	1.17	6.40
29 Zntpp	5.15	0.620	59 $CH_3C_4H_4N$	1.26	6.47
30 Cotpp	4.44	0.58	60 N-methyl imidazole	0.934	8.96

[a]These parameters yield enthalpy values in kcal/mole adduct.
[b]The abbreviations used are as follows: For acid 20 hfac is hexafluoroacetylacetonate, for 22 dmg is dimethylglyoximate, for 24 tfaccam is trifluoroacetylcamphorato, for 29 and 30 tpp is tetraphenylporphine, for base 52 $C_9H_{18}NO$ is the free radical 2,2,6,6-tetramethylpiperidine-*N*-oxyl, for 54 $C_6H_{10}O$ is 7-oxabicyclo[2.2.2]heptane.
[c]For systems in which these acids are involved, the following equation should be used $- \Delta H + W = E_A E_B + C_A C_B$, where W is 1.10 for acid 12, 3.1 for acid 26, 6.3 for acid 27, and 11.3 for acid 28.
[d]From R. S. Drago, *Coord. Chem. Rev.* 1980, *33*, 251.

GENERAL REFERENCES

R. P. Bell, *The Proton in Chemistry,* 2nd ed., Cornell University Press, Ithaca, New York, 1973.

J. E. Bartmess and R. T. McIver, Jr., "The Gas Phase Acidity Scale" in *Gas-Phase Ion Chemistry,* M. T. Bowers, Ed., Academic Press, New York, 1978.

R. S. Drago and N. A. Matwiyoff, *Acids and Bases,* Heath, Boston, 1968.

W. B. Jensen, *The Lewis Acid–Base Concepts,* Wiley, New York, 1980.

J. E. Huheey, *Inorganic Chemistry,* 2nd ed Chap. 7, Harper and Row, New York, 1978.

R. G. Pearson, *Survey of Progress in Chemistry,* Vol. 1, Chap. 1, A. Scott, Ed., Academic Press, New York, 1969.

C. S. G. Phillips and R. J. P. Williams, *Inorganic Chemistry,* Vol. II, Chapter 31, Oxford, Oxford, 1966.

PROBLEMS

12.1   *a.* Indicate the conjugate bases of the following: NH_3, NH_2^-, NH^{2-}, H_2O, HI.

   b. Indicate the conjugate acids of the above species.

   c. What relationship exists between the strength of a conjugate acid and a conjugate base for a neutral substance, such as NH_3 in parts *a.* and *b.* above.

12.2   Construct a table similar to Table 12.3 in which the system contains excess $HI(g)$. Would you characterize this "solvent" as giving a more acidic or more basic system than $H_2O(g)$? Explain.

12.3   The heat of neutralization of $H^+(aq)$ and $OH^-(aq)$ is considerably higher in concentrated salt solution ($\Delta H_{neut.} = 85.4$ kJ/mole at $\mu = 16$) than in dilute aqueous solution ($\Delta H_{neut.} = 56.5$ kJ/mole as $\mu \rightarrow 0$). Explain what might cause the increase in the heat of neutralization when a high concentration of NaCl is present.

12.4   Write a thermochemical cycle from which you can obtain the hydride affinity of a positive ion from the bond dissociation energy of EH and other suitable data.

12.5   Give the approximate pK_a values for the following acids.

   a. H_3PO_3 $pK_1 =$ $pK_2 =$

   b. HNO_3 $pK_1 =$

   c. $HClO_4$ $pK_1 =$

   d. H_5IO_6 $pK_1 =$ $pK_2 =$

12.6   Select the best answer and give the basis for your selection.

   a. Thermally most stable: PH_4Cl PH_4Br PH_4I.

   b. Strongest acid: H_2O H_2S H_2Se H_2Te.

   c. Acidic oxide: Ag_2O V_2O_5 CO Ce_2O_3.

   d. Strongest acid: MgF_2 $MgCl_2$ $MgBr_2$.

   e. Stronger base (toward a proton): PH_2^- NH_2^-.

12.7   Give equations to explain why adding ammonium acetate to either zinc amide(s) in liquid ammonia or zinc acetate(s) in acetic acid causes the solid to dissolve.

12.8   Would the following *increase, decrease,* or have *no effect* on the acidity of the solution?

   a. Addition of Li_3N to liquid NH_3.

   b. Addition of HgO to an aqueous KI solution.

   c. Addition of SiO_2 to molten $Fe + FeO$.

   d. Addition of $CuSO_4$ to aqueous $(NH_4)_2SO_4$.

     e. Addition of $Al(OH)_3$ to aqueous NaOH.
     f. Addition of $KHSO_4$ to H_2SO_4.
     g. Addition of CH_3CO_2K to liquid NH_3.

12.9   Select the best response within each horizontal group and indicate the major factor governing your choice.

     Strongest protonic acid

a. SnH_4	SbH_3	H_2Te
b. NH_3	PH_3	SbH_3
c. H_5IO_6	H_6TeO_6	HIO
d. $Fe(H_2O)_6^{3+}$	$Fe(H_2O)_6^{2+}$	H_2O
e. $Na(H_2O)_x^+$	$K(H_2O)_x^+$	

     Strongest Lewis acid

f. BF_3	BCl_3	BI_3
g. $BeCl_2$	BCl_3	
h. $B(nBu)_3$	$B(tBu)_3$	

     More basic toward BMe_3

i. Me_3N	Et_3N		
j. 2-MePy	4-MePy	Py	(Py = pyridine)
k. 2-MeC_6H_4CN	C_6H_5CN		

12.10   In general, the anhydrous nitrites of a given metal ion have a lower thermal stability than the nitrates, and similarly for the sulfites and the sulfates. How may this be explained? (For a more general survey of such data, and generalizations about the area, see R. T. Sanderson, *Chemical Periodicity*, Chap. 7, Reinhold, New York, 1960.)

12.11   Use the data given in Table 12.6 (p. 531) to sketch a field distribution diagram for a 1 *M* solution of V(V) as a function of pH. What pH range is suitable for precipitation of V(V) from solution? (*Hint:* First determine the log [conc] ratios as a function of pH.)

12.12   Compared with the spectra of the halopentaamminecobalt(III), the spectra of the comparable rhodium compounds show shifts in the *d-d* bands toward the *uv* end of the spectrum, and in the charge transfer bands toward the visible. Offer an explanation for these shifts.

12.13   Tetrahydrofuran has an ionization energy of 11.1 eV and diethyl ether has an ionization energy of 9.6 eV. How can we account for this difference? What difference is expected in their *PA*'s? Which is higher? THF is much more basic toward $MgCl_2$ and Grignard reagents than Et_2O—how might this observation be rationalized?

12.14   Indicate which of the following ligands should show ambidentate character, and indicate for such ligands the hard and soft ends: NO_3^-, SO_3^{2-}, $S_2O_3^{2-}$, $S_2O_7^{2-}$, NO_2^-, $(CH_3)_2SO$, CN^-, $SeCN^-$. Why is glycine not classified as an ambidentate ligand?

12.15   Use Table 12.12 to decide which of the following in each pair is the stronger base:
     a. Acetone or dimethylsulfoxide.
     b. Dimethylsulfide or dimethylsulfoxide.
     Comment on your conclusions regarding possible ambiguity.

XIII
The Chemistry of Some Nonmetals

Although this is the only chapter dealing exclusively with nonmetals, the chemistry of nonmetals appears in most of the book. The stereochemistry and bonding of compounds of nonmetals are treated in Chapters 2, 4, and 5. The solids treated in Chapter 6 include metal compounds with nonmetals, covalent crystals, and metal silicates. Coordination chemistry (Chapters 7 and 8) and Chapter 10 on organometallic compounds deal with the chemistry of metals with nonmetal molecules or groups as ligands or substituents. Such compounds are important in bioinorganic chemistry (Chapter 16). Chapter 12, on acid–base chemistry, is concerned largely with compounds of nonmetals. The chemistry of boranes and carboranes (Chapter 15) is covered extensively. Here we examine most of the nonmetals of Groups VB, VIB, VIIB, and 0 in the framework of the periodic table in order to observe and understand the similarities and differences in their physical and chemical properties. The coverage is not comprehensive.

13.1 NITROGEN AND PHOSPHORUS

13.1.1 Family Trends for Group VB

The elements N, P, and As are nonmetals, although the chemistry of As shows many characteristics of metals. Antimony and bismuth are metals. Except for N, the family trends are fairly regular. Whereas negative oxidation states are very important in the chemistry of N—NH_3, RNH_2, R_2NH, for example—the positive oxidation states are more important for P (and still more so for As). Nitrogen forms compounds involving very stable multiple bonds of the p-p-π type, as in N_2, $C\equiv N^-$, and $RC\equiv NR$, whereas multiple bonds to P usually involve p-d-π bonding, as in oxo anions. The availability of only s and p orbitals in the valence shell

Table 13.1   Some properties of the group VB elements

	N	P	As	Sb	Bi
Crystal radius (pm)	$132(N^{3-})$	$212(P^{3-})$	$72(As^{3+})$	$94(Sb^{3+})$	$117(Bi^{3+})$
Covalent radius (pm)	72.5	111	121	145	152
Electronegativity	3.0	2.2	2.2	2.1	2.0
First ionization (MJ/mole)	1.4023	1.0118	0.947	0.8337	0.7033
energy (eV)	14.534	10.486	9.81	8.641	7.289
Electron affinity (kJ)	≤ 0	71.7	77	101	91.4
Outer electron configuration	$2s^2 2p^3$	$3s^2 3p^3$	$3d^{10} 4s^2 4p^3$	$4d^{10} 5s^2 5p^3$	$4f^{14} 5d^{10} 6s^2 6p^3$

of N limits compound formation to those satisfying the octet rule, whereas the octet is often exceeded in P compounds such as PCl_5 and $M^+PCl_6^-$.

The ionization energies change by small amounts from one element to the next for $P \rightarrow Bi$, but the ionization energy of N is much higher than that of P (Table 13.1). The electron affinities in general for the Group VB elements are lower than those of the neighboring elements (see Table 1.7, p. 39) in the same period, because an electron must be added to the half-filled p^3 configuration. Unlike the trend for most main-group families (see Tables 13.1, 13.3, and 13.5), the electron affinities for Group VB *increase* with increasing atomic number (Figure 13.1), except for Bi. The increase might be expected because of the decrease in pairing energy with increasing atomic radius (the p orbitals become larger and more diffuse with increasing n value). The electron affinity of N is zero or slightly negative as a result of the combination of three trends: (1) it is expected to be lower than that of P because of the high pairing energy of N; (2) it is expected to be lower than the values of C or O because of the half-filled configuration of N; and (3) the first members of Groups VB–VIIB are anomalous with respect to further increase in electron density for the small compact atoms, as seen in their low electron affinities and single-bond energies. From Table 1.7, the electron affinity of N appears to deviate more from the family trend than does O or F. However, because of the opposite trends in electron affinities for Groups VB and VIB, plots of ionization energy versus electron affinity (Figure 13.1) reveal that N comes closer to the straight line determined by the rest of the family than does O.

13.1.2   The Elements and Their Occurrence

Nitrogen occurs in organic matter as proteins and amino acids. As the main constituent of air (78.09 vol %), it supplies the major source for preparing nitrogen compounds. The deposits of sodium nitrate found in a few arid regions, such as Chile, are important for fertilizers, but most of the nitrogen for fertilizers comes from the fixation of N_2. Dinitrogen (N_2) has accumulated in the atmosphere because the very stable triple-bond makes it quite chemically inert. N_2 is recovered on a large scale by the fractional distillation of liquid air, which produces O_2 and also the noble gases. For small-scale laboratory uses, pure N_2 can be obtained by the thermal decomposition of NaN_3 at about 275°C.

$$2NaN_3 \xrightarrow{\text{heat}} 2Na + 3N_2$$

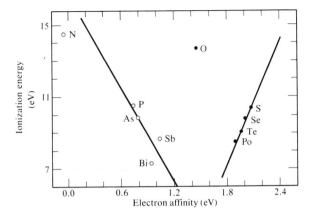

Figure 13.1 Ionization energy versus electron affinity plots for Groups VB and VIB. (Adapted from P. Politzer in *Homoatomic Rings, Chains and Macromolecules of Main Group Elements*, A. L. Rheingold, Ed., Elsevier, Amsterdam, 1977, p. 95.)

The thermal decomposition of sodium azide also is used to inflate the air bags used as safety devices in cars. With moderate heating, dinitrogen reacts with Li to form lithium nitride, Li_3N, and with some transition metals. Nitrogen-fixing bacteria, using the enzyme nitrogenase (Section 16.8.5), can reduce N_2 to NH_3. At elevated temperature, N_2 is more reactive. The Haber process is used in large-scale production of ammonia from the elements.

$$N_2 + 3H_2 \underset{\underset{\sim 500 \text{ atm}}{\sim 500°C}}{\overset{\text{Fe catalyst}}{\rightleftharpoons}} 2NH_3$$

Most other N compounds are obtained from NH_3. Lightning converts N_2 to N oxides, and rainfall supplies the resulting HNO_3 to the soil.

Phosphorus occurs in sedimentary rocks (see p. 610) as phosphates—primarily apatite, $Ca_5(PO_4)_3F$; and phosphorite, $Ca_5(PO_4)_3(OH,F,Cl)$, where there are variable relative proportions of OH^-, F^-, and Cl^-. Elemental phosphorus is obtained from phosphate rock by reduction with coke at 1300 to 1450°C and with sand added to form slag.

$$Ca_3(PO_4)_2 + 5C + 3SiO_2 \rightarrow 3CaSiO_3 + 5CO + P_2(g)$$

The P_2 vapor is condensed under water to form white phosphorus, P_4 (m.p. 44°C). Most phosphorus compounds are produced from the free element obtained in this way, with the notable exception of phosphate fertilizer. The phosphate rock can be used directly in pulverized form, or it can be converted to the more soluble *superphosphate*.

$$2\ Ca_5(PO_4)_3F(s) + 7\ H_2SO_4(aq) + 10\ H_2O$$
$$\rightarrow \underbrace{3\ Ca(H_2PO_4)_2 \cdot H_2O(s) + 7\ CaSO_4 \cdot H_2O(s)}_{\text{superphosphate}} + 2HF(g)$$

Treatment of phosphate rock with aqueous phosphoric acid gives *triple superphosphate*, which has a higher P content.

$$Ca_5(PO_4)_3F(s) + 7\ H_3PO_4(aq) + 5\ H_2O \rightarrow \underset{\substack{\text{triple}\\\text{superphosphate}}}{5\ Ca(H_2PO_4)_2 \cdot H_2O(s)} + HF(g)$$

White phosphorus is a soft, waxy, highly toxic solid that, because it ignites spontaneously in air, is stored under water. It exists as P_4 molecules, with each P forming three single bonds.

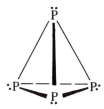

When white P is heated to about 250°C, or to lower temperature in the presence of light, it is converted to red phosphorus (m.p. *ca.* 600°C). Red P is polymeric and does not dissolve in solvents (diethyl ether and benzene) in which white P is soluble; it also is much less reactive than the white allotrope and does not ignite in air below about 400°C.

13.1.3 Hydrogen Compounds

Nitrogen forms the following volatile hydrogen compounds: ammonia, NH_3; hydrazine, H_2NNH_2; diimine, HNNH; tetrazene, N_4H_4; and hydrogen azide (hydrazoic acid), HN_3. Ammonia is the most important of these. Liquid ammonia can be used directly as a fertilizer, or it can be converted to ammonium salts (nitrate, sulfate, or phosphate) for fertilizer. The catalytic oxidation of NH_3 produces NO *(Ostwald process)*.

$$4 \ NH_3 \ + \ 5 \ O_2 \ \xrightarrow{\text{Pt catalyst}} \ 4 \ NO \ + \ 6 \ H_2O$$

Heat is required to initiate the reaction, but then it becomes self-sustaining. The NO reacts with more O_2 to form NO_2. Nitric acid is produced by dissolving NO_2 in water—NO_2 disproportionates to form HNO_3 and HNO_2, which then reacts with O_2 to form more HNO_3.

$$2 \ NO_2 \ + \ H_2O \rightarrow HNO_3 \ + \ HNO_2$$

Interestingly, although NH_3 and amines, including $N(CH_3)_3$, are pyramidal, $N(SiH_3)_3$ is planar. This suggests p-d-π bonding involving the empty d orbitals on Si and the filled p orbital on N (sp^2 hybridization). As this description implies, the basicity of $N(SiH_3)_3$ is very low. Since $P(SiH_3)_3$ is pyramidal, it appears that p-d-π bonding is less favorable for P, because of poorer overlap with the more diffuse $3p$ orbital of P.

> As noted earlier (p. 164), there is still controversy over the extent of participation of the outer d orbitals in bonding, including p-d-π bonding. Glidewell *(Inorg. Chim. Acta* 1975, *12*, 219) prefers to explain the planarity of $N(SiH_3)_3$ in terms of stronger nonbonding interactions (for the larger Si substituents). Although his arguments are more profound, this is basically steric repulsion.

Hydrazine, with a weak N—N single bond (bond dissociation energy 247 kJ/mole), is a stronger reducing agent and a weaker base than ammonia. The *Raschig synthesis* of N_2H_4 uses NaOCl for oxidation of NH_3 to NH_2Cl, which then reacts with NH_3.

$$NH_3 \ + \ NaOCl(aq) \rightarrow NH_2Cl \ + \ NaOH$$

$$NH_2Cl + NH_3 + NaOH \rightarrow H_2NNH_2 + NaCl + H_2O$$

Competing reactions are catalyzed by traces of transition metal ions. Gelatin or glue is added to sequester the metal ions, providing yields of ~70%.

Hydrogen azide, the only acidic uncharged N—H compound, is a weak acid (Table 13.2). Anhydrous HN_3 (b.p. 36°C) decomposes explosively, but dilute aqueous solutions (up to ~20 wt %) can be handled safely. The N_3^- ion is linear and isoelectronic with CO_2. Ionic azides, such as NaN_3, are stable and decompose thermally (but controllably) to the elements. Covalent azides such as HN_3 and those of heavy metals (highly polarizing cations) are detonators. Heavy metal azides, such as those of lead and mercury, can be used as primers for explosives.

Derivatives of Ammonia

Amines have the general formulas RNH_2, R_2NH, and R_3N. There are also derivatives in which R = OH, NH_2OH (hydroxylamine) and R = X—for example, NH_2Cl (chloroamine) and $NHCl_2$ (dichloroamine). The amide, NH_2^-; imide, NH^{2-}; and nitride, N^{3-}, ions yield metal salts. The formation of N^{3-} from N(g) is highly endothermic (~2200 kJ/mole), because the addition of a second and then a third electron is very unfavorable. Consequently, ionic nitrides are obtained only for very electropositive metals.

Phosphine and Diphosphine

Phosphine, PH_3, is a toxic gas formed from the hydrolysis of metal phosphides or, on a larger scale, by the reaction of P_4 with NaOH solution.

$$Ca_3P_2 + 6 H_2O \rightarrow 3 Ca(OH)_2 + 2 PH_3(g)$$

$$P_4 + 3 NaOH + 3 H_2O \rightarrow PH_3(g) + 3 NaH_2PO_2$$

Phosphine is not very soluble in water and is very weakly basic, but phosphonium salts, PH_4X, can be obtained. Pure phosphine is stable in air, but traces of P_2H_4 (diphosphine) commonly present cause the material prepared directly by the reactions cited above to burn spontaneously in air. Diphosphine can be separated from PH_3 by fractional distillation.

The fact that the bond angle for PH_3 (93.3°) is much smaller than that for NH_3 (107.3°) can be rationalized by the VSEPR theory (Section 2.3.3) or interpreted as indicating greater *p* character (and less *s* character) for bonding in PH_3.

Table 13.2 Acid dissociation constants for some NH compounds

Acid	K_a (25°C)
NH_4^+	5.5×10^{-10}
$N_2H_5^+$	1.2×10^{-8}
$N_2H_6^{2+}$	11.2
HN_3	1.8×10^{-5}

13.1.4 Halides

The halides of N and P are shown in Figure 13.2. In addition,

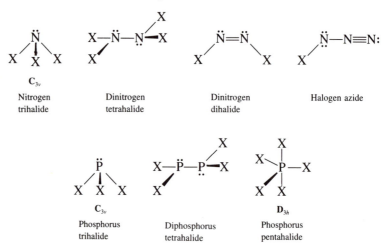

Figure 13.2   Nitrogen and phosphorus halides.

phosphorus forms PX_4^+ ($\mathbf{T}_d$) and PX_6^- ($\mathbf{O}_h$) ions. All the various N halides listed are encountered only for fluorine, since the stabilities of the N halides decrease with increasing atomic weight of the halogen. NF_3 is stable, but the other trihalides are explosive compounds. Of the N_2X_4 compounds, only N_2F_4 is known. Gaseous N_2F_4 contains both *gauche-* and *trans*-conformers. Difluorodiazene (dinitrogen difluoride, N_2F_2) is a gaseous substance consisting of an equilibrium mixture of *cis-* and *trans*-isomers, the *cis*-isomer predominating. The azide ion, considered to be a pseudohalide ion (see Section 13.3.7), forms halogen azides that are similar to the interhalogen compounds (Section 13.3.6).

All phosphorus trihalides are known. They hydrolyze easily and react with ammonia to produce $P(NH_2)_3$.

$$PCl_3 + 3\ H_2O \rightarrow P(OH)_3 + 3\ HCl$$

$$PCl_3 + 6\ NH_3 \rightarrow P(NH_2)_3 + 3\ NH_4Cl$$

In addition, they react with oxygen to produce phosphoryl halides, POX_3 (also called phosphorus oxohalides). All P pentahalides are known except for PI_5, where steric factors probably are critical. PF_5 and PCl_5 have $\mathbf{D}_{3h}$ symmetry in the vapor phase, but PCl_5 exists as $[PCl_4]^+$-$[PCl_6]^-$ in the solid. Phosphorus pentabromide exists as $[PBr_4]^+Br^-$ in the solid and dissociates into $PBr_3 + Br_2$ in the vapor phase.

Compounds of the type PR_5 are named as substituted phosphoranes. The fully hydrogenated hydrides are named as methane, CH_4; silane, SiH_4; and phosphorane, PH_5. The parent compound of the phosphoranes is unknown. Chemists show no inclination to replace the names for water and ammonia by systematic names.

13.1.5 Oxygen Compounds

Oxides of Nitrogen

Nitrogen forms an extensive series of oxides: N_2O (nitrous oxide), NO (nitric oxide), N_2O_3, NO_2, N_2O_4, N_2O_5, and NO_3. Scheme 1 demonstrates the complexity of the chemistry of the oxo compounds of N. The higher oxides are named systematically—dinitrogen trioxide, dinitrogen pentaoxide, etc. Nitrous oxide is obtained by moderate heating of NH_4NO_3,

$$NH_4NO_3 \xrightarrow{\sim 250°C} N_2O + 2\,H_2O \quad \text{(DANGER: Explosive!)}$$

or by heating an acidic solution of NH_4NO_3 containing some Cl^-. Nitrous oxide is not very reactive. Along with serving as an anesthetic (laughing gas), it is used as an aerosol propellant in whipped cream, because it is soluble in fats.

Nitric oxide is obtained by reducing dilute HNO_3 by Cu.

$$8\,HNO_3 + 3\,Cu \rightarrow 3\,Cu(NO_3)_2 + 4\,H_2O + 2\,NO$$

The Ostwald process (p. 558) for the catalytic oxidation of NH_3 in the production of HNO_3 has been mentioned. NO, an odd-electron molecule, dimerizes in the liquid or solid state to form ONNO; but in the presence of HCl or a Lewis acid, a red solid shown to be the asymmetric dimer, ONON, is formed. Nitric oxide is oxidized easily to NO^+, the nitrosyl ion. Note that in the electron-counting scheme for metal carbonyl nitrosyl compounds (Section 10.2.3), NO is regarded as a three-electron donor.

Dinitrogen trioxide, $ONNO_2$, is blue as a liquid or solid. Obtained from the proper combination of NO and O_2 or NO and N_2O_4, it dissociates above $\sim -30°C$ to form NO + NO_2. The asymmetric $ONNO_2$ is converted into a symmetrical isomer $O{=}N\diagdown{}_O\diagup N{=}O$ by irradiation at 720 nm and is reconverted to the asymmetric isomer by irradiation at 380 nm. Nitrogen dioxide and dinitrogen tetraoxide exist in equilibrium.

$$2\,NO_2 \rightleftharpoons N_2O_4$$

brown   colorless

The equilibrium often is used for lecture demonstrations or laboratory experiments, since it is strongly dependent on temperature and pressure, and only one species is colored. Liquid N_2O_4

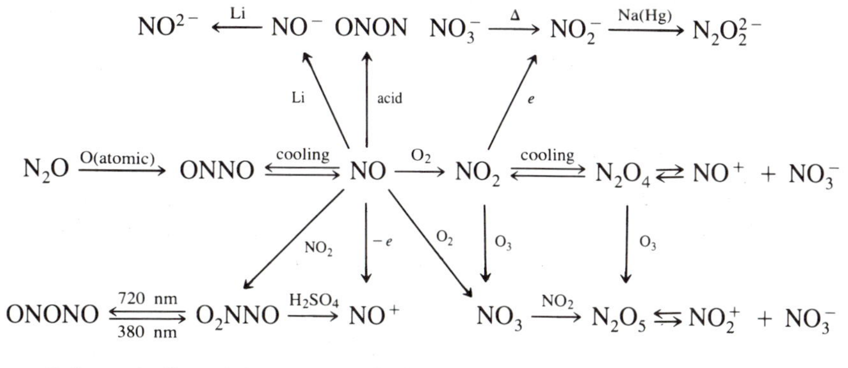

Scheme 1 From J. Laane and J. R. Ohlsen, *Prog. Inorg. Chem.* 1980, 27, 465.

is useful as a nonaqueous solvent[1] in which NO^+ salts are acids and NO_3^- salts are bases. NO_2 is easily oxidized to NO_2^+ (nitryl ion) and reduced to NO_2^- (nitrite ion).

Dinitrogen pentaoxide, the anhydride of HNO_3 can be obtained by dehydrating HNO_3 with P_4O_{10}. As a solid (m.p. 30°C, decomposes ~47°C) that dissolves in H_2O to form HNO_3, it is a true anhydride. Nitrogen trioxide is formed when N_2O_5 decomposes or by its reaction with ozone.

Oxoacids of Nitrogen

Nitric acid, the most important acid of nitrogen, is synthesized from NH_3 (p. 558) on a large scale as one of the common strong acids and for the production of N fertilizers. Concentrated HNO_3 is 15.7 M. It can be dehydrated and vacuum-distilled to give anhydrous HNO_3. Nitric acid is a strong oxidizing agent, commonly being reduced to NO in dilute solution and NO_2 in concentrated solution. The E^0 values are given for comparison but these refer to unit

$$NO_3^- + 4\,H^+ + 3\,e \rightleftarrows NO + 2\,H_2O \qquad E_A^0 = 0.96 \text{ V}$$

$$NO_3^- + 2\,H^+ + e \;\; \rightleftarrows NO_2 + H_2O \qquad E_A^0 = 0.803 \text{ V}$$

activities. NO_2 is the major reduction product only for concentrated HNO_3. A strong reducing agent such as Zn reduces NO_3^- in dilute acidic solution to NH_4^+. Nitrate salts are used frequently, because most metal nitrates are water-soluble and the nitrate ion is a weak ligand for complex formation.

Nitrous acid is unstable except in dilute solution, which can be obtained from H_2SO_4 + $Ba(NO_2)_2$ after removal of the $BaSO_4$ precipitate. Nitrous acid oxidizes Fe^{2+} or I^- and reduces MnO_4^-. Alkali-metal nitrite salts can be prepared by thermal decomposition of the nitrates or by reducing the nitrates with carbon, lead, or iron. Sodium nitrite is used as an additive in cured meats (hot dogs, hams, bacon, and cold cuts), to which the presence of some NO from decomposition of NO_2^- imparts an appealing red color, because of the formation of a bright red compound with hemoglobin. Also, the NO_2^- ion inhibits bacterial growth—particularly, that of *Clostridium botulinum,* which causes botulism. Nitrites are controversial food additives, since they can be converted during cooking to nitrosoamines ($R_2NN{=}O$), compounds considered to be carcinogens. Recently, however, nitrites have been approved as food additives.

$$HNO_2 + H^+ + e \;\; \rightleftarrows NO + H_2O \qquad E^0 = 1.0 \text{ V}$$

$$NO_3^- + 3\,H^+ + 2\,e \rightleftarrows HNO_2 + H_2O \qquad E^0 = 0.94 \text{ V}$$

Hyponitrous Acid

The hyponitrite ion, $N_2O_2^{2-}$, exists as *cis-* and *trans-*isomers. The more stable *trans-*isomer is obtained by reducing NO in organic solvents, or by reducing aqueous $NaNO_2$ with sodium amalgam. The reaction of NO with Na or K in liquid ammonia yields the *cis-*isomer. A solution

---

[1]C. C. Addison, "Dinitrogen Tetraoxide, Nitric Acid and Their Mixtures as Media for Inorganic Reactions," *Chem. Rev.* 1980, *80,* 21.

of the free acid can be obtained from the silver salt of the *trans*-isomer plus HCl. A variety of redox reactions is possible for hyponitrites, but they usually serve as reducing agents.

trans-hyponitrite ion (C_{2h})

Oxides and Oxoacids of Phosphorus

The reaction of P_4 with a limited supply of O_2 yields P_4O_6; with an excess of O_2, P_4O_{10} is formed. Oxygen atoms are inserted into the P—P bonds of P_4 to form P_4O_6, and then terminal oxygen atoms are added to each P to form P_4O_{10}, with the retention of T_d symmetry (Figure 13.3). For the terminal P—O bonds we can write either single-bonded structures, corresponding to the addition of an O atom to the lone pair of P, or double-bonded structures. Figure 13.3 shows the formal charges. Taking into account the polarization effects that displace electron density toward the more electronegative O, the double bonded structure is more favorable for charge distribution. There are also the intermediate oxides P_4O_7, P_4O_8, and P_4O_9, which correspond to the stepwise addition of terminal oxygens to P_4O_6.

Phosphorus(III) oxide is the anhydride of phosphorous acid, H_3PO_3, which features four covalent bonds to P. Only the H attached to O are acidic, so $H_2(HPO_3)$ is a dibasic acid, forming salts of $H_2PO_3^-$ and HPO_3^{2-}. Correspondingly, hypophosphorous acid is a monobasic

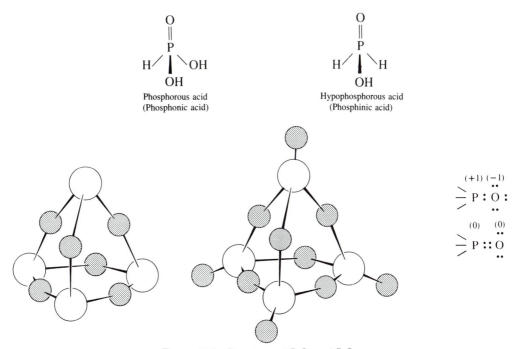

Phosphorous acid
(Phosphonic acid)

Hypophosphorous acid
(Phosphinic acid)

Figure 13.3 Structures of P_4O_6 and P_4O_{10}.

acid. It is obtained by reaction of P_4 with $Ba(OH)_2$ and subsequent treatment with a stoichiometric amount of H_2SO_4.

Phosphorus(V) oxide reacts with water stepwise to produce cyclic metaphosphoric acid, $(HPO_3)_4$, diphosphoric acid ($H_4P_2O_7$, also called pyrophosphoric acid), and finally, orthophosphoric acid (H_3PO_4—the ortho prefix is usually omitted unless the distinction is important). Other polyphosphates are formed by dehydration of H_3PO_4.

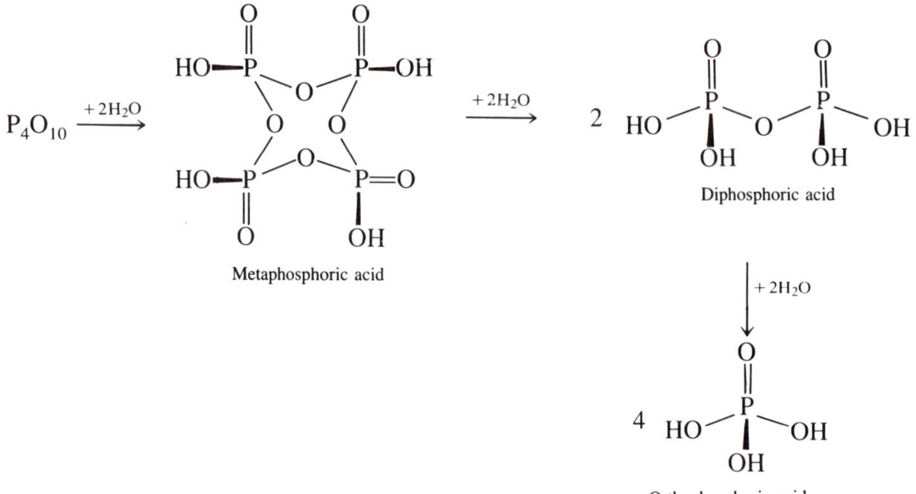

Metaphosphoric acid

Diphosphoric acid

Orthophosphoric acid

P_4O_{10} is used as a desiccant when a more powerful desiccant than $CaCl_2$ or other low-cost agent is needed. Reaction of P_4O_{10} with aqueous hydrogen peroxide gives H_3PO_5, peroxophosphoric acid. Salts of peroxodiphosphoric acid, $H_4P_2O_8$, are obtained by anodic oxidation of phosphate ion. Sometimes called (improperly) *per*phosphoric acids, these acids contain the O—O linkage, as H—O—O—P in H_3PO_5 and as P—O—O—P in peroxodiphosphates.

13.1.6 Redox Chemistry of N and P

As the potential diagrams for N reveal, the stable species are NH_4^+ (NH_3), N_2, and NO_3^-. For P the stable species are PH_3, H_3PO_3 (HPO_3^{2-}),

$$E_A^0 \quad NO_3^- \xrightarrow{0.94} HNO_2 \xrightarrow{1.45} N_2 \xrightarrow{0.27} NH_4^+$$

$$H_3PO_4 \xrightarrow{-0.276} H_3PO_3 \xrightarrow{-0.50} P_4 \xrightarrow{-0.11} PH_3$$

$$E_B^0 \quad NO_3^- \xrightarrow{0.01} NO_2^- \xrightarrow{0.41} N_2 \xrightarrow{-0.74} NH_3(aq)$$

$$PO_4^{3-} \xrightarrow{-1.12} HPO_3^{2-} \xrightarrow{-1.73} P_4 \xrightarrow{-0.89} PH_3$$

and H_3PO_4 (PO_4^{3-}). Elemental P disproportionates. Although many oxides and oxoacids of N are kinetically stable, in aqueous solution the intermediate oxidation states are thermodynamically unstable. Note that HNO_3 is strongly oxidizing, but H_3PO_4 is not. Whereas HNO_2 is an even stronger oxidizing agent than HNO_3, H_3PO_3 is easily oxidized but not easily reduced.

> Families of Pourbaix (E/pH) diagrams (in the form of the periodic table as given by Campbell and Whiteker in *J. Chem. Educ.* 1969, *46*, 90) are useful for comparing the range of stabilities of various species. See Section 11.7 for a discussion of Pourbaix diagrams. In this section the species referred to as stable are those that do not oxidize or reduce water (or H^+ or OH^-) or undergo disproportionation. These reactions limit redox stability in the absence of added oxidizing or reducing agents.

13.1.7 Phosphazenes

The phosphazenes contain the —P≡N— linkage, most commonly as the chlorophosphazenes, $(-PCl_2=N-)_n$. There are cyclic chlorophosphazenes,[2] with $n = 3$ to 8, and chain phosphazenes. The cyclic phosphazenes are formed by the reaction of PCl_5 with NH_4Cl at 120 to 135°C.

$$n\ PCl_5 + n\ NH_4Cl \rightarrow (PCl_2N)_n + 4n\ HCl$$

The oligomers can be separated by vacuum distillation. The Cl can be replaced partially or fully by other groups, such as F or organic groups, R. Hexachlorotriphosphazene (Figure 13.4) has a considerable amount of π delocalization, although there is a sign mismatch for p-d-π bonding going around the ring for the trimer (or for any case where n is odd). Puckering occurs for larger rings.

Polyphosphazenes[3]

Chemists have known since 1897 that hexachlorotriphosphazene could be treated to form a rubbery material now known to be a cross-linked polymer that is hydrolytically unstable. Allcock and coworkers, however, found the means to control the polymerization and substitute alcohol, phenol, and amine groups for Cl^-. By careful purification and control of conditions, they obtained soluble uncrosslinked polymers from $(PNCl_2)_3$. The Cl of the soluble polymers are replaced by organic groups and the polymers are cross-linked.

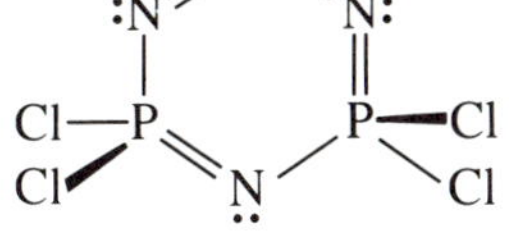

Figure 13.4 Hexachlorotriphosphazene.

[2]S. S. Krishnamurthy and A. C. Sau, *Adv. Inorg. Chem. Radiochem.* 1978, *21*, 41.

[3]H. R. Allcock, "Poly(organophosphazenes)—Unusual New High Polymers," *Angew. Chem. Int. Ed.* 1977, *16*, 147.

The polyphosphazenes include polymers that are flame retardants for other polymers, glasses, elastomers that retain their elasticity at low temperature and are oil-resistant, film-forming polymers, and polymers that can be fabricated into fibers. Some of the polymers are water-repellant and compatible with living tissue for use in biomedical devices. Those derived from amino acid esters are biodegradable into harmless products through hydrolysis.

13.2 OXYGEN, SULFUR, AND SELENIUM

13.2.1 Family Trends for Group VIB

In Table 13.3 are listed some properties of the elements of the oxygen family. Oxygen stands apart in many physical and chemical properties with close similarities among S, Se, and Te. A major difference is that third- (and later) period elements have d orbitals available for p-d-π bonding or for the expansion of the valence shell beyond the octet. There are no oxygen counterparts of SF_4 and SF_6. Thus we have a big change in properties from the first member of each group to the next, but a very small change from the second member to the third. Within each main-group family to the right of the transition series, the pairs Al and Ga, Si and Ge, P and As, S and Se, and Cl and Br display the greatest similarities, because of their equivalent sets of valence shell orbitals and similar radii. The corresponding similarity between Zr and Hf and between Nb and Ta is attributed to the lanthanide contraction (p. 631). Actually, the decrease in atomic radius is greater for the addition of d electrons than for f electrons, so

Table 13.3 Some properties of group VIB elements

	O	*S*	*Se*	*Te*	*Po*
Crystal radius (X^{2-}, pm)	126	170	184	207	—
Covalent radius (pm)	74	113	118.7	141.7	—
Electronegativity	3.4	2.6	2.6	2.1	2.0
First ionization (MJ/mole)	1.3140	0.99960	0.9409	0.8693	0.812
energy (eV)	13.61	10.360	9.752	9.009	8.42
Electron affinity (kJ)	141.1	200.43	194.97	190.16	183
Outer electron configuration	$2s^2 2p^4$	$3s^2 3p^4$	$3d^{10}4s^2 4p^4$	$4d^{10}5s^2 5p^4$	$4f^{14}5d^{10}6s^2 6p^4$

that the atomic radius of Ga (136 pm from the average Ga–Ga distance in the solid), following the first transition series, is *smaller* than that of Al (143 pm). This is the consequence of the "transition series contraction." As for the lanthanide contraction, the effect diminishes for each succeeding group, but the chemical consequences are profound.

The ionization energy and electronegativity of O are much higher than those of other members of the family. The electron affinity is the lowest of the group (see Table 13.3 and Figure 13.1). Nitrogen, oxygen, and fluorine have much lower electron affinities than expected from the group trends, because of their small size and very high electron densities. Removal of electrons from the neutral atoms requires a great deal of energy, but the addition of an electron is less favorable than for less compact atoms (see Section 1.6).

Oxygen shows negative oxidation states except in combination with F. The oxide ion is a hard Lewis base, whereas S^{2-} and Se^{2-} are very soft bases. Metal oxides commonly have typical ionic-type lattice structures, whereas metal sulfides and selenides more likely will have layer-type (p. 213) structures or other structures, such as NiAs (p. 204), encountered only where polarization effects are great.

Family trends are quite regular for S, Se, and Te, but with greater importance of positive oxidation states and an increasing similarity to metals. The last member of the group, Po, is a radioactive metal.

13.2.2 The Elements and Their Occurrence

Oxygen

Oxygen is the most abundant element (46.6 wt %) in the earth's crust, which consists primarily of Si, H, and metal oxides. We obtain the element most readily from the atmosphere (20.9 vol %) by fractional distillation of liquid air. Liquid-air plants are located at major usage sites, particularly in the steel industry. Liquid and solid O_2 are pale blue. Molecular orbital theory gives a description of bonding in O_2 consistent with a bond order of 2 and two unpaired electrons.

The earth is believed to have lost most of its volatile constituents during formation. The original atmosphere was reducing in character (H_2 and H compounds), the present O_2 atmosphere being produced by photosynthesis. Accordingly, elements in igneous rocks (rocks formed originally, see p. 610) usually are in relatively low oxidation states and elements in sedimentary rocks (those subjected to the weathering process) commonly are in higher oxidation states. Dioxygen is a rather strong oxidizing agent, as indicated by the reduction potentials, but the reactions are slow except at elevated temperature. In our remarkable environment, living organisms are dependent on O_2, but the organic molecules comprising the organisms are thermodynamically unstable in an O_2 atmosphere. These molecules eventually wind up as CO_2 and H_2O, and other oxides, but the process is slow except at elevated temperature. In living organisms, cells die, but they are replaced continuously as energy from metabolism brings about the thermodynamically unfavorable reactions. When an organism dies, the life-sustaining functions cease. The usual reduction sequence for O_2 is to O_2^- (superoxide ion) by a one-electron gain and then to O_2^{2-} (or HO_2^-). The active oxidant is usually O_2^{2-} or HO_2^-, except in the combustion process.

$$O_2 + 4\,H^+ + 4\,e \rightleftarrows 2\,H_2O \qquad E_A^0 = +1.229 \text{ V}$$

$$O_2 + 2 H_2O + 4 e \rightleftarrows 4 OH^- \qquad E_B^0 = +0.401 \text{ V}$$

Ozone, O_3, is an allotropic form of oxygen formed by the absorption of ultraviolet light (sunlight) or electrical energy (lightning or a spark source). The ozone layer in the upper atmosphere (15 to 25 km altitude) absorbs most ultraviolet radiation (200 to 360 nm) harmful to life on earth. Great concern has arisen about the depletion of the ozone layer by catalytic decomposition of O_3 to O_2 caused by chlorofluorocarbons used for refrigerants and aerosol propellants (hair sprays, etc.) and by NO and NO_2 produced by supersonic aircraft flying in the ozone layer. The chlorofluorocarbons (for example, $CFCl_3$ and CF_2Cl_2) decompose photochemically to form atomic Cl, which reacts with O_3 to form O_2.

The bond distance in O_2 is 121 pm and in O_3, 127.8 pm. The Lewis structure of O_3 is similar to that of NO_2^- and SO_2. Ozone is a powerful and highly reactive oxidant.

$$O_3 + 2 H^+ + 2 e \rightleftarrows O_2 + H_2O \qquad E^0 = +2.076 \text{ V}$$

Sulfur and Selenium

Sulfur occurs in the crust of the earth as metal sulfides, many of which are important ores for metals. Sulfur is recovered in the smelting process for some of the metal sulfides. Gaseous H_2S and SO_2 are produced in volcanic activity, which is the source of some S deposits. Sulfates, such as anhydrite ($CaSO_4$), occur in sedimentary rocks (see p. 610) as a result of complete oxidation of S. The most important sources of S are the remarkably pure deposits of elemental S in North and South America, Japan, Poland, the U.S.S.R., and Sicily. Many of these underground deposits were formed by the leaching of sulfide minerals, the eventual oxidation to SO_4^{2-}, and deposition of metal sulfates. Bacteria reduced the sulfates to H_2S, which reacted with more SO_4^{2-} to produce elemental S. Underground S is recovered by the Frasch process, in which superheated water melts the S and an admixture of air brings the molten S to the surface. The major use of S is in the production of H_2SO_4.

Sulfur in coal and petroleum is a source of pollution, because of combustion to SO_2; on the other hand, removal of SO_2 from flue gases is an increasingly important source of the element. The effective removal of S at reasonable cost is still a major problem in increasing the use of coal with high S content.

High-performance batteries that are smaller, cheaper, and have higher capacity than the lead storage cell are of interest for vehicles and for energy storage in off-peak hours in electric utility networks. The sodium-sulfur and lithium-sulfur batteries being developed for such applications[4] operate at high temperature and use molten or solid electrolytes (see p. 231).

Selenium and tellurium occur as metal selenides and tellurides, usually accompanying metal sulfides. Se and Te, as the dioxides, concentrate in fly ash from the roasting of sulfide ores.

---

[4]J. R. West, Ed., "New Uses of Sulfur," *Adv. Chem. Ser.* 1975, *140*, 186, 203, 216.

The many allotropic forms of sulfur include S_6, S_8, S_{12}, and S_n rings, where n can be as high as 20.[5] Rapid cooling of molten S yields metastable plastic S, consisting of zigzag chains. The S_8 rings are most important, as encountered in rhombic sulfur, the stable room-temperature form. Monoclinic sulfur crystallizes from the melt (119°C), but converts only slowly to the stable rhombic form below the transition temperature of 95.5°C. The stable metallic form of Se contains helical chains.

13.2.3 Oxides, Peroxides, and Superoxides

Although oxygen reacts with most elements to form oxides, the reactions are usually slow except at high temperature. The active metals and others in low oxidation states give ionic oxides. Nonmetals and metals in high oxidation states give covalent compounds—for example, CO_2, SO_3, CrO_3, Mn_2O_7, and OsO_4. Ionic oxides commonly have high heats of formation, although energy is required to form the metal atom (heat of vaporization) and ion (ionization energy), dissociate O_2, and add two electrons to O. The favorable heats of formation result from the high lattice energies of ionic oxides:

$$\Delta H$$

$$O_2(g) \rightarrow 2\,O \qquad\qquad 493.59 \text{ kJ/mole}$$
$$O(g) + e \rightarrow O^-(g) \qquad -141.1 \;\; \text{kJ/mole}$$
$$O^-(g) + e \rightarrow O^{2-}(g) \qquad 764 \quad\;\; \text{kJ/mole}$$
$$O(g) + 2e \rightarrow O^{2-}(g) \qquad 623 \quad\;\; \text{kJ/mole}$$

The formation of the peroxide ion is more favorable, because it is not necessary to break the O—O bond, and the addition of a second electron is not so unfavorable, because of delocalization of charge over two atoms.

$$\Delta H$$

$$O_2(g) + e \rightarrow O_2^-(g) \qquad -42.5 \text{ kJ/mole}$$
$$O_2^-(g) + e \rightarrow O_2^{2-}(g) \qquad +512 \text{ kJ/mole}$$
$$O_2^{2-}(g) \rightarrow 2\,O^-(g) \qquad +126 \text{ kJ/mole}$$

However, the lattice energies are smaller for peroxides (see Table 6.14). The peroxides are highly reactive, because of the low O—O bond energy. The two electrons added to O_2 to form O_2^{2-} are antibonding.

Potassium superoxide is used in a self-contained breathing apparatus. The moisture from the user's breath reacts with KO_2 to release oxygen at a steady rate, and CO_2 is absorbed to form K_2CO_3.

The superoxide ion, as a radical anion, is highly reactive. The alkali metal hydroxides react with ozone to produce metal ozonides, MO_3. As with superoxides, the ozonides are more stable for larger cations.

[5]B. Meyer, *Adv. Inorg. Chem. Radiochem.* 1976, *18*, 287.

13.2.4 The Dioxygenyl Cation

The discovery that oxygen gas reacts with PtF_6 to form an orange solid, O_2PtF_6, together with recognition of the similar ionization energies of O_2 and Xe, led to the discovery of Xe compounds (p. 601). The O_2^+ ion can be obtained by oxidation of O_2 with F_2, if a large anion[6] is available for stabilization (see Section 6.3.5 for a discussion of the stabilizing effects of large counter ions).

$$O_2 + \tfrac{1}{2} F_2 + GeF_4 \xrightarrow[-78°C]{\text{Photolysis}} O_2^+ GeF_5^-$$

13.2.5 Hydrogen Compounds of Oxygen

The hydrogen compounds of oxygen are H_2O; OH^-; H_2O_2; HO_2^-; the unstable HO_2; and H_2O_3 and H_2O_4 formed in the decomposition of H_2O_2 in a glow discharge. Water, the most familiar of these compounds, is the most common reaction medium. Since it is so familiar, we tend to think of it as a typical solvent. In fact, it is an extraordinary solvent, because of its high dielectric constant, large dipole moment, and great tendency to form hydrogen bonds—both with itself and with other species capable of participating as H donors or acceptors (see Chapter 5).

Hydrogen peroxide is produced by the hydrolysis of the peroxodisulfate ion, which is obtained by electrolytic oxidation (p. 577). The more important methods produce H_2O_2 as one of the products of oxidation of organic compounds such as a substituted 9,10-anthracenediol.

The quinone is reduced catalytically with H_2 to the diol, and H_2O_2 is extracted into water.

Pure H_2O_2 has a greater liquidus range (m.p. $-0.43°C$, b.p. $150.2°C$) and greater density (1.44, 25°C) than H_2O. That it is more acidic ($K_{20°} = 1.5 \times 10^{-12}$) than H_2O as expected, since the negative charge on HO_2^- can be delocalized to a greater extent. H_2O_2 is thermodynamically unstable with respect to disproportionation.

$$2\,H_2O_2 \rightarrow 2\,H_2O + O_2 \qquad \Delta G^0 = -210 \text{ kJ/mole}$$

The decomposition is catalyzed by traces of transition or heavy metal ions, so complexing agents are added to sequester the metal ions. Usually, it is used at concentrations of 30% or less.

The long O—O bond (148 pm) in H_2O_2 indicates a weak single bond. The H–O–O bond angle is 96.8°. If you imagine the O—O bond at the intersection of this page and the one opposite, with one O—H bond drawn in each page, you would have to open the pages to an angle of 93.8°. The barrier to rotation about the O—O bond is small.

[6]K. O. Christe, R. D. Wilson, and I. B. Goldberg, *Inorg. Chem.* 1976, *15*, 1271.

The standard electrode potentials for H_2O_2 indicate that it is a strong oxidizing agent, particularly in acidic solution, and a mild reducing agent. It oxidizes NO_2^- to NO_3^- and Cr(III) to chromate and is reduced by strong oxidizing agents such as MnO_4^-, Cl_2, and Ce(IV) salts.

$$H_2O_2 + 2\,H^+ + 2\,e \rightleftarrows 2\,H_2O \qquad E^0 = 1.77\text{ V}$$

$$O_2 + 2\,H^+ + 2\,e \rightleftarrows H_2O_2 \qquad E^0 = 0.68\text{ V}$$

$$HO_2^- + H_2O + 2\,e \rightleftarrows 3\,OH^- \qquad E_B^0 = 0.87\text{ V}$$

Example What are the reactions expected of H_2O_2 in $1M$ acid with U^{4+}? Write the overall equation and calculate E_{cell}^0.

$$UO_2^{2+} \xrightarrow{0.163} UO_2^+ \xrightarrow{0.380} U^{4+} \xrightarrow{-0.520} U^{3+} \qquad E_A^0$$

Solution First consider whether H_2O_2 can reduce U^{4+} or oxidize U^{4+} to the next higher oxidation state (UO_2^+).

$$U^{4+} + e \rightarrow U^{3+} \qquad E^0 = -0.520\text{ V}$$

$$H_2O_2 \xrightarrow{\rightharpoonup} O_2 + 2\,H^+ + 2\,e \qquad E^0 = -0.68\text{ V as a reducing agent}$$

Since both have negative E^0 values, the overall reaction cannot occur. The potential for oxidation of U^{4+} to UO_2^+ is -0.38 V, and that for reduction of H_2O_2 ($+1.77$ V) is very high, so this reaction will occur. However, note that the next oxidation step is even easier (-0.16 V), so UO_2^+ disproportionates also; and if enough H_2O_2 is present, U^{4+} will be oxidized to UO_2^{2+}. See Section 11.6.1 for calculation of the potential for a third half-reaction (**3**) from (**1**) and (**2**).

1. $U^{4+} + 2\,H_2O$	$\rightarrow UO_2^+ + 4\,H^+ + e$	E^0	$=$	-0.38 V
2. UO_2^+	$\rightarrow UO_2^{2+} + e$	E^0	$=$	-0.16
3. $U^{4+} + 2\,H_2O$	$\rightarrow UO_2^{2+} + 4\,H^+ + 2\,e$	E^0	$=$	-0.27
$H_2O_2 + 2\,H^+ + 2\,e \rightarrow 2\,H_2O$		E^0	$=$	1.77
$H_2O_2 + U^{4+}$	$\rightarrow UO_2^{2+} + 2\,H^+$	E_{cell}^0	$=$	1.50 V

Note that for an intermediate oxidation state such as Mn^{3+}, the reactions for oxidation *and* reduction of H_2O_2 are favorable and should occur, decomposing H_2O_2 to H_2O and O_2 catalytically.

$$MnO_2 \xrightarrow{+0.95} Mn^{3+} \xrightarrow{1.51} Mn^{2+}$$

13.2.6 Oxygen Fluorides

Since O is less electronegative than F, the OF compounds are regarded as fluorides rather than oxides. These include OF_2, and the unstable HOF, O_2F_2, O_4F_2, and the radical O_2F obtained by matrix isolation. Oxygen difluoride is prepared by the reaction of F_2 with dilute (2%) NaOH solution.

$$2\,F_2 + 2\,NaOH \rightarrow OF_2 + H_2O + 2\,NaF$$

The product can be handled in the presence of reducing agents such as H_2 or CH_4, but it reacts

Table 13.4   Bond distances for some oxygen and fluorine compounds

Molecule	O–H distance	O–O distance	O–F distance
H_2O	96 pm		
HOF	96		144 pm
OF_2			142
H_2O_2	96	148 pm	
O_2F_2		122	158
O_2F (dist. est.)		(122)	(158)
O_2		121	
Sum of covalent radii		148	145

violently with the other halogens and oxidizes or fluorinates many elements. It hydrolyzes in basic solution, and more slowly in neutral solution.

$$OF_2 + 2\ OH^- \rightarrow O_2 + 2\ F^- + H_2O$$

The compound HOF is obtained in low yield from the reaction of F_2 and H_2O at low temperature (p. 594). For this unusual oxygen compound, assigning the most electronegative element (F) a $-I$ oxidation number and H $+I$ gives oxygen an oxidation number of zero.

Dioxygen difluoride, O_2F_2, is obtained as a yellow-orange solid from mixtures of O_2 and F_2 in a high-voltage electric discharge at low pressure and temperature. It decomposes into the elements, even at $-50°C$.

Earlier we noted that single bonds are weak between highly electronegative atoms: N—N, O—O, and F—F (see p. 556). The same effect is apparent for O—F bonds. The compound O_2F_2 is interesting because of the unusually short O–O distance and long O–F distances (Table 13.4). The structure resembles that of H_2O_2, but with O—O—F bond angles of 109.5° and an 87.5° angle between the planes containing the O—F bonds. The O–O bond distance corresponds to a bond order of 2 and the O—F bond to a fractional bond order. This suggests that the O—O bond is much like that of O_2, with a singly occupied σ orbital of each F atom overlapping one of the singly occupied π^* orbitals of oxygen to form a weak three-center electron-pair bond. Because of the high electronegativity of F, little electron density is transferred into the π^* orbital to weaken the O—O bond. In the case of H_2O_2, the π^* orbital is occupied fully, decreasing the bond order of the O—O bond to 1. Spratley and Pimentel discuss these cases and the related FNO, HNO, and LiNO.[7]

13.2.7   Chemistry of Sulfur and Selenium Compounds

The chemistry of S is very similar to that of Se, and both differ greatly from oxygen. One unique feature of S is its great tendency toward self-linkage in the free element—for example, S_8 and S_x—in the polysulfides and in the thionates.

---

[7] R. D. Spratley and G. C. Pimentel, *J. Am. Chem. Soc.* 1966, *88*, 2394.

Chalcogen Cations[8]

Sulfur, selenium, and tellurium dissolve in oleum (H_2SO_4/SO_3) to form colored cyclic cations. All three elements form square planar cyclic X_4^{2+} cations, but S also gives S_8^{2+} (blue) and S_{16}^{2+} (red), and Se gives Se_8^{2+} (green). The X_8^{2+} rings are folded (Figure 13.5) with one short transannular distance, suggesting a weak bond to give a bicylic ion. A unique trigonal prismatic cluster cation[9] is found for Te_6^{4+} (Figure 13.5).

Hydrogen Compounds of Sulfur and Selenium

The hydrides H_2X can be obtained from the elements at elevated temperature or by reaction of metal sulfides and selenides with acid. They are highly toxic, and the odor of H_2Se is even worse than that of H_2S. The strengths of the acids H_2X increase in the order $H_2S < H_2Se < H_2Te$. Hydrogen sulfide dissolves in molten sulfur to form polysulfanes, H_2S_n. We also can produce polysulfanes by acidifying solutions of polysulfide salts at low temperature or by reaction of H_2S or H_2S_2 with SCl_2 or S_2Cl_2.

$$2\ H_2S + SCl_2 \rightarrow H_2S_3 + 2\ HCl$$

$$2\ H_2S + S_2Cl_2 \rightarrow H_2S_4 + 2\ HCl$$

The polysulfanes, H_2S_n, form zigzag chains up to $n = 8$.

Interest in the molten Na–S battery as a high-energy battery has led to the thorough study of the Na–S phase diagram.[10] The species identified are Na_2S, Na_2S_2, Na_2S_3, Na_2S_4, and Na_2S_5; in addition, the potassium salts include K_2S_6.

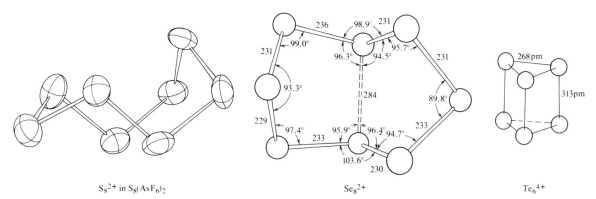

S_8^{2+} in $S_8(AsF_6)_2$ Se_8^{2+} Te_6^{4+}

Figure 13.5 Structures of S_8^{2+}, Se_8^{2+}, and Te_6^{4+} ions. Se_8^{2+} structure is reproduced with permission from R. K. McMullan, D. J. Prince, and J. D. Corbett, *Inorg. Chem.* 1971, *10*, 1749. Copyright 1971, American Chemical Society.

[8]R. J. Gillespie and J. Passmore, *Adv. Inorg. Chem. Radiochem.* 1975, *17*, 49; J. D. Corbett, *Prog. Inorg. Chem.* 1976, *21*, 129.

[9]R. C. Burns, R. J. Gillespie, W.-C. Luk, and D. R. Slim, *Inorg. Chem.* 1979, *18*, 3086.

[10]D.-G Oei, *Inorg. Chem.* 1973, *12*, 435, 438.

Sulfur-Nitrogen Compounds

Disulfur dichloride reacts with NH_3 in CCl_4 to form S_4N_4 (m.p. 185°C). The crystals are thermochromic, changing from almost colorless at liquid N_2 temperature to yellow at room temperature and red above 100°. Although the compound is the starting material for many S–N compounds, it explodes on grinding, by percussion, or by rapid heating. The S_4N_4 molecule consists of a cage (p. 714) with an average S–N bond length of 161.6 pm and a long S–S distance (258 pm) that indicates weak bonding (Figure 13.6). Treatment of S_4N_4 with an Ag_2F in CCl_4 gives $(NSF)_4$, breaking the weak S—S bond and forming an $(S—N)_4$ ring. Tetrasulfur tetranitride can be vaporized at about 80°C at low pressure and passed through silver wool to form S_2N_2. Disulfur dinitride, which is very nearly square-planar, can be obtained as a crystalline product that explodes on heating.

Disulfur dinitride vapor or single crystals can be polymerized to form crystals of $(SN)_x$, a shiny metallic solid. Polymeric sulfur nitride (polythiazyl)[11] crystallizes as oriented fibers, each consisting of parallel zigzag $(SN)_x$ chains. Crystals of $(SN)_x$ are metallic conductors along the long axis of the crystal—a one-dimensional ''metal'' containing only S and N. There is great interest in $(SN)_x$ not only because of its one-dimensional conductivity, but also because it becomes superconducting at 0.26 K.

Crystals of high purity $(SN)_x$ are stable indefinitely *in vacuo* at room temperature. The crystals are relatively inert toward O_2 and H_2O, but heating in air causes explosions at *ca.* 240°C. In a vacuum, $(SN)_x$ sublimes at *ca.* 135°C to form a red-purple linear tetramer, $(SN)_4$.

Sulfur and Selenium Halides

Sulfur forms S_2X_2 with all of the halogens: SF_2 (unstable), S_2F_4 (stable to $-75°C$), SCl_2, SF_4, SCl_4 (dissociates above $-31°C$), SF_6, and S_2F_{10}. Selenium forms SeF_4, SeF_6, Se_2Cl_2, $SeCl_4$ (dissociates in the vapor), Se_2Br_2 and $SeBr_2$ (both dissociate in the vapor), and $SeBr_4$. The important fluorides are SF_4 and SF_6. Sulfur dichloride is used as a solvent for S to form dichlorosulfanes, S_nCl_2, in the vulcanization of rubber. Chlorination of molten S gives S_2Cl_2, but in the presence of $FeCl_3$ or I_2 as a catalyst, SCl_2 is formed.

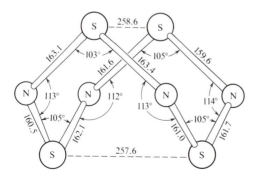

Figure 13.6 Structure of S_4N_4. (From B. D. Sharma and J. Donohue, *Acta Cryst.* 1963, *16*, 891.)

[11]M. M. Labes, P. Love, and L. F. Nichols, *Chem. Rev.* 1979, *79*, 1; A. G. MacDiarmid *et al.*, *Adv. Chem. Ser.* 1976, *150*, 63.

The reaction of SCl_2 and NaF in acetonitrile yields sulfur tetrafluoride,

$$3 \, SCl_2 \, + \, 4 \, NaF \xrightarrow[75°C]{CH_3CN} SF_4 \, + \, S_2Cl_2 \, + \, 4 \, NaCl$$

which is hydrolyzed readily to SO_2 and HF. It serves as a selective fluorinating agent, converting $\diagdown\overset{\diagup}{C}{=}O$ into $\diagdown\overset{\diagup}{C}F_2$, $-CO_2H$ into $-CF_3$, $\overset{\diagup}{-}P{=}O$ into $\overset{\diagup}{-}PF_2$, and $\diagdown\overset{\diagup}{P}{-}OH$ into $\diagup$

$\diagdown\overset{\diagup}{P}F_3$. The structure of the SF_4 molecule, as expected from the VSEPR treatment, is derived from a trigonal bipyramid with a lone pair in an equatorial position.

Direct fluorination of S or SO_2 yields SF_6, a colorless, odorless, nontoxic gas (b.p. $-64°C$) that is quite inert. The hydrolysis reaction is thermodynamically favorable, so the stability toward hydrolysis must arise from kinetic factors—the effective screening of S from nucleophilic attack.

The S–F bond length (156 pm) in the octahedral SF_6 is shorter than expected for an S—F single bond. Using a valence bond description, we would assume sp^3d^2 hybridization, with the possibility of some degree of S—F p-d-π bonding to account for the short bond length. An equivalent MO description can be given also. There has been controversy[12] for some time over the participation of outer d orbitals in bonding, because of their high energy and diffuse character. The latter objection is met, in part, by the contraction of the d orbitals about a positive center, such as S bonded to very electronegative atoms. Kiang and Zare[13] determined the stepwise bond-dissociation energies of SF_6 and SF_4, finding that they are relatively high ($\sim$360 kJ/mole) for SF_6, SF_4, and SF_2 and considerably lower ($\sim$240 kJ/mole) for SF_5 and SF_3. The alternation was already known and had been explained in terms of a valence bond interpretation. Zare prefers the MO description using p orbitals on S for three-center bonding, as described for XeF_2 (Figure 4.24).

Sulfur difluoride is isoelectronic with OF_2, so bonding is localized with only two-center electron-pair bonds. An F is added to form $F^-(SF_2)^+$ such that the F^- (on a line perpendicular to SF_2^+) is bonded by a two-center, three-electron bond: that is, the third electron is antibonding in the σ^* orbital. Another F is added axially to $(SF_2)^+F^-$ to form a three-center bond in SF_4 (Figure 13.7). The bonding three-center MO is occupied by an electron pair, and the second electron pair occupies the nonbonding orbital localized on the axial F atoms. Similarly, a fifth F is added to SF_4 to form a weak two-center, three-electron bond. SF_6 has three equivalent three-center, four-electron bonds. This description explains the low bond-dissociation energies of SF_3 and SF_5. Zare describes three-center, four-electron bonds as strong—but for SF_4, weaker than the two-center equatorial S—F bonds. Although the bond order is 1/2 for each S—F bond of the three-center bond, there is an ionic contribution for SF_4 and SF_6.[14] Presumably, this is sufficiently important to enhance the strength of the S—F bond, with a covalent bond order of 1/2, to account for a bond length somewhat *shorter* than that expected for a single bond.

[12]K. A. R. Mitchell, "The Use of Outer d Orbitals in Bonding," *Chem. Rev.* 1969, *69*, 157.

[13]T. Kiang and R. N. Zare, *J. Am. Chem. Soc.* 1980, *102*, 4024.

[14]R. Maclagan, "Symmetry, Ionic Structures and d Orbitals in SF_6", *J. Chem. Educ.* 1980, *57*, 428.

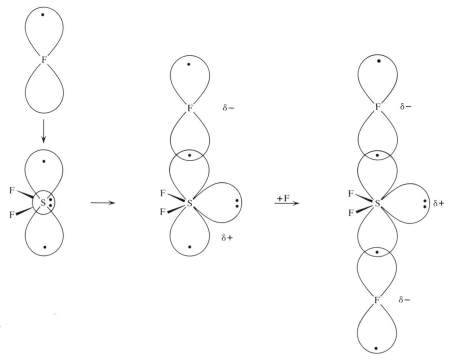

Figure 13.7   S—F bonding using only *p* orbitals.

Sulfur and Selenium Oxides

Unstable SO and (SO)$_2$ are known, but the important oxides are SO$_2$, SO$_3$, and SeO$_2$. Selenium trioxide is obtained as the anhydride of H$_2$SeO$_4$. The burning of S and the roasting of many sulfides in air produce SO$_2$ (b.p. $-10.1°$C). Many deposits of coal and oil contain enough S that their use as fuels produces objectionable amounts of SO$_2$. Selenium dioxide, a white volatile solid, can be obtained by burning Se in air or by the reaction of Se with HNO$_3$, followed by dehydration of the H$_2$SeO$_3$ formed.

Sulfur dioxide is oxidized slowly in air, but the reaction is catalyzed by V$_2$O$_5$, Pt sponge, or NO. In the gas phase, SO$_3$ is planar (**D**$_{3h}$) with a bond order of about 1 1/3. In one of the forms of solid SO$_3$, γ-SO$_3$, there are cyclic (SO$_3$)$_3$ molecules, involving three oxygen-bridged SO$_4$ tetrahedra. A more stable asbestoslike form consists of helical chains of SO$_4$ tetrahedra. Other forms also have been reported.

Oxoacids of Sulfur and Selenium

Sulfur forms an extensive series of oxoacids. Sulfur dioxide dissolves in water to produce solutions containing sulfurous acid. Although the free H_2SO_3 is not known, salts of HSO_3^- and SO_3^{2-} are stable. Sulfites are oxidized easily to sulfates.

$$SO_4^{2-} + 4\,H^+ + 2\,e \rightleftharpoons H_2SO_3 \text{ (or } SO_2 \cdot xH_2O) + H_2O \qquad E^0 = 0.17\text{ V}$$

Adding an excess of SO_2 to $NaHSO_3$ solution gives the disulfite ion, which has an S—S bond and the unsymmetrical structure shown in Figure 13.8 for the acid.

Sulfur trioxide dissolves in water to produce sulfuric acid; when dissolved in sulfuric acid, it produces disulfuric acid, $H_2S_2O_7$, or higher polysulfuric acids, $H_2S_nO_{3n+1}$, involving O-bridged SO_4 tetrahedra. Replacing one oxygen of sulfuric acid by S gives thiosulfuric acid, known in the form of its salts. In addition, there are polyacids involving S—S linkages with neither S in a terminal position, including dithionous acid, $H_2S_2O_4$; disulfurous acid, $H_2S_2O_5$; dithionic acid, $H_2S_2O_6$; and polythionic acids, $H_2S_{n+2}O_6$ (Figure 13.8).

Anodic oxidation of solutions of HSO_4^- salts produces $S_2O_8^{2-}$, peroxodisulfate ion. The electrolysis is carried out at low temperature and high current density to diminish the extent of oxidation of H_2O to O_2. Hydrolysis of peroxodisulfate ion is one method for preparing H_2O_2. Peroxodisulfate salts are useful strong oxidizing agents, oxidizing Mn^{2+}

$$S_2O_8^{2-} + 2\,e \rightleftharpoons 2\,SO_4^{2-} \qquad E^0 = 2.10\text{ V}$$

to MnO_4^-. Oxidations are often slow unless a catalyst, such as Ag^+, is added. Presumably Ag^{II} is the active oxidant in this case. Hydrolysis of $H_2S_2O_8$ produces

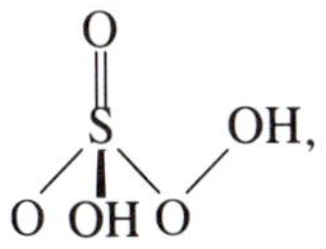

peroxomonosulfuric acid or Caro's acid, which is obtained also by the reaction of H_2O_2 with H_2SO_4 or with HSO_3Cl (chlorosulfuric acid).

Selenium and tellurium do not form such extensive series of oxoacids; those containing Se—Se or Te—Te bonds are unknown. Selenium dioxide dissolves in H_2O to form selenous

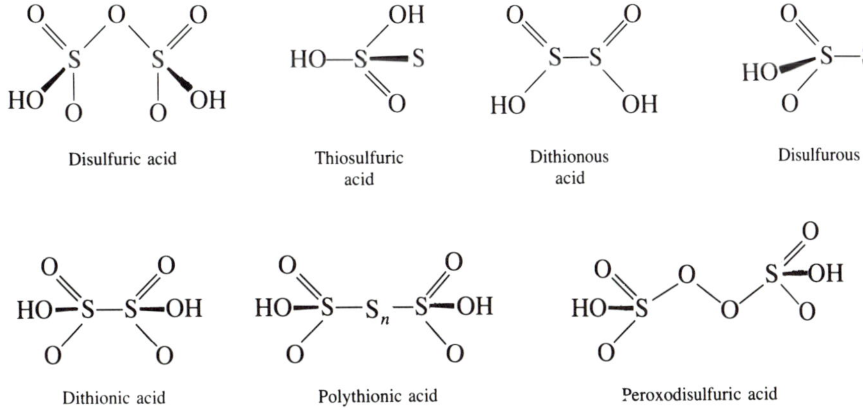

Disulfuric acid   Thiosulfuric acid   Dithionous acid   Disulfurous acid

Dithionic acid   Polythionic acid   Peroxodisulfuric acid

Figure 13.8 Oxoacids of sulfur

acid, H_2SeO_3, which can be obtained as colorless crystals by evaporation of the solution. It decomposes into SeO_2 and water on warming. Selenous acid is reduced easily to Se by mild reducing agents such as HI, SO_2, and N_2H_4. Concentrated solutions of $HSeO_3^-$ contain the $Se_2O_5^{2-}$ ion, with a symmetrical structure, unlike $S_2O_5^{2-}$ (Figure 13.8), with Se—O—Se bonding. Very strong oxidizing agents (O_3, MnO_4^-, and H_2O_2) oxidize H_2SeO_3 to H_2SeO_4. Crystalline selenic acid (m.p. 58°C) can be obtained from aqueous solution by evaporation at low pressure. Selenic acid is a strong acid and a rather strong oxidizing agent.

$$SeO_4^{2-} + 4\,H^+ + 2\,e \rightleftarrows H_2SeO_3 + H_2O \qquad E^0 = 1.15\ \text{V}$$

Sulfuric acid, the chemical produced in largest quantity (over 40 million tons annually), is obtained by the catalytic oxidation of SO_2 to SO_3 (contact process). Sulfur trioxide is water-soluble but tends to produce aerosols, so it is dissolved in concentrated H_2SO_4 to produce fuming sulfuric acid, oleum, which is diluted to produce H_2SO_4. In the manufacture of fertilizers and other chemicals, sulfuric acid is used whenever a strong acid is needed, except when its oxidizing character is a limitation. H_2SO_4 is a good drying agent—particularly for non-reactive gases, which can be bubbled through the liquid.

Thiosulfate salts are obtained by boiling aqueous solutions of sulfites with either H_2S or S.

$$8\,Na_2SO_3 + S_8 \rightarrow 8\,Na_2S_2O_3$$

Sodium thiosulfate is used as a "fixer" in developing photographic film or prints, since it dissolves the AgBr not reduced by the developer.

$$AgBr + 2\,S_2O_3^{2-} \rightarrow [Ag(S_2O_3)_2]^{3-} + Br^-$$

Thiosulfates are useful as analytical reducing agents, particularly for the titration of I_2. The oxidation product is the tetrathionate ion.

$$2\,S_2O_3^{2-} + I_2 \rightarrow S_4O_6^{2-} + 2\,I^-$$

Dithionite ($S_2O_4^{2-}$) salts are obtained by using Zn to reduce solutions of sulfites containing an excess of SO_2; they are oxidized easily by O_2. Oxidation of sulfites by a mechanism not involving oxygen transfer produces dithionates.

$$MnO_2 + 2\,SO_3^{2-} + 4\,H^+ \rightleftarrows Mn^{2+} + S_2O_6^{2-} + 2\,H_2O$$

Dithionate salts are stable with respect to reactions with water and with most oxidizing and reducing agents (kinetic, not thermodynamic, stability).

The polythionic acids, $H_2S_{n+2}O_6$, are unstable with respect to decomposition to form S and a variety of oxides and oxoacids of S. Polythionate salts are well characterized for $n = 1$ to 4, with much higher values of n reported. Polythionates are related to the dithionates only in the formal sense that the formula for dithionate ion, $S_2O_6^{2-}$, corresponds to $S_{n+2}O_6^{2-}$ with $n = 0$. Preparation of tetrathionate ion from $S_2O_3^{2-}$ and I_2 was described above. Oxidation of thiosulfate salts by H_2O_2 produces trithionate salts.

$$2\,S_2O_3^{2-} + 4\,H_2O_2 \rightarrow S_3O_6^{2-} + SO_4^{2-} + 4\,H_2O$$

The reactions of $S_2O_3^{2-}$(aq) with H_2S and SO_2 produces polythionates of varying chain length.

The rather complex chemistry of S provides many examples of the importance of reaction mechanisms in determining the stability of a compound (for example, SF_6 and $S_2O_6^{2-}$) and the reaction products (for example, $S_2O_3^{2-}$ plus an oxidizing agent such as I_2 or H_2O_2).

13.2.8 Redox Chemistry of Sulfur and Selenium

Figures 13.9 and 13.10 show the Pourbaix (E/pH) diagrams for S and Se, respectively.

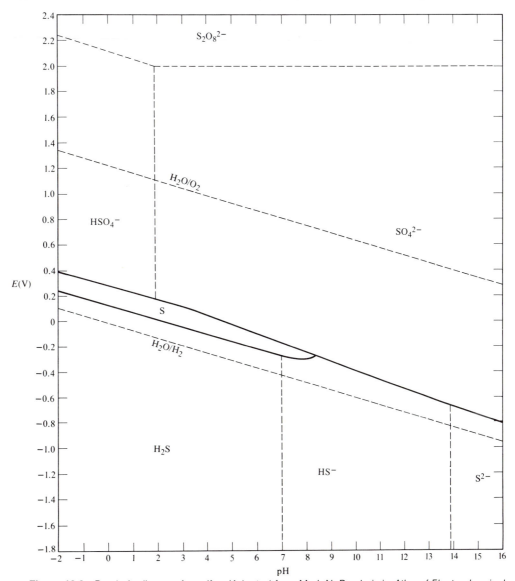

Figure 13.9 Pourbaix diagram for sulfur. (Adapted from M. J. N. Pourbaix in *Atlas of Electrochemical Equilibria in Aqueous Solution* (English Translation), Pergamon, New York, 1966.)

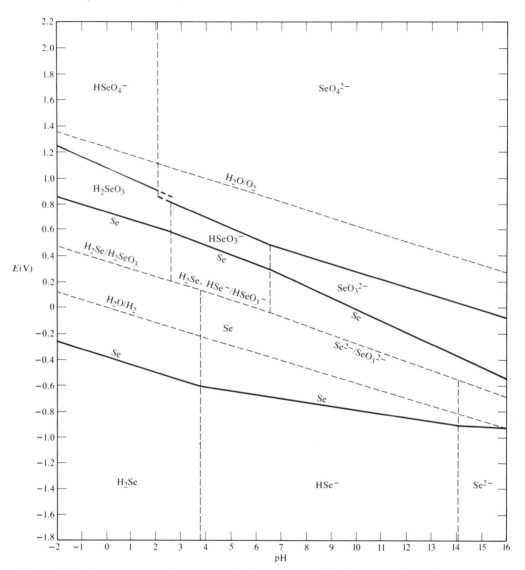

Figure 13.10 Pourbaix diagram for selenium. (Adapted from M. J. N. Pourbaix in *Atlas of Electrochemical Equilibria in Aqueous Solution* (English Translation), Pergamon, New York, 1966.)

$$E_A^0 \quad SO_4^{2-} \xrightarrow{0.17} H_2SO_3 \xrightarrow{0.45} S \xrightarrow{0.14} H_2S$$

$$SeO_4^{2-} \xrightarrow{1.15} H_2SeO_3 \xrightarrow{0.74} Se \xrightarrow{-0.40} H_2Se$$

$$E_B^0 \quad SO_4^{2-} \xrightarrow{-0.91} SO_3^{2-} \xrightarrow{-0.61} S \xrightarrow{-0.48} S^{2-}$$

$$SeO_4^{2-} \xrightarrow{0.05} SeO_3^{2-} \xrightarrow{-0.37} Se \xrightarrow{-0.92} Se^{2-}$$

For the discussion of Pourbaix diagrams, see Section 11.7. In these diagrams a horizontal or sloping line represents the potential for two species in equilibrium as a function of pH. Solid lines separate thermodynamically stable species. The dotted lines are for the oxidation of H_2O (to O_2) and the reduction of H_2O (to H_2). The vertical lines separate the predominant species involved in acid–base equilibria. Figure 13.9 demonstrates that under reducing conditions, H_2S, HS^-, and S^{2-} are the stable species. H_2 is a strong enough reducing agent (thermodynamically) to reduce S or any other S compound. Under oxidizing conditions, H_2SO_4 and its salts are stable. $S_2O_8^{2-}$ is produced only at such high potentials that H_2O is oxidized. The element has a narrow range of stability, and H_2SO_3 does not appear because it disproportionates (see the potential diagram above).

The diagram for Se (Figure 13.10) is more complex. H_2SeO_3 and its salts are stable, and elemental Se has a wide range of stability. A dotted line represents the reduction of H_2SeO_3 to H_2Se, because this is within the region of stability for Se and H_2SeO_3 will oxidize H_2Se to Se (see Problem 13.21). H_2Se is a strong enough reducing agent to reduce water, but the reaction should be slow, because of the hydrogen overvoltage.

13.3 THE HALOGENS

13.3.1 Family Trends

Table 13.5 lists some properties of the halogens.

Table 13.5 Some properties of the halogens

	F	*Cl*	*Br*	*I*	*At*
Crystal radius (X^-, pm)	119	167	182	206	
Covalent radius (pm)	70.9	99.4	114.1	133.3	
Electronegativity	4.0	3.0	2.8	2.5	2.2
First ionization (MJ/mole)	1.6810	1.2511	1.1399	1.0084	0.92
energy (eV)	17.422	12.967	11.814	10.451	9.5
Outer electron configuration	$2s^2 2p^5$	$3s^2 3p^5$	$3d^{10} 4s^2 4p^5$	$4d^{10} 5s^2 5p^5$	$4f^{14} 5d^{10} 6s^2 6p^5$
$E^0 \frac{1}{2}X_2 + e \rightarrow X^-$ (V)	+2.87	+1.36	+1.07	+0.535	+0.3
$E_B^0\ XO^- + e \rightarrow \frac{1}{2}X_2$ (V)	—	+0.40	+0.45	+0.45	0.0
Dissociation energy (kJ/mole)	154.6	239.23	190.15	148.82	115.9
Electron affinity (kJ)	328.0	348.8	324.6	295.4	270
Polarizability (cm^3/atom)	1.04×10^{-24}	3.66×10^{-24}	4.77×10^{-24}	7.10×10^{-24}	

Note that the difference between F and Cl generally is greater than the differences between the succeeding heavier halogens. Of the halogens, chlorine and bromine are the most similar in chemical properties. That the crystal radius of the Br^- ion is only slightly greater than that of the Cl^- ion may be attributed to the incomplete nuclear shielding that occurs as the $3d$ electrons are added (see also p. 32). The small difference in electronegativity between Br and Cl indicates there should be little difference in the polarity of the bonds formed by these halogens with any particular element. The large differences between the effect of F and that of Cl, and the similarity between that of Cl and Br, are illustrated in the pK_a values of HX (Table 13.7), the oxidizing ability of X_2 (Table 13.5), the solubilities of the silver halides (p. 236) and the alkali metal halides (p. 234), the oxidation states of known binary compounds (p. 637), the structures of compounds, and the hydrogen bond strength of XHX^- (Table 5.3).

The high oxidizing ability of fluorine, together with its small size, permits it to form many compounds in which the elements exhibit their highest oxidation state. Thus there are 16 known binary hexafluorides, but no known hexaiodides. On the other hand, the lack of low-lying empty *d* orbitals prevents F from forming more than one normal covalent bond.

The crystal structures of the halides depend not only on the radius ratio of cation to anion, but also on the electronegativity difference of the metal and the halogen and on the polarizabilities of each. For compounds of the types MX, MX_2, or MX_3, the fluorides have structures anticipated from the radius ratio, since F is sufficiently electronegative to form essentially ionic bonds with metals in low oxidation states, and the F^- ion is not very polarizable. The chlorides, bromides, and iodides of MX_2- and MX_3-type compounds show the effect of lower electronegativities and greater polarizabilities by the formation of layer structures (p. 213). Covalent bond formation or strong polarization of the halide ion is necessary to account for these layer structures.

We can also relate the colors of many of the halides and halide complexes to the polarizability of the anions. Thus, the colors of the TiX_6^{2-} ions vary from colorless for the fluoro complex to yellow for the chloro complex, red for the bromo complex, and deep red for the iodo complex. These spectra result from the excitation of an electron essentially in a halogen orbital to an orbital belonging to the metal. The energy of this transition is least for the iodo complex (see p. 288).

13.3.2 Occurrence

The halogens occur in combined form in the earth's crust in the approximate amounts F, 0.08%; Cl, 0.05%; Br, 0.001%; and I, 0.001%. Chlorides and bromides are concentrated in the oceans by the leaching processes of natural waters (see Table 16.1, p. 721). Large deposits of NaCl and, to a lesser extent, calcium, potassium, and magnesium chlorides are found in the dried-up beds of landlocked lakes, where calcium and magnesium sulfates and mixed salts of KBr and $MgBr_2$ also occur. Because of differences in concentrations and temperature during deposition, the different salts often are well separated. Limiting the fluoride content of sea water is the large concentration of calcium ion present and the very low solubility of CaF_2. CaF_2's low solubility compared with the other calcium halides stems from the high lattice energy of the fluoride, which in turn results from the small radius of F^-. Natural deposits of calcium fluoride, called fluorspar or fluorite, serve as the primary souce of fluorine. Cryolyte, Na_3AlF_6, another commercially important fluoride mineral, is used as the electrolyte in the production of Al.

Although the concentration of iodine in the ocean is small, it is absorbed selectively by and can be obtained from seaweed. A more important source is sodium iodate and sodium periodate, which occur in deposits in Chile. The natural occurrence of these oxo anions for iodine contrasts sharply with the lack of natural deposits of oxo salts of Cl and Br and illustrates the greater ease with which I attains the higher oxidation states, compared with the lighter halogens.

Astatine occurs naturally in very small amounts as the beta decay product of ^{215}Po, ^{216}Po, and ^{218}Po. As polonium itself is rather rare, and α decay is the usual decay route, At has not been isolated from natural sources; it has been prepared by α bombardment of bismuth, $^{209}Bi(\alpha,2n)^{211}At$. Astatine itself is radioactive: its longest-lived isotope has a half-life of 8.3 hr.

13.3.3 Preparation of the Elements

Since F, Cl, and Br occur in nature as halides, preparing the elements involves suitable oxidation reactions. Electrolytic oxidation of fused KF-HF adducts (p. 183) is used to produce F_2. HF is removed from the F_2 by reaction with NaF. The cells are constructed of copper, monel, or steel with amorphous carbon anodes. Graphite electrodes cannot be used because of the graphite compounds formed. Chlorine is produced by electrolysis of either fused NaCl or concentrated aqueous solutions of NaCl, the latter being the more commonly used method.

$$Na^+ + Cl^- + H_2O \xrightarrow{electr.} Na^+ + OH^- + \tfrac{1}{2} Cl_2 + \tfrac{1}{2} H_2$$

In the laboratory, Cl_2 can be made by the action of oxidizing agents such as MnO_2 on HCl.

Bromine is produced commercially from bromides in sea water by the oxidizing action of chlorine. After the pH of the sea water is adjusted to a value between 1 and 4 with sulfuric acid and then treated with chlorine, the liberated Br_2 is blown out by an air stream and the bromine concentrated by a sequence of reactions, such as absorption in a carbonate solution and subsequent acidification.

$$Cl_2 + 2Br^- \rightarrow Br_2 + 2Cl^-$$

$$3Br_2 + 3CO_3^{2-} \rightarrow 5Br^- + BrO_3^- + 3CO_2$$

$$5Br^- + BrO_3^- + 6H^+ \rightarrow 3Br_2 + 3H_2O$$

Cl_2 is removed from the Br_2 produced by reaction with a bromide such as iron(III) bromide. Sulfite reduction of iodates yields I_2 which is purified by sublimation.

$$2NaIO_3 + 5NaHSO_3 \xrightarrow{aq.\ soln.} 3NaHSO_4 + 2Na_2SO_4 + H_2O + I_2$$

Table 13.6 lists some physical properties of the halogens.

Table 13.6 Properties of the halogens

	Fluorine	*Chlorine*	*Bromine*	*Iodine*
Isotopes	19	35,37	79,81	127
Melting point (°C)	−223	−102.4	−7.3	+113.7
Boiling point (°C)	−187.9	−34.0	+58.8	+184.5
Density (g/cm³)	1.108[a]	1.57[a]	3.14	4.94
Color of gas	light yellow	greenish yellow	reddish brown	violet

[a]For the liquid at the boiling point.

13.3.4   The Hydrogen Halides

The hydrogen halides can be prepared by the action of a nonvolatile, nonoxidizing acid on a halide salt. Sulfuric acid serves well for the preparation of HF and HCl but is too strong an oxidizing agent to be used with bromides and iodides.

$$CaF_2 + H_2SO_4 \rightarrow CaSO_4 + 2HF$$

$$NaCl + H_2SO_4 \rightarrow NaHSO_4 + HCl$$

Phosphoric acid serves for the similar preparation of HBr and HI. Commercially, the direct reaction of Cl_2 with H_2 is most important. Although all of the hydrogen halides may be obtained commercially, they usually contain appreciable amounts of impurities. HF, HCl, and HI usually contain H_2 from reaction with the metal cylinder. Hydrogen iodide usually contains I_2 in addition to H_2, from thermal decomposition. For laboratory use the best source of HF is the thermal decomposition of dry KHF_2. Hydrogen iodide is usually prepared by dropping the aqueous solution on phosphorus pentaoxide.

Table 13.7 gives some of the properties of the hydrogen halides and their aqueous solutions.

The following topics relating to the hydrogen halides have already been discussed: the effect of hydrogen bonding on the boiling point of HF (p. 174), the ability to form hydrogen bonds with halide ions (p. 183), and the acid strength of the hydrohalic acids (p. 515). Further specific discussion will be limited to HF as a solvent system.

Hydrogen fluoride is highly toxic and should be handled only in a good fume hood. Since it reacts with glass, it is handled in metal systems, especially copper or monel. It is characterized by its dehydrating ability and its strong acidity. Because of its high dielectric constant and low viscosity, most of its solutions show high conductivity although the pure liquid itself has a conductivity comparable with that of distilled water. Many of the reaction products formed upon dissolving substances in HF have been inferred from conductivities. Some typical reactions are

$$HNO_3 + HF \rightleftharpoons H_2NO_3^+ + F^-$$

$$H_2SO_4 + HF \rightarrow HSO_3F + H_3O^+ + F^-$$

Table 13.7   Properties of the hydrogen halides

	HF	HCl	HBr	HI
Melting point (°C)	−83.07	−114.19	−86.86	−50.79
Boiling point (°C)	19.9	−85.03	−66.72	−35.35
Density at b.p. (g/cm³)	0.991	1.187	2.160	2.799
Dielectric constant of liquid	66	9	6	3
Percent dissociation at 1000°C	—	3×10^{-7}	0.003	19
Composition of azeotrope with water at 1 atm (wt-%)	35.37	20.24	47	57.0
pK_a values (aqueous)	3.17	−7	−9	−10

$$NaCl + HF \rightarrow NaF + HCl$$

$$ZnO + HF \rightarrow ZnF_2 + H_3O^+ + F^-$$

Anhydrous HF compares in strength with anhydrous H_2SO_4. The only known acids in anhydrous HF are fluoride acceptors such as NbF_5, SbF_5, AsF_5, and BF_3, the strongest of which is SbF_5. Adding metal fluorides to such solutions precipitates the metal hexafluoroantimonate(V).

13.3.5 Reactions of the Halogens

When halogens react with other elements, the products often depend on the temperature and pressure at which the reaction is carried out. We can grasp the effect of these variables most readily by means of idealized phase diagrams.

Figure 13.11 shows generalized pressure-composition isotherms expected for a halogen-metal system, in which the MX_y compounds are assumed to be nonvolatile and no solid solutions are formed.

The vertical sections of the isotherms indicate the existence of a single solid compound (with the stoichiometry indicated on the abscissa) in equilibrium with gaseous X_2. The horizontal sections or plateaus indicate the coexistence of two solid phases in equilibrium with gaseous X_2. As the pressure of X_2 is increased, for a given equilibrium temperature a higher halide will be produced. As the equilibrium temperature is increased, for a given pressure of X_2 a lower halide will be produced. From the phase diagram in Figure 13.11, heating X_2 and M together at a pressure of P_2 produces MX_2 at the higher temperature T_2 and MX_4 at the lower temperature T_1. MX_2 may also be formed at the lower temperature T_1 by reducing the pressure of X_2 to P_1. Since the metal can coexist in equilibrium with only the lowest halide, heating the metal with an excess of a particular metal halide with no additional halogen will produce some of the next-lower halides at equilibrium. If equilibrium is attained rapidly, it may be necessary to quench the products to keep other phases from forming.

The temperature range within which the various platinum chlorides may exist at 1 atmosphere pressure of chlorine is as follows: $PtCl_4$, below 370°; $PtCl_3$, 370-435°; $PtCl_2$, 435–481°; PtCl, 481–482°; Pt, above 482°.

Although the discussion here has used the systems $M–X_2$ as an example, similar considerations apply to other gas-solid systems in which the products are nonvolatile: $O_2(g)–M(s)$,

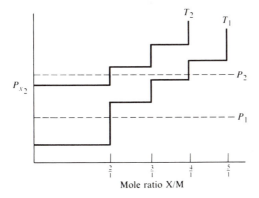

Figure 13.11 Generalized pressure-composition isotherms for the system $M(s)—X_2(g)$.

$NH_3(g)$–$MX_y(s)$, $H_2O(g)$–$MO_y(s)$, etc. When manganese oxides are ignited in a Bunsen flame in air, Mn_3O_4 is produced. If the partial pressure of oxygen is increased from 0.2 to 1 atm by using pure oxygen, Mn_2O_3 is formed; at still higher pressures, MnO_2 is formed.

We can make some predictions as to whether a compound might be stable with respect to disproportionation from known and/or estimated heats of formation of the halides.

The heat of the reaction

$$2MX \rightarrow M + MX_2$$

may be found from

$$\Delta H = \Delta H_f(MX_2) - 2\Delta H_f(MX)$$

If ΔH is negative, then ΔG probably will also be negative and MX will be unstable with respect to disproportionation. More generally, if we plot ΔH_f against the number of halogen atoms in MX_y, as in Figure 13.12 we find that only the highest halide is stable if the curve has an increasing rate of ascent (*A*); that all halides are stable if the curve has a decreasing rate of ascent but no maximum (*B*); and that if a maximum does occur, only those halides at and preceding the maximum are stable (*C*). These situations are often approximated by the transition metal fluorides (*A*), the chlorides (*B*), and the bromides and iodides (*C*). (See Table 14.10.)

Preparation of the Binary Metal Halides

The binary metal halides of the transition elements are listed in Table 14.10. The metal halides formed in the solid state for the Group IA, IIA, and IIIA elements are those expected: that is, MX, MX_2, and MX_3, respectively.

The preparation of MX or MX_2 usually may be carried out by treating an oxide or carbonate of M with aqueous HX, evaporating the solution to obtain the metal halide hydrate, and then removing the water by means of a suitable desiccant—or if necessary to prevent hydrolysis, in a stream of HX. The dehydration step often is troublesome, and researchers have developed many specific procedures for individual preparations not involving aqueous solutions. Several preparations use the metal hydride as starting material.

$$2LiH(xs.) + I_2 \xrightarrow{\text{ether}} 2LiI + H_2$$

$$BaH_2(xs.) + 2NH_4I \xrightarrow{\text{pyridine}} BaI_2 + 2NH_3 + H_2$$

The excess hydride is insoluble in the solvent and may be removed by filtering.

The uranium(III) halides are made conveniently by the reaction of the hydrogen halide with uranium hydride.

$$2U + 3H_2 \xrightarrow{250°} 2UH_3$$

$$UH_3 + 3HX \xrightarrow{250-300°} UX_3 + 3H_2 \qquad X = F, Cl$$

Anhydrous $FeCl_2$ has been obtained by the reaction

$$C_6H_5Cl + FeCl_3 \rightarrow C_6H_4Cl_2 + FeCl_2 + HCl$$

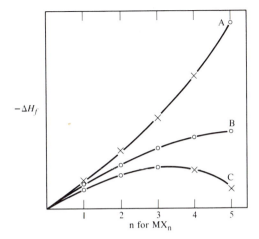

Figure 13.12 Stability curves for halides: ×, unstable compositions; ○, stable compositions. (From L. H. Long, *Quart. Rev.* 1953, 7, 134–174.)

Compounds of the type MX_3, MX_4, and MX_5 usually are obtained by one of the following reactions.

$$M + X_2 \rightarrow MX_n$$

$$M + HX \rightarrow MX_n + H_2$$

$$MO + C + X_2 \rightarrow MX_n + CO$$

$$MO + CCl_4 \rightarrow MCl_n + CO + COCl_2$$

$$MO + S_2Cl_2 \rightarrow MCl_n + SO_2$$

Since the direct reaction of the metal with the halogen usually is vigorous, an inert solvent medium often is used with Br_2 or I_2, the solvent serving as a diluent. Often, the vapor of the halogen is used, and for further safety a noble gas sometimes is used as a diluent. The van Arkel process for purifying metals (Ti, Hf, Zr, V, W, etc.) takes advantage of (1) the formation of a volatile halide by direct reaction of the halogen with the metal; (2) the decomposition of the metal halide to the metal and halogen at a higher temperature.

The reaction of the metal directly with anhydrous HX is useful for metals above hydrogen in the activity series. $AlCl_3$ and $CrCl_2$ are made in this way. In the reactions involving C, S_2Cl_2, or CCl_4, we can presume that reduction to metal occurs, followed by halogenation.

Where a series of halides exists with the metal in different oxidation states, a lower halide can be produced by hydrogen reduction of the higher halide.

13.3.6 Interhalogen Compounds and Ions

The halogens form all combinations of XX′ compounds, of which only IF is unstable. It is detectable spectroscopically, but it disproportionates to form I_2 and IF_5. Most of the XX′ compounds can be oxidized further to form XX'_n compounds, as shown in Table 13.8. Valence bond descriptions have been given (p. 45), and these compounds were used as examples for

Table 13.8   Interhalogen compounds and ions (boiling or melting points given for the neutral compounds)

	Interhalogens	*Cations*	*Anions*
ClF	− 101°C (b.p.)		ClF_2^-
ClF_3	11.8 (b.p.)	ClF_2^+	ClF_4^-
ClF_5	− 14 (b.p.)	ClF_4^+	
BrF	20 (b.p.)		BrF_2^-
BrF_3	126 (b.p.)	BrF_2^+	BrF_4^-
BrF_5	41 (b.p.)	BrF_4^+	BrF_6^-
(IF)	—		IF_2^-
IF_3	—	IF_2^+	IF_4^-
IF_5	101 (b.p.)	IF_4^+	IF_6^-
IF_7	subl. 40	IF_6^+	IF_8^-
BrCl	5 (b.p.)		$BrCl_2^-$
ICl	27 (m.p.)		ICl_2^-
$(ICl_3)_2$	—	ICl_2^+	ICl_4^-
IBr	41 (m.p.)		IBr_2^-

the VSEPR approach. ICl_3 exists as a dimer, giving planar ICl_4 units. The planar structure can be predicted from VSEPR once we know that there are two bridging Cl.

The interhalogen compounds can serve as halogenating agents. The oxidizing power increases with increasing oxidation number in a series such as $BrF < BrF_3 < BrF_5$. Among the commercially available fluorinating agents the oxidizing power may be rated roughly as

$$F_2 > ClF_5 > ClF_3 > BrF_3 > IF_5 > SF_4 > AsF_5 > SbF_5 > AsF_3 > SbF_3$$

This is the thermodynamic order of oxidizing power, not the order of the rate of reaction. Thus SF_6 would probably be better as an oxidizing agent than IF_5, based on thermodynamic considerations, but SF_6 is inert, because of the lack of a low-energy path for reaction to take place. In similar fashion, ClF_3 is more reactive (greater *rate* of reaction) than F_2 with most substances, despite its lower oxidizing power. This may be attributed to the lower energy needed to break the F_2Cl—F bond compared with the F—F bond. ClF_3 reacts explosively with substances such as cotton, paper, picien wax, and stopcock grease! ClF_3 has been used as a fuel for short-range rockets. Where hydrazine, N_2H_4, is the fuel to be oxidized, both reactants may be stored without refrigeration. This mixture is a hypergolic fuel: that is, it ignites spontaneously on being mixed.

Interhalogen Halides

Almost all of the interhalogen compounds react with metal halides to form the XX'^-_n ions (Table 13.8), as well as "mixed" anions such as I_2Cl^-, I_4Cl^-, ICl_3F^-, etc. The less stable anions can be stabilized in the solid as salts of large cations (p. 229). Since ions such as IF_2^- and IF_4^- have been known for a long time, and are isoelectronic with XeF_2 and XeF_4, respectively, it seems surprising, in retrospect, that the Xe compounds were not sought earlier. We have considered bonding in XeF_2 in some detail (p. 162). The XX'^-_2 ions are linear and XX'^-_4 ions are planar (p. 69). The IF_6^- ion shows the expected distortion from octahedral symmetry, although for the smaller Br the lone pair is stereochemically inactive, giving a centrosymmetric BrF_6^- (see the discussion of the isoelectronic XeF_6, p. 602).

In addition to the interhalogen halides, where the more electropositive halogen serves as the central atom, the halogens form polyhalide ions, X_n^-. The triiodide ion, I_3^-, is familiar from the use of KI to increase the amount of I_2 that can be dissolved in aqueous solution. Ions Cl_3^- and Br_3^- are known, but iodine gives MI_n salts with n as high as 9. I_n^- ions involve bonding between I_2 molecules and I^-. The I_5^- ion consists of a zigzag chain of two I_2 molecules bonded to I^- ($\mathbf{C}_{2v}$).

Cations

Oxidation of Br_2 and I_2 by $S_2O_6F_2$ gives the salts $X_2^+(SO_3F^-)$. Also known[15] are the cations Cl_3^+, Br_3^+, and I_3^+, which can be obtained in oleum (H_2SO_4/SO_3). Table 13.8 lists known interhalogen cations.

13.3.7 Pseudohalogens

Analogy plays an important role in descriptive chemistry. The ionic groups given in Table 13.9 behave chemically very much like halide ions and are often called "pseudohalides."

The approximate order of electronegativities of the pseudohalogens and the halogens is $F— > N_3— > NC— > Cl— > NCS— > Br— > I—$.[16]

The order of strength as oxidizing agents is approximately $F_2 > Cl_2 > Br_2 > (SCN)_2 > (CN)_2 > I_2 > (SeCN)_2$.

Electrolysis or chemical oxidation may be used to prepare many of the pseudohalogens from their ions.

$$4HSCN + MnO_2 \rightarrow (SCN)_2 + Mn(SCN)_2 + 2H_2O$$

Similarities to the halogens appear in reactions such as hydrolysis, addition to carbon-carbon double bonds, formation of complex ions and pseudointerhalogen compounds, and insolubility of silver, lead and Hg(I) salts.

$$(CN)_2 + 2OH^- \rightarrow OCN^- + CN^- + H_2O$$

$$CH_2CH_2 + (SCN)_2 \rightarrow NCSCH_2CH_2SCN$$

[15]J. D. Corbett, *Prog. Inorg. Chem.* 1976, *21*, 129.

[16]D. H. McDaniel and A. Yingst, *J. Am. Chem. Soc.* 1964, *86*, 1334.

Table 13.9   The pseudohalogens

Ion	Dimeric Molecule	m.p. of X_2, °C	pK_a of HX
N_3^- azide	—	—	4.55
CN^- cyanide	$(CN)_2$ cyanogen	m.p. -27.9 b.p. -21.17	9.21
OCN^- cyanate	$(OCN)_2$ oxycyanogen	—	3.92
SCN^- thiocyanate	$(SCN)_2$ thiocyanogen	-2 to $-3°$	(-0.74)
$SeCN^-$ selenocyanate	$(SeCN)_2$ selenocyanogen	Yellow powder	—

$$Ag^+ + CN^- \rightarrow AgCN(s)$$

$$AgCN(s) + CN^- \rightarrow Ag(CN)_2^-$$

Since the pseuodhalides, like the halides themselves, have more than one pair of electrons with which coordinate bonding can take place, they can serve as bridging groups. However, the bridging may result in a different degree of polymerization of the parent compound, because of the different geometric orientation of the available electron pairs. Thus, although dialkylgold chloride and bromide are dimeric, the cyanide is tetrameric.

In each case the geometry about the gold atom is essentially square planar.

Other differences between the halogens and the pseudohalogens occur in the polymerization of many of the pseudohalogens under heating.

$$x(CN)_2 \xrightarrow[400°]{} \tfrac{1}{2}(CN)_x$$

$$x(SCN)_2 \xrightarrow[\text{room temp.}]{} \tfrac{1}{2}(SCN)_x$$

13.3.8   Chlorine Oxofluorides

The known chlorine oxofluorides form a nearly complete series of netural, cationic, and anionic species (Figure 13.13). The Cl(III) species ClF_2^+, ClF_3, and ClF_4^-, although not oxofluorides, are included to complete the series. Similarly, ClF, ClF_2^-, and Cl^+ (not known as a simple

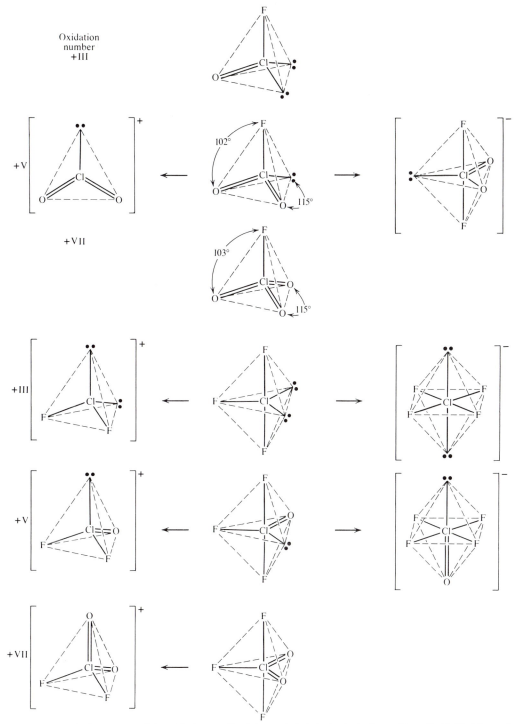

Figure 13.13 Structures of the chlorine oxofluorides and their ionic derivatives. (From K. O. Christe and C. J. Schack. *Adv. Inorg. Chem. Radiochem.* 1976. *18*, 319.)

cation) could be added to complete the FClO, FClO$_2$, FClO$_3$ series. Note that here the formulas are written with the central atom (Cl) between the atoms bonded to it. This avoids possible confusion with isomers—for example, FClO and ClOF, sometimes called chlorine hypofluorite.

The geometries of the species in Figure 13.13 are those expected from VSEPR considerations (see Section 2.3.3), with mutual repulsion decreasing in the order lone pair > double-bonded oxygen > fluorine. Detailed structures are known only for FClO$_2$ and FClO$_3$, but vibrational spectra are available for most of the compounds. Force constants reveal that the Cl—O bonds have varying degrees of double-bond character, whereas the Cl—F bonds appear to be of three types: (*a*) ordinary two-center covalent bonds, (*b*) three-center four-electron bonds (as described for XeF$_2$, p. 164) for F$_2$ClO$_2^-$, F$_4$ClO$^-$, and for the axial F atoms in F$_3$ClO and F$_3$ClO$_2$; and (*c*) polar (*p*-π*) σ three-center two-electron bonding (as described for O$_2$F$_2$, p. 572) for FClO$_2$ and FClO.

Table 13.10 summarizes the properties of and syntheses for the chlorine oxofluorides and their ionic derivatives. The oxofluorides are rather similar to the chlorine fluorides. As liquids, they undergo little self-ionization. They form stable adducts with strong Lewis acids (primarily fluorides such as BF$_3$, PF$_5$, and MF$_5$) or Lewis bases such as CsF (a fluoride salt of a large nonoxidizable cation). The neutral molecules based upon a trigonal bipyramid form adducts easily, giving stable pseudotetrahedral cations and pseudooctahedral anions. The tetrahedral FClO$_3$ is kinetically stable and does not form adducts.

13.3.9 Halogen Oxides

Oxygen fluorides (p. 571) and ClO$_2$ (p. 505) have been discussed already. The other oxides of chlorine are Cl$_2$O, Cl$_2$O$_4$, Cl$_2$O$_6$, and Cl$_2$O$_7$. Two can be considered perchlorates: the unstable Cl$_2$O$_4$, with the structure Cl–O–ClO$_3$, is commonly called chlorine perchlorate; Cl$_2$O$_6$ has an ionic structure, ClO$_2^+$ClO$_4^-$ in the solid. Chlorine monooxide, a yellow-red gas, is obtained by the reaction of Cl$_2$ with freshly prepared HgO. An angular molecule (**C$_{2v}$**), it decomposes explosively when heated. It reacts with water to produce HOCl. The most stable of the chlorine oxides, Cl$_2$O$_7$, is obtained by dehydration of HClO$_4$ with P$_4$O$_{10}$. It is isolated as a colorless liquid by low-pressure distillation. Even Cl$_2$O$_7$ is explosive.

All of the bromine oxides are unstable thermally: Br$_2$O (dark brown liquid), Br$_3$O$_8$ (white solid), and BrO$_2$ (yellow solid). Two of the oxides of I, I$_2$O$_4$ and I$_4$O$_9$, are not well characterized. The important oxide is I$_2$O$_5$, a white crystalline product yielded by dehydration of HIO$_4$ at 250°C.

13.3.10 Oxoacids and Oxo anions of the Halogens

We discussed the preparation of the oxo anions of Cl and some of their redox chemistry in Section 11.10. The electrode potentials allow us to predict thermodynamically favorable reactions, but we cannot rely on a correlation of emf values with rates. In particular, the reactions of oxo anions (or acids) often involve oxygen transfer mechanisms. The processes of breaking and reforming covalent bonds and the necessary molecular rearrangements often are slow.

Table 13.10 Synthesis and properties of chlorine oxofluorides and their cationic and anionic derivatives

Species	Properties	Synthesis
FClO	Unstable	Observed as intermediate and by matrix isolation
F_3ClO	Colorless m.p. -42 to $-44.2°C$ b.p. 27 to 29°C Stable at room T	From $F_2 + Cl_2 + O_2$ and $2F_2 + ClONO_2 \xrightarrow{-35°C} F_3ClO + FNO_2$
$FClO_2$ Chloryl fluoride	Colorless m.p. $\sim -120°C$ b.p. $\sim -6°C$ Stable at room T	$6NaClO_3 + 4ClF_3 \xrightarrow{-196°C} 6NaF$ $+ 2Cl_2 + 3O_2 + 6FClO_2$
F_3ClO_2	Colorless m.p. $-81.2°C$ b.p. $-21.58°C$ Marginally stable at room T	From $FClO_2 + PtF_6$ followed by treatment with FNO_2
$FClO_3$ Perchloryl fluoride	Colorless m.p. $-147.75°C$ b.p. $-46.67°C$ High thermal stability Not easily hydrolyzed	From $KClO_4 + SbF_5$
F_2ClO^+ Difluorooxochloronium cation	Stable white solids with anions such as HF_2^-, BF_4^-, PF_6^-, and most MF_6^-	F_3ClO + Lewis acid
ClO_2^+ Chloryl cation	White solids unless anion is colored Moderately stable	$FClO_2$ + Lewis acid
$F_2ClO_2^+$ Difluoroperchloryl cation	White stable solids React violently with H_2O or organic materials	$F_3ClO_2 + BF_3 \rightarrow [F_2ClO_2]BF_4$
F_4ClO^- Tetrafluorooxochlorate ion	Stable white solids	$MF + F_3ClO \rightarrow M[F_4ClO]$ $M = Cs^+, Rb^+,$ or K^+
$F_2ClO_2^-$ Difluorochlorate ion	White solid Stable at room T Reacts explosively with H_2O	$CsF + FClO_2 \rightarrow Cs[F_2ClO_2]$

Thus the reduction potential for ClO_4^- indicates that it should be a strong oxidizing agent, and it is—in hot and/or concentrated solutions. Yet cold dilute $HClO_4$ does not oxidize I^-, even though the E_{cell}^0 is 0.65 V. This does not make the potential diagrams useless; knowing which reactions are thermodynamically impossible is very valuable. With experience, you acquire a feeling for which reactions are likely to be slow, even though thermodynamically favorable. Below, we examine some of the intercomparisons among and the chemistry of the halogen oxo anions.

The Hypohalites

The compound HOF, which has been prepared by the reaction of F_2 and H_2O at 0°C,[17] is unstable, even oxidizing H_2O further to form H_2O_2. Although HOF has the same formula type as HOCl, F is assigned its usual $-I$ oxidation number, so it is hydrogen oxygen fluoride, or hydroxyl fluoride, not hypofluorous acid. Nevertheless, HOF should be acidic, and probably a stronger acid than HOCl. The $-OF$ group occurs in CF_3OF, O_2NOF, F_5SOF, and O_3ClOF—all of which are strongly oxidizing.

$$F_2 + H_2O \xrightarrow{0°C} HOF + HF$$

For the other halogens, the relative increase in ease of oxidation of the halide ion with increasing atomic number is demonstrated in the following potential diagrams for basic solution.

$$E_B^0 \qquad ClO^- \xrightarrow[]{+0.40} Cl_2 \xrightarrow{+1.36} Cl^-$$

(with $+0.88$ spanning ClO^- to Cl^-)

$$BrO^- \xrightarrow{+0.45} Br_2 \xrightarrow{+1.07} Br^-$$

(with $+0.76$ spanning BrO^- to Br^-)

$$IO^- \xrightarrow{+0.45} I_2 \xrightarrow{+0.535} I^-$$

(with $+0.49$ spanning IO^- to I^-)

All of the above hypohalite ions may be formed from the disproportionation of the corresponding halogen in base. The stability of the hypohalite ions with regard to further disproportionation to halite ions and halide ions decreases with increasing atomic number. Thus although hypoiodite solutions decompose within a few hours, hypochlorite solutions decompose only slightly over a period of weeks, relative rates being ClO^- 1, BrO^- 100, and IO^- 3×10^6. Accordingly, hypobromite is generated for analytical use from hypochlorite solutions at a pH of around 10.

$$ClO^- + Br^- \rightarrow BrO^- + Cl^-$$

Despite its lower emf, hypobromite is often a faster oxidizing agent than hypochlorite.

Hypoiodite ion is an oxidant used to determine β-keto functions through the "iodoform" reaction, so-called because of one of the reaction products.

$$\underset{\underset{O}{\|}}{RCCH_3} + 3I_2 + 4OH^- \rightarrow HCI_3 + RCO_2^- + 3I^- + 3H_2O$$

The anhydrous hypohalite acids cannot be isolated. Their pK_a values are given in Table 12.4 (p. 526). Hypoiodous acid also undergoes dissociation to produce hydroxide ion in solution, its pK_b being 10.

[17]E. H. Appelman, *Acc. Chem. Res.* 1973, *6*, 113.

The Halites

Chlorite salts and chlorous acid, $HClO_2$, have been known for a long time. The acid cannot be isolated in anhydrous form. Earlier efforts to prepare bromites and iodites were unsuccessful, but bromite ion has been prepared by anodic oxidation of Br^- with very high current efficiency.[18] The bromite ion was isolated as a tetraalkylammonium salt, because of its insolubility in water. The salt was recrystallized from chloroform to give orange crystals.

The Halates

The oxidizing strengths of chlorate and bromate solutions are similar, but the iodate is much weaker.

$$E_B^0 \qquad ClO_3^- \xrightarrow{+0.63} Cl^-$$

$$BrO_3^- \xrightarrow{+0.61} Br^-$$

$$IO_3^- \xrightarrow{+0.29} I^-$$

All the standard emf values are more positive by 0.83 V in acid solution.

As oxidizing reagents, these ions have rates of reaction in the order $IO_3^- > BrO_3^- > ClO_3^-$. This order has been explained as stemming from the decreasing multiple-bond character in the O—X bond as the atomic number increases. The iodate thus has more negative charge at the oxygen surface than the bromate, which in turn has more surface charge than the chlorate. The increase in surface charge would allow oxygen to coordinate more readily with other ions, and the low multiple-bond character also would allow the iodine to expand its coordination sphere more readily through dative bonding to empty d orbitals.

Moreover, the low double-bond character and consequent higher negative charge on oxygen would account for the generally lower solubility of transition metal iodates compared with chlorates or bromates, and also for the greater viscosity of comparable solutions of iodates compared with chlorates. It might also help explain why acid salts containing the hydrogen-bonded units $(IO_3 \cdot HIO_3)^-$ and $(IO_3 \cdot 2HIO_3)^-$ are formed with iodic acid, but not with bromic or chloric acids. The high molecular weight of $KH(IO_3)_2$ and ease of purification through recrystallization make it suitable for standardizing base solutions for analytical work. Iodic acid also can be isolated in anhydrous form, in contrast to chloric and bromic acids.

We can make both iodic and bromic acids in the laboratory by oxidizing the halogen or halide ion with Cl_2, removing the Cl^- with silver oxide.

$$X_2 + 5Cl_2 + 6H_2O \rightarrow 2HXO_3 + 10HCl$$
$$2HCl + Ag_2O \rightarrow 2AgCl + H_2O$$

Commercially, iodates and bromates are produced by electrolytic oxidation. As with Cl_2, reaction with hot basic solution will produce the halate.

---

[18]T. Kageyama and T. Yamamoto *Bull. Chem. Soc. Japan* 1980, *53*, 1175; *Chem. Lett. 1980*, 671.

$$3X_2 + 6OH^- \rightarrow XO_3^- + 5X^- + 3H_2O$$
$$X = Cl, Br, I$$

In acidic solution, the above reaction is reversed for X = Br and I, and this serves as a convenient method of preparing Br_2 and I_2 in solution, usually using an excess of the halide ion.

$$6H^+ + 5Br^- + BrO_3^- \rightarrow 3Br_2 + 3H_2O$$

The extent of unsaturation in olefins may be determined by the addition of the Br_2 produced to the double bond.

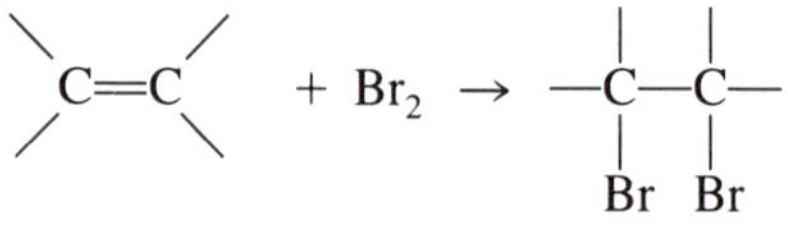

$$\underset{/}{\overset{\backslash}{}}C=C\underset{\backslash}{\overset{/}{}} \quad + Br_2 \quad \rightarrow \quad -\underset{|}{\overset{|}{C}}-\underset{|}{\overset{|}{C}}-$$
$$ Br \; Br$$

We can determine the excess of Br_2 by adding KI and titrating with thiosulfate solution.

$$2I^- + Br_2 \rightarrow 2Br^- + I_2$$
$$I_2 + S_2O_3^{2-} \rightarrow 2I^- + S_4O_6^{2-}$$

Unlike the reactions of iodates and bromates with their halide ions, chlorates react with Cl^- in the presence of acid to produce ClO_2 as well as Cl_2.

$$2HClO_3 + 2HCl \rightarrow 2ClO_2 + Cl_2 + 2H_2O$$

The Perhalates

The marked differences among the perhalic acids and their anions are not peculiar to this family, as can be illustrated by examining the properties of the following acids.

H_3PO_4	H_2SO_4	$HClO_4$
H_3AsO_4	H_2SeO_4	$HBrO_4$
$HSb(OH)_6{}^a$	H_6TeO_6	H_5IO_6
^{a}(in solution only)		

Acid strength and oxidizing ability increase regularly on going from phosphoric acid to sulfuric acid to perchloric acid; the melting points decrease and the tendency toward formation of polyacid units decreases from phosphoric acid to perchloric acid, which forms no polyacids. Likewise, the ability to form metal ion complexes decreases from phosphate through perchlorate. In acid strength, solubility of salts, and structure of the salts, arsenic and selenic acid resemble phosphoric and sulfuric acids more than antimonic and telluric acids. Perbromic acid resembles perchloric acid more than it does periodic acid, but it is a stronger oxidizing agent than $HClO_4$. Arsenic and selenic acids also are much stronger oxidizing agents than H_3PO_4 and H_2SO_4. Unlike H_2SO_4, H_2SeO_4 upon heating is unstable toward decomposition to O_2, H_2O, and Se. Perbromic acid also is less stable thermally than $HClO_4$.

The coordination number of the central atom increases for antimonates and tellurates, which show only six-coordination. The smaller size of I^{7+} compared with Te^{6+} and Sb^{5+} permits I to show C.N. 4 as well as 6 in periodates. Typical salts formed by these acids are $NaSb(OH)_6$, $LiSb(OH)_6$, $Na_3H_3TeO_6$, Ag_6TeO_6, Ag_5IO_6, $Na_3H_2IO_6$, and KIO_4.

$$E_A^0 \qquad ClO_4^- \xrightarrow{\;1.19\;} ClO_3^-$$
$$BrO_4^- \xrightarrow{\;1.74\;} BrO_3^-$$
$$H_5IO_6 \xrightarrow{\;1.60\;} IO_3^-$$

Perchloric Acid. Perchloric acid is the only oxoacid of Cl that can be isolated in an anhydrous state. It is prepared most readily through the dehydration of the dihydrate by fuming sulfuric acid and removal of the anhydrous $HClO_4$ by vacuum distillation in a grease-free system.

$$HClO_4 \cdot 2H_2O + 2H_2S_2O_7 \xrightarrow{25-80°} HClO_4 + 2H_2SO_4$$

The anhydrous acid melts at $-102°C$. The $HClO_4$ molecular units exist in the gas, liquid, and solid. Contact with organic materials such as wood, paper, and rubber produces violent explosions, but it may be stored without explosive decomposition for 30 to 60 days at liquid air temperature ($-190°$) or for 10 to 30 days at room temperature. Normally colorless, it develops an amber color prior to detonation.

Perchloric acid and water form an azeotrope (72.5% $HClO_4$) that boils at approximately 203°C at 1 atm. Under these conditions, appreciable decomposition of the acid occurs. The commercially available concentrated acid is 70% $HClO_4$ (11.6 M).

Cold concentrated $HClO_4$ (70%) is a weak oxidizing agent. Hot concentrated $HClO_4$ has been used as a rapid "wet ashing" procedure for the complete oxidation of the organic matter in samples, leaving the inorganic components for analysis—the so-called liquid fire reaction. Alcohols should not be present, as the perchlorate esters formed may explode, particularly ethyl perchlorate. This hazard can be eliminated by using a mixture of concentrated nitric and perchloric acids. The HNO_3 oxidizes the alcohols and other readily oxidizable groups, and is itself displaced on heating. The temperature of the reaction mixture slowly increases as the perchloric acid is concentrated, approaching the azeotropic conditions. As the temperature and perchloric acid concentration increase, the oxidizing ability of the solution increases. Completion of the reaction may often be determined by adding a small amount of $K_2Cr_2O_7$ to the reaction mixture. The dichromate reacts rapidly with the organic material, giving colorless or green solutions containing Cr(III) ion. At the end of the reaction, the Cr(III) ion is oxidized to orange CrO_3. Wet-ashing in this fashion, which may be carried out in as little as 10 to 15 minutes, has been used for analysis of samples of organic origin such as leather, wood, grain, and coal.

Perchlorates. Ammonium perchlorate was used formerly as a nonfreezing blasting compound in mining operations. Today it is used as an oxidant in solid-fuel missiles.

The perchlorate ion is less extensively hydrated than the other oxo halogen anions and also shows very little tendency to form complexes with metal ions. Accordingly, it is used often as an inert anion in studies of metal ion complexes in aqueous solution; but in the absence of other ligands, ClO_4^- can function as a unidentate or bidentate ligand.[19]

Magnesium perchlorate is a very efficient desiccant. In this case the Mg^{2+} ions behave as if they were isolated in an inert matrix and accordingly form a very stable hexahydrate when in contact with water. Ammonium perchlorate will absorb sufficient ammonia to liquify at

[19]K. O. Christe and C. J. Schack, *Inorg. Chem.* 1974, *13*, 1452.

room temperatures, giving solutions resembling Diver's solution (NH_4NO_3, NH_3). Many perchlorate salts are isomorphous with permanganate, perrhenate, and tetrafluoroborate salts.

Perbromic Acid and Perbromates. Efforts to prepare the perbromate ion were unsuccessful until it was obtained from the β-decay of $^{83}SeO_4^{2-}$. This led to the discovery that perbromates can be prepared by the oxidation of bromates electrolytically, by XeF_2, or by oxidation by F_2 in basic solution. Solid perbromates are stable: for example, $KBrO_4$ is stable up to 275°C. Perbromate ion is a stronger oxidizing agent than ClO_4^- or IO_4^-, but in dilute solution reduction occurs slowly—it does not oxidize Cl^-. The free acid is stable in solutions up to 6 M. In concentrated solution, it is a vigorous oxidizing agent.

Periodic Acid and Periodates. Periodates may be made in the laboratory by the oxidation of an iodate in basic solution using Cl_2. Commercially, periodates are prepared by electrolytic oxidation of iodates.

The equilibria involved in the dissociation of periodic acid in water are as follows.

$$H_5IO_6 \rightleftharpoons H^+(aq) + H_4IO_6^- \qquad K_1 = (10 \pm 4) \times 10^{-4}$$
$$H_4IO_6^- \rightleftharpoons H^+(aq) + H_3IO_6^{2-} \qquad K_2 = (3 \pm 1) \times 10^{-7}$$
$$H_4IO_6^- \rightleftharpoons 2H_2O + IO_4^- \qquad K = 43 \pm 17$$

Periodic acid is thus a much weaker acid than $HClO_4$. The H_5IO_6 form is called "paraperiodic acid"—the form of the acid stable as a solid in contact with water. At 100°C, this form loses water to convert into *meta*-periodic acid, HIO_4.

Periodic acid is especially useful as an analytical reagent for the quantitative determination of α,β-dihydroxyorganic compounds.

$$-\overset{|}{\underset{\underset{OH}{|}}{C}}-\overset{|}{\underset{\underset{OH}{|}}{C}}- \;+\; H_5IO_6 \;\rightarrow\; -\overset{|}{\underset{\underset{O}{\|}}{C}} \;+\; \overset{|}{\underset{\underset{O}{\|}}{C}}- \;+\; HIO_3 \;+\; 3H_2O$$

An intermediate in this highly selective reaction is the complex

Mercury(II) ion may be determined by precipitating it as $Hg_5(IO_6)_2$.

13.4 THE NOBLE GASES

13.4.1 Family Trends

The elements of Group 0 (or Group VIIIB) have been called the rare gases or the inert gases, but these are misnomers. Argon constitutes almost 1% of air by volume, and helium is available readily from some natural gas deposits; moreover, the discovery of the xenon fluorides demonstrated that Xe is not truly inert. The term *noble gases* seems most appropriate, since we classify the least reactive metals as noble metals. The noble gases have ''closed'' eight-electron outer-shell configurations, ns^2np^6—except for He, for which the first shell is complete, $1s^2$. The ionization energies of the noble gases follow the regular trend of increasing across each period. This corresponds to the increase in Z, with each electron added to the same outer shell not shielding the nuclear charge by a full unit (see p. 32). The nonmetals become increasingly more reactive from left to right within a period until we reach the noble gases. The abrupt change in reactivity for the noble gases results from their very low electron affinities, since an added electron would occupy a new shell. Actually, the electron affinities are zero or slightly negative, indicating that no bound state for the electron is achieved.

13.4.2 Discovery, Occurrence, and Recovery

Cavendish (1784) removed the nitrogen from air by sparking with added O_2 to form NO_2. The NO_2 was absorbed in aqueous alkali and the excess O_2 was removed by burning S. After removal of the SO_2 in aqueous alkali, a small volume of unidentified, unreactive gas remained. Rayleigh (1892) prepared nitrogen from ammonia and by removing the other known constituents of air (O_2, CO_2, and H_2O). The N_2 from ammonia was always about 0.5% less dense than that from air. Rayleigh and Ramsay soon isolated a new gas, with an emission spectrum different from that of any known element, that they named argon (''inert''). Ramsay and Travers then isolated neon (''new''), krypton (''hidden''), and xenon (''stranger'') by the fractional distillation of liquid air.

During an 1868 eclipse, Jansen identified a new yellow line in the spectrum of the sun. Since this line did not appear in the spectrum of any known element, Frankland and Lockyer concluded that the sun contained a new element, which they named helium (Greek *helios*, ''the sun''). By heating cleveite (a radioactive mineral related to uranite) Hillebrand (1889) obtained an unreactive gas that he believed to be N_2, but Ramsay (1895) showed it to be a sample of Lockyer's helium. The study of the radioactive series led to the discovery of radon, the ''emanation'' from the radioactive decay of radium. Isotopes of radon from other radioactive decay series were called actinon and thoron.

The earth's atmosphere contains about 5 ppm He, but it would be very expensive to recover the He. Some natural gas deposits contain 0.5 to 0.8% He trapped from radioactive decay. These extensive deposits, which occur almost entirely in the southwestern United States account for the present world production of He. They will be depleted in 20 to 30 years. Helium-lean (0.03 to 0.3%) natural gas deposits are found primarily in the Soviet Union, the United States, Algeria, and Canada. There are extensive deposits in the Soviet Union, but the Algerian deposits are more important, because of their higher helium content.[20]

The U.S. government has had an on-again/off-again helium conservation program. Some He is stored in depleted gas fields, but most of the He in natural gas now being used is lost to the atmosphere. Helium can be obtained now at low cost from helium-rich gas, and the potential supply far exceeds the demand. The demand for He for the use of superconductors operating at liquid He temperature for efficient power transmission could increase greatly just as the important supplies in natural gas are exhausted.

[20]E. Cook, ''The Helium Question,'' *Science* 1979, *206*, 1141.

Table 13.11   Properties of the noble gases

	He	*Ne*	*Ar*	*Kr*	*Xe*	*Rn*
Atomic radius[a] (van der Waals)	150 pm	160	190	200	220	—
First ionization (eV) energy (MJ/mole)	24.587 2.3723	21.564 2.0806	15.759 1.5205	13.999 1.3507	12.130 1.1704	10.748 1.0370
Outer shell configuration	$1s^2$	$2s^2 2p^6$	$3s^2 3p^6$	$4s^2 4p^6$	$5s^2 5p^6$	$6s^2 6p^6$
B.p. (K)	4.18	27.13	87.29	120.26	166.06	208.16
% by volume in atm.	5.2×10^{-4}	1.82×10^{-3}	0.934	1.14×10^{-3}	8.7×10^{-6}	—

[a]Calculated from the cubic close-packed lattice parameters extrapolated to 0 K. Helium also has hexagonal close-packed and body-centered cubic phases. Lattice parameters are from J. Donohue, *The Structures of the Elements*, Wiley, New York, 1974, Chapter 2.

The other nonradioactive noble gases are recovered from the fractional distillation of liquid air. Only Ar is abundant in air (see Table 13.11). The greater abundance of Ar than Ne is taken as evidence that the earth lost most of its atmosphere during its formation. There are three isotopes of Ar, but the one of overwhelming abundance (99.63%) is ^{40}Ar, formed from the β-decay of ^{40}K. Argon in the atmosphere has accumulated from this source. The decay of ^{40}K also is believed to have provided a major source of heat in bringing the earth to its present temperature.

13.4.3   Physical Properties

Some of the physical properties of the noble gases are given in Table 13.11. The gases are all low-boiling, because of their spherical closed shells and the lack of any mutual interaction other than the very weak van der Waals attraction. As we might expect in the absence of any preferred bonding directions, the solids have close-packed structures.

Helium, the lowest-boiling substance known, is the only element that forms a solid only at high pressure (~25 atm). Liquid He (helium I) is a normal liquid down to a λ-point (~2 K, depending on pressure), where it is converted to helium II—a most extraordinary liquid, with zero viscosity and very high heat conductivity, among other unusual physical properties.

The major uses of He are for cryogenic work and for lighter-than-air vehicles. Neon is used for display signs and lights. Argon is used to provide an inert atmosphere, as in incandescent bulbs, and for chemical applications where N_2 is too reactive.

13.4.4   Chemical Properties

Introductions to molecular orbital theory commonly deal with molecular ions such as He_2^+ and HeH^+. These ions, along with those of other noble gases—Ar_2^+, $HeNe^+$, etc.—are formed in gaseous discharge tubes, but the neutral molecules are unstable. Also, clathrate compounds (see Section 5.4) of the noble gases have been known for a long time. Unlike ordinary compounds displaying typical chemical bonding, these compounds involve the trapping of atoms or molecules of the proper dimensions in cavities in a crystal. Aqueous hydroquinone solution under several atmospheres pressure of a noble gas (G = Ar, Kr, or Xe) can be cooled to give crystalline solids with the approximate composition $[C_6H_4(OH)_2]_3G$. Noble gas hydrates are formed as clathrates, with the noble gas atoms occupying the cavities of regular pentagonal dodecahedra. The fact that the anesthetic effect of substances such as chloroform has been attributed to the formation of clathrate hydrate crystals in the brain might explain the anesthetic effect of xenon.

The first real noble-gas compound was obtained when Bartlett and Lohman oxidized Xe with platinum hexafluoride. Prior to this experiment, they had prepared $O_2^+PtF_6^-$. Since the ionization energy of both O_2 and Xe is 12.1 eV (1.17 MJ/mole), only lattice energy effects would prevent formation of the Xe derivative. They formulated the product obtained as $Xe^+[PtF_6]^-$, but it is now known to be more complex. Bartlett described the experiment as follows:

> The predicted interaction of Xe and PtF_6 was confirmed in a simple and visually dramatic experiment. The deep red PtF_6 vapor, of known pressure, was mixed, by breaking a glass diaphragm, with the same volume of xenon, the pressure of which was greater than that of the hexafluoride. Combination, to produce a yellow solid, was immediate at room temperature, and the quantity of Xe which remained was commensurate with a combining ratio of 1:1.
>
> (From N. Bartlett, *Am. Sci.* 1963, *51,* 114.)

Soon after publication of the preparation of $XePtF_6$, XeF_4 was obtained by the direct reaction of Xe and F_2 at 400°C. The XeF_4 is a colorless volatile solid that is isoelectronic (in terms of valence electrons) and isostructural with ICl_4^-. The parallel with interhalogen halides was apparent, and groups all over the world began the study of noble gas compounds.

Table 13.12 lists the xenon compounds. Krypton forms a linear KrF_2, but the ionization energies of He and Ar are too high to expect similar compounds. Radon gives compounds similar to those of Xe and should have even more varied chemistry; the high level of radioactivity has limited work on radon compounds, however, there is some evidence for the formation of $XeCl_2$, and $XeCl_4$ has been characterized by Mössbauer spectroscopy from the β-decay of ^{129}I in $KICl_4 \cdot 4H_2O$ to form $^{129}Xe^*Cl_4$.

Xenon Difluoride

This compound is obtained by the direct reaction of F_2 with Xe at high pressure. The XeF_2 is linear, as expected from VSEPR considerations. It is soluble in water, undergoing hydrolysis that is slow except in the presence of base.

$$XeF_2 + 4 OH^- \rightarrow Xe + O_2 + 2 F^- + 2 H_2O$$

It is a very powerful oxidizing agent.

$$XeF_2(aq) + 2 H^+ + 2 e^- \rightarrow Xe + 2 HF \qquad E^0 = +2.64 \text{ V}$$

Table 13.12 Xenon compounds

Oxidation State	Compound	M.p. (°C)	Structure	Properties
II	XeF_2	129	Linear	Hydrolyzed to Xe + O_2
	$XeF^+AsF_6^-$		Nearly linear XeF_2 with 1 bridging F	
IV	XeF_4	117	Square planar	Stable
	$XeOF_2$	31		Unstable
VI	XeF_6	49.6	Distorted octahedral	Stable
	$CsXeF_7$			dec. >50°C
	Cs_2XeF_8		Archimedes' antiprism	Stable to 400°C
	$XeOF_4$		Square pyramidal	Stable
	XeO_2F_2			Unstable
	XeO_3		Pyramidal	Explosive
VIII	XeO_4		Tetrahedral	Explosive
	XeO_6^{4-}		Octahedral	Anion of a very weak acid

Estimated Values of E^0

$$E_A^0 \quad H_4XeO_6 \xrightarrow{2.36} XeO_3 \xrightarrow{2.10} Xe$$

$$XeF_2 \xrightarrow{2.64} Xe$$

$$E_B^0 \quad HXeO_6^{3-} \xrightarrow{0.94} HXeO_4^- \xrightarrow{1.26} Xe$$

Xenon Tetrafluoride

The compound is easily prepared from the elements, but accompanied by XeF_2 at low F_2/Xe ratios and by XeF_6 at high F_2/Xe ratios. It is a strong oxidizing agent and reacts violently with water to give XeO_3.

$$3\ XeF_4 + 6\ H_2O \rightarrow XeO_3 + 2\ Xe + \tfrac{3}{2}\ O_2 + 12\ HF$$

Xenon Hexafluoride

The reaction of Xe with excess F_2 at high pressure at about 250°C yields XeF_6, which is hydrolyzed rapidly to form XeO_3 and reacts with quartz.

$$2\ XeF_6 + SiO_2 \rightarrow 2\ XeOF_4 + SiF_4$$

There has been controversy about the structure of XeF_6, in which the Xe has one lone pair of electrons. Although it now seems clear that the molecule is *not* octahedral, it does not correspond to a structure expected where the lone pair plays a dominant role. In the solid, XeF_6 exists as XeF_5^+ (square pyramidal) and F^-, but the F^- ions form bridges. Electron diffraction studies indicate that XeF_6 is slightly distorted from an octahedron in the gas phase.[21] It might

[21]L. S. Bartell and R. M. Gavin, *J. Chem. Phys.* 1968, *48*, 2460, 2466.

involve fluxional behavior, passing from one nonoctahedral structure, such as one with the lone pair directed through an octahedral face or edge, to another. The isoelectronic ions $SbBr_6^{3-}$, $TeCl_6^{2-}$, and $TeBr_6^{2-}$ are octahedral. The lone pair is "stereochemically inactive," possibly in an unhybridized s orbital.[22] BrF_6^- is centrosymmetric, whereas the lone pair lowers the symmetry of IF_6^- to $\mathbf{C}_{2v}$ for the larger central atom.[23]

XeF$_6$ is a fluoride donor, forming complexes such as $2XeF_6 \cdot SbF_5$ and an adduct with PtF_5, $XeF_5^+ PtF_6^-$. The latter compound has a sixth, bridging F^- with an Xe–F distance longer than the others. Similarly, XeF_2 forms $XeF^+ AsF_6^-$ with AsF_5.[24] An attempt to prepare the low-spin d^6 Au(V) ion, AuF_6^-, yielded the compound $Xe_2F_{11}^+ AuF_6^-$. The cation consists of 2 XeF_5 bridged by a shared F^- at greater distance (223 pm) than for the other Xe—F bonds (184 pm).[25] The geometry about each Xe suggests that the lone pair is stereochemically active. C.N.'s of 7 and 8 are achieved in reactions of XeF$_6$ with alkali metal fluorides (M = Na, K, Rb, or Cs) to form $MXeF_7$ and M_2XeF_8. The $MXeF_7$ salts decompose above room temperature to form M_2XeF_8 and XeF$_6$. The Cs^+ and Rb^+ salts of XeF_8^{2-} are stable to *ca.* 400°C.

The oxofluorides of Xe are $XeOF_2$, an unstable partial hydrolysis product of XeF_4; $XeOF_4$, from the partial hydrolysis of XeF_6; and XeO_2F_2, from the reaction of $XeOF_4$ and XeO_3.

$$XeOF_4 + XeO_3 \rightarrow 2\ XeO_2F_2.$$

Bonding in Xenon Fluorides

Because XeF$_2$ and XeF$_4$ closely parallel ICl_2^- and ICl_4^-, respectively, and because of the success of the VSEPR approach in predicting their structures, these are commonly dealt with in terms of an expanded octet. Whether the $5d$ orbitals are sufficiently low in energy for effective bonding is questionable however (p. 164).

Oxides

Xenon trioxide is an explosive white solid that appears to be present as XeO_3 molecules in water but forms $HXeO_4^-$ in basic solution. The $HXeO_4^-$ disproportionates slowly to produce perxenate ion, XeO_6^{4-}.

$$2\ HXeO_4^- + 2\ OH^- \rightarrow XeO_6^{4-} + Xe + O_2 + 2\ H_2O.$$

The strongly oxidizing perxenate ion also can be made by oxidizing $HXeO_4^-$ with O_3. The alkali metal and barium salts of XeO_6^{4-} are stable. Xenon tetraoxide is formed as an explosive gas by the reaction of Ba_2XeO_6 with concentrated sulfuric acid.

GENERAL REFERENCES

F. A. Cotton and G. Wilkinson, *Advanced Inorganic Chemistry*, 4th ed., Wiley-Interscience, New York, 1980. The best single-volume source for up-to-date inorganic descriptive chemistry.

[22]R. J. Gillespie, *J. Chem. Educ.* 1970, *47*, 18.

[23]K. O. Christe, *Inorg. Chem.* 1972, *11*, 1215.

[24]A. Zalkin, D. L. Ward, R. N. Biagioni, D. H. Templeton, and N. Bartlett, *Inorg. Chem.* 1978, *17*, 1318.

[25]K. Leary, A. Zalkin, and N. Bartlett, *Inorg. Chem.* 1974, *13*, 775.

L. Gmelin, *Handbuch der anorganischen Chemie,* Verlag Chemie, Weinheim, Germany, 1924–1971.

J. W. Mellor, *Comprehensive Treatise of Inorganic and Theoretical Chemistry,* Longmans Green, London, 1922–1937; Supplements, 1950–1967.

M. C. Sneed, J. L. Maynard, and R. C. Brasted, Eds., *Comprehensive Inorganic Chemistry,* Van Nostrand-Reinhold, New York, 1953–1961.

A. F. Wells, *Structural Inorganic Chemistry,* 4th ed., Oxford, Oxford, 1975. The best single-volume source for inorganic structures.

Inorganic Chemistry of the Main-Group Elements, A Specialist Periodical Report, Chemical Society, London, Vol. 5, 1978, and other volumes.

Nitrogen

C. B. Colburn, Ed., *Developments in Inorganic Nitrogen Chemistry,* 2 Vols., Elsevier, Amsterdam, 1966 and 1973.

J. Jander, "Nitrogen Triiodide, Tribromide, Trichloride, and Related Compounds," *Adv. Inorg. Chem. Radiochem* 1976, *19* 1.

J. Laane and J. R. Ohlsen, "Characterization of Nitrogen Oxides by Vibrational Spectroscopy," *Prog. Inorg. Chem.* 1980, *27,* 465.

Phosphorus

J. Emsley and D. Hall, *The Chemistry of Phosphorus,* Harper & Row, New York, 1976. A thorough source for P.

A. D. F. Toy, *Chemistry of Phosphorus,* Pergamon, New York, 1975.

Group VIB

G. Nickless, Ed., *Inorganic Sulfur Chemistry,* Elsevier, Amsterdam, 1960.

R. Stendel, "Properties of Sulfur-sulfur Bonds," *Angew. Chem. Int. Ed.* 1975, *14,* 655.

J. R. West, "New Uses of Sulfur," ACS *Advances in Chemistry* series, No. 140, 1975.

Group VIIB, the Halogens

A. J. Downs and C. J. Adams, *The Chemistry of Chlorine, Bromine, Iodine, and Astatine,* Pergamon, New York, 1975.

H. J. Eméleus, *The Chemistry of Fluorine and Its Compounds,* Academic Press, New York, 1969.

V. Gutman, Ed., *Halogen Chemistry,* Academic Press, New York, 1967. Three authoritative volumes.

P. Tarrant, Ed., *Fluorine Chemistry Reviews,* Dekker, New York. A continuing series of volumes.

J. C. Tatlow *et al.,* Eds., *Advances in Fluorine Chemistry,* CRC Press, Cleveland. A continuing series of volumes.

Noble Gases

N. Bartlett, *The Chemistry of the Noble Gases,* Elsevier, Amsterdam, 1971.

D. T. Hawkins, W. E. Falconer, and N. Bartlett, *Noble Gas Compounds,* Plenum Press, 1978. A bibliography covering 1962–1976.

G. J. Moody, "A Decade of Xenon Chemistry," *J. Chem. Educ.* 1974, *51,* 628.

PROBLEMS

13.1   The electron affinity of N(≤ 0) seems anomalous. Explain the order of electronegativities for Group VB (Table 13.1) and why one can conclude that N is really more "regular" than O.

13.2   What compound is used for inflating airbags in cars?

13.3   Write balanced equations for the preparation of HNO_3, starting with N_2.

13.4   When dilute nitric acid reacts with Cu turnings in a test tube, a colorless gas is formed that turns brown near the mouth of the tube. Explain the observations and write equations for the reactions involved.

13.5   Write equations for the preparation of five nitrogen oxides.

13.6   Sketch the *cis*- and *trans*-isomers of hyponitrous acid.

13.7   What properties of polymers of $(SN)_x$ and $(N\text{-}PR_2)_x$ make them of interest for practical uses?

13.8   Give the formula of a biodegradable polyphosphazene polymer that could be used for medical devices.

13.9 PI_5 is not known. What would be its likely structure in the solid?

13.10   *a.* What N species are compatible with the H_3PO_4-PO_4^{3-} species from the Pourbaix diagrams given by J. A. Campbell and R. A. Whiteker (*J. Chem. Educ.* 1969, *46*, 90).
      b. What P species are compatible with NH_3(aq.)?

13.11   How can white P be separated from red P?

13.12   How are P_4, P_4O_6, and P_4O_{10} related structurally?

13.13   In what way are phosphine ligands in metal complexes similar to CO?

13.14   Write the formulas for the diethylester of phosphorous acid and the monoethylester of hypophosphorous acid. Would these be protonic acids?

13.15   Sketch the *p-d-*π bonding in hexachlorotriphosphazene (Figure 13.4) and in cyclic $(PCl_2N)_4$ assuming it to be planar.

13.16   Write separate balanced equations and calculate E^0_{cell} for the reaction of H_2O_2 and NO_2^- and with MnO_4^- in acid solution.

13.17   What thermodynamic factors are important in determining the relative stabilities of solid metal oxides, peroxides, and superoxides? What cation characteristics (size and charge) would you choose to prepare metal oxides, peroxides, superoxides, and ozonides?

13.18   Give a description of bonding in O_2F_2 to account for the very short O—O bond and very long O—F bonds.

13.19   Write (separate) equations for the oxidation of $S_2O_3^{2-}$ by I_2 and by H_2O_2.

13.20   In the Pourbaix diagram for oxygen, H_2O_2 does not appear. Explain. (If you care to check the diagram, see reference in Problem 13.10.)

13.21   From Figures 13.10 and 13.11, *a.* What S and Se species are stable in contact with O_2? *b.* What S and Se species are stable in contact with H_2? *c.* What S species are stable in contact with Se?

13.22   How are dithionates and polythionates alike, and in what respects do they differ?

13.23   What are the products of reaction of SF_4 with alkylcarbonyls, carboxylic acids, and phosphonic acids [$R(RO)PO(OH)$]?

13.24   Give valence bond descriptions for SF_3 and SF_5 and indicate why they have low stability.

13.25   Write balanced equations for the following preparations:
Cl_2 (from NaCl).
Br_2 (recovery from seawater).
I_2 (from $NaIO_3$).
HF (from CaF_2).
HCl (from NaCl).

13.26   Draw the structures of the following species, indicating the approximate bond angles: IF_2^+, IF_4^+, and IF_6^+.

13.27   Indicate reactions which might be suitable for the preparation of:
a. Anhydrous tetramethylammonium fluoride.
b. Aluminum bromide.
c. Barium iodide.

13.28   The ΔH_f^0 of BrF(g), $BrF_3(l)$, and $BrF_5(l)$ are -61.5, -314, and -533 kJ/mole, respectively. Which of these species should be predominant on reacting Br_2 and F_2 under standard conditions?

13.29   Bromites and perbromates have been obtained only recently. Give the preparation of each.

13.30   What are the expected structures of ClO_3^+, $F_2ClO_3^-$, and $F_4ClO_2^-$? Sketch the regular figure and indicate deviations from idealized bond angles.

13.31   Use the emf data in the appendix to predict the results of mixing the following: *a.* Cl^- and BrO_3^- and 1 M acid; *b.* Cl_2 and IO_3^- in 1 M base; *c.* At_2 and Cl_2 in 1 M base.

13.32   Compare the expected *rate* of reaction of AtO_3^- and IO_3^- as oxidizing agents. What formula is expected for perastatinate? Why?

13.33   Give practical uses of He, Ne, and Ar, and give the sources of each.

13.34   What known compounds related in bonding and structure to the Xe halides should have prompted a search for the Xe halides earlier?

13.35   Give the expected shape and approximate bond angles for ClF_3O, considering the effects of the lone pair on Cl and the directional effects of the Cl=O π bond. (See K. O. Christe and H. Oberhammer, *Inorg. Chem.* 1981, *20*, 296.)

13.36   Draw the structure expected for $XeOF_4$ and indicate the approximate bond lengths (the Xe—F bond length in XeF_4 is 195 pm).

13.37 XeF_6 gives solutions in HF which conduct electricity. How might one distinguish between the following possible modes of dissociation:

$$XeF_6 + HF \rightleftharpoons XeF_5^+ + HF_2^-$$

and

$$XeF_6 + 2HF \rightleftharpoons XeF_7^- + H_2F^+$$

13.38   How might one distinguish between Xe^+ ions and Xe_2^{2+} ions in the compound $XePtF_6$?

XIV

Chemistry and Periodic Trends of Compounds of Metals (Groups IA–IIB)

We dealt with the structures of and the bonding in metals in Chapter 6, on inorganic solids. Compounds of metals range from those of the most extreme ionic type to covalent volatile substances such as $TiCl_4$ and OsO_4. The periodic variations among the metals and their compounds provide the basis for the use of size, charge, and electron configuration to remember, and even predict, chemical and physical properties. Recently, highly interesting research has revealed numerous examples of compounds involving metal–metal bonding. Simple compounds involving M—M bonding are included here, but metal cluster compounds are considered in Chapter 15, on cage and cluster compounds.

14.1 GENERAL PERIODIC TRENDS AMONG METALS

Periodic trends among the metals are rather regular. Melting points and hardness of the metals increase from Group IA to the middle of the transition series (Group VI) and decrease to the soft, low-melting metals of Group IIB. The other post–transition series metals are also rather soft and low-melting. The metals Sn, Pb, Sb, and Bi are used for low-melting alloys. In looking for periodic trends among melting points, only large differences and broad trends are significant, since melting points depend on specific interactions in the solid. Thus Ga has the widest liquidus range of any element (m.p. 29.8°C, b.p. *ca.* 2250°C), indicating that the attraction between Ga atoms is great (high b.p.), but the lattice energy is low. Ga is used in high-temperature thermometers.

A general view of the activities of metals is provided by ionization energies. For most metals, however, stable oxidation states can be achieved by removing more than a single electron, and we must consider also the stabilization of the ion through solvation and/or lattice effects. These stabilization effects increase with increasing oxidation states, so a balance must be achieved with the large increases in successive ionization energies (see examples for CaCl and NaF_2—p. 228). As a result, there is only one important oxidation state for Groups IA, IIA, and IIIA, except for some inner transition elements. Most of the other transition metals have $+2$ oxidation states of varying stability and higher oxidation states achieved by the stepwise removal of d electrons. For all but the largest metal atoms, such as Th, the removal of four electrons requires too much energy, so ionic compounds are limited to those of M^+, M^{2+}, M^{3+}, and in very few cases, M^{4+}. In addition to the high ionization energies involved, ions of very high charge density are strongly polarizing, so compounds of metals in high oxidation states are covalent. The insistence on balancing redox equations with CrO_4^{2-} and MnO_4^-, rather than the nonexistent Cr^{6+} and Mn^{7+} ions, is valid. These are covalent anions. The ease of formation of the highest oxidation states (same as the group number) increases generally with increasing atomic radius within a transition metal group.

Group IB metals show variation in oxidation states because the d orbitals participate. The $+1$ state for Group IIB involves the dimeric M_2^{2+} ions. Main-group metals to the right of the transition series all show the positive oxidation number corresponding to the group number (giving 18 electron or pseudo–noble-gas configurations). This highest oxidation state becomes more strongly oxidizing with increasing atomic number in each family (the reverse is true for transition metal families). With the decrease in the stability of the highest oxidation state, the oxidation state two units lower than the group number becomes increasingly more important (Tl^+, Pb^{2+}, Bi^{3+}). These ions have the $18 + 2$ configurations, corresponding to the retention of the ns^2 "inert" pair. This tendency to retain the "inert" pair (not necessarily stereochemically inert) increases with the greater penetration of the ns orbitals with increasing n.

Size effects can be treated in a straightforward manner for ionic substances and in a qualitative way using Fajans' rules where polarization (covalence) is important. The Mn^{2+} and Fe^{2+} ions are comparable in size with Mg^{2+} and substitute for the more abundant Mg^{2+} ion in minerals. Such isomorphous substitution of transition metal ions for others of similar size is very common. Radii used for the ions of the transition metals are usually those of the high-spin ions, unless the metal ion is coordinated to strong-field ligands. Size trends are rather regular, except for significant contraction following the first transition series and the lanthanide series. These contractions have important effects in the chemistry of Ga–Br and the members of the third transition series.

The chemistry of metals is largely that of their complexes, with the metal ions coordinated by anions in the solid and solvent molecules or other ligands in solution. Metals to the far left of the periodic table with low ionic potential and no ligand field stabilization effects tend to form labile complexes. In solution, they form solvated ions, except in the presence of ligands that are stronger hard Lewis bases than the solvent or multidentate ligands such as $edta^{4-}$ and cryptand ligands. Main-group metals to the right of the transition series also have no ligand field stabilization effects, but they are more highly polarizing and complexes of soft Lewis bases, particularly anionic ligands, are important—for example, complexes of Cl^- are more important than those of F^-. Transition metal complexes show the full range from labile to substitution inert and from hard acid–base interaction to the complexes of zero-valent metals that are very soft Lewis acids. The colors of transition metal complexes make us very much aware of the presence of complexes. Often, we can observe striking color changes as substitution or redox reactions occur.

The most common ligand is H_2O, since metal ions in solution are aqua complexes (in the absence of other ligands). Hydrolysis of metal ions such as Fe^{3+} in H_2O does not involve OH^- addition, but simple ionization of the acid $Fe(H_2O)_6^{3+}$. Loss of H^+ decreases the positive charge of hydrated metal ions, and the acidity increases with increasing ionic potential of the metal ion. For high oxidation states, proton loss is complete and oxo anions result.

14.2 GROUP IA—THE ALKALI METALS

14.2.1 Group Trends

The alkali metals are soft and low-melting because of weak bonding in the solid state. Only one electron beyond the noble-gas core is available for bonding. The melting points, boiling points, and hardnesses decrease with increasing atomic number, indicating a weakening in the bonding between the atoms. The atoms are partly associated as diatomic molecules at temperatures just above the boiling points; the dissociation energies of the molecules decrease from 100.9 kJ/mole for Li_2 to 38.0 kJ/mole for Cs_2. The metals' densities, which are low because of the large radii, increase with increasing atomic number—except for K, which has a lower density than Na. This irregularity results from differences in the rate of change in atomic weights as compared with the changes in the atomic radii. See Table 14.1 for a summary of some properties of the alkali metals.

Table 14.1 The alkali metals

ns^1	$_3Li$	$_{11}Na$	$_{19}K$	$_{37}Rb$	$_{55}Cs$
Abundance (% of earth's crust)	0.0065	2.83	2.59	0.028	3.2×10^{-4}
Density (g/cm³)	0.534	0.97	0.87	1.53	1.873
Melting point (°C)	179	97.9	63.7	38.5	28.5
Boiling point (°C)	1317	883	760	668	705
Sublimation energy (kJ/mole 25°C)	155.1	108.7	90.00	85.81	78.78
Ionization energy (eV)					
1st	5.392	5.139	4.341	4.177	3.894
2nd	75.638	47.286	31.625	27.28	25.1
3rd	122.451	71.64	45.72	40	35
Atomic radius (pm)	152	185	231	246	263
Ionic radius (pm), C.N. 6	90	116	152	166	188 (C.N. 8)
Heat of hydration of $M^+(g)$ (kJ/mole)	515	406	322	293	264
E^0 for $M(aq)^+ + e \rightleftharpoons M(s)$	-3.040	-2.714	-2.931	-2.925	-3.08

Low ionization energies make the alkali metals the most active family of metals. They have bright luster but, except for Li, tarnish so rapidly in air that they are stored under hydrocarbons.

Since the alkali metals give only 1+ ions with noble-gas configurations, the properties of their compounds vary more systematically than those of any other family in the periodic table. The very high second-ionization energies indicate that oxidation numbers higher than I are ruled out for the alkali metals. Most of the compounds are predominantly ionic in character, and since the ionic radii increase regularly down through the family, the Born treatment (p. 220) is particularly successful in dealing with trends. The cations have low charge and large radii—each has the largest radius of any cation from the same period—so that the lattice energies of their salts are relatively low. Consequently, most of the simple salts of the alkali metals are water-soluble. The low interionic attraction also results in high conductance of the salts in solution or in the molten state. Most of the salts are dissociated completely in aqueous solution, and the hydroxides are among the strongest bases available.

14.2.2 Occurrence

The only members of the alkali family abundant in the earth's crust are Na and K. The abundance of light elements generally is high, but that of Li, like that of Be and B, is quite low. Not only is the terrestrial abundance of these elements low, but also the cosmic abundance. According to all current theories of the origin of the elements, these elements, with low nuclear charge, would be expected to undergo thermonuclear reactions to produce heavier elements and/or helium. The low nuclear charge gives a low potential barrier for proton- or alpha-capture reactions. These elements could not accumulate during the formation of the elements, because they were used up in such reactions.

Geochemically, the most important occurrence of the alkali metals is in the aluminosilicate minerals, which make up the bulk of the earth's crust. Since the abundances of Rb and Cs are low, these elements rarely form independent minerals, but often are found in K minerals. In spite of its low abundance, Li forms independent minerals, because Li^+ is too small to replace the more abundant Na^+ or K^+ in their minerals. Lithium occurs in minerals along with Mg in aluminosilicates, which would be expected to separate in the very late stages of crystallization of a *magma*. In addition to the lithium ores obtained from *pegmatite* minerals, Searles Lake in California is an important source, furnishing about half of the world's supply of lithium. Searles Lake, an almost dry lake in the Mohave Desert, is strongly alkaline (pH 9.48).

A *magma* is the parent molten mass from which igneous rocks can be considered to separate.

Pegmatites are formed during the last stages of cooling and solidification of a deep-seated plutonic igneous rock. Pegmatites are characterized by large and irregular grain size. Since pegmatites represent the end product of the crystallization of a magma, rare elements become concentrated and the conditions are right for further differentiation and the growth of large crystals. The overall composition of pegmatites does not differ greatly from that of granites, but they are coarser-grained and are important sources of many rare elements.

The *sedimentary rocks* are secondary in origin, formed by such processes as the weathering of other rocks. The ores found in sedimentary rocks have been concentrated to a great extent by the chemical and physical changes caused by weathering. In the geological sense, a rock is any bed, layer, or mass of the material of the earth's crust. Natural waters are rocks composed mainly of the mineral water.

Although the bulk of the Na and K present in the earth's crust occurs in the aluminosilicate minerals, the important ores are found in the *sedimentary rocks*. Sodium and potassium are leached from the parent rocks by weathering. The Na^+ tends to concentrate in the seas, but K^+ is absorbed strongly by clays. Some NaCl is recovered from sea water, but the most important sources are the extensive deposits of rock salt left by the evaporation of isolated bodies of water. Deposits of K^+ salts of marine origin are also important sources of K.

14.2.3 Preparation of Metals

The most important general process for the preparation of the alkali metals is the electrolysis of fused salts or hydroxides. Most Na, the metal of this family produced in the greatest quantity, is obtained by the electrolysis of fused NaCl. Lithium also is obtained by electrolysis of the fused chloride, although KCl usually is added to lower the melting point of the mixture and permit the electrolysis to be carried out at lower temperature (about 450°C). The heavier alkali metals also are obtained by electrolysis, but thermal reduction processes are important as well. Potassium is obtained by reducing KF with CaC_2. Reduction of KCl with Na produces a K-Na alloy suitable for use as a heat exchanger. The heavier alkali metals also are obtained by reduction of the oxides with Al, Mg, Ca, Zr, or Fe.

$$2KF + CaC_2 \rightarrow CaF_2 + 2K + 2C$$

$$3Cs_2O + 2Al \rightarrow Al_2O_3 + 6Cs(g)$$

This process takes advantage of the high lattice energy of the oxides of the metals used as reducing agents and of the volatility of the alkali metals, which are vaporized in the process. Cesium can be obtained by reduction of Cs_2CO_3 with carbon or by the thermal decomposition of cesium tartrate. Thermal decomposition of the azides yields all alkali metals except Li in a high state of purity. Lithium forms the very stable nitride, Li_3N, and cannot be prepared in this way.

$$Cs_2CO_3 + 2C \rightarrow 2Cs + 3CO$$

$$2NaN_3 \rightarrow 2Na + 3N_2$$

The strong reducing power of the alkali metals accounts for their extensive use. Sodium is the metal usually used for this purpose, because it is easily available and inexpensive. Lithium imparts toughness to certain alloys and acts as a scavenger in steel. Cesium is used in photoelectric cells, since the absorption of radiant energy in the visible region of the spectrum can remove an electron because of its low ionization energy. Rb and Cs serve as ''getters'' in vacuum tubes, removing the last traces of corrosive gases in the tube when it is first put into use.

14.2.4 Cryptates[1]

Because of their large size and low charge density, the alkali metals show minimal tendency to form complexes. They do form complexes with β-diketones, and some of the fluorinated derivatives can be sublimed. The bonding should be largely ionic, and the chelates are insoluble in nonpolar solvents, except for some of their solvates.

An important development in the chemistry of the alkali metals has been the recent work on so-called *cryptate* complexes of polyethers (Figure 14.1) and nitrogen–oxygen macrocycles. (The name *cryptate* comes from the Greek word for "hidden.") The polyether macrocycles show selectivity among the alkali metal ions, depending on the size of the ring opening in the "crown." A cyclic polyether of four oxygens (crown-4) is selective for Li^+, whereas Na^+ prefers a crown-5 and K^+ a crown-6 macrocycle. The organic linkages joining the oxygens are puckered to give the "crown" arrangement, whereas the oxygens, with their lone pairs, are arranged in a nearly planar fashion about the metal ion at the center of the ring. The ligand $N[CH_2CH_2OCH_2CH_2OCH_2CH_2]_3N$ (a cryptand, abbreviated C222) can enclose Rb^+ completely in a "cage." These cryptates owe their stabilities largely to entropy effects in displacing several solvent molecules. They are important because the complex presents a hydrocarbon exterior, and they are soluble in organic solvents. Also, they are interesting models of those natural compounds that can transport Na^+ or K^+ selectively across membranes in order to maintain the remarkable electrolyte balance inside and outside cells.

14.2.5 Solutions of the Alkali Metals in Ammonia[2]

In the presence of catalysts, such as Fe, the alkali metals react with ammonia to form the metal amide and hydrogen. If no impurities are present, the metals dissolve without the liberation of hydrogen and can be recovered by evaporating the ammonia. All of the alkali metals dissolve to give solutions that are bronze if concentrated and blue if dilute. The bronze solutions conduct

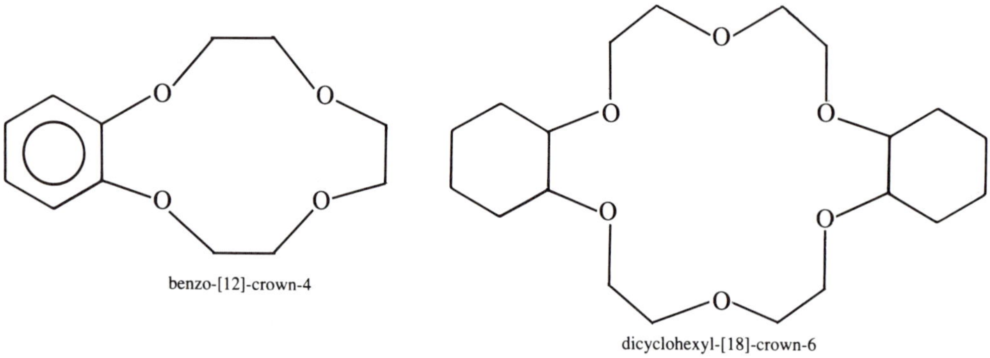

benzo-[12]-crown-4

dicyclohexyl-[18]-crown-6

Figure 14.1 Examples of crown ethers, using a simplified nomenclature that indicates the size of the macrocycle and the number of ligand atoms.

[1]C. J. Pedersen, *J. Am. Chem. Soc.* 1967, *89*, 2459, 7017; M. R. Truter, *Chem. in Britain 1971*, 203; C. J. Pedersen and H. K. Frensdorff, *Angew. Chem. Int. Ed.* 1972, *11*, 16.
[2]W. L. Jolly, *Metal Ammonia Solutions*, Dowden, Hutchinson and Row, Stroudsburg, Pa., 1972.

electricity about as well as many metals, and the blue solutions conduct electricity somewhat better than solutions of strong electrolytes (see Figure 14.2).

The characteristics of the solutions are essentially the same for all of the alkali metals, for the more active alkaline earth metals, and for europium and ytterbium. Thus the metals that dissolve in liquid ammonia are those with high negative electrode potentials—in other words, metals with low ionization energies, low sublimation energies, and high solvation energies. The solubilities of the alkaline earth metals are much lower than those of the alkali metals. The solubilities of the alkaline earth metals are limited, in part, by a tendency to form ammoniates. A saturated solution of Cs contains 25.1 moles of metal per 1000 g of ammonia at −50°C.

The very dilute blue solutions are regarded as solutions of electrolytes with the metal atoms dissociated to give ammoniated cations and ammoniated electrons. Since there is a very marked increase in volume when the metals dissolve in ammonia, the electrons are thought to occupy cavities with the surrounding NH_3 molecules oriented such that the hydrogens are directed inward. The unusually high conductivity for an electrolyte results from the high mobility of the solvated electron.

In the polarographic reduction of the alkali metal ions in liquid ammonia, amalgams are formed. In effect, the alkali metal ions are captured by the electron-rich mercury electrode. A cathodic wave is observed polarographically for a solution of tetrabutylammonium iodide in ammonia, but this corresponds to the dissolution of electrons. Instead of the tetraalkylammon-

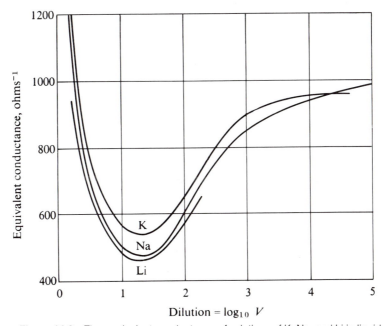

Figure 14.2 The equivalent conductance of solutions of K, Na, and Li in liquid ammonia at −33.5°C. V = liters of ammonia per gram-atom of metal. (From W. L. Jolly, *The Inorganic Chemistry of Nitrogen,* Benjamin, New York, 1964.)

ium ions dissolving in the mercury, the electrons are released as ammoniated electrons. The process is described by

$$e + xNH_3 = e(NH_3)_x \qquad E^0 = -1.89 \text{ V at } -34°C$$

When the electrolysis is carried out on a large scale, a blue solution results.

The decrease in conductance with increasing concentration apparently results from the metal ions becoming bound together in clusters by the electrons. At high concentrations the clusters become large enough to resemble liquid metals, with the cationic sites occupied by ammoniated metal ions.

14.2.6 Metal Anions[3]

Since its electron affinity is higher than that of any element other than the halogens, gold might be expected to form Au^- compounds. CsAu is an ionic compound containing Au^-, and the dissolution of metallic Au in ammonia solutions by K, Rb, or Cs yields solvated Au^-.[4] Table 14.2 provides thermodynamic estimates to judge the stability of M^-_{solv} in ammonia relative to the solvated electron and the solid metal. Since estimates are involved, negative or even small positive ΔG^0 values indicate that the formation of ammoniated metal anions might be expected for all of the alkali metals, Au, Ag, Pb, Bi, Tl, and Te. Because of the stabilization of metal–metal bonding, several main group metals to the right of the periodic table form polyatomic anions such as Sn_5^{2-}, Sn_9^{4-}, Pb_5^{2-}, Pb_9^{4-}, Sb_3^{3-}, Bi_4^{2-}, Te_2^{2-}, and Te_3^{2-}. Corbett and coworkers[5] have characterized several of these and determined their structures. Even though Pt has a very high electron affinity, its high sublimation energy makes the formation of Pt^-_{solv} very unfavorable. This does not preclude the formation of anionic clusters.

Early studies of metal–ammonia solutions might have led to the discovery of M^-, but the high solvation energies of M^+ and e in liquid ammonia favor the reaction

$$M^-_{solv} \rightleftharpoons M^+_{solv} + 2e_{solv}$$

in dilute solution. Solutions of Na in ethylenediamine and in methylamine contain Na^+ and Na^-, with electrical conductance of a normal 1:1 electrolyte. In these solvents the very high conductivity of dilute solutions of Cs indicates that the important species are Cs^+ and e_{solv}, but at higher concentrations the conductivity approaches that of Cs^+Cs^-.

The low solubility of alkali metals in amines has limited studies of M^- species. The crown ether (Figure 14.1) and cryptand ligands increase the solubilities of the metals and shift the equilibrium to the right, forming more M^-.

$$2M(s) \stackrel{\text{cry}}{\rightleftharpoons} M(cry)^+ + M^-$$

Sodium and potassium form $M^+(C222)M^-$ in tetrahydrofuran and in simple amines. The crystal structure of $Na^+(C222)Na^-$ confirms the existence of the simple Na^- anion. Solids containing K^-, Rb^-, and Cs^- also have been obtained.

[3] J. L. Dye, *Angew. Chem. Int. Ed.* 1979, *18*, 587.

[4] W. J. Peer and J. J. Lagowski, *J. Am. Chem. Soc.* 1978, *100*, 6260.

[5] J. D. Corbett *et al.*, *J. Am. Chem. Soc.* 1975, *97*, 6267; 1976, *98*, 7234; 1977, *99*, 3313; *Inorg. Chem.* 1977, *16*, 632.

Table 14.2 Thermodynamic estimates to judge the stability of M_{solv}^- in ammonia (kJ/mole)[a]

Metal	*Radius M^-, est. (pm)*	*Electron Affinity*[b]	$\Delta H^{0\,c}$	$\Delta G^{0\,c}$	$\Delta H^{0\,d}$	$\Delta G^{0\,d}$	
						298 K	238 K
Li	235	59.8	96.7	67.7	−65.8	1.8	−11.8
Na	272	52.9	53.4	30.6	−78.1	−5.6	−20.2
K	327	48.4	39.1	18.2	−59.3	13.5	−1.1
Rb	339	46.9	36.4	14.4	−56.3	15.2	0.8
Cs	355	45.5	30.7	10.9	−54.9	18.5	3.8
Au	200	222.7	140.9	109.0	−61.5	5.1	−8.3
Ag	200	125.7	156.4	125.4	−46.0	21.4	7.9
Cu	193	118.3	217.5	185.7	5.5	72.5	59.0
Ba	350	−52.1	225.4	202.3	137.7	207.9	193.7
Pt	355	205.3	357.5	320.6	271.9	328.2	316.9
Te	230	183.3	11.0	−20.8	−156.5	−91.4	−104.5
Pb	220	101.3	91.2	66.0	−86.8	−14.7	−29.2
Bi	220	101.3	103.3	72.3	−74.7	−8.4	−21.8
Tl	210	48.2	131.5	104.6	−58.1	12.8	−1.4
Sb	220	101.3	158.6	126.3	−19.5	45.6	32.5
Sn	220	120.6	179.0	152.2	1.0	71.4	57.3

[a]Adapted from J. L. Dye, *Angew. Chem. Int. Ed. Engl.* 1979, *18*, 587.
[b]Electron affinities are from the reference and differ slightly from those in Table 1.7.
[c]For $M_s + e_g \rightarrow M_g^-$ (Sum of E. A. and sublimation energy).
[d]For $M_s + e_{solv}^- \rightarrow M_{solv}^-$.

14.2.7 Francium

Mlle. Perey in 1939 found Element 87 in nature as a product of the decay of ^{227}Ac. The major product is ^{227}Th by β decay, but about 1% of the ^{227}Ac undergoes α decay to produce ^{223}Fr. This nuclide undergoes β decay with a half-life of 21 min.

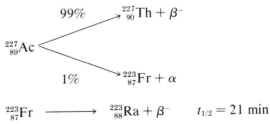

$$^{223}_{87}\text{Fr} \longrightarrow \quad ^{223}_{88}\text{Ra} + \beta^- \qquad t_{1/2} = 21 \text{ min}$$

Enough tracer work has been done to confirm that the chemistry of Fr resembles that of Cs. Considering the great similarity between Rb and Cs, the properties of Fr should be easily predictable.

14.2.8 Diagonal Relationships—The Anomalous Behavior of Lithium

In many of its properties, Li is quite different from the other alkali metals. This behavior is not unusual, in that the first member of each main group of the periodic table shows marked deviations from the regular trends for the group as a whole. The deviations shown by Li can be explained on the basis of the small radius of Li^+ and its high charge density. The nuclear charge of the Li^+ ion is screened only by a shell of two electrons. Since in ionic radius Li^+ (90 pm) is closer to Mg^{2+} (86 pm) than to Na^+ (116 pm), Li is more similar to Mg than to Na in a number of respects. Such *diagonal relationships* are of great importance among the active metals—for example, Na^+ and Ca^{2+}, K^+ and Sr^{2+}, Be^{2+} and Al^{3+}.

We can illustrate the diagonal relationship well by noting that the solubilities of Li compounds in water resemble those of Mg compounds to a greater extent than those of the other alkali metal compounds. Thus the fluoride, carbonate, and phosphate of Li or Mg are relatively insoluble and those of the other alkali metals are reasonably soluble. Adding acid increases the solubility of the Li or Mg salts mentioned through the formation of acid salts (FHF^-, HCO_3^-, $H_2PO_4^-$). However, adding acid to a solution of sodium carbonate precipitates $NaHCO_3$.

The low solubility of $NaHCO_3$ in water serves as the basis for the production of Na_2CO_3 by the Solvay process. In this remarkably efficient process, CO_2 (from limestone, $CaCO_3$) is passed through a water solution of ammonia and salt.

$$NaCl + NH_4HCO_3 \rightleftharpoons NaHCO_3 + NH_4Cl$$

The sodium hydrogen carbonate that precipitates is converted to sodium carbonate by heating.

$$2NaHCO_3 \rightarrow Na_2CO_3 + H_2O + CO_2$$

The CO_2 is recycled. After decomposing the NH_4HCO_3 in the mother liquor by warming, the NH_3 (from NH_4Cl) is displaced by adding $Ca(OH)_2$ (from the CaO produced on heating limestone). A solution of sodium carbonate is strongly basic, because of hydrolysis. For many large-scale uses calling for a strong base, sodium carbonate can be substituted for more expensive bases such as NaOH. Because of the efficiency of the Solvay process and the low-cost raw materials, sodium carbonate is the least expensive strong base and consequently is produced in very large quantities.

Lithium is the only alkali metal that reacts with N_2 gas. Presumably, the expected high lattice energy of Li_3N is responsible for the great stability of Li_3N. On burning Mg in air, some Mg_3N_2 is formed along with MgO. Lithium carbide is the only alkali metal carbide formed readily by direct reaction.

Lithium salts of small anions, like those of magnesium, have exceptionally high lattice energies, accounting for their high stability and low solubility. The anomalous behavior of lithium salts formed by large anions was discussed in the treatment of radius ratio effects (p. 217). For these salts the lattice energies are smaller than expected because of anion–anion repulsion.

The hydration energy of lithium ion exceeds that of any other alkali metal ion, because of the small radius of Li^+. Even though Li^+ is the smallest of the alkali metal ions, its ionic mobility is the lowest of the group, because of the extensive hydration resulting from its high charge density. Lithium is the only alkali metal for which hydrolysis of salts is of any importance. In this respect also it resembles Mg. Several Li compounds are useful because of properties appreciably different from those of Na compounds. Lithium alkyls or aryls are used in organic syntheses, where they undergo reactions similar to Grignard reagents (RMgX). $LiAlH_4$ and $LiBH_4$ are employed extensively as reducing agents in organic reactions. The organolithium compounds and the lithium hydride salts are used because of their solubility in nonpolar solvents. These compounds react with oxygen, water, or water vapor and are best handled in the absence of air.

The ionization and sublimation energies decrease regularly down through the alkali metal family. However, although the *IE* and *S* are greatest for Li, the electrode potential of Li is as negative as that of Cs. Lithium is a highly active reducing agent because of its very high hydration energy. The energy released in the hydration of the small Li^+ ion more than compensates for its higher ionization and sublimation energies. On the basis of heat effects alone we would conclude that Li should be much more active than Cs, but we have neglected the unfavorable entropy change for the formation of a well-ordered sheath of water molecules in $Li(aq)^+$. Although there are no outstanding anomalies among the electrode potentials of the remaining alkali metals, the order is not perfectly regular. Heat and entropy effects are rather delicately balanced, so the differences between the E^0 values are small and somewhat irregular.

14.2.9 Compounds of the Other Alkali Metals

The trends in the properties of the alkali metal compounds do not become really regular until we get to K, Rb, and Cs. Sodium does not deviate from the regular trends nearly so much as Li does, but it does differ from the heavier alkali metals in properties such as the solubilities and the extent of hydration of salts. Except for a few Li salts (mentioned above) and $NaHCO_3$, the slightly soluble salts of the alkali metals are those of K, Rb, and Cs—for example, the perchlorates and hexachloroplatinates. Since the ammonium ion is comparable in size (ionic radius 161 pm) to K^+ (152 pm), the solubilities of ammonium salts frequently parallel those of potassium. Actually, the radius of the NH_4^+ is the same as that of Rb^+, but the comparison is usually made to potassium salts because they are more common. We should expect some differences between ammonium and alkali metal salts, because of specific interactions between NH_4^+ ion and a solvent such as water through hydrogen bonding.

The similarities among the alkali metal salts often leave no reason to prefer the salt of one alkali metal over another. Consequently when a soluble salt with a cation not likely to cause interference is needed, the Na salt is selected because of its lower cost and availability. Potassium salts sometimes are preferred, especially for analytical purposes, because they are less often hydrated and usually are not hygroscopic. The more expensive Rb and Cs salts rarely offer advantages over those of Na or K. Organosodium and organopotassium compounds are ionic and not soluble in nonpolar solvents. They usually are not isolated, but prepared *in situ* for carbanionic reactions.

14.3   GROUP IIA—THE ALKALINE EARTH METALS

14.3.1   Group Trends

We conveniently can extend the original definition of the alkaline earth metals to include Be, Mg, and Ra. The metals of this family are harder, more dense, and higher melting than the corresponding members of the group of alkali metals. Each of the metal atoms has two valence electrons beyond the noble-gas configuration. Hence the alkaline earth metal atoms are more strongly bonded in the solid state than are the alkali metals. The trends in properties dependent on the bonding between the atoms indicate a decrease in the strength of bonding with increasing atomic radius, as is true for the alkali metals.

The first ionization energies of the alkaline earth metals are higher than those of the corresponding alkali metals because of the smaller radii and higher nuclear charge (compare Tables 14.1 and 14.3). Since the second ionization energies are considerably greater than the first, compounds of the metals with oxidation number I might be expected. Indeed, the formation of a compound such as MgF is an exothermic process ($\Delta H = -84$ kJ/mole) for $MgF(g)$, and such molecules exist at high temperatures in the gaseous state. Although the heat of formation of $MgF(s)$ would be considerably larger than -84 kJ/mole, the much greater lattice energy of MgF_2 favors the disproportionation to give Mg and MgF_2 in the solid state (see p. 228). $MgCl(g)$, $CaCl(g)$, and $CaF(g)$ have been reported also. In solution the monohalides are unstable, because of the much higher hydration energy of the divalent cations. Since the removal of a third electron would break up a noble-gas configuration and the energy required is prohibitively high, oxidation numbers higher than II are not encountered.

The large amount of energy necessary to sublime the metals and remove two electrons is compensated for by the high hydration energy of the divalent cations (or high lattice energy of the solids), so that the more active alkaline earth metals are comparable in activity to the alkali metals. Beryllium and magnesium are reasonably stable in air in spite of their high electrode potentials, presumably because oxidation is inhibited by the formation of a thin, adherent layer of metal oxide on the surface. The electrode potentials increase in magnitude with increasing atomic radius, following the order of decreasing ionization energies. The high hydration energy of Be^{2+} is not great enough to offset the trend established by the ionization and sublimation energies, because of the very large amount of energy required for these processes. Consequently, Be follows more regular trends than does Li.

The salts of the larger alkaline earth cations are typically ionic and conduct electricity well in aqueous solution and in the molten state. The conductances of fused salts of beryllium are low. Magnesium salts containing large, polarizable anions also are poor conductors of electricity in the fused state.

14.3.2   Occurrence

Magnesium and calcium are the only abundant alkaline earth metals in the earth's crust. The cosmic, as well as terrestrial, abundance of Be is low, as is true for Li for the same reasons (p. 610). Most of the Be in the earth's crust is found in silicate minerals, with Be replacing Si in tetrahedral SiO_4 units. Additional cations must be incorporated in the crystal lattice to prevent an imbalance of charge brought about by the replacement of Si(IV) by Be(II). The

Table 14.3 The alkaline earth metals

ns^2	$_4Be$	$_{12}Mg$	$_{20}Ca$	$_{38}Sr$	$_{56}Ba$	$_{88}Ra$
Abundance (% of earth's crust)	6×10^{-4}	2.09	3.63	0.015	0.040	1.3×10^{-10}
Density (g/cm³)	1.845	1.74	1.54	2.6	3.5	~5
Melting point (°C)	1284	651	851	770	710	(960)
Boiling point (°C)	2507	1103	1440	1380	1500	(1140)
Sublimation energy (kJ/mole, 25°C)	319.2	150	192.6	164	175.6	130
Ionization energy (eV)						
1st	9.322	7.646	6.113	5.695	5.212	5.279
2nd	18.211	15.035	11.871	11.030	10.004	10.147
3rd	153.893	80.143	50.908	43.6	35.5	—
Atomic radius (pm)	111	160	197	215	217	—
Ionic radius (pm) C.N. 6	41[a]	86	114	132	149	162[b]
Heat of hydration of $M^{2+}(g)$ (kJ/mole)	2385	1940	1600	1460	1320	—
E^0 for $M(aq)^{2+} + 2e = M(s)$	−1.85	−2.37	−2.87	−2.89	−2.90	−2.92

[a]C.N. 4.
[b]C.N. 8.

most important Be mineral is beryl, $Be_3Al_2[Si_6O_{18}]$. Emerald is crystalline beryl colored by a small amount of Cr^{3+}; aquamarine is also beryl of gem quality. Magnesium's abundant minerals might be considered to separate in the early stages of the crystallization of a magma. The *mantle* is believed to consist largely of forsterite, Mg_2SiO_4, and olivine, $(Mg,Fe)_2SiO_4$. The earth has a thin *crust* (17 km) covering the *mantle* (thickness 2880 km) and inner *core* (radial thickness 3471 km). The important magnesium ores are magnesite, $MgCO_3$; dolomite, $(Mg,Ca)CO_3$; and brucite, $Mg(OH)_2$—all found primarily in sedimentary rocks. Seawater is another important source of Mg.

Calcium is concentrated in minerals expected to separate during the main stage of crystallization of a magma. The bulk of the Ca in the earth's crust is found in the feldspars (p. 240), which make up about two thirds of the crust. Ca^{2+}, leached from rocks by weathering, does not concentrate in the sea, because of the low solubility of $CaCO_3$. The $Ca^{2+} + CO_2 + H_2O \rightleftharpoons CaCO_3 + 2H^+$ equilibrium is important in regulating the acidity of the sea and the CO_2 in the atmosphere. The most important Ca ore is $CaCO_3$ as limestone, marble, or sea shells. There are also extensive deposits of gypsum, $CaSO_4 \cdot 2H_2O$.

Strontium and barium form few independent minerals. Strontium frequently accompanies Ca and sometimes accompanies K, whereas barium often replaces K in minerals such as the potash feldspars. Heavy spar, $BaSO_4$, is an important Ba ore. Radium is found only as one of the daughters in the radioactive decay of heavier elements and occurs in all U minerals.

14.3.3 Preparation of the Metals

We can obtain Be metal by electrolyzing a fused BeF_2-NaF mixture; adding NaF permits the bath to operate at a lower temperature and improves the conductivity of the melt. Used in alloys with Cu for nonsparking tools, beryllium is also added to alloys to improve the resilience and fatigue-resisting properties of springs.

Magnesium, the most commercially important metal of the alkaline earth family, is obtained by electrolyzing a fused KCl-MgCl$_2$ mixture. The MgCl$_2$ can be obtained from ores or from seawater: after the magnesium is precipitated from seawater as the hydroxide by the addition of Ca(OH)$_2$, the Mg(OH)$_2$ is converted to the chloride for use in the electrolytic cell. We can also get Mg by reducing MgO with CaC$_2$, C, or Si. The reduction must be carried out at high temperature followed by quenching of the products to prevent reversal of the reaction. Mg's low density makes it useful as a structural metal. Mg reacts rapidly with water at elevated temperatures, but slowly at ordinary temperatures. Some of its low-density alloys (Mg-Al and Mg-Mn) are much more resistant to air oxidation than is the pure metal, presumably because the alloys form protective oxide layers to a greater extent.

The remaining alkaline earth metals can be obtained also by electrolyzing fused salts or by reducing the alkaline earth metal oxide or chloride with Al. Ca, Sr, and Ba are oxidized rapidly in air. Compared with Mg, they are produced in small amounts.

14.3.4 Anomalous Behavior of Beryllium

Like Li, Be stands apart from the remainder of the family in many respects. The hardness of Be (6–7 on Mohs' scale) is much greater than that of Mg (2.6) or Ca (2.2–2.5). The melting points of all alkaline earth metals except for Be fall within a narrow range (Table 14.3).

Since many properties depend greatly on charge density, there is some validity to the statement that Be is more similar to Al (diagonal relationship) than to Mg. The ionic potential (ionic charge/radius) of Be^{2+} [0.039, using r (C.N. 6) = 58 pm for comparison to Al^{3+} and Mg^{2+}] is closer to that for Al^{3+} (0.044) than to that for Mg^{2+} (0.023). Beryllium compounds are largely covalent. Thus, fused BeCl$_2$ is a rather poor electrolyte. Beryllium oxide is hard and refractory, like Al$_2$O$_3$; when fired, it is practically insoluble in acids. Beryllium nitrate, sulfate, and the halides are all soluble, but the hydroxide, oxide, and phosphate are all insoluble. The normal beryllium carbonate, like that of Al, is unstable except in the presence of CO$_2$. Basic beryllium carbonate of variable composition is precipitated by the addition of sodium carbonate to a solution of a soluble Be salt. Be$_2$C and Al$_4$C apparently are the only common saltlike metal carbides that yield methane on hydrolysis.

The extent of hydrolysis of Be salts in solution parallels much more that of Al salts than that of the salts of the rest of the alkaline earth family. Beryllium has a great tendency to form stable complexes with F$^-$ or oxoanions. Tetrahedral complexes are formed usually. Salts of the type [NaBeF$_3$]$_n$ contain long chains of repeating BeF$_3^-$ units, which are analogous to the metasilicates (SiO$_3^{2-}$)$_n$ and metaphosphates (PO$_3^-$)$_n$. We can think of fluorides as weakened models of oxides of the same structural type—for example, BeF$_2$ for SiO$_2$, MgF$_2$ for TiO$_2$, and NaF for CaO. The fluorides have lower lattice energies than the corresponding oxides, because of the lower charges. Because of the charge effects, salts of (BeF$_3^-$)$_n$ in solubilities and melting points resemble more closely (PO$_3^-$)$_n$ salts than (SiO$_3^{2-}$)$_n$ salts.

The remaining members of the family show little tendency to form complexes, except with rather exceptional ligands such as edta (ethylenediaminetetraacetic acid). The complexes of the alkaline earth metals other than Be usually are octahedral.

Beryllium forms an unusual compound by combining with acetic acid (and also with other carboxylic acids). The compound is basic beryllium acetate, $Be_4O(O_2CCH_3)_6$, in which the four Be^{2+} are arranged tetrahedrally around a central oxide ion, with a carboxylate ion spanning each of the six edges of the tetrahedron (Figure 14.3). The molecule contains interlocking six-membered rings, satisfying the tetrahedral bonding requirements of Be^{2+}. Basic beryllium acetate is insoluble in water but soluble in organic solvents such as chloroform. It can be distilled under reduced pressure. Similar, though less stable, compounds of Zn^{2+} and ZrO^{2+} are known.

The fact that the solubilities of both BeO and $BeSO_4$ are increased when both are present in the same solution further demonstrates beryllium's great tendency to form complexes involving oxide bridges. Individually, $BeSO_4$ is moderately soluble and BeO is virtually insoluble. Together they form a complex in which the Be^{2+} is "solvated" with four BeO molecules (Figure 14.4). The coordination requirements of these BeO molecules can be satisified by hydration.

The characteristic coordination number of Be^{2+} in crystals is four. The oxide, BeO, has the wurtzite structure (*4PT*) and BeS has the zinc blende structure (*6PT*). Most metal ions that do not replace Si^{4+} in silicates have C.N. 6, but in beryl, $Be_3Al_2[Si_6O_{18}]$, and phenacite, $Be_2[SiO_4]$, the Be^{2+} is in tetrahedral sites. The structure of BeF_2 is that of cristobalite (a polymorph of SiO_2—see Figure 6.29). Crystals of the more covalent $BeCl_2$ contain chains of $BeCl_4$ tetrahedra (distorted). The structure of solid $Be(CH_3)_2$ is similar, although the $Be(CH_3)_2$ molecules are monomeric and linear in the vapor.

It is easy to separate Be^{2+} analytically from the remaining alkaline earth metals, but separation from Al^{3+} is difficult. We can take advantage of the formation of basic beryllium acetate, which can be extracted into chloroform, leaving Al^{3+} in the aqueous phase. Also, Al^{3+} can be precipitated as the hydroxide by adding an excess of ammonium carbonate, but Be^{2+} is soluble, probably because of the formation of a carbonato complex. We can precipitate Al^{3+} as $AlCl_3 \cdot 6H_2O$ by adding ether to a solution saturated with HCl gas. Beryllium chloride remains in solution.

Beryllium compounds are *extremely* toxic. Do *not* try to verify the sweet taste of beryllium compounds (which led to the old name glucinium for Be). Be^{2+} can displace Mg^{2+} from important Mg-activated enzymes, blocking enzymatic activity.

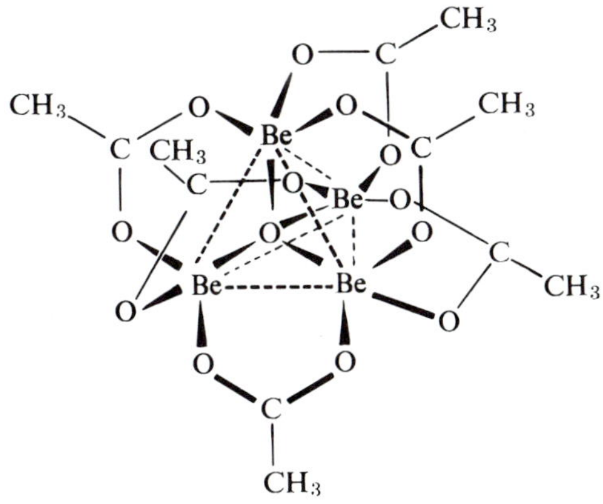

Figure 14.3 Basic beryllium acetate.

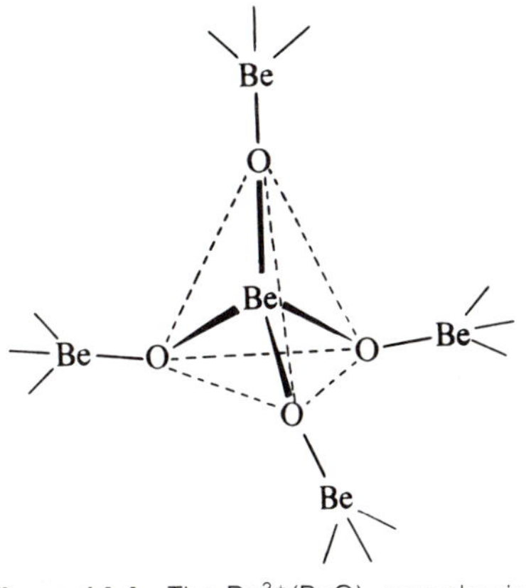

Figure 14.4 The $Be^{2+}(BeO)_4$ complex ion.

14.3.5 Compounds of the Remaining Alkaline Earth Metals

Most of the compounds of the remaining members of Group IIA are rather typical ionic compounds for which the trends are quite regular. The hydroxides are not very soluble, except for $Ba(OH)_2$ (a saturated solution is about 0.05 M). Barium hydroxide is a strong base and although slaked lime, $Ca(OH)_2$, is quite caustic, milk of magnesia, $Mg(OH)_2$, is mild enough to be swallowed as a laxative.

Magnesium bridges the gap between Be and the heavier members of the family in its slight tendency both to hydrolyze and to form complex ions. The compounds of Ca, Sr, and Ba are quite similar to one another in many respects. In an analytical separation scheme, these three metals usually are precipitated as the carbonates and then separated from one another utilizing differences in the solubilities of salts such as $MCrO_4$ at different acidities.

All of the alkaline earth metal compounds of the oxygen family except those of Be (see above and Table 6.4) have the NaCl structure. The fluorite structure ($9PTT$ C.N. 8) appears in the halides with the largest radius ratios: CaF_2, SrF_2, BaF_2, and $SrCl_2$. The rutile structure ($4PO_{1/2}$, C.N. 6) occurs for MgF_2, but layer structures (p. 213) are encountered for $MgCl_2$ ($CdCl_2$ structure), CaI_2, $MgBr_2$, and MgI_2 (CdI_2 structure). A distorted rutile structure characterizes $CaCl_2$ and $CaBr_2$. The complex structure of $PbCl_2$ with C.N. 9 appears in $BaCl_2$, $BaBr_2$, and BaI_2.

The Grignard reagents, usually given the general formula RMgX, have been widely used in organic reactions. The reagents are prepared by the reaction of Mg metal with the alkyl or aryl halide: for example, C_2H_5Br yields C_2H_5MgBr in anhydrous diethyl ether. See Section 10.6 for discussion of metal alkyls and Grignard reagents.

14.4 GROUP IB—THE COINAGE METALS

14.4.1 Group Trends

Although K and Cu each contains a single electron in the fourth shell, the differences in properties are very great. The Cu atom has a smaller radius, because of contraction through the transition series, and a higher nuclear charge. Both factors favor an increase in the ionization energy and sublimation energy of Cu, resulting in a decrease in electropositive character of reactivity as a metal (see Table 14.4). The increases in sublimation and ionization energies are both quite large, and each contributes significantly to the nobility of the coinage metals. However, the differences in the activities of the alkali and coinage metals are even greater than would be expected on the basis of the decrease in size and increase in nuclear charge through the transition series. The additional factor is the difference in the electronic configuration. Beneath the outer shell, each of the alkali metals has a noble-gas core; but each of the coinage metals has an 18-electron shell—a pseudo–noble-gas configuration. The 18-electron shell does not screen the positive nuclear charge as effectively as an 8-electron shell. The contrast in reactivity is obvious from the fact that the alkali metals include the most active metals and the coinage metals include Au, the most noble metal.

The coinage metals offer a striking contrast to the alkali metals in another respect: they decrease in reactivity with increasing atomic number. Because of the less effective shielding

Table 14.4 The coinage metals

$(n-1)d^{10}ns^1$	$_{29}Cu$	$_{47}Ag$	$_{79}Au$
Abundance (% of earth's crust)	7×10^{-3}	1×10^{-5}	5×10^{-7}
Density (g/cm³)	8.94	10.49	19.32
Melting point (°C)	1083	960.5	1063
Boiling point (°C)	2595	2212	2966
Sublimation energy (kJ/mole, 25°C)	341.1	299.2	344.3
Ionization energy (eV)			
1st	7.726	7.576	9.225
2nd	20.292	21.49	20.5
3rd	36.83	34.83	—
Atomic radius (pm)	128	144	144
Ionic radius (pm) C.N. 2	Cu^+ 60	Ag^+ 81	Au^+ 151 (C.N. 6)
C.N. 6	Cu^{2+} 87	Ag^{2+} 93 (C.N. 4)^{sq}	Au^{III} 82 (C.N. 4)^{sq}
Heat of hydration of M^{n+} (kJ/mole)	Cu^+ 581 Cu^{2+} 2,120	Ag^+ 485	Au^+ 644
E^0 for $M^+(aq) + e \rightarrow M(s)$	+0.521	+0.799	*ca.* +1.68
$M^{2+}(aq) + 2e \rightarrow M(s)$	+0.337	+1.39	—

of the nuclear charge by the 18-electron shell, the increase in nuclear charge is more important than the increase in atomic radius in determining the attraction for electrons. The formation of Ag^+ requires more energy than for Cu^+, as indicated by the emf values, even though the first ionization energy of Ag is slightly *lower* than that of Cu and the sublimation energy is appreciably lower. Only the higher hydration energy of the smaller Cu^+ ion makes the oxidation process more favorable for Cu (see Table 14.4). The fact that the lanthanide series comes between Ag and Au causes the atomic radii to be the same. The resulting increase in the first ionization energy and the much higher sublimation energy account for the much greater nobility of Au.

The coinage metals are among the most dense metals (Table 14.4), and their melting points are very similar. There is a surprisingly large minimum in the sublimation energy for Ag. As well as having the highest electrical conductivities of any element, they are the most malleable and ductile structural metals. Gold can be rolled into sheets as thin as 10^{-5} cm.

14.4.2 Occurrence

In spite of their low abundances, the coinage metals were among the earliest discovered metals. All three metals occur in the native state. Native Au usually is found in placer deposits, such as in the bed of a stream, left by the weathering of the original rocks. Since Au is dense and resistant to oxidation, it remains after most of the other minerals present have been removed by chemical attack or by the mechanical forces of weathering. The coinage metals also commonly occur as sulfide minerals, from which early man could have obtained the metals in his campfires—as the minerals are easily reduced.

14.4.3   **Preparation of the Metals**

The copper sulfide ores are converted to the oxide by roasting in air. The oxide is reduced to the metal by carbon. If the supply of air is cut off before the conversion to the oxide is complete, the Cu can be obtained directly.

$$2Cu_2S + 3O_2 \rightleftharpoons 2Cu_2O + 2SO_2$$

$$2Cu_2O + Cu_2S \rightleftharpoons SO_2 + 6Cu$$

The Cu obtained directly from the ores is generally of low purity. Purification can be accomplished by heating the crude Cu in a furnace to form some Cu_2O, which serves to oxidize the more reactive metals present. The remaining Cu_2O is reduced to the metal by wood charcoal or by stirring with wooden poles. The Cu obtained is more than 99% pure.

Electrolytic refining is the more important process for obtaining pure Cu. The crude blister Cu cast as a large anode is suspended in an acidified copper sulfate solution containing a small amount of chloride ion. Electrolysis oxidizes Cu and the more active metals to their ions and deposits pure Cu at the cathode. Noble metals present in the blister Cu are not oxidized and fall to the bottom of the cell along with other impurities as the anode *mud*. Any Pb that is oxidized precipitates as lead sulfate; any Ag that might be oxidized would precipitate as AgCl. The recovery of precious metals from the anode mud is profitable enough to pay a large share of the cost of refining Cu. More active metals remain in solution. Because of the low operating voltage (a few tenths of a volt), many cells normally are operated in series.

Silver is obtained as a by-product in the recovery of Pb, Zn and Cu from their ores. Silver ores (Ag_2S, AgCl, or Ag) commonly are treated by the cyanide process. In the presence of oxygen, Ag dissolves in aqueous NaCN solution as the complex ion $[Ag(CN)_2]^-$. Sulfide ion is oxidized to SO_4^{2-} or SCN^-. The Ag is oxidized by oxygen, because of the great stability of $Ag(CN)_2^-$. The emf for the oxidation is shifted by over a volt [compared with the formation of $Ag(aq)^+$].

$$Ag + 2CN^- \rightarrow Ag(CN)_2^- + e \qquad E^0 = +0.31 \text{ V (as oxidation)}$$

Silver is recovered from the solution by the addition of Zn dust, or in some applications, by electrolysis.

Gold-containing ore can be enriched by a washing process in which the ore is broken up and suspended in water in order to separate the less dense impurities from the Au, which settles rapidly because of its great density. More complete recovery of Au is possible if the concentrated ore is treated by an amalgamation process. The Au dissolves in the Hg to form an amalgam and then is recovered by distilling the Hg.

More important for the extraction of Au is the cyanide process, which is applicable even to very low grade ores. The ore is ground finely and treated with a dilute cyanide solution (0.1 to 0.2% KCN or NaCN). Oxygen in the air oxidizes the Au to a soluble cyanide complex, to which Zn is added to yield the Au. Gold(I) is stabilized even more than silver(I) by coordination with CN^-, so that Au becomes a strong reducing agent. The emf is shifted by over two volts [compared with the formation of $Au(aq)^+$].

$$Au + 2CN^- \rightarrow Au(CN)_2^- + e \qquad E^0 = +0.60 \text{ V (as oxidation)}$$

From this shift in emf, the formation constant for the complex is 2×10^{38}. (See Section 11.10.1 for the calculation of equilibrium constants from E^0 values.)

14.4.4 Compounds of the Coinage Metals

Only one oxidation state is encountered for the alkali metals, because the first ionization energies are very low and the energy needed to remove additional electrons from the underlying noble-gas shell is prohibitively high (see Tables 14.1 and 14.4). The differences among the first three ionization energies are not so great for the coinage metals, and the formation of compounds in the higher oxidation states is more favorable because of complex formation.

A number of subhalides of the type M_2X have been reported for the coinage metals, but most of them probably are mixtures. The only verified subhalide of the group is Ag_2F, a bronze-colored solid obtained by the cathodic reduction of AgF in aqueous solution. The solid has double layers of Ag "atoms" separated by layers of F^- ions. With the Ag "atoms" held together by metal–metal bonds, no discrete Ag^+ or Ag_2^+ ions can be identified.

The *oxidation number I* is common to all three of the coinage metals. Copper(I) is stable only in solid compounds or in a few complexes formed by ligands such as CN^-, SCN^-, $S_2O_3^{2-}$, or thiourea. In solid $K[Cu(CN)_2]$, each Cu^I is bonded to three CN^- in an almost planar trigonal arrangement. Two of the CN^- are bound through C, and the third (bridging) CN^- is bound through C and N. Water molecules doubtless are coordinated in aqueous solution. Compounds of the type $M_2^I CuX_3$ and $M^I Cu_2 X_3$ contain single and double chains, respectively, of CuX_4 tetrahedra through sharing of X^- ions. The polynuclear complexes of the type $Cu_4 I_4 L_4$ (where L is a tertiary phosphine or arsine) contain tetrahedra of four Cu with an I above each face to complete a cube of four Cu and four I with one phosphine or arsine attached to each Cu.

Copper(II) is reduced by I^- or CN^- with the formation of insoluble CuI or CuCN, but CuCN dissolves in the presence of excess CN^- with complex formation.

$$2Cu^{2+} + 4I^- \rightarrow 2CuI(s) + I_2$$

$$2Cu^{2+} + 4CN^- \rightarrow 2CuCN(s) + (CN)_2$$

$$CuCN(s) + CN^- \rightarrow Cu(CN)_2^-$$

$$2Cu^+ \rightarrow Cu + Cu^{2+} \qquad E^0 = +0.37 \text{ V}$$

The solids CuCl, CuBr, and CuI have the zinc blende structure (*6PT*, C.N. 4). The hydrated copper(I) ion and solid CuF are unstable with respect to disproportionation. Solutions of Cu(I) complexes absorb olefins by forming Cu(I) olefin complexes. This process has been used to separate olefins from other hydrocarbons and monoolefins from polyolefins. As is generally true, lower oxidation states are more stable at high temperatures. CuO is converted to Cu_2O at elevated temperatures.

$$4CuO \xrightarrow{\text{heat}} 2Cu_2O + O_2$$

Silver(I) is the most stable oxidation state of Ag. Silver nitrate, silver fluoride, and silver perchlorate are very soluble in water. The acetate, sulfate, chlorate, and bromate range from slightly soluble to moderately soluble, but most of the rest of the common salts are quite insoluble. The solubilities of the silver halides decrease in the order AgCl (K_{sp} = 1.8 $\times$ 10^{-10}), AgBr(K_{sp} = 3.3 $\times$ 10^{-13}), AgI(K_{sp} = 8.5 $\times$ 10^{-17}). The NaCl structure (6*PO*) characterizes all the halides except AgI (wurtzite, 4*PT*), whose interesting polymorphism has been discussed (see Section 6.4.3). Chain structures are encountered for the covalent AgCN and AgSCN. Many silver compounds containing colorless anions are themselves colored: for example, Ag_3PO_4, Ag_2CO_3, and AgI are yellow, and Ag_2S is black. Salts containing polarizable anions and highly polarizing cations—for example, those with 18 or 18 + 2 electronic configurations—often are colored (see p. 198). Silver(I) forms many stable complexes, particularly with N and S donors. Although in complexes Ag^+ generally shows C.N. 2, in solution, solvation perhaps increases its effective C.N. to 4.

The silver(I) halides (except the covalent AgI) show a higher C.N. in the solid than the copper(I) halides, but in its complexes, Ag(I) demonstrates a pronounced preference, compared with Cu(I), for the lower C.N. 2. This indicates a greater tendency of Ag(I) to form linear bonds. Whereas C.N. 2 seems predominant for Au(I), in Group IIB only Hg(II) shows a preference for C.N. 2. Linear bonding involving *s-p* hybridization can be enhanced, as a result of more directional character of the hybrid orbitals, by some mixing of the d_{z^2} orbital also. The energies of the *ns* and (*n* − 1)*d* levels, as indicated by 2nd *IE* − 1st *IE*, are about the same—12.6 ± 1.3 eV for Cu, Ag, and Au, although lowest for Au. These energy differences (3rd *IE* − 2nd *IE*) are much greater for Zn (21.7 eV) and Cd (20.6 eV), but comparable for Hg (15.4 eV). Planar trigonal coordination is encountered for Au(I) in chlorobis(triphenylphosphine)gold(I).

Gold(I) chloride and bromide are obtained by gently heating the corresponding gold(III) halides. These gold(I) halides disproportionate in aqueous solution to give the metal and the corresponding gold(III) halide.

$$AuCl_3 \xrightarrow{\text{heat}} AuCl + Cl_2$$

$$3AuCl + H_2O \longrightarrow AuCl + 2Au$$

$$AuCl_3 + 3I^- \longrightarrow AuI(s) + I_2 + 3Cl^-$$

Gold(I) iodide is precipitated from solutions containing gold(III) compounds and iodide ion. Because of its low solubility, AuI is decomposed more slowly by water. Gold(I) cyanide, so insoluble that it is stable in the presence of water, dissolves in the presence of a cyanide salt to give a soluble complex, such as K[Au(CN)$_2$]. Gold(I) also gives soluble complexes with $S_2O_3^{2-}$ and SO_3^{2-}. The insoluble Au_2S can be precipitated from an aqueous solution of any soluble gold(I) complex by saturating the solution with H_2S.

Copper(II) occurs in most of the stable copper salts and complexes in the solid state and in solution. The copper(II) complexes are usually square planar, although there are often two additional ligands or solvent molecules at slightly greater distances, one above and one below the plane of the four closer ligands. Copper(II) fluoride has the fluorite structure typical of ionic salts. Copper(II) chloride and bromide have chain structures commonly encountered among salts with an appreciable amount of covalent character.

Silver(II) compounds are uncommon. Silver(II) fluoride, formed by the action of F_2 on AgF, is used as a fluorinating agent for organic compounds. An oxide, AgO, is obtained by the anodic oxidation of aqueous solutions of silver salts. Since it is diamagnetic, it is probably a silver(I)-silver(III) compound, $Ag^I[Ag^{III}O_2]$. The divalent state is stabilized by coordination with pyridine, $[Ag(py)_4](NO_3)_2$; pyridine derivatives such as picolinic acid; 1,10-phenanthroline; bipyridine; and a few other heterocyclic nitrogen ligands. The complexes presumably are planar, like those of copper(II).

Of the several gold(II) compounds reported, most have been shown to be mixed Au(I)–Au(III) compounds. The simple salts $AuSO_4$, AuO, and AuS have been reported, but whether they are authentic gold(II) compounds is questionable. No gold(II) complexes have been verified.

Copper(III) and silver(III) complexes with periodate and tellurate ions, such as $K_7[Cu(IO_6)_2]\cdot nH_2O$ and $K_9[Cu(TeO_6)_2]\cdot nH_2O$, have been reported, and the fluorocomplex $K[AgF_4]$ has been prepared. Most commonly encountered are gold(III) compounds. Gold is usually dissolved in *aqua regia* to obtain chloroauric acid, $H[AuCl_4]\cdot 4H_2O$. Neither hydrochloric nor nitric acid alone will dissolve Au. Gold(III) is stabilized sufficiently by coordination with chloride ion to lower the emf required such that the oxidation can be accomplished by HNO_3. Copper(III) can be stabilized by coordination to peptides, and perhaps it is biologically important (p. 744).

All gold compounds are decomposed easily by heating. $AuCl_3$ and $AuBr_3$ give the monohalides upon gentle heating, but the metal is formed when the halides are heated more strongly. AuI_3 is unstable with respect to the loss of I_2, even without heating. Except for a few stable solids, the gold(III) compounds encountered are planar complexes. Fluorination of $Cs[AuF_4]$ yields $Cs[AuF_6]$ containing gold(V) (d^6) as the octahedral AuF_6^- ion.[6]

14.5 GROUP IIB—THE ZINC SUBGROUP

14.5.1 Group Trends

The zinc subgroup elements are more dense and less active than their counterparts in the alkaline earth family, as would be expected from their smaller radii and higher nuclear charge. Sublimation energies of the Group IIB metals are lower than for any other group of metals except the alkalis (Table 14.5). Correspondingly, the metals have low melting points and boiling points and can be distilled. The 18 + 2 configuration characteristic of this family practically constitutes a closed configuration from the standpoint of metal–metal bonding in the free state. The increased activity of the IIB family compared with the IB family results partly from the low sublimation energies of the metals and partly from the high hydration energy of the divalent ions. Obviously, it is not caused by a decrease in ionization energies. The radii of the IIB metals are slightly larger than those of the corresponding IB metals.

The IIB metals decrease in activity with increasing atomic number, like the IB metals. The increase in nuclear charge is more important than the increase in atomic radius. Zinc and cadmium are very active metals, and Hg is a distinctly noble metal—in fact, it has the highest first ionization energy of any metal. Group IIB is the first group after Group IVA in which the members of the first and second transition series (Zn and Cd) are more similar than the members of the second and third transition series (Cd and Hg).

[6]K. Leary and N. Bartlett, *J. Chem. Soc. Chem. Comm.* 1972, 903; K. Leary, A. Zaldin, and N. Bartlett, *Inorg. Chem.* 1974, *13*, 775.

Table 14.5   The zinc family

$(n-1)d^{10}ns^2$	$_{30}Zn$	$_{48}Cd$	$_{80}Hg$
Abundance (% of earth's crust)	8×10^{-3}	1.8×10^{-5}	5×10^{-5}
Density (g/cm³)	7.133	8.65	13.55
Melting point (°C)	419.5	320.9	-38.87
Boiling point (°C)	906	767	357
Vaporization energy (kJ/mole 25°C)	130.5	112.8	60.84
Ionization energy (eV)			
1st	9.394	8.993	10.437
2nd	17.964	16.908	18.756
3rd	39.722	37.48	34.2
Atomic radius (pm)	133	149	150
Ionic radius (pm) M(II) C.N. 6	88	109	116
Heat of hydration of M^{2+} (kJ/mole)	2,060	1,830	1,840
E^0 for $M^{2+}(aq) + 2e \rightarrow M$	-0.763	-0.403	$+0.849$

14.5.2   Occurrence

The most important zinc ores are smithsonite, $ZnCO_3$, and zinc blende, ZnS. Cadmium occurs along with zinc in the same minerals. Independent cadmium minerals are rare. The principal Hg ore is cinnabar, HgS. Often, some native Hg is associated with the cinnabar.

14.5.3   Preparation of the Metals

Roasting zinc ores (ZnS or $ZnCO_3$) yields the oxide, which is then reduced with C. The Zn is distilled from the furnace. In the electrolytic process, which is becoming increasingly more important than C reduction, the ore is leached with sulfuric acid to give a solution of zinc sulfate, electrolysis of which produces Zn. Zinc is deposited even from the acidic solution, because of the high hydrogen overvoltage at the Zn cathode.

Cadmium can be recovered from the Zn produced by C reduction, because of the greater volatility of Cd. The Cd is obtained by repeated fractional distillation; it also can be separated from Zn in the electrolytic preparation, because of the lower deposition potential of Cd.

Mercury can be recovered by heating the sulfide in air—the Hg vapor distills. Metals more active than Hg can be removed by leaching with dilute nitric acid.

14.5.4   Compounds of the Group IIB Metals

The characteristic oxidation number of the Group IIB metals is II. Their salts are associated extensively in solution. The Group IIB metals resemble Be and Mg in the tendency to form stable complexes. Zinc forms four-coordinate tetrahedral complexes, with few examples of higher C.N. Most of the Cd complexes are similar to those of Zn, except for the OH^- complex—$Cd(OH)_2$ is not amphoteric.

Mercury(II) usually gives tetrahedral complexes. The formation constants for the addition of the first two ligands usually are considerably greater than for the last two, suggesting a definite tendency to form linear complexes with C.N. 2. Mercury(II) shows a very marked preference for large, polarizable ligands. Although complexes containing nitrogen ligands are very stable, there is a great preference for S-ligands over O-ligands. The order of increasing stability of the halide complexes is $Cl^- < Br^- < I^-$, with little or no tendency to form fluoride complexes. Mercury(II) oxide is basic and does not tend to form hydroxide complexes.

Except for CdO and HgO, the MX compounds of O, S, Se, and Te have the zinc blende or wurtzite structures (C.N. 4—see Table 6.4). Cadmium oxide has the NaCl structure (C.N. 6), whereas the orthorhombic form of HgO contains zigzag chains of essentially linear O–Hg–O units. A second, hexagonal, form of HgO is isostructural with a second (and more common) form of HgS, cinnabar. Cinnabar is built of helical chains of S–Hg–S units (S–Hg–S angle 172°). Two Hg–S distances are 236 pm, two are 310 pm, and two are 330 pm.

ZnF_2 (rutile structure, C.N. 6) and CdF_2 (fluorite structure, C.N. 8) are the only ionic-type structures of the halides of Zn and Cd. Of the three polymorphic forms of $ZnCl_2$, two have structures built of $ZnCl_4$ tetrahedra and the third has the HgI_2 layer structure (p. 214). Cadmium chloride and zinc bromide have a layer structure containing a cubic close-packed array of X^- with M^{2+} in octahedral holes in alternate layers (p. 214). The CdI_2, $CdBr_2$, and ZnI_2 structures are similar except that the X^- ions are hexagonally close-packed.

The mercury(II) halides show interesting variations in lattice type (see p. 215). The fluoride has the fluorite structure common for ionic salts. Mercury(II) chloride has a molecular lattice consisting of discrete linear $HgCl_2$ molecules. (See Section 8.4 for discussion of the low force constants of linear molecules of the type HgX_2.) Although both $HgBr_2$ and red HgI_2 have layer lattices, $HgBr_2$ represents a transition between the molecular and layer lattice, since each Hg^{2+} has two nearest-neighbor Br atoms and two Br atoms at a greater distance. Red HgI_2 undergoes a reversible phase transition at 126°C to a yellow form that contains linear HgI_2 molecules.

Organozinc compounds, such as $(C_2H_5)_2Zn$, were used in organic reactions before the discovery of the Grignard reagents, which can be prepared and handled more conveniently. The organometallic compounds of Li, Mg, Zn, and Cd all decompose rapidly upon contact with oxygen or moisture. The organomercury compounds are formed easily by the reaction of Grignard reagents (organomagnesium compounds) with $HgBr_2$. The organomercury compounds are resistant to attack by oxygen or moisture and can be handled in contact with air (see Section 10.6.1).

Zinc is an essential element that occurs at the active site of many enzymes (Section 16.8.2). At high levels, zinc is toxic, but its toxicity is obscured by the inevitable presence of the much more toxic Cd(II) in Zn(II) compounds. The amount of highly toxic Cd in cigarette smoke is great enough for concern, and problems with Cd poisoning have surfaced in Sweden, in connection with the manufacture of alkaline Ni–Cd batteries, and in Japan, where the painful *itai itai* disease has been attributed to chronic Cd poisoning. Mercury also is highly toxic. The vapor pressure of Hg metal is sufficiently high so that spilled Hg is a laboratory hazard. Hg(II) has a high affinity for S and inhibits metal enzymes, particularly those containing thiol groups. Microorganisms can transform Hg compounds into CH_3Hg^+, which concentrates in the food chain and poses a major threat of mercury poisoning in fish and other seafood. Cd, Hg, Pb, Sb, and Sn seem to have a cumulative effect, since they have no apparent biological function and hence there is no mechanism to regulate the levels of these toxic metals.[7]

Compounds of zinc(I) and cadmium(I) have been reported. The existence of cadmium(I) compounds has been demonstrated clearly,[8] but the compounds decompose in contact with oxygen or in water. Several mercury(I) compounds are stable. The compounds are unusual in that they contain the only common diatomic metal cation, Hg_2^{2+}. The compounds are diamagnetic and hence cannot contain monomeric Hg^+. X-ray structure analysis has revealed that solid mercury(I) compounds contain Hg—Hg bonds. Mercury(I) compounds can be prepared generally by the reaction of Hg with the corresponding Hg(II) compound. The equilibrium constant for the reaction of Hg with Hg^{2+}, $Hg + Hg^{2+} \rightleftharpoons Hg_2^{2+}$ is 166. Mercury(I) gives few stable complexes, so adding a complexing agent to a mercury(I) salt usually results in disproportionation to give Hg and the Hg(II) complex. The cadmium(I) and zinc(I) ions have been shown to be M_2^{2+} from Raman spectra of melts containing MCl_2 and excess M. A yellow diamagnetic solid containing $Cd_2(AlCl_4)_2$ was isolated from melts containing excess $AlCl_3$. Unusual low oxidation states of metals are often obtained in a system in which the anion is $AlCl_4^-$ or some other large anion. Force constants for the M—M bonds demonstrate that the bond strengths increase in the order $Zn_2^{2+} < Cd_2^{2+} << Hg_2^{2+}$.

14.6   GROUP IIIA—THE SCANDIUM FAMILY AND RARE EARTHS

14.6.1   Group Trends

The trends in the family Sc, Y, La, and Ac are quite regular and follow the pattern expected from the properties of the Group IA and IIA metals. The metals are very active in spite of the large amount of energy required to remove three electrons and the fact that the sublimation energies are also appreciably higher than those of the alkaline earth metals. The lattice energies of the compounds containing triply charged cations or the hydration energies of the ions are sufficiently high to account for the great activity of the metals, as shown by the high heats of formation of their solid compounds and the high oxidation emf's of the metals (see Table 14.6). The trivalent state is most important for these metals.

Scandium, much more than other members of the family, tends to form complexes. Its salts hydrolyze extensively in solution, and the tendency to hydrolyze decreases with increasing ionic radius of the metal ion through the family.

Discovery of the rare earth metals.   The minerals gadolinite and cerite were discovered in Sweden in the latter part of the 18th century. The minerals were found to contain new "earths" or metal oxides, later shown to be mixtures and separated into other metal oxides. Berzelius characterized the mineral cerite as containing the new earth ceria. Later, ceria was resolved into lanthanum, cerium, and didymium oxides. The name *didymium* ("twin") refers to the supposed new element as a twin

---

[7]D. E. Carter and Q. Fernando, "Chemical Toxicology II: Metal Toxicity," *J. Chem. Educ.* 1979, *56*, 490.

[8]J. D. Corbett, *Prog. Inorg. Chem.* 1976, *21*, 135.

of lanthanum. Didymium oxide later was resolved into the oxides of praseodymium (''green twin'') and neodymium (''new twin''). More careful separation and characterization of the earths obtained from these minerals ultimately yielded Sc, Y, La, and the 14 lanthanide metals. (The term *lanthanoid* is recommended by the I.U.P.A.C., to reserve the *-ide* ending for anions, but the well established term *lanthanide* persists.)

The rare earth metals are always found together in nature, because they occur as compounds of the 3+ ions, which have very similar radii. The chemical similarities are so great that separating them is very difficult. The classical separation method involved the fractional crystallization of salts from aqueous solution—a slow, tedious process not suitable for the production of appreciable amounts of the rare earths of high purity. Since the rare earths are among the fission products of U, their separation became a practical problem in the study of uranium fission. We now can separate the rare earths though using ion exchange procedures or a combination of solvent extraction and ion exchange procedures, so they currently are available in a high state of purity at only a small fraction of earlier costs.

Although the decrease in radius from one element to the next is very small within the rare earth series, the cumulative decrease is quite significant (the *lanthanide contraction*). The atomic radii decrease fairly regularly through the series, except for significant increases at Eu and Yb. Since the latter elements have only two electrons beyond the respective half-filled and

Table 14.6 Group IIIA metals and the rare earths

	Abundance (% of earth's crust)	Density (g/cm³)	Melting Point (°C)	Atomic Radius (pm)	Ionic Radius[a] M^{3+} (pm)	E^0 M^{3+}, M	E^0 M^{3+}, M^{2+}	E^0 M^{4+}, M^{3+}
$_{21}$Sc	5×10^{-4}	2.992	1539	164.1	88.5	-2.02		
$_{39}$Y	2.8×10^{-3}	4.472	1509	180.1	104.0	-2.37		
$_{57}$La	1.8×10^{-3}	6.174	920	187.7	117.2	-2.36		
$_{58}$Ce	4×10^{-3}	6.66	795	182	115	-2.34		$+1.76$
$_{59}$Pr	5.5×10^{-4}	6.782	935	182.8	113	-2.35		$+3.9$
$_{60}$Nd	2.4×10^{-3}	7.004	1024	182.1	112.3	-2.32		
$_{61}$Pm	—	—	1035	—	111	-2.29		
$_{62}$Sm	6.5×10^{-4}	7.536	1072	180.2	109.8	-2.30	-1.57	
$_{63}$Eu	1×10^{-4}	5.259	826	204.2	108.7	-1.99	-0.35	
$_{64}$Gd	6.5×10^{-4}	7.895	1312	180.2	107.8	-2.29		
$_{65}$Tb	9×10^{-5}	8.272	1356	178.2	106.3	-2.30		
$_{66}$Dy	4.5×10^{-4}	8.536	1407	177.3	105.2	-2.29		
$_{67}$Ho	1.1×10^{-4}	8.803	1461	176.6	104.1	-2.33		
$_{68}$Er	2.5×10^{-4}	9.051	1497	175.7	103.0	-2.31		
$_{69}$Tm	2×10^{-5}	9.332	1545	174.6	102.0	-2.31		
$_{70}$Yb	2.7×10^{-4}	6.977	824	194.0	100.8	-2.22	-1.04	
$_{71}$Lu	7.5×10^{-5}	9.843	1652	173.4	100.1	-2.30		

[a]C.N. 6.

completely filled f orbitals, apparently they have only two electrons available for metallic bonding, rather than the three available in the other rare earth metals. These irregularities disappear for the ionic radii (M^{3+}) where the three outer electrons have been removed. Although the ionic radius of La^{3+} is appreciably larger than that of Y^{3+}, the decrease in size through the rare earth series proceeds steadily until Ho^{3+} is only slightly larger than Y^{3+}. In the separation of the rare earths Y concentrates with Dy and Ho.

The rare earth metals are very strong reducing agents, because of the high lattice energies of their compounds or their high hydration energies. Consequently, they are useful for thermite-type reactions, taking advantage of the high ΔH_f of M_2O_3. (The thermite reaction uses Al to reduce metal oxides.) A mixture containing mostly Ce and La and smaller amounts of other rare earths and iron, known as mischmetall, is produced for technical uses. Mischmetall is pyrophoric when finely divided. When alloyed with more Fe to increase the hardness, it is used for flints in lighters.

We obtain the rare earth metals by reducing the anhydrous fluorides by Ca at 800–1000°C. The reduction process proceeds only to the difluorides for Sm, Eu, and Yb. These metals can be obtained by reduction of their oxides by La metal *in vacuo,* since they are more volatile than La. The anhydrous chlorides of La–Gd can be reduced to the metals with Ca, Mg, Li, or Na. The chlorides are hygroscopic and those of the heavier rare earth metals are too volatile, so the fluorides are used to obtain the metals.

$$2LaF_3 + 3Ca \rightarrow 3CaF_2 + 2La$$

$$2EuF_3 + Ca \rightarrow 2EuF_2 + CaF_2$$

$$Eu_2O_3 + 2La \rightarrow 2Eu(g) + La_2O_3$$

14.6.2 Crystal Structures

Of the various crystal structures that characterize lanthanide compounds, many are less common than the structures considered in Chapter 6 and are not easily visualized from plane figures or descriptions. One striking characteristic of the lanthanide metal ions in crystal structures and in complexes is their range of coordination numbers—6, 7, 8, 9, 10, and 12. C.N. 6 is encountered in some crystals of divalent ions, in a few octahedral complexes of anionic ligands, and in the MCl_3 structures of the smaller ions forming the latter part of the series (Dy–Lu). C.N.'s 10 and 12 are encountered for the complexes of the bidentate nitrate ligand $Ce(NO_3)_5^{2-}$ and $Ce(NO_3)_6^{3-}$ (p. 325).

The more common coordination numbers are seven and, in particular, eight and nine. C.N. 8 is represented by $La(acac)_3(H_2O)_2$ (distorted square antiprism), $Ce^{IV}(acac)_4$ (Archimedes square antiprism), $Cs[Y(CF_3COCHCOCF_3)_4]$ (dodecahedron), CeO_2 (cube, CaF_2 structure), and LaI_3 (the $PuBr_3$ layer structure containing bicapped trigonal prisms). We see C.N. 9 in $Nd(H_2O)_9^{3+}$ (a tricapped trigonal prism—see Figure 14.5) and in the complex structures of $La(OH)_3$, LaF_3, and the trichlorides of the larger lanthanide ions (La–Gd).

14.6.3 Oxidation States Other Than III

The oxidation number III is characteristic of the rare earths. Other oxidation numbers occur just about as expected on the basis of the stability of a group of orbitals that is empty, half-

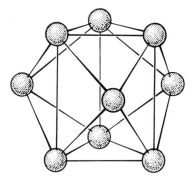

Figure 14.5 Positions of oxygen in $Nd(H_2O)_9^{3+}$.

filled, or completely filled (Table 14.7). Thus Ce^{4+} achieves a noble-gas configuration. Tb^{4+} and Eu^{2+} have the same configuration as Gd^{3+}—seven electrons in the $4f$ orbitals. Yb^{2+} has completely filled $4f$ orbitals, as does Lu^{3+}. There are a few other examples for metals that do not quite achieve one of the stable configurations of the f orbitals. Pr(IV) is encountered in the mixed oxide Pr_6O_{11}, which can be obtained by heating Pr_2O_3 in the presence of O_2 at high pressure. The compounds PrF_4 and TbF_4 have been prepared. Terbium(IV) has been reported to form the oxide Tb_4O_7, but the composition does not agree with this formulation. Samarium(II) salts can be obtained, but the blood-red Sm(II) ion is unstable in aqueous solution, because it reduces water. Although thulium(II) can be obtained as the solid TmI_2, the ion is very unstable in water.

All of the rare earths have been obtained as M^{2+} ions in CaF_2 as the host lattice. When a rare earth fluoride, MF_3, is incorporated in solid CaF_2, the M^{3+} ions should occupy the cation sites, with the "extra" F^- ions in interstitial sites. If electrolysis is carried out with graphite electrodes at each end of a crystal, electrons enter from the cathode to reduce M^{3+} to M^{2+} and the "extra" F^- ions migrate into the anode.

We know of quite a variety of rare earth compounds that show formal oxidation states other than III. The oxidation number of Eu in EuS clearly is II, since the magnetic and spectroscopic properties are those of $4f^7$ ion. The situation is not so clear for GdS, for which

Table 14.7 Electronic configurations of some rare earth metals and ions

Metal	Electronic Configuration	Configuration of M^{3+}	Configuration of M^{4+}	Configuration of M^{2+}
$_{57}$La	[Xe] $5d^1 6s^2$	[Xe]	—	—
$_{58}$Ce	[Xe] $4f^2 6s^2$ | add $4f$ electrons	[Xe] $4f^1$	[Xe]	—
$_{63}$Eu	[Xe] $4f^7 6s^2$	[Xe] $4f^6$	—	[Xe] $4f^7$
$_{64}$Gd	[Xe] $4f^7 5d^1 6s^2$	[Xe] $4f^7$	—	—
$_{65}$Tb	[Xe] $4f^9 6s^2$ | add $4f$ electrons	[Xe] $4f^8$	[Xe] $4f^7$	—
$_{70}$Yb	[Xe] $4f^{14} 6s^2$	[Xe] $4f^{13}$	—	[Xe] $4f^{14}$
$_{71}$Lu	[Xe] $4f^{14} 5d^1 6s^2$	[Xe] $4f^{14}$	—	—

the magnetic properties are also those of a $4f^7$ cation, corresponding to Gd^{3+}. This can be interpreted as a Gd(III) sulfide with the "extra" electron in a conduction band.[9]

14.7 THE ACTINIDE METALS

The metals Ac, Th, Pa, and U appeared to be members of the III, IV, V, and VI transition element families, respectively, until the discovery of the transuranium metals. Each additional element discovered made it more obvious that these metals were members not of a transition series, but of an inner-transition series. The widely varying oxidation numbers of these metals made it difficult to determine which element was the first member of the series. Greater variation in oxidation numbers (compared with the rare earths) is to be expected among the actinide metals, because the atomic radii are larger and the energy levels of the valence electrons are closer together. Several stable oxidation numbers are encountered for most of the actinide metals.

It has been suggested that the second inner-transition series begins with thorium. Most of the metals following Th give dioxides that are isomorphous with ThO_2 and there are other similarities to Th. On proceeding through the series, however, the trivalent state becomes increasingly more important (Table 14.8). This is one of the important oxidation numbers of Pu and the most stable state for americium and the remainder of the series. It is now clear that the second inner-transition series begins with actinium, to give a series similar to the rare earths. Americium is the actinide counterpart of europium in the lanthanide series and, as expected, it exhibits the divalent state. However, the $5f$ electrons are removed more easily than the $4f$ electrons, and consequently, for the actinide metals the lower oxidation numbers are less important and the higher oxidation numbers are more important, compared with the lanthanide metals. Even curium gives Cm(IV) in solid CmO_2 and CmF_4, whereas only Gd(III) compounds are known. The Cm^{4+} ion is not stable in solution.

With very short half-lives, the last few members of this series have been produced in very small quantities (a few atoms, in some cases). Studying elements that show a very high level

Table 14.8 Oxidation numbers of the actinide elements (very stable states in boldface type; unstable states enclosed in parentheses)

Ac	Th	Pa	U	Np	Pu	Am	Cm	Bk	Cf	Es	Fm	Md	No	Lr
				(II)		(II)			(II)	(II)	(II)	(II)	(II)	
III	(III)	(III)	(III)	III	**III**	**III**	**III**	**III**	**III**	**III**	**III**	**III**	**III**	**III**
	IV	IV	**IV**	**IV**	**IV**	IV	IV	IV	(IV)					
		V	V	V	V	V								
			VI	**VI**	**VI**	VI								
				(VII)										

[9]L. B. Asprey and B. B. Cunningham, "Unusual Oxidation States of Some Actinide and Lanthanide Elements," *Prog. Inorg. Chem.* 1960, *2*, 267; D. A. Johnson, "Recent Advances in the Chemistry of the Less Common Oxidation States of the Lanthanide Elements," *Adv. Inorg. Chem. Radiochem.* 1977, *20*, 1.

of activity is difficult, because of the drastic chemical changes which occur in the sample. Even the water in which a compound of the element is dissolved is decomposed by the radioactive decay to give H_2 and O_2. Consequently, the chemical properties of the last few actinides have not been investigated in great detail. The last member of the actinide series should be element 103 (Lawrencium) which seems to have only III as an important oxidation number. Element 104 should be a member of the transition metal family IVA. Russian claims for the discovery of this element have not been substantiated.

14.8 TRANSITION METALS, GROUPS IVA–VIIA

According to the Bohr classification (p. 23) the metals in the Cu and Zn families are representative elements, not transition elements. However, the chemical similarities throughout a series of metal ions such as Fe^{2+}, Co^{2+}, Ni^{2+}, Cu^{2+}, and Zn^{2+} make the exclusion of Cu and Zn from the transition series seem artificial. Hence, these metals and the others in their families are commonly included with the transition metals.

The metals of Groups IB and IIB have been discussed along with Groups IA and IIA in order to emphasize the periodic relationships among these groups. Group IIIA has been discussed separately because these metals, unlike most of the transition metals, vary little in oxidation numbers and the presence of the lanthanide and actinide metals makes this group unique. Only some of the overall trends will be discussed for Groups IVA–VIIA. Group VIII's unique character mandates separate treatment.

14.8.1 Size Effects

The cumulative effect of the lanthanide contraction makes the radii of the members of the third transition series very similar to those of the corresponding members of the second transition series. The effect is great enough that Hf, which follows immediately after the rare earths, has a slightly smaller atomic radius than Zr (see Table 14.9). The almost identical ionic radii of Zr^{4+} and Hf^{4+} account for the fact that these metals always occur together in nature and are difficult to separate. There are no hafnium minerals. The less abundant Hf is too similar to Zr to become enriched. For most uses, a few percent of Hf in Zr does not matter. Separating them became important when Zr was found to be a good structural metal for atomic piles, because of its low neutron cross section (it does not absorb neutrons greatly). Hf has a very high cross section and can be used as a moderator to absorb neutrons. Early separation methods involved fractional crystallization or precipitation procedures, but solvent-extraction procedures are more effective.

The atomic radius of Ta is the same as that of Nb, but the two elements are not as similar to one another as are Zr and Hf. They occur together in nature, but not always in the same proportions. The mineral $Fe(MO_3)_2$ is called columbite if niobium (formerly called columbium in America) predominates and tantalite if tantalum predominates. We can separate the two metals by crystallizing the less soluble K_2TaF_7 from a concentrated HF solution. In hot water the Nb complex dissolves as $K_2[NbOF_5]$, and the Ta complex is converted to a basic salt of low solubility. The chemical similarity between Mo and W is somewhat less—they form

Table 14.9 Properties of some transition metals

	$_{22}Ti$	$_{23}V$	$_{24}Cr$	$_{25}Mn$
Atomic radius (pm)	132	122	118	117
Ionic radius, C.N. 6 (pm)	M^{III} 81.0	M^{II} 93	M^{II} 94	M^{II} 97.0, 81[a]
	M^{IV} 74.5	M^{III} 78	M^{III} 75.5	M^{III} 78.5, 72[a]
Ionization 1st	6.82	6.74	6.766	7.435
energy (eV) 2nd	13.58	14.65	16.50	15.640
3rd	27.491	29.310	30.96	33.667
4th	43.266	46.707	49.1	51.2

	$_{40}Zr$	$_{41}Nb$	$_{42}Mo$	$_{43}Tc$
Atomic radius (pm)	145	134	130	127
Ionic radius, C.N. 6 (pm)	M^{IV} 86	M^{IV} 82	M^{IV} 79	M^{IV} 78.5
		M^{V} 78	M^{VI} 73	M^{VII} 70
Ionization 1st	6.84	6.88	7.099	7.28
energy (eV) 2nd	13.13	14.32	16.15	15.26
3rd	22.99	25.04	27.16	29.54
4th	34.34	38.3	46.4	

	$_{72}Hf$	$_{73}Ta$	$_{74}W$	$_{75}Re$
Atomic radius (pm)	144	134	130	128
Ionic radius, C.N. 6 (pm)	M^{IV} 85	M^{V} 78	M^{IV} 80	M^{IV} 77
			M^{VI} 74	M^{VII} 67
Ionization 1st	7.0	7.89	7.98	7.88
energy (eV) 2nd	14.9	16.2	17.7	16.6

[a]Low spin.

independent minerals and their chemical separations are accomplished much more easily. The effect óf the *lanthanide contraction* diminishes for elements beyond W, but for each of the families of transition metals beyond Group IIIA, the second and third members are much more similar in properties than are the first and second members.

14.8.2 Oxidation States

Varying oxidation numbers characterize most of the transition metals. The exclusive oxidation number III for the members of Group IIIA arises because the energy required for the removal of all three outer electrons does not much exceed that needed for the removal of two electrons. The energy for the removal of three electrons is supplied by the high hydration energies of the 3^{+} ions or the high lattice energies of their compounds.

The energy required for the removal of four or more electrons is prohibitively high for the formation of simple ions, except in the case of very large atoms with low ionization energies. We expect an appreciable amount of covalent character in the higher oxidation states.

Except for Group VIII, the highest oxidation number for each transition metal corresponds to the group number. The lowest state usually encountered in simple salts is II, corresponding to the removal of the two *s* electrons in the outermost shell (Table 14.10). Compounds of second and third transition-series metals containing metal–metal bonds and metal atom clusters (see p. 704) cloud comparisons of oxidation states. Thus, we can compare the oxidation number II for $CrCl_2$ with that of Mo in Mo_6Cl_{12} only in the most formal sense.

Although the first ionization energies do not differ greatly within most families of the transition metals, the third and higher ionization energies generally are highest for the first member of each family. The differences between successive ionization energies also are greatest for the first member of each family. Consequently, the highest oxidation state is most stable, and the corresponding compounds are poor oxidizing agents, for the second or third member of each group (for example, Zr^{4+} and Hf^{4+}). The lower oxidation states are relatively more stable for the first member of each group (for example, Ti^{2+} and Ti^{3+}) as compared with the second and third members. Solid $ZrCl_2$, $ZrCl_3$, and $HfCl_3$ can be obtained by reducing MCl_4 with the free metal, but the lower oxidation states are not stable in solution. A variety of Ti(II) and Ti(III) compounds is known (see Table 14.10). Titanium(III) compounds can be obtained readily in solution by reducing Ti(IV) salts with Zn and acid, or electrolytically. Titanium(III)

Table 14.10 Oxidation states of some binary compounds of transition and post-transition metals

	IVA	VA	VIA	VIIA	VIII			IB	IIB	IIIB	IVB
	Ti	V	Cr	Mn	Fe	Co	Ni	Cu	Zn	Ga[a]	Ge
F	3,4	3,4,5	2,3,4,5	2,3,4	2,3	2,3	2	2	2	3	2,4
Cl	2,3,4	2,3,4	2,3	2	2,3	2	2	1,2	2	1,3	2,4
Br	2,3,4	2,3,4	2,3	2	2,3	2	2	1,2	2	3	2,4
I	2,3,4	2,3	2,3	2	2	2	2	1	2	3	2,4
O	2,3,4	2,3,4,5	2,3,4, 6	2,3,4, 7	2,3	2,3	2,3,4	1,2	2	1,3	2,4
	Zr	Nb	Mo	Tc	Ru	Rh	Pd	Ag	Cd	In[a]	Sn
F	4	3,4,5	3,4,5,6	6	3,4,5,6	3,4,5,6	2, 4	1,2	2	3	2,4
Cl	1,2,3,4	3,4,5	2,3,4,5,6	4, 6	3,4	3	2	1	1,2	1,3	2,4
Br	2,3,4	2,3,4,5	2,3,4	—	3	3	2	1	2	1,3	2,4
I	2,3,4	2,3,4,5	3,4	—	3	3	2	1	2	1,3	2,4
O	4	2, 4,5	4,5,6	4, 6,7	3,4, 8	3,4	2	1, 3	2	1,3	2,4
	Hf	Ta	W	Re	Os	Ir	Pt	Au	Hg	Tl	Pb
F	4	5	4,5,6	4,5,6,7	4,5,6,7	3,4,5,6	4,5,6	3	1,2	1,3	2,4
Cl	1, 3,4	3,4,5	2,3,4,5,6	3,4,5,6	3,4	3,4	2,3,4	1, 3	1,2	1,3	2,4
Br	3,4	3,4,5	2,3,4,5,6	3,4,5,6	3	3,4	2,3,4	1, 3	1,2	1,3	2
I	3,4	3,4,5	2,3,4	2,3,4	3	3,4	2,3,4	1, 3	1,2	1	2
O	4	4,5	4, 6	3,4, 6,7	4, 8	3,4, 6	4	1, 3	2	1,3	2,4

[a]Gallium and indium dihalides are diamagnetic compounds, which can be formulated as $M^IM^{III}X_4$.

chloride, which can be obtained by dissolving Ti metal in hydrochloric acid, finds some application as a reducing agent in analytical work.

Vanadium(II) compounds are strongly reducing, while vanadium(III) salts are stable in solution (see Section 11.7 for discussion of the Pourbaix diagram for V). Niobium and tantalum most commonly show oxidation number V. Chromium(II) compounds also are strongly reducing, and most of the common salts are those of Cr(III). Until recently little-known, the oxide CrO_2 now is used as a magnetic oxide for high-quality recording tape. The oxidation number VI is encountered in CrO_3, K_2CrO_4, $K_2Cr_2O_7$, etc. These compounds are strong oxidizing agents, being reduced to compounds of chromium(III) in solution. As is generally true, oxidation to higher oxidation states encountered as oxo anions is accomplished much more easily in basic solution and the oxo anions are poorer oxidizing agents in basic solution. Molybdenum and its compounds are oxidized to MoO_3 by heating in air or by reaction with HNO_3. Molybdenum trioxide dissolves in alkali hydroxides or ammonium hydroxide to form molybdates, $M_2^I[MoO_4]$. The occurrence of Mo as a sulfide, MoS_2 (molybdenite), is unusual among the heavier metals in Groups III–VII, although ReS_2 sometimes occurs with molybdenite. The layer structure of MoS_2 accounts for its use as a solid lubricant, similar to graphite.

Tungsten has an even greater tendency than Mo to form compounds with an oxidation number VI. It forms a hexachloride, WCl_6, by heating freshly reduced W with Cl_2. The corresponding fluoride and even the bromide can be obtained. The hexahalides are hydrolyzed completely in water to give tungstic acid, H_2WO_4.

Manganese(I) occurs in the cyano complex, $K_5[Mn(CN)_6]$. The characteristic oxidation number of Mn in solution is II, although the III state is stabilized in complexes such as $K_3[Mn(CN)_6]$, $K_3[Mn(C_2O_4)_3]$, and $K_3[MnF_6]$. Manganese occurs in the igneous rocks as the Mn^{2+} ion, replacing Fe^{2+} and Mg^{2+}. During the weathering process Mn^{2+} is oxidized and finally deposited as MnO_2, the important Mn ore. Potassium permanganate, $KMnO_4$, the most important compound of Mn(VII), can be obtained by electrolytic oxidation or Cl_2 oxidation of K_2MnO_4, which is produced by strong oxidation of $Mn(OH)_2$ or $Mn(OH)_3$ in a basic medium. Potassium permanganate is a strong oxidizing agent that finds wide analytical application because of the sharp color change on reduction. It usually is used in acidic solution, where it is reduced to the nearly colorless Mn^{2+} ion. In the absence of an excess of strong acid or strong base, MnO_2 is formed. The manganate(VI) ion is formed by reduction of MnO_4^- in concentrated NaOH or KOH solution. The bright green MnO_4^{2-} salts are unstable except in the solid or in strongly basic solution (see Section 11.7 for a detailed discussion of the redox chemistry of Mn in terms of Pourbaix diagrams). The oxide, Mn_2O_7, can be obtained by the reaction of $KMnO_4$ and concentrated H_2SO_4. A dark, oily liquid that decomposes explosively when heated, it is an acid anhydride, since it reacts with an excess of water to produce $HMnO_4$, a very strong acid.

The characteristic oxidation number of technetium and rhenium is VII. The compounds, such as $KTcO_4$ and $KReO_4$, are not strong oxidizing agents. The volatile oxides, Tc_2O_7 and Re_2O_7, and sulfides, Tc_2S_7 and Re_2S_7, are stable. In properties, Tc is much closer to Re than to Mn.

The second most important oxidation number of Re is IV. Reducing Re_2O_7 in H_2 at about 300°C yields ReO_2, which combines with alkali hydroxides on fusion to produce the alkali metal rhenites, $M_2^I ReO_3$. TcO_2 also is stable. The reported Re(−I) turned out to be a hydride compound (ReH_9^{2-}), rather than a simple solvated Re^- ion.

The very high oxidation states of the transition metals usually are brought out in combination with F or O. The size of the anion is rarely the limiting factor. More important are the oxidizing power of the cation in higher oxidation states and the increasing ease of oxidation of the anions with increasing size within a group (for example, F^-, Cl^-, Br^-, I^-). The oxides usually display the greatest number of oxidation states of a given metal, although this may be true in part because the oxides have been investigated most extensively (see Table 14.10). The emf diagrams given in Appendix C, show the relative stabilities of the various oxidation states of the metals as hydrated cations or oxo anions. The relative stabilities of the various oxidation states can be altered greatly by complex formation—for example, the stabilization of high oxidation states in oxo anions—or by lattice forces in the solid state.

The basicity of the metal ions in various oxidation states can be treated on the basis of charge/size relationships. For the same oxidation number—for example, IV for Group IVA—basicity increases markedly with increasing ionic radius and the covalent character of the resulting compounds decreases accordingly. The Ti(IV) compounds are much more covalent than those of Zr(IV) or Hf(IV) and are hydrolyzed much more extensively. Thus $TiCl_4$ is a liquid while $ZrCl_4$ is a solid, although it sublimes at about 300°C. In Group VIA, Cr(III) is amphoteric and can be obtained as Cr^{3+} salts or salts of the chromite ion, $Cr(OH)_4^-$, whereas Mo(III) and W(III) are basic in character. The Cr^{2+} ion is basic because of the lower charge and larger radius (compared with Cr^{3+}). In general, the members of the first transition series give ions that are predominately basic only in the lower oxidation states. The oxidation numbers greater than IV for the first members of Groups V–VIII are generally encountered in oxides or oxo anions. The second and third members of these families give ions that are much more basic, showing a much greater tendency than the first members to form cationic species in higher oxidation states. Chromium trioxide is quite acidic and is very soluble in water, giving H_2CrO_4. Molybdenum trioxide is less acidic and is only slightly soluble in water, and WO_3 is very insoluble. All three oxides dissolve readily in bases. Because the difference between the radii of Mo and W is small, the basicities of the compounds of Mo and W are very similar.

Usually, zero and negative oxidation numbers are encountered for metal carbonyls (see Section 10.2).

14.8.3 Ligand Field Effects

The hydration energies of the first-transition-series metal ions should increase with decreasing ionic radius (and increasing atomic number) through the series. We see from plots of hydration energies for M^{2+} and M^{3+} (Figure 14.6) that there are two maxima with minima for ions with the configurations d^0 (Ca^{2+} and Sc^{3+}), d^5 (Mn^{2+} and Fe^{3+}), and d^{10} (Zn^{2+} and Ga^{3+}). These are cases with zero ligand field stabilization energy (LFSE). For the other ions the hydration energy is increased by the LFSE (see Section 7.5.1), giving the two maxima. Subtracting the calculated LFSE from the observed hydration energies gives the "corrected" points, which fall close to the straight lines joining the d^0, d^5, and d^{10} ions.

Although the increasing positive charge density through the transition series primarily determines the "natural order of stability" of metal complexes ($Mn^{2+} < Fe^{2+} < Co^{2+} < Ni^{2+} < Cu^{2+} > Zn^{2+}$—see section 12.3.3), the LFSE is also a factor, particularly for the decrease in stability of Zn^{2+} complexes as compared with those of Cu^{2+}. Figure 14.7 shows a plot of ΔH_f for the formation of $[M(en)_2]^{2+}$ and $[M(en)_3]^{2+}$ complexes in aqueous solution.

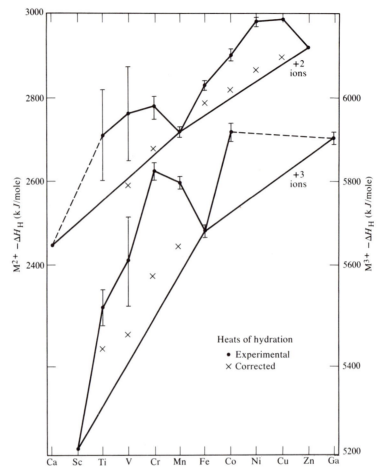

Figure 14.6 The hydration energies of the divalent and trivalent ions of the first transition series. (From P. George and D. S. McClure, *Prog. Inorg. Chem.* 1959, *1*, 418.)

Subtracting the LFSE for $[M(en)_3]^{2+}$ gives values close to the straight line joining Mn^{2+} and Zn^{2+}. No value is plotted for $[Cu(en)_3]^{2+}$, because the C.N. of Cu^{2+} is uncertain in this case. Cu^{2+} shows strong tetragonal distortion in six-coordinate complexes (Jahn-Teller effect—see Section 7.5.2), lowering the stability of tris(bidentate)copper(II) complexes. The stability trend is normal for $[M(en)_2]^{2+}$.

The "natural order of stability" applies to high-spin complexes and to cases in which the ligand does not impose some fixed stereochemistry, as do porphyrin or phthalocyanine ligands. Low-spin complexes have additional LFSE, but at the expense of the pairing energy. The ΔH_f values given in Figure 14.7 are for the formation of the complexes in aqueous solution and involve replacement of coordinated H_2O by N of ethylenediamine. The LFSE from the plots is actually the difference between the coordination values of N and O, but this additional stabilization is proportional to the calculated LFSE.

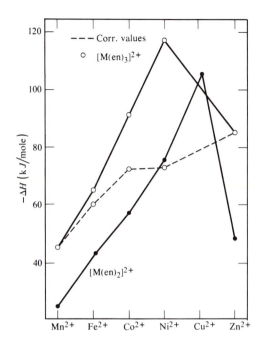

Figure 14.7 Heats of formation of $[M(en)_2]^{2+}$ and $[M(en)_3]^{2+}$ and the heats of formation of $[M(en)_3]^{2+}$ corrected for LFSE. (Adapted from M. Ciampolini, P. Paoletti, and L. Sacconi, "Heats and Entropies of Reaction of Transition Metal Ions with Polyamines" in *Advances in the Chemistry of the Coordination Compounds*, S. Kirschner, Ed., Macmillan, New York, 1961, pp. 305 and 307.)

14.8.4 Isopoly and Heteropoly Anions

Isopoly Anions

One of the important characteristics of Mo(VI), W(VI), and to a lesser extent, V(V), Nb(V), and Ta(V) is the tendency to form polymeric oxo anions containing MO_6 octahedra. For the tetrahedral oxo anions of the nonmetals, XO_4^{n-}, the tendency to form polyanions is greatest for Si(IV) where the oxidation number is the same as the coordination number. The oxidation number is also the same as the C.N. for the Mo(VI) and W(VI) polyanions. The chromate ion CrO_4^{2-} forms $Cr_2O_7^{2-}$ in acid solution; in concentrated H_2SO_4 or $HClO_4$, CrO_3 precipitates. The tetrahedral CrO_4 units share corners to form chains in solid CrO_3. Simple tetrahedral anions of MoO_4^{2-}, WO_4^{2-}, VO_4^{3-}, NbO_4^{3-}, amd TaO_4^{3-} are encountered, but in forming the polyacids or polyanions the C.N. increases to six. It seems likely that the small size of Cr(VI) prevents the increase in C.N. Solid MoO_3 has a layer structure in which each MoO_6 octahedron shares two edges and one corner. The WO_3 structure involves WO_6 octahedra sharing all corners in a distorted version of the ReO_3 structure. Vanadium in V_2O_5 has a distorted octahedral environment with one V–O distance much longer than the others.

The isopoly anions can be regarded as fragments of oxide structures. Since discrete MO_6^{n-} ions have unfavorably high charges, the MO_4^{n-} ions are encountered instead. It is striking that we encounter only aggregates of well-defined size and that their sizes are different for different metals. The fact that the stable isopoly anions are compact rather than extended aggregates indicates that polymerization is favored by charge reduction and the highly favorable entropy change accompanying elimination of water molecules with the formation of compact structures.[10]

[10]D. L. Kepert, *Inorg. Chem.* 1969, *8*, 1556; K. -H. Tytoko and O. Glemser, *Adv. Inorg. Chem. Radiochem.* 1976, *19*, 239.

The sharing of edges between MO_6 octahedra causes repulsion between the metal ions. This repulsion is reduced partially by displacement of M^{n+} from the exact center of the octahedra. Repulsion should increase with increasing ionic radius ($V^{5+} < Mo^{6+} \leqslant W^{6+} < Nb^{5+} \leqslant Ta^{5+}$), in agreement with the degree of distortion observed and the sizes of the edge-shared isopoly anions: $V_{10}O_{28}^{6-}$, $Mo_8O_{26}^{4-}$, $Mo_7O_{24}^{6-}$, $HW_6O_{21}^{5-}$, $Nb_6O_{19}^{8-}$, and $Ta_6O_{19}^{8-}$. The choice of edges to be shared—and consequently, the shape of the aggregate—is determined also by the tendency to minimize charge repulsion. Larger aggregates require some extent of apex sharing, in order to reduce cation–cation repulsion.

As fragments of oxide structures, the isopoly anions contain cubic close-packed arrangements of oxide ions with metal ions in octahedral holes. The *hcp* arrangement involves face- and edge-shared octahedra. A basic unit is the "super" octahedron, M_6O_{19} (Figure 14.8). The $M_{10}O_{28}$ unit contains two "super" octahedra sharing an edge. Fragments of this unit encountered include M_8O_{26}, from the removal of octahedra *a* and *d* in Figure 14.8; and M_7O_{24}, from the removal of octahedra *b, c,* and *d*.

Heteropoly Anions

As a qualitative test, phosphate ion is detected by adding $(NH_4)_2MoO_4$ to form a precipitate of ammonium 12-molybdophosphate, $(NH_4)_3[P(Mo_3O_{10})_4] \cdot 6H_2O$. This heteropoly anion has four groups of three edge-shared MoO_6 octahedra arranged tetrahedrally about P (Figure 14.9). A 2:18 ratio occurs in $K_6[P_2W_{18}O_{62}] \cdot 14H_2O$, where two PO_4 tetrahedra pointing in opposite directions are enclosed in a cage formed by two groups of three edge-shared octahedra (sharing the oxygens of the opposed apices of the PO_4 units) and six groups of two edge-shared octahedra. The edge-shared octahedral groupings are joined by shared apices. This structure would result from removing three octahedra from the 1:1 structure and then joining two of these fragments.

A second class of heteropoly anions involves edge-shared MO_6 units clustered about an octahedral hetero atom. A 1:6 ratio—for example, $[Te^{VI}Mo_6O_{24}]^{6-}$ and $[M^{III}Mo_6O_{24}]^{9-}$—results from a ring of six octahedra about the hetero atom (Figure 14.10). Adding three edge-shared octahedra above and below this arrangement gives us a plausible 1:12 structure. Re-

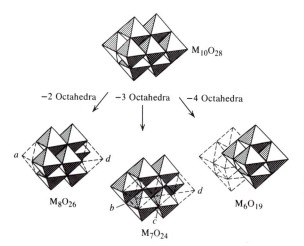

Figure 14.8 The structure of isopoly anions, showing their relationship to the $M_{10}O_{28}$ structure. (Reproduced with permission from D. L. Kepert, *Inorg. Chem.* 1969, 8, 1557. Copyright 1969, American Chemical Society.)

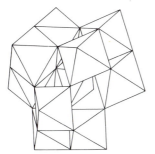

Figure 14.9 Structure of the 12-heteropoly anion $[P(Mo_3O_{10})_4]^{3-}$

moving alternate octahedra from the ring of the 1:12 structure gives us the 1:9 ratio for example, $[Ni^{IV}Mo_9O_{32}]^{6-}$. Other fragments are possible, but more structural work is needed. Many hetero atoms capable of forming tetrahedral or octahedral oxo anions occur in heteropoly anions. The compound $(NH_4)_6[H_4Co_2Mo_{10}O_{38}] \cdot 7H_2O$ can be pictured as resulting from removal of an MoO_2 unit from the ring of each of two 1:6 heteropoly anions so as to expose the central CoO_6 groups, which are then joined with the elimination of six more oxygens along the shared edges. The dissymmetric $[H_4Co_2Mo_{10}O_{38}]^{6-}$ ion has been resolved to give stable optical isomers.[11]

The $[CeMo_{12}O_{42}]^{8-}$ ion does not have the 1:12 structure shown in Figure 14.10. There are six pairs of face-shared octahedra coordinated to Ce^{IV} to form an icosahedron.

We have taken only a broad structural view of isopoly and heteropoly anions. Many structures are unknown, and others do not fit into the simple pattern presented. Important questions remain to be answered in this not-fully-explored field.

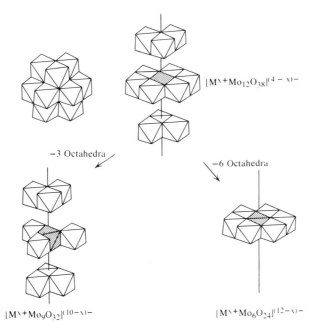

$[M^{x+}Mo_{12}O_{38}]^{(4-x)-}$

−3 Octahedra

−6 Octahedra

$[M^{x+}Mo_9O_{32}]^{(10-x)-}$

$[M^{x+}Mo_6O_{24}]^{(12-x)-}$

Figure 14.10 The structures of heteropolymolybdates and their relationship to the hypothetical $M^{x+}Mo_{12}O_{38}^{(4-x)-}$. (Reproduced with permission from D. L. Kepert, *Inorg. Chem.* 1969, *8*, 1557. Copyright 1969, American Chemical Society.)

[11]T. Ama, J. Hidaka, and Y. Shimura, *Bull. Chem. Soc. Japan* 1970, *43*, 2654.

14.9   GROUP VIII

14.9.1   Periodic Classification

The Group VIII metals logically belong to a single periodic group, since they exhibit close similarities in the horizontal triads as well as in the vertical groups (Table 14.11). Properties of complexes are most closely related to the number of d electrons, which is similar for the ions of the metals in each of the vertical columns. Not only are the radii very similar in each horizontal triad, but also the radii of the second- and third-transition-series metals in each vertical column are very similar, because of the lanthanide contraction. Hence, the most useful classification places Fe, Co, and Ni in one group, known as the iron group or triad. The platinum metals might be divided into the light platinum metals (Ru, Rh, and Pd) and the heavy platinum metals (Os, Ir, and Pt), but the chemical similarities are more apparent in the vertical groups of two: Ru and Os, Rh and Ir, and Pd and Pt. In each of these pairs the metals have similar radii and similar electronic configurations.

Table 14.11   Properties of Group VIII metals

		$_{26}Fe$		$_{27}Co$		$_{28}Ni$
Atomic radius (pm)		126		125		125
Ionic radius^a (pm)	Fe^{II}	92, 75^b	Co^{II}	88.5, 79^b	Ni^{II}	83, 71^c
				72^c		
	Fe^{III}	78.5, 69^b	Co^{III}	75, 70^b	Ni^{III}	68^b
Ionization 1st		7.870		7.86		7.635
energy (eV) 2nd		16.18		17.06		18.168
3rd		30.651		33.50		35.17

		$_{44}Ru$		$_{45}Rh$		$_{46}Pd$
Atomic radius (pm)		134		134		137
Ionic radius^a (pm)	Ru^{III}	82	Rh^{III}	80.5	Pd^{II}	78^c
					Pd^{IV}	75.5
Ionization 1st		7.37		7.46		8.34
energy (eV) 2nd		16.76		18.08		19.43
3rd		28.47		31.06		32.93

		$_{76}Os$		$_{77}Ir$		$_{78}Pt$
Atomic radius (pm)		135		136		139
Ionic radius^a (pm)	Os^{IV}	77.0	Ir^{III}	82	Pt^{II}	74^c
			Ir^{IV}	76.5	Pt^{IV}	76.5
Ionization 1st		8.7		9.1		9.0
energy (eV) 2nd		17		17		18.563

^{a}Ionic radii for C.N. 6, except as noted.
^{b}Low-spin.
^{c}Square-planar.

14.9.2 The Iron Triad (Fe, Co, and Ni)

The metals of the iron triad often occur together in nature, and the simple salts of their M^{2+} ions are very similar. Only iron, as Fe(III), gives stable simple salts with an oxidation number other than II. Iron(II) salts are oxidized easily to those of iron(III), whereas the hydrated Co^{3+} ion oxidizes water and the only simple salts of nickel are those of Ni(II). Strong oxidation of $Fe(OH)_3$ in strongly alkaline solution produces the ferrate(VI) ion—as in K_2FeO_4, a very strong oxidizing agent that forms lustrous black crystals. Although the highest oxidation state encountered increases steadily through the first transition series for Groups III–VII (see Table 14.9, particularly for the oxides), there is an abrupt change for Fe, Co, and Ni.

The Fe^{2+} ion is a d^6 ion, which gives very stable low-spin octahedral complexes of the type $[Fe(CN)_6]^{4-}$ and $[Fe(phen)_3]^{2+}$. Most ligands other than 1,10-phenanthroline (phen) and bipyridine give more stable complexes with Fe^{3+}, a d^5 ion—presumably because of the higher charge on the cation. The $[Fe(CN)_6]^{3-}$ ion is a low-spin complex, but most others, such as FeF_6^{3-}, $[Fe(C_2O_4)_3]^{3-}$, and Fe(acetylacetonate)$_3$, have high spin.

The cobalt(II) ion forms octahedral complexes, such as $[Co(NH_3)_6]^{2+}$; planar complexes, such as,

and tetrahedral complexes, such as $[CoCl_4]^{2-}$. However, in the presence of strong ligands, cobalt(II) is oxidized easily to cobalt(III) to give low-spin octahedral complexes such as $[Co(NH_3)_6]Cl_3$ and $K_3[Co(CN)_6]$. The cobalt(III) complexes are exceptionally stable and give stable geometrical isomers, such as *cis*- and *trans*-[Co(en)$_2$Cl$_2$]Cl, and optical isomers, such as (+)- and (−)-[Co(en)$_3$]$^{3+}$ and (+)- and (−)-*cis*-[Co(en)$_2$Cl$_2$]$^+$. There are very few high-spin octahedral Co(III) complexes—for example, $[CoF_6]^{3-}$. The trigonal bipyramidal structure of $[Co(Et_3P)_2Cl_3]$ is even more unusual.

Most of the nickel complexes are those of nickel(II). Planar complexes formed by very strong field ligands—$[Ni(CN)_4]^{2-}$, for example—are diamagnetic. Octahedral complexes, such as $[Ni(H_2O)_6]^{2+}$, $[Ni(NH_3)_6]^{2+}$, and $[Ni(en)_3]^{2+}$, have two unpaired electrons. Halide complexes, such as $[NiCl_4]^{2-}$, or complexes containing halide ligands and bulky ligands, such as triphenylphosphine or triphenylphosphine oxide, are tetrahedral and paramagnetic. The compound with the stoichiometry Ni(*acac*)$_2$ (*acac* = acetylacetonate ion) is a trimer with bridging oxygens to make Ni six-coordinate. Yellow diamagnetic bis(NH$_2$CHRCHRNH$_2$) planar complexes are formed, but they add solvent molecules or other ligands to form blue octahedral complexes. Complexes of some Schiff bases and a few other ligands have shown planar-tetrahedral equilibria. Several phosphine and arsine ligands form five-coordinate trigonal bipyramidal complexes. Since quadridentate "tripod" ligands, such as N[C$_2$H$_4$N(CH$_3$)$_2$]$_3$ and N[C$_2$H$_4$PPh$_2$]$_3$, cannot span the apices of a regular tetrahedron, one face is left open for a fifth ligand, such as a halide ion, to form a trigonal bipyramidal complex. Nickel(III) complexes of the type $[Ni(PR_3)_2Br_3]$ also are trigonal bipyramidal.

14.9.3   Ruthenium and Osmium

Ruthenium and osmium form complexes similar to those of iron(II) and iron(III). In addition, many of their complexes can be oxidized to M^{IV}. Oxidation of ruthenium metal produces RuO_2. Strong oxidation of ruthenium salts in alkaline solution produces RuO_4, an orange liquid above 25°C. The stable oxo anions encountered for Ru are the ruthenate(VI) ion, RuO_4^{2-}, and ruthenate(VII), RuO_4^-. We obtain osmium tetraoxide, OsO_4, by direct oxidation of the metal or the compounds of Os. Like RuO_4, OsO_4 is volatile (m.p., 40°C; b.p., *ca.* 100°C) and highly toxic. The oxo anions of osmium are less stable than those of ruthenium, because of the ease of formation of OsO_4. Mild reduction of OsO_4 in alkaline solution produces osmate salts such as K_2OsO_4.

14.9.4   Rhodium and Iridium

Most of the simple and complex salts of Rh are those of rhodium(III). Iridium(III) compounds are most stable, although some octahedral complex halides can be oxidized to M^{IV}. Numerous examples of Rh(I) and Ir(I) compounds have been prepared in recent years.

14.9.5   Palladium and Platinum

The important compounds of Pd and Pt are those of oxidation numbers II and IV, and primarily their complexes. The Pd compounds generally are planar complexes of Pd(II), such as $K_2[PdCl_4]$ and $[Pd(NH_3)_4]Cl_2$. Even $PdCl_2$ in the solid state contains chains of planar $PdCl_4$ units. Oxidation of $K_2[PdCl_4]$ with Cl_2 or *aqua regia* produces $K_2[PdCl_6]$. One of the polymorphs of $PtCl_2$ is soluble in benzene, with Pt_6Cl_{12} molecular units having the structure of the M_6Cl_{12} metal atom clusters (Figure 15.38), containing planar $PtCl_4$ units. Since the Pt orbitals are filled, the bond order is zero, and the Pt_6Cl_{12} unit is held together primarily by the Cl bridges. Pt complexes closely resemble those of Pd. Platinum(II) is the metal ion that gives the most stable planar complexes, giving rise to geometrical isomerism as exemplified by *cis*- and *trans*-$[Pt(NH_3)_2Cl_2]$. Pt(IV) is much more stable than Pd(IV). Octahedral complexes of Pt(IV) give particularly stable geometrical isomers of the type mentioned for Co(III).

14.10   COMPOUNDS CONTAINING METAL–METAL BONDS[12]

The dimeric cation Hg_2^{2+} was long considered to be unique. The existence of Cd_2^{2+} and Zn_2^{2+} has been demonstrated (p. 630), although the bonding is much weaker than in Hg_2^{2+}.

---

[12]F. A. Cotton, "Quadruple Bonds and other Multiple Metal to Metal Bonds," *Chem. Soc. Rev.* 1975, *4*, 27; F. A. Cotton, "Discovering and Understanding Multiple Metal-to-Metal Bonds," *Acc. Chem. Res.* 1978 *11*, 226; J. L. Templeton, "Metal–Metal Bonds of Order Four," *Prog. Inorg. Chem.* 1979, *26*, 211; W. C. Trogler and H. B. Gray, "Electronic Spectra and Photochemistry of Complexes Containing Quadruple Metal–Metal Bonds," *Acc. Chem. Res.* 1978, *11*, 232.

The compound $K_3W_2Cl_9$ has long been known to contain discrete $W_2Cl_9^{3-}$ ions consisting of two octahedra of Cl^- ions with one face in common (Figure 14.11). The octahedra are distorted because of the very short W–W distance (241 pm)—shorter than in the metal. Strong W–W bonding draws the positive ions together, and the compound is diamagnetic. The opposite distortion occurs in $Cr_2Cl_9^{3-}$, because of repulsion between the Cr^{3+} ions in the face-shared octahedra. The Cr–Cr distance is 312 pm, and there is no spin pairing between the d^3 ions here. The bridging M–Cl–M bond angle is *less* than that for two regular face-shared octahedra for $W_2Cl_9^{3-}$, and *greater* for $Cr_2Cl_9^{3-}$. The isoelectronic $Re_2Cl_9^-$ has a structure similar to that of $W_2Cl_9^{3-}$, with a short Re–Re distance ($\sim$271 pm).

Metal–metal bonding occurs in several metal carbonyls. The structure of $Fe_2(CO)_9$ is similar to that of $W_2Cl_9^{3-}$, with three bridging carbonyls and an Fe—Fe bond. The dimeric carbonyls $Mn_2(CO)_{10}$ and $Re_2(CO)_{10}$ (Figure 10.3) contain metal–metal bonds without the aid of bridging carbonyls.

The high selectivity of dimethylglyoxime for the precipitation of nickel(II) does not result from extraordinarily great stability of the discrete complex, but from the unusual structure in the solid, which leads to low solubility. The planar units are stacked above one another so that the nickel "atoms" are bonded. Because of similar stacking in $K_{1.75}[Pt(CN)_4]\cdot1.5H_2O$, the solid is a one-dimensional metallic conductor (p. 233).

Metal–metal bonding is unmistakable from the structure of diamagnetic $Mn_2(CO)_{10}$, in which there are no bridging groups. The presence of M—M bonds is often inferred from low magnetic moments such as for the diamagnetic dimer $Cu_2(C_2H_3O_2)_4\cdot2H_2O$, which contains two d^9 Cu(II) ions (Figure 14.12). The long Cu—Cu distance (264 pm) indicates very weak bonding interaction in this case. The corresponding Cr—Cr compound has a much shorter Cr—Cr distance (236 pm).[13] The strong M—M bonding corresponds to a quadruple bond. A bond length of only 184.7 pm has been found for $Cr_2[2,6-C_6H_3(OMe)_2]_4$. Magnetic interactions between metal ions might occur through overlap of diffuse orbitals without significant bonding, or through bridging groups. Lower-than-expected magnetic moments might also occur because of distortions, with the resulting removal of degeneracy of sets of orbitals in the lower symmetry; so magnetic properties must be interpreted with caution.

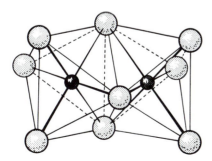

Figure 14.11 Structure of $W_2Cl_9^{3-}$

[13]J. Catterick and P. Thornton, "Structures and Physical Properties of Polynuclear Carboxylates", *Adv. Inorg. Chem. Radiochem.* 1977, *20*, 291.

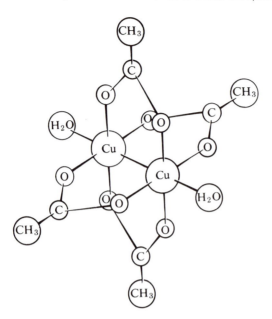

Figure 14.12 Structure of $[Cu(CH_3CO_2)_2 \cdot H_2O]_2$.

14.10.1 Metal–Metal Quadruple Bonds

Whereas the rhenium(III) compounds such as $CsReCl_4$ were presumed to contain low-spin tetrahedral $ReCl_4^-$ ions, Cotton and coworkers found instead a dimer, $Re_2Cl_8^{2-}$ (Figure 14.13), and a trimer, $Re_3Cl_{12}^{3-}$ (Figure 14.14). The dimer has a surprising structure containing two *eclipsed* planar $ReCl_4^-$ units joined only by a Re—Re bond. The eclipsed configuration is unexpected, because of strong repulsion between adjacent Cl^- ions. This repulsion is evident from the distortion from planar $ReCl_4$ units with Cl–Cl distances between units much longer than the Re–Re distance. The unusually short Re–Re distance (224 pm, compared with 275 pm in the metal) and the eclipsed configuration led Cotton *et al.* to propose a *quadruple* bond. The nearly planar $ReCl_4$ units are assumed to use the $d_{x^2-y^2}sp^2$ orbitals of Re for Re—Cl

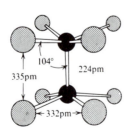

Figure 14.13 The structure of the $Re_2Cl_8^{2-}$ ion. (Reproduced with permission from F. A. Cotton, *Acc. Chem. Res.* 1969, *2*, 242. Copyright 1969, American Chemical Society.)

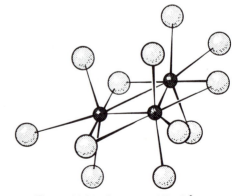

Figure 14.14 Structure of $Re_3Cl_{12}^{3-}$

bonding. A *sigma* Re—Re bond can be formed using a p_z–d_{z^2} hybrid orbital on each Re. The d_{xz} orbitals on each Re would overlap to form a *pi* bond, and an equivalent *pi* bond would result from the d_{yz} orbitals. The d_{xy} orbitals of the two Re atoms are parallel and form a *delta* bond. This bond scheme *requires* the eclipsed configuration.

The very short Re–Re distance is retained in $Re_2Br_8^{2-}$, or by replacement of one halide ion on each Re by a tertiary phosphine ligand. The halide ions of $Re_2X_8^{2-}$ can be replaced by reaction with carboxylic acids to give dimeric carboxylates, $Re_2(O_2CR)_4X_2$, with the structure of the dimeric Cu(II) acetate (Figure 14.12). The intermediate case $Re_2(O_2CR)_2X_4(H_2O)_2$ is also obtained with bridging carboxylates on opposite sides. Presumably the rhenium–rhenium quadruple bond is retained in these compounds.

The isoelectronic Mo^{II} compounds $Mo_2Cl_8^{4-}$ and $Mo_2(O_2CCH_3)_4$ (without axial ligands) also have the eclipsed configuration (required by the bridging acetates) and exceedingly short Mo–Mo distances—214 and 211 pm, respectively.

Although many examples of compounds containing metal–metal bonds have been known for some time, they have been considered anomalies. Recent structural investigations indicate that this type of interaction is much more common than previously supposed. Quadruple bonds occur for Cr, Mo, W, Re, and Tc. Although W, as compared with Mo and Cr, forms few compounds containing quadruple bonds, some of these compounds are very stable. An air-stable compound, $W_2(C_8H_8)_3$ (Figure 14.15), has a bond length of only 237.5 pm. The bond length of the corresponding Cr compound is 221.4 pm. Sharp and Schrock[14] suggested that few W≡W compounds have been obtained because of the lack of a convenient starting material corresponding to $Mo_2(O_2CCH_3)_4$, which is used for preparing many of numerous Mo≡Mo compounds. They prepared a class of compounds $W_2Cl_4(PR_3)_4$ by reducing $(WCl_4)_x$ in the presence of PR_3 with sodium amalgam. The premise is that once an M—M multiple bond is present, the metal likely can be reduced to give a bond of higher order. The structures of the four compounds obtained were determined.[15] The average W—W distance (227.6 pm) for the

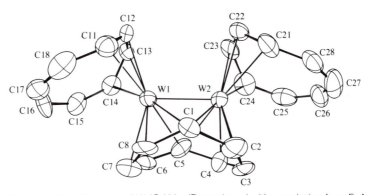

Figure 14.15 Structure of $W_2(C_8H_8)_3$. (Reproduced with permission from F. A. Cotton, *Acc. Chem. Res.* 1978, *11*, 227. Copyright 1978, American Chemical Society.)

[14]P. R. Sharp and R. R. Schrock, *J. Am. Chem. Soc.* 1980, *102*, 1430.

[15]F. A. Cotton, T. R. Felthouse, and D. G. Lay, *J. Am. Chem. Soc.* 1980, *102*, 1431.

presumably quadruply-bonded compounds is similar to distances for $W\equiv W$ compounds, but only slightly longer than for the quadruply bonded $[W_2(CH_3)_8]^{4-}$ and $[W_2(CH_3)_{8-n}Cl_n]^{4-}$ ions (226 pm).

14.10.2   Bond Length and Bond Order

Electron count is used as a check in determining that the assignment of bond order of an M—M compound is reasonable. Bond distance alone is not a good criterion of the bond order in such compounds. The quadruple bond lengths do not vary greatly for Re—Re (218–226 pm) and Mo—Mo (206.5–218 pm), but the range is very great for Cr—Cr (185–251 pm). Cotton, in introducing a basis for comparing bond shortening for different metals, calculated a *formal shortness ratio* as the observed bond length divided by the sum of the single bond radii (Pauling's value) for the atoms. In Figure 14.16's plot of formal shortness versus observed bond length, we see that the widest range is spanned by Cr. The effect of changing bond order on bond length can be seen for $Mo_2(SO_4)_4^{4-}$ (d^4, 211.1 pm for B.O. 4) and $Mo_2(SO_4)_4^{3-}$ (d^3, 216.4 pm for B.O. 3.5). This is an ideal case in which the two ions differ only in charge, so there is only one δ electron from $Mo_2(SO_4)_4^{3-}$. A bond order of 3.5 also can be achieved by adding an electron to a δ^* orbital. Although this appears to be the case for Re_2^{5+} compounds,[16] no identical pairs are known for these cases.

Bond lengths for triple-bonded M—M species vary for Mo—Mo from 216.7 pm $[Mo_2(CH_2SiMe_3)_6]$ to 224.2 $[Mo_2(OSiMe_3)_6(NHMe_2)_2]$, for W—W from 225.4 $[W_2(CH_2SiMe_3)_6]$ to 230.1 $[W_2(NEt_2)_4Cl_2]$, and for Re—Re from 223.2 $[Re_2Cl_4(PEt_3)_4]$ to 229.3 $[Cl_4ReRe(MeSC_2H_4SMe)_2Cl]$. We see that there is some overlap in the bond-length ranges for

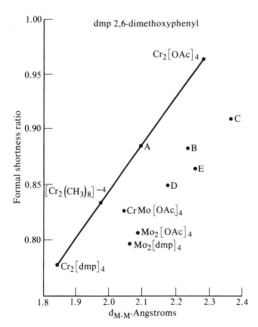

Figure 14.16 A plot of the "formal shortness" ratio versus M—M bond lengths. Points not identified by formula are for (*A*) $Cr_2(C_8H_8)_3$, (*B*) $Re_2Cl_8^{2-}$, (*C*) $W_2(C_8H_8)_3$, (*D*) $Re_2(CH_3)_8^{2-}$, (*E*) $W_2(CH_2SiMe_3)_6$; (Reproduced with permission from F. A. Cotton, *Acc. Chem. Res.* 1978, *11*, 228. Copyright 1978, American Chemical Society.)

[16]J. L. Templeton, *Prog. Inorg. Chem.* 1979, *26*, 211.

bond orders 3, 3.5, and 4. Other factors affecting the bond length include the presence or absence of and the nature of bridging groups, as well as the presence or absence of axial ligands. A theoretical consideration of the reasons for having or not having bridging ligands has been given.[17]

Since the correlation between bond length and bond order is not straightforward, the correlation with bond strength should be even more elusive. A wide range of estimates of bond strengths has been given, with recent thermochemical estimates of $\sim$500 kJ/mole for quadruple Mo—Mo bonds and $\sim$400 kJ/mole for quadruple Re—Re bonds.[18] Estimates from thermochemical data have been higher for Mo—Mo triple bonds ($\sim$590 kJ/mole) and for W—W triple bonds ($\sim$775 kJ/mole).[19] By comparison, the bond dissociation energy for N_2 is 946 kJ/mole.

14.10.3 Some Reactions of Compounds Involving M—M Multiple Bonds

Substitution reactions occur readily for some Re≡Re and Mo≡Mo compounds.

$$Re_2Cl_8^{2-} + 8SCN^- \xrightarrow{CH_3OH} Re_2(NCS)_8^{2-} + 8Cl^-$$

$$Mo_2(O_2CCH_3)_4 + 8\ HCl(aq) \xrightarrow{0°C} Mo_2Cl_8^{4-} + 4CH_3CO_2H + 4H^+$$

At 60°C, adding HCl to $Mo_2(O_2CCH_3)_4$ results in oxidative addition to the Mo—Mo quadruple bond,

$$Mo_2^{II}(O_2CCH_3)_4 + 8HCl(aq) \xrightarrow{60°C} [Mo_2^{III}Cl_8H]^{3-} + 3H^+ + 4CH_3CO_2H$$

to form a symmetrical compound (C_{2v}) containing an M—M bond (238 pm) with two bridging Cl and one symmetrical bridging H (Figure 14.17). We might expect easy oxidation (removal

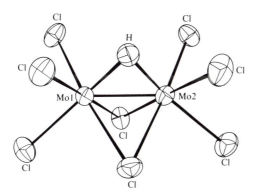

Figure 14.17 The structure of the $[Cl_3Mo(\mu\text{-}Cl)_2(\mu\text{-}H)MoCl_3]^{3-}$ ion as it occurs in $[C_5NH_6]_3[Mo_2Cl_8H]$. (From A. Bino and F. A. Cotton, *Angew. Chem. Int. Ed. Engl.* 1979, *18*, 332.)

[17]S. Shaik and R. Hoffmann, *J. Am. Chem. Soc.* 1980, *102*, 1194.

[18]L. R. Morss, R. J. Porcja, J. W. Nicoletti, J. SanFilippo, Jr., and H. D. B. Jenkins, *J. Am. Chem. Soc.* 1980, *102*, 1923.

[19]J. A. Conner *et al. J. Am. Chem. Soc.* 1978, *100*, 7738.

of δ *e*) or reduction (addition of δ* *e*) to produce compounds of lower bond order, but redox processes usually involve structural changes.[20] Oxidation of $[Re_2Cl_8]^{2-}$ by Cl_2 gives $[Re_2Cl_9]^-$ with the structure of $W_2Cl_9^{3-}$ (Figure 14.11). The $Mo_2(SO_4)_4^{4-}/Mo_2(SO_4)_4^{3-}$ case is an exception: the oxidation of $Mo_2(SO_4)_4^{4-}$ causes little structural change in the dimeric unit other than that expected for the decrease in bond order.

The unbridged triple-bonded $Mo_2(O\text{-}t\text{-}Bu)_6$ adds a molecule of CO reversibly to form $Mo_2(O\text{-}t\text{-}Bu)_6CO$, which contains two bridging alkoxides, a bridging CO, and a rare Mo=Mo bond (249.8 pm) (Figure 14.18). $Mo_2(OR)_6$ compounds (R = Me_3Si, Me_3C, Me_2CH, and Me_3CCH_2) add two moles of CO_2 reversibly in hydrocarbon solvents to produce $Mo_2(OR)_4(O_2COR)_2$. The two carboxylates produced by the insertion reaction serve to bridge the Mo—Mo bond.[21] Carbon monoxide can be inserted reversibly as a bridging carbonyl in $Pd_2(dmp)_2X_2$ (X = Cl, Br; dmp = $Ph_2PCH_2PPh_2$) in dichloromethane[22] to form an "A-frame" complex.

The A-frame complexes have a vacant bridging site (X) that might serve as an active site for the addition of a small molecule for possible catalytic activation (Figure 14.19). The complex ion $[CORh(\mu\text{-}Cl)(Ph_2PCH_2PPh_2)_2RhCO]^+$ adds SO_2 reversibly at the active site.[23] Carbon monoxide adds reversibly also, to give a compound with a bridging CO, but labeling studies show that CO attack occurs at a terminal position and then one of the coordinated CO molecules swings into the bridging site.

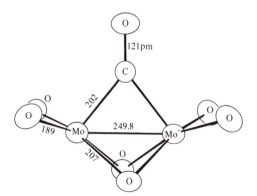

Figure 14.18 A view of the coordination geometry of $Mo_2(O\text{-}t\text{-}Bu)_6(CO)$, showing the main internuclear distances. The tertiary butyl groups are omitted for clarity. (Reproduced with permission from M. H. Chisholm, R. L. Kelly, F. A. Cotton, and M. W. Extine, *J. Am. Chem. Soc.* 1978, *100*, 2256. Copyright 1978, American Chemical Society.)

[20]T. J. Meyer, "Oxidation–Reduction of Metal–Metal Bonds," *Prog. Inorg. Chem.* 1975, *19*, 1.

[21]M. H. Chisholm, F. A. Cotton, M. W. Extine, and W. W. Reichert, *J. Am. Chem. Soc.* 1978, *100*, 1727.

[22]M. M. Olmstead, H. Hope, L. S. Benner, and A. L. Balch, *J. Am. Chem. Soc.* 1977, *99*, 5502.

[23]M. Cowie and S. K. Dwight, *Inorg. Chem.* 1980, *19*, 209 and references cited.

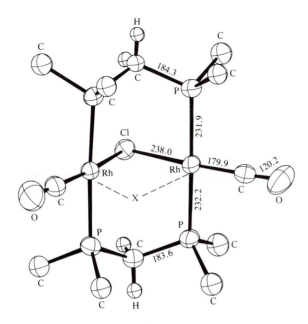

Figure 14.19 The inner coordination sphere of the $[Rh_2(CO)_2(\mu\text{-}Cl)$-$(Ph_2PCH_2PPh_2)]^+$ ion, showing some relevant bond lengths. Only the first carbon atom of each phenyl ring is shown. (Reproduced with permission from M. Cowie and S. K. Dwight, *Inorg. Chem.* 1979, *18*, 2700. Copyright 1979, American Chemical Society.)

GENERAL REFERENCES

K. W. Bagnall, *The Actinide Elements*, Elsevier, Amsterdam, 1972.

K. W. Bagnall, Ed., *Lanthanides and Actinides*, Butterworths, London, 1972.

J. C. Bailar, Jr., H. J. Emeléus, R. Nyholm, and A. F. Trotman-Dickerson, Eds., *Comprehensive Inorganic Chemistry*, Pergamon, Oxford, 1953. Five volumes.

F. A. Cotton and G. Wilkinson, *Advanced Inorganic Chemistry*, 4th ed., Wiley, New York, 1980. The best and most up-to-date inorganic reference book in a single volume.

S. A. Cotton and F. A. Hart, *The Heavy Transition Elements*, Wiley, New York, 1975. Good coverage of the second and third transition series, including the lanthanides and actinides.

P. R. Fields and T. Moeller, Eds., "Lanthanide-Actinide Chemistry," *Adv. Chem. Ser.*, 1967, *71*. Symposium papers.

L. Gmelin, *Handbuch der anorganischen Chemie*, Verlag, Weinheim, 1924–71.

W. L. Jolly, Ed., *Metal-Ammonia Solutions*, Dowden, Hutchinson and Ross, Stroudsburg, Pa., 1972. Reprints of 63 important papers.

C. Keller, *The Chemistry of the Transuranium Elements*, Verlag, Weinheim, 1971.

D. L. Kepert, *The Early Transition Metals*, Academic Press, London, 1972.

A. F. Wells, *Structural Inorganic Chemistry*, 4th ed., Oxford, Oxford, 1975. The best single source for structural inorganic chemistry.

PROBLEMS

14.1 Write balanced equations for the preparation of

a. Na_2CO_3.
b. MgF.
c. $TiBr_3$.
d. $ZrCl_3$.
e. $VOCl_3$.
f. WCl_6.

14.2 The bond dissociation energies for the alkali metal M_2 molecules decrease regularly from 100.9 kJ/mole for Li_2 to 38.0 kJ/mole for Cs_2. The bond dissociation energies are *greater* for the M_2^+

ions, decreasing from 138.9 kJ/mole for Li_2^+ to 59 kJ/mole for Cs_2^+. Explain why the dissociation energies are greater for the M_2^+ ions. What is the bond order for M_2^+?

14.3 Predict the following for Fr.
a. The product of the burning of Fr in air.
b. An insoluble compound of Fr.
c. The structure of FrCl.
d. The relative heats of formation of FrF and FrI.

14.4 How could you remove unreacted Na (metal) in liquid ammonia safely?

14.5 Write equations for the preparation of the following metals.
a. Na.   *c.* Cs.
b. K.   *d.* Mg.

14.6 How can you stabilize solutions containing Na^-?

14.7 What are laboratory or everyday uses of Li, K, and Cs?

14.8 Cite several properties that show the diagonal relationships between Li and Mg, and between Be and Al.

14.9 The elements Li-Mg, Na-Ca, and Be-Al are closely related because of the diagonal relationship. Would you expect Mg^{2+} to be more closely related to Sc^{3+} or to Ga^{3+}? Why?

14.10 Why is Au expected to form Au^-?

14.11 Write equations for the reduction of Cu^{2+} with a limited amount of CN^- and with an excess of CN

14.12 Why is *aqua regia* (HCl—HNO_3) effective in oxidizing noble metals when neither HCl nor HNO_3 is effective alone?

14.13 Why are large anions effective in stabilizing unusually low oxidation states, such as Cd(I)?

14.14 Sketch a linear combination of s, p_z, and d_{z^2} that would give very favorable overlap for bonding in a linear MX_2 molecule.

14.15 The structures of ZnO, CdO, and HgO are quite different. Describe the structures.

14.16 The structures of HgF_2, $HgCl_2$, $HgBr_2$, and HgI_2 show interesting variations. Describe the structures.

14.17 Sketch the structure of the silicate anion in beryl.

14.18 Give equations for reactions that could be used to separate Zn^{2+}, Cd^{2+}, and Hg^{2+} present in solution.

14.19 Can one obtain
a. Hg^{2+} salts free of Hg_2^{2+}?
b. Hg_2^{2+} salts free of Hg^{2+}?
Explain.

14.20 Discuss the factors involved in determining the following solubility patterns: LiF is much less soluble than LiCl, but AgF is much more soluble than AgCl.

14.21 The metal perchlorates have been referred to as "universal solutes." What properties are important in causing most metal perchlorates to be quite soluble in water and several other solvents?

14.22 For which elements in the rare earth series are M(II) and M(IV) oxidation numbers expected?

14.23 Why should it have been expected that there would be more uncertainty concerning the identity of the first member of the second inner transition series compared with the lanthanide series?

14.24 One of the common M_2O_3 structures is that of α-alumina (Al_2O_3). Another is the La_2O_3 structure (C.N.7). Check the La_2O_3 structure in Wells and describe the coordination about La^{3+}.

14.25 Ti is the ninth most abundant element in the earth's crust, and its minerals are reasonably concentrated in nature. Why is it less commonly used than rarer metals?

14.26 Some elements are known as dispersed elements, forming no independent minerals, even though their abundances are not exceptionally low; whereas others of comparable or lesser abundance are highly concentrated in nature. Explain the following cases:
Dispersed: Rb, Ga, Ge, Hf.
Concentrated: Li, Be, Au.

14.27 The stabilities of the oxidation states of the lanthanides other than $+$ III can be explained in terms of empty, half-filled, and filled f orbitals. Attempt to apply a similar approach to the series Sc–Zn. Where does it work well? Why is it much less useful? Is this approach effective for the actinide elements?

14.28 Compare the syntheses of the highest and lowest stable oxidation states of Mn and Re? Which halide ions can be oxidized by MnO_4^- and by ReO_4^-?

14.29 Would the removal of Hf from Zr be important in most applications of zirconium compounds? In the use of Zr metal in flash bulbs?

14.30 What properties of tungsten make it so suitable for filaments for light bulbs?

14.31 The isopoly anions containing MO_6 can be considered as fragments of oxide structures. Why are the arrangements usually *ccp*?

14.32 The simple anions CrO_4^{2-}, MoO_4^{2-} and WO_4^{2-} are tetrahedral. Why do the polyacids and poly-anions of Cr differ structurally from those of Mo and W?

14.33 Give an example of
a. An acidic oxide of a metal.
b. An amphoteric oxide of a transition metal.
c. A diamagnetic rare earth metal ion.
d. A compound of a metal in the $+8$ oxidation state.
e. A liquid metal chloride.
f. A compound of a transition metal in a negative oxidation state.

14.34 Describe the quadruple bonding in $Re_2Cl_8^{2-}$ in terms of the bond types (σ, π, etc.) and the atomic orbitals involved. What is significant about the eclipsed configuration?

14.35 Determine the bond order and oxidation number for $Mo_2(O_2CCH_3)_4$ (bridging acetate ions), $W_2Cl_4(PR_3)_4$ (no bridging ligands), and $[Mo_2(O\text{-}t\text{-Bu})_6CO]$ (Figure 14.18).

14.36 Give an example of each type of Co complex.
a. Co(II) tetrahedral.
b. Co(II) square planar.
c. Co(III) octahedral, high-spin.
d. Co(III) optically active.

14.37 Give one example of a nickel complex illustrating square planar, tetrahedral, and octahedral coordination. What type of ligands favor each of these cases?

14.38 The structures of complexes with high coordination number involving bidentate groups can be described in terms of the "average" positions of the bidentate groups. Describe $Ce(NO_3)_5^{2-}$ and $Ce(NO_3)_6^{3-}$ in this way.

14.39 Give syntheses for *cis-* and *trans-*[$Pt(NH_3)_2Cl_2$], starting with [$Pt(NH_3)_4$]$^{2+}$ and [$PtCl_4$]$^{2-}$.

14.40 How can one account for the color of the following.
a. Fe_3O_4.   *d.* $KMnO_4$.
b. Ag_2S.   *e.* $Ti(H_2O)_6^{3+}$.
c. $KFeFe(CN)_6$.   *f.* $Cu(NH_3)_4^{2+}$.

XV

Boron Hydrides, Cluster and Cage Compounds

The preceding chapters showed how to predict the formulas and structures of several kinds of compounds with the aid of valence bond theory, molecular orbital theory, the octet rule, and the 18-electron rule. But the electronic and molecular structures of one large class of compounds are not understandable in these terms. This chapter discusses a number of such compounds, starting with the boron hydrides—their structures, synthesis, and chemistry. Subsequently, we study related compounds containing transition metals, noting their remarkable behavior and structures as well as the possible relevance of these cluster compounds to heterogeneous catalysis. Finally, we deal with cage compounds containing several metals but featuring fewer metal–metal bonds than seen in clusters.

15.1 INTRODUCTION

The valence bond model dominated bonding theory for some time. Couper's lines between symbols (1858) were interpreted as electron pair bonds by Lewis (1916). The stereochemistry postulated by van't Hoff and LeBel for organic compounds and by Werner for inorganic complexes was justified on a quantum-mechanical basis through hybridization (Pauling, 1931). Along with VSEPR of Nyholm and Gillespie (1957), these considerations accommodated the x-ray structures of thousands of ionic and molecular species. Using the valence bond approach, Pauling related bond lengths to bond order, as well as electronegativity to bond polarity. Organic chemists developed "electronic" theories of reactivity that helped greatly to systematize the field.

Yet at the very time G. N. Lewis proposed the electron-pair bond, Alfred Stock was preparing a series of compounds whose formulas gave no hint as to their structures and whose structures, once determined, could not be accommodated by a simple valence-bond model. These remarkable compounds are the boron hydrides or boranes. Because they are so reactive toward O_2 (boron is found in nature as oxides), Stock and his coworkers devised vacuum line techniques to handle the volatile boron hydrides in an O_2- and moisture-free environment. Even so, the experimental difficulties to be overcome by the early workers were enormous. Yields of the hydrides typically were only 4 to 5%. The mixtures produced often required tedious and laborious procedures for fractionation. The quantities of pure compounds attained were millimolar at maximum. In spite of these problems, Stock was able to prepare and characterize B_2H_6, B_4H_{10}, B_5H_9, B_5H_{11}, B_6H_{10}, and $B_{10}H_{14}$. It is possible to divide these compounds into two groups—a hydrogen-"rich," of general formula B_pH_{p+6}; and a hydrogen-"poor," of formula B_pH_{p+4}. Subsequently, other members of both series have been discovered. A third series of very stable anions, $B_pH_p^{2-}$ (which can be thought of as derived via deprotonation of B_pH_{p+2}), has been prepared more recently (for $p = 6$–12).

Figure 15.1 depicts the structures of the hydrides discovered by Stock. Each B is bonded to at least one H, with some H's bridging two boron atoms. These hydrogens are called terminal and bridging, respectively. The physical properties of the boron hydrides are reported in Table 15.1. A very important chemical property of most of the boranes is their vigorous (often explosive) reactivity with O_2. Of the neutral boranes, only $B_{10}H_{14}$ can be handled in air. Table 15.2 records some known boron hydride species, ionic as well as neutral. Several potential members of each series are missing—perhaps because no one has yet discovered a workable synthesis, or because the compounds themselves are not stable. A likely example of the latter

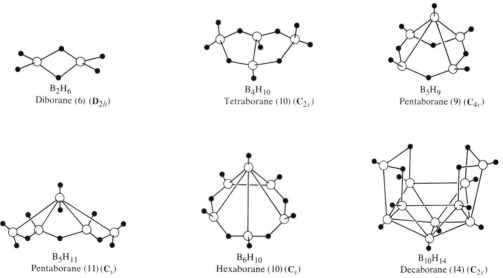

| B_2H_6 | B_4H_{10} | B_5H_9 |
| Diborane (6) (D_{2h}) | Tetraborane (10) (C_{2v}) | Pentaborane (9) (C_{4v}) |

| B_5H_{11} | B_6H_{10} | $B_{10}H_{14}$ |
| Pentaborane (11) (C_s) | Hexaborane (10) (C_s) | Decaborane (14) (C_{2v}) |

Figure 15.1 Structures of Stock's boron hydrides. (From S. G. Shore in *Boron Hydride Chemistry*, E. L. Muetterties, Ed., Academic Press, New York, 1975.)

Table 15.1   Physical properties[a] of boron hydrides prepared by Stock

Formula	Name	Melting Point (°C)	Boiling Point (°C)
B_2H_6	Diborane	-165.5	-92.5
B_4H_{10}	Tetraborane	-120	16
B_5H_9	Pentaborane(9)	-46.8	58.4
B_5H_{11}	Pentaborane(11)	-123.3	65
B_6H_{10}	Hexaborane(10)	-65.1	108
$B_{10}H_{14}$	Decaborane	99.5	213

[a]From S. G. Shore in *Boron Hydride Chemistry*, E. Muetterties, Ed., Academic Press, New York, 1975.

would be B_3H_9, which would have three very crowded bridging hydrogens.[1] The isoelectronic species resulting from its (hypothetical) deprotonation, $B_3H_8^-$, is known to be stable. The different series of boron hydrides, or boranes, are named *closo* (for species isoelectronic with B_pH_{p+2}), *nido* (isoelectronic with B_pH_{p+4}), *arachno* (isoelectronic with B_pH_{p+6}), and *hypho* (isoelectronic with B_pH_{p+8}).

Table 15.2   Some boron hydride species containing 12 or fewer borons

Closo (B_pH_{p+2})	Nido (B_pH_{p+4})	Arachno (B_pH_{p+6})	Hypho (B_pH_{p+8})
	B_2H_6	$B_2H_7^-$	
		$B_3H_8^-$	
	$B_4H_7^-$	B_4H_{10}	
		$B_4H_9^-$	
	B_5H_9	B_5H_{11}	$B_5H_{12}^-$
	$B_5H_8^-$		
$B_6H_6^{2-}$	B_6H_{10}	B_6H_{12}	
	$B_6H_9^-$	$B_6H_{11}^-$	
	$B_6H_{11}^+$		
$B_7H_7^{2-}$		$B_7H_{12}^-$	
$B_8H_8^{2-}$	B_8H_{12}	B_8H_{14}	
$B_9H_9^{2-}$	B_9H_{13}	$n\text{-}B_9H_{15}$	
	$B_9H_{12}^-$	$i\text{-}B_9H_{15}$	
		$i\text{-}B_9H_{14}^-$	
$B_{10}H_{10}^{2-}$	$B_{10}H_{14}$	$B_{10}H_{15}^-$	
	$B_{10}H_{13}^-$	$B_{10}H_{14}^{2-}$	
$B_{11}H_{11}^{2-}$	$B_{11}H_{15}$ (?)		
	$B_{11}H_{14}^-$		
	$B_{11}H_{13}^{2-}$		
$B_{12}H_{12}^{2-}$			

[1]R. E. Williams, *Adv. Inorg. Radiochem.* 1976, *18*, 67.

The progressively more open structures of these series (see Figure 15.2) are reflected in their names: *closo* = "closed"; *nido* is Greek "nest"; *arachno* and *hypho* mean "web" and "net," respectively.

Nomenclature for Boranes[2]

Neutral boron hydrides are all named *borane;* a Greek prefix indicates the number of B atoms, and an Arabic number in parentheses gives the number of H atoms. So B_5H_{11} is pentaborane(11). The number is omitted if only one borane containing a particular B count is known. B_2H_6 is usually referred to as diborane.

   Anionic species are named as hydroborates. Greek prefixes separately indicate the numbers of H and B; the charge on the anion is given in parentheses following. So $B_5H_8^-$ is octahydropentaborate(1−). The structural type sometimes is specified when naming anions. So $B_5H_8^-$ is also octahydro-*nido*-pentaborate(1−).

15.2 THE BONDING PROBLEM IN BORANES

15.2.1 Localized Bonding Picture

Retaining the valence bond concept of the relationship between bond distance and bond order, we immediately encounter a problem on examining the known structures of boron hydrides such as those shown in Figure 15.1: the coordination number of each boron (and of some of the hydrogens) exceeds the number of low-energy orbitals—not to mention the number of electron pairs. Obviously, the lines in the structures of Figure 15.1 cannot represent electron-pair bonds: they serve only to show which atoms are connected. Such structures are called topological structures. As you can see, the boron hydrides are an example of the so-called electron-deficient compounds, in which the number of valence orbitals exceeds the number of valence electrons.

   Ideally, a bonding picture for electron-deficient compounds would allow the same straightforward prediction of geometry, reactivity, stoichiometry, redox properties, acidity, etc. that the valence bond approach permits for "regular" compounds. Early attempts to account for the electronic structure of diborane, the simplest member of the class, included the observation that B_2H_6 is isoelectronic with ethylene, C_2H_4. In this view, we could regard the two bridging H's in the structure as protonating the double bond of $B_2H_4^{2-}$. Subsequent research has confirmed the acidic nature of bridging H's in the boranes; however, this bonding picture is difficult to extend to the higher boranes. A straightforward application of valence bond theory to the electronic structure of diborane requires some 20 resonance structures (see Figure 15.3). Plainly, as the task of describing more complex molecules begins to require unmanageable numbers of canonical structures, the simple valence bond approach loses its utility.

   A valence bond approach for simple molecules such as H_2O and NH_3 differs from a localized MO approach mainly in terminology. A simple extension to include three-center, two-electron bonds accommodates many electron-deficient compounds. If each B were to form at least one three-center bond, it could achieve an octet of electrons. Figure 15.4 shows possible ways of forming three-center B—B—B and B—H—B bonds, along with diagrammatic representations of each. The difference between open and closed B—B—B bonds lies in the

[2]*Inorg. Chem.* 1968, *7*, 1945; *Pure Appl. Chem.* 1972, *30*, 683.

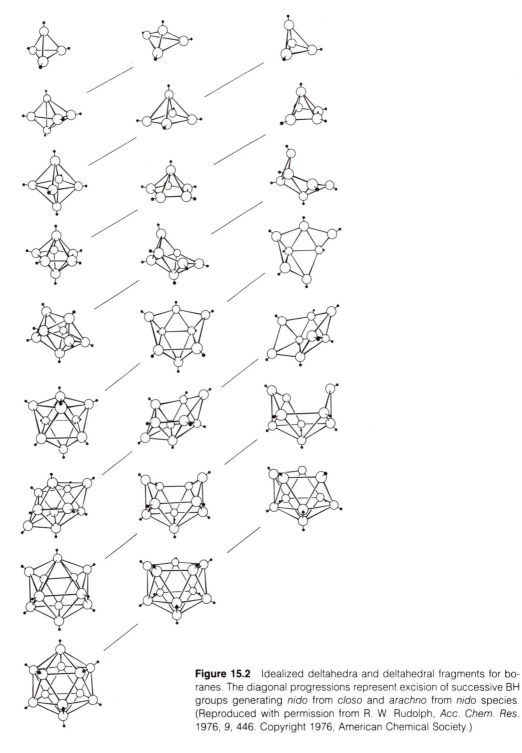

Figure 15.2 Idealized deltahedra and deltahedral fragments for boranes. The diagonal progressions represent excision of successive BH groups generating *nido* from *closo* and *arachno* from *nido* species. (Reproduced with permission from R. W. Rudolph, *Acc. Chem. Res.* 1976, *9*, 446. Copyright 1976, American Chemical Society.)

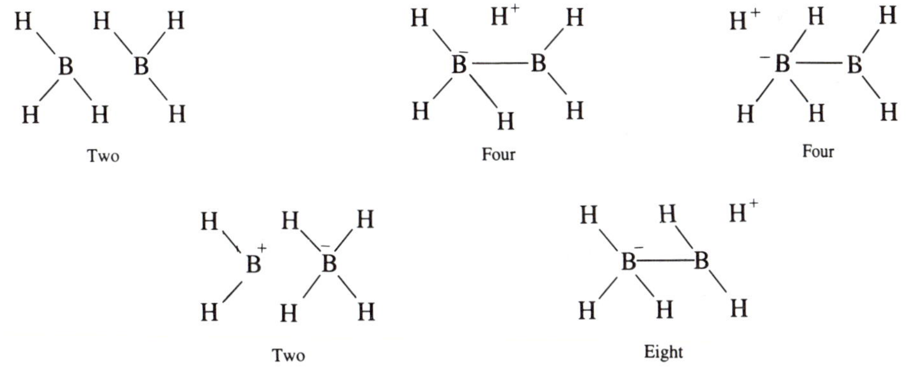

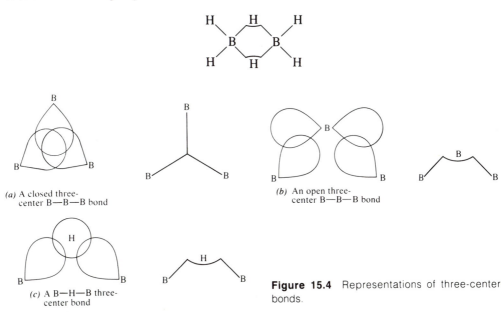

Figure 15.3 Resonance structures for B_2H_6.

choice of hybrid orbitals used. Since theoretical studies have not revealed evidence in favor of the open three-center bond, all valence structures in this book depict only closed ones.

We can provide a simple picture of the electronic structure of B_2H_6 by employing the notion of the three-center bond. The B atoms may be regarded as sp^3 hybridized. Each B forms 2 two-center, two-electron bonds with H atoms, thus accounting for eight of the 12 available electrons.

In terms of molecular orbitals, we can consider the three-center B—H—B bonds as resulting from the combination of one sp^3 orbital from each boron and the s orbital of a bridging hydrogen. The two bonding orbitals are occupied by the remaining four electrons. The non-bonding and antibonding orbitals, which also arise, would be unfilled. The observed bond angles in B_2H_6 are far from tetrahedral angles, so the molecular orbitals probably have some of the character of sp^2 hybrids also. A valence structure for diborane is

Figure 15.4 Representations of three-center bonds.

(a) A closed three-center B—B—B bond

(b) An open three-center B—B—B bond

(c) A B—H—B three-center bond

15.2.2 *styx* Numbers

Lipscomb and coworkers established a systematic procedure for obtaining the valence structures of more complex boron hydrides, incorporating three-center bonding. The valence structures drawn by Lipscomb's procedure may be employed in two ways: given the molecular formula, to predict a molecular structure that conforms to a reasonable electronic structure; or, given the molecular structure, to describe the electronic structure. The procedure consists essentially of determining the total number of orbitals and electrons available for bonding. The number of B—H bonds and B—H—B three-center bonds is then counted and the requisite orbitals and electrons assigned. The remaining orbitals and electrons, considered to be available for framework bonding, are distributed among two-center B—B bonds and three-center B—B—B bonds. A systematic prescription for accomplishing this is outlined.

Consider a neutral borane whose formula can be written as B_pH_{p+q}. The molecule consists of p (BH) groups and q "extra" hydrogens distributed between bridging positions and BH groups (converting them to BH_2 groups). If we let

$$s = \text{number of B—H—B bonds}$$
$$t = \text{number of B—B—B bonds}$$
$$y = \text{number of B—B bonds}$$
$$x = \text{number of } BH_2 \text{ groups}$$

then we can formulate several relations between structural features and available orbitals and electrons, called Equations of Balance.

Hydrogen Balance

$$q = s + x$$

All the "extra" hydrogens must be in B—H—B or BH_2 units.

Orbital Balance

$$p = s + t$$

The structure contains p boron atoms, each of which must participate in 1 three-center bond if it is to attain a completed octet. This can be either a B—H—B or B—B—B bond.

Electron Balance

The total number of electron pairs available for framework bonding is p from the BH groups (since each B uses one electron in bonding to the terminal H) plus $\frac{1}{2} q$ from the "extra" H's. These must be just enough to occupy the $s + t + y$ framework bonds and the x BH_2 bonds. Hence,

$$p + \tfrac{1}{2} q = s + t + y + x$$

Substituting from the hydrogen balance equation,

$$p - \tfrac{1}{2} q = t + y$$

Applying the Equations of Balance to a compound of given composition allows us to determine a set of *styx* numbers that specify a valence structure. For example, for B_2H_6

$[(BH)_2H_4]$, $p = 2$, $q = 4$, and we have $4 = s + x$, $2 = s + t$, $0 = t + y$. Since all numbers must be ≥ 0, the only possible solution is $s = 2$, $t = 0$, $y = 0$, $x = 2$ (written "2002"), and the structure corresponds to that already given for diborane.

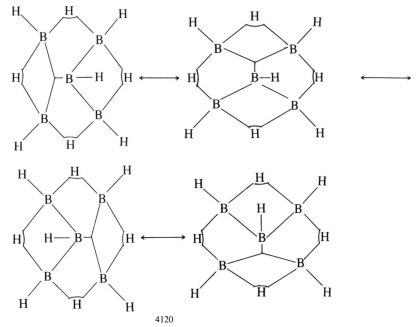

2002

You can write structures for higher boranes with this same procedure. For B_5H_9 $[(BH)_5H_4]$, the *nido* species with a square pyramidal framework shown in Figure 15.1: $4 = s + x$, $5 = s + t$; $5 - 2 = 3 = t + y$. Three sets of *styx* numbers are possible solutions: 2302, 3211, and 4120. Obviously, the Equations of Balance do not always give unequivocal answers, but they do aid us by limiting the structures to be considered. In choosing the best structure, keep in mind these additional considerations.

1. Every pair of adjacent B's must be bonded to each other through a B—B, B—H—B, or B—B—B bond.

2. Pairs of B atoms bonded by a B—B bond may not also be bonded to one another by B—B—B or B—H—B bonds.

3. Nonadjacent pairs of B atoms may not be bonded by framework bonds.

4. Other things being equal, the preferred structure is the one with the highest symmetry.

These considerations eliminate the first two solutions, leaving the 4120 structure (see Figure 15.5). Although no one of the structures shown exhibits the $\mathbf{C}_{4v}$ symmetry of the B_5 framework,

4120

Figure 15.5 Symmetry-equivalent 4120 structures for B_5H_9.

the entire group of symmetry-equivalent structures taken together does. A C_4 operation yields each successive structure from the preceding one. Of course, the existence of several symmetry-equivalent structures corresponds to electron delocalization.

We can modify the Equations of Balance to treat species of charge c. If we imagine that positive charge is created by protonation of neutral species, then the formulas can be written in the form $(BH)_p H^c_{q+c}$, and the Equations of Balance become

$$q + c = s + x$$

$$p + c = s + t$$

$$p - \frac{q}{2} - c = t + y$$

So for $B_3H_8^-$, $p = 3$, $q = 6$, $c = -1$, and two solutions are 1104 and 2013. The first cannot be physically meaningful, since it is impossible to have four BH_2 groups with only three B's. The structure is

2013

Lipscomb[3] also has formulated rules for estimating from these structures charges on individual atoms.

The number of valence structures that can be written for boron hydride species increases rapidly with the complexity of the species; this indicates considerable electron delocalization in the higher boron hydrides. Hence, another fruitful approach to the electronic structures lies in molecular orbital methods (also useful in treating delocalization in ''regular'' species). As usual, it is necessary to know something about the molecular geometry before the MO approach can be applied usefully.

15.3 STRUCTURES OF THE BORON HYDRIDES

X-ray diffraction, electron diffraction and ^{11}B NMR techniques have allowed elucidation of the structures of many boron hydride species besides those prepared by Stock. Figure 15.2 shows idealized structures for boron hydrides. With the exception of B_2H_6 and $B_3H_8^-$ (not shown), these structures are based on polyhedra (Figure 15.18) or polyhedral fragments having triangular faces with BH groups at vertices. Such polyhedra are often called deltahedra. Boron hydrides undoubtedly adopt such structures because of the favorable possibilities for three-center bonding by borons sharing a triangular face.

[3] W. N. Lipscomb, *Boron Hydrides*, Benjamin, New York, 1963.

> The structures of several other known hydrides result from the joining of deltahedral frameworks or fragments of them by sharing edges or vertices or by formation of B—B bonds with elimination of H_2. The known species of this type are listed in Table 15.3.

Conceptually, it is simplest to regard the structures of the B_pH_{p+4} and B_pH_{p+6} *nido* and *arachno* series as derived from those of the $B_pH_p^{2-}$ *closo* anions. The *closo* species (known for $p = 6$ to 12) adopt structures based on closed deltahedra having all vertices occupied by BH units. If one BH unit is removed from the most highly connected vertex and replaced by two electrons (the number contributed by the BH unit to the framework), a series of hypothetical anions $B_pH_p^{4-}$ results. If these are protonated by distributing $4H^+$ in the most symmetrical fashion, each bridging two B's around the open face, the series of neutral *nido* compounds B_pH_{p+4} results. Removing a second BH unit of high connectivity on the open face of a B_pH_{p+4} molecule and replacing it with a pair of electrons gives the hypothetical $B_pH_{p+4}^{2-}$ series of anions. If these species are protonated, the *arachno* compounds B_pH_{p+6}, with a still more open structure, result. The placement of ''extra'' protons in the *arachno* species is not always easy to predict. At least one is always attached to a boron adjacent to the BH unit formally removed. This creates a BH_2 unit, a structural feature appearing for the first time in the *arachno* series. (A lone exception to this statement is B_2H_6.)

In summary, we can think of the known boron hydrides as derived from the parent *closo* anions $B_pH_p^{2-}$ by successive excision of BH units, replacement by electron pairs, and protonation of the resulting hypothetical anions to give B_pH_{p+4} and B_pH_{p+6} compounds featuring progressively more open structures. In the *nido* series, the ''extra'' H's (that is, the number

Table 15.3 Boron hydride species having fused or linked deltahedral structures

Species	*Comments*
$2,2'\text{-}B_8H_{18}$	Made of two B_4H_{10} frameworks joined via one B—B bond.
$1,1'\text{-}B_{10}H_{16}$	Two B_5H_9 frameworks joined at apical B's via B—B bond.
$2,2'\text{-}B_{10}H_{16}$	Two B_5H_9 frameworks joined at basal B's via B—B bond.
$1,2'\text{-}B_{10}H_{16}$	Two B_5H_9 frameworks joined at one apical and one basal B via B—B bond.
$B_{13}H_{19}$	Fusion of $n\text{-}B_9H_{15}$ and B_6H_{10} frameworks sharing common edge.
$B_{13}H_{18}^-$	
$B_{14}H_{18}$	Fusion of $B_{10}H_{14}$ and B_6H_{10} frameworks sharing common edge.
$B_{14}H_{20}$	Two B_8H_{12} frameworks fused at B(7)–B(12) positions.
$B_{14}H_{22}$	Dissociated to B_6H_{10} and B_8H_{12} at $-41°$.
$B_{15}H_{23}$	B_6H_{10} framework bonded via three-center interaction of non–H-bridged B—B bond with B_9H_{13} framework.
$B_{16}H_{20}$	$B_{11}H_{17}$ and B_6H_{10} frameworks share a common vertex.
$n\text{-}B_{18}H_{22}$	Fusion of two $B_{10}H_{14}$ units, both opening in same direction sharing two B's.
$i\text{-}B_{18}H_{22}$	Fusion of two $B_{10}H_{14}$ units opening in opposite directions sharing two B's.
$B_{20}H_{16}$	Face-to-face fusion of two $B_{10}H_{14}$ units; four shared B's have no H bonded.
$B_{20}H_{26}$	Fusion of two $B_{10}H_{14}$ units with one B—B bond. Several isomers possible.

in addition to the ones attached to the B atoms at each polyhedral vertex) are distributed around the open face in bridging positions. In the *arachno* series, at least one is placed so as to give a terminal BH_2 unit. The diagonal series in Figure 15.2 show this process. Clearly, the formulas of the hydrides have structural consequences—the *closo*, *nido*, and *arachno* series having increasing numbers of H as well as increasingly more open structures.

15.4   MOLECULAR ORBITAL DESCRIPTION OF BONDING IN BORON HYDRIDES

A molecular orbital treatment of the bonding in boron hydrides describes electron delocalization more satisfactorily than the valence bond approach. Although the MO description is to some extent an *ad hoc* case for each species, regularities become apparent with experience. A good illustration is a treatment of the *closo* species $B_6H_6^{2-}$.[4] The description starts with the radially-directed *sp* hybrid and two tangential *p* orbitals contributed by each B to framework bonding (Figure 15.6); each B also contributes two electrons. The other B *sp* hybrid and electron go to form the B—H bond to the exopolyhedral *(exo)* hydrogen. Figure 15.7 shows the framework molecular orbitals from radial *sp* hybrids (a_{1g}, e_g, and t_{1u}) and tangential *p*'s (t_{1g}, t_{2g}, t_{1u}, t_{2u}). The angular overlap model suggests that the MO energy diagram will be as shown in Figure 15.8. The a_{1g}, t_{2g}, and $1t_{1u}$ (the in-phase combination of the radial and tangential t_{1u}'s) orbitals are bonding, whereas the e_g, t_{1g}, $2t_{1u}$ (out-of-phase combination), and t_{2u} are antibonding. The total of seven electron pairs (one from each BH unit and one from the charge) to be accommodated exactly fills the bonding MO's. This *closo* species with six vertices has an MO energy diagram which accommodates seven pairs in bonding orbitals. In general, graph theoretical methods show that the *closo* anions $B_pH_p^{2-}$ can accommodate $p + 1$ electron pairs in bonding framework MO's. Thus any species having p BH groups and $p + 1$ electron pairs for framework bonding should assume a *closo* structure.

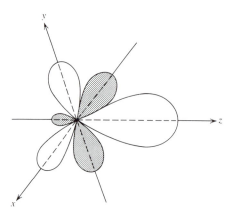

Figure 15.6   The radially directed *sp* hybrid and the tangentially directed p_x and p_y orbitals used for framework bonding in the boron hydrides.

[4]See Section 4.4 for B_2H_6.

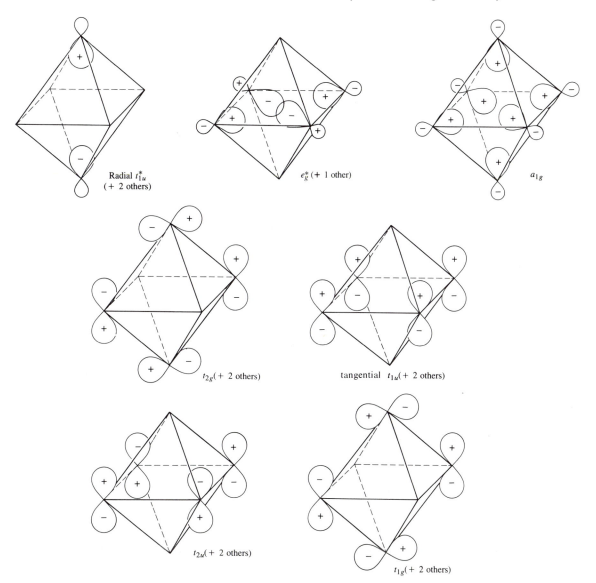

Figure 15.7   Framework molecular orbitals for $B_6H_6^{2-}$.

Electron-deficient molecules usually have only bonding orbitals filled, whereas most other stable molecules have both bonding and nonbonding orbitals filled.

In constructing MO descriptions of other boranes, you may find it convenient to use other hybrids whose geometry corresponds to bond directions in the structure—for example, sp^3 for B_4H_{10} and for basal borons in B_5H_9.

You should note that the framework orbitals for $B_pH_p^{2-}$ are the same as the ligand group orbitals for a complex ML_p with the same symmetry. Thus, the combinations of the six sp orbitals for $B_6H_6^{2-}$ are the same as the σ LGO for ML_6, and the others correspond to the π LGO for ML_6.

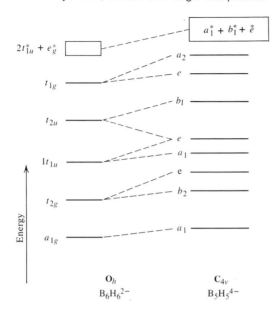

Figure 15.8 Energy-level diagrams for framework bonding orbitals in $B_6H_6^{2-}$ and $B_5H_5^{4-}$. The "exo" hydrogens are involved in localized two-center bonds and are not included here.

Suppose a BH unit is now removed from $B_6H_6^{2-}$ and an electron pair added to generate $B_5H_5^{4-}$. The five BH groups occupy all but one of the six vertices of an octahedron, giving a C_{4v} square pyramid. Figure 15.8 provides the energy-level diagram for the $B_6H_6^{2-}$ bonding orbitals and depicts their fate on formation of $B_5H_5^{4-}$. The main change in the energies of bonding MO's is a splitting of the triply degenerate orbitals and the fact that the second e level becomes less bonding. The seven skeletal electron pairs (one from each BH unit and two from the overall charge) exactly fill the bonding MO's. A general feature[5] of *nido* compounds is that their energy-level diagrams provide stable levels for $p+2$ electron pairs when the p BH units occupy all but one of the vertices of the $p+1$ *closo* polyhedron. Thus we can expect any species having p BH units and $p+2$ skeletal electron pairs to adopt a *nido* structure, with the BH units occupying all but one vertex of the $p+1$ *closo* polyhedron.

We can extend the same procedure to produce *arachno* $B_4H_4^{6-}$. A general result[5] is that the energy diagrams of $B_pH_p^{6-}$ anions provide stable orbitals for $p+3$ skeletal electron pairs when the BH units occupy all but two adjacent vertices of the $p+2$ *closo* polyhedron.

Note that in this view all the "extra" hydrogens required to give neutral species from the (hypothetical) anions are thought of as arising by protonation, usually in the most symmetrical way, of B—B bonds around the open face (in *nido* and *arachno* compounds) or of BH groups to give BH_2 units (in *arachno* compounds). In other words, the electrons contributed by these hydrogens are incorporated into the framework electron count.

A significant result of a generalized MO approach is the classification of boron hydride species as *closo*, *nido*, or *arachno* via framework electron count as well as by molecular formula. Thus a link between electronic structure and molecular geometry is established for electron-deficient species.

[5]R. B. King and D. H. Rouvray, *J. Am. Chem. Soc.* 1977, *99*, 7834.

Framework electron counting can be done from molecular formulas for species of the type $[(BH)_p H_{q+c}]^c$ by assuming the following contributions.

1 electron pair from each BH unit
½ electron pair from each "extra" H
$-$½ electron pair from each $+$ charge

Hence the total number of electron pairs is

$$p + \tfrac{1}{2}(q+c) - \tfrac{1}{2}c = p + \tfrac{1}{2}q$$

where p is the number of vertices of the polyhedron. Thus, $B_9 H_{14}^-$ has $p = 9$, $c = -1$, and $q+c = q-1 = 5$—so that $q = 6$. The species has $p + \tfrac{1}{2}q = 9 + 3$ skeletal electron pairs and is expected to exhibit an *arachno* structure.

We also can use *styx* numbers to count framework electron pairs.[6] The total number of bonding pairs is exactly that required to fill the $s + t + y + x$ orbitals, and from the Equations of Balance,

$$s + t + y + x = q + c + p - \frac{q}{2} - c = p + \tfrac{1}{2}q$$

For hydrides having p vertices of formula $[(BH)_p H_{q+c}^c]^c$ the relation between structure and skeletal electron count may be summarized as follows.

Structural Type	Skeletal electron pairs
	$s + t + y + x = p + \dfrac{q}{2}$
closo	$p + 1$
nido	$p + 2$
arachno	$p + 3$
hypho	$p + 4$

15.5 HETEROBORANES

In many known compounds one or more BH groups in a deltahedral framework are replaced by atoms of a different element. Structures of these species are related simply to those of the parent boron hydrides, except that atoms of differing size may cause some distortion of the polyhedral framework. For example, CH^+, P^+, and S^{2+} are all equivalent to BH (having two electrons for framework bonding plus two more in a bond to an *exo*-H or as a lone pair). Hence a series of neutral *closo* compounds exists of formula $C_2 B_p H_{p+2}$ derived by substitution of CH^+ for BH in the $B_p H_p^{2-}$ series. These carbon-containing compounds are called carboranes. Although $B_5 H_5^{2-}$ has not yet been prepared, the isoelectronic $C_2 B_3 H_5$ having the anticipated trigonal bipyramidal structure is known. Likewise, other generic types of *closo* compounds

[6]W. N. Lipscomb, *Inorg. Chem.* 1979, *18*, 2328.

include SB_pH_p, B_pCH_{p+2}, $B_pH_{p+1}P$. Similar series of *nido* and *arachno* heteroboranes exist, although all members of any series are not always known. For example, $C_2B_4H_8$ is isoelectronic with B_6H_{10}, and $C_2B_7H_{13}$ is isoelectronic with $B_{19}H_{15}$: that is, both pairs of species have the same skeletal electron count.

A more frequently encountered problem is the reverse of the one above: namely, given the molecular formula, to predict the structure of the heteroborane. We can extend previous methods in a straightforward way to carboranes $[(CH)_a(BH)_pH_{q+c}]^c$. Assuming that the number of electron pairs contributed by CH is $\frac{3}{2}$, the total number of skeletal electron pairs is

$$\frac{3}{2}a + p + \frac{1}{2}(q+c) - \frac{1}{2}c = (a+p) + \frac{1}{2}(a+q) = n + \frac{1}{2}(a+q)$$

where n is the number of polyhedral vertices. Hence for $C_2B_8H_{12} \equiv [(CH)_2(BH)_8H_2]$, $a = 2$, $p = 8$, $q = 2$, $c = 0$, and the number of electron pairs is $10 + \frac{1}{2}(2+2) = 10 + 2$. A *nido* structure is expected. The "extra" electrons in bridging positions in carboranes normally are not bonded to C.

Nomenclature for Heteroboranes

Vertices of *closo-*, *nido-*, and *arachno-*polyhedra are given numbers based by convention on planar projections of polyhedral structures. The numbering is by zones (planes) perpendicular to the major axis (see Figure 15.9). Interior vertices on the projection are numbered first, then peripheral ones. This corresponds to numbering apical vertices with lowest numbers. The numbering proceeds clockwise starting from the twelve o'clock position or at the first position clockwise. The location of heteroatoms can be specified by numbers. Thus the isomer of $C_2B_3H_7$, shown in Figure 15.9, is 1,2-dicarba-*nido*-pentaborane(7). $PB_{11}H_{12}$ is phospha-*closo*-dodecaborane(12). There is no need to specify the P position since all icosahedral vertices are equivalent.

Table 15.4, which gives several isoelectronic equivalents of various groups, can be used to predict and rationalize the structures of heteroboranes. Note especially that B:L (where L is a Lewis base) is isoelectronic with CH, and thus we expect the structures of Lewis base adducts of the boron hydrides to be related to those of the carboranes.

15.6 SYNTHESIS OF THE BORON HYDRIDES

Stock prepared boron hydrides from the hydrolysis of magnesium boride with aqueous acid. Because many boranes are susceptible to hydrolysis, the boron hydrides isolated by Stock were those whose reactions with water were slowest—especially tetraborane. This compound then was pyrolyzed to give diborane and other boron hydrides.

$$MgB_2 + Mg + HCl \rightarrow B_4H_{10} + H_2 + MgCl_2 \tag{15.1}$$

$$B_4H_{10} \xrightarrow{\Delta} B_2H_6 + \text{higher boranes} \tag{15.2}$$

As Stock realized, protonic reagents are not good choices for preparing hydrides, since pro-

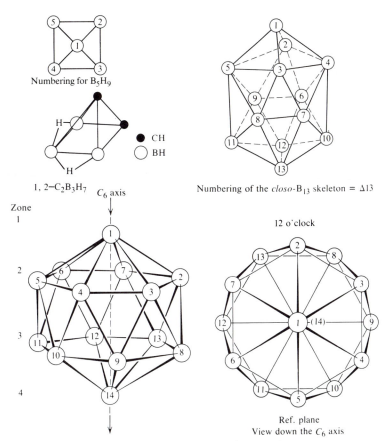

Numbering for B_5H_9

$1, 2-C_2B_3H_7$ C_6 axis

CH ●
BH ○

Numbering of the *closo*-B_{13} skeleton = $\Delta13$

12 o'clock

Zone

Ref. plane
View down the C_6 axis

Figure 15.9 Numbering systems for some boranes.

trolysis occurs with liberation of H_2. The metal hydrides now available are usually employed— for example,

$$3LiAlH_4 + 4BCl_3 \xrightarrow{Et_2O} 2B_2H_6(g) + 3LiCl(s) + 3AlCl_3 \tag{15.3}$$

Historically, boranes often have been prepared by pyrolysis of B_2H_6. We can write reaction schemes involving successive additions of BH_3 and H_2 eliminations to give most of the known *nido* and *arachno* boranes. In recent years, Shore has developed systematic procedures for making boron hydrides in good yield, taking advantage of the acidity of μ-H. The latter can be abstracted (Section 15.7) to afford anions that may then react to add BH_3, yielding species subject to protonation and H_2 elimination.

$$B_4H_{10} + KH \xrightarrow[Et_2O]{-78°C} K[B_4H_9] + H_2$$

$$K[B_4H_9] + \tfrac{1}{2}B_2H_6 \xrightarrow[Et_2O]{-35°C} K[B_5H_{12}] \tag{15.4}$$

$$K[B_5H_{12}] + HCl(l) \xrightarrow{-110°C} B_5H_{11} + H_2 + KCl$$

Table 15.4 Isoelectronic equivalents of various groups

		1e	2e	3e	BH 4e	CH 5e	(BH + 2e) 6e	(BH + 3e) 7e
II	A	Be^+	Be/BeH^+	BeH	BeH^- $BeNR_3$			
			Mg					
	B		Zn Cd		ZnH^- CdH^-	ZnH_2^{2-} CdH_2^{2-}		
III	B		B^+	B/BH^+	BH	BH^- $B(NR_3)$		
			Al^+	Al/AlH^+	AlH	AlH^-		
			Ga^+	Ga/GaH^+	GaH	GaH^-		
			In^+	In/InH^+	InH	InH^-		
IV	B			C^+	C/CH^+	CH	CH^- (CNR_3)	
				Si^+	Si	SiH	SiH^-	
				Ge^+	Ge	GeH	GeH^-	
				Sn^+	Sn	SnH	SnH^-	
V	B				N^+	N	NH	NH^-
					P^+	P/PH^+	PH	PH^-
					As^+	As	AsH	AsH^-
					Sb^+	Sb	SbH	SbH^-
VI	B					S^+	S/SH^+	SH
						Se^+	Se	SeH
						Te^+	Te	TeH

$$1\text{-}BrB_5H_8 + KH \xrightarrow[Me_2O]{-78°C} K[1\text{-}BrB_5H_7] + H_2$$

$$K[1\text{-}BrB_5H_7] + \tfrac{1}{2}B_2H_6 \xrightarrow[Me_2O]{-78°C} K[B_6H_{10}Br]$$

(15.5)

$$K[B_6H_{10}Br] \xrightarrow[Me_2O]{-35°C} B_6H_{10} + KBr$$

In another approach, H^- may be abstracted from anions with BX_3, giving species that transfer BH_3, sometimes with accompanying H_2 elimination.

$$[n\text{-}Bu_4N][BH_4] + BCl_3 \xrightarrow[CH_2Cl_2]{25°C} \tfrac{1}{2}B_2H_6 + [n\text{-}Bu_4N][HBCl_3]$$

(15.6)

$$[n\text{-}Bu_4N][B_3H_8] + BBr_3 \xrightarrow{0°C} \tfrac{1}{2}B_4H_{10} + [n\text{-}Bu_4N][HBBr_3] + \frac{1}{x}(BH_2)_x \qquad (15.7)$$

$$[Me_4N][B_9H_{14}] + BCl_3 \xrightarrow{25°C} \tfrac{1}{2}B_{10}H_{14} + [Me_4N][HBCl_3] + \tfrac{1}{2}H_2 + \frac{1}{2x}(B_8H_{10})_x \qquad (15.8)$$

Because a mixture of $B_5H_8^-$ and B_5H_9 thermally decomposes to $B_9H_{14}^-$, Reaction 15.8 represents a means of converting commerically available B_5H_9 to $B_{10}H_{14}$. Reactions 15.6 and 15.7 also take advantage of easily available starting materials. Solvent, temperature, and cation choice are critical in these preparations.[7]

Among the *closo* boranes, $B_{10}H_{10}^{2-}$ and $B_{12}H_{12}^{2-}$ are the most thermally stable and easiest to prepare. The condensation and closure reaction involving a *nido* boron hydride gives a nearly quantitative yield of $B_{12}H_{12}^{2-}$.

$$B_{10}H_{14} + 2(C_2H_5)_3NBH_3 \xrightarrow[\text{solution}]{190°C \atop \text{hydrocarbon}} [(C_2H_5)_3NH]_2B_{12}H_{12} + 3H_2 \qquad (15.9)$$

$B_{10}H_{10}^{2-}$ can be prepared nearly quantitatively by a simple base-promoted closure reaction.

$$B_{10}H_{14} + 2(C_2H_5)_3N \rightarrow [(C_2H_5)_3NH]_2B_{10}H_{10} + H_2 \qquad (15.10)$$

Other *closo* hydrides can be synthesized by related processes involving condensations and closures (both base-induced and pyrolytic) or degradation of larger *closo* species to smaller ones. The *closo* hydrides ordinarily are isolated as their salts. By using cationic exchange columns, the crystalline acid hydrate forms of more hydrolytically stable *closo* ions $(H_3O)_2(B_pH_p)(H_2O)_m$ can be prepared. These are strong acids in aqueous solution with acidities comparable to H_2SO_4.

15.7 CHEMISTRY OF THE BORANES

Boron hydrides are extremely reactive with air, often explosively so. Their heats of combustion are quite large—up to 1.5 times greater than those of the most nearly comparable hydrocarbons. Accordingly, it is not surprising that they have been of interest as potential rocket fuels (a fact which provided considerable impetus for the investigation of their chemistry).

The thermal stability of the boron hydrides varies widely. Whereas B_8H_{12} decomposes within a few minutes at room temperature, B_5H_9 can be handled at temperatures above 250°C. $Cs_2B_{12}H_{12}$ is stable on heating to 810°C in a sealed, evacuated quartz tube, whereas $Cs_2B_{10}H_{10}$ is stable to 600°C. This behavior reflects the low reactivity of closed structures.

All the boron hydrides are thermodynamically unstable to hydrolysis that affords $B(OH)_3$ and H_2. Some species, especially $B_{10}H_{10}^{2-}$ and $B_{12}H_{12}^{2-}$, are very kinetically stable, however. In general, the closed structures of $B_pH_p^{2-}$ species lead to lower reactivity than is the case for *nido* and *arachno* boranes.

> Reactions of $B_{12}H_{12}^{2-}$ can be handled in open beakers in aqueous solutions.

[7]J. B. Leach, M. A. Toft, F. L. Himpsl and S. G. Shore, *J. Am. Chem. Soc.* 1981, *103*, 988.

15.7.1 Reactions with Lewis Bases

Not surprisingly, the electron-deficient boron hydrides react with reagents having unshared electron pairs (Lewis bases). Depending on the base and the reaction conditions, several types of reactions have been observed: cleavage (symmetric and unsymmetric), addition, and bridge proton abstraction.

Base Cleavage

Diborane undergoes two different types of cleavage reactions with Lewis bases: symmetric and unsymmetric bridge cleavage.

$$
\begin{array}{c}
\text{H} \quad \text{H} \diagup \text{H} \\
\diagdown \diagup \diagdown \\
\text{B} \diagup \diagdown \text{B} \qquad + \ 2L \rightarrow 2LBH_3 \quad \text{Symmetric} \qquad (15.11) \\
\diagup \diagup \times \diagdown \diagdown \\
\text{H} \diagup \text{H} \qquad \text{H}
\end{array}
$$

$$
\begin{array}{c}
\text{H} \ \vert \ \text{H} \qquad \text{H} \\
\diagdown \diagup \diagup \\
\text{B} \ \vert \quad \text{B} \qquad + \ 2L \rightarrow [L_2BH_2]^+ BH_4^- \quad \text{Unsymmetric} \qquad (15.12) \\
\diagup \ \vert \ \diagdown \\
\text{H} \ \vert \ \text{H} \qquad \text{H}
\end{array}
$$

where L = amines, ethers, phosphines, H^-, NCS^-, CN^-. Some symmetric cleavage reactions of diborane are

$$\tfrac{1}{2}B_2H_6 + (CH_3)_2O \rightarrow (CH_3)_2\ddot{O}:BH_3 \qquad (15.13)$$

$$\tfrac{1}{2}B_2H_6 + CO \rightarrow H_3B:CO \qquad (15.14)$$

$$\tfrac{1}{2}B_2H_6 + C_5H_5N \rightarrow C_5H_5N:BH_3 \qquad (15.15)$$

$$\tfrac{1}{2}B_2H_6 + (CH_3)_3N \rightarrow (CH_3)_3N:BH_3 \qquad (15.16)$$

$$\tfrac{1}{2}B_2H_6 + MH \rightarrow MBH_4 \qquad (15.17)$$

In all of the above reactions, the diborane is split into BH_3 units, each of which accepts an electron pair from the attacking reagent, thereby completing its octet of electrons. The order of thermal stability of the products is generally that expected on the basis of strength of the Lewis base (see Chapter 12), dimethyl ether-borane being the least stable of the above and the borohydride the most stable.

Reaction 15.13 has been used in the purification of diborane. When dimethyl ether is used, the unstable addition compound with diborane is formed at −78.5°C. Other volatile compounds that may be present are removed by pumping on the addition compound at −100°C. On warming to −78.5°, the addition compound is dissociated sufficiently to allow vacuum-line fractionation of the dimethyl ether and the regenerated diborane. The technique of purifying a volatile compound through the formation of an adduct, which is stable at one temperature but dissociates at a higher temperature, is a fairly general one. Other examples include the purification of HF through the KHF_2 adduct, of BF_3 through the anisole adduct, of low-molecular-weight ethers through the $LiBH_4$ adducts, etc.

Reaction 15.17 proceeds only in the presence of a solvent, diethyl ether being satisfactory for the formation of $LiBH_4$ and diglyme (that is, diethylene glycol dimethyl ether) for $NaBH_4$. These compounds are quite stable thermally; $LiBH_4$ loses hydrogen at 275°C, and $NaBH_4$ loses hydrogen above 400°C. The borohydride ion, BH_4^-, is reasonably stable in water at high pH values. Lithium borohydride reacts slowly with water, apparently because of the hydrolysis of the Li^+ ion, which increases the hydrogen ion concentration (see Section 14.2.8). Other borohydrides may be prepared by metathetical reactions with $LiBH_4$ and $NaBH_4$.

Ammonia reacts with diborane to afford the ionic product $[(NH_3)_2BH_2]^+BH_4^-$ of unsymmetrical cleavage, known as the "diammoniate of diborane." It is not completely clear which factors promote one type of cleavage over the other. Probably the first step is the displacement of a bridging H, giving a singly-bridged species

$$(15.18)$$

When $L = H^-$, we can prepare salts of the monobridged $B_3H_7^-$ anion that are stable in vacuum at room temperature. The second step involves another bridge displacement and leads to symmetric cleavage (via attack on the BH_3 boron) or unsymmetric cleavage (via attack on the BH_2L boron). When L is bulky, symmetric cleavage is more likely. Electronic factors may also be important. As in much of boron chemistry, solvent plays an important role in cleavage, more basic solvents giving a higher yield of symmetric cleavage products. Many adducts BH_3L can be prepared via displacement of weaker Lewis bases by stronger ones, especially using $BH_3(THF)$.

The reactions of higher boron hydrides often resemble those of diborane, especially when a BH_2 group is present. Otherwise, fragmentation by Lewis bases occurs less readily. Tetraborane reacts with L [where L may be $(CH_3)_3N$, R_2O, $(CH_3)_3P$ or H^-] to give adducts of BH_3 and B_3H_7, with the exact structure of B_3H_7L dependent on L.

$$(15.19)$$

B_3H_7L is isoelectronic with the unknown B_3H_9. When an excess of $L = N(CH_3)_3$ is used, $(CH_3)_3N:BH_3$ and $[HB:N(CH_3)_3]_x$ result from further attack.

The reaction of tetraborane with ammonia proceeds by unsymmetrical cleavage, as in the case of diborane.

$$2NH_3 + B_4H_{10} \rightarrow [H_2B(NH_3)_2]^+[B_3H_8]^- \tag{15.20}$$

THF affords an unsymmetric cleavage product at low temperature, $[(THF)_2BH_2]^+B_3H_8^-$, which rearranges to $BH_3(THF)$ and $B_3H_7(THF)$ on warming.

We can obtain an ammonia-triborane adduct in the same fashion as the ammonia-borane adduct: that is, by displacing the ether from an etherate with ammonia.

$$H_3N + R_2O:B_3H_7 \rightarrow H_3N:B_3H_7 + R_2O \tag{15.21}$$

Even though B_5H_9 has no BH_2 group, it is cleaved unsymmetrically by NH_3.

$$B_5H_9 + 2NH_3 \rightarrow [(NH_3)_2BH_2]^+[B_4H_7]^- \tag{15.22}$$

Other higher hydrides are also cleaved by Lewis bases, NH_3 usually yielding ionic products.

$$B_5H_{11} + 2NH_3 \rightarrow [(NH_3)_2BH_2][B_4H_9] \tag{15.23}$$

$$B_5H_{11} + 2CO \rightarrow BH_3(CO) + B_4H_8(CO) \tag{15.24}$$

$$B_6H_{12} + 2NH_3 \rightarrow [(NH_3)_2BH_2][B_5H_{10}] \tag{15.25}$$

$$B_6H_{12} + P(CH_3)_3 \rightarrow BH_3[P(CH_3)_3] + B_5H_9 \tag{15.26}$$

The amine boranes usually are much more stable in air than boron hydrides. Thus a common way of disposing of reaction wastes is to condense a volatile amine into the reaction vessel and allow it to react on warming before exposing the contents to air.

Base Addition

Not all Lewis bases react to give cleavage products. The formation of the adduct $B_2H_7^-$ from B_2H_6 has been mentioned already. Evidence also exists for the monobridged $(NH_3)H_2B\overset{H}{\diagup}{\diagdown}BH_3$, which has not been isolated.

$P(CH_3)_3$ reacts with B_5H_9 to form the adduct $B_5H_9[P(CH_3)_3]_2$, whose structure is shown in Figure 15.10. The electron count of $p + 4$ electron pairs makes this the first well-characterized member of the *hypho* class of boron hydrides. Other adducts of Lewis bases with boron hydrides have been prepared, but they are not well characterized.

Because $B_{10}H_{14}$ is an air-stable, crystalline, and easily handled compound, its chemistry has been investigated extensively. Lewis bases such as CH_3CN, C_5H_5N, and $(CH_3)_2S$ react with decaborane to liberate H_2.

$$2CH_3CN + B_{10}H_{14} \rightarrow B_{10}H_{12} \cdot 2CH_3CN + H_2 \tag{15.27}$$

The structures of these species are shown in Figure 15.11.

Triethylamine reacts with the bis-acetonitrile decaborane to give two compounds of the

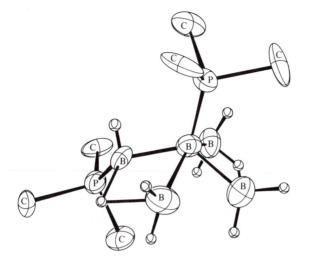

Figure 15.10 Structure of the *hypho* compound $B_5H_9(PMe_3)_2$. (Reproduced with permission from A. V. Fratini, G. W. Sullivan, M. L. Denniston, R. K. Hertz, and S. G. Shore, *J. Am. Chem. Soc.* 1974, *96*, 3013. Copyright 1974, American Chemical Society.)

formula $B_{10}H_{12} \cdot 2(C_2H_5)_3N$. One is assumed to be a covalent compound similar to those discussed above, since it is soluble in benzene and the amine readily may be displaced by triphenylphosphine. The other compound is ionic, containing the triethylammonium ions and the *closo* ion $B_{10}H_{10}^{2-}$.

We also can obtain the ionic compound by the direct reaction of triethylamine with decaborane (Equation 15.10). An interesting reaction of $B_{10}H_{12} \cdot 2(CH_3)_2S$ is that with $HC \equiv CH$, which affords 1,2-dicarbadodecahydro-*closo*-dodecaborane $1,2-C_2B_{10}H_{12}$. This carborane also can be prepared directly from the reaction of Lewis base and acetylene with $B_{10}H_{14}$ without isolation of the intermediate (see Section 15.8).

Bridge Proton Abstraction

As early as 1956 it was shown that $B_{10}H_{14}$ behaved as an acid in aqueous media and could be titrated with NaOH. Exchange experiments revealed the exchange of the four bridging protons in decaborane and pointed to the acidic character of μ-H.

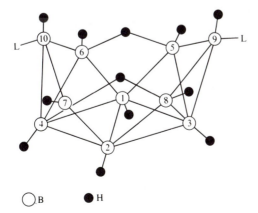

○ B   ● H

Figure 15.11 Structure of $B_{10}H_{12} \cdot 2L$.

The bridging protons in *nido* and *arachno* boron hydrides may be abstracted by (usually) very strong bases in ethers. See Reaction 15.4; also:

$$B_5H_9 + NaH \rightarrow Na[B_5H_8] + H_2 \qquad (15.28)$$

$$B_6H_{10} + KH \rightarrow K[B_6H_9] + H_2 \qquad (15.29)$$

$$B_{10}H_{14} + NH_3 \rightarrow [NH_4][B_{10}H_{13}] \qquad (15.30)$$

Metathesis to the tetraalkylammonium salts gives significantly more stable products.

The relative acidities of boron hydrides can be established by competition reactions such as

$$B_4H_9^- + B_{10}H_{14} \rightarrow B_4H_{10} + B_{10}H_{13}^- \qquad (15.31)$$

showing that the acidity ordering is $B_4H_{10} < B_{10}H_{14}$. The ordering is

$$\textit{nido} \qquad B_5H_9 < B_6H_{10} < B_{10}H_{14} < B_{16}H_{20}$$

$$\textit{arachno} \qquad B_4H_{10} < B_5H_{11} < B_6H_{12}$$

The experimental orderings $B_4H_{10} < B_{10}H_{14}$ and $B_5H_9 < B_6H_{12}$ provide a connection between the two series. Some idea of absolute acidity can be gained from the fact that NH_3 deprotonates B_4H_{10}, showing that $NH_4^+ < B_4H_{10}$ in ethereal solution. In general, the larger boranes are more acidic than the smaller ones, and *arachno* hydrides more acidic than *nido* hydrides of comparable size.[8]

The conjugate bases contain electron-rich B—B bonds that can be reprotonated to give the parent boranes. Figure 15.12 shows the structures of $B_5H_8^-$ and $B_6H_9^-$ as elucidated by ^{11}B NMR.

These electron-rich sites react with electrophiles to insert them into a bridging position. An isomer of $B_6H_8(CH_3)_2$ having a bridging $B(CH_3)_2$ group has been prepared from the reaction of LiB_5H_8 with $(CH_3)_2BCl$ at low temperature; its structure is shown in Figure 15.13. On warming, the compound rearranges to the more thermally stable 4,5-dimethylhexaborane(10).

15.7.2 Electrophilic Substitution

Terminal H atoms on all classes of boranes can be substituted by electrophiles. The most studied examples involve $B_{10}H_{10}^{2-}$, $B_{12}H_{12}^{2-}$, B_5H_9, and $B_{10}H_{14}$. Typical of these reactions are

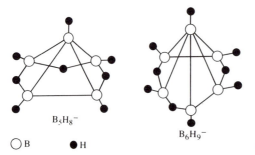

$B_5H_8^-$

$B_6H_9^-$

○ B   ● H

Figure 15.12 Static structures of borane anions.

[8]R. J. Remmel, H. D. Johnson II, I. S. Jaworiwsky, and S. G. Shore, *J. Am. Chem. Soc.* 1975, *97*, 5395.

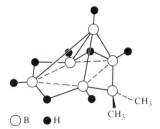

○ B   ● H   CH$_3$ CH$_3$

Figure 15.13 The low-temperature isomer of $B_6H_8(CH_3)_2$.

halogenations. $B_{10}H_{10}^{2-}$ has two different kinds of H, apical and equatorial, whereas in icosahedral $B_{12}H_{12}^{2-}$ all positions are equivalent. Molecular orbital calculations indicate that negative charge on B atoms decreases in the order apical $B_{10}H_{10}^{2-}$ > equatorial $B_{10}H_{10}^{2-} \geq B_{12}H_{12}^{2-}$. As expected for electrophilic attack, the reactivity order is $B_{10}H_{10}^{2-} > B_{12}H_{12}^{2-}$ and $Cl_2 > Br_2 > I_2$. Halogenations proceed easily in the dark and multiply-substituted products are obtained. The fully substituted $B_{10}X_{10}^{2-}$ and $B_{12}X_{12}^{2-}$ (X = Cl, Br, I) can be prepared.

B_5H_9 reacts with elemental halogens to give $1\text{-}XB_5H_8$ in the presence of a Friedel-Crafts catalyst. In the absence of a catalyst, both 1- and $2\text{-}XB_5H_8$ are obtained. The formation of the apical(1-) isomer is consistent with the assignment of greater negative charge to the apical boron and with the assumption that the charge distribution in the ground state parallels that in the transition state for electrophilic attack. The following rearrangement occurs quantitatively, indicating that apical substitution is a kinetically controlled process.

$$1\text{-}ClB_5H_8 \xrightarrow{\text{Et}_2O} 2\text{-}ClB_5H_8 \tag{15.32}$$

Repeated substitution followed by isomerization allows the preparation of multiply-substituted species.

Halogenation of $B_{10}H_{14}$ gives both 1- and $2\text{-}XB_{10}H_{13}$, with the latter predominating. Long treatment with Br_2 or I_2 gives 2,4- and 1,2-disubstituted derivatives. The reactivity toward electrophilic substitutions by several different reagents appears to be 2,4 > 1,3, paralleling the ground-state charge distribution.

Boranes undergo reactions with other electrophiles (alkyl halides, for example) that parallel those with halogens. Acids bring about H exchange except when an electron-rich B—B bond is available for protonation. An example is B_6H_{10}, which reacts with HCl in BCl_3 to yield $B_6H_{11}^+$.

15.8 CARBORANES

The most important heteroboranes are those containing carbon—the carboranes. Electron-counting rules for these species were given in Section 15.5. The carboranes are much more stable to air and moisture than the boranes. An interesting feature of their structures is the appearance of C with C.N. up to 6. C's are the most electropositive centers in the frameworks. For species containing two or more C (many are C_2 species), isomers are possible. The most thermodynamically stable are the ones in which the electropositive C's have maximum separation. The most common preparative route is the reaction of boranes with acetylenes.

$$B_5H_9 + C_2H_2 \xrightarrow{500\text{-}600°C} 2,4\text{-}C_2B_5H_7 + 1,6\text{-}C_2B_4H_6 + 1,5\text{-}C_2B_3H_5 \tag{15.33}$$

$$B_8H_{12} + MeC\equiv CMe \rightarrow (MeC)_2B_7H_9 \qquad (15.34)$$

$$B_{10}H_{14} + C_2H_2 \rightarrow 1,2\text{-}C_2B_{10}H_{12} \qquad (15.35)$$

The isomers of the *closo* compound $C_2B_{10}H_{12}$ have icosahedral geometry and exhibit extremely high kinetic and thermodynamic stability. The 1,2-, 1,7-, and 1,12- isomers have the common names *o*-, *m*-, and *p*-carborane, respectively. *o*-carborane has been given the following symbol in the literature.

$$\begin{array}{c} HC\!-\!CH \\ \diagdown\!\bigcirc\!\diagup \\ B_{10}H_{10} \end{array}$$

Reactions at B centers in carboranes parallel those of the boranes: bridge proton abstraction and electrophilic substitution, including halogenation. The terminal H attached to electropositive C are relatively acidic. Hence these C centers can be metallated.

$$1,6\text{-}C_2B_4H_6 + 2\,n\text{-}BuLi \rightarrow Li_2[1,6\text{-}C_2B_4H_4] + 2C_4H_{10} \qquad (15.36)$$

$$\begin{array}{c} HC\!-\!CH \\ \diagdown\!\bigcirc\!\diagup \\ B_{10}H_{10} \end{array} + RMgX \rightarrow \begin{array}{c} HC\!-\!CMgX \\ \diagdown\!\bigcirc\!\diagup \\ B_{10}H_{10} \end{array} + RH \qquad (15.37)$$

The metallated products retain structural integrity and can react as nucleophiles to produce a large number of C-substituted derivatives. *o*-carborane has a very extensive organic chemistry.

Thermal isomerizations occur. One mechanism that often seems to operate is the diamond-square-diamond rearrangement proposed by Lipscomb. A pair of triangular faces at an angle open into a square and rejoin with a different pair of vertices connected. Figure 15.14 shows this, as well as its application to the isomerization of *o*- to *m*-carborane. For further information on the carboranes, the references should be consulted.[9]

o-carborane is attacked by base that excises a BH group, generating a *nido* anion that retains structural integrity. The 1,7-isomer can be obtained by thermal rearrangement of the anion or by starting with *m*-carborane.

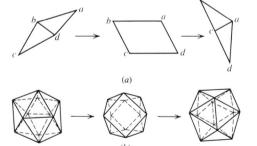

Figure 15.14 (a) The diamond-square-diamond mechanism. (b) Rearrangement of o- to m-carborane via a cuboctahedral intermediate. (From H. Beall, in *Boron Hydride Chemistry*, E. L. Muetterties, Ed., Academic Press, New York, 1975.)

[9]R. N. Grimes, *Carboranes*, Academic Press, New York, 1970; T. Onak in *Boron Hydride Chemistry*, E. L. Muetterties, Ed., Academic Press, New York, 1975, p. 349; H. Beal, *ibid.*, p. 301; G. B. Dunks and M. F. Hawthorne, *Acc. Chem. Res.* 1973, 6, 124.

$$1,2\text{-B}_{10}\text{C}_2\text{H}_{12} + \text{OEt}^- + 2\text{EtOH} \xrightarrow{85°\text{C}} 1,2\text{-B}_9\text{C}_2\text{H}_{12}^- + \text{H}_2 + \text{B(OEt)}_3 \qquad (15.38)$$

NaH in tetrahydrofuran deprotonates the anion, giving a dianion $\text{B}_9\text{C}_2\text{H}_{11}^{2-}$ whose structure is shown in Figure 15.15a. Assuming for convenience sp^3 hybridization of the five atoms on the open face, a set of MO's reminiscent of the Cp^- anion may be constructed. These MO's are occupied by six electrons. Hawthorne[10] has exploited the analogy between $\text{B}_9\text{C}_2\text{H}_{11}^{2-}$ (the

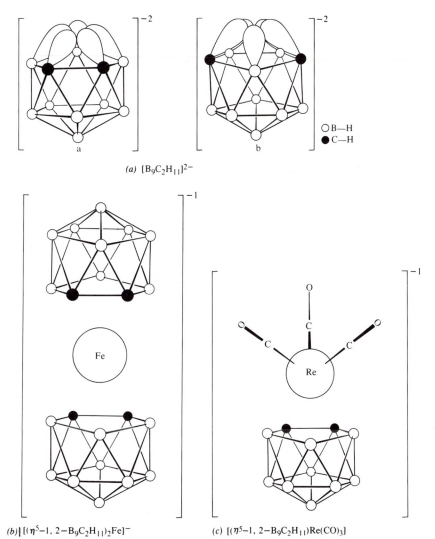

(a) $[\text{B}_9\text{C}_2\text{H}_{11}]^{2-}$

○ B—H
● C—H

(b) $[(\eta^5\text{-}1, 2\text{-B}_9\text{C}_2\text{H}_{11})_2\text{Fe}]^-$ (c) $[(\eta^5\text{-}1, 2\text{-B}_9\text{C}_2\text{H}_{11})\text{Re(CO)}_3]$

Figure 15.15 Structures of dicarbollides. (Reproduced with permission from M. F. Hawthorne, *et al.*, *J. Am. Chem. Soc.* 1968, *90*, 879. Copyright 1968, American Chemical Society.)

[10]M. F. Hawthorne, *Pure Appl. Chem.* 1972, *29*, 547. L. F. Warren and M. F. Hawthorne, *J. Am. Chem. Soc.*, 1970, *92*, 1157.

dicarbollide ion) and Cp^- to prepare metal dicarbollide complexes related to the metallocenes.

$$2 [1,2\text{-}B_9C_2H_{11}]^{2-} + Fe^{2+} \xrightarrow[THF]{} [\eta^5\text{-}(1,2\text{-}B_9C_2H_{11})_2Fe]^{2-} \qquad (15.39)$$

$$2 [1,7\text{-}B_9C_2H_{11}]^{2-} + Re(CO)_5Br \rightarrow [(\eta^5\text{-}B_9C_2H_{11})Re(CO)_3]^- + Br^- + 3CO \quad (15.40)$$

$$6 [1,2\text{-}B_9C_2H_{11}]^{2-} + 3CoCl_2 \rightarrow 2[(\eta^5\text{-}B_9C_2H_{11})_2Co]^- + Co(0) + 3 Cl^- \quad (15.41)$$

Some representative structures appear in Figures 15.15*b* and 15.15*c*. Dicarbollide complexes undergo one-electron electron transfer processes, giving products having a range of metal oxidation numbers. For example, complexes of Ni (II), (III), and (IV) can be prepared.

The structures in Figures 15.15*b* and 15.15*c* can be regarded as metallocarboranes having a framework vertex occupied by a metal with its attached ligands. A general technique for preparing such complexes involves excision of a BH group via base degradation and reaction of the resulting anion with a metal halide.

15.9 SOME CHEMISTRY OF GROUP IIIB

15.9.1 Borazine[11] and Boron Nitride

Reaction of alkali metal borohydrides with NH_4Cl produces $B_3N_3H_6$, borazine (formerly called borazole). This compound is isoelectronic with benzene, and because of its very similar physical properties has been referred to as "inorganic benzene." A few of these properties are compared in Table 15.5.

The possible analogy with benzene is apparent from examination of its molecular structure parameters. In particular, the molecule exhibits $\mathbf{D}_{3h}$ symmetry and displays B–N distances of 143.5 pm—significantly shorter than the distance in H_3BNH_3 (156 pm) and thus suggestive of a bond order considerably greater than one. This information can be represented by the resonance structures showing aromatic character

^{1}H NMR studies of substituted borazines show that the influence of Br, Cl, and Me is about equal on the chemical shifts of *ortho* and *para* NH protons. The *para* N–H chemical shifts can only be influenced by π effects, whereas the *ortho* shifts result from both σ and π effects. This behavior indicates the aromatic character of borazines.

[11]D. F. Gaines and J. Borlin in *Boron Hydride Chemistry*, E. L. Muetterties, Ed., Academic Press, New York, 1975, p. 241.

Table 15.5 Physical properties of borazine and benzene

	Borazine	*Benzene*
Molecular weight	80.5	78.1
b.p. (°C)	55.0	80.10
m.p. (°C)	−56.2	5.51
Critical temperature (°C)	252	288.0
Liquid density at b.p. (g/cm³)	0.81	0.81
Crystal density at m.p. (g/cm³)	1.00	1.01
Trouton constant (J/K mole)	89.5	88.2

The chemical behavior of the borazines, however, seems to be dominated by the polarity of the B—N bonds. Both of the electrons required for aromaticity must come from the N. Whereas Lewis bases always add to the B's, Lewis acids add to N, suggesting an important contribution from the structure

For example, *B*-substituted alkyl borazines can be prepared from borazine and nucleophilic reagents such as Li alkyls or Grignard reagents. *N*-substitution can be achieved by reaction sequences involving *B*-alkylated species which have acidic *N*-hydrogens.

$$R_3B_3N_3H_3 + MeLi \rightarrow R_3B_3N_3H_2(Li) + CH_4 \tag{15.42}$$

$$R_3B_3N_3H_2Li + R'X \rightarrow R_3B_3N_3H_2R' + LiX$$

(The first-listed substituents are those bound to B, and the last-listed are bound to N.) This reactivity contrasts with the alkylation of benzene, which is an electrophilic substitution process. Also in contrast to benzene chemistry is the facile addition of polar reagents across the B—N bonds. For example,

$$H_3B_3N_3H_3 + 3HCl \rightarrow H_3Cl_3B_3N_3H_6 \tag{15.43}$$

The saturated products are formal analogues of cyclohexanes. Thus, the chemistry of substituted borazines differs quite substantially from that of benzene. Figure 15.16 summarizes the chemistry of borazines.

Pure samples of borazine are light-sensitive and have been known to explode unless stored in the dark.

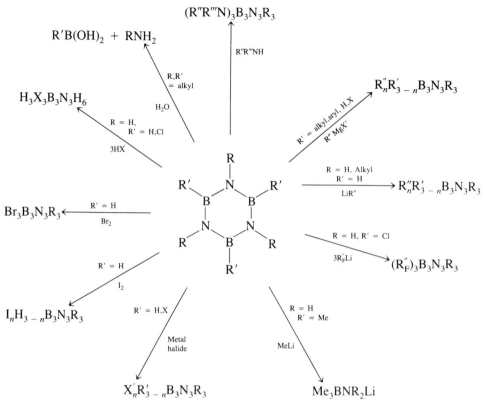

Figure 15.16   Some chemistry of the borazines.

Substituted borazines can be made directly instead of from the parent. Metal borohydrides react with monosubstituted alkylammonium salts to afford *N*-alkylated borazines.

$$3MBH_4 + 3RNH_3^+Cl^- \rightarrow H_3B_3N_3R_3 + 3MCl + 9H_2 \qquad (15.44)$$

Using a mixture of ammonium salts produces unsymmetrically substituted borazines. *B*-halogenated borazines result from the reaction of boron trihalides with primary amines.

$$BX_3 + RNH_2 \rightarrow RNH_2BX_3 \xrightarrow[\Delta]{-HX} RNHBX_2 \xrightarrow[\Delta]{-HX} \frac{1}{3}(X_3B_3N_3R_3) \qquad (15.45)$$

Complete pyrolysis of borazine, or any compound containing a 1:1 ratio of B to N, yields boron nitride, sometimes referred to as "inorganic graphite." The crystal structure of the hexagonal form of BN is shown in Figure 15.17. The stacking of the layers differs from that in graphite in that B atoms in one layer lie directly over N atoms in another, and *vice versa*. This suggests the importance of polar interactions between the layers; however, interlayer bonding energy is only about 16 kJ/mole. Both graphite and BN are good lubricants, because of the ability of layers to slide over one another. In contrast to graphite, BN is a white insulator. High pressures convert BN to a cubic form displaying the diamond structure. The hardness of the cubic form surpasses that of diamond. Cubic BN can be used as an abrasive with metals

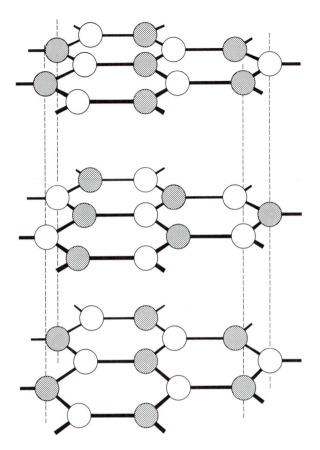

Figure 15.17 The layer structure of hexagonal BN. (From K. Niedenzu and J. W. Dawson, *Boron-Nitrogen Chemistry*, Academic Press, New York, 1965.)

prone to carbide formation. Like diamond, cubic BN is thermodynamically unstable at low pressures, but very stable kinetically.

15.9.2 The Borohydrides and Lithium Aluminum Hydride

Lithium borohydride was synthesized by applying the Lewis acid–base concept directly to the metal hydrides and boron compounds. The hydride ion is a very strong Lewis base, and will react even with the weak Lewis acid methyl borate.

$$NaH(s) + B(OCH_3)_3(l) \underset{68°}{\rightarrow} NaBH(OCH_3)_3 \qquad (15.46)$$

The relative strengths of Lewis acids in the sequence $BF_3 > (BH_3)_2 > (CH_3)_3B > B(OCH_3)_3$ suggested that diborane might displace methyl borate from the trimethoxyborohydride ion.

$$2Na[HB(OCH_3)_3] + B_2H_6 \rightarrow 2NaBH_4 + 2B(OCH_3)_3 \qquad (15.47)$$

This reaction does take place—so rapidly that it may be used to absorb diborane from a gas stream. The commercial synthesis of sodium borohydride is carried out without having to use diborane. The reaction used is

$$4NaH + B(OCH_3)_3 \xrightarrow[225-275°C]{} NaBH_4 + 3NaOCH_3 \tag{15.48}$$

The reaction of lithium hydride with aluminum chloride in dry ether yields lithium aluminum hydride.

$$4LiH + AlCl_3 \xrightarrow[(C_2H_5)_2O]{} LiAlH_4 + 3LiCl \tag{15.49}$$

It has also been found possible to prepare the alkali-metal aluminum hydrides directly from the elements in tetrahydrofuran as solvent.

$$Na + Al + H_2(5000 \text{ psi}) \xrightarrow[THF]{140°C} NaAlH_4 \tag{15.50}$$

Reactions of the Borohydrides

The borohydride ion serves as a starting material for most of the boron hydrides and the hydroborate ions. The reactions between diborane and the borohydride ion provide an excellent example of product control by variations in pressure, temperature, solvent, or "inert cation."

$$MBH_4 + B_2H_6 \xrightarrow[no\ solvent]{} no\ reaction \quad M = Li, Na, K \tag{15.51}$$

$$[(C_2H_5)_4N]BH_4 + \tfrac{1}{2}B_2H_6 \xrightarrow[-80°C]{1\ atm} [(C_2H_5)_4N]B_2H_7 \tag{15.52}$$

$$NaBH_4 + \tfrac{1}{2}B_2H_6 \xrightarrow[polyethers]{0°C} NaB_2H_7 \tag{15.53}$$

$$NaBH_4 + B_2H_6 \xrightarrow[\substack{3\ atm \\ polyethers}]{25\ to\ 50°C} NaB_3H_8 \ (90\%\ yield) + H_2 \tag{15.54}$$

$$NaBH_4 + B_2H_6 \xrightarrow[polyethers]{120°C} NaB_{11}H_{14} \ (50\%) + Na_2B_{12}H_{12} \ (40\%) + H_2 \tag{15.55}$$

$$NaBH_4 + 5B_2H_6 \xrightarrow[dioxane]{120°C} NaB_{11}H_{14} + 10H_2 \tag{15.56}$$

$$2NaBH_4 + 5B_2H_6 \xrightarrow[(C_2H_5)_3N]{100\ to\ 180°C} Na_2B_{12}H_{12} + 13H_2 \tag{15.57}$$

$$2NaBH_4 + 2B_2H_6 \xrightarrow[polyethers]{162°C} Na_2B_6H_6 \ (5-10\%\ yield) + 7H_2 \tag{15.58}$$

We can offer the following, somewhat speculative, explanations for the differences in behavior displayed above. Reactions 15.52 and 15.53 are analogous to the base cleavages of Section 15.7.1. The energy released in forming the $B_2H_7^-$ ion must be greater than that needed to expand the MBH_4 lattice. Thus reaction occurs with tetraethylammonium borohydride where the tetraethylammonium ion dominates the lattice, and only slight expansion is needed to accommodate the new ion formed. With the smaller cations Li^+, Na^+, and K^+, no reaction occurs in the absence of solvent. Polyethers form relatively stable complexes with Na^+, and this large solvated cation loses relatively little ion-pair energy in solution (or lattice energy in the solid) when the $B_2H_7^-$ species is formed from BH_4^- (Reaction 15.53). Potassium ion, because of its larger size, does not solvate as readily, and hence KBH_4 does not react at low temperatures with B_2H_6 in polyethers.

The higher pressures of Reaction 15.54 probably will lead to formation of a $B_3H_{10}^-$ species, which splits out hydrogen to form the $B_3H_8^-$ ion.

$$(15.59)$$

Reactions 15.55, 15.56, and 15.57 apparently occur as a result of the formation of decaborane and subsequent reaction of the decaborane. Thus Reaction 15.56 may be carried out by allowing sodium borohydride to react with decaborane at 90°C.

$$MBH_4 + B_{10}H_{14} \xrightarrow[R_2O]{90°C} MB_{11}H_{14} + 2H_2 \qquad (15.60)$$

At 25°C, the BH_4^- abstracts a proton from $B_{10}H_{14}$ to give $B_{10}H_{13}^-$, H_2, and B_2H_6. This is thought to be a step in Reaction 15.56, followed by the addition of BH_3; that is,

$$R_2O:BH_3 + MB_{10}H_{13} \rightarrow MB_{11}H_{14} + H_2 + R_2O \qquad (15.61)$$

The last reaction has been shown to occur at 90°C.

The $B_{11}H_{14}^-$ ion is postulated to resemble $B_{12}H_{12}^{2-}$, but with a triangular H_3^+ group replacing a BH group.

In Reaction 15.57 the triethylamine solvent would form a relatively stable borane adduct, thereby stopping the reaction producing $B_{11}H_{14}^-$. The route to $B_{12}H_{12}^{2-}$ has not been established, but may involve solvent coordination to the $B_{10}H_{14}$ species. The solvent used in Reaction 15.55 is probably intermediate in base strength between dioxane and triethylamine.

The emf values for boron half-reactions are given in Appendix C.

The BH_4^- is stable in water above a pH of 9 and is a mild reducing agent, useful for selective organic reductions in water or alcohol solvents. It reduces aqueous solutions of arsenites, antimonites, germanates, or stannates to the corresponding hydrides.

$$3BH_4^- + 4H_3AsO_3 + 3H^+ \rightarrow 3H_3BO_3 + 4AsH_3 + 3H_2O \qquad (15.62)$$

Many transition-metal ions are reduced quantitatively to lower oxidation states in aqueous solution—the products being dependent on their half-cell emf values.

As indicated earlier, strong acids, protonic or Lewis, displace B_2H_6 from borohydrides. If the solutions are dilute aqueous, complete hydrolysis of the borohydride yields boric acid and hydrogen.

Reactions of Lithium Aluminum Hydride

Lithium aluminum hydride, a much stronger reducing agent than the borohydrides, reacts violently with water. Its solutions in diethyl ether or tetrahydrofuran are used widely for organic reductions, and its reaction with binary halides provides a general method for preparing the volatile hydrides.

$$SiCl_4 + LiAlH_4 \xrightarrow{(C_2H_5)_2O} SiH_4 + LiCl + AlCl_3 \tag{15.63}$$

$$4PCl_3 + 3LiAlH_4 \xrightarrow{(C_2H_5)_2O} 4PH_3 + 3LiCl + 3AlCl_3 \tag{15.64}$$

The solvent plays an essential role in the above reactions—thus, the reduction of PCl_3 does not take place in the absence of ether.

Lithium aluminum hydride reacts with diborane to produce a liquid aluminum borohydride.

$$LiAlH_4 + 2B_2H_6 \rightarrow LiBH_4 + Al(BH_4)_3 \tag{15.65}$$

This borohydride has six bridge bonds about each aluminum. It is extremely hazardous, detonating violently on contact with air containing trace amounts of moisture, and slowly evolves hydrogen on standing at room temperature. $Al(BH_4)_3$ is the most volatile aluminum compound known.

15.9.3 Hydroboration

A new area of organic chemistry involving boron hydrides (a thorough treatment of which is beyond the scope of this text) has been developed in recent years by H. C. Brown,[12] who was awarded the 1979 Nobel Prize in Chemistry for this work.

Diborane attacks the double bonds in olefins and has the further unique property of undergoing reversible addition and dissociation, during which time the double bond "walks" to the end of the chain! Alkaline peroxide oxidation then produces predominately terminal alcohols from mixed olefins.

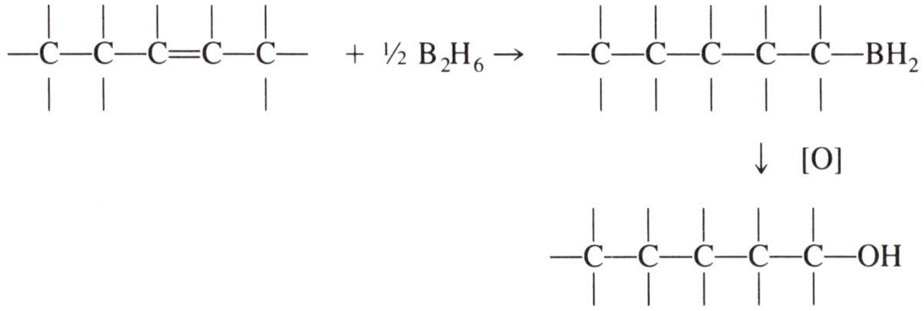

$$-\overset{|}{\underset{|}{C}}-\overset{|}{\underset{|}{C}}-\overset{|}{C}=\overset{|}{C}-\overset{|}{\underset{|}{C}}- \ + \ \tfrac{1}{2}\,B_2H_6 \rightarrow \ -\overset{|}{\underset{|}{C}}-\overset{|}{\underset{|}{C}}-\overset{|}{\underset{|}{C}}-\overset{|}{\underset{|}{C}}-\overset{|}{\underset{|}{C}}-BH_2$$

$$\downarrow [O] \tag{15.66}$$

$$-\overset{|}{\underset{|}{C}}-\overset{|}{\underset{|}{C}}-\overset{|}{\underset{|}{C}}-\overset{|}{\underset{|}{C}}-\overset{|}{\underset{|}{C}}-OH$$

The diborane is prepared *in situ* from the reaction of $LiBH_4$ or $NaBH_4$ with acids such as boron trifluoride etherate, hydrogen chloride, or sulfuric acid. An ether solvent, such as diethyl ether, tetrahydrofuran, or polyethers, must be used. Also, the oxidation usually is carried out *in situ*.

[12]H. C. Brown, *Organic Synthesis via Boranes,* Wiley-Interscience, New York, 1975.

Addition reactions with 1-alkenes give ~94% B placement on the terminal position. Branching of the alkyl on the 2-position does not change the distribution significantly.

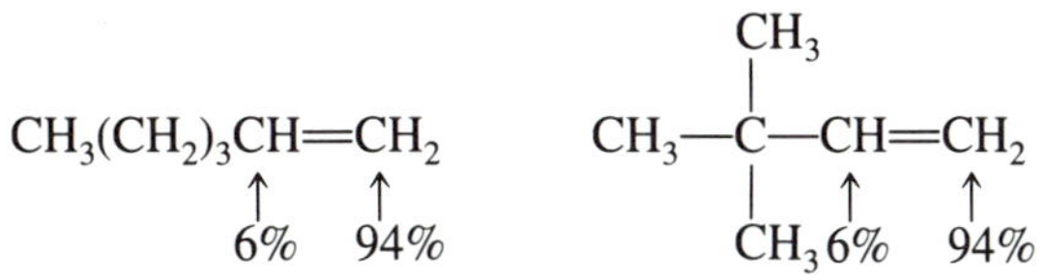

$$CH_3(CH_2)_3CH=CH_2 \qquad CH_3-\overset{\overset{\displaystyle CH_3}{|}}{\underset{\underset{\displaystyle CH_3}{|}}{C}}-CH=CH_2$$
$$\quad\; 6\% \quad 94\% \qquad\qquad\qquad 6\% \quad 94\%$$

With internal olefins, B shows a similar preference to add at the less-hindered site.

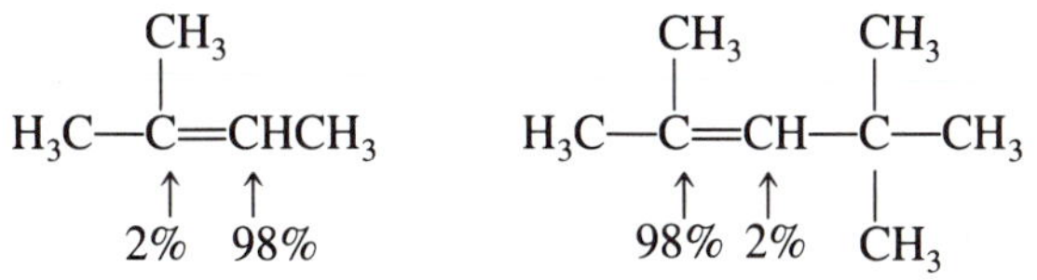

$$H_3C-\overset{\overset{\displaystyle CH_3}{|}}{C}=CHCH_3 \qquad H_3C-\overset{\overset{\displaystyle CH_3}{|}}{C}=CH-\overset{\overset{\displaystyle CH_3}{|}}{\underset{\underset{\displaystyle CH_3}{|}}{C}}-CH_3$$
$$2\% \quad 98\% \qquad\qquad 98\% \; 2\%$$

However, no great preference exists for substituted positions containing alkyl groups of very different steric requirements.

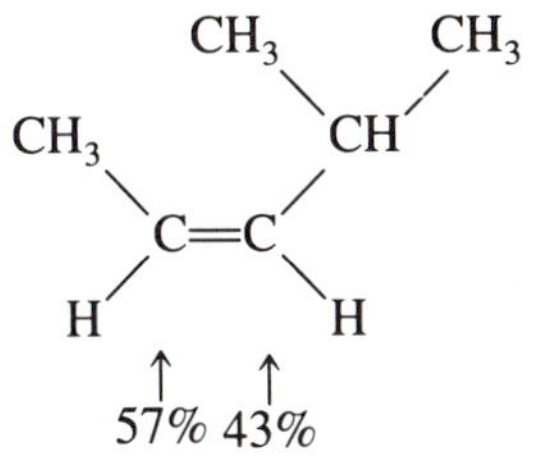

$$57\% \; 43\%$$

Steric effects do have some influence on the extent of hydroboration, as the following reactions show.

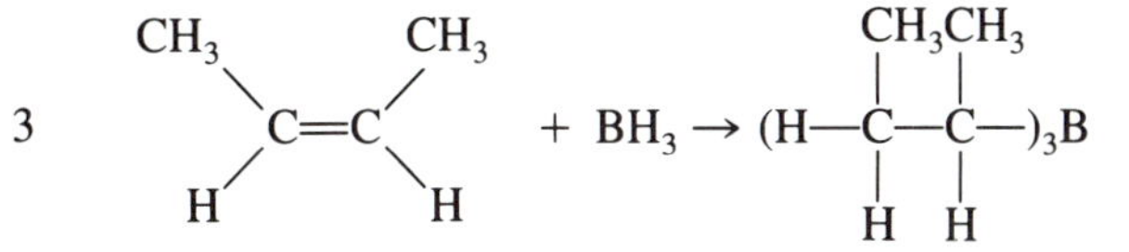

$$3 \quad \overset{CH_3}{\diagdown}C=C\overset{CH_3}{\diagup} \; + \; BH_3 \rightarrow (H-\overset{CH_3CH_3}{\underset{H\;H}{C-C}}-)_3B \qquad (15.67)$$

$$2 \quad \overset{CH_3}{\diagdown}C=C\overset{CH_3}{\diagup} \; + \; BH_3 \rightarrow (H-\overset{CH_3CH_3}{\underset{CH_3H}{C-C}}-)_2BH \qquad (15.68)$$

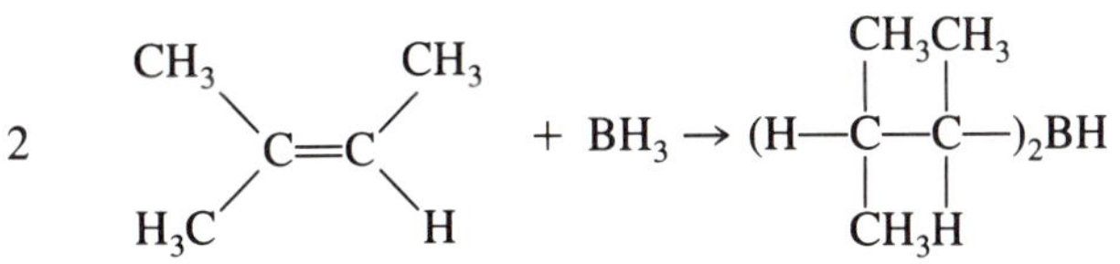

$$\overset{CH_3}{\diagdown}C=C\overset{CH_3}{\diagup} \; + \; BH_3 \rightarrow (H-\overset{CH_3CH_3}{\underset{CH_3CH_3}{C-C}}-)BH_2 \qquad (15.69)$$

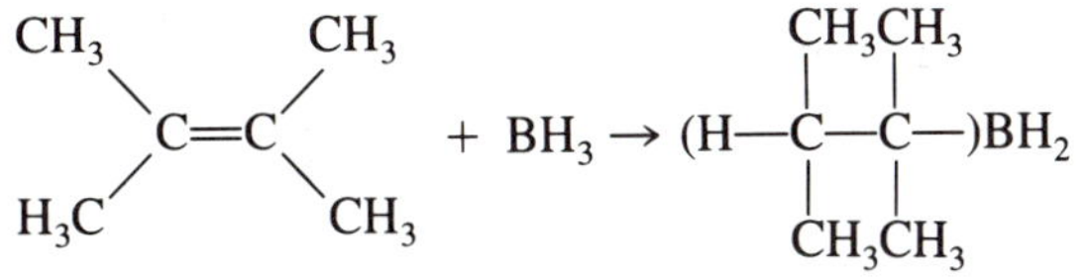

Under forcing conditions, both Reaction 15.68 and Reaction 15.69 can proceed to the next stage, giving trialkyl and dialkyl boranes, respectively.

Electronic rather than steric effects seem to predominate in directing the site for B substitution. For example, in crotyl chloride, $CH_3CH=CHCH_2Cl$, the 2-position is substituted ≈90% of the time with B.

The *cis* stereochemistry of hydroboration permits stereoselective reactions such as the following.

(15.70)

15.10   CLUSTER COMPOUNDS

Inorganic chemistry contains numerous examples of compounds such as the boron hydrides, in which atoms are bonded together in cage structures. We already have seen several examples in this book, including P_4, P_4O_6, P_4O_{10} (Section 13.1.7), and the trigonal prismatic Te_6^{2+} (Section 13.2.7). To be bonded in cage structures, elements must have sufficient valence orbitals properly directed for three-dimensional bonding. Among the elements satisfying this requirement are metals, which often form metal *cluster compounds*—structures displaying polyhedral frameworks held together by two or more metal–metal bonds per metal (Section 14.10 discusses the ability of metals to bond to each other). Figure 15.18 shows the triangulated polyhedra on which cluster structures are based.

Some cluster compounds contain only metals. Examples include Bi_9^{5+} (a tricapped trigonal prism—Figure 15.18*h*), Sn_9^{4-} (a capped Archimedian antiprism—Figure 15.18*i*, with bottom vertex missing), and Pb_5^{2-} (a trigonal bipyramid—Figure 15.18*b*). These species are present in ammonia solutions of so-called Zintl alloys consisting of intermetallic phases of Na and the metal. Recently, the use of cryptate ligands (Section 14.1.4) to complex the Na^+ ions has permitted crystallization of some of these strange species.[13]

Of more widespread recent interest are transition-metal cluster compounds. Each metal generally is also bonded to other ligands that are π acids, halides, hydrides, and/or organic π donors. These are related to the boron hydrides in that a transition metal–containing group can often replace a BH unit in a borane framework. A significant apparent difference between most metal clusters and the boron hydrides is that clusters, even though structurally similar, are not electron-deficient. An example of such a compound is $H_2Ru_6(CO)_{18}$ (Figure 15.19), which contains an octahedron of Ru atoms, each bonded to three terminal CO's. Opposite octahedral faces are triply H-bridged. Regarding the compound as the diprotonated form of $[Ru_6(CO)_{18}]^{2-}$

---

[13]J. D. Corbett and P. A. Edwards, *J. Am. Chem. Soc.* 1977, *99*, 3313 and refs. therein.

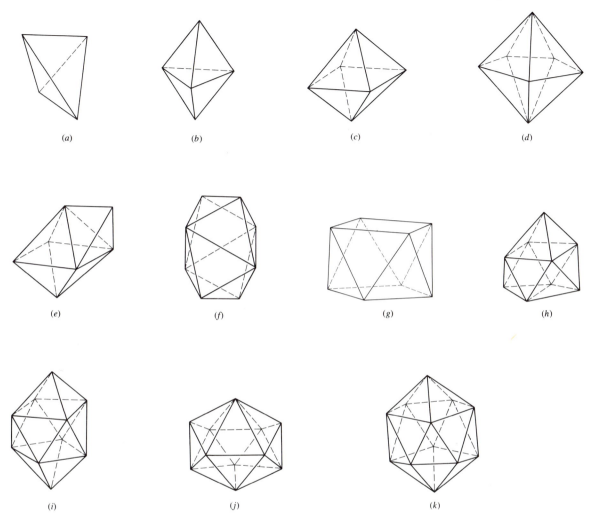

Figure 15.18 Closed triangulated polyhedra that form the structural basis for cluster compounds. *a.* Tetrahedron ($\mathbf{T}_d$). *b.* Trigonal bipyramid ($\mathbf{D}_{3h}$). *c.* Octahedron ($\mathbf{O}_h$). *d.* Pentagonal bipyramid ($\mathbf{D}_{5h}$). *e.* Capped octahedron ($\mathbf{C}_s$). *f.* Dodecahedron ($\mathbf{D}_{2d}$). *g.* Archimedian (square) antiprism ($\mathbf{D}_{4d}$). *h.* Tricapped trigonal prism ($\mathbf{C}_{3v}$). *i.* Bicapped Archimedian antiprism ($\mathbf{D}_{4d}$). *j.* Octadecahedron ($\mathbf{C}_{2v}$). *k.* Icosahedron ($\mathbf{I}_h$).

gives a total of 54 Ru valence orbitals (the *d, s,* and *p* orbitals of each Ru). Allotting 18 of these for Ru–CO bonding, the remaining 36 orbitals and 50 electrons could be used to bond the Ru_6 octahedron. The situation seems just the reverse of that in the boron hydrides: there are more electrons than valence orbitals. Moreover, there is no way to rationalize the electron count via the EAN rule: there are two electrons too many. Thus important questions emerge regarding transition-metal clusters: *(a)* how their molecular structures are related to their electronic structures, and *(b)* why the molecular structures are related to those of the boron hydrides.

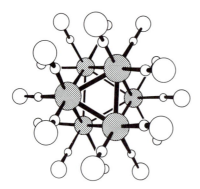

Figure 15.19 The structure of $H_2Ru_6(CO)_{18}$ projected onto one of the open faces. (From M. R. Churchill, J. Wormald, J. Knight and M. Mays, *Chem. Comm.* 1970, 458.)

15.10.1 Clusters and Catalysis

A principal reason for the extensive research on cluster compounds is their possible relevance to the study of heterogeneous catalysis.[14] Industrially important syntheses of organic compounds rely heavily on catalysis (see Section 10.8).

Most industrial reactions are catalyzed heterogeneously. Catalysts usually are transition metals themselves or their solid oxides, and reactions occur at the solid surface with reactants and products in solution. Practical considerations often dictate a preference for heterogeneous catalysts. Ease of product separation is increased, since the solid can be filtered off from a solution or suspended in a gaseous mixture. These solids also are more stable at the high temperatures at which many reactions are conducted.

Directly combining the elements of synthesis gas could afford such industrially useful products as CH_3OH, $HOCH_2CH_2OH$, etc. It is desirable to develop suitable catalysts for the combination of CO and H_2 with high yield and selectivity. The Fischer-Tropsch process is now known to catalyze such reactions heterogeneously. However, very high temperatures with attendant energy expenditure are required.

Investigation of reactions of these organic building blocks catalyzed at solid surfaces presents a formidable problem—in particular, usual solution spectroscopic techniques no longer are applicable.[15] Nevertheless, some knowledge of the course of reactions at surfaces would be highly desirable as an aid in designing improved heterogeneous catalysts.

Some features of the interaction of small molecules with metal surfaces have been investigated. We know that many small molecules are chemisorbed on metallic surfaces: that is, a bond is formed between metal atoms and the molecules that may even disrupt intramolecular bonding. Figure 15.20 indicates several modes of attachment for small molecules of catalytic interest to surfaces containing many metal atoms. None of these bonding modes is available for mononuclear species. The types of molecules bonded and the extent of surface coverage depend critically on the nature of the metal. Transition metals display by far the greatest activity. Moreover, chemisorbed species can migrate across the surfaces. We can visualize the course of a chemical reaction that occurs at a metal surface as involving chemisorption of different molecules at different sites, possibly with bond breaking, migration of reactive species

[14]E. L. Muetterties, *Bull. Soc. Chim. Belg.* 1975, *84*, 959.

[15]J. C. Slater and K. H. Johnson, *Physics Today,* October 1974, 34.

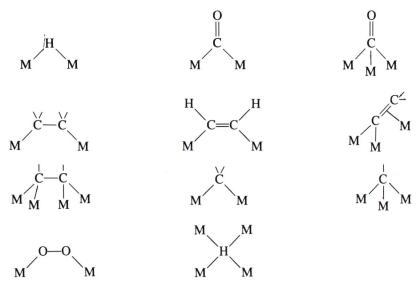

Figure 15.20 Possible modes of attachment of small molecules to metal surfaces or clusters. (After E. L. Muetterties in *Bull. Soc. Chem. Belg.* 1975, *84*, 959.)

to adjacent sites, reaction, and finally, product desorption. In addition, surface metal atoms themselves are known to move.

Two modeling approaches currently are employed for investigating the interaction of small molecules with metal atoms under experimentally accessible conditions. The first[16] (which will not be discussed further) involves vaporization of metal atoms and their condensation at cryogenic temperatures in a matrix of CO, $CH_2=CH_2$, or some other small molecule of interest on the window of an infrared cell. The IR spectra of the mixture, as relative numbers of metal atoms and small molecules are varied, give clues as to the nature of the species present. The second modeling approach, and the one of interest here, involves transition metal–containing clusters as models for metal surfaces with their chemisorbed species. The next sections point out several analogies between cluster compounds and metal surfaces.

It may be that cluster compounds ultimately will not prove very relevant to heterogeneous catalysis. However, this possible relation is an example of why the scientific community focuses on a particular area. Whatever the ultimate result, much interesting chemistry has been and continues to be found.

15.10.2 Molecular Structures of Metal Clusters

Although bare metal skeletons generally are derived from the polyhedra of Figure 15.18, in some known examples the arrangement of metal atoms is related to that in bulk metals. The Rh framework in $[Rh_{13}(CO)_{24}H_3]^{2-}$ corresponds to an *hcp* structure (Figure 15.21), whereas in $[Rh_{14}(CO)_{25}]^{4-}$ a *bcc* arrangement is evident (Figure 15.22). These metal frameworks can

[16]G. A. Ozin, *Acc. Chem. Res.* 1977, *10*, 21; K. J. Klabunde, *ibid.* 1975, *8*, 393.

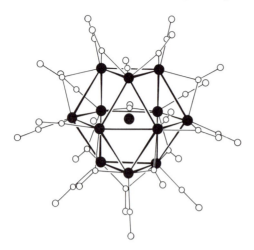

Figure 15.21 The structure of $[Rh_{13}(CO)_{24}H_3]^{2-}$ (H omitted). (Reproduced with permission from S. Martinengo, G. Ciani, A. Sironi, and P. Chini, *J. Am. Chem. Soc.* 1978, *100*, 7096. Copyright 1978, American Chemical Society.)

be regarded as "round surfaces." It is also significant that catalytically active metals have *hcp* or *ccp* structures displaying the same triangular arrangements of metal atoms seen on the faces of cluster polyhedra.

In contrast to single metal centers, metal surfaces and clusters present possibilities for binding small molecules to more than one atom simultaneously. $[Ni_5(CO)_{12}]^{2-}$ (Figure 15.23) shows both terminal and μ_2-bridging carbonyls, and $[Rh_6(CO)_{15}I]^-$ (Figure 15.24) [derived from $Rh_6(CO)_{16}$ by carbonyl substitution] displays four μ_3-carbonyls [as does $Rh_6(CO)_{16}$]. As more negative charge accumulates in isoelectronic species, more of the carbonyls tend to be bridging. A case in point is $[Co_6(CO)_{15}]^{2-}$ (Figure 15.25), which is isoelectronic with $[Rh_6(CO)_{15}I]^-$ and has two more bridging CO's, whereas $[Co_6(CO)_{14}]^{4-}$ (Figure 15.26), which is also isoelectronic, has eight μ_3-CO's. Other examples of coordination to multiple metal sites

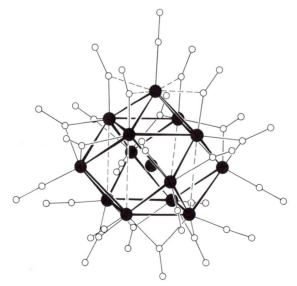

Figure 15.22 The structure of $[Rh_{14}(CO)_{25}]^{4-}$ (Reproduced with permission from S. Martinengo, G. Ciani, A. Sironi, and P. Chini, *J. Am. Chem. Soc.* 1978, *100*, 7096. Copyright 1978, American Chemical Society.)

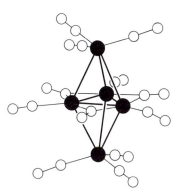

Figure 15.23 The structure of $[Ni_5(CO)_{12}]^{2-}$. (J. K. Ruff, R. P. White and L. F. Dahl, *J. Am. Chem. Soc.* 1971, *93*, 2159. Reproduced from P. Chini, G. Longoni, and V. G. Albano, *Adv. Organomet. Chem.* 1976, *14*, 285.)

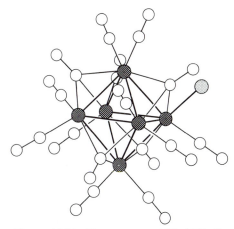

Figure 15.24 The structure of $[Rh_6(CO)_{15}I]^-$. (V. G. Albano, P. L. Bellon, and M. Sansoni, *J. Chem. Soc. A* 1971, 678. Reproduced from P. Chini, G. Longoni, and V. G. Albano, *Adv. Organomet. Chem.* 1976, *14*, 285.)

include μ_2-H and μ_2-CH_2 in $H_2Os_3(CO)_{10}CH_2$ (Figure 15.27). μ_3-H in $H_2Ru_6(CO)_{12}$ was mentioned previously. In $Co_3(CO)_9CCH_3$ (Figure 15.28), μ_3-CCH_3 is found, whereas μ_4-(Et_2C_2) appears in $Co_4(CO)_{10}(EtCCEt)$ (Figure 15.29). Coordination of a ligand to more than one metal may modify the reactivity of small molecules acting as cluster ligands in ways paralleling their reactivity on metal surfaces. Indeed, researchers recently have found that cluster compounds catalyze reactions that mononuclear species do not catalyze. A case in point is the demonstration that $Ir_4(CO)_{12}$ catalyzes the reaction of CO with H_2 to produce CH_3OH.[17] Coordination of CO to more than one metal possibly may weaken the triple bond and permit

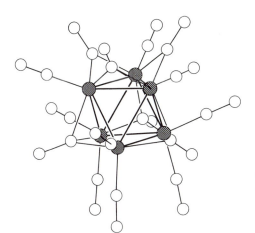

Figure 15.25 The structure of $[Co_6(CO)_{15}]^{2-}$. (V. G. Albano, P. Chini, and V. Scatturin, *J. Organomet. Chem.* 1968, *15*, 423. Reproduced from P. Chini, G. Longoni, and V. G. Albano, *Adv. Organomet. Chem.* 1976, *14*, 285.)

[17]M. G. Thomas, B. F. Beier, and E. L. Muetterties, *J. Am. Chem. Soc.* 1976, *98*, 1296.

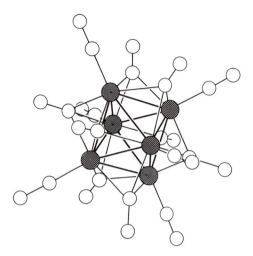

Figure 15.26   The structure of $[Co_6(CO)_{14}]^{4-}$. (V. G. Albano, P. L. Bellon, P. Chini, and V. Scatturin, *J. Organomet. Chem.* 1969, *16*, 461. Reproduced from P. Chini, G. Longoni, and V. G. Albano, *Adv. Organomet. Chem.* 1976, *14*, 285.)

hydrogenation. A lengthening and bond weakening of terminal CO has recently been observed in $Cp_2Mo_2(CO)_4$,[18] because of donation from the π-system to a second Mo atom.

Another similarity between the structures of clusters and those of metals is the incorporation of nonmetallic atoms in the framework. $Fe_5(CO)_{15}C$ is analogous to the interstitial carbides (see Figure 15.30).

15.10.3   Stereochemical Nonrigidity in Clusters

Rearrangements of cluster-bound ligands can occur via suitable vibrational motions and are tantamount to migration over the metal surface of the cluster. Vibrational modes allow per-

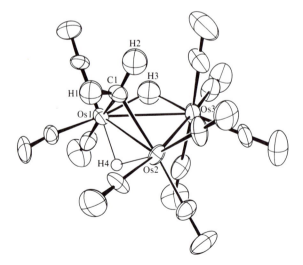

Figure 15.27   The structure of $H_2Os_3(CO)_{10}CH_2$. (Reproduced with permission from R. B. Calvert, J. R. Shapley, A. J. Schultz, J. M. Williams, S. L. Suib, and G. D. Stuckey, *J. Am. Chem. Soc.* 1978, *100*, 6240. Copyright 1978, American Chemical Society.)

[18]R. J. Klingler, W. M. Butler, and M. D. Curtis, *J. Am. Chem. Soc.* 1978, *100*, 5034.

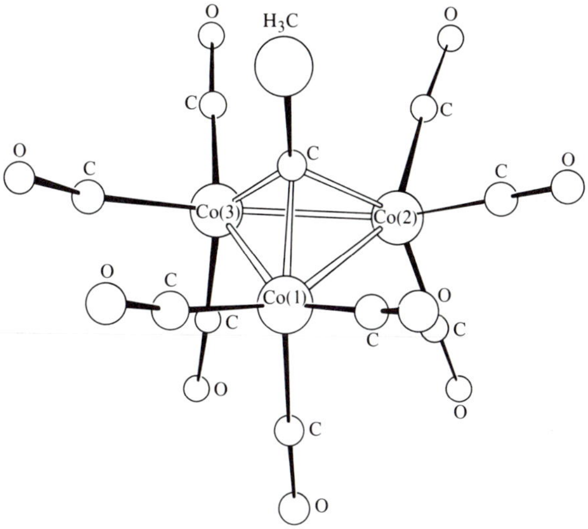

Figure 15.28 The structure of $Co_3(CO)_9CCH_3$. (Reproduced with permission from P. W. Sutton and L. F. Dahl, *J. Am. Chem. Soc.* 1967, *89*, 261. Copyright 1967, American Chemical Society.)

mutation of atomic positions. Structural techniques (for example, NMR) that have a slower time scale than the vibrations permuting nuclear positions "see" a time-averaged structure.

The time required to make an observation (its time scale) is related to the frequency of exciting electromagnetic radiation by the Uncertainty Principle:

$$\Delta E \, \Delta t \sim h$$
$$\Delta(h\nu)\Delta t = h\Delta\nu\Delta t \sim h \qquad (15.71)$$
$$\Delta\nu\Delta t \sim 1$$

Since we adopt the most pessimistic view in uncertainty calculations, $\Delta\nu$ can be set $\approx \nu$. Then $\Delta t \approx \frac{1}{\nu}$. All other things being equal, the time scale is inversely proportional to the frequency of the exciting radiation. Other considerations connected with the sample (for example, relaxation times) may effectively lengthen Δt. Table 15.6 gives time scales for several common structural techniques.

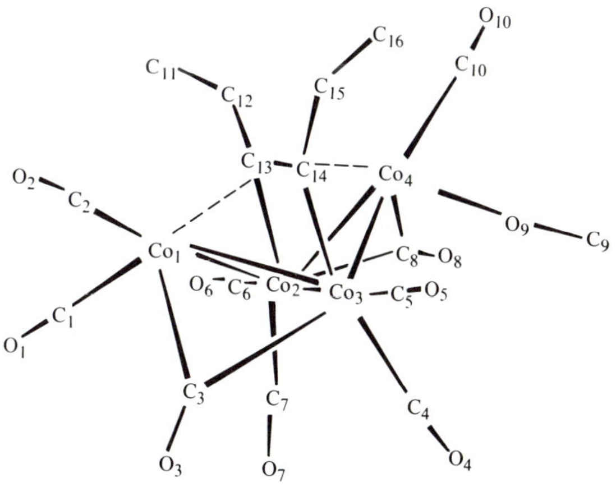

Figure 15.29 The structure of $Co_4(CO)_{10}(EtC\equiv CEt)$. (Reproduced with permission from L. F. Dahl and D. L. Smith, *J. Am. Chem. Soc.* 1962, *84*, 2450. Copyright 1965, American Chemical Society.)

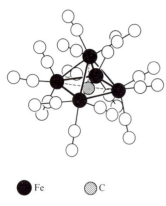

Fe   C

Figure 15.30   The structure of $Fe_5(CO)_{15}C$. (E. H. Braye, L. F. Dahl, W. Hubel, and D. Wampler, *J. Am. Chem. Soc.* 1962, *84*, 4633. Reproduced from P. Chini, G. Longoni, and V. G. Albano, *Adv. Organomet. Chem.* 1976, *14*, 285.)

NMR spectroscopy is suited to investigating stereochemical nonrigidity because the time scales for molecular rotations and vibrations that permute nuclei often are much greater than the time scale for nuclear spin transitions at low temperature. At high temperature they are often smaller. Figure 15.31 shows an example. At 273 K, sufficient thermal energy is available for rotations of 1,2-dibromo-1,1-dichloro-2,2-difluoroethane to occur rapidly enough relative to the NMR time scale that the F's see an averaged environment over the time required for observation. On lowering the temperature to 193 K, the available thermal energy is less than the rotational activation energy. Hence rotations are slowed sufficiently to detect the presence of rotational isomers in solution. (Since I and III are enantiomers, their NMR spectra are the same in a nonchiral solvent.) Rotational barriers rarely are so high. More common is the situation in which vibrational barriers result in the detection of rigid structures at low temper-

Table 15.6   Time scale for structural techniques[a]

Technique	Approx. Time Scale, sec
Electron diffraction	10^{-20}
Neutron diffraction	10^{-18}
X-ray diffraction[b]	10^{-18}
Ultraviolet	10^{-15}
Visible	10^{-14}
Infrared-Raman	10^{-13}
Electron spin resonance[c]	$10^{-4} - 10^{-8}$
Nuclear magnetic resonance[c]	$10^{-1} - 10^{-9}$
Quadrupole resonance[c]	$10^{-1} - 10^{-8}$
Mössbauer (iron)	10^{-7}
Molecular beam	10^{-6}
Exptl. sepn. of isomers	$>10^2$

[a]From E. L. Muetterties, *Inorg. Chem.* 1965, *4*, 769.
[b]Individual measurements of this duration are taken over a long time span. Hence a time-averaged structure is obtained.
[c]Time scale sensitivity defined by chemical systems under investigation.

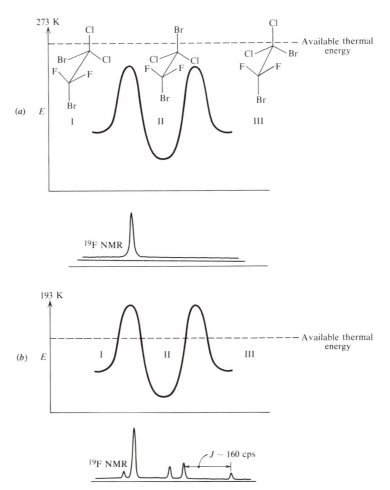

Figure 15.31 NMR spectra and energy diagrams for a halogenated ethane. (Spectra from *Nuclear Magnetic Resonance* by J. D. Roberts. Copyright 1959, McGraw-Hill Book Company. Used with permission of McGraw-Hill Book Company.)

atures. In cluster compounds, the vibrations may include ligand motions or motions of the polyhedral metal structure.

An interesting case involving ligand mobility is $Cp_3Rh_3(CO)_3$, whose ^{13}C NMR spectrum in the CO region at $-65°C$ and $+26°C$ appears in Figure 15.32. Since Rh has nuclear spin ½, each equivalent bridging CO gives a triplet at low temperatures. At high T each CO visits each Rh and is thus coupled to all three; a ^{13}C quartet $[2(\frac{3}{2})+1]$ results. Figure 15.32 also shows a possible idealized intermediate in the process that permutes CO's; the proposed species involves conversion of μ_2-CO's to terminal CO's.[19] The permutation of CO positions in the cluster is analogous to the migration of chemisorbed species on surfaces.

[19] R. J. Lawson and J. R. Shapley, *J. Am. Chem. Soc.* 1976, *98*, 7433.

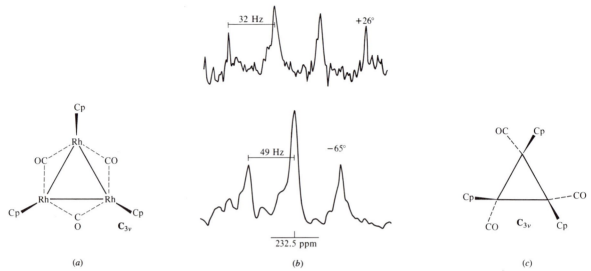

Figure 15.32 *(a)* Molecular structure of $Cp_3Rh_3(CO)_3$. *(b)* Carbonyl ^{13}C NMR spectra for $Cp_3Rh_3(\mu_2\text{-}CO)_3$ in the slow-exchange ($-65°C$) and fast-exchange ($26°C$) limits. *(c)* Proposed intermediate in CO exchange process. (Reproduced with permission from R. J. Lawson and J. R. Shapley, *J. Am. Chem. Soc.* 1976, *96*, 7433. Copyright 1976, American Chemical Society.)

The variable temperature ^{13}C NMR spectrum of $Rh_4(CO)_{12}$[20] appears in Figure 15.33. The low-temperature spectrum indicates a static structure containing four different kinds of CO's coupled to one or two Rh atoms. At low temperatures, the time spent by each CO in a given chemical environment is long compared with the time for a nuclear transition. As a result, the measuring technique "sees" separate resonances for each kind. When the sample is warmed, nuclei begin to permute, and each one spends less and less time in a particular environment. As the environment becomes less and less well defined over the time period required for a nuclear transition, the lines begin to broaden. When the time of residence is comparable to the time required for nuclear transitions (at about 7.5°C), the signal is broadened into the base line. On raising the temperature further, the exchange processes become sufficiently rapid as to reduce the residence time in any environment far below the time scale of the measurement. The radiation then "sees" the CO ligands in their time-averaged environment. That the ligand exchange process is intramolecular is shown by the persistence of Rh—C coupling at high temperatures. Again, a bridge-terminal interchange of CO's provides a possible mechanism for the CO exchange. Such a mechanism has been established firmly for some dinuclear complexes.[21] $[Rh_6(CO)_{15}C]^{2-}$ (Figure 15.34), on the other hand, does not show carbonyl exchange below 298 K, perhaps because all edges are occupied by CO's and concerted motion of several CO's would be required to effect bridge-terminal interchange.[22]

An example of H migration around a cluster framework is provided by the anion

[20]F. A. Cotton, L. Kruczynski, B. L. Shapiro, and L. F. Johnson, *J. Am. Chem. Soc.* 1972, *94*, 6191; R. D. Adams and F. A. Cotton, in *Dynamic Nuclear Resonance Spectroscopy*, L. M. Jackman and F. A. Cotton, Eds., Academic Press, New York, 1975, p. 520.

[21]R. D. Adams and F. A. Cotton, *J. Am. Chem. Soc.* 1973, *95*, 6589.

[22]V. G. Albano, P. Chini, S. Martinengo, D. J. A. McCaffrey, D. Strumolo, and B. T. Heaton, *J. Am. Chem. Soc.* 1974, *96*, 8106.

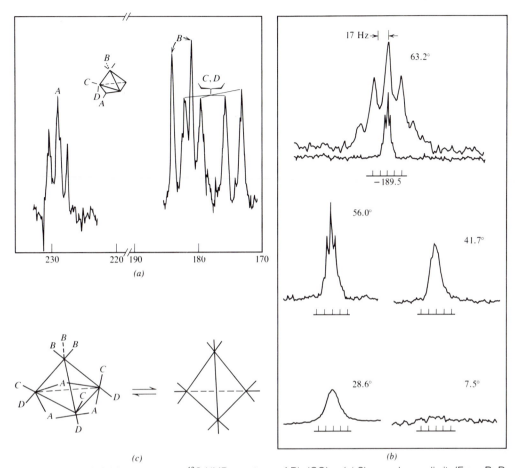

Figure 15.33 Variable-temperature ^{13}C NMR spectrum of $Rh_4(CO)_{12}$. *(a)* Slow exchange limit. (From R. D. Adams and F. A. Cotton in *Dynamic Nuclear Resonance Spectroscopy*, L. M. Jackman and F. A. Cotton, Eds. Academic Press, New York, 1975). *(b)* At higher temperatures. *(c)* Proposed mechanism for CO exchange. (*b* and *c* reproduced with permission from F. A. Cotton, L. Kruczynski, B. L. Shapiro, and L. F. Johnson, *J. Am. Chem. Soc.* 1972, *94*, 6191. Copyright 1972, American Chemical Society.)

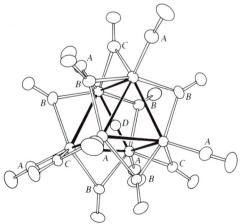

Figure 15.34 Structure of $[Rh_6(CO)_{15}C]^{2-}$. (Reproduced with permission from V. G. Albano, P. Chini, S. Martinengo, D. J. A. McCaffrey, D. Strumolo, and B. T. Heaton, *J. Am. Chem. Soc.* 1974, *96*, 8106. Copyright 1974, American Chemical Society.)

$[H_3Ru_4(CO)_{12}]^-$, which exists in solution as a mixture of two isomers (Figure 15.35)[23] thought to have structures A and B_1 or B_2. On warming, the three signals broaden, collapse, and reform into a single peak at the fast exchange limit, indicating the equivalence of all the H's on the

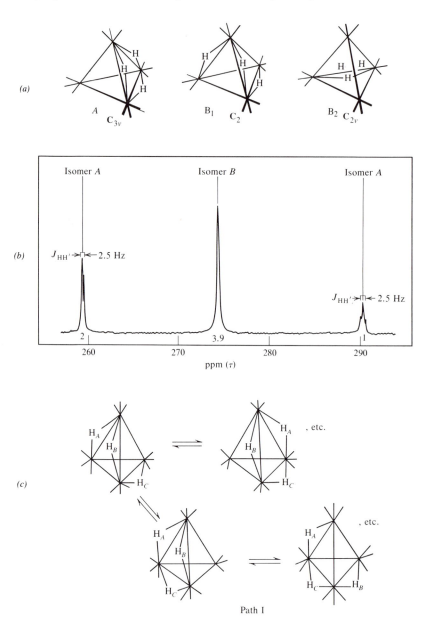

Figure 15.35 (a) Isomers of $[H_3Rh_4(CO)_{12}]^-$. (b) Low temperature limiting ${}^1H$ NMR spectrum of $[H_3Rh_4(CO)_{12}]^-$. (c) Possible pathways for H exchange in $[H_3Rh_4(CO)_{12}]^-$.

(Continued)

[23]J. W. Koepke, J. R. Johnson, S. A. R. Knox, and H. D. Kaesz, *J. Am. Chem. Soc.* 1975, *97*, 3947.

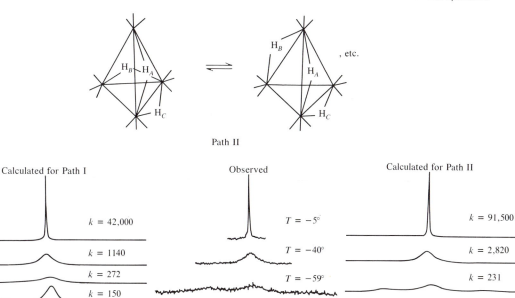

Figure 15.35 (*Continued*). (*d*) Computed and observed ¹H NMR spectra at various temperatures. (Reproduced with permission from J. W. Koepke, J. R. Johnson, S. A. R. Knox, and H. D. Kaesz, *J. Am. Chem. Soc.* 1975, *97*, 3947. Copyright 1975, American Chemical Society.)

NMR time scale. Two possible mechanisms may lead to H exchange: an intramolecular shift coupled with isomer exchange (Figure 15.35*c*, path I) or isomer exchange (Figure 15.35*c*, path II). In principle, separate pathways leading to different permutation patterns can often be distinguished by computing the line shapes expected on the basis of each and comparing with experimental line shapes. As Figure 15.35*d* shows, however, such a choice is not possible for this system.

Mobility in clusters is not confined to ligands. Relatively small motions can deform polyhedral structures with seven or more vertices into alternative structures. Figure 15.36 shows how stretching or compression of edges easily can lead to the traverse of three different polytopal forms in eight-atom clusters. Recent examples of metal motion have been found in the mixed-metal four-atom clusters $H_2FeRu_3(CO)_{13}$ and $H_2FeRuOs_2(CO)_{13}$[24] and in Rh clus-

[24]G. L. Geoffroy and W. L. Gladfelter, *J. Am. Chem. Soc.* 1977, *99*, 6775.

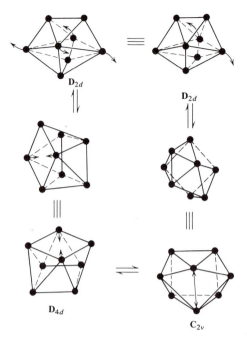

Figure 15.36 Interconversion of eight-vertex struc-tues.

ters.[25] Undoubtedly, more such examples will be forthcoming.[26] NMR investigations of ligand and metal mobility in clusters have revealed aspects that should apply to species adsorbed on metal surfaces.

Stereochemical nonrigidity also is exhibited by the ligands of mononuclear species. Such species can be related to the polyhedra of metal clusters if we think of the ligands as occupying vertices of a polyhedron circumscribed around a central atom. For example, the CO's in $Fe(CO)_5$ can be considered to lie at the vertices of a trigonal bipyramid circumscribed around Fe. Down to the lowest experimentally accessible temperatures, only a single ^{13}C line is observed for this compound. This obviously indicates a process with extremely low activation energy that permutes all the CO groups. Ligand exchange in such C.N.5 species was discussed in Chapter 9.

The structures of boron hydrides also are very commonly nonrigid and temperature-dependent. ^{11}B and ^{1}H NMR are important structural techniques in boron chemistry.

15.10.4 Electronic Structures

The relation between electron count and molecular structure for metal clusters displays an important analogy with that of boron hydrides, as pointed out by Wade.[27]

Returning to the problem of $H_2Ru_6(CO)_{18}$, recall that after reserving 18 Ru orbitals for

[25]O. A. Gansow, D. S. Gill, F. J. Bennis, J. R. Hutchinson, J. L. Vidal, and R. C. Schoening, *J. Am. Chem. Soc.* 1980, *102*, 2449.

[26]B. F. G. Johnson and R. E. Benfield, *J. Chem. Soc., Dalton Trans.* 1978, 1554.

[27]K. Wade, *Chem. in Britain* 1975, *11*, 177; *Adv. Inorg. Radiochem.* 1976, *18*, 1.

carbonyl bonding, 36 Ru orbitals and 50 electrons remain. Since transition metal compounds usually contain nonbonding electrons, it is plausible to assign some electrons as nonbonding on Ru atoms. Assuming six nonbonding electrons for each Ru and reserving three atomic orbitals per Ru to accommodate them, leaves a total of 18 orbitals [three per $Ru(CO)_3$ unit] and 14 electrons (seven pairs) for bonding the Ru_6 framework. This same situation prevailed in $B_6H_6^{2-}$. The group at each vertex contributes three orbitals and two electrons to framework bonding. Seven electron pairs are available for bonding of the six-vertex framework, and by analogy with the $B_6H_6^{2-}$ case, we expect the octahedral *closo* structure.

Another interesting compound is $Fe_5(CO)_{15}C$ (Figure 15.30). The total number of framework electrons (the ones in Fe—CO bonds are not counted) is

$$5 \text{ Fe} \quad + \quad C$$
$$5 \times 8e + 4e = 44e$$

Reserving three orbitals on each Fe for bonding to CO's leaves us 30 Fe framework valence orbitals. If we assign six electrons to occupy three orbitals as nonbonding on each Fe, this leaves 14 electrons (seven pairs) and 15 Fe orbitals for framework bonding. The structure is the expected *nido* arrangement of five $Fe(CO)_3$ groups at vertices bonded by $7(5 + 2)$ electron pairs.

The electron counting procedure consists of assigning sufficient metal electrons and orbitals as nonbonding so that each metal group also has enough orbitals to bond to its ligands and still contribute three orbitals plus any remaining electrons to framework bonding. We actually must assume that some electrons are nonbonding; otherwise, the number of skeletal bonding electrons would be more than the number of orbitals remaining after accounting for metal–ligand bonding. Plainly, accommodating all these electrons in bonding MO's would be impossible.

Table 15.7 gives the number of framework electrons contributed by various groups. Each transition metal has nine valence orbitals. We want to reserve three of these, along with whatever electrons they contain, for framework bonding. This means that six metal orbitals must be either nonbonding or used in bonding to ligands. Twelve electrons are required to fill these orbitals. Since the total number of electrons for each group is equal to the v valence shell electrons of the metal plus the x electrons supplied by the ligands (see Tables 10.2 and 10.4), the number of electrons left to contribute to framework bonding is $(v + x - 12)$.

Table 15.7 Number of skeletal bonding electrons $(= v + x - 12)$ contributed by some transition-metal units

Metal Group	M	$M(CO)_2$	$M(\eta^5\text{-}C_5H_5)$	$M(CO)_3$	$M(CO)_4$
			Cluster Unit		
VI	Cr, Mo, W	-2	-1	0	2
VII	Mn, Tc, Re	-1	0	1	3
VIII	Fe, Ru, Os	0	1	2	4
	Co, Rh, Ir	1	2	3	5
	Ni, Pd, Pt	2	3	4	6

Example Calculate the number of electrons contributed by CpCo to framework bonding.

Solution Co(I) has eight valence electrons; hence, $v = 8$.
Cp$^-$ contributes six electrons; hence $x = 6$.
$(v + x - 12) = 2$ electrons contributed.

Example Predict the molecular geometry of $[Ni_6(CO)_{12}]^{2-}$.

Solution According to Table 15.7, each $Ni(CO)_2$ unit contributes $2e$ to framework bonding. The framework electron count is

$$6 \; Ni(CO)_2 \; + \; charge$$
$$6 \times 2e \; + \; 2e \; = 14e$$

Seven electron pairs are available for framework bonding. Hence an octahedral arrangement of the Ni atoms [the *closo* structure for n vertices and $(n + 1)$ electron pairs] is both expected and observed. (The actual structure is distorted toward a trigonal prism.)

Example What is the predicted structure of $Fe_5(CO)_{15}C$?

Solution Since carbides generally occupy an interstitial position, the vertices will all be occupied by $Fe(CO)_3$ units. The electron count is

$$5 \; Fe(CO)_3 \; + \; C$$
$$5 \times 2e \; + 4e = 14e$$

The seven electron pairs represent two more than the number of vertices. A *nido* structure like that in Figure 15.30 is predicted.

Example What structure is to be expected for $Rh_6(CO)_{16}$?

Solution Six $Rh(CO)_3$ units obviously is not possible. We then suppose that there are six $Rh(CO)_2$ units and four "extra" CO's. The electron count will be

$$6 \; Rh(CO)_2 \; + \; 4 \; CO$$
$$6 \times 1e \; + 4 \times 2e = 14e$$

The seven framework bonding pairs and six Rh-containing groups predict a *closo* structure. In the actual structure, alternate octahedral faces are triply bridged by the "extra" CO's. We cannot predict the distribution of CO between bridging and terminal positions, because the electron count will be the same in either case.

The analogy between transition metal fragments and BH units makes it reasonable that there should be compounds in which some BH units are replaced by transition metal fragments. Their structures can be predicted using these considerations. For example, B_5H_9, $B_4H_8Fe(CO)_3$,[28] and $B_3H_7[Fe(CO)_3]_2$[29] all have five groups and seven skeletal electron pairs leading to *nido* structures, as does the six-group, eight-skeletal-pair species $BC_4H_5Fe(CO)_3$.[30] Of course, Table 15.4 fulfills the same function for main-group heteroatoms as Table 15.7 does for transition metal groups. Structural predictions based on Wade's approach do not always distinguish between terminal and bridging CO ligands or locate "extra" hydrogens.

[28]N. N. Greenwood, C. G. Savory, R. N. Grimes, L. G. Sneddon, A. Davison, and S. S. Wreford, *Chem. Comm.* *1974*, 718.

[29]E. L. Anderson, K. J. Haller and T. P. Fehlner, *J. Am. Chem. Soc.* 1979, *101*, 4390.

[30]T. P. Fehlner, *J. Am. Chem. Soc.* 1978, *100*, 3250.

15.10.5 Structures Not Rationalized by Wade's Approach

Although the analogy with boron hydrides pointed out by Wade correctly predicts a large number of cluster structures, many remain unrationalized. This is especially the case for five-membered clusters. An example with eight vertices is $(CpCo)_4B_4H_4$, which has only eight skeletal bonding pairs but adopts the *closo* dodecahedral structure (Figure 15.37).

Several known "electron-hyperdeficient" compounds seem to contain only $2p$ electron pairs and adopt the *closo* structure of $p-1$ vertices with the "extra" group capping one face. Examples include $Os_6(CO)_{18}$, a capped trigonal bipyramid;[31] $[Rh_7(CO)_{16}]^{3-}$, a capped octahedron,[32] and $1,6$-Cp_2-$1,6,2,3$-$Fe_2C_2B_6H_8$, a capped tricapped trigonal prism.[33] A recent suggestion[34] is that this latter compound (and presumably others) has a structure best described as the bicapped Archimedean antiprism expected for a 10-vertex, 11-electron-pair *closo* compound distorted by formation of an additional polyhedral metal–metal bond to relieve electron hyperdeficiency.

We have adopted the viewpoint of seeking to predict structural type from electron count in metal clusters. Although such an approach works fairly well for boron compounds, a number of important exceptions are known—such as B_4Cl_4, which is tetrahedral. A possible alternative approach starts with the geometry given and develops a qualitative MO scheme for a metal framework.[35] The framework electron count that will just fill the bonding and nonbonding levels is determined and used to rationalize the geometry adopted. For example, $[Ni_5(CO)_{12}]^{2-}$ has a framework electron count of 16, leading to the prediction of an *arachno* structure derived from a dodecahedron. Instead, the complex is an elongated trigonal bipyramid (Figure 15.23). A regular trigonal bipyramid would be predicted for a complex having six skeletal electron

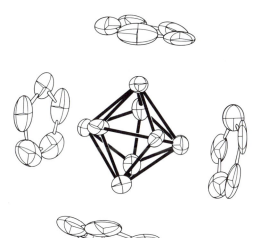

Figure 15.37 The structure of $(CpCo)_4B_4H_4$. (Reproduced with permission from J. R. Pipal and R. N. Grimes, *Inorg. Chem.* 1979, *18*, 257. Copyright 1979, American Chemical Society.)

[31]R. Mason, K. M. Thomas, and D. M. P. Mingos, *J. Am. Chem. Soc.* 1973, *95*, 3802.

[32]V. G. Albano, P. L. Bellon, and G. F. Ciani, *Chem. Comm.* 1969, 1024.

[33]K. P. Callahan, *et al.*, *J. Am. Chem. Soc.* 1975, *97*, 296.

[34]E. K. Nishimura, *Chem. Comm. 1978*, 858.

[35]J. W. Lauher, *J. Am. Chem. Soc.* 1978, *100*, 5305.

pairs. If this regular structure is distorted by lengthening the axial distances, however, calculations show that two antibonding orbitals become stable enough to accommodate the two "extra" electron pairs. This procedure serves to rationalize the structures of "nonconforming" complexes, but leaves open the question of why the geometry exhibited ought to be the preferred one.

Two long-known cluster compounds representative of halide clusters having the formulas $[Mo_6Cl_8]^{4+}$ and $[Ta_6Cl_{12}]^{2+}$ are depicted in Figure 15.38. As with other clusters of the types M_6X_8 and M_6X_{12}, both have octahedra of metal atoms. Each metal is surrounded by a square planar array of halide ligands. Their electronic structures can be rationalized in terms of localized bonds. Each metal uses dsp^2 hybrids to bond with its four square-planar ligands. A pd hybrid directed outward is used for bonding the *exo* ligands usually present. Four atomic orbitals for each metal are available for framework bonding. The Mo_6^{12+} and Ta_6^{14+} cores have 12 and 8 electron pairs, respectively. These could be used for M—M two-center bonds along the (12) octahedral edges or three-center bonds on the (eight) faces.[36]

Additional nonconforming structures likely will be discovered, and the designations *closo, nido, arachno,* and *hypo* may well lose structural significance for metal clusters and might refer only to electron count.

Clusters containing four metals are usually tetrahedral electron-precise species that can be viewed as having two-electron metal–metal bonds directed along tetrahedral edges. Consequently, there is no need to employ Wade's approach for rationalization of their electronic structures.

15.10.6 Synthesis of Metal Clusters

The most active area in metal cluster chemistry currently is that of carbonyl-containing clusters. The synthesis of these species is presently in a rather unsatisfactory state. Unexpected products are obtained—for example, $H_2Ru_6(CO)_{18}$ is the product of the reaction in THF of $Ru_3(CO)_{12}$ with $[CpFe(CO)_2]^-$. Yields often are low. Single products seldom result, and separations are

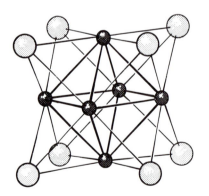 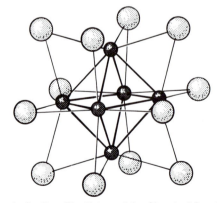

Figure 15.38 Structures of $[Mo_6Cl_8]^{4+}$ and $[Ta_6Cl_{12}]^{2+}$. (From L. Pauling, *The Nature of the Chemical Bond,* Third edition, © 1960 by Cornell University, p. 440. Used by permission of Cornell University Press.)

[36]F. A. Cotton and T. E. Haas, *Inorg. Chem.* 1964, *3,* 10; S. F. A. Kettle, *Theor. Chim. Acta* 1965, *3,* 211; R. F. Schneider and R. A. Mackay, *J. Chem. Phys.* 1968, *48,* 843.

difficult. Syntheses are hard to plan and product structures often are hard to rationalize, since the EAN rule seldom applies. However, the redox condensation method of Chini does represent a considerable advance in systematic preparative strategies. Because of the large numbers of atoms, IR and NMR techniques are not always very informative for structure determination. Most progress depends on x-ray structure determinations (and to a lesser extent, on mass spectra).

Thermochemical Considerations

The ability to prepare clusters and the conditions under which they are stable are related to the strengths of metal–metal and metal–ligand bonds. Table 15.8 gives bond energies determined from thermochemical experiments (on the assumption that the M—CO terminal bond energy is invariant for all neutral carbonyls of the same metal). Also reported (in columns labeled 2) are values calculated on the assumption that metal–metal bond energies are proportional to r^{-k}, where r is the metal–metal separation in the compound and k is a constant that can be evaluated from atomization enthalpies of the pure metals.

Both methods indicate that the strengths of metal–metal bonds increase on descending columns in the periodic table. Even so, the contribution of metal–metal bonds to the total bond energy is not very large. It accounts for only ~6% of the total bond energy in dinuclear

Table 15.8 Bond energies for some carbonyl complexes (kJ/mole)

Compound	$d_{M—M}(pm)$	$E_{M—M}$		E_{CO}	
		Set 1[a]	*Set 2*[b]	*Set 1*[a]	*Set 2*[b]
$Fe(CO)_5$				117	117
$Fe_2(CO)_9$	252	82	70	$117t^c$	123
				$64b^c$	
$Fe_3(CO)_{12}$	256	82	65	$117t^c$	126
				$64b^c$	
$Ru_3(CO)_{12}$	285	117	78	172	182
$Os_3(CO)_{12}$	268	130	94	190	201
$Co_2(CO)_8$	252	83	70	$136t^c$	136
				$68b^c$	
$Co_4(CO)_{12}$	249	83	74	136	140
$Rh_4(CO)_{12}$	273	114	86	166	178
$Rh_6(CO)_{16}$	278	114^d	80^d	166	182
$Ir_4(CO)_{12}$	268	130	117	190	196

[a]Assuming constant $E_{M—M}$ and E_{CO} in series of compounds. From J. A. Connor, *Topics in Current Chemistry* 1977, *71*, 71.
[b]Assuming $E_{M—M} = Ad_{M—M}^{-k}$. From C. E. Housecroft, *et al.*, *J. Chem. Soc., Chem. Comm.* 1978, 765; C. E. Housecroft, *et al.*, *J. Organomet. Chem.* 1981, *213*, 35.
[c]t = terminal; b = bridging
[d]Assuming 11 two-center, two-electron M—M bonds resonating among 12 octahedral edges.

clusters, ~10% in trinuclear clusters, ~20% in tetranuclear clusters, and ~25% in hexanuclear clusters. Metal–CO bonds are energetically dominant.

Metal–ligand bond energies also increase on descending the periodic table (at about the same rate as M—M bond energies). In the calculations that do not assume constancy for M—CO bond energies, we can see a significant increase in the average M—CO bond energy with increasing nuclearity of the clusters (as the number of CO's per metal decreases). The more electron-rich the metal portion, or the more negative the charge on the cluster, the more CO functions as a π acid and the stronger the bonds become. In the set of calculations that differentiates bond energies for the different bonding modes, bond energies for metal-terminal CO bonds are larger than those for metal-bridging CO bonds.

Thermal Condensation

A very important route to the synthesis of metal clusters is condensation reactions. Carbonyl ligands are removed, creating coordinatively unsaturated species that condense.

$$2Co_2(CO)_8 \underset{25° \text{ C, 1 atm CO}}{\overset{60°C}{\rightleftharpoons}} Co_4(CO)_{12} + 4CO \qquad \Delta H \sim 7.8 \text{ kJ} \qquad (15.72)$$

$$3Rh_4(CO)_{12} \underset{-19°C, 490 \text{ atm CO}}{\overset{25° \text{ C, 490 atm CO}}{\rightleftharpoons}} 2Rh_6(CO)_{16} + 4CO \qquad \Delta H \sim 15 \text{ kJ} \qquad (15.73)$$

Both the above reactions are endothermic, as we would expect for processes that break strong M—CO bonds and substitute weaker M—M bonds.

The greater strength of M—CO bonds as compared with M—M bonds for first-row transition elements requires that cluster synthesis be carried out in the absence of CO, to avoid degrading the cluster. This dictates that a preformed carbonyl be used. In working with second- and third-row transition elements, reduction to carbonyls and condensation to clusters can be carried out in a single step: for example,

$$7K_3RhCl_6 + 48KOH + 28CO \xrightarrow[\text{MeOH}]{25°C, \text{ 1 atm CO}} K_3[Rh_7(CO)_{16}]$$
$$+ 42KCl + 12K_2CO_3 + 24H_2O \qquad (15.74)$$

The buildup of negative charge renders even some third-row clusters unstable to CO. Although $Rh_6(CO)_{16}$ and $[Rh_6(CO)_{15}]^{2-}$ are stable to CO at 25°C, the isoelectronic $[Rh_6(CO)_{14}]^{4-}$ is not.

Redox Condensation

The second method of general utility in preparing carbonyl clusters is redox condensation. These reactions often take place under very mild conditions. Examples are

$$[Fe_3(CO)_{11}]^{2-} + Fe(CO)_5 \rightarrow [Fe_4(CO)_{13}]^{2-} + 3CO \qquad (15.75)$$

$$Ru_2Os(CO)_{12} + [Fe(CO)_4]^{2-} \xrightarrow{H^+} H_2OsRu_2Fe(CO)_{13} + 3CO \qquad (15.76)$$

Redox condensations offer particularly attractive possibilities for making mixed-metal clusters.

$$2Co_2(CO)_8 + Rh_2(CO)_4Cl_2 \rightarrow Co_2Rh_2(CO)_{12} + 2CoCl_2 + 8CO \qquad (15.77)$$

15.10.7 Reactivity of Cluster Compounds

Comparatively little is known of the reactivity of cluster species, and practically nothing is known of reaction mechanisms. Only some rather general reaction types can be mentioned here.

Reduction

Although in several isoelectronic series, carbonyl ligands are replaced by electron pairs (for example, $Co_6(CO)_{16}$, $[Co_6(CO)_{15}]^{2-}$, $[Co_6(CO)_{14}]^{4-}$), these are usually not prepared cleanly by reduction of the first member with release of CO. The released carbon monoxide often leads to cluster degradation.

$$11[Co_6(CO)_{15}]^{2-} + 22Na \rightarrow 11[Co_6(CO)_{14}]^{4-} + 22Na^+ + 11CO$$
$$\underline{2[Co_6(CO)_{14}]^{4-} + 11CO \rightarrow [Co_6(CO)_{15}]^{2-} + 6[Co(CO)_4]^-} \quad (15.78)$$
$$10[Co_6(CO)_{15}]^{2-} + 22Na \rightarrow 9[Co_6(CO)_{14}]^{4-} + 6[Co(CO)_4]^- + 22Na^+$$

Use of strong base reduces CO, thus providing a better yield of reduction product.

$$Rh_4(CO)_{12} + OCH_3^- \xrightarrow[CH_3OH]{25°} [Rh_4(CO)_{11}(COOCH_3)]^- \xrightarrow{3OH^-}$$
$$[Rh_4(CO)_{11}]^{2-} + CH_3OH + CO_3^{2-} + H_2O \quad (15.79)$$

In this reaction, the intermediate carboalkoxy complex has been isolated.

Oxidation

Electrochemical oxidation of $Cp_4Fe_4(CO)_4$ to the $+1$ and $+2$ species is known; it probably indicates that the HOMO of the parent compound is slightly antibonding or nonbonding.

A second method is by protonation of anionic clusters to give intermediate hydrido compounds, which then react with an additional proton, eliminating H_2.

$$[Ir_6(CO)_{15}]^{2-} + 2H^+ \xrightarrow[CH_3CO_2H]{CO} Ir_6(CO)_{16} + H_2 \quad (15.80)$$

$$[Ir_4(CO)_{11}H]^- + H^+ \xrightarrow{CO} Ir_4(CO)_{12} + H_2 \quad (15.81)$$

Some chemical oxidations can be effected with addition of the oxidant.

$$[Rh_6(CO)_{15}]^{2-} + I_2 \xrightarrow{THF} [Rh_6(CO)_{15}I]^- + I^- \quad (15.82)$$

Ligand Substitution

Carbonyl ligands often can be displaced by other Lewis bases. However, cluster fragmentation often occurs with first-row transition metal species, because of their relatively weak metal–metal bonds.

$$Co_4(CO)_{12} + CH_3NC \rightarrow Co_4(CO)_{11}(CNCH_3)$$

$$\xrightarrow{CH_3NC} Co_4(CO)_{10}(CNCH_3)_2 \xrightarrow{CH_3NC} Co_4(CO)_9(CNCH_3)_3 \qquad (15.83)$$

$$\xrightarrow{CH_3NC} Co_4(CO)_8(CNCH_3)_4$$

$$Ru_6(CO)_{17}C + PPh_3 \rightarrow Ru_6(CO)_{16}(PPh_3)C + CO \qquad (15.84)$$

$$Fe_3(CO)_{12} + 3PPh_3 \rightarrow Fe(CO)_4(PPh_3) + Fe(CO)_3(PPh_3)_2 \qquad (15.85)$$

Clusters containing nonmetal atoms may display reactivity at these centers, as well. Figure 15.39 summarizes several reactions of $ClCCo_3(CO)_9$.

15.11 CAGE COMPOUNDS AND THEIR RELATION TO CLUSTERS

Many known compounds have structures related to those of metal clusters but not held together by two or more metal–metal bonds per metal. Atoms in these cage compounds are located at the vertices of triangulated or other polyhedra. Boron hydrides and metal clusters generally conform to a molecular orbital approach for electron-deficient species. This approach is not necessary in dealing with electron-precise compounds such as tetrahedral clusters.

Many cage compounds may be considered as electron-rich species derived from electron-precise ones. We can imagine each "extra" pair of electrons as occupying a localized M—M antibonding orbital, thereby breaking a metal–metal bond.[37] Thus $Co_4(CO)_{12}Sb_4$ containing a tetrahedron of Co atoms has no Co—Co bonds and is held together by μ_3-Sb's (see Figure 15.40). The structure is a distorted cube with Co and Sb at alternate corners of two

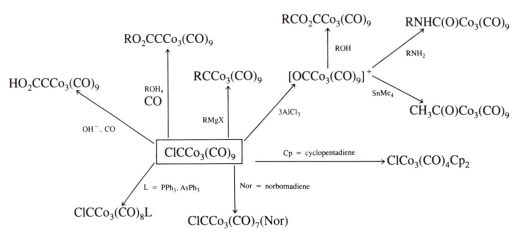

Figure 15.39 Reactions of $ClCCo_3(CO)_9$. (Adapted with permission from W. R. Penfold and B. H. Robinson, *Acc. Chem. Res.* 1976, *6*, 73. Copyright 1976, American Chemical Society.)

[37]D. M. P. Mingos, *Nature, Phys. Sci.* 1972, *236*, 99.

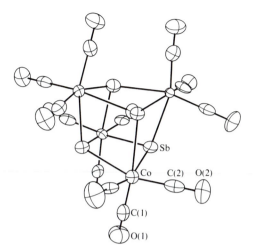

Figure 15.40 The structure of $Co_4(CO)_{12}Sb_4$. (Reproduced with permission from A. S. Foust and L. F. Dahl, *J. Am. Chem. Soc.* 1970, *92*, 7337. Copyright 1970, American Chemical Society.)

interpenetrating tetrahedra. The relation of electron count to structure may be thought of as follows: each $Co(CO)_3$ unit contributes three orbitals and 3 e to skeletal bonding (Table 15.7), for a total of 12 e. Each Sb donates 3 e (Table 15.4), for a total of 12. Altogether there are 24 framework e. If the Sb and $Co(CO)_3$ units are placed at alternate corners of a cube, exactly 24 e are required to form a bond along each of the 12 cube edges. The distortion results from differing sizes of the groups.

Example   Predict the structure of $[Re_4(CO)_{16}]^{2-}$.

Solution   The total number of framework electrons is

$$4 \ Re(CO)_4 + \text{charge}$$
$$4 \times 3 \ e + 2 \ e = 14 \ e$$

If we arrange the four Re atoms at the vertices of a tetrahedron, 12 e (six pairs) are required to form a Re—Re bond along each edge. We imagine the extra pair of electrons as occupying an antibonding orbital, breaking one Re—Re bond and giving the structure shown in Figure 15.41.

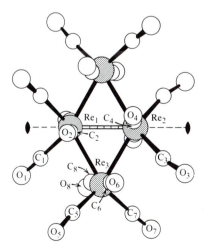

Figure 15.41 The structure of $[Re_4(CO)_{16}]^{2-}$. (Reproduced with permission from M. R. Churchill and R. Bau, *Inorg. Chem.* 1968, *7*, 2606. Copyright 1968, American Chemical Society.)

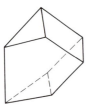

Figure 15.42   The structure of cuneane C_8H_8, an isomer of cubane.

Example   Predict the structure of S_4N_4.

Solution   We extend the approach here to cage compounds of the nonmetals. The total number of framework electrons can be determined from Table 15.4 as 28. If we start by arranging the eight atoms at the corners of a cube, 24 *e* would be required to form S—N bonds along all the edges. The two "extra" pairs of electrons would break two of the bonds. The actual structure of S_4N_4 shown in Figure 13.6 does not comform to this prediction. Instead, it results from breaking two of the 12 bonds in the cage structure of cuneane (Figure 15.42), an isomer of the eight-carbon cubane. Hence, with main group compounds, the straightforward choice of the cube as the starting eight-vertex polyhedron must be modified.

A detailed MO treatment for molecules such as $Co_4Sb_4(CO)_{12}$ related to cubane has been worked out by Dahl and co-workers.[38]

The relation between molecular and electronic structures for metal-containing cages and clusters is, as yet, incompletely worked out. Broadly speaking, four-membered species are electron-precise or electron-rich, whereas six-membered ones are electron-deficient. Pentanuclear clusters present a variety of situations. Structures of higher nuclearity are so far too few in number for the systematics to be apparent.

GENERAL REFERENCES

H. C. Brown, *Boranes in Organic Chemistry,* Cornell University Press, Ithaca, N.Y., 1972.

P. Chini and B. T. Heaton, *Top. Curr. Chem.* 1977, *71,* 1. Molecular structures and synthesis of tetrahedral clusters.

P. Chini, G. Longoni, and V. G. Albano, *Adv. Organomet. Chem.* 1976, *14,* 285. A discussion of high nuclearity clusters, their structures and syntheses.

J. Evans, *Adv. Orgmet. Chem.* 1977, *16,* 319. Molecular rearrangements in clusters.

R. N. Grimes, *Acc. Chem. Res.* 1978, *11,* 420. Structure and bonding in borane and metalloborane clusters.

A. P. Humphries and H. D. Kaesz, *Prog. Inorg. Chem.* 1979, *25,* 145. A discussion of hydride-containing clusters.

L. M. Jackman and F. A. Cotton, Eds., *Dynamic Nuclear Magnetic Resonance Spectroscopy,* Academic Press, New York, 1975. A comprehensive treatment of stereochemical nonrigidity.

B. F. G. Johnson, Ed., *Transition Metal Clusters,* Wiley, New York, 1980. Chapters on cluster compounds by leading investigators in the field.

B. F. G. Johnson and J. Lewis, *Adv. Inorg. Radiochem.* 1981, *24,* 225. An interesting review, of manageable size, of cluster compounds.

[38]M. A. Neuman, Trinh-Toan, and L. F. Dahl, *J. Am. Chem. Soc.* 1972, *94,* 3383; Trinh-Toan, W. P. Fehlhammer, and L. F. Dahl., *J. Am. Chem. Soc.* 1972, *94,* 3389.

E. L. Muetterties, Ed., *Boron Hydride Chemistry*, Academic Press, New York, 1975. Excellent chapters on the chemistry of boranes.

D. Seyferth, *Adv. Organomet. Chem.* 1976, *14*, 98. Chemistry of $RCo_3(CO)_9$ compounds.

PROBLEMS

15.1   Classify the following species as *closo, nido, arachno,* or *hypho.*

C_8H_8 B_9H_{15} B_4H_8

B_6H_{12} B_4H_{10} $B_6H_{11}^+$

15.2   What structure do you expect for each species in Problem 15.1?

15.3   How may $(CH_3)_2B_2H_4$ be prepared? Draw structural formulas for all isomers expected for $(CH_3)_2B_2H_4$. How might one identify these isomers if they were all separated?

15.4   Calculate *styx* numbers and draw valence structures for the following.

B_5H_{11} $B_5H_5^{2-}$

B_6H_{10} B_8H_{12}

Which of the structures is the preferred one for each species?

15.5   Solve the Equations of Balance for $B_3H_6^+$ (not known) and write a "reasonable" structure for such a hydride.

15.6   Show that the number of framework electrons contributed by the groups in Table 15.4 is given by the formula $(v + x - 2)$. v is the number of valence electrons in the cage atom and x is the number contributed by exo ligands or lone pairs. For example, for BH, $v = 3$ and $x = 1$.

15.7   Classify the following species as *closo, nido,* or *arachno.*

$2\text{-}CB_5H_9$ $B_{11}SH_{10}Ph$ (phenyl is attached to B)

$5\text{-}CH_3\text{-}2,3\text{-}C_2B_4H_7$ $B_9H_{12}NH^-$

$1,2\text{-}C_2B_9H_{11}$ $C_2B_6H_{10}$

$B_9H_{11}S$ $B_3H_4(PF_3)_2$

$1,7\text{-}B_{10}CPH_{11}$

Name and sketch the structures of the above species.

15.8   Use the Equations of Balance to obtain a reasonable bonding picture of 1,5-dicarba-*closo*-pentaborane(7).

15.9   Predict the products of the following reactions.

a. $B_5H_{11} + KH$ *e.* $B_6H_{10} + Br_2$ *i.* $2,3\text{-}C_2B_4H_8 + NaH$

b. $B_5H_9 + NMe_3$ *f.* $Li_2o\text{-}C_2B_{10}H_{10} + CH_3I$ *j.* $ZrCl_4 + 4LiBH_4$

c. $B_{10}H_{14} + SMe_2$ *g.* $Li_2o\text{-}C_2B_{10}H_{10} + R_3SiCl$ *k.* $(C_6H_5)_2PCl + LiAlH_4$

d. $B_5H_9 + HCl(l)$ *h.* $B_6H_5^- + Me_2SiCl_2$

15.10   Give a reasonable method for preparing and purifying B_2H_6. How might the purity of the sample be determined? How could one dispose of the diborane?

15.11   *a.* Show that the diamond-square-diamond mechanism (Figure 15.14) cannot account for the known thermal rearrangement of *o*-carborane to *p*-carborane.

         b. What experiments might be helpful in shedding light on the *o*- to *p*-rearrangement?

15.12   Write an essay on the possible relevance of transition cluster compounds to heterogeneous catalysis. (You will want to consult some of the references.)

15.13   Classify the following as *closo, nido,* or *arachno.*

$[Co_4Ni_2(CO)_{14}]^{2-}$ $(Et_3P)_2Pt(H)B_9H_{10}S$

$[Fe(CO)_3]B_4H_8$ $Rh_6(CO)_{16}$

$(CpCo)C_2B_7H_{11}$ $[Rh_9P(CO)_{21}]^{2-}$ (P is at cage center)

                                $Os_5(CO)_{16}$

15.14 Sketch the predicted geometry of the species in Problem 15.13.

15.15 Some known cluster compounds contain *n* vertices and *n* pairs of framework bonding electrons. The structures of these species are often $(n - 1)$-vertex *closo* structures with one of the triangular faces capped. One example of such a structure is $Os_6(CO)_{18}$ (Figure 15.43). Contrary to the prediction by Wade's rules, this compound is a trigonal bipyramid with one triangular face capped by an $Os(CO)_3$ group. Show that this capping arrangement has the effect of contributing two more electrons and no more orbitals to the framework bonding, thus rationalizing the basic TBP geometry.

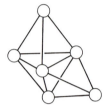

Figure 15.43 The structure of $Os_6(CO)_{18}$.

15.16 Figure 15.44 shows the ^{13}C NMR spectrum of $(cot)Fe(CO)_3$ at several temperatures. The signals at 214 and 212 ppm are attributed to CO, and the others to cot. Explain the appearance of the spectrum at $-134°$ C, and its change with *T*. (See F. A. Cotton, and D. L. Hunter, *J. Am. Chem. Soc.* 1976, *98*, 1413).

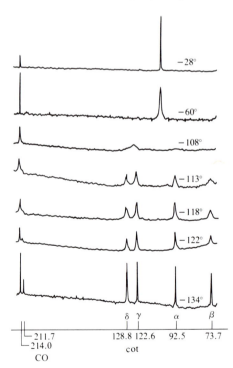

	δ	γ		α	β
		128.8	122.6	92.5	73.7
211.7			cot		
214.0					
CO					

Figure 15.44 ^{13}C NMR spectrum of $Fe(CO)_3$ at several temperatures. (Reproduced with permission from F. A. Cotton, and D. L. Hunter, *J. Am. Chem. Soc.* 1976, *98*, 1413. Copyright 1968, American Chemical Society.)

15.17 *a.* The ^1H NMR spectrum $CpFe(CO)_2(\eta^1\text{-}C_5H_5)$ at several temperatures is reproduced in Figure 15.45*a*. Account for the appearance of the spectrum as the temperature changes.

b. (i) The compound $(C_5H_5)_4Ti$ has been prepared. What does the EAN rule suggest about the attachment of the C_5H_5 rings to Ti?

(ii) From the results in ***a*** and ***b***(i), rationalize the appearance of the ^1H NMR spectrum (Figure 15.45*b*) as the temperature changes.

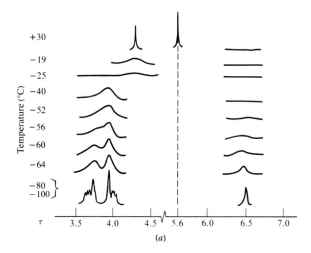

(*a*)

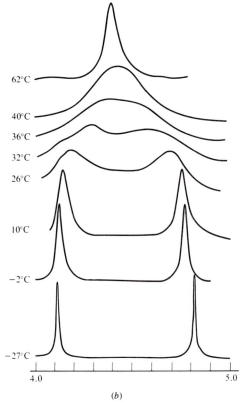

(*b*)

Figure 15.45 *a.* ^1H NMR spectrum of $CpFe(CO)_2(\eta^1\text{-}C_5H_5)$ at several temperatures. (Reproduced with permission from M. J. Bennett, *et. al.*, *J. Am. Chem. Soc.* 1966, *88*, 4371. Copyright 1966, American Chemical Society.) *b.* ^1H NMR spectrum of $(C_5H_5)_4$ Ti at several temperatures. (Reproduced with permission from J. L. Calderon, F. A. Cotton, and J. Takats, *J. Am. Chem. Soc.* 1971, *93*, 3587; J. L. Calderon, *et. al.*, *J. Am. Chem. Soc.* 1971, *93*, 3592. Copyright 1971, American Chemical Society.)

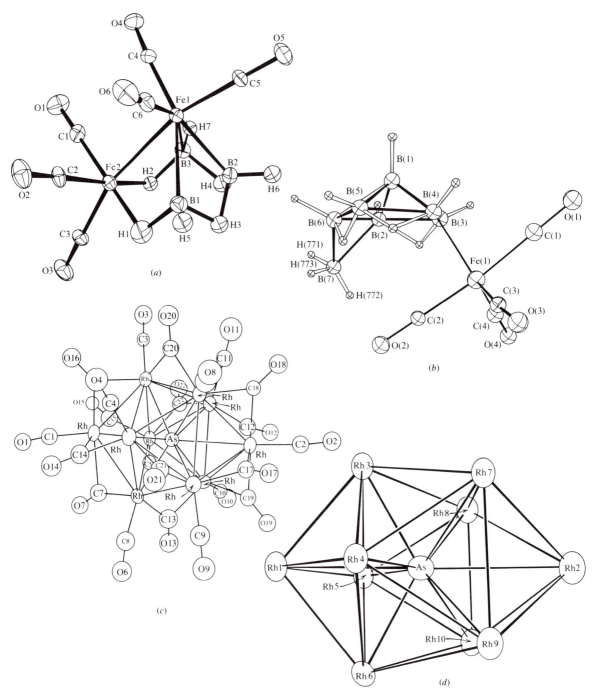

Figure 15.46 *a.* The structure of $Fe_2(CO)_6B_3H_7$. (Reproduced with permission from K. J. Haller, E. L. Andersen, and T. P. Fehlner, *Inorg. Chem.* 1981, *20*, 309. Copyright 1981, American Chemical Society.) *b.* The structure of $Fe(CO)_4B_7H_{12}^-$. (Reproduced with permission from M. M. Mangion, W. R. Clayton, O. Hollander, and S. G. Shore, *Inorg. Chem.* 1977, *16*, 2110. Copyright 1977, American Chemical Society.) *c* and *d.* The structure of $Rh_{10}As(CO)_{22}^{3-}$ with (*c*) and without (*d*) the carbonyl ligands. (Reproduced with permission from J. L. Vidal, *Inorg. Chem.* 1981, *20*, 243. Copyright 1981, American Chemical Society.)

15.18 The compound $Co_3Rh(CO)_{12}$ is a tetrahedral cluster.

    a. At $-85°$ C, its ^{13}C NMR spectrum displays seven signals in intensity ratio 1:2:2:2:3:1:1. The second and the last two signals are coupled to ^{103}Rh (I = 1/2). What is the structure of the species "frozen out" at this temperature?

    b. On warming to $+10°$ C, two signals appear—a single line of relative intensity 2 showing coupling to ^{103}Rh and one of relative intensity 10 which is somewhat broadened. At $+30°$ C, only a single broad resonance is visible. Account for these observations. (See B. F. G. Johnson, J. Lewis, and T. W. Matheson, *Chem. Comm.* 1974, 441.)

15.19 *a.* Give the electron count and sketch the geometry for $1,2\text{-}B_9C_2H_{11}^{2-}$ and $1,7\text{-}B_9C_2H_{11}^{2-}$.

    b. Both of these ligands form a number of compounds with transition metals. Their common names are 1,2- and 1,7-dicarbollide, respectively. Give the electron count and predict the structure for $CpCo(1,2\text{-}C_2B_9H_{11})$.

    c. MO calculations indicate that six electrons are available on the open faces of the dicarbollide ligands for donation to transition metals. This makes them formally equivalent to Cp ligands. Show that the following species obey the EAN for the metal.

        $(\eta^4\text{-}Ph_4C_4)Pd(1,2\text{-}Me_2C_2B_9H_9)$, $[(1,2\text{-}B_9C_2H_{11})Re(CO_3]^-$, $[(1,2\text{-}B_9C_2H_{11})Mo(CO)_3W(CO)_5]^{2-}$

15.20 Using the approach developed by Mingos for cage and ring compounds (Section 15.11), predict plausible structures for the following.

    S_8 P_4

    $P_4(C_6H_{11})_4$ $[Fe(NO)_2]_2(SEt)_2$

15.21 Several recently prepared clusters are depicted in Figures 15.46*a* to *d*. Rationalize these structures according to Wade's rules. With what neutral boranes are $Fe_2(CO)_6B_3H_7$ and $Fe(CO)_4B_7H_{12}^-$ isoelectronic?

XVI

Some Aspects of Bioinorganic Chemistry

We now understand many biological processes at the molecular level. Complete sequences of amino acids, x-ray structures, and total syntheses have been achieved for large complex molecules of biological importance. The more complete our understanding of biological processes, the more we have recognized the key roles of metal ions in the function of enzymes, oxidation–reduction reactions, and many other processes, in addition to their long-known roles in chlorophyll and hemoglobin. Researchers have found an increasing number of elements to be essential for life processes. The hybrid field of bioinorganic chemistry is developing rapidly.

Obviously, life processes involve much more than just the reactions and syntheses of complex organic molecules. In model studies, which have been very important in bioinorganic chemistry, the goal is not to improve on nature, but to focus on a simple system that we can understand and to evaluate alterations of the model.

16.1 ESSENTIAL ELEMENTS

16.1.1 Introduction

The essential biological roles of a few elements have been recognized for many years. Delicate electrolyte balances are achieved by Na^+ and Ca^{2+} as the major extracellular cations, by K^+ and Mg^{2+} as the major cellular cations, and by Cl^- as the major anion commonly found within and without the cell. The importance of calcium in bones was recognized before the development of biochemistry. Quite early, iodine deficiences were known to cause abnormal functioning of the thyroid gland. Iodine is an esential constituent of the thyroid hormones—for example, thyroxine.

Thyroxine

The insecticide DDT [1,1,1-trichloro-2,2-bis(*p*-chlorophenyl)ethane, $HC(CCl_3)(\phi Cl)_2$] has been banned in the United States because of harmful effects on birds and other animals. These effects stem from biological amplification in the food chains, caused when ultraviolet light— sunlight, for example—converts DDT in the vapor phase into polychlorinated biphenyls that are known to be toxic. The polychlorinated biphenyls are structurally similar to thyroxine, and their toxicity might result from interference with the functions of thyroid hormones.

The 27 elements now recognized as essential for life[1] (Table 16.1) include most of the elements up to atomic number 34, except for Li, Be, Al, Sc, Ti, Ga, Ge, and the noble gases. Selenium, molybdenum, and iodine also are essential. In testing to determine if an element is essential at low concentrations, it is difficult to exclude an abundant element such as aluminum so completely that symptoms will appear as a result of the deficiency. Strikingly, all of the

Table 16.1 Chemical selectivity in the human body[a] (percent of total number of atoms)

Composition of Earth's Crust		Composition of Seawater		Composition of Human Body[b]	
O	46.6%	H	66%	H	63%
Si	27.1	O	33	O	25.5
Al	8.1	Cl	0.33	C	9.5
Fe	5.0	Na	0.28	N	1.4
Ca	3.6	Mg	0.033	Ca	0.31
Na	2.8	S	0.017	P	0.22
K	2.6	Ca	0.006	Cl	0.03
Mg	2.1	K	0.006	K	0.06
Ti	0.46	C	0.0014	S	0.05
H	0.22	Br	0.0005	Na	0.03
C	0.19			Mg	0.01
All others < 0.1		All others < 0.1		All others < 0.1	

[a]E. Frieden, "The Chemical Elements of Life," *Sci. Am.*, July 1972, 52.
[b]Other elements, commonly referred to as trace elements, shown to be essential to some form of life include B, F, Si, V, Cr, Mn, Fe, Co, Ni, Cu, Zn, As, Se, Mo, and I. See H. J. Saunders, "Nutrition and Health," *Chem. and Eng. News* 1979, p. 27.

[1]H. J. Bowen, *Trace Elements in Biochemistry*, Academic Press, New York, 1966; E. W Ainscough, "The Role of Metal Ions in Proteins and other Biological Molecules," *J. Chem. Educ.* 1976, *53*, 156.

elements appreciably abundant in the human body, except for phosphorus, also are abundant in seawater—suggesting that our family tree is rooted in the sea. The fact that F, Si, Sn, V, Ni, B, and As are essential for some forms of life has been established only recently. It is not at all unusual that elements essential at low concentrations, such as F, Se, As, and even Fe, are toxic at higher concentrations.[2] More of the "toxic" elements probably will be found to be essential at low concentrations. Even NaCl is toxic at high concentrations, because it upsets the electrolyte balance.

16.1.2   Cobalt Deficiency

Sheep raised in certain regions of Australia were found to suffer from an illness that was traced to cobalt deficiency resulting from cobalt-deficient soil. Adding cobalt salts to the soil remedies the problem, but it is expensive and must be repeated periodically. A more economical remedy is to force each sheep to swallow a cobalt pellet and a small screw.[3] The pellet and screw stay in the rumen, adding sufficient cobalt salts to the diet while the mechanical action between them removes any coatings that might form. The pellet can be recovered when the animal dies, and used again. But if a little cobalt is good, more is not better. A higher rate of congestive heart failure among heavy beer drinkers was linked to the addition of small amounts of cobalt salts (1.2–1.5 ppm) to beer, to improve the foaming properties. It seems that the deleterious effect occurred only if there were a high alcohol level and a dietary deficiency of protein or thiamine, as often occurs among heavy drinkers.

16.1.3   Availability of Elements

An organism must adapt to its environment, using the "raw materials" available to it and coping with unwanted or even toxic substances. For metal ions, control is exercised through the formation of metal complexes. In basic soils, iron, as $Fe(OH)_3$, is quite insoluble. Plants can synthesize chelating agents to form soluble iron complexes (see Section 16.5). Other complexes can be important for the transport of iron through cell membranes. Where the soil is deficient in iron, a synthetic soluble iron complex of edta (ethylenediaminetetraacetic acid) is added. If the concentration of a metal ion is too high, the ions can be tied up as metal complexes or "sequestered." The treatment for toxic metals, such as lead, is to inject a chelating agent (edta) to form a soluble complex that can be excreted.[4] The alkali metal salts are very soluble in water, but solubility in a nonpolar medium can be important in cell transport. So nature synthesizes chelating agents that contain oxygens as ligand atoms, and a hydrocarbon exterior of the complex formed to impart solubility in nonpolar solvents. These complexes resemble the crown ether complexes (see Section 14.2.4).

Although phosphorus is not abundant in the earth's crust or in seawater (Table 16.1), it serves important functions in plant and animal life. A few years ago, controversy embroiled detergents containing phosphates, which helped pollute lakes and streams by fostering the runaway growth of algae, often followed by its decay. In the normal balance of nutrients, P

---

[2]D. E. Carter and Q. Fernando, "Chemical Toxicology. Part II: Metal Toxicity," *J. Chem. Educ.* 1979, *56*, 490.

[3]J. C. Bailar, Jr., "Some Coordination Compounds in Biochemistry," *Am. Scientist* 1971, *59*, 586.

[4]M. M. Jones and T. H. Pratt, "Therapeutic Chelating Agents," *J. Chem. Educ.*, 1976, *53*, 342.

is likely to be the limiting nutrient. The mean life of inorganic phosphate added to lake water is only a few minutes—it is used rapidly. The phrase "eutrophication of lakes" has been used to describe this situation, which results in excessive growth. (Actually, *eutrophication* means the process of becoming well fed.[5]) The phosphate detergents thus upset the natural balance.

The primary means for energy storage and release in cells involves the formation and hydrolysis of polyphosphates (see Figure 16.14). The low abundance of phosphorus makes its compounds good candidates for energy carriers: the molecules storing energy are less likely to get lost in a multitude. The *nucleic acids* DNA (deoxyribonucleic acid) and RNA (ribonucleic acid) consist of long chains of cyclic five-carbon sugars joined by phosphate linkages. Attached to the sugars along the chain are heterocyclic nitrogenous bases, the sequence of which constitutes the genetic code. The intertwined strands of DNA forming the double helix permits the replication of the code.

16.1.4   Heme and Chlorophyll

Most early research into the role of metal ions in biological processes focused on magnesium in chlorophyll and iron in hemoglobin, two striking examples of the evolution of complex molecules for highly specialized roles. The heme of hemoglobin is the iron complex of a substituted porphine (a porphyrin). The substituents differ for chlorophyll and the C_7—C_8 double bond of porphine is reduced. Chlorophyll *a*, shown in Figure 16.1, is the most abundant of the group of closely related pigments. The chlorophylls absorb light in the red region, making the energy available for photosynthesis (see Section 16.7). The appearance of hemoglobin in the evolutionary process might have depended on the availability of chlorophyll, or

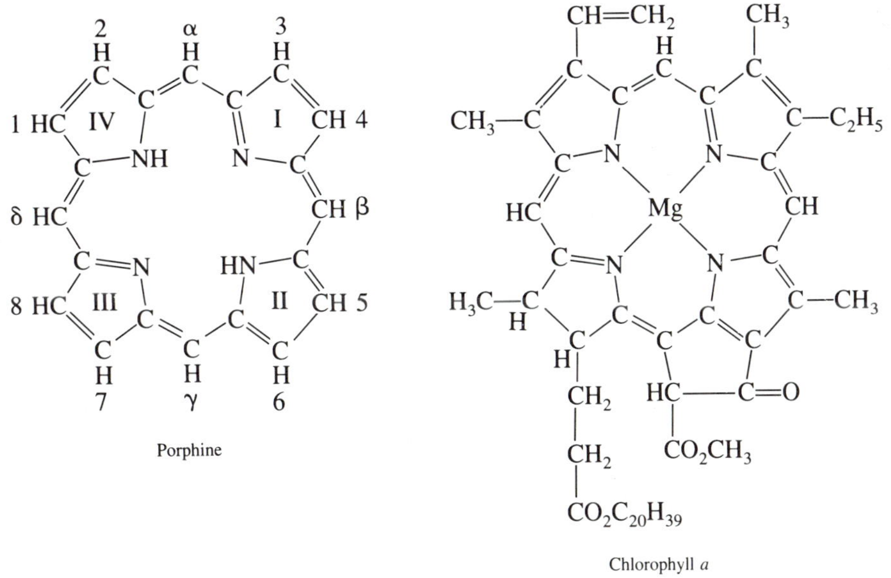

Figure 16.1 Porphine and Chlorophyll *a*.

[5]G. E. Hutchinson, "Eutrophication," *Am. Scientist*, 1973, *61*, 267.

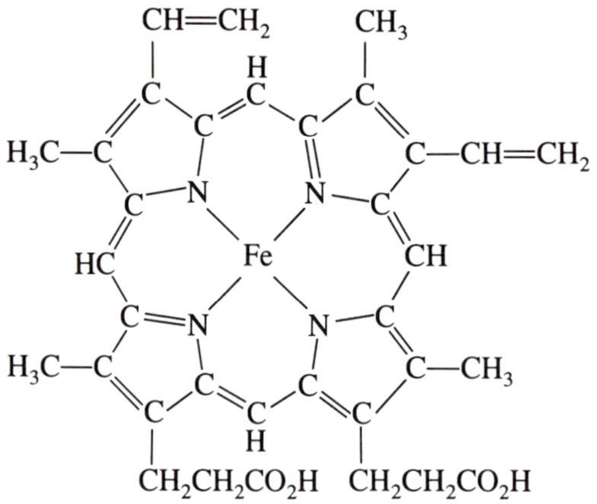

Figure 16.2 Heme *b* (protoheme).

at least on the processes for its synthesis. Hemoglobin contains an iron porphyrin with a large protein molecule, globin, coordinated to the iron on one side of the plane. The sixth coordination site is available for coordination of an O_2 molecule. One of the well-characterized hemes, heme *b* (shown in Figure 16.2), is isolated from hemoglobin of beef blood, or from other sources. All higher animals use iron porphyrins for oxygen transport (hemoglobins) and storage (myoglobins).

16.2 BLOOD

16.2.1 Oxygen Transport and Storage

The best known function of iron in biological systems is as an oxygen carrier in hemoglobin. The molecular weight of hemoglobin is about 64,500. There are four subunits, each of which contains one heme group, an iron(II) complex of protoporphyrin IX (protoheme—Figure 16.2), associated with the protein globin. In the most common form of hemoglobin, two of the subunit proteins form *alpha* chains of 141 amino acids and two form *beta* chains of 146 amino acids. The chains are coiled so that a histidine side chain is coordinated to the iron on one side of the porphyrin ring. The sixth site is occupied by O_2 in oxyhemoglobin (Figure 16.3); in deoxyhemoglobin, it is vacant or a water molecule is weakly bonded.

The iron(II) is high-spin in deoxyhemoglobin and has a larger radius than low-spin iron(II). High-spin iron(II) has the larger radius, because one electron occupies the $d_{x^2 - y^2}$ orbital directed toward the N ligand atoms of the porphyrin. Since high-spin iron(II) is too large for the "hole" of the porphyrin, the Fe is above the plane of the nitrogens by about 70 pm, giving a square pyramidal arrangement with the pyrrole rings tipped out-of-plane. When O_2 is coordinated, the ligand field is strong enough to cause spin-pairing, giving a low-spin d^6 (t_{2g}^6) complex. Release of the strain energy of the pyramidal complex helps balance the loss in spin-exchange energy (pairing energy) on going to the oxygenated hemoglobin, in much the same way as an overhead garage door spring and gravity counterbalance each other so that little energy is expended in changing states. If too stable an oxygen complex were formed, too much

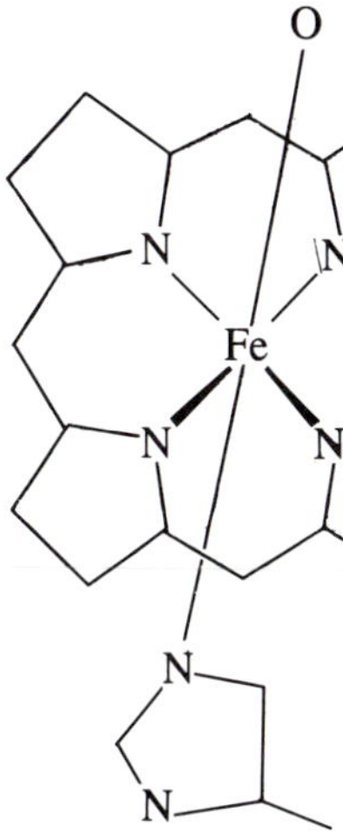

Figure 16.3 The coordination of Fe(II) in oxyhemoglobin.

energy would be released in the lungs and less energy would be available when oxygen is released for use in the muscles. In the planar conjugated porphyrin ring, stable π and low-lying π^* orbitals are responsible for the characteristic charge-transfer electronic absorption spectrum of red blood. As the heme adds oxygen, the iron slips into the hole of the planar porphyrin ring, shifting with it the coordinated side chain and causing important conformational changes.

It is remarkable that O_2 does not oxidize hemoglobin, considering the electrode potentials for the reduction of O_2 and oxidation of Fe^{2+}(aq). Fortunately, the reversible adduct is stabilized by the unique bonding features of the porphyrin ring system and the hydrophobic blocking by the large protein (globin). As we will see from model studies, an essential feature for reversible oxygen addition in iron porphyrins is the blocking of the irreversible formation of an oxidized dimer.

Cooperative binding of O_2 by the subunits of hemoglobin permits each successive O_2 molecule to be bound more strongly than the one before. In solution the tetrameric hemoglobin molecules $(\alpha_2\beta_2)$ are in equilibrium with dimers $(\alpha\beta)$. It has been proposed that upon addition of O_2 movement of the histidine in one chain of a dimer changes the conformation of the other chain, increasing its affinity for O_2 and affecting the dimer-tetramer equilibrium. Oxygenation of the second chain of the tetramer causes conformational changes in the other two chains, promoting addition of O_2. Although the four heme sites are well separated, movement of one chain affects the conformations of the others significantly and changes the hydrophilic and hydrophobic surface area. This plausible explanation accounts for the cooperative binding of O_2 and for the differences in the oxygen saturation curves (Figure 16.4) of hemoglobin and myoglobin (see below), for which there is no cooperative effect. However, the effect is by no means completely understood, and we can anticipate further interesting developments. The affinity of hemoglobin for O_2 decreases with decreasing pH (Bohr effect), but blood is well buffered, so the pH decreases only slightly with the accumulation of CO_2 in muscles (venous blood—see Figure 16.4).

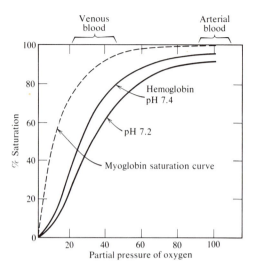

Figure 16.4 The oxygen saturation curves for myoglobin and hemoglobin. In muscles (venous blood, pH = 7.2), the partial pressure of oxygen varies considerably at work or at rest; but it varies little for arterial blood (pH = 7.4).

The skeletal carbons in the peptide linkages, $\overset{O}{\underset{C}{\overset{\|}{C}}-\overset{C}{\underset{H}{N}}$, of globin are *trans* with planar geometry about N, because of the high degree of double-bond character of the $\overset{O}{\underset{}{\overset{\backslash\backslash}{C}}}$═N bonds. Consideration of the optimum hydrogen bonding within the chain led Pauling to postulate the α-helix structure with 3.7 peptide units per turn (the secondary structure of proteins). This unusual screw axis provided a key element in Watson and Crick's unlocking of the DNA structure.

There has been controversy over whether the O_2 molecule is attached to Fe in a linear Fe—O—O group, an angular Fe—O$\overset{\diagup O}{}$, or a symmetrical group with sidewise interaction, Fe$\overset{O}{\underset{O}{\big<}}$. Recent studies indicate end-on angular binding for model compounds and for oxyhemoglobin.[6]

After being separated from air and transported from the lungs to the muscles by hemoglobin, O_2 is transferred to myoglobin for storage until needed for energetic processes. Myoglobin (Figure 16.5), although similar to one of the subunits of hemoglobin, binds O_2 more strongly than does hemoglobin, particularly at the low concentrations of O_2 and high concentrations of CO_2 (low pH) that exist in active muscles (Figure 16.4). After the first O_2 molecule is transferred by hemoglobin, the others are released even more easily, because of the coop-

[6]L. L. Duff, E. H. Appelman, D. F. Shriver, and I. M. Klotz, *Biochem. Biophys. Research Comm.* 1979, *90*, 1098.

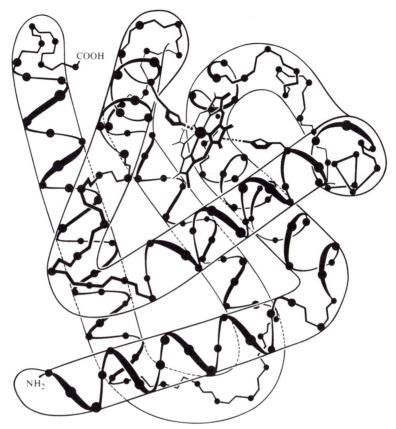

Figure 16.5 Structure of myoglobin. (From R. E. Dickerson in *The Proteins*, Vol. 2, H. Neurath, Ed., Academic Press, New York, p. 634.)

erative effect (see above) in reverse. These effects result in efficient transfer of O_2 to myoglobin. Terminal amine groups bind CO_2 produced in muscles for transport to the lungs.

16.2.2 Hemerythrin

Hemerythrin,[7] a nonheme iron protein, is the oxygen carrier for some marine worms. The typical and widely studied hemerythrin from the worm *Goldfingia* has a molecular weight of 108,000, with eight subunits. The myohemerythrin in the muscles of the worm *Thermiste pyroides* consists of just one of these subunits—as is the case for myoglobin in the heme series. Complete amino-acid sequences have been established for some of the hemerythrins. The long chain protein has high helix content, with two coordinated high-spin Fe(II) close together (344 pm) in deoxyhemerythrin. A coordinated water molecule in deoxyhemerythrin is replaced by

[7]D. M. Kurtz, Jr., D. F. Shriver, and I. M. Klotz, "Structural Chemistry of Hemerythrin," *Coord. Chem. Rev.* 1977, *24*, 145; R. E. Stenkamp and L. H. Jensen, "Hemerythrin and Myohemerythrin," *Adv. Inorg. Biochem.* 1979, *1*, 219; J. S. Loehr and T. M. Loehr, "Hemerythrin," *Adv. Inorg. Biochem.* 1979, *1*, 235.

O_2, presumably bridging the two irons in oxyhemerythrin. The irons in oxyhemerythrin are antiferromagnetically-coupled Fe(III). Oxygen is bound as O_2^{2-}.

It is interesting that the blue blood of crabs, lobsters, snails, scorpions, and octopuses contains oxygenated hemocyanins, copper-containing proteins that bind one O_2 molecule for two Cu atoms. The name means "blue blood"; hemocyanin contains no heme. It appears that deoxygenated hemocyanins contain copper(I). The blood cells of sea squirts contain a vanadium-containing protein, hemovanadin, that probably is an oxidation-reduction enzyme but might serve also as an O_2 carrier.

16.2.3   Models for Oxygen Carriers

Model Compounds

The question of central interest about hemoglobin is how it can bind O_2 reversibly. Unfortunately, the complexity of the hemoglobin molecule makes identification of the relevant factors very difficult—a not-uncommon situation for biological (and even nonbiological) systems. For example, even isolating hemoglobin from natural sources and purifying it is a tremendously difficult process. Preparing solutions of even moderate molarity for substances with very high molecular weight is expensive, and the high molecular weights often limit solubilities. (*Exercise:* What is the maximum molarity of a substance with mol. wt. = 100,000 and density = 1 g/mL? *Answer:* 0.01 M neat–no solvent!) The tremendous number of atoms makes it impossible to determine by x-ray methods the mode of coordination—whether unidentate

bent Fe $\overset{\displaystyle O—O}{\diagup}$ or bidentate $Fe\underset{\diagdown O}{\overset{\diagup O}{|}}$ —in oxyhemoglobin.

Inorganic chemists have learned a good deal about the mode of functioning of hemoglobin by studying model compounds. These are simpler Fe-containing species that function like the hemoglobin prototype. Factors relevent to the biological functioning of hemoglobin can be understood by studying changes in chemical behavior of model compounds with planned structural differences.

The fact that free heme is oxidized irreversibly by O_2 in aqueous solution to hematin— the Fe(III) form of heme, which does not bind oxygen—suggests that the globin part of the molecule somehow provides a structural feature required for reversibility. X-ray structure of hemoglobin reveals that the heme portion of the molecule is buried in the globin in such a way as to provide a hydrophobic "pocket" in the region of the Fe coordination site available for O_2 bonding. Studies on model compounds helped elucidate the role of this hydrophobic region. Possibilities include the importance of the low dielectric constant in the region in retarding possible oxidation processes, such as

$$Fe(II)heme + O_2 \rightarrow Fe(III)heme + O_2^-$$

or

$$O_2—Fe(II)heme \rightarrow Fe(III)heme + O_2^-$$

(16.1)

the low acidity of the medium, which would retard

$$\text{Fe(II)heme} + O_2 \xrightarrow{\text{HA}} \text{Fe(III)heme} + HO_2$$

or (16.2)

$$O_2\text{—Fe(II)heme} \xrightarrow{\text{HA}} \text{Fe(III)heme} + HO_2$$

and the steric hindrance provided by the protein, which would retard oxidation with dimerization,

$$O_2\text{—Fe(II)heme} \rightarrow \text{Fe(III)—O—Fe(III)} + \ldots \qquad (16.3)$$

a process known for many simple Fe(II) complexes.

In general, a good model for the active site should reproduce biological activity and approach or reproduce the bio unit in terms of composition, types of ligand, oxidation states, and physiochemical properties. The study of such models, by analogy, provides information on the role of proteins and the structure of the inside of the bio molecule. With hemoglobin, we are interested in clearing away much of the overwhelming mass of the molecule, in order to focus our attention on the active sites where the action is.

Iron Porphyrins

Early studies of hemoglobin models encountered problems with irreversible oxidation, but recently,[8] success has been achieved with synthetic iron(II) oxygen carriers. An iron porphyrin with a coordinated imidazole side chain adds O_2 reversibly at $-45°C$. The iron(II) complex of $\alpha,\beta,\gamma,\delta$-tetraphenylporphine with two pyridine ligands coordinated adds O_2 reversibly at $-79°C$ in methylene chloride solution. One pyridine is displaced by O_2. At higher temperature, these and other simple models, lacking the protective protein, form dimers, leading to the irreversible oxidation of Fe(II).

Because model studies have shown that irreversible oxidation can occur through dimerization, two approaches to obtaining oxygen carriers by preventing the formation of dimers have been devised. One involves attaching the iron porphyrin to a surface so that the units cannot come together to form dimers. The other approach is illustrated by the "picket fence" porphyrins, in which substituents on tetraphenylporphine form a "fence" around the O_2 binding site (Figure 16.6). The O_2 is sigma bonded unsymmetrically, $\text{Fe—O}^{\diagup O}$. Cobalt(II) derivatives of the "picket fence" porphyrins add oxygen reversibly at 25°C. Without globin or the "picket fence" protection, the cobalt(II) porphyrins have little tendency to bind O_2, except at very low temperature.

Baldwin achieved reversible O_2 addition with a macrocyclic Fe(II) complex involving bulky ring substituents. He also prepared a porphyrin with a bridge involving a benzene ring as a "canopy" over the center of the porphyrin ring.[9] With another ligand such as l-methyl-imidazole coordinated below the porphyrin ring, O_2 could be added reversibly "beneath the

[8]F. Basolo, B. M. Hoffman, and J. A. Ibers, "Synthetic Oxygen Carriers of Biological Interest," *Acc. Chem. Res.* 1975, *8*, 384.

[9]J. E. Baldwin *et al., J. Am. Chem. Soc.* 1975, *97*, 226, 227.

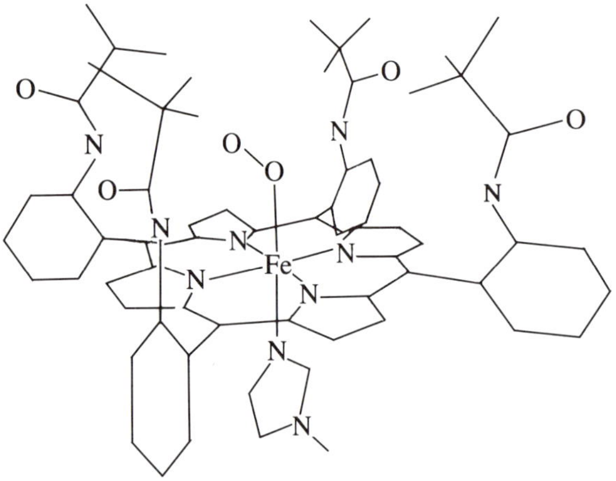

Figure 16.6 Dioxygen bonding to a "picket fence" porphyrin. (Adapted with permission from J. P. Collman *et al., J. Am. Chem. Soc.* 1975, *97*, 1427. Copyright 1975, American Chemical Society.)

canopy.'' After several hours, irreversible oxidation of the oxygenated capped porphyrin formed the oxo-bridged dimer.

Other Synthetic Oxygen Carriers

Metals usually react irreversibly with O_2 to give metal oxides or other oxocompounds. Knowing which metals, in what oxidation states, and combined with what ligands can add O_2 reversibly is important not only for the production of "synthetic blood," but also for understanding the function of hemoglobin. The controlled activation of O_2 as achieved so delicately in biological systems is an important practical goal. During World War II, a considerable effort was made to try to synthesize oxygen carriers for storage of O_2, since shipment of O_2 at high pressure in heavy metal cylinders is expensive and hazardous.

The most successful models for oxygen carriers have been those of cobalt(II), and iron(II). Vaska, however, discovered that the complex $Ir(CO)Cl[P(C_6H_5)_3]_2$ can add O_2 (Figure 16.7). The complex formed involves symmetrical $M\begin{smallmatrix}O\\|\\O\end{smallmatrix}$ bonding. The coordination number is five if we consider the O_2 as a single ligand, perhaps π-bonded; it is six if the oxygen is regarded as a bidentate ligand. The diamagnetic complex yields H_2O_2 on treatment with acid. Model compounds that more closely resemble natural oxygen carriers have been studied more extensively.

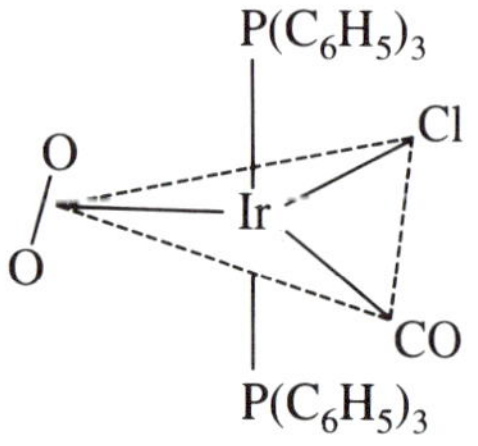

Figure 16.7 Vaska's oxygenated iridium complex.

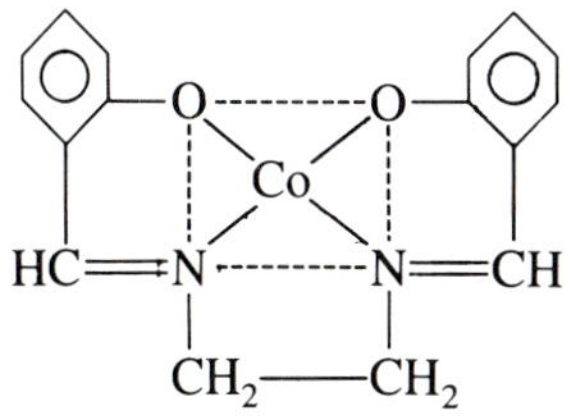

Figure 16.8 *N,N'*-ethylenebis(salicylaldiminato)cobalt(II) [Co(salen)].

Until recently, the greatest success had been achieved in the study of cobalt(II) complexes.[10] A ligand containing $-NH_2$ groups, such as ethylenediamine, can be condensed easily with salicylaldehyde through the formation of Schiff-base linkages (C=N). The resulting ligand (salen) gives a square planar complex with cobalt(II). The complex [Co(salen)] (Figure 16.8) and some related complexes react with O_2 in solution containing a base such as pyridine to give a diamagnetic species involving an O_2^{2-} bridge, [(py)(salen)Co^{III}—O_2—Co^{III}(salen)(py)]. The bis(3-methoxy) derivative of the [Co(salen)] complex gives a monomeric complex with O_2 (1:1). The bis(dimethylglyoximato)cobalt(II) complex with a base such as pyridine coordinated (see Figure 16.12) can add O_2. Two molecules of the complex are bridged by O_2. The cobalt complex of the quadridentate macrocycle cyclam with an anion (X = Cl$^-$, NO$_2^-$, or NCS$^-$) coordinated also gives a bridged (2:1) complex with O_2 (Figure 16.9).

The cobalt(II) complex formed by *N,N'*-ethylenebis(acetylacetonimine) (the Schiff base from acetylacetone and ethylenediamine) adds O_2 reversibly below 0°C to give a 1:1 complex. Oxidation of the ligand occurs at higher temperature. An x-ray crystallographic study reveals that the Co(III)–O_2^- bonding is bent, end-on, $Co-O_{125°}^{O}$. The five-coordinate complex formed by the ligand derived from $NH_2C_3H_6NHC_3H_6NH_2$ and salicylaldehyde adds O_2 to give a 1:1 complex (Figure 16.10). The similar cationic five-coordinate complex of cobalt(II) with $(NH_2C_2H_4NHC_2H_4)_2NH$ gives a dimeric (2:1) complex with O_2. The 1:1 complex that would be formed initially would contain Co(III) and superoxide ion. The superoxide ion easily could oxidize another Co(II) coordinated to a strong field ligand, to produce the dimer.

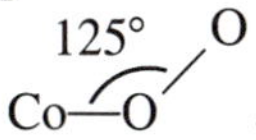

Synthetic *coboglobins,* prepared from cobalt(II) protoporphyrin IX (Figure 16.2) and globin from hemoglobin or myoglobin, function as oxygen carriers. The coordination of the Co(II) and the orientation of the porphyrin in deoxy cobalt hemoglobin or myoglobin is the same as in hemoglobin or myoglobin, respectively.

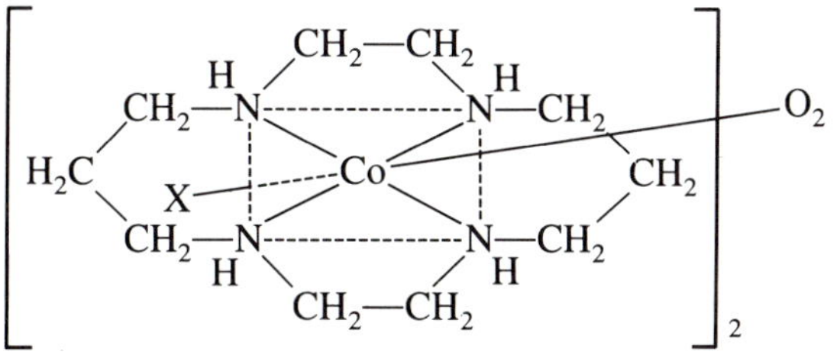

Figure 16.9 The oxygen complex of [Co(cyclam)X]$^+$

[10]R. G. Wilkins, "Uptake of Oxygen by Cobalt(II) Complexes in Solution," in *Bioinorganic Chemistry*, R. Dessy, J. Dillard, and L. Taylor, Eds., *Adv. Chem. Ser.* 1971, *100*, 111.

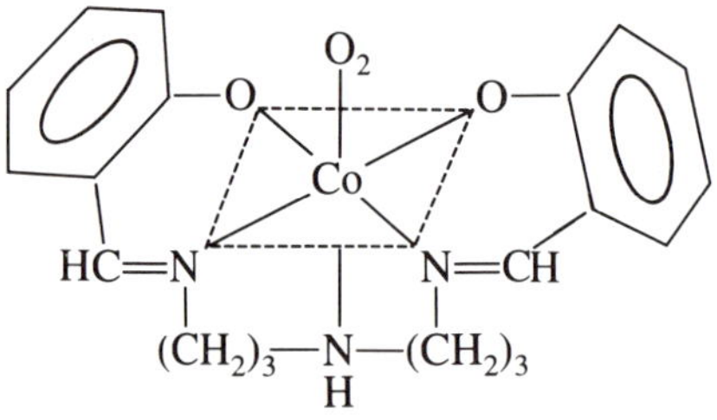

Figure 16.10 The 1:1 salicylaldiminecobalt(II)-dioxygen complex.

16.2.4 Supply and Storage of Iron in Animals

Animals absorb ingested iron in the gastrointestinal tract as Fe(II), but unlike bacteria, they do not require significant new supplies of iron continually. Iron is recycled; little is absorbed from the diet or excreted. In humans, about 70% of Fe (4g total) is present in hemoglobin (0.8 kg for an average person) and myoglobin. Most of the rest of the iron is stored as *ferritin*.[11] Red-brown, water-soluble ferritin consists of a shell of protein (apoferritin, an apoprotein—the protein with the prosthetic group removed—see p. 736) surrounding a micelle (an aggregate whose surface bears a charge) of iron(III) hydroxide oxide phosphate. The micelle contains about 2000–4500 iron atoms, or 12 to 20% iron. The apoferritin consists of about 24 subunits, each of molecular weight about 18,500.

The micelle for horse ferritin is approximately spherical, with an outer diameter of 12,200 pm (122 Å) and an inner diameter of 7300 pm (73Å). Presumably the core is primarily iron(III) oxide, so we expect a close-packed oxide matrix with Fe^{3+} in octahedral holes. The hydroxide and phosphate groups probably serve only to balance charge and complete bonding at the surface. This view is supported by studies showing that careful neutralization of iron(III) nitrate solution yields a polymer with the formula $[Fe_4O_3(OH)_4(NO_3)_2 \cdot 3/2H_2O]_n$. The micelles obtained have the same size as and properties similar to the ferritin core. This polymer and apoferritin yield a synthetic ferritin.

Ferritin stores iron in spleen, liver, and bone marrow. Ingestion of iron stimulates the synthesis of apoferritin. Iron salts are toxic at moderate concentrations. How ferritin releases iron is not known, but it can be achieved by reduction to Fe(II) and/or by chelating agents. The iron is released to an iron-binding protein to form *transferrin*.[12] Serum transferrin is a single-chain polypeptide with a molecular weight of 76,000–80,000. There are two Fe(III) binding sites involving two or three tyrosyl residues (tyrosine is

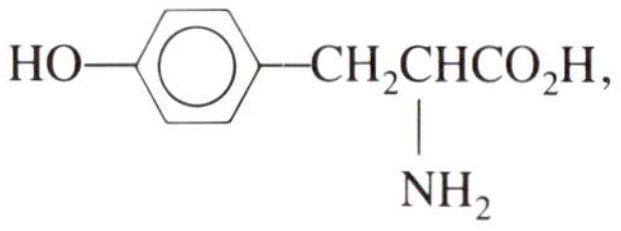

$$HO-\bigcirc-CH_2CHCO_2H,$$
$$NH_2$$

providing a phenolic oxygen for coordination) and one or two nitrogen atoms, perhaps from imidazole rings of histidyl residues. The iron is bound only if a suitable anion also is bound. Several anions can promote the binding of iron, but carbonate (or HCO_3^-) seems to be the

[11] R. R. Crichton, "Structure and Function of Ferritin," *Angew. Chem. Internat. Edit.* 1973, *12*, 57.

[12] P. Aisen and A. Leibman, "Transport by Transferrin," in *Bioinorganic Chemistry II*, K. N. Raymond, Ed., *Adv. Chem. Ser.* 1977, *162*, 104.

physiologically active one. Under physiological conditions the Fe(III) is very strongly bound (binding constant is $\sim 5 \times 10^{23}$ M^{-1}), leaving less than one free Fe(III) per liter of blood plasma. Carbonate ion could be displaced easily by acid, with the release of Fe(III), which has little affinity for the protein without the bound anion. The iron is transported to the bone marrow by transferrin, which can specifically recognize reticulocytes (immature red blood cells). Since the iron (*ca.* 30 mg) used each day in building red cells in an adult is about 10 times the amount of iron bound in transferrin, many cycles are required during the lifetime of a protein molecule (half-life *ca.* 7–8 days). Individuals with atransferrinemia, the genetic inability to synthesize transferrin, suffer from iron-deficiency anemia *and* an overload of iron.

16.3 VITAMIN B$_{12}$[13]

16.3.1 Structure

It has long been recognized that cobalt is one of the essential trace metals, but the discovery that vitamin B$_{12}$ is a cobalt complex came as a great surprise. The vitamin was isolated from liver after it was found that eating raw liver would alleviate pernicious anemia. Vitamin B$_{12}$ is now used for treatment (monthly 1 mg injections).

Figure 16.11 depicts the structure of vitamin B$_{12}$. The cobalt is coordinated to a corrin ring. This macrocycle is similar to the porphyrin ring, except that rings *A* and *D* are joined directly in the corrin ring and there is less conjugation. On one side of the corrin ring the ligand is α-5,6-dimethylbenzimidazole nucleotide, which also is joined to the corrin ring. The sixth ligand is CN$^-$, but actually the CN$^-$ is introduced in the isolation procedure. The active form of the vitamin, called coenzyme B$_{12}$, contains an adenosine group (Figure 16.11) as the sixth ligand. Solving the structure of B$_{12}$ coenzyme required both chemical information and x-ray crystallography. Dorothy Crowfoot Hodgkin received the Nobel Prize in Chemistry in 1964 for her crystallographic work. The coenzyme has been synthesized.

Vitamin B$_{12}$ without the CN$^-$ is called cobalamin, so B$_{12}$ is cyanocobalamin, the form used for treating pernicious anemia. Other ligands can substitute for CN$^-$. It was considered very unusual that the coenzyme contains a Co—C sigma bond, but –CH$_3$ or other organic groups can be so bonded. Studies of B$_{12}$ derivatives have demonstrated that transition metal–carbon sigma bonding is not so rare. Methylcobalamin reacts with Hg^{2+} to give methylmercury, whose accumulation in biological systems poses a problem. The vitamin and coenzyme are treated as formally cobalt(III) compounds. Hydroxocobalamin (and other derivatives) can be reduced (H$_2$/Pt) to Co(II) cobalamins, designated B$_{12r}$, and by stronger reduction (NaBH$_4$) to Co(I) cobalamins, designated B$_{12s}$.

16.3.2 Reactions

Coenzyme B$_{12}$ serves as a *prosthetic group* (see Section 16.8.1) to various enzymes in catalyzing reactions. In the case of enzymes the prosthetic group (see Section 16.4.1) is a coenzyme

[13]R. H. Abeles and D. Dolphin, ''The Vitamin B$_{12}$ Coenzyme,'' *Acc. Chem. Res.* 1976, *9*, 114; A. Eschenmoser and C. E. Wintner, ''Natural Product Synthesis and Vitamin B$_{12}$,'' *Science* 1977, *196*, 1410; J. M. Pratt, *Inorganic Chemistry of Vitamin B$_{12}$*, Academic Press, London, 1972; J. M. Wood and D. G. Brown, ''Vitamin B$_{12}$ Enzymes,'' *Str. Bonding* 1972, *11*, 47.

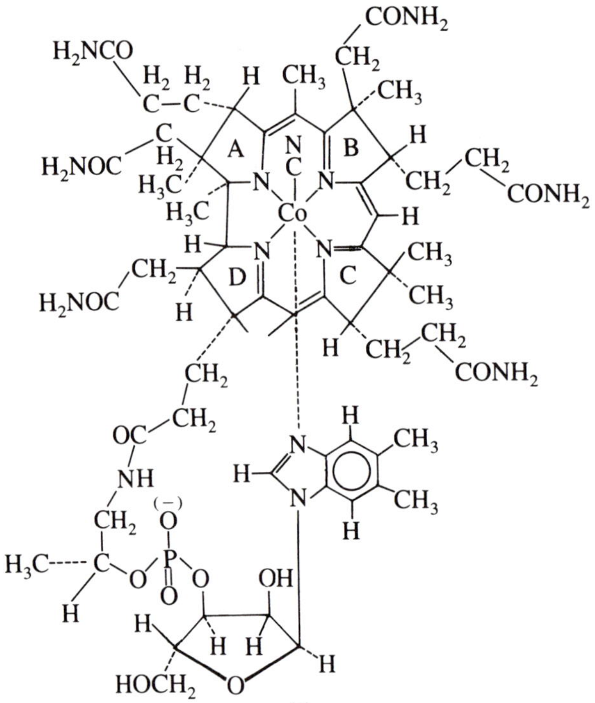

Figure 16.11 (a) Vitamin B_{12}. (b) The group present in place of CN^- in coenzyme B_{12}.

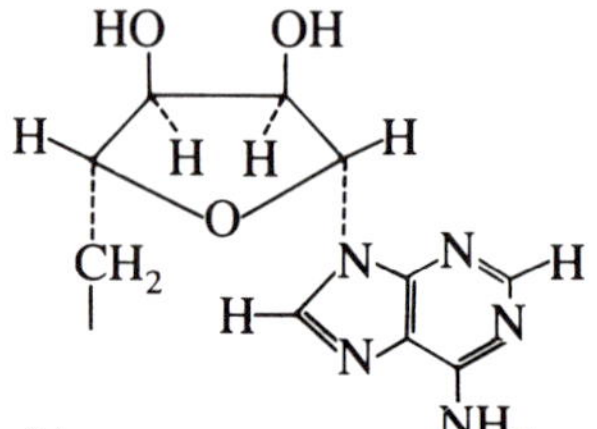

tightly bound to the enzyme. Many remarkable reactions have been discovered.[14] We would expect a variety of reactions from the three ways the Co—C bond of alkylcobalamines cleaves.

$$\text{Co—R} \begin{cases} \text{Co(I)} + \text{R}^+ \\ \text{Co(II)} + \text{:R} \\ \text{Co(III)} + \text{:R}^- \end{cases} \qquad (16.4)$$

Among the very impressive reactions of the enzyme are:

One-Carbon Transfer. The introduction or transfer of a one carbon unit can be achieved.

$$\underset{\substack{| \\ \text{NH}_2}}{\text{HO}_2\text{CCHCH}_2\text{CH}_2\text{SH}} \xrightarrow[\text{+ enzyme}]{\text{coenzyme B}_{12}} \underset{\substack{| \\ \text{NH}_2}}{\text{HO}_2\text{CCHCH}_2\text{CH}_2\text{SCH}_3} \qquad (16.5)$$

$$\underset{\text{homocysteine}}{} \qquad \qquad \underset{\text{methionine}}{}$$

$$\text{H}_2\text{N—CH}_2\text{—CO}_2\text{H} \xrightarrow[\text{+ enzyme}]{\text{coenzyme B}_{12}} \underset{\substack{| \\ \text{CH}_2\text{OH}}}{\text{H}_2\text{N—CH—CO}_2\text{H}} \qquad (16.6)$$

$$\underset{\text{glycine}}{} \qquad \qquad \underset{\text{serine}}{}$$

[14]R. H. Abeles, "Current Status of the Mechanism of Action of B_{12}-Coenzyme" in *Biological Aspects of Inorganic Chemistry*, A. W. Addison, W. R. Cullen, D. Dolphin, and B. R. James, Eds. Wiley-Interscience, New York, 1977, p. 245.

Isomerization. The isomerization reactions that occur involve moving a substituent along a carbon chain.

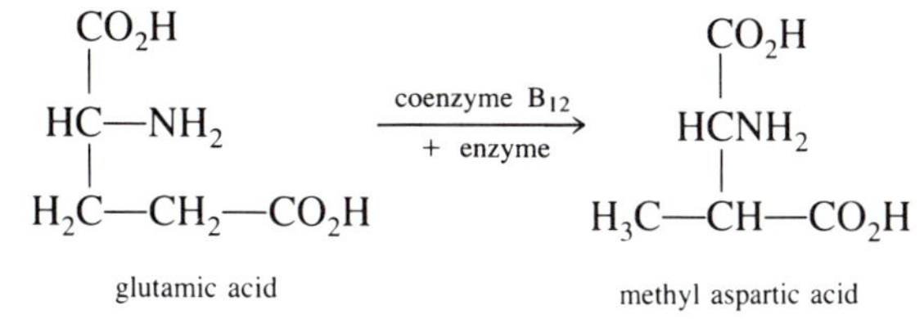

$$
\begin{array}{ccc}
\text{CO}_2\text{H} & & \text{CO}_2\text{H} \\
| & \xrightarrow[\text{+ enzyme}]{\text{coenzyme B}_{12}} & | \\
\text{HC}-\text{NH}_2 & & \text{HCNH}_2 \\
| & & | \\
\text{H}_2\text{C}-\text{CH}_2-\text{CO}_2\text{H} & & \text{H}_3\text{C}-\text{CH}-\text{CO}_2\text{H}
\end{array}
\tag{16.7}
$$

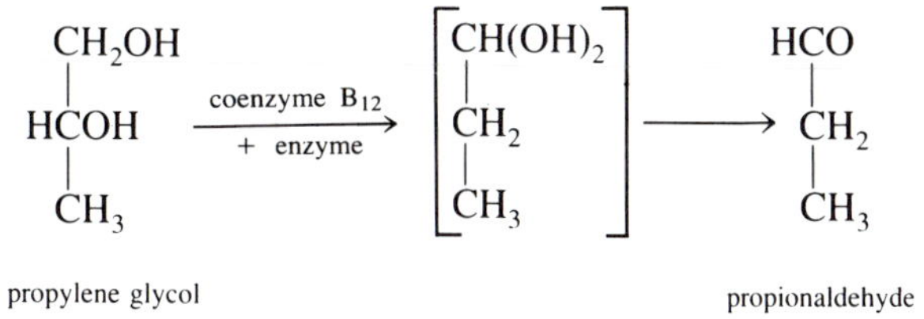

glutamic acid   methyl aspartic acid

$$
\begin{array}{cccc}
\text{CH}_2\text{OH} & & \left[\begin{array}{c}\text{CH(OH)}_2\end{array}\right. & \text{HCO} \\
| & \xrightarrow[\text{+ enzyme}]{\text{coenzyme B}_{12}} & \text{CH}_2 & | \\
\text{HCOH} & & \text{CH}_2 & \longrightarrow \quad \text{CH}_2 \\
| & & \left.\text{CH}_3\right] & | \\
\text{CH}_3 & & & \text{CH}_3
\end{array}
\tag{16.8}
$$

propylene glycol   propionaldehyde

16.3.3 Models of B$_{12}$

The cobalt complex of dimethylglyoxime (cobaloxime—see Figure 16.12) is an effective model for B$_{12}$. The cobaloximes can be reduced to cobaloximes(II) and cobaloximes(I). The reactions shown by adenosine and alkyl derivatives of cobaloximes(I), which resemble those of B$_{12s}$, include methyl group transfer, reduction, and rearrangements.

16.4 OXIDATION–REDUCTION PROCESSES—CONTROLLED ENERGY RELEASE

16.4.1 Cytochromes[15]

The oxygen transported to cells is utilized for the reversible oxidation of cytochromes as a source of energy in the mitochondria (energy-producing granules of aerobic cells). *Cyto-*

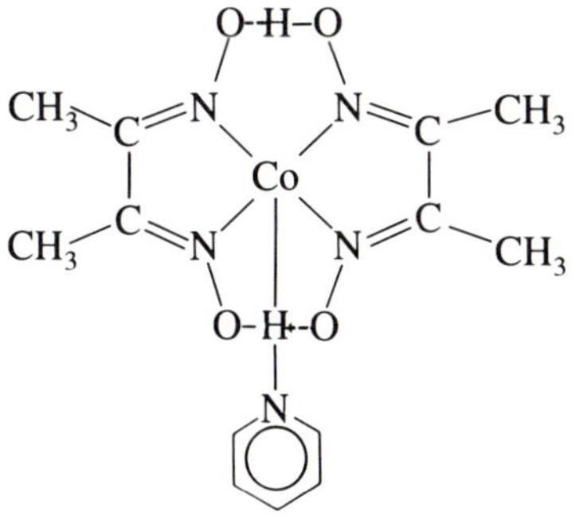

Figure 16.12 Pyridine cobaloxime.

[15]P. Cloud and A. Gibor, "The Oxygen Cycle," *Sci. Amer.* Sept. 1970, 111; R. J. P. Williams, G. R. Moore, and P. E. Wright, "Oxidation–reduction Properties of Cytochromes and Peroxidases," in *Biological Aspects of Inorganic Chemistry,* A. W. Addison, W. R. Cullen, D. Dolphin, and B. R. James, Eds., Wiley-Interscience, New York, 1977, p. 369.

chromes, which also are utilized in energy transfer in photosynthesis, are electron-transferring proteins that act in sequence to transfer electrons to O_2. The *prosthetic* group in all cytochromes is heme, which undergoes reversible Fe(II)–Fe(III) oxidation. For cases other than enzymes, the prosthetic group is the non–amino acid portion of the protein. The cytochromes in a sequence differ from one another in electrode potentials by about 0.2 volt or less, with a total potential difference of about 1 volt.[16] Differences in potentials for the Fe(II)–Fe(III) oxidation result from changes in the porphyrin substituents, changes in the protein, and in some cases, changes in axial ligands. Cytochrome a_3, the terminal member of a cytochrome chain, has water (instead of S) as a ligand. The H_2O can be replaced by O_2 to initiate the electron transfer. Cytochromes a and a_3 together are called cytochrome oxidase (an oxidase catalyzes oxidation of another species). The complex $a + a_3$ system (mol. wt. about 240,000) consists of six subunits, each containing a single heme a group and a copper atom. Two of the units form cytochrome a, and only the remaining four units of cytochrome a_3 react directly with O_2. Oxidation of copper ($Cu^I \rightarrow Cu^{II}$) and iron($Fe^{II} \rightarrow Fe^{III}$) occur in the reaction of a_3 with oxygen. Cytochrome c receives an electron from cytochrome c_1 and passes it along to cytochrome $a + a_3$. The structures of horse-heart and tuna-heart cytochrome c reveal that the protein chain is coordinated to the iron on one side of the porphyrin ring through a histidine, and on the other side through S of methionine [$H_3CSC_2H_4CH(NH_2)CO_2H$]. Higher resolution (200 pm) structure determinations of tuna-heart cytochrome c show that the structures of the oxidized and reduced forms are very similar, with no great conformational changes.

The sequence of reversible oxidations in a cytochrome chain is shown in Figure 16.13. At two stages in the cytochrome chain where there are large differences in electrode potentials, enough energy is supplied for the combination of inorganic PO_4^{3-} with adenosine diphosphate (ADP) to form adenosine triphosphate (ATP)—a conversion called *oxidative phosphorylation*. The reverse conversion of ATP to ADP (involving a Mg^{2+} complex) provides energy for muscle activity (Figure 16.14). The oxidized form of cytochrome b oxidizes the hydroquinone form of coenzyme Q (named ubiquinone, because it is ubiquitous in cells) to the quinone form (Figure 16.15). The quinone form oxidizes flavin adenine dinucleotide hydride ($FADH_2$) to flavin adenine dinucleotide (FAD). The last step in the cycle involves the oxidation of nicotinamide adenine dinucleotide hydride (NADH) to nicotinamide adenine dinucleotide (NAD^+) (Figure 16.16), with the release of more energy for conversion of ADP to ATP. Each of the oxidations is reversed in the next step, except for the last. The nicotinamide adenine dinucleotide hydride is restored from the citric acid cycle. The complete sequence, bringing about the oxidation of NADH indirectly by O_2, is known as the *respiratory chain*.

The respiratory chain for the simple net oxidation of NADH by O_2 might seem unnecessarily complex, but the many steps serve to break down the large amount of energy involved in the reduction of O_2 into smaller units that can be stored as ATP and keep the reactants (NADH and O_2) well separated. The chain achieves high biological specificity and H^+ is produced or consumed where needed. Photosynthesis accomplishes the reverse reactions, with O_2 as an end product, by a similar cytochrome chain.

The electron-transfer proteins commonly are membrane-bound. This has the advantage of holding the links of the chain in the correct juxtaposition for reaction, and thus avoiding side

[16]D. F. Wilson, P. L. Dutton, M. Erecinska, J. G. Lindsay, and N. Sato, *Acc. Chem. Res.* 1972, *5*, 234—a discussion of the energetics of the mitochondrial respiratory chain, including *in vivo* determinations of potentials in pigeon-heart mitochondria.

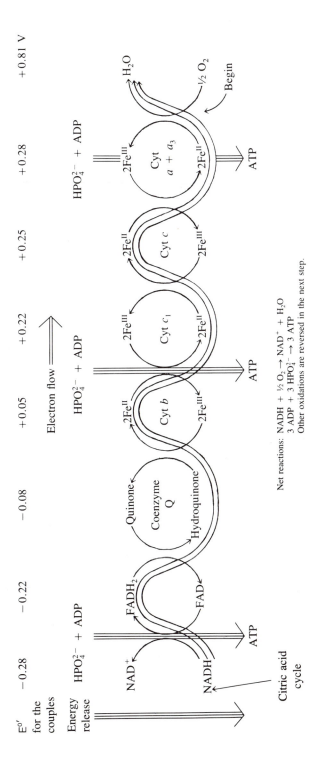

Figure 16.13 The respiratory chain, showing electron transport and oxidative phosphorylation. The $E^{0'}$ values for couples are shown above the figure. The potential ($+0.81$ V) is that for the reduction of gaseous (1 atm) O_2 in neutral solution. The potential for reduction of O_2 bound to cytochrome would differ.

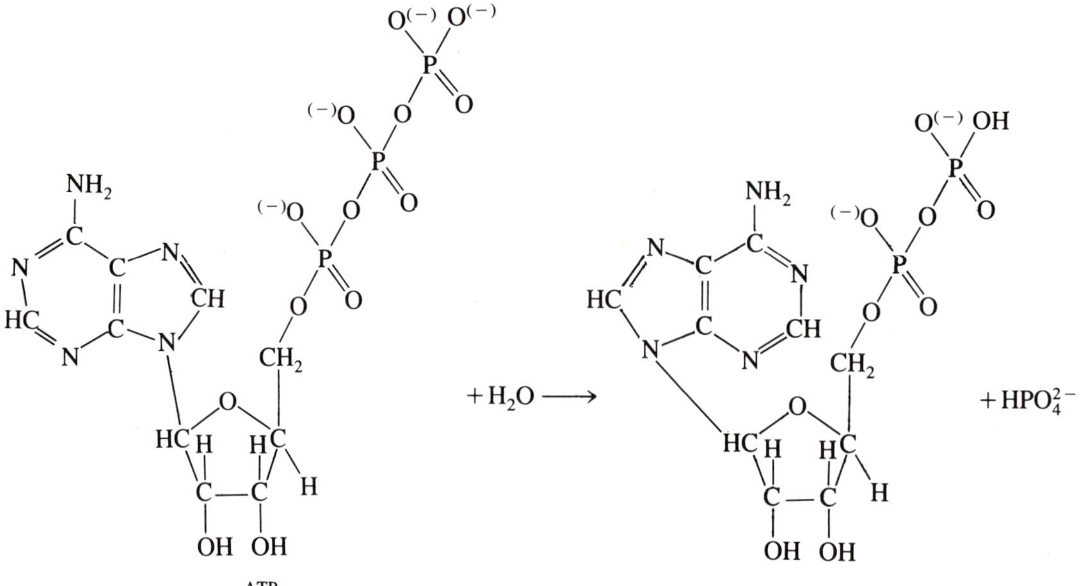

Figure 16.14 The energy-producing hydrolysis of adenosine triphosphate (ATP) to adenosine diphosphate (ADP).

reactions or short-circuiting of the chain. In particular, the coupled oxidative phosphorylation centers must not be bypassed. Membrane binding also provides a nonaqueous environment for dehydration in the synthesis of ATP. Peter Mitchell received the Nobel Prize in Chemistry in 1978 for formulating the chemiosmotic theory of oxidative phosphorylation.[17] Although the cytochromes have been characterized rather well, the links between them that provide efficient electron transfer along the chain offer an intriguing challenge.

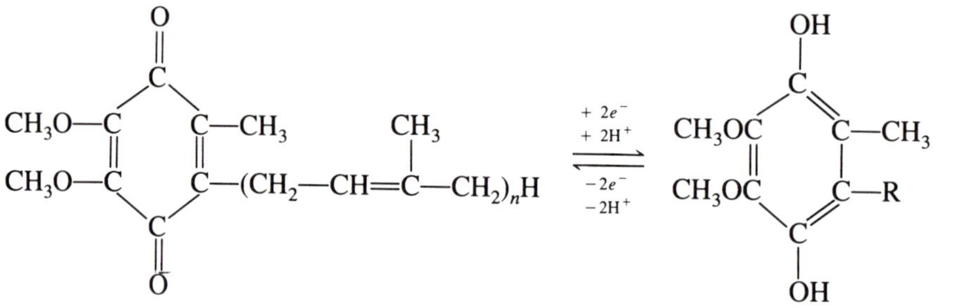

Conenzyme Q, Ubiquinone
(For most mammals, $n = 10$, for some microorganisms $n = 6$.)

Figure 16.15 Ubiquinone.

[17]P. Mitchell, Nobel Address, *Science* 1979, *206*, 1148.

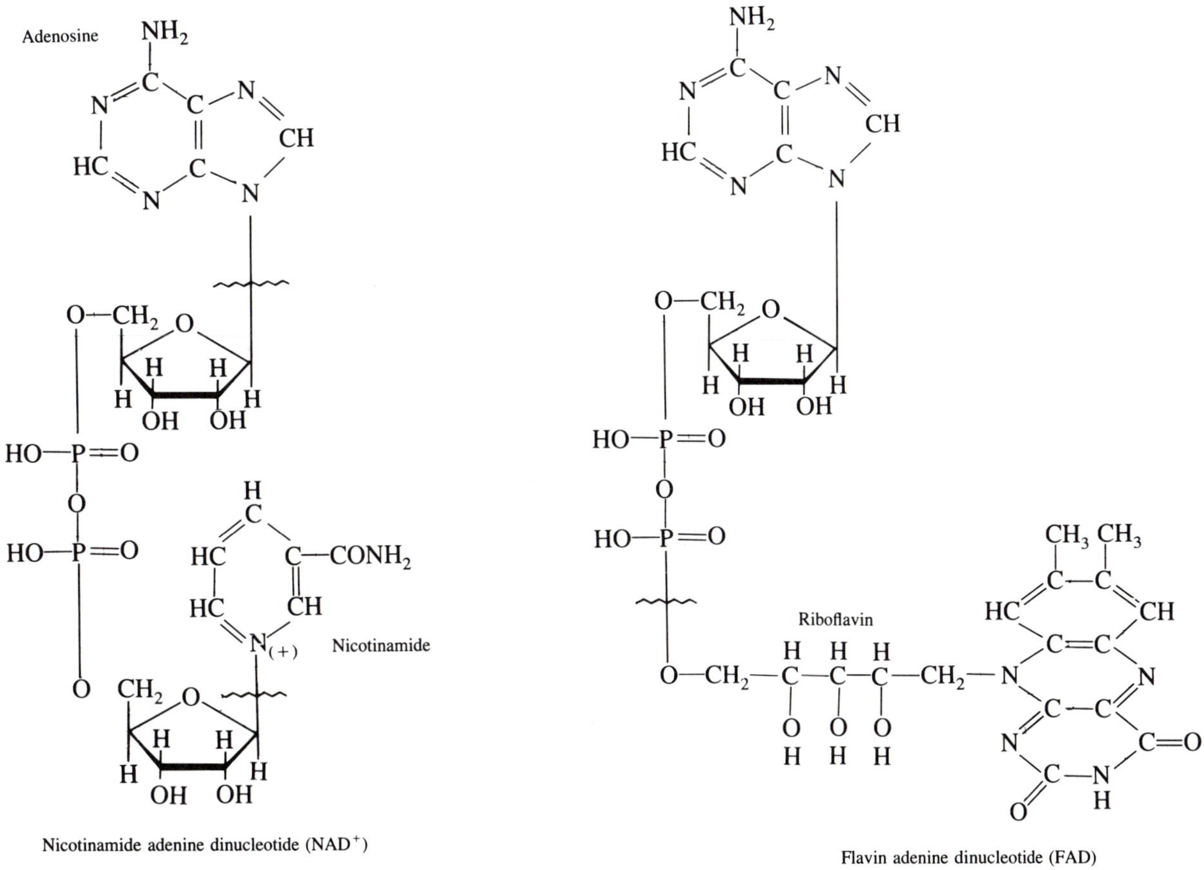

Figure 16.16 Structures of NAD$^+$ and FAD.

16.4.2 Rubredoxins and Ferredoxins[18]

These are nonheme iron proteins that participate in electron-transfer processes. They both contain iron bonded to sulfur. Rubredoxin was first isolated from the bacterium *Clostridium pasturianum,* but it also occurs in other anaerobic bacteria. The iron is bonded to four S atoms of cysteine [HSCH$_2$CH(NH$_2$)CO$_2$H] as part of a protein of about 55 amino acid residues (mol. wt. about 6000). The sulfur atoms form a distorted tetrahedron around iron, with one very short Fe—S bond (Figure 16.17). The electrode potential of rubredoxin in neutral solution is close to 0 V.

[18]R. H. Holm, "Iron–sulfur Clusters in Natural and Synthetic Systems," *Endeavour* 1975, *34*, 38; R. H. Holm, "Identification of Active Sites in Iron–sulfur Proteins," in *Biological Aspects of Inorganic Chemistry,* A. W. Addison, W. R. Cullen, D. Dolphin, and B. R. James, Eds., Wiley-Interscience, New York, 1977, p. 71; R. Mason and J. A. Zubieta, "Iron-sulfur Proteins" *Angew. Chem. Internat. Ed.* 1973, *12*, 390; G. R. Moore and R. J. P. Williams, "Electron-Transfer Proteins," *Coord. Chem. Rev.* 1976, *18*, 126.

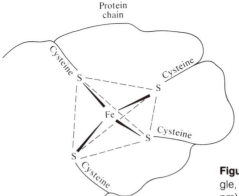

Figure 16.17 The FeS$_4$ flattened tetrahedron (smallest angle, 101°) in rubredoxin. There is one short Fe—S bond (205 pm), with others at 224–235 pm.

There are at least three types of ferredoxin, one containing two Fe and two sulfide ions per unit, one containing four Fe and four sulfide ions per unit, and a third containing eight Fe and eight sulfide ions. The 2Fe:2S^{2-} proteins also contain cysteine groups bound to Fe, but the structure is not known. These proteins are important in photosynthesis and gain one electron per iron upon reduction. Both the Fe(III) and Fe(II) are high-spin. The 2Fe:2S^{2-} ferredoxins were believed to occur only in plants, but Holm's core-extrusion technique, which permits the active Fe—S clusters to be removed by adding simple thiols to replace the S attached to the peptide chain, has shown that xanthine oxidase (easily available from milk) contains 2Fe:2S^{2-} units.

Xanthine oxidase also contains Mo. In man, this enzyme oxidizes purines to uric acid, which is excreted through the kidneys. An excess of uric-acid accumulation leads to gout, which can be treated with inhibitors of xanthine oxidase. See E. I. Stiefel, *Bioinorganic Chemistry II*, K. N. Raymond, Ed., *Adv. Chem. Ser.* 1977, *162*, p. 356.

The ferredoxin from *Peptococcus aerogenes* (8Fe:8S^{2-}), whose structure is known, has a molecular weight of 6000. It contains two well-separated Fe$_4$S$_4$ cubes with a cysteine group attached to each Fe, completing a tetrahedron of 4S about each Fe (Figure 16.18). The 4Fe:4S^{2-}

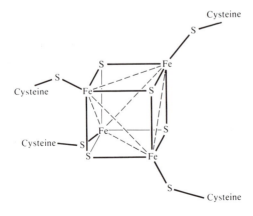

Figure 16.18 The cubane unit Fe$_4$S$_4$(cysteine)$_4$ found in *Peptococcus aerogenes*. The Fe$_4$ tetrahedron outlined is smaller than the S$_4$ tetrahedron.

proteins presumably contain just one cubane unit. Another group of 4Fe:4S^{2-} proteins are known as *high-potential iron proteins*. They also contain the cubane cluster, but the high-potential protein from *Chromatrium* has an electrode potential of $+0.3$ volt, whereas ferre-doxins such as the one from *P. aerogenes* have potentials of about -0.4 volt.

16.5 THE MEANS FOR SUPPLYING IRON IN LOWER ORGANISMS

Since iron plays a vital role in most organisms, several mechanisms have evolved for its storage and transport, depending on the availability and form of the iron. Many organisms can use Fe^{3+} if it is available directly; if not, high-affinity mechanisms are activated.

In our oxidizing environment, iron is usually found as Fe(III). The iron is present as very insoluble Fe(OH)$_3$ in basic soils, and thus is not utilized easily by plants. Such iron deficiency can be remedied by supplying the soluble iron(II) complex of ethylenediaminetetraacetic acid, as is done on a large scale in Florida citrus groves.

Siderophores[19]

Red-brown iron-containing complexes for the transport of iron in lower organisms are known as *siderophores*. The chelating agents are hydroxamates in fungi and yeast, and hydroxamates or substituted catechols in bacteria (Figure 16.19). Some bacteria can produce both types of chelating ligands, and other ligands combine these functional groups. The production of the ligands is activated when the concentration of iron is low. The siderophores that promote growth have been called sideramines. Those that are antibiotics have been called sideromycins—these function as antibiotics by making iron unavailable for bacterial growth.

Siderophores also are known as siderochromes, because some of them are intensely colored. The classifications as sideramines and sideromycins are not clear-cut, and they are not recommended by those active in the field.

These ligands give very stable Fe(III) complexes (high formation constants), and they are rather specific for Fe(III). The formation constants for the Fe(II) complexes are much lower, so that reduction of the iron provides a mechanism for its release. Presumably, the ligands

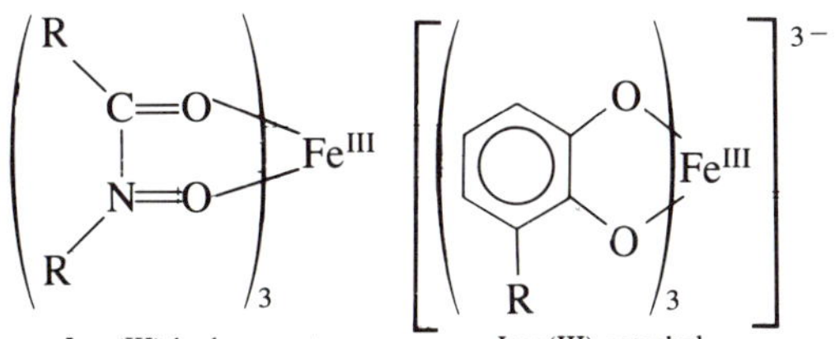

Iron(III) hydroxamate complex

Iron(III) catechol complex

Figure 16.19 Siderophores.

[19]J. Leong and K. N. Raymond, *J. Am. Chem. Soc.* 1974, *96*, 1757.

capture Fe^{3+} at the cell membrane (the mechanism of transfer through the cell membrane depends on the charge type of the complex) and transport it within the cell. In cases of very low concentrations of Fe(III), overproduction of the ligands can result in their release to the surrounding medium to dissolve Fe(III).

Ferrichrome (Figure 16.20) is an hydroxamate complex in which the hydroxamate groups are side chains to a peptide ring. Crystallographic studies have shown that ferrichrome A, which differs in substituents, has the Λ absolute configuration, a left spiral of chelate rings about Fe(III). Other ferrichromes differ in ring substituents. The *ferrioxamines* contain the hydroxamates as part of the peptide chain. Ferrioxamine D_1 (Figure 16.21) has a linear chain; there are cyclic ferrioxamines where the chain is closed. *Iron(III) enterobactin* (Figure 16.22) provides an example of a catechol-type siderophore in which the catechols are side chains to a cyclic ester. This siderophore has the Δ absolute configuration, a right spiral of chelate rings.

The iron is released from hydroxamate siderophores by reduction of Fe(III) to Fe(II), which forms much less stable hydroxamate complexes. Then the ligand is available for reuse. The iron(III) enterobactin requires a potential for reduction (*ca.* -0.75 V at pH 7) that is too low for physiological reductants.[20] The iron is released by the destruction of the ligand, a result of the hydrolysis of the ester linkages by an enzyme specific for the complex. The hydrolysis products are not used for the ligand synthesis. The synthesis of a complex ligand for the transport of a single Fe(III) seems very inefficient. However, the very great stability of the Fe(III) complex (log K_f *ca.* 56) allows the bacteria using enterobactin to acquire iron present at very low concentrations, even at the expense of microorganisms using other ligands. Raymond refers to this as the "American approach"—use it once and throw it away.

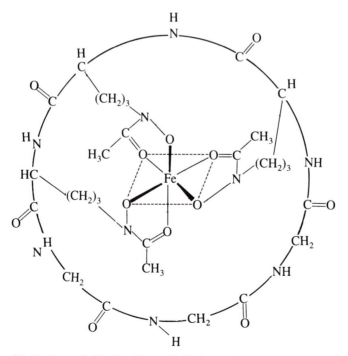

Figure 16.20 Ferrichrome.

[20]S. R. Cooper, J. V. McArdle, and K. N. Raymond, *Proc. Nat. Acad. Sci., U.S.A.* 1978, *75*, 3551.

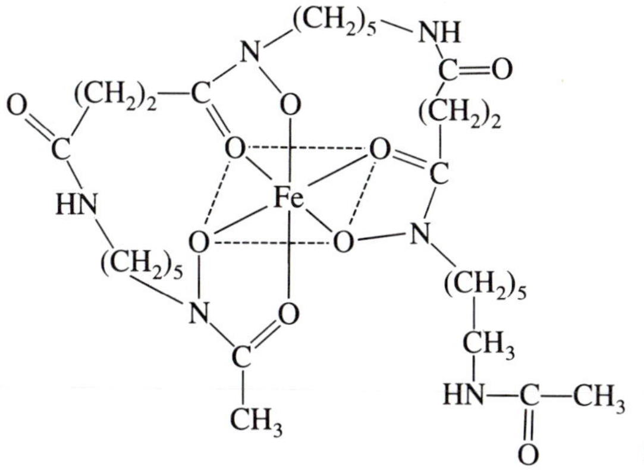

Figure 16.21 Ferrioxamine D_1.

16.6 THE ROLE OF COPPER

Copper is vital in plants and animals, but some of its roles are less well understood than are those of iron. The copper protein hemocyanin has been mentioned as the oxygen-carrier for a number of invertebrates. Cytochrome oxidase, an enzyme containing copper and iron (heme), has been discussed as the terminal member of the respiratory chain. Some other copper enzymes will be discussed later (Section 16.8.4).

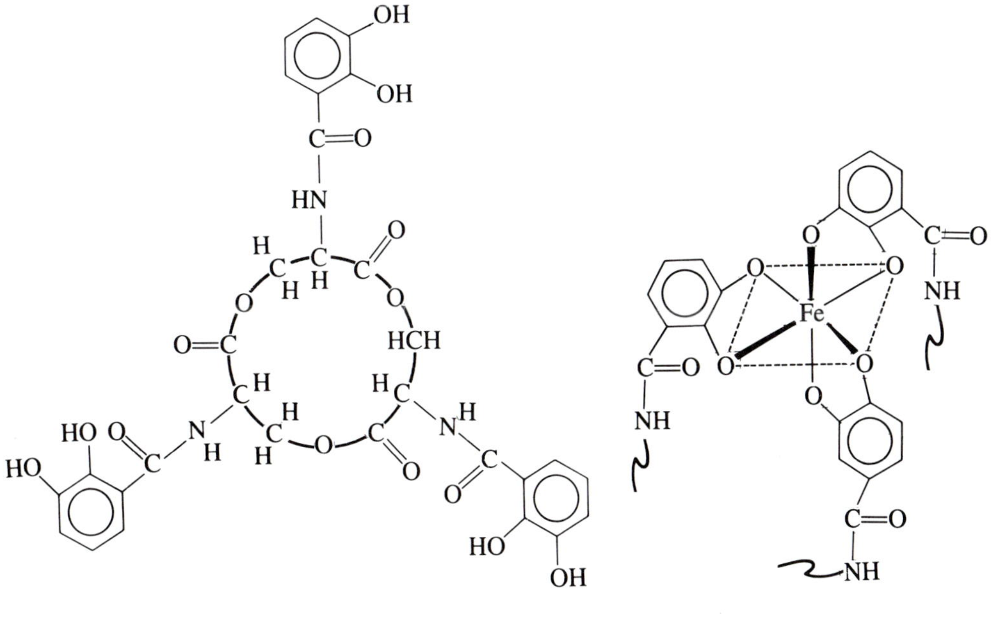

Enterobactin

Chelated portion of Fe(III) enterobactin

Figure 16.22 Enterobactin and the chelated portion of the Fe(III) complex.

16.6.1   Ceruloplasmin

Ceruloplasmin from human plasma is an intensely blue protein containing 0.3% copper. Its molecular weight is about 134,000, with six or seven copper atoms per molecule, and it has oxidase (enzymic oxidation) activity toward polyamines, polyphenols, and Fe(II). The protein is found in the plasma of most animals. Interestingly, it occurs in the frog only after metamorphosis, accompanying changes in hemoglobin synthesis. Ceruloplasmin apparently plays a role in iron metabolism. Copper deficiency in animals results in reduced iron mobilization and eventually, anemia, even though there is still abundant iron storage in the liver. Presumably, ceruloplasmin has ferroxidase activity (catalyzes $Fe^{II} \rightarrow Fe^{III}$) in the uptake of iron by transferrin. Although iron is stored and transported as Fe(III), the transfer requires reduction to Fe(II) and then reoxidation. Other biological functions of ceruloplasmin have been suggested, including the storage and transport of copper for other needs.

Several transition metals at trace levels can regulate the synthesis and degradation of heme. They inhibit δ-aminolevulinate ($H_2NCH_2\overset{\displaystyle O}{\overset{\|}{C}}C_2H_4CO_2H$) synthetase (the rate-limiting enzyme in heme synthesis) and induce heme oxygenase (the rate-limiting enzyme in heme degradation). M. A. Maines and A. Kappas, *Science*, 1977, *198*, 1215.

A deficiency of ceruloplasmin occurs in individuals afflicted with the hereditary Wilson's disease. Copper accumulates in the liver, brain, and kidneys. The symptoms can be relieved by the use of the chelating agents edta [as $Na_2Ca(edta)$] or penicillamine for the mobilization of the accumulated copper.

16.6.2   Blue Copper Proteins

The intensely blue copper proteins[21] *azurin* (from bacteria; mol. wt., 16,000; 1 Cu/molecule), *stellacyanin* (from the lacquer tree; mol. wt., 20,000; 1 Cu/molecule), and *plastocyanin* (from the chloroplasts of plants; mol. wt., 10,500; 1 Cu/molecule) are electron carriers. Plastocyanin is involved in electron transfer in photosynthesis. The absorption bands of these blue proteins are much more intense ($\epsilon \sim 4000$) than those associated with *d-d* transitions. The presumption that these are charge-transfer bands was confirmed by H. Gray's substitution of Co^{2+} for Cu^{2+}, causing an expected great shift in the energy of the band. Azurin and plastocyanin are structurally similar. The copper atom is coordinated to a cysteine thiol group, a methionine thioether group, and two histidine imidazole groups in a highly distorted tetrahedral configuration.

16.6.3   Iron and Copper Enzymes

Presumably, iron and copper play many vital roles in life processes, because of the availability of two oxidation states under biological conditions. Recent work has shown that Cu(III) is stabilized in deprotonated amide or deprotonated peptide complexes to the extent that Cu(II) can be oxidized to Cu(III) by O_2. It was suggested that the biological role of Cu(III), involving

[21]J. A. Fee, "Copper Proteins: Systems Containing the 'Blue' Copper Center," *Str. Bonding* 1975, *23*, 1.

a two-electron reduction, $Cu^{III} \rightarrow Cu^I$, could be important in eliminating high-energy free radical intermediates.[22] Evidence for this suggestion has been found in the galactose oxidation reaction.

16.7 PHOTOSYNTHESIS

Except for nuclear fuel and geothermal energy in the form of geysers, our energy sources ultimately are solar. Wood and fossil fuels (oil and coal) represent solar energy stored through photosynthesis. The oxygen in our atmosphere is a by-product of photosynthesis. We associate photosynthesis with plants, but perhaps half or more of the photosynthesis on earth occurs in algae, diatoms, and in certain red, green, brown, and purple bacteria.

The usual net reaction converts H_2O and CO_2 to carbohydrates and O_2.

$$n\ H_2O\ +\ n\ CO_2\ \xrightarrow{\text{light}}\ (CH_2O)_n\ +\ n\ O_2. \tag{16.9}$$

In photosynthetic bacteria, however, the species oxidized is not H_2O, but an organic molecule or H_2S. In the latter case, elemental S is produced instead of O_2 by anaerobic bacteria that are poisoned by O_2. Plants can use NO_3^- as the electron acceptor instead of CO_2 (to form NH_3), and some organisms can use H^+ to form H_2, or N_2 to form NH_3.

Chlorophyll a is just one of the many pigments involved in photosynthesis. A variety of colors of photosynthetic cells is possible from combinations of chlorophylls, yellow carotenoids, and blue or red pigments (phycobilins). Green plant cells contain chloroplasts, the membrane-surrounded units for photosynthesis, as well as mitochondria (p. 735) for respiration—the reverse of the photosynthetic reactions. These oxygen-producing chloroplasts contain two kinds of chlorophyll, one of which is chlorophyll a. Oxygen is produced from H_2O accompanying the reduction of nicotinamide adenine dinucleotide phosphate ($NADP^+$). $NADP^+$ is NAD^+ (Figure 16.16) with one of the OH groups esterified with phosphate. This reaction is driven by the light energy absorbed and also produces ATP (Figure 16.14). The reduction of CO_2 to carbohydrates occurs in a dark reaction utilizing the energy stored by NADPH and ATP.

$$2H_2O\ +\ 2NADP^+\ \xrightarrow{\text{light}}\ 2NADPH\ +\ 2H^+\ +\ O_2 \tag{16.10}$$

There are two light-reactions in oxygen-producing plants. The first of these (Photosystem I) involves chlorophyll a along with other pigments and does not produce O_2. This is the only photosystem for anaerobic photosynthetic bacteria and is believed to have been the first such system to evolve. Chlorophyll a absorbs at about 680 nm in cells. Photosystem II, which is believed to bring about the formation of O_2, involves chlorophyll a; another type of chlorophyll, which absorbs at shorter wavelengths; and other associated pigments. Typically, Photosystem I contains about 200 chlorophyll a molecules and about 50 carotenoid pigment molecules. Several light quanta are needed for Reaction 16.10, so the energy must be accumulated. Photosystem I contains an energy trap, P700, that serves to accumulate the energy absorbed

[22]D. W. Margerum *et al.*, ''Copper(II) and Copper(III)-Peptide Complexes'' in *Bioinorganic Chemistry II*, K. N. Raymond, Ed., *Adv. Chem. Ser. 162* 1977, p. 281.

by the individual chlorophyll *a* molecules. P700 is thought to be a specialized chlorophyll that in the excited state can lose electrons to bring about the reduction of $NADP^+$. Photosystem II is believed to restore the electrons (from H_2O) to oxidized P700 and form O_2 in the process. Manganese(II) is involved in this process in an unknown manner. There are also cyclic dark reactions that do not bring about Reaction 16.10 but do produce ATP. These processes are represented in Figure 16.23.

The electron transfer processes in Photosystem I utilize ferredoxin (Section 16.4.2). The electron transport between the photosystems involves a cytochrome chain similar to that for the respiratory chain (Figure 16.13). A blue copper protein, *plastocyanin* (Sec. 16.6.2), also is involved in this electron transfer process.

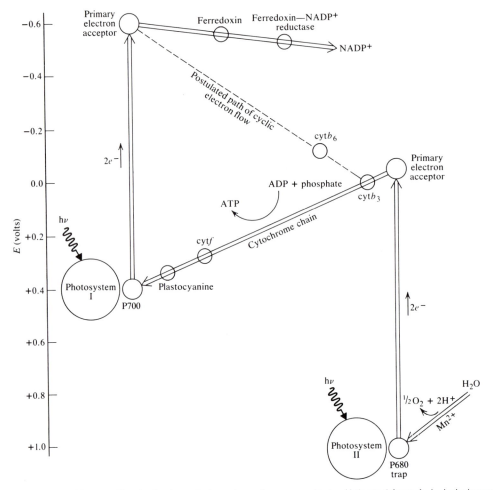

Figure 16.23 Electron flow in the two photosystems for green plants. (Adapted from A. L. Lehninger, *Biochemistry*, 2nd ed., Worth, New York, 1975, p. 605.)

16.8 METALLOENZYMES

16.8.1 Enzymes

Enzymes are catalysts that greatly enhance the rates of specific reactions. The yields are often 100%, and reactions occur under the mildest conditions. High yields are achieved because the rate of the reaction catalyzed might be increased by a factor of 10^{12}, making side reactions unimportant. Enzymes are proteins built up by the cells from amino acids. The names of the over 1500 enzymes identified are obtained by adding the suffix *ase* to the name of the process catalyzed or to the name of the molecule on which the enzyme acts (the *substrate*). Some enzymes function alone, whereas others require the cooperation of a *cofactor,* a metal ion or an organic molecule. Some of the enzymes requiring metal ions as cofactors *(metalloenzymes)* are given in Table 16.2. Coenzymes that are tightly bound to the enzyme are called *prosthetic groups*—for example, coenzyme B_{12} (Section 16.3). We shall consider only a few of the metalloenzymes.

Catalysis is vitally important to the chemical industry, but the usual catalysts are very poor in comparison to enzymes. Imagine the frustration a research director who is asked to devise a reaction for the synthesis of an organometallic compound with a molecular weight of 50,000. The reaction is to be carried out in about 30 minutes total time, at 1 atm and at a temperature of less than 50°C, and to give a 100% yield! Such reactions occur with enzymes at low concentrations of reactants—in fact, the reactants are scavenged from the environment! What a problem for the body of a student who subsists on snack foods for an extended period! We have a lot to learn about synthesis.

Table 16.2   Some metalloenzymes

Mg^{II}	Phosphohydrolases
	Phosphotransferases
Mn^{II}	Arginase
	Phosphotransferases
Fe^{II} or Fe^{III}	Cytochromes
	Peroxidase
	Catalase
	Ferredoxin
Cu^{II} or Cu^{I}	Tyrosinase
	Amine oxidases
	Cytochrome oxidase
	Ascorbate oxidase
	Galactose oxidase
	Dopamine-β-hydroxylase
Zn^{II}	Alcohol dehydrogenase
	Alkaline phosphatase
	Carbonic anhydrase
	Carboxypeptidase
Fe and Mo	Nitrogenase

16.8.2   Carbonic Anhydrase and Carboxypeptidase (Zinc Enzymes)[23]

Carbonic anhydrase is an enzyme that catalyzes the dehydration of carbonic acid (and the hydrolysis of certain esters). The uncatalyzed dehydration of carbonic acid is too slow for respiration in animals. The enzyme (mol. wt., 30,000) occurs in the blood. The one Zn^{2+} per molecule lies inside a cavity associated with the active site. The zinc has a tetrahedral environment in carbonic anhydrase C.

Carboxypeptidase A is a zinc enzyme (see front cover) that has been studied more thoroughly. The enzyme hydrolyzes (cleaves) the terminal peptide (amide) bond of a peptide chain for the carboxyl terminal end. The enzyme is selective

$$-CH-\underset{R}{\overset{}{}}\;\overset{O}{\underset{}{\overset{\parallel}{C}}}-NH-\underset{R'}{\overset{}{CH}}-\overset{O}{\overset{\parallel}{C}}\Big\}-NH-\underset{R''}{\overset{}{CH}}-CO_2^-$$

for terminal peptides where the terminal amino acid has an aromatic or branched aliphatic substituent (R'') and the L configuration.

The structure of carboxypeptidase A has been studied extensively by x-ray crystallography. The one Zn^{2+} per molecule (mol. wt., 34,600) is bonded to two imidazoles of histidine, one glutamic acid group, and an H_2O molecule that can be replaced by the substrate. The Zn^{2+} sits in a depression in the surface of the molecule. This is the active site, from which extends a pocket into the interior of the molecule, for accommodating the substrate. The enzyme protects its own terminal end against self-digestion by tucking it inside.

The zinc ion can be replaced by cobalt(II) in carboxypeptidase, and in several other zinc enzymes, with retention of activity of the enzyme. In some cases there is even enhanced activity—perhaps cobalt was not available, and nature had to settle for zinc. Replacement of Zn^{2+} by Co^{2+} has been particularly useful because Zn^{2+} (d^{10}) is colorless whereas Co^{2+} (d^7) absorbs in the visible region. Thus spectral studies (absorption, circular dichroism, and magnetic circular dichroism) of the Co-substituted enzyme provide valuable information about the metal ion environment at the active site.

16.8.3   Models

Cobalt(III) complexes of the type cis-$[Co(en)_2(H_2O)(OH)]^{2+}$ and cis-β-$[Co(trien)(H_2O)(OH)]^{2+}$ (en = $NH_2C_2H_4NH_2$ and trien = $NH_2C_2H_4NHC_2H_4NHC_2H_4NH_2$) promote the hydrolytic cleavage of N-terminal amino acids of a peptide chain. The reaction is not catalytic, since a stable complex of the chelated amino acid is formed. The trien complex is the more effective. The mechanism of enzymatic hydrolysis resembles that of the hydrolysis of esters by such cobalt complexes. The first step is the coordination of the terminal NH_2 in place of H_2O (Figure 16.24). By one path, the carbonyl oxygen coordinates next, replacing OH^-. Hydroxide attack on the carbonyl group leads to the chelated amino acid complex. By the other path, the coordinated OH^- directly attacks the peptide carboxyl carbon, hydrolyzing the peptide bond and giving the amino acid complex. This case involves cleavage of the N-terminal amino acid,

[23]R. H. Prince, "Some Aspects of the Bioinorganic Chemistry of Zinc," *Adv. Inorg. Chem. Radiochem.* 1979, *22*, 349.

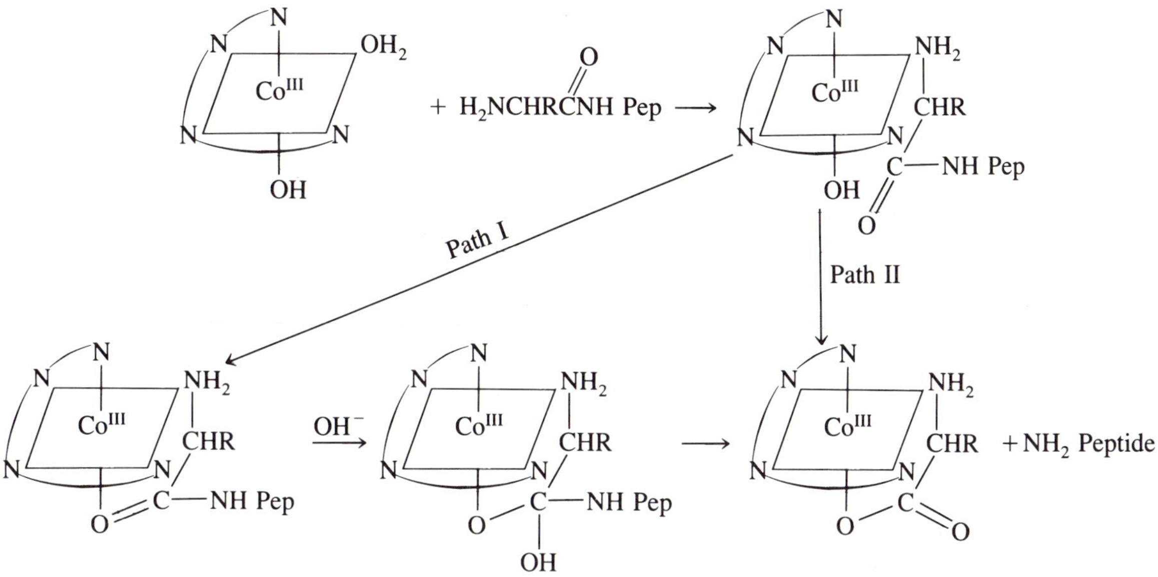

Figure 16.24 Mechanisms for peptide cleavage by *cis*-β-[Co(trien)(H₂O)OH]²⁺

rather than the carboxyl-terminal amino acid cleaved by carboxypeptidase. However, metalloenzymes, such as leucine aminopeptidase [activated by Mg(II) or Mn(II)], catalyze the hydrolysis of *N*-terminal peptide bonds through chelation.

16.8.4 Enzymes of Iron and Copper— Oxidase, Peroxidase, and Catalase

The first forms of life probably emerged when the earth had a reducing atmosphere. The present oxygen atmosphere is believed to have evolved later, from photosynthesis. Anaerobic bacteria still survive, but the dominant forms of life utilize oxygen for releasing energy stored primarily through photosynthesis. A crucial step in the evolution of aerobic life was the development of a means of coping with the highly toxic and reactive products of the reduction of O_2—peroxide or superoxide. A strain of *micrococcus* that dies on contact with oxygen was found to survive in air if heme was added to their nutrient medium. The heme was taken up by the cells to provide the bacteria with respiratory enzymes and a catalyst for decomposing H_2O_2.

Life forms on earth are scientifically amazing, since they are made up of easily oxidizable organic molecules but depend on O_2 for their existence. The thermodynamically favored reaction gives CO_2 and H_2O (plus oxides of N, etc.)—the products formed by decay or at high temperature (burning of fossil fuels). However, living organisms can use O_2 for controlled oxidation to supply the energy they need. Systems have evolved for the remarkable four-electron reduction of O_2 to form H_2O (cytochrome chain) or for dealing with the even stronger oxidizing agents produced by the more likely one-electron reduction (O_2^-) or two-electron reduction (O_2^{2-}).

The earliest iron proteins utilized apparently were the ferredoxins and rubredoxins (non-heme Fe-S proteins—Section 16.4.2) found in anaerobic bacteria. Some scientists have speculated that the porphyrins were formed geochemically and utilized before enzymes were developed for their synthesis.

The close link between the biological activity of Fe and Cu is apparent in *cytochrome oxidase* (Section 16.4.1) and in the role of *ceruloplasmin* in iron metabolism (Section 16.6.1). The Cu proteins *erythocuprein* and *hepatocuprein* (mol. wt., 35,000; 2 Cu/molecule), which occur in mammalian blood and liver, are believed to catalyze the conversion of superoxide to H_2O_2.

$$2H^+ + 2O_2^- \xrightarrow[\text{dismutase}]{\text{superoxide}} H_2O_2 + O_2 \tag{16.11}$$

(*Dismutation* is the term used by biochemists for disproportionation.)

The H_2O_2 can be handled by enzymes known as *peroxidases*. Bovine erthrocyte superoxide dismutase is one of the most thoroughly studied Cu proteins. The complete amino-acid sequence and x-ray crystal structure are known. There are two subunits with molecular weight 15,700, each containing one Cu and one Zn. At the active site a deprotonated imidazole ring bridges Cu and Zn. Lippard has studied some imidazolate-bridged model compounds of Cu(II) as an aid to understanding the essential features of the active site of the enzyme.[24] Model studies usually involve simple systems that can be studied in solution. Structural changes are made to evaluate their effects on the catalytic activity.

The peroxidases occur widely in plants and animals. The sap of the fig tree and the root of the horseradish are the richest known sources. Horseradish peroxidase (mol. wt., *ca.* 40,000) contains the iron(III) derivative of heme *b*—(see Figure 16.2) as the prosthetic group. A fifth coordination site is occupied by a protein of some 16 amino-acid residues. The sixth coordination site can be occupied by H_2O or other ligands that exchange as the peroxidase functions. No single mechanism can explain all peroxidase oxidations. Horseradish peroxidase is oxidized by H_2O_2 to the extent of two equivalents. There is evidence for Fe(IV), suggesting that the second equivalent of oxidation involves the porphyrin or the protein. The oxidized enzyme oxidizes a substrate. The net reaction is

$$H_2O_2 + \text{SubstrateH}_2 \xrightarrow{\text{peroxidase}} 2H_2O + \text{Substrate} \tag{16.12}$$
$$\text{(reduced} \qquad\qquad \text{(oxidized}$$
$$\text{form)} \qquad\qquad \text{form)}$$

The *catalases* are enzymes (mol. wt., *ca.* 240,000) with four iron(III) heme *b* groups per molecule. Each heme group has a separate peptide chain coordinated. Catalases promote the disproportionation of H_2O_2.

$$2H_2O_2 \xrightarrow{\text{catalase}} 2H_2O + O_2 \tag{16.13}$$

Catalases can also catalyze peroxide oxidations of substrates.

---

[24]S. J. Lippard *et al.*, ''Physical and Chemical Studies of Bovine Erythrocyte Superoxide Dismutase,'' in *Bioinorganic Chemistry II*, K. N. Raymond, Ed., *Adv. Chem. Ser.* 1977, *162*, p. 251.

A Model Compound

The iron(III) complex of triethylenetetraamine, $[Fe(trien)(OH)_2]^+$, has catalase activity. In the proposed mechanism (Figure 16.25), the hydroxide ions are replaced by HO_2^- coordinated as a bidentate group, stretching and weakening the O—O bond. A second HO_2^- can react to produce O_2 and OH^-, regenerating the original complex. Inorganic promoted reactions of H_2O_2 always seem to require substitution into the first coordination sphere of metal ions.[25]

The intensely blue copper proteins azurin, stellacyanin, and plastocyanin (Section 16.6.2) do not have oxidase activity. Two related groups of copper proteins do have oxidase activity, catalyzing the reduction of O_2 to H_2O and, in turn, oxidizing another substrate. These so-called "blue" copper oxidases contain four Cu atoms per molecule; "nonblue" Cu oxidases contain only one or two Cu atoms per molecule. The "blue" oxidases contain one Cu(II), which is responsible for the very intense color, as in azurin. The other Cu atoms are referred to as nonblue, although one of them is believed to be Cu(II), but without the intense charge-transfer absorption. Two of the Cu atoms are diamagnetic. It has been shown for *laccase* that the removal of only the nonblue Cu(II) causes loss of catalytic activity, even though three atoms of Cu per molecule remain. Adding Cu to the sample restores the original Cu content and the oxidase activity. The ions N_3^-, CN^-, and F^- inhibit oxidase activity through the coordination to the nonblue Cu(II). It has been proposed that the two diamagnetic Cu atoms are a spin-paired Cu(II)–Cu(II) pair capable of acting as a two-electron acceptor site in oxidizing a substrate. Only the Cu(II) $\rightleftharpoons$ Cu(I) conversion has been considered for Cu enzymes involved in oxidation–reduction processes. Since Cu(III) has been shown to exist in biological systems, scientists will have to consider its possible role in redox processes.

16.8.5 Nitrogenase

In recent years, nitrogen fertilizers have been in short supply, because they are expensive and manufacturing them requires high energy consumption. Over the period 1949–1974, consumption of N fertilizers increased more than tenfold, to 40×10^6 metric tons annually. Usage is expected to increase still more, as underdeveloped countries strive to improve their agriculture to meet the needs of population increases.[26]

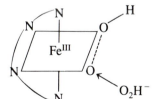

Figure 16.25 Activation of HO_2^- by coordination.

[25]M. L. Bowers, D. Kovacs, and R. E. Shepherd, *J. Am. Chem. Soc.* 1977, *99*, 6555.

[26]R. W. F. Hardy and U. D. Havelka, "Nitrogen Fixation Research: A Key to World Food?" *Science* 1975, *188*, 633.

The N_2 molecule is so unreactive that N_2 commonly is used to provide an inert atmosphere. The commercial Haber process for *nitrogen fixation* (conversion to N compounds) requires high temperature and pressure.

$$N_2 + 3H_2 \xrightarrow[\substack{200-600 \text{ atm} \\ 450-600°C}]{\text{catalyst}} 2NH_3 \qquad (16.14)$$

Nevertheless, blue-green algae and some bacteria are able to fix N_2 at ambient temperature and pressure. *Rhizobium* bacteria living in the root nodules of certain legumes are an important source of fixed N.

A nitrogen-fixing enzyme, *nitrogenase*, was isolated from the anaerobic bacterium *Clostridium pasteurianum* in 1960. It now appears that there is a single nitrogenase from any of several different sources. Nitrogenase has been found to contain two components, an Mo–Fe-containing protein and an Fe-containing protein. The Mo–Fe protein contains two Mo, 24–32 atoms each of Fe and acid-labile sulfide per molecule (mol. wt., 220,000). The Fe protein contains two identical subunits; there are four Fe and four S per molecule (mol. wt., 60,000).

X-ray absorption edge spectroscopic (EXAFS) studies of the Mo–Fe protein from two sources of nitrogenase indicate the presence of a Mo, Fe, S cluster, thought to be a constituent of all nitrogenase systems. The cluster $[Mo_2Fe_6S_9(SC_2H_5)_8]^{3-}$ has been prepared by Holm *et al.*, and the structure determined (Figure 16.26). EXAFS studies of this cluster indicate that it is the closest model yet achieved for the active site of nitrogenase. It could differ from the core of the Mo–Fe site of nitrogenase only by the additional ligands to Mo.

Nitrogenase reduces N_2 to NH_3, with no evidence for intermediates not bound by the enzyme. Presumably, N_2 is complexed by the molybdenum and the iron protein, probably of the ferredoxin type, bringing about the reduction through electron transfer. Adenosine tri-

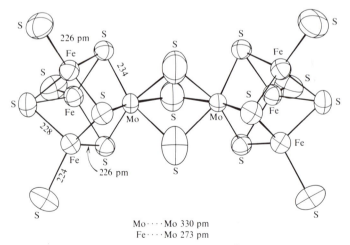

Mo····Mo 330 pm
Fe····Mo 273 pm

Figure 16.26 Structure of $[Mo_2Fe_6S_9(SC_2H_5)_8]^{3-}$. The ethyl groups of the six terminal thiolate ligands and the two bridging thiolates are omitted. (Reprinted with permission from T. E. Wolff, J. M. Berg, C. Warrick, K. O. Hodgson, and R. H. Holm, *J. Am. Chem. Soc.* 1978, *100*, 4630. Copyright 1978, American Chemical Society.)

phosphate (ATP) is essential for nitrogenase activity. Nitrogenase is effective in reducing N_3^-, N_2O, RCN, RNC, and HCCH. The reduction of acetylene to ethylene has been used for monitoring nitrogenase activity, since the process is more efficient than the reduction of N_2. The discovery of stable complexes of N_2 led to much more intense study of model compounds[27] and a possible new nitrogen fixation process. Ammonia has been obtained from metal complexes of N_2, and intermediate stages of reduction and N substitution have been observed.[28] A study of the nature of bonding in N_2 complexes[29] favored end-on coordination and suggested that σ donation is more important than π back donation (from the metal to the π* orbitals of N).

16.9 ROLES OF Na⁺, K⁺, Mg²⁺, AND Ca²⁺

So far we have dealt with the biological roles of trace elements (Table 16.1), except for phosphorus in the phosphorylation process. The pairs of major elements Na^+ and K^+, and Ca^{2+} and Mg^{2+} are so similar chemically that it is surprising that they differ so greatly in their biological functions. The ions Ca^{2+} and Na^+ are concentrated in body fluids outside of cells. Calcium forms solid skeletal materials such as bones, stabilizes conformations of proteins, and triggers muscle contraction and the release of hormones. The ions Mg^{2+} and K^+ are concentrated in cells. Magnesium ion forms a complex with ATP and is required for most enzymatic reactions involving ATP within the cell.

In most animal cells the concentration of K^+ is about 0.15 M and that of Na^+ is about 0.01 M. In the fluids outside the cells the concentration of Na^+ is about 0.15 M and that of K^+ is less than 0.004 M (concentrations rather close to those of seawater—see Table 16.1). Maintenance of these large-concentration gradients requires a "sodium pump" (really, an Na^+–K^+ pump). The energy of transport of the ions is provided by the hydrolysis of ATP. Kidney and brain cells use about 70% of the energy from ATP for this transport. In some cells, each ATP molecule hydrolyzed transports 3 Na^+ out of the cell and 2 K^+ ($+H^+$) into the cell. The K^+ is required in the cell for glucose metabolism, protein synthesis, and activation of some enzymes. The transport of glucose and amino acids into the cell is coupled with Na^+ transport, which is favored by the great concentration gradient. The Na^+ entering the cell in this way must be pumped out again.

GENERAL REFERENCES

A. W. Addison, W. R. Cullen, D. Dolphin, and B. R. James, Eds., *Biological Aspects of Inorganic Chemistry*, Wiley-Interscience, New York, 1977.

[27]J. Chatt, "The Activation of Molecular Nitrogen," p. 229, and A. E. Shilov, "Dinitrogen Fixation in Protic Media: A Comparison of Biological Dinitrogen Fixation with its Chemical Analogues," p. 197, in *Biological Aspects of Inorganic Chemistry*, A. W. Addison, W. R. Cullen, D. Dolphin, and B. R. James, Eds., Wiley-Interscience, New York, 1977.

[28]J. Chatt et al., *J. Chem. Soc., Dalton* 1977, 688, 1852, 2139; D. Sellmann and W. Weiss, *Angew. Chem. Int. Ed.* 1977, *16*, 880; 1978, *17*, 269.

[29]T. Yamabe, K. Hori, T. Minato, and K. Fukui, *Inorg. Chem.* 1980, *19*, 2154.

S. P. Cramer and K. O. Hodgson, "X-Ray Absorption Spectroscopy: a New Structural Method and Its Applications to Bioinorganic Chemistry," *Prog. Inorg. Chem.* 1979, *25*, 1.

R. Dessy, J. Dillard, and L. Taylor, Eds., *Bioinorganic Chemistry, Adv. Chem. Ser.* 1971, *100*. Symposium papers.

G. L. Eichorn, Ed., *Inorganic Biochemistry*, Elsevier Scientific Publishing Co., New York, 1973. A two-volume comprehensive treatise; there are chapters on the topics covered here (and others) written by experts.

G. L. Eichorn and L. G. Marzilli, Eds., *Advances in Inorganic Biochemistry*. A series of volumes updating the treatise.

M. N. Hughes, *The Inorganic Chemistry of Biological Processes*, 2nd ed., Wiley-Interscience, New York, 1981.

J. A. Ibers and R. H. Holm, "Modeling Coordination Sites in Metallobiomolecules," *Science*, 1980, *209*, 223. An excellent article dealing with model studies of oxygen transport and storage, and electron carriers.

E. Ochiai, *Bioinorganic Chemistry*, Allyn and Bacon, Boston, 1977.

E. Ochiai, "Principles in Bioinorganic Chemistry," *J. Chem. Educ.* 1978, *55*, 631.

S. J. Lippard, Ed., *Current Research Topics in Bioinorganic Chemistry*, 1973, *18*.

K. N. Raymond, Ed., *Bioinorganic Chemistry II, Adv. Chem. Ser.* 1977, *162*. An excellent collection of symposium papers.

L. Vaska, "Dioxygen–Metal Complexes: Toward a Unified View," *Acc. Chem. Res.* 1976, *9*, 175.

R. A. D. Wentworth, "Mechanisms of Molybdenum in Enzymes," *Coord. Chem. Rev.* 1976, *18*, 1.

PROBLEMS

16.1   For each of the following elements, identify one significant role in biological processes: Fe, Mn, Cu, Zn, I, Mg, Co, Ca, and K.

16.2   What prevents simple iron porphyrins from functioning as O_2 carriers? How has this problem been avoided in successful models of Fe–porphyrin O_2 carriers?

16.3   How is iron stored and transported in mammals? What is the oxidation state of iron for storage and for transfer?

16.4   Give an example of each of two types of reactions brought about by vitamin B_{12}.

16.5   What is the cytochrome chain? What are the advantages of such a complex system?

16.6   What are the prosthetic groups of cytochromes and hemoglobin?

16.7   What are the two important systems for biological electron-transfer processes?

16.8   Identify two chemical types of siderophores.

16.9   The formation constant of Fe(III) enterobactin is about 10^{52}. Calculate the concentration of Fe^{3+} in equilibrium with $10^{-4}\,M$ Fe(III) enterobactin and $10^{-4}\,M$ ent^{6-}. This corresponds to how many liters per Fe^{3+}? The volume of the hydrosphere (all bodies of water, snow, and ice) is *ca.* 1.37×10^{21} L (See K.N. Raymond *et al.*, *J. AM. Chem. Soc.* 1979, *101*, 6097).

16.10   Wilson's disease causes the accumulation of what element in the body? How can symptoms be relieved?

16.11   What chemical properties of Fe and Cu make them suitable for redox processes in biological systems?

16.12   What electron transport systems are used in photosynthesis?

16.13 The conversion of carbonic acid to $CO_2 + H_2O$ is a natural process; why is carbonic anhydrase needed?

16.14 Give an example of the substitution of Co(II) for Zn(II) in an enzyme to provide a ''spectral probe'' for study of the enzyme.

16.15 The direct-reduction products of water, H_2O_2 (or HO_2^-) and O_2^-, are toxic. How are these handled in biological systems?

16.16 What metals are at the active centers of nitrogenase? Name some reduction processes other than that of N_2 that are accomplished by nitrogenase?

16.17 High-spin iron(II) is too large for the opening of the porphyrin ring, but low-spin iron(II) can be accommodated in the opening. Why does the high-spin ion have a larger radius?

APPENDIX A

Units and Physical Constants

By international convention the units employed for scientific measurement are those of the International System of Units (SI).[1] We use SI units in this book, except in a few cases (for example, we express pressure in atm). Seven base units regarded as dimensionally independent are recognized. Other units are derived by combining base units according to algebraic relations. Some of these derived units have special names. The SI base units are as follows.

	Name	*Symbol*
Mass	kilogram	kg
Length	meter	m
Time	second	s
Electric current	ampere	A
Thermodynamic temperature	kelvin	K
Amount of substance	mole	mol
Luminous intensity	candela	cd

[1]See NBS Special Publication 330, 1977 Edition, U.S. Government Printing Office, Washington, 1977.

Some derived units recognized in SI are listed below.

Quantity	Name	Symbol	Expression in Terms of Other Units	Expression in Terms of SI Base Units
Frequency	hertz	Hz		s^{-1}
Force	newton	N		$m \cdot kg \cdot s^{-2}$
Pressure, stress	pascal	Pa	N/m^2	$m^{-1} \cdot kg \cdot s^{-2}$
Energy, work, quantity of heat	joule	J	$N \cdot m$	$m^2 \cdot kg \cdot s^{-2}$
Power, radiant flux	watt	W	J/s	$m^2 \cdot kg \cdot s^{-3}$
Quantity of electricity, electric charge[a]	coulomb	C	$A \cdot s$	$s \cdot A$
Electric potential, potential difference, electromotive force	volt	V	W/A	$m^2 \cdot kg \cdot s^{-3} \cdot A^{-1}$
Magnetic flux	weber	Wb	$V \cdot s$	$m^2 \cdot kg \cdot s^{-2} \cdot A^{-1}$
Magnetic flux density	tesla	T	Wb/m^2	$kg \cdot s^{-2} \cdot A^{-1}$
Celsius temperature	degree Celsius	°C		K
Heat capacity, entropy	joule per kelvin		J/K	$m^2 \cdot kg \cdot s^{-2} \cdot K^{-1}$
Specific heat capacity, specific entropy	joule per kilogram kelvin		$J/(kg \cdot K)$	$m^2 \cdot s^{-2} \cdot K^{-1}$
Thermal conductivity	watt per meter kelvin		$W/(m \cdot K)$	$m \cdot kg \cdot s^{-3} \cdot K^{-1}$
Energy density	joule per cubic meter		J/m^3	$m^{-1} \cdot kg \cdot s^{-2}$
Electric field strength	volt per meter		V/m	$m \cdot kg \cdot s^{-3} \cdot A^{-1}$
Electric charge density	coulomb per cubic meter		C/m^3	$m^{-3} \cdot s \cdot A$
Electric flux density	coulomb per square meter		C/m^2	$m^{-2} \cdot s \cdot A$
Molar energy	joule per mole		J/mol	$m^2 \cdot kg \cdot s^{-2} \cdot mol^{-1}$
Molar entropy, molar heat capacity	joule per mole kelvin		$J/(mol \cdot K)$	$m^2 \cdot kg \cdot s^{-2} \cdot K^{-1} \cdot mol$

[a]In lattice energy calculations (see Section 6.3), dimensional analysis usually is simplified by expressing electrical charge in esu whose units are $dyne^{1/2}$ cm.

Multiples of both base and derived units are indicated by one of the following prefixes.

	SI Prefixes				
Factor	Prefix	Symbol	Factor	Prefix	Symbol
10^{18}	exa	E	10^{-1}	deci	d
10^{15}	peta	P	10^{-2}	centi	c
10^{12}	tera	T	10^{-3}	milli	m
10^{9}	giga	G	10^{-6}	micro	μ
10^{6}	mega	M	10^{-9}	nano	n
10^{3}	kilo	k	10^{-12}	pico	p
10^{2}	hecto	h	10^{-15}	femto	f
10^{1}	deka	da	10^{-18}	atto	a

Note that the kilogram is the only base unit that contains one of these multiplicative prefixes. Physical constants[2] are given here in SI and other units. Conversion factors are given inside the back cover. In this country L is used for liter to avoid the confusion between l (printed ℓ) and 1.

Physical Constants

Velocity of light	$c = 2.997925 \times 10^8$ m/s
Planck's constant	$h = 6.626176 \times 10^{-34}$ J s
	$\hbar = 1.054589 \times 10^{-34}$ J s
	$= h/2\pi$
Boltzmann constant (gas constant per molecule)	$k = 1.380662 \times 10^{-23}$ J/K
Gas constant	$R = 8.31441$ J/mol K
	$= 1.98719$ cal/mol K
	$= 0.08205$ L atm/mol K
Avogadro's number	$N_A = 6.0220 \times 10^{23}$/mol
Electron charge	$e = 1.60219 \times 10^{-19}$ C
	$= 4.80324 \times 10^{-10}$ esu
Faraday	$F = 9.648456 \times 10^4$ C/mol
Atomic mass unit	amu $= 1.6605665 \times 10^{-27}$ kg
Electron rest mass	$m_e = 9.109534 \times 10^{-31}$ kg
	$= 5.4858026 \times 10^{-4}$ amu
Proton rest mass	$m_p = 1.6726485 \times 10^{-27}$ kg
	$= 1.007276470$ amu
Neutron rest mass	$m_n = 1.6749543 \times 10^{-27}$ kg
	$= 1.00866501$ amu
Bohr magneton	$\mu_B = 9.274078 \times 10^{-24}$ J/T
Nuclear magneton	$\mu_N = 5.050824 \times 10^{-27}$ J/T
Debye	$D = 1.0 \times 10^{-18}$ esu cm
	$= 3.33 \times 10^{-32}$ C cm
Gravitational constant	$G = 6.6720 \times 10^{-11}$ N m^3/sec kg
Acceleration of gravity	$g = 0.980665$ m/sec^2
Rydberg constant	$R = 1.097373177 \times 10^7$/m

[2]Values from E. R. Cohen and B. N. Taylor, *J. Phys. Chem. Ref. Data* 1973, *2*, 663.

APPENDIX B

Nomenclature of Inorganic Chemistry*

ATOMIC SYMBOLS, MASS, ATOMIC NUMBER, ETC.

The approved symbols and names of the elements are given in Table B.1. The mass number, atomic number, number of atoms, and atomic charge are to be represented as follows:

left upper index	mass number
left lower index	atomic number
right lower index	number of atoms
right upper index	ionic charge

Example $^{200}_{80}Hg_2^{2+}$ represents the doubly charged ion containing two mercury atoms, each of which has the mass number 200. The charge is to be written as Hg_2^{2+}, not Hg_2^{+2}.

In the periodic table subgroups A are to the left and subgroups B to the right. [Freshman texts commonly designate the main group elements as A and transition groups (Sc group-Zn group) as B. Confusion is avoided if we talk about the C family or the N family, *etc*. An interesting solution is to number the columns in the long form of the Table 1–18. Check it.]

*Abbreviated version of the 1970 Report of the Commission on the Nomenclature of Inorganic Chemistry, International Union of Pure and Applied Chemistry; published with the permission of the International Union of Pure and Applied Chemistry and Pergamon Press, Oxford; also published in *Pure and Applied Chemistry,* Vol. 28, No. 1 (1971).

FORMULAS

The molecular formula is used for compounds that exist as discrete molecules, for example, S_2Cl_2, not SCl; $Co_2(CO)_8$, not $Co(CO)_4$. If the molecular weight varies with changes in conditions, the simplest formula may be used unless it is desired to indicate the molecular complexity for given conditions, for example, S, P, and NO_2 may be used instead of S_8, P_4, and N_2O_4.

The electropositive constituent is placed first in the formula, for example, NaCl, $MgCO_3$. If the compound contains more than one electropositive or more than one electronegative constituent, the sequence within each class is in alphabetical order of their symbols. Acids are treated as hydrogen salts. In the case of binary compounds between nonmetals, that constituent should be placed first which appears earlier in the sequence.

$$\text{Rn, Xe, Kr, B, Si, C, Sb, As, P, N, H, Te, Se, S, At, I, Br, Cl, O, F} \tag{1}$$

Ths sequence roughly follows the order of electronegativities without overlap of periodic groups.

Examples NH_3, H_2Te, $BrCl$, Cl_2O, OF_2, XeF_2.

Exceptions to the above order are encountered in compounds in which the sequence of symbols is used to indicate the order in which the atoms are bonded in the molecule or ion, for example, HOCN (cyanic acid), HONC (fulminic acid), HNCO (isocyanic acid).

Table B.1   The elements

Name	Symbol	Atomic Number	Name	Symbol	Atomic Number
Actinium	Ac	89	Carbon	C	6
Aluminum	Al	13	Cerium	Ce	58
Americium	Am	95	Cesium	Cs	55
Antimony	Sb	51	Chlorine	Cl	17
Argon	Ar	18	Chromium	Cr	24
Arsenic	As	33	Cobalt	Co	27
Astatine	At	85	Copper	Cu	29
Barium	Ba	56	Curium	Cm	96
Berkelium	Bk	97	Dysprosium	Dy	66
Beryllium	Be	4	Einsteinium	Es	99
Bismuth	Bi	83	Erbium	Er	68
Boron	B	5	Europium	Eu	63
Bromine	Br	35	Fermium	Fm	100
Cadmium	Cd	48	Fluorine	F	9
Calcium	Ca	20	Francium	Fr	87
Californium	Cf	98	Gadolinium	Gd	64

Table B.1 *(Continued)*

Name	Symbol	Atomic Number	Name	Symbol	Atomic Number
Gallium	Ga	31	Potassium	K	19
Germanium	Ge	32	Praseodymium	Pr	59
Gold	Au	79	Promethium	Pm	61
Hafnium	Hf	72	Protactinium	Pa	91
Helium	He	2	Radium	Ra	88
Holmium	Ho	67	Radon	Rn	86
Hydrogen	H	1	Rhenium	Re	75
Indium	In	49	Rhodium	Rh	45
Iodine	I	53	Rubidium	Rb	37
Iridium	Ir	77	Ruthenium	Ru	44
Iron	Fe	26	Samarium	Sm	62
Krypton	Kr	36	Scandium	Sc	21
Lanthanum	La	57	Selenium	Se	34
Lawrencium	Lr	103	Silicon	Si	14
Lead	Pb	82	Silver	Ag	47
Lithium	Li	3	Sodium	Na	11
Lutetium	Lu	71	Strontium	Sr	38
Magnesium	Mg	12	Sulfur	S	16
Manganese	Mn	25	Tantalum	Ta	73
Mendelevium	Md	101	Technetium	Tc	43
Mercury	Hg	80	Tellurium	Te	52
Molybdenum	Mo	42	Terbium	Tb	65
Neodymium	Nd	60	Thallium	Tl	81
Neon	Ne	10	Thorium	Th	90
Neptunium	Np	93	Thulium	Tm	69
Nickel	Ni	28	Tin	Sn	50
Niobium	Nb	41	Titanium	Ti	22
Nitrogen	N	7	Tungsten	W	74
Nobelium	No	102	Uranium	U	92
Osmium	Os	76	Vanadium	V	23
Oxygen	O	8	Xenon	Xe	54
Palladium	Pd	46	Ytterbium	Yb	70
Phosphorus	P	15	Yttrium	Y	39
Platinum	Pt	78	Zinc	Zn	30
Plutonium	Pu	94	Zirconium	Zr	40
Polonium	Po	84			

If two or more different atoms or groups are attached to a single central atom, the symbol of the central atom is placed first followed by the symbols of the remaining atoms or groups in alphabetical order.

Parentheses () or brackets [] are used to improve clarity, for example $[Co(NH_3)_6]_2(SO_4)_3$. Hydrates are written as follows: $Na_2SO_4 \cdot 10H_2O$.

The prefixes *cis* and *trans* are italicized and separated from the formula by a hyphen, for example, *cis*-$[PtCl_2(NH_3)_2]$.

SYSTEMATIC NAMES

The name of the electropositive constitutent is not modified and is placed first. The name of the electronegative constituent [or the element later in sequence (1) for compounds of non-metals] is modified to end in -ide if it is monoatomic or homopolyatomic. Elements other than those in sequence (1) are taken in order of the elements in the long form of the periodic table starting with Rn (Group 0) and going upward in each family (from Group I on the left) before starting the next, thus ending with fluorine, sequence (2).

Examples   Sodium chloride, magnesium sulfide, lithium nitride, nickel arsenide, silicon carbide, sulfur hexafluoride, oxygen difluoride, potassium triiodide, and sodium plumbide.

If the electronegative constituent is heteronuclear it should be designated by the termination -ate. Exceptions include OH^-, hydroxide ion; NH^{2-}, imide ion; NH_2^-, amide ion; and CN^-, cyanide ion.

In case of two or more electronegative constituents their sequence should be in alphabetical order. This order might differ in names and formulas.

Complex anions can be named using the name of the characteristic or central atom modified to end in -ate. Ligands attached to the central atom are indicated by the termination -o. The oxidation number of the central atom should be indicated by a Roman numeral (Stock system) or by making the charge on the anion clear by the use of prefixes.

Examples

$Na_2[SO_4]$	sodium tetraoxosulfate(VI) or disodium tetraoxosulfate	$Na[SO_3F]$	sodium trioxofluorosulfate(VI)
$Na_2[SO_3]$	sodium trioxosulfate(IV) or disodium trioxosulfate	$Na[ICl_4]$	sodium tetrachloroiodate(III)
$Na_2[S_2O_3]$	disodium trioxothiosulfate	$Na[PCl_6]$	sodium hexachlorophosphate(V)

Common names of oxo anions and oxo acids may be used (see p. 766).

Stoichiometric proportions may be denoted by the use of Greek numerical prefixes (mono, di, tri, tetra, penta, hexa, hepta, octa, ennea, deca, hendeca, dodeca) preceding without hyphen the names of the elements to which they refer. The Latin prefixes nona (9) and undeca (11) are also used. The prefix mono may be omitted. Beyond 10, Arabic numerals may be used. The end vowels of numerical prefixes should not be elided, for example, tetraoxide.

Examples

N_2O	dinitrogen oxide	Fe_3O_4	triiron tetraoxide
N_2O_4	dinitrogen tetraoxide	U_3O_8	triuranium octaoxide
S_2Cl_2	disulfur dichloride		

The proportions of the constitutents may also be indicated indirectly by the Stock system, in which Roman numerals are used to represent the oxidation number of an element or central atom. For zero the Arabic 0 is used. Latin names of the elements or Latin stems may be used with the Stock system; such usage is common for complex anions. The charge on the aggregate can be shown by an Arabic numeral followed by the charge in parentheses (Ewens and Bassett system) instead of giving the oxidation number of the central atom.

$FeCl_2$	iron(II) chloride
$FeCl_3$	iron(III) chloride
$SnSO_4$	tin(II) sulfate
MnO_2	manganese(IV) oxide
BaO_2	barium(II) peroxide
$Pb_2^{II}Pb^{IV}O_4$	dilead(II) lead(IV) oxide or trilead tetraoxide
$K_4[Ni(CN)_4]$	potassium tetracyanonickelate(0) or potassium tetracyanonickelate(4 −) or potassium tetracyano*niccolate*(4 −)
$K_4[Fe(CN)_6]$	potassium hexacyanoferrate(II)
$K_2[Fe(CO)_4]$	potassium tetracarbonylferrate(-II) or potassium tetracarbonylferrate(2 −)

The use of the endings -ous and -ic for cations is not recommended and should not be used for elements exhibiting more than two oxidation numbers.

Hydrides

The following names are acceptable:

H_2O	water	PH_3	phosphine
NH_3	ammonia	P_2H_4	diphosphane
N_2H_4	hydrazine	AsH_3	arsine
BH_3	borane	SbH_3	stibine
B_2H_6	diborane	BiH_3	bismuthine
SiH_4	silane (Si_2H_6, disilane, etc.)	H_2S_5	pentasulfane
GeH_4	germane		

NAMES FOR IONS AND RADICALS

Cations

Names of monatomic cations are the same as the names of the elements. Oxidation numbers are designated by use of the Stock system.

Examples Cu^+, copper(I) ion; Cu^{2+}, copper(II) ion.

Polyatomic cations formed from radicals which have special names use those names without change. Complex cations are discussed later (pp. 768–771).

Examples

NO^+	nitrosyl cation
NO_2^+	nitryl cation

UO_2^{2+} uranyl(VI) ion

Polyatomic cations derived by the addition of protons to monatomic anions are named by adding the ending -onium to the root of the name of the anion.

Examples Phosphonium, arsonium, stibonium, oxonium, sulfonium, selenonium, telluronium, and iodonium ions.

Exceptions NH_4^+, ammonium ion; $HONH_3^+$, hydroxylammonium ion; $NH_2NH_3^+$, hydrazinium ion; $C_6H_5NH_3^+$, anilinium ion; $C_5H_5NH^+$, pyridinium; $HO_2CCH_2NH_3^+$, glycinium, etc.

The H_3O^+ ion is the oxonium ion. Hydrogen ion may be used for the indefinitely solvated proton or when the hydration is of no particular importance to the matter under consideration.

Anions

Monoatomic anions are named by adding the ending -ide to the stem of the name of the element.

Examples H^-, hydride ion; F^-, fluoride ion; Cl^-, chloride ion; Te^{2-}, telluride ion; N^{3-}, nitride ion; Sb^{3-}, antimonide ion, etc.

Certain polyatomic anions have names ending in -ide. These are:

OH^-	hydroxide	CN^-	cyanide ion	N_3^-	azide ion
O_2^{2-}	peroxide ion	C_2^{2-}	acetylide ion	NH^{2-}	imide ion
O_2^-	hyperoxide ion*	I_3^-	triiodide ion	NH_2^-	amide ion
O_3^-	ozonide ion	HF_2^-	hydrogendifluoride	$NHOH^-$	hydroxylamide ion
S_2^{2-}	disulfide ion		ion	$N_2H_3^-$	hydrazide ion

Ions such as HS^- and HO_2^- are named hydrogensulfide ion and hydrogenperoxide ion, respectively. The names of other polyatomic anions consist of the name of the central atom with the termination -ate in accordance with the naming of complex anions.

Example $[Sb(OH)_6]^-$ hexahydroxoantimonate(V) ion.

Certain anions have names using prefixes (hypo-, per-, etc.) that are well established. These are in accord with the names of the corresponding acids (see p. 765). The termination -ite has been used to denote a lower oxidation state and may be retained in trivial names in the following cases:

*The name superoxide ion is in common use.

NO_2^-	nitrite	SO_3^{2-}	sulfite	ClO_2^-	chlorite
$N_2O_2^{2-}$	hyponitrite	$S_2O_5^{2-}$	disulfite	ClO^-	hypochlorite
NOO_2^-	peroxonitrite	$S_2O_4^{2-}$	dithionite	(and correspondingly for
AsO_3^{3-}	arsenite	$S_2O_2^{2-}$	thiosulfite	the other halogens)
		SeO_3^{2-}	selenite	

Other anions that have used the -ite ending, for example, antimonite, should be named according to the general rule, that is, antimonate(III).

Radicals

The names ending in -yl of the following radicals are approved:

HO	hydroxyl	SO	sulfinyl (thionyl)	ClO	chloroxyl
CO	carbonyl	SO_2	sulfonyl (sulfuryl)	ClO_2	chloryl
NO	nitrosyl	S_2O_5	disulfuryl	ClO_3	perchloryl
NO_2	nitryl	SeO	seleninyl	(and similarly for
PO	phosphoryl	SeO_2	selenonyl	other halogens)
		CrO_2	chromyl	NpO_2	neptunyl
		UO_2	uranyl	PuO_2	plutonyl, etc.

The prefixes thio-, seleno-, etc., are used for other chalcogens in place of oxygen.

Examples PS, thiophosphoryl; CSe, selenocarbonyl, etc. The oxidation number of the characteristic element is denoted by the Stock system or the charge on the ion is indicated by the Ewens-Bassett system. UO_2^{2+} may be uranyl(VI) or uranyl(2+) and UO_2^+ may be uranyl(V) or uranyl(1+).

Radicals are treated as the positive part of a compound.

Examples $COCl_2$, carbonyl chloride; NOS, nitrosyl sulfide; POCl, phosphoryl(III) chloride, IO_2F, iodyl(V) fluoride, etc.

Acids

Acids giving rise to -ide anions are named as binary and pseudobinary compounds of hydrogen, for example, hydrogen chloride, hydrogen cyanide, hydrogen azide, etc.

Other acids may be named as pseudobinary compounds of hydrogen, for example, H_2SO_4, hydrogen sulfate; $H_4Fe(CN)_6$, hydrogen hexacyanoferrate(II).

Oxoacids

Most of the common acids are oxoacids commonly named using the -ic and -ous endings in place of the anion endings -ate and -ite, respectively. The acids using the -ous ending should be restricted to those listed above for the -ite anions.

The prefix hypo- is used to denote a lower oxidation state and the prefix per- is used to denote a higher oxidation state. These prefixes should be limited to the following cases:

$H_2N_2O_2$	Hyponitrous acid	HClO	Hypochlorous acid
			(Also HBrO and HIO)
$H_4P_2O_6$	Hypophosphoric acid	$HClO_4$	Perchloric acid
			(Also $HBrO_4$ and HIO_4)

$(HO)_2PHO$ is phosphonic acid rather than phosphorous acid in order to reserve the name phosphite for organic derivatives of $P(OH)_3$. Phosphinic acid is the name for $HOPH_2O$, rather than hypophosphorous acid.

The prefixes ortho- and meta- may be used to distinguish acids differing in the "content of water" in the following cases:

H_3BO_3	orthoboric acid	$(HBO_2)_n$	metaboric acids
H_4SiO_4	orthosilicic acid	$(H_2SiO_3)_n$	metasilicic acids
H_3PO_4	orthophosphoric acid	$(HPO_3)_n$	metaphosphoric acids
H_5IO_6	orthoperiodic acid		
H_6TeO_6	orthotelluric acid		

Acids obtained by removing water from H_5IO_6 or H_6TeO_6 and other acids not covered by specific names should be given systematic names, for example, H_2ReO_4, tetraoxorhenic(VI) acid; H_2NO_2, dioxonitric(II) acid; and H_2MnO_4, tetraoxomanganic(VI) acid.

The prefix di- is preferred to the prefix pyro- for $H_2S_2O_7$, disulfuric acid; $H_2S_2O_5$, disulfurous acid; and $H_4P_2O_7$, diphosphoric acid.

The names germanic acid, stannic acid, molybdic acid, etc., may be used for substances with indefinite "water content" and degree of polymerization.

Peroxoacids

The prefix peroxo- indicates the substitution of —O— by —O—O— (see coordination compounds).

Examples

HNO_4	dioxoperoxonitric acid or peroxonitric acid
H_3PO_5	trioxoperoxophosphoric acid or peroxophosphoric acid
$H_4P_2O_8$	μ-peroxo-bis-trioxophosphoric acid or peroxodiphosphoric acid
H_2SO_5	trioxoperoxosulfuric acid or peroxosulfuric acid
$H_2S_2O_8$	μ-peroxo-bis-trioxosulfuric acid or peroxodisulfuric acid

Thioacids

The prefix thio- indicates the replacement of oxygen by sulfur. The prefixes seleno- and telluro- may be used in a similar manner.

Examples

$H_2S_2O_2$	thiosulfurous acid	H_3PO_3S	monothiophosphoric acid
$H_2S_2O_3$	thiosulfuric acid	$H_3PO_2S_2$	dithiophosphoric acid
HSCN	thiocyanic acid	H_2CS_3	trithiocarbonic acid

Acids containing ligands other than O and S are generally named as complexes.

Salts and Saltlike Compounds

Simple salts are named as binary compounds using the names as prescribed for ions.

Acid Salts

Salts that contain acidic hydrogens are named by treating hydrogen as a positive constituent.

Examples

$NaHCO_3$ sodium hydrogencarbonate
NaH_2PO_4 sodium dihydrogenphosphate

Double Salts, etc.

Cations

Cations other than hydrogen are cited in alphabetical order, which may be different in formulas and names.

Examples

$KMgF_3$	magnesium potassium fluoride
$NaTl(NO_3)_2$	sodium thallium(I) nitrate
$KNaCO_3$	potassium sodium carbonate
$MgNH_4PO_4 \cdot 6H_2O$	ammonium magnesium phosophate hexahydrate (or 6-water)
$Na(UO_2)_3[Zn(H_2O)_6](C_2H_3O_2)_9$	hexaaquazinc sodium triuranyl(VI) nonaacetate
$NaNH_4HPO_4 \cdot 4H_2O$	ammonium sodium hydrogenphosphate tetrahydrate
$AlK(SO_4)_2 \cdot 12H_2O$	aluminum potassium sulfate 12-water

Anions

Anions are to be cited in alphabetical order which may be different in names and formulas.

Examples

$NaCl \cdot NaF \cdot 2Na_2SO_4$ or $Na_6ClF(SO_4)_2$	hexasodium chloride fluoride bis(sulfate)*
$Ca_5F(PO_4)_3$	pentacalcium fluoride tris(phosphate)*

Basic salts should be treated as double salts, not as oxo or hydroxo salts.

*The prefixes bis, tris, tetrakis, etc., are used for anions to avoid confusion with disulfate, etc., and for complex expressions to avoid ambiguity.

Examples

MgCl(OH)	magnesium chloride hydroxide
BiClO	bismuth chloride oxide
$ZrCl_2O \cdot 8H_2O$	zirconium dichloride oxide octahydrate
$CuCl_2 \cdot 3Cu(OH)_2$ or $Cu_2Cl(OH)_3$	dicopper chloride trihydroxide
$VO(SO_4)$	vanadium(IV) oxide sulfate

Coordination Compounds

The symbol for the central atom is placed first in the formula of coordination compounds. The ligands are written in the alphabetical order of the symbols of the ligating atoms. The formula for the complex molecule or ion is enclosed in square brackets []. In names the central atom is placed after the ligands. The ligands are listed in alphabetical order regardless of the number of each. Thus diammine is listed under "a" and dimethylamine under "d."

The oxidation number of the central atom is indicated by the Stock notation. Alternatively, the proportion of constituents may be given by means of stoichiometric prefixes, or the charge on the entire ion can be designated by the Ewens-Bassett number. Formulas and names may be supplemented by italicized prefixes *cis, trans, fac, mer,* etc. Names of complex anions end in -ate. Complex cations and neutral molecules are given no distinguishing ending.

Names of Ligands

The names for anionic ligands end in -o (-ido, -ito, and -ato commonly).

Examples

$Li[AlH_4]$	lithium tetrahydridoaluminate
$K_2[OsCl_5N]$	potassium pentachloronitridoosmate(VI) or potassium pentachloronitridoosmate(2−)
$Na_3[Ag(S_2O_3)_2]$	sodium bis(thiosulfato)argentate(I) or (3−)
$[Ni(C_4H_7O_2N_2)_2]$	bis(2,3-butanedione dioximato)nickel(II) or omit (II)
$[Cu(C_5H_7O_2)_2]$	bis(2,4-pentanedionato)copper(II) or omit (II)
$K_2[Cr(CN)_2O_2(O_2)NH_3]$	potassium amminedicyanodioxoperoxochromate(VI) or (2−)

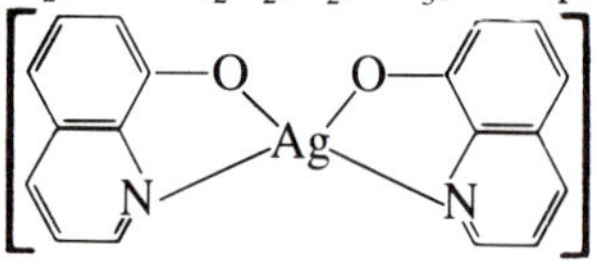

bis(8-quinolinolato)silver(II) or omit (II)

The following exceptions are recognized:

H^-	hydrido (or hydro)		HS^-	mercapto
F^-	fluoro		S^{2-}	thio
Cl^-	chloro		CN^-	cyano
Br^-	bromo		CH_3O^-	methoxo or methanolato
I^-	iodo		$C_6H_5^-$	phenyl
O^{2-}	oxo		$C_5H_5^-$	cyclopentadienyl

OH^-	hydroxo	Other hydrocarbon anions are also given
O_2^{2-}	peroxo	radical names without the -o ending.

Neutral and cationic ligands are given no special endings. Water and ammonia are called aqua and ammine, respectively, in complexes. Groups such as NO and CO are named as radicals and treated as neutral ligands.

Examples

$Ba[BrF_4]_2$	barium tetrafluorobromate(III) or $(1-)$
$K[CrF_4O]$	potassium tetrafluorooxochromate(V) or $(1-)$
$Na[BH(OCH_3)_3]$	sodium hydrotrimethoxoborate(III) or $(1-)$
$[CuCl_2(CH_3NH_2)_2]$	dichlorobis(methylamine)copper(II) or omit (II)
$[Pt(py)_4][PtCl_4]$	tetrakis(pyridine)platinum(II) tetrachloroplatinate(II) or $(2+)$ and $(2-)$
$[Co(en)_3]_2(SO_4)_3$	tris(ethylenediamine)cobalt(III) sulfate or $(3+)$
$K[PtCl_3(C_2H_4)]$	potassium trichloro(ethylene)platinate(II) or $(1-)$
$[Al(OH)(H_2O)_5]^{2+}$	pentaaquahydroxoaluminum(III) ion or $(2+)$
$K_3[Fe(CN)_5NO]$	potassium pentacyanonitrosylferrate(II) or $(3-)$
$[CoCl_3(NH_3)_2\{(CH_3)_2NH\}]$	diamminetrichloro(dimethylamine)cobalt(III) or omit (III)
$K[SbCl_5C_6H_5]$	potassium pentachloro(phenyl)antimonate(V) or $(1-)$
$Fe(C_5H_5)_2$	bis(cyclopentadienyl)iron(II) or omit (II)
$[Cr(C_6H_6)_2]$	bis(benzene)chromium(0) or omit (0)
$[Ru(NH_3)_5N_2]Cl_2$	pentaammine(dinitrogen)ruthenium(II) chloride or $(2+)$

Alternative Modes of Linkage

A ligand that may be attached through different atoms, for example SCN^-, may be distinguished as follows:

$$M—SCN \quad \text{thiocyanato-S} \quad or \quad M—NCS \quad \text{thiocyanato-N}$$

$$K_2\left[Ni\left(\begin{array}{c} S—C=O \\ | \\ S—C=O \end{array}\right)_2\right] \quad \text{potassium bis(dithiooxalato-S,S')nickelate(II) or (2-)}$$

$$\left[\begin{array}{c} Cl \quad H_2N—CH_2 \\ \diagdown Pt \diagup \\ Cl \quad S—CH_2 \\ | \\ C_2H_4N(CH_3)_2 \end{array}\right]$$

dichloro [*N*,*N*-dimethyl-2,2'-thiobis(ethylamine)-S,N']platinum(II) or omit (II), or dichloro[2-methyl-2,8-diaza-5-thia octane-S, 8-N]-platinum

Where special names are recognized for alternative modes of linkage, these may be used, for example, thiocyanato (—SCN), isothiocyanato (—NCS), nitro (—NO_2), and nitrito (—ONO).

Examples

$[Co(NO_2)_3(NH_3)]$	triamminetrinitrocobalt(III) or omit (III)
$[Co(ONO)(NH_3)_5]SO_4$	pentaamminenitritocobalt(III) sulfate or $(2+)$
$[Co(NCS)(NH_3)_5]Cl_2$	pentaammineisothiocyanatocobalt(III) chloride or $(2+)$

Di- and Polynuclear Compounds

Bridging groups are indicated by adding the Greek letter μ immediately before the names of the groups. Two or more bridging groups of the same kind are indicated by di-μ-, etc. If a bridging group bridges more than two metals, use μ_3, μ_4, etc. Bridging groups are listed with other groups in alphabetical order unless the symmetry of the molecule permits a simpler name (first example). If the same ligand is present in bridging and nonbridging roles, it is cited first as a bridging ligand.

Examples

$[(NH_3)_5Cr—OH—Cr(NH_3)_5]Cl_5$	μ-hydroxo-bis[pentaamminechromium(III)] chloride or (5+)
$[(CO)_3Fe(CO)_3Fe(CO)_3]$	tri-μ-carbonyl-bis(tricarbonyliron)
$[\{Au(CN)(C_3H_7)_2\}_4]$	cyclo-tetra-μ-cyano-tetrakis(dipropylgold)
$[Be_4O(CH_3COO)_6]$	hexa-μ-acetato(O,O')-μ_4-oxo-tetraberyllium(II) or omit (II)
$[(CH_3Hg)_4S]^{2+}$	μ_4-thio-tetrakis[methylmercury(II)] ion or (2+)

Extended structures may be indicated by the prefix *catena-μ*, for example, $CsCuCl_3$ contains the anion

$$\cdots Cl—\underset{\underset{Cl}{\overset{Cl}{|}}}{Cu}—Cl—\underset{\underset{Cl}{\overset{Cl}{|}}}{Cu}—Cl—\underset{\underset{Cl}{\overset{Cl}{|}}}{Cu}—Cl—\underset{\underset{Cl}{\overset{Cl}{|}}}{Cu}\cdots$$

The compound may be named cesium *catena-μ*-chloro-dichlorocuprate(II). If the structure were in doubt, however, the substance would be called cesium copper(II) chloride (as a double salt).

Compounds Without Bridging Groups

Symmetrical compounds such as those containing metal-metal bonds are named by the use of multiplicative prefixes. When unsymmetrical, one central atom and its attached ligands are treated as a ligand on the other central atom. The one considered as the primary central atom is the first encountered in sequence (2). The ligating atom is named as a radical.

Examples

$[(CO)_5Mn—Mn(CO)_5]$	bis(pentacarbonylmanganese)
$[(CO)_4Co—Re(CO)_5]$	pentacarbonyl(tetracarbonylcobaltio)rhenium
$[(C_6H_5)_3AsAuMn(CO)_5]$	pentacarbonyl[(triphenylarsine)aurio]manganese

Metal Cluster Compounds

The geometrical shape of the cluster is designated by *triangulo, quadro, tetrahedro, octahedro*, etc.

Examples

$Os_3(CO)_{12}$	dodecacarbonyl-*triangulo*-triosmium
$Cs_3[Re_3Cl_{12}]$	cesium dodecachloro-*triangulo*-trirhenate(3−)

B_4Cl_4	tetrachloro-*tetrahedro*-tetraboron
$[Nb_6Cl_{12}]^{2+}$	dodeca-μ-chloro-*octahedro*-hexaniobium(2+) ion
$[Mo_6Cl_8]^{4+}$	octa-μ_3-chloro-*octahedro*-hexamolybdenum(4+) ion
$Cu_4I_4(PEt_3)_4$	tetra-μ_3-iodo-tetrakis(triethylphosphine)-*tetrahedro*-tetracopper

Some aspects of the designation of geometrical and optical isomers are covered in Chapter 8, and the naming of boron compounds is discussed briefly in Chapter 15. For more details and for rules dealing with isopolyanions, heteropolyanions, and nonstoichiometric phases, and for tables of names of ions and radicals, see the original 106-page report.

APPENDIX C

Standard Half-Cell emf Data

ACID SOLUTION (E_A^0)

$$H^+ \xrightarrow{\ 0.00\ } H_2 \xrightarrow{\ -2.25\ } H^-$$

$$O_2 \xrightarrow{\ 0.682\ } H_2O_2 \xrightarrow{\ 1.776\ } H_2O$$

with $O_2 \xrightarrow{\ 1.229\ } H_2O$

Group 0

$$H_4XeO_6 \xrightarrow{\ 2.36\ } XeO_3 \xrightarrow{\ 2.10\ } Xe$$

Group IA		*Group IIA*		*Group III*	
$M^+ \longrightarrow M$		$M^{2+} \longrightarrow M$		$M^{3+} \longrightarrow M$	
Li	−3.040	Be	−1.85	Al	−1.66
Na	−2.714	Mg	−2.37	Sc	−2.02
K	−2.931	Ca	−2.87	Y	−2.37
Rb	−2.925	Sr	−2.89	La	−2.36
Cs	−3.08	Ba	−2.90	Ac	−2.13
		Ra	−2.92		

Group III (con't)

$$
\begin{array}{c}
\overset{-0.47}{\underset{}{\longrightarrow}} \\
\overset{-0.69}{\longrightarrow} \\
H_3BO_3 \xrightarrow{-0.90} B \xrightarrow{-0.08} B_{12}H_{12}^{2-} \xrightarrow{-0.32} B_{10}H_{10}^{2-} \xrightarrow{0.50} B_{10}H_{14} \xrightarrow{-0.16} B_2H_6 \xrightarrow{-0.36} BH_4^- \\
\underset{-0.67}{\longrightarrow} \qquad \underset{-0.24}{\longrightarrow}
\end{array}
$$

Lanthanides

	IV/III	III/II	III/0
La	—	-3.74	-2.36
Ce	1.76	-3.76	-2.34
Pr	3.9	-3.03	-2.35
Nd	4.9	-2.62	-2.32
Pm	5.4	-2.67	-2.29
Sm	5.2	-1.57	-2.30
Eu	6.2	-0.35	-1.99
Gd	7.4	-3.82	-2.29
Tb	3.1	-3.47	-2.30
Dy	4.5	-2.42	-2.29
Ho	5.7	-2.80	-2.33
Er	5.7	-2.96	-2.31
Tm	5.6	-2.27	-2.31
Yb	6.8	-1.04	-2.22
Lu	8.1	—	-2.30

Actinides

$$
\begin{array}{c}
\overset{\text{III/0}}{\overline{\qquad\qquad\qquad}} \\
MO_2^{2+} \xrightarrow{\text{VI/V}} MO_2^+ \xrightarrow{\text{V/IV}} M^{4+} \xrightarrow{\text{IV/III}} M^{3+} \xrightarrow{\text{III/II}} M^{2+} \xrightarrow{\text{II/0}} M
\end{array}
$$

	V/III	VI/V	V/IV	IV/III	III/II	III/0	II/0
Ac					(-4.9)	-2.13	(-0.8)
Th				(-3.7)	(-4.9)	-1.17	(0.7)
Pa	-1.0		(0.0)	(-2.0)	(-4.7)	-1.49	(0.1)
U		0.163	0.380	-0.520	(-4.7)	-1.66	(-0.1)
Np		1.236	0.74	0.155	(-4.7)	-1.79	(1.3)
Pu		1.013	1.17	0.982	(-3.5)	-2.00	(1.2)

Actinides (con't)

	V/III	VI/V	V/IV	IV/III	III/II	III/0	II/0
Am		1.70	0.86	2.62	(−2.3)	−2.07	(2.0)
Cm		(2.0)	(2.7)	3.1	(−4.4)	−2.06	(−0.8)
Bk		(2.8)	(3.4)	1.64	(−2.8)	−1.97	(−1.6)
Cf		(3.9)	(0.9)	(3.2)	−1.6	−2.01	(−2.2)
Es		(1.8)	(2.1)	(4.5)	−1.2	−1.98	(−2.4)
Fm		(3.3)	(2.9)	(4.9)	(−1.1)	−1.95	(−2.4)
Md		(4.4)	(2.8)	(5.4)	−0.15	−1.66	−2.42
No		(4.4)	(2.9)	(6.5)	1.45	−1.18	−1.04
Lr		(4.8)	(3.5)	(7.9)		−2.06	

Group IVA

$$TiO^{2+} \xrightarrow{\text{0.10}} Ti^{3+} \xrightarrow{\text{−0.37}} Ti^{2+} \xrightarrow{\text{−1.63}} Ti$$

$$ZrO_2 \xrightarrow{\text{−1.46}} Zr \qquad Zr^{4+} \xrightarrow{\text{−1.54}} Zr \qquad HfO_2 \xrightarrow{\text{−1.51}} Hf$$

$$Hf^{4+} \xrightarrow{\text{−1.70}} Hf$$

Group VA

$$V(OH)_4^+ \xrightarrow{\text{1.00}} VO^{2+} \xrightarrow{\text{0.34}} V^{3+} \xrightarrow{\text{−0.26}} V^{2+} \xrightarrow{\text{−1.18}} V$$

$$Nb_2O_5 \xrightarrow{\text{0.05}} Nb^{3+} \xrightarrow{\text{−1.10}} Nb$$

$$Ta_2O_5 \xrightarrow{\text{−0.75}} Ta$$

Group VIA

$$Cr_2O_7^{2-} \xrightarrow{\text{1.33}} Cr^{3+} \xrightarrow{\text{−0.41}} Cr^{2+} \xrightarrow{\text{−0.91}} Cr$$

$$MoO_2^{2+} \xrightarrow{\text{0.48}} MoO_2^+ \xrightarrow{\text{0.311}} Mo^{3+} \xrightarrow{\text{−0.20}} Mo$$

$$WO_3 \xrightarrow{\text{−0.03}} W_2O_5 \xrightarrow{\text{−0.04}} WO_2 \xrightarrow{\text{−0.15}} W^{3+} \xrightarrow{\text{−0.11}} W$$

Group VIIA

$$MnO_4^- \xrightarrow{\text{0.564}} MnO_4^{2-} \xrightarrow{\text{0.274}} MnO_4^{3-} \xrightarrow{\text{4.246}} MnO_2 \xrightarrow{\text{0.95}} Mn^{3+} \xrightarrow{\text{1.51}} Mn^{2+} \xrightarrow{\text{−1.19}} Mn$$

(1.507)

(1.695)

$$TcO_4^- \xrightarrow{0.65} TcO_3 \xrightarrow{0.83} TcO_2 \xrightarrow{0.281} Tc^{2+} \xrightarrow{-0.5} Tc \xrightarrow{-1} TcH_9^{2-}$$

$$ReO_4^- \xrightarrow{0.768} ReO_3 \xrightarrow{0.385} ReO_2 \xrightarrow{0.26} Re \xrightarrow{-0.5} ReH_9^{2-}$$

Group VIII

$$FeO_4^{2-} \xrightarrow{2.20} Fe^{3+} \xrightarrow{0.771} Fe^{2+} \xrightarrow{-0.473} Fe$$

$$RuO_4 \xrightarrow{1.40} RuO_2 \cdot xH_2O \xrightarrow{1.99} Ru^{3+} \xrightarrow{0.25} Ru^{2+} \xrightarrow{0.45} Ru$$
with 0.68 branch from $RuO_2 \cdot xH_2O$ to Ru^{2+}, and 0.84 branch

$$OsO_4 \xrightarrow{0.964} OsO_2 \cdot 2H_2O \xrightarrow{0.72} Os$$
(0.84 branch from OsO_4 to Os)

$$CoO_2 \xrightarrow{1.416} Co^{3+} \xrightarrow{1.82} Co^{2+} \xrightarrow{-0.277} Co$$

$$RhO_4^{2-} \xrightarrow{1.5} RhO^{2+} \xrightarrow{1.4} Rh^{3+} \xrightarrow{1.1} Rh^{2+} \xrightarrow{0.6} Rh^+ \xrightarrow{0.6} Rh$$

$$IrO_4^{2-} \xrightarrow{2.056} IrO_2 \xrightarrow{0.233} Ir^{3+} \xrightarrow{1.73} IrO \xrightarrow{0.87} Ir$$
(1.448 branch; 1.156 branch; 0.926 branch)

$$NiO_4^{2-} \longrightarrow NiO_2 \xrightarrow{1.68} Ni^{2+} \xrightarrow{-0.232} Ni$$
(>1.8 branch)

$$Pd^{2+} \xrightarrow{0.92} Pd$$

$$PtO_3 \xrightarrow{2.00} PtO_2 \xrightarrow{0.84} Pt^{2+} \xrightarrow{1.2} Pt$$

$$PtO_3 \xrightarrow{2.00} PtO_2 \xrightarrow{1.05} PtO \xrightarrow{0.98} Pt$$

Group IB

$$CuO^+ \xrightarrow{1.8} Cu^{2+} \xrightarrow{0.153} Cu^+ \xrightarrow{0.521} Cu$$

$$AgO^+ \xrightarrow{2.1} Ag^{2+} \xrightarrow{1.98} Ag^+ \xrightarrow{0.799} Ag$$

$$Au^{3+} \xrightarrow{<1.29} Au^{2+} \xrightarrow{>1.29} Au^+ \xrightarrow{1.68} Au$$
(1.50 branch from Au^{3+} to Au)

Group IIB

$$Zn^{2+} \xrightarrow{-0.76} Zn$$

$$Cd^{2+} \xrightarrow{-0.403} Cd$$

$$Hg^{2+} \xrightarrow{0.91} Hg_2^{2+} \xrightarrow{0.79} Hg$$

Group IIIB

$$Ga^{3+} \xrightarrow{-0.65} Ga^{2+} \xrightarrow{-0.45} Ga$$

with an overarching -0.56 from Ga^{3+} to Ga

$$In^{3+} \xrightarrow{-0.49} In^{2+} \xrightarrow{-0.40} In^{+} \xrightarrow{-0.14} In$$

$$Tl^{3+} \xrightarrow{0.30} Tl^{2+} \xrightarrow{2.22} Tl^{+} \xrightarrow{-0.34} Tl$$

Group IVB

$$CO_2 \xrightarrow{-0.12} CO \xrightarrow{0.51} C \xrightarrow{0.13} CH_4$$

$$SiO_2 \xrightarrow{-0.86} Si \xrightarrow{0.10} SiH_4$$

$$GeO_2 \xrightarrow{-0.50} Ge^{2+} \xrightarrow{0.00} Ge \xrightarrow{-0.867} GeH_4$$

with an overarching -0.25 from GeO_2 to Ge

$$Sn^{4+} \xrightarrow{0.15} Sn^{2+} \xrightarrow{-0.14} Sn$$

$$PbO_2 \xrightarrow{1.46} Pb^{2+} \xrightarrow{-0.13} Pb$$

Group VB

$$NO_3 \xrightarrow{0.79} N_2O_4 \xrightarrow{1.07} HNO_2 \xrightarrow{0.996} NO \xrightarrow{1.59} N_2O \xrightarrow{1.77} N_2 \xrightarrow{-1.87} NH_3OH^{+} \xrightarrow{1.41} N_2H_5^{+} \xrightarrow{1.275} NH_4^{+}$$

with overarching values: 0.95 (NO_3 to HNO_2), 0.71 to $H_2N_2O_2$ $\xrightarrow{2.65}$, 0.27, and -0.11 (NH_3OH^{+} to NH_4^{+})

$$H_3PO_4 \xrightarrow{-0.276} H_3PO_3 \xrightarrow{-0.50} H_3PO_2 \xrightarrow{-0.51} P_4 \xrightarrow{-0.17} P_2H_4 \xrightarrow{0.006} PH_3$$

$$H_3AsO_4 \xrightarrow{0.56} H_3AsO_3 \xrightarrow{0.247} As \xrightarrow{-0.60} AsH_3$$

$$Sb_2O_5 \xrightarrow{0.581} SbO^{+} \xrightarrow{0.212} Sb \xrightarrow{-0.51} SbH_3$$

$$Bi_2O_5 \xrightarrow{1.60} BiO^{+} \xrightarrow{0.32} Bi \xrightarrow{-0.80} BiH_3$$

Group VIB

$$O_3 \xrightarrow{2.076} O_2\ (+H_2O) \qquad O_2 \xrightarrow{-0.131} HO_2 \xrightarrow{1.495} H_2O_2 \xrightarrow{1.776} H_2O$$

$$O_2 \xrightarrow{0.682} H_2O_2$$

$$O_2 \xrightarrow{1.229} H_2O$$

$$SO_4^{2-} \xrightarrow{0.17} SO_2 \qquad SO_2 \xrightarrow{0.45} S_2O_3^{2-}$$

$$SO_4^{2-} \xrightarrow{-0.22} S_2O_6^{2-} \xrightarrow{0.57} SO_2 \xrightarrow{0.51} S_4O_6^{2-} \xrightarrow{0.08} S_2O_3^{2-} \xrightarrow{0.50} S \xrightarrow{0.14} H_2S$$

$$SeO_4^{2-} \xrightarrow{1.15} H_2SeO_3 \xrightarrow{0.74} Se \xrightarrow{-0.40} H_2Se$$

$$H_6TeO_6(s) \xrightarrow{1.02} TeO_2(s) \xrightarrow{0.53} Te \xrightarrow{-0.72} H_2Te$$

$$PoO_2 \xrightarrow{0.73} Po$$

$$PoO_3 \xrightarrow{1.52} PoO_2 \xrightarrow{0.80} Po^{2+} \xrightarrow{0.65} Po \xrightarrow{-1.00} H_2Po$$

Group VIIB

$$F_2 \xrightarrow{2.87} F^-$$

$$ClO_4^- \xrightarrow{1.19} ClO_3^- \xrightarrow{1.21} HClO_2 \xrightarrow{1.645} HClO \xrightarrow{1.63} Cl_2 \xrightarrow{1.36} Cl^-$$

$$BrO_4^- \xrightarrow{1.743} BrO_3^- \xrightarrow{1.49} HBrO \xrightarrow{1.59} Br_2(1) \xrightarrow{1.07} Br^-$$

$$H_5IO_6 \xrightarrow{1.60} IO_3^- \xrightarrow{1.14} HIO \xrightarrow{1.45} I_2(s) \xrightarrow{0.54} I^-$$

$$H_5AtO_6 \xrightarrow{1.6} AtO_3^- \xrightarrow{1.5} HAtO \xrightarrow{1.0} At \xrightarrow{0.3} At^-$$

BASE SOLUTION (E$_B^0$)

$$O_3 \xrightarrow{1.24} O_2 + OH^- \qquad\qquad O_2 \xrightarrow{0.076} HO_2^- \xrightarrow{0.878} OH^-$$

$$H_2O \xrightarrow{-0.828} H_2 + OH^- \qquad\qquad O_2 \xrightarrow{0.401} OH^-$$

Group 0

$$HXeO_6^{3-} \xrightarrow{0.94} HXeO_4^- \xrightarrow{1.24} Xe$$

Group IIA

$$Be_2O_3^{2-} \xrightarrow{-2.62} Be \qquad\qquad Sr(OH)_2 \cdot 8H_2O \xrightarrow{-2.99} Sr$$

Group IIA (con't)

$$Mg(OH)_2 \xrightarrow{-2.69} Mg \qquad\qquad Ba(OH)_2\cdot8H_2O \xrightarrow{-2.97} Ba$$

$$Ca(OH)_2 \xrightarrow{-3.03} Ca$$

Group III

$$B_4O_7^{2-} \xrightarrow{-1.76} B \xrightarrow{-0.81} B_{12}H_{12}^{2-} \xrightarrow{-0.32} B_{10}H_{10}^{2-} \xrightarrow{-1.15} B_{10}H_{14}(s) \xrightarrow{-0.98} B_2H_6 \xrightarrow{-0.78} BH_4^-$$

$$H_2AlO_3^- \xrightarrow{-2.35} Al \qquad\qquad Y(OH)_3 \xrightarrow{-2.8} Y$$

$$Sc(OH)_3 \xrightarrow{-2.6} Sc \qquad\qquad La(OH)_3 \xrightarrow{-2.9} La$$

Lanthanides

$$M(OH)_3 \longrightarrow M$$

La	-2.90	Sm	-2.83	Ho	-2.77
Ce	-2.87	Eu	-2.83	Er	-2.75
Pr	-2.85	Gd	-2.82	Tm	-2.74
Nd	-2.84	Tb	-2.79	Yb	-2.73
Pm	-2.84	Dy	-2.78	Lu	-2.72

Actinides*

$$Th(OH)_4 \xrightarrow{-2.42} Th$$

$$MO_5^{3-} \xrightarrow{\text{VII/VI}} MO_4^{2-} \qquad MO_2(OH)_3^- \xrightarrow{\text{VI/V}} MO_2(OH)_2^- \xrightarrow{\text{V/IV}} M(OH)_5^- \xrightarrow{\text{IV/III}} M(OH)_4^-$$

	VII/VI	VI/V	V/IV	IV/III
U		-0.69	-0.03	-2.78
Np	0.538	0.38	-0.09	-1.88
Pu	0.857	0.16	0.52	-1.04

$$UO_2(OH)_2 \xrightarrow{-0.62} U(OH)_4 \xrightarrow{-2.14} U(OH)_3 \xrightarrow{-2.17} U$$

$$NpO_2(OH)_2 \xrightarrow{0.48} NpO_2OH \xrightarrow{0.39} Np(OH)_4 \xrightarrow{-1.76} Np(OH)_3 \xrightarrow{-2.25} Np$$

$$PuO_2(OH)_2 \xrightarrow{0.26} PuO_2OH \xrightarrow{0.76} Pu(OH)_4 \xrightarrow{-0.95} Pu(OH)_3 \xrightarrow{-2.42} Pu$$

$$AmO_2(OH)_2 \xrightarrow{1.1} AmO_2OH \xrightarrow{0.7} Am(OH)_4 \xrightarrow{-0.5} Am(OH)_3 \xrightarrow{-2.71} Am$$

*The charged species of U, Np, and Pu in the potential diagrams represent the predominant solution species in dilute basic media (see Allard).

Group IVA

$$TiO_2 \xrightarrow{-1.69} Ti \qquad H_2ZrO_3 \xrightarrow{-2.36} Zr \qquad HfO(OH)_2 \xrightarrow{-2.50} Hf$$

Group VA

$$HV_6O_{17}^{3-} \xrightarrow{-1.15} V$$

Group VIA

$$CrO_4^{2-} \xrightarrow{-0.13} Cr(OH)_3 \xrightarrow{-1.1} Cr(OH)_2 \xrightarrow{-1.4} Cr$$
$$CrO_2^- \xrightarrow{\hspace{2cm} -1.2 \hspace{2cm}}$$

$$MoO_4^{2-} \xrightarrow{-0.96} MoO_2 \xrightarrow{-0.91} Mo$$

$$WO_4^{2-} \xrightarrow{-1.007} W$$

Group VIIA

$$MnO_4^- \xrightarrow{0.564} MnO_4^{2-} \xrightarrow{0.27} MnO_4^{3-} \xrightarrow{0.93} MnO_2 \xrightarrow{-0.2} Mn(OH)_3 \xrightarrow{0.1} Mn(OH)_2 \xrightarrow{-1.55} Mn$$

$$TcO_4^- \xrightarrow{-0.322} TcO_2 \xrightarrow{-0.55} Tc$$

$$ReO_4^- \xrightarrow{-0.7} ReO_4^{3-} \xrightarrow{-0.5} ReO_2 \xrightarrow{-0.53} Re(OH)_3 \xrightarrow{-0.6} Re \xrightarrow{-0.4} ReH_9^{2-}$$

Group VIII

$$FeO_4^{2-} \xrightarrow{0.72} Fe(OH)_3 \xrightarrow{-0.56} Fe(OH)_2 \xrightarrow{-0.887} Fe$$

$$RuO_4 \xrightarrow{1.00} RuO_4^- \xrightarrow{0.59} RuO_4^{2-} \xrightarrow{0.35} RuO_2 \cdot xH_2O \xrightarrow{-0.15} Ru$$

$$HOsO_5^- \xrightarrow{0.1} OsO_2 \cdot 2H_2O \xrightarrow{-0.12} Os$$

$$CoO_2 \xrightarrow{0.62} Co(OH)_3 \xrightarrow{0.17} Co(OH)_2 \xrightarrow{-0.72} Co$$

$$RhO_4^{2-} \xrightarrow{-0.1} Rh(OH)_3 \xrightarrow{0.0} Rh$$

$$IrO_4^{2-} \xrightarrow{0.4} IrO_2 \xrightarrow{0.1} Ir_2O_3 \xrightarrow{0.1} Ir$$

$$Ni(OH)_4 \xrightarrow{0.6} Ni(OH)_3 \xrightarrow{0.48} Ni(OH)_2 \xrightarrow{-0.72} Ni$$

Group VIII (con't)

$$Pd(OH)_4 \xrightarrow{0.7} Pd(OH)_2 \xrightarrow{0.07} Pd$$

$$PtO_4^{2-} \xrightarrow{>0.4} Pt(OH)_6^{2-} \xrightarrow{\sim 0.2} Pt(OH)_2 \xrightarrow{0.15} Pt$$

Group IB

$$Cu(OH)_2 \xrightarrow{-0.08} Cu_2O \xrightarrow{-0.36} Cu$$

$$Ag_2O_3 \xrightarrow{0.74} AgO \xrightarrow{0.60} Ag_2O \xrightarrow{0.34} Ag$$

$$HAuO_3^{2-} \xrightarrow{0.8} Au(OH)_2^{-} \xrightarrow{0.4} Au$$

Group IIB

$$ZnO_2^{2-} \xrightarrow{-1.216} Zn$$

$$Cd(OH)_2 \xrightarrow{-0.809} Cd$$

$$HgO \xrightarrow{0.0984} Hg$$

Group IIIB

$$H_2GaO_3^{-} \xrightarrow{-1.22} Ga \qquad In(OH)_3 \xrightarrow{-1.0} In$$

$$Tl(OH)_3 \xrightarrow{-0.05} Tl(OH) \xrightarrow{-0.34} Tl$$

Group IVB

$$CO_3^{2-} \xrightarrow{-1.02} HCO_2^{-} \xrightarrow{-0.52} C \xrightarrow{-0.70} CH_4$$

$$SiO_3^{2-} \xrightarrow{-1.70} Si \xrightarrow{-0.93} SiH_4$$

$$HGeO_3^{-} \xrightarrow{-1.0} Ge \xrightarrow{<-1.1} GeH_4$$

$$Sn(OH)_6^{2-} \xrightarrow{-0.90} HSnO_2^{-} \xrightarrow{-0.91} Sn$$

$$PbO_2 \xrightarrow{0.25} PbO \xrightarrow{-0.58} Pb$$

Group VB

$$NO_3^{-} \xrightarrow{0.86} N_2O_4 \xrightarrow{0.88} NO_2^{-} \xrightarrow{-0.46} NO \xrightarrow{0.76} N_2O \xrightarrow{0.94} N_2 \xrightarrow{-3.04} NH_2OH \xrightarrow{0.73} N_2H_4 \xrightarrow{0.1} NH_3$$

Group VB (con't)

$$PO_4^{3-} \xrightarrow{-1.12} HPO_3^{2-} \xrightarrow{-1.57} H_2PO_2^- \xrightarrow{-2.05} P_4 \xrightarrow{-0.89} PH_3$$

$$AsO_4^{3-} \xrightarrow{-0.67} H_2AsO_3^- \xrightarrow{-0.68} As \xrightarrow{-1.21} AsH_3$$

$$Sb(OH)_6^- \xrightarrow{-0.40} SbO_2^- \xrightarrow{-0.66} Sb \xrightarrow{-1.34} SbH_3$$

$$Bi_2O_5 \xrightarrow{0.78} Bi_2O_4 \xrightarrow{0.56} Bi_2O_3 \xrightarrow{-0.46} Bi \xrightarrow{<-1.6} BiH_3$$

Group VIB

$$O_2 \xrightarrow{-0.08} HO_2^- \xrightarrow{0.88} OH^-$$

$$SO_4^{2-} \xrightarrow{-0.93} SO_3^{2-} \xrightarrow{0.79} S_4O_6^{2-} \xrightarrow{0.08} S_2O_3^{2-} \xrightarrow{-0.74} S \xrightarrow{-0.48} S^{2-}$$

$$SeO_4^{2-} \xrightarrow{0.05} SeO_3^{2-} \xrightarrow{-0.37} Se \xrightarrow{-0.92} Se^{2-}$$

$$TeO_2(OH)_4^{2-} \xrightarrow{0.4} TeO_3^{2-} \xrightarrow{-0.57} Te \xrightarrow{-1.14} Te^{2-}$$

$$Po \xrightarrow{<-1.4} Po^{2-}$$

Group VIIB

$$F_2 \xrightarrow{2.87} F^-$$

$$ClO_3^- \xrightarrow{-0.50} ClO_2 \xrightarrow{1.16} ClO_2^-$$

$$ClO_4^- \xrightarrow{0.36} ClO_3^- \xrightarrow{0.33} ClO_2^- \xrightarrow{0.66} ClO^- \xrightarrow{0.40} Cl_2 \xrightarrow{1.36} Cl^-$$

$$BrO_4^- \xrightarrow{0.915} BrO_3^- \xrightarrow{0.54} BrO \xrightarrow{-0.45} Br_2 \xrightarrow{1.07} Br^-$$

$$H_3IO_6^{2-} \xrightarrow{0.7} IO_3^- \xrightarrow{0.14} IO^- \xrightarrow{0.45} I_2 \xrightarrow{0.535} I^-$$

$$AtO_3^- \xrightarrow{0.6} AtO^- \xrightarrow{0.3} At_2 \xrightarrow{0.3} At^-$$

REFERENCES

Except as noted below the emf half-cell data were taken from W. M. Latimer, *The Oxidation States of the Elements and Their Potentials in Aqueous Solution,* 2nd Ed., Prentice-Hall, New York, 1952.

Part or all of the data for the elements O, V, Mo, Mn, Ru, Ag, Hg, Ga, and Tl were taken from G. Charlot, *Selected Constants Tables No. 8 Oxydo-Reduction Potentials,* Pergamon, New York, 1958.

Part or all of the data for the elements Li, K, Ti, Hf, Ta, Nb, W, Tc, Ni, Au, Pb, Bi, S were taken from G. Milazzo and S. Caroli, *Tables of Standard Electrode Potentials,* Wiley-Interscience, New York, 1978.

In addition to these sources, data for individual elements (listed in the order of their appearance in the appendix) were taken from the literature cited below.

Xe   E. Wiberg, *Lehrbuch der Anorganischen Chemie,* Holleman-Wiberg, de Gruyter, Berlin, 1976.

Cs   H. L. Friedman and M. Kahlweit, *J. Am. Chem. Soc.* 1956, *78,* 2141.

B   A. Kaczmarczyk, W. C. Nichols, W. H. Stockmayer, and T. B. Ames, *Inorg. Chem.* 1968, *7,* 1057.

Sc   J. G. Travers, I. Dellien, and L. G. Hepler, *Thermochimica Acta* 1976, *15,* 89.

La   L. G. Hepler and P. P. Singh, *Thermochimica Acta* 1976, *16,* 95.

Lanthanides   L. R. Morss, *Chem. Rev.* 1976, *76,* 827.

Actinides   B. Allard, H. Kipatsi, and J. O. Liljenzin *J. Inorg. Nucl. Chem.* 1980, *42,* 1015;

S. Ahrland, J. O. Liljenzin, and J. Rydberg in *Comprehensive Inorganic Chemistry,* J. C. Bailar, Jr., *et al.,* Eds., Pergamon, Oxford, 1973;

J. R. Brand and J. W. Cobble, *Inorg. Chem.* 1970, *9,* 912;

F. David, K. Samhoun, R. Guillaumont, and N. Edelstein, *J. Inorg. Nucl. Chem.* 1978, *40,* 69;

D. Langmuir, *Geochimica et Cosmochimica Acta* 1978, *42,* 547;

D. Langmuir and J. S. Herman, *Geochimica et Cosmochimica Acta* 1980, *44,* 1753;

L. R. Morss and J. Fuger, *J. Inorg. Nucl. Chem.* 1981, *43,* 2059;

L. J. Nugent, "Chemical Oxidation States of the Lanthanides and Actinides," in *International Review of Science, Inorganic Chemistry, Series 2, Vol. 7, Lanthanides and Actinides,* K. W. Bagnall, Ed., Buttersworth, London, 1975.

Mo   Th. Heumann and N. D. Stolica in *Encyclopedia of Electrochemistry,* Vol. V, A. J. Bard, Ed., Dekker, New York, 1976.

Mn   E. Wiberg (see Xe ref.).

Tc   A. P. Ginsberg, *Inorg. Chem.* 1964, *3,* 567; R. Colton and R. D. Peacock, *Quart. Rev.* 1962, *16,* 299.

Re   J. P. King and J. W. Cobble, *J. Am. Chem. Soc.* 1957, *79,* 1559; J. W. Cobble, *J. Phys. Chem.* 1957, *61,* 727; S. C. Abrahams, A. P. Ginsberg, and K. Knox, *Inorg. Chem.* 1964, *3,* 567.

Group VIII   R. N. Goldberg and L. G. Hepler, *Chem. Rev.* 1968, *68,* 239;

J. W. Larson, P. Cerrutti, H. K. Garber, and L. G. Hepler, *J. Chem. Phys.* 1968, *72,* 2902; J. F. Llopis and F. Colom in *Encyclopedia of Electrochemistry,* Vol. VI, A. J. Bard, Ed., Dekker, New York, 1976;

M. Pourbaix, *Atlas of Electrochemical Equilibria in Aqueous Solution,* Pergamon, Oxford, 1966.

Ag   J. A. Macmillan, *Chem. Rev.* 1962, *62,* 65.

Tl   H. A. Schwartz, D. Comstock, J. K. Yandell, and R. W. Dodson, *J. Phys. Chem.* 1974, *78,* 488;

B. Falcinella, P. D. Felgate, and G. S. Laurence, *J. C. S. Dalton 1974,* 1367.

Si, Ge   M. Pourbaix (see under Group VIII).

N   W. L. Jolly, *The Inorganic Chemistry of Nitrogen,* Benjamin, New York, 1964.

As   W. L. Jolly, *The Principles of Inorganic Chemistry,* McGraw-Hill, New York, 1976.

Br   F. Schreiner, D. Osborne, A. Pocius, and E. Appelman, *Inorg. Chem.* 1970, *9,* 2320.

At   E. H. Appelman, *J. Am. Chem. Soc.* 1961, *83,* 805.

Index